Stedman's

OB-GYN & GENETICS

WORDS

THIRD EDITION

Stedman's

OB-GYN & GENETICS WORDS

THIRD EDITION

LIPPINCOTT
WILLIAMS
& WILKINS

Series Editor: Beverly J. Wolpert
Associate Managing Editor: Trista A. DiPaula
Associate Managing Editor: William A. Howard
Art Director: Jennifer Clements
Production Manager: Julie K. Stegman
Production Coordinator: Kevin Iarossi
Typesetter: Peirce Graphic Services, Inc.
Printer & Binder: Data Reproductions Corporation.

Third Edition, 2001

Library of Congress Cataloging-in-Publication Data

01
1 2 3 4 5 6 7 8 9 10

Stedman's OB-GYN & genetics words.— 3rd ed.
        p. ; cm.
    Rev. ed. of: Stedman's OB-GYN words. 2nd ed. c1995.
    Includes bibliographical references.
    ISBN 0-7817-2704-9
        1. Gynecology—Terminology. 2. Obstetrics—Terminology. I. Title: OB-GYN &
    genetics words. II. Title: Stedman's OB-GYN and genetics words.
    III. Title: OB-GYN and genetics words. IV. Stedman, Thomas Lathrop,
    1853-1938. V. Stedman's OB-GYN words.
        [DNLM: 1. Obstetrics—Dictionary—English. 2. Genetics—Dictionary—English.
    3. Gynecology—Dictionary—English. WQ 13 S812 2001]
RG45 .S74 2001
618'.03—dc21   00-052022

Contents

Acknowledgments

An important part of our editorial process is the involvement of medical transcriptionists—as advisors, reviewers, and editors.

We extend special thanks to Ellen Atwood, as well as to Diane LeMieux Zook, CMT, and Jeanne Bock, CSR, MT, for editing the manuscript, helping to resolve many difficult content questions, and contributing material for the appendix sections. We also extend special thanks to Helen Littrell, CMT, for performing the final prepublication review.

We are grateful, as well, to our MT Editorial Advisory Board members, including Natasha Brown; Marty Cantu, CMT; Patricia Gibson; Nancy Hill, MT; Darcy Johnson; Heather Little, CMT; Wendy Ryan, ART; and Sandra Wideburg, CMT. These medical transcriptionists and medical language specialists served as important editors and advisors.

Other important contributors to this edition include Kathryn Mason, CMT, who focused on the appendix sections, and Rose M. Berry; Susan Carkin, CMT; Sherry Crawford, CMT; Shemah Fletcher; Deborah Hahn, CMT; Sandy Kovacs, CMT; Robin Koza; Diana Rezac, CMT; Cheri Sawyer, CMT; Patricia White, CMT; and DéBorah Wiggins.

Barb Ferretti played an integral role in the process by reviewing the content files for format, updating the database, and providing a final quality check. Also integral to the development of this reference were the advice and assistance of our Lippincott Williams & Wilkins colleagues, including Tim Hiscock, Lisa McAllister, Rebecca Roberson, and Heather Whary.

As with all our *Stedman's* word references, this resource incorporates the suggestions and expertise of our many contacts in the medical transcriptionist community. Thanks to all of our advisory board participants, reviewers, and editors; AAMT meeting attendees; and others who have written us with requests and comments—keep talking, and we'll keep listening.

Editor's Preface

Not so very long ago (my grandmother's generation) photographs of new babies were taken at about one year of age; infant mortality was high, even expected, and it wasn't safe to take little ones out and expose them to germs. The next generation photographed their newborns at about six weeks or so—when mom came home from her two- or three-week (or longer) stay in the hospital and recovered from the ordeal of childbirth. My own babies were photographed moments after birth, when a hospital stay of a few days was normal. My daughter's generation now has keepsakes of ultrasound pictures as early as a few weeks in utero, and many new moms and babies go home within 24 hours of birth. More recently, couples undergoing fertility treatments have photographs taken of the actual egg and sperm at the instant of fertilization. Thinking about this simple timing of photographs can help anchor us in the vast sea of change and innovation that have deluged gynecology, obstetrics, and reproductive health in the 20th century.

Thousands of couples who, in generations past, would not have been able to conceive now produce their own genetic offspring via an increasingly complex variety of methods. The miracles of assisted reproductive techniques (ART) are not without challenges:
- increasing numbers of higher order multiple births
- keeping a mother, who is carrying 20 or more pounds of babies, healthy
- maintaining a healthy, nurturing environment in the womb for three, four, five, or more fetuses
- advancing maternal ages for even singleton pregnancies

Nor have the scientific advances been without surprises, such as the discovery that azoospermia, a primary cause of infertility, is now known to be a heritable genetic defect. This leads us to wonder—shades of *Brave New World*—will all reproduction eventually take place outside the body?

A new law was recently passed to establish guidelines for the scientific, moral, and ethical handling of frozen embryos fertilized during ART procedures yet not selected for implantation; these wisps of new life may unlock mysteries surrounding not just reproduction but serious disease as well. Appropriate study of embryos may even sustain the long-held hope that some of the degenerative or chronic diseases like Alzheimer, amyotrophic lateral sclerosis, and especially the heart-rending disorders of the unborn and newborn can be treated, cured, and even eliminated via understanding embryonic stem cells and tailoring them in the lab. Perhaps we need to thank the infertile couples who desired so strongly to have their own babies that they inspired the reproductive scientists. Discoveries from deeply delving into the secrets locked within germ cells have fulfilled not only parents' dreams of conceiving and giving birth to their babies, but also the understanding of pregnancy and reproduction in myriad ways to benefit all women, all couples, their babies, and, eventually, humankind.

Other advances in fetal and maternal health include almost routinely saving the lives of extremely premature infants; operations and transfusions in utero can repair damage that would otherwise result in the death of babies before or soon after birth. Triple screening, early sonography, and ever more sophisticated tests and techniques result in detection of anomalies not compatible with life and severe physical or mental retardation ever earlier in the pregnancy, giving parents the information needed to make life-altering decisions. These methods and many others help to soothe, early on, what has always been an underlying worrisome aspect of pregnancy: Is my baby normal? Adding a simple vitamin component, folic acid, has been shown to markedly decrease the incidence of some common neural tube defects, one of the leading causes of fetal and infant mortality and deformity.

Today we are shocked to hear of a woman dying in childbirth, rather than taking such news as normal or acceptable. Life-threatening events of labor and delivery—for example, disseminated intravascular coagulation (DIC)—can frequently be anticipated and immediately treated, while prebirth imaging and diagnosis of the fetus can allow specialty teams to gather and participate in surgical or medical intervention, thus saving lives of both babies and their moms.

Unraveling of the human genome, recently accomplished, holds great promise for the future of the health of women and their babies. Even now there are techniques and procedures in place to detect or protect against gene changes and incompatibilities. On the horizon are more sophisticated and distinct tests to monitor and predict the health status of parents and children. While these efforts are not yet absolute, and may never be, the more scientists understand the little bits of DNA and RNA that make us tick, the more we can understand why nature sometimes goes awry.

Although no longer headline news, HIV and other viral infections, particularly the herpesviruses, still wreak havoc, especially for women and their offspring. While antiviral and antiretroviral therapies are still in their own infancies, great strides have been made in understanding viruses and protecting women and children during their most vulnerable periods. Other women-specific health problems—breast and uterine cancers, effects of menopause, osteoporosis—now receive intense scrutiny and substantial research funding; no longer are they relegated to that obscure category of "women's problems."

Previously excluded from general medical research studies because of the risk of pregnancy (thus damage to a developing fetus) and because of hormone fluctuations, women are now welcomed as volunteers because birth control and other hormonal therapies help standardize fluctuations, and pregnancies can be detected practically the "next morning." Women's bodies are being studied appropriately so that ranges of normal for hormones, blood components, and other bodily indicators of health status reflect not just men's standards, but separate out the women's specifics and variations where appropriate, providing more precise understanding of the distinct differences between the sexes in areas not directly related to reproduction.

As briefly sketched above, so much has happened in reproductive, gynecological, and genetic science in the last few years that it is difficult to keep pace. *Stedman's OB-GYN & Genetics Words, Third Edition,* offers a comprehensive look at the terms important to medical transcription for these specialties, encompassing conception, reproduction, birth, women's cyclical and general health, and life's instructions previously secreted within the genes.

Thanks to Diane LeMieux Zook, for her contributions to the extensive genetics section that make this new word book both timely and comprehensive. Special thanks to Jeanne Bock, for her perseverance and high quality standards for each aspect of the content this reference provides.

Ellen Atwood

Publisher's Preface

Stedman's OB-GYN & Genetics Words, Third Edition, offers an authoritative assurance of quality and exactness to the wordsmiths of the healthcare professions—medical transcriptionists, medical editors and copyeditors, health information management personnel, court reporters, and the many other users and producers of medical documentation.

We received many requests for updates to *Stedman's OB-GYN Words, Second Edition,* including expanded coverage of medical genetics. As the requests continued to accumulate, we realized that medical language professionals needed a current, comprehensive reference for these specialties.

In *Stedman's OB-GYN & Genetics Words, Third Edition,* users will find thousands of words related to gynecologic oncology, maternal-fetal medicine, endocrinology, infertility, ART, neonatology, and medical genetics. Users will also find terms for diagnostic and therapeutic procedures, new techniques, and lab tests, as well as equipment names and abbreviations with their expansions. The appendix sections provide anatomical illustrations with useful captions and labels; sample reports, including genetic counseling notes; common terms by procedure; and drugs listed by indication. For quick reference, we have also included tables of genetic symbols, amino acids, storage diseases, biophysical profile scores, and normal lab values.

This compilation of more than 65,000 entries, fully cross-indexed for quick access, was built from a base vocabulary of approximately 38,000 medical words, phrases, abbreviations and acronyms. The extensive A-Z list was developed from the database of *Stedman's Medical Dictionary, 27th Edition,* and supplemented by terminology found in current medical literature (please see list of References on page xix).

We at Lippincott Williams & Wilkins strive to provide you with the most up-to-date and accurate word references available. Your use of this word book will prompt new editions, which we will publish as often as updates and revisions justify. We welcome your suggestions for improvements, changes, corrections, and additions—whatever will make this *Stedman's* product more useful to you. Please complete the postpaid card at the back of this book, and send your recommendations care of "Stedman's" at Lippincott Williams & Wilkins.

Explanatory Notes

Medical transcription is an art as well as a science. Both approaches are needed to correctly interpret the dictation of a physician, whose language is a product of education, training, and experience. This variety in medical language means that there are several acceptable ways to express certain terms, including jargon. *Stedman's OB-GYN & Genetics Words, Third Edition,* provides variant spellings and phrasings for many terms. These elements, in addition to complete cross-indexing, make *Stedman's OB-GYN & Genetics Words, Third Edition,* a valuable resource for determining the validity of terms as they are encountered.

Alphabetical Organization

Alphabetization of main entries is letter by letter as spelled, ignoring punctuation, spaces, prefixed numbers, or other special characters. For example:

hydroxyphenyluria
17-hydroxypregnenolone

Terms beginning with Greek letters show the Greek letter spelled out and listed alphabetically. For example:

alpha, α
    a. antitripsin level
    estrogen receptor a. (Er-alpha)
    transforming growth factor a. (TGF-alpha)

In subentry alphabetization, the abbreviated singular form or the spelled-out plural form of the noun main entry word is ignored.

Format and Style

All main entries are in **boldface** to expedite locating a sought-after term, to enhance distinction between main entries and subentries, and to relieve the textual density of the pages.

Irregular plurals and variant spellings are shown on the same line as the singular or preferred form of the word. For example:

os, pl. ora

hygroma, hydroma

Hyphenation

As a rule of style, multiple eponyms (e.g., Counsellor-Davis artificial vagina operation) are hyphenated. Also, hyphens have been added between a manufacturer and one or more eponyms (e.g., Vital-Metzenbaum dissecting scissors). Please note that in many cases, hyphenation is a question of style, not of accuracy, and thus is a matter of choice.

Possessives

Possessive forms have been dropped in this reference for the sake of consistency and conformance with the guidelines of the American Association for Medical Transcription (AAMT) and other groups. Please note, however, that in many cases, retaining the possessive, like hyphenating, is a question of style, not of accuracy, and thus is a matter of choice. To form the possessive of a word, simply add the apostrophe or apostrophe "s" to the end of the word.

Cross-indexing

The word list is in an index-like main entry-subentry format that contains two combined alphabetical listings:

(1) A *noun* main entry-subentry organization, which is typical of the A-Z section of medical dictionaries like *Stedman's*:

chromosome
    accessory c.
    c. aberration

syndrome
    Down s.
    s. of inappropriate secretion of antidiuretic hormone (SIADH)

(2) An *adjective* main entry-subentry organization, which lists words and phrases as you hear them. The main entries are the adjectives or modifiers in a multiword term. The subentries are the nouns around which the terms are constructed and to which the adjectives or modifiers pertain:

uterine
    u. artery
    u. contraction

twins
    twin birth
    diovular t.

This format provides the user with more than one way to locate and identify a multiword term. For example:

estradiol
 e. transdermal system

system
 estradiol transdermal s.

gene
 tumor suppressor g.

tumor
 t. suppressor gene

It also allows the user to see together all terms that contain a particular descriptor, as well as all types, kinds, or variations of a noun entity. For example:

antigen
 ABO a.
 a. activity
 allogeneic a.

transfer
 blastocyst t.
 direct oocyte t. (DOT)
 direct oocyte sperm t.
  (DOST)

Wherever possible, abbreviations are separately defined and cross-referenced. For example:

GASA
 growth-adjusted sonographic age

growth-adjusted
 g.-a. sonographic age (GASA)

age
 growth-adjusted sonographic a.

Genetics Terminology

The first 22 human chromosomes are described with Arabic numerals and the 23rd by an uppercase letter, usually X or Y. Any of these designators can also be delineated with a lowercase letter, usually p or q, to represent the short or long "arm" of the chromosome, respectively, and with other descriptive words. In this word book, to avoid lengthy repetitive lists, terms for chromosomes 1 to 22 and their variations are represented in a single entry with a range of numbers. For example:

chromosome
    c. 1–22
    c. ring 1–22
    c. 1–22 trisomy
    c. 8p mosaic tetrasomy

Subentries for the X and Y chromosomes are listed separately.
For the chromosome terms "deletion" and "duplication," the abbreviations are the first three letters with the word "chromosome" understood and the chromosome description in parentheses:

del (deletion)
    d. (Xq) syndrome

dup (duplicate)
    d. (Yp) syndrome

For chromosomes 1 to 22, these terms are listed in a single entry with a range of numbers:

del
    d. (1p)-(22p) syndrome

syndrome
    del (1p)-(22p) s.

References

In addition to the manufacturers' literature we gather at various medical meetings, scientific reports from hospitals, and the lists created by our MT Editorial Advisory Board members from their daily transcription work, we used the following sources for new terms in *Stedman's OB-GYN & Genetics Words, Third Edition*:

Books

Berek JS, Adashi E, Hillard PA, eds. Novak's Gynecology, 12th Edition. Baltimore: Williams & Wilkins, 1996.

Gabbe SG, Niebyl JR, Simpson JL, eds. Pocket Companion to Obstetrics: Normal & Problem Pregnancies, 3rd Edition. Philadelphia: Churchill Livingstone, 1999.

Jacobs DS, DeMott WR, Grady HJ, Horvat RT, Huestis DW, Kasten BL, eds. Laboratory Test Handbook, 4th Edition. Hudson, OH: Lexi-Comp, 1996.

Lambrou NC, Morse AN, Wallach EE, eds. The Johns Hopkins Manual of Gynecology and Obstetrics. Philadelphia: Lippincott Williams & Wilkins, 1999.

Lance LL. Quick Look Drug Book 2000. Baltimore: Lippincott Williams & Wilkins, 2000.

McKusick VA. Mendelian Inheritance in Man: A Catalog of Human Genes and Genetic Disorders, 12th Edition. Baltimore: The Johns Hopkins University Press, 1998.

Pyle V. Current Medical Terminology, 7th Edition. Modesto, CA: Health Professions Institute, 1998.

Reece EA, Hobbins JC, eds. Medicine of the Fetus and Mother, 2nd Edition. Philadelphia: Lippincott-Raven Publishers, 1999.

Reiss HE, ed. Reproductive Medicine: From A to Z. Oxford: Oxford University Press, 1998.

Rivlin, ME, Martin RW, eds. Manual of Clinical Problems in Obstetrics and Gynecology, 5th Edition. Philadelphia: Lippincott Williams & Wilkins, 2000.

Scott JR, Di Saia PJ, Hammond CB, Spellacy WN, eds. Danforth's Obstetrics and Gynecology, 8th Edition. Philadelphia: Lippincott Williams & Wilkins, 1999.

Stedman's Medical Dictionary, 27th Edition. Baltimore: Lippincott Williams & Wilkins, 2000.

CDs

Briggs GB, Freeman RK, Yaffe S, eds. Drugs in Pregnancy and Lactation: A Reference Guide to Fetal and Neonatal Risk, on CD-ROM. Philadelphia: Lippincott Williams & Wilkins, 1998.

Sciarra JJ, ed. Gynecology and Obstetrics Looseleaf CD-ROM, Volumes 1–6. Philadelphia: Lippincott Williams & Wilkins, 1999.

Journals

ADVANCE for Health Information Professionals. King of Prussia, PA: Merion, 1999.

American Journal of Human Genetics. Chicago: University of Chicago Press, 1995.

American Journal of Obstetrics and Gynecology. St. Louis: Mosby, 1999.

Clinical Dysmorphology. Philadelphia: Lippincott Williams & Wilkins, 1999–2000.

Contemporary OB/GYN. Montvale, NJ: Medical Economics, 1995–1997, 1999.

The Female Patient. Chatham, NJ: Quadrant HealthCom, 1999–2000.

Genetics in Medicine. Baltimore: Lippincott Williams & Wilkins/American College of Medical Genetics, 1998–1999.

Infertility and Reproductive Medicine Clinics of North America. Philadelphia: WB Saunders, 1995.

Internal Medicine. Montvale, NJ: Medical Economics, 1995–1997.

Journal of Reproductive Medicine. St. Louis: Journal of Reproductive Medicine, 1995.

The Latest Word. Philadelphia: WB Saunders, 1995.

OB-GYN Clinical Alert. Atlanta: American Health Consultants, 2000.

OB-GYN News. Morristown, NJ: International Medical News Group, 1995–1997, 1999–2000.

Obstetrical and Gynecological Survey. Baltimore: Lippincott Williams & Wilkins, 1995, 1997, 1999–2000.

Obstetrics and Gynecology. New York: Elsevier/The American College of Obstetricians and Gynecologists, 1995–1997, 1999–2000.

Stedman's WordWatcher. Baltimore: Lippincott Williams & Wilkins, 1995–2000.

Websites

http://asrm.abstracts.org/1997TOC.HTM

http://genomics.phrma.org

http://news.bmn.com/hmsbeagle

http://www.acog.com

http://www.ama-assn.org/insight/h_focus/wom_hlth/wom_hlth.htm

http://www.ama-assn.org/special/womh/womh.htm

http://www.amwa-doc.org

http://www.cdc.gov/genetics/update/current.htm

http://www.centerwatch.com/studies/LISTING.HTM

http://www.geneticalliance.org

http://www.hpisum.com

http://www.hum-molgen.de

http://www.mayohealth.org/mayo/0003/htm/hrt.htm

http://www.mtdesk.com

http://www.mtmonthly.com

http://www.nhgri.nih.gov

http://www.nlm.nih.gov/mesh/jablonski/syndrome_title.html

http://www.obgyn.net/medical.asp

http://www.sph.uth.tmc.edu/retnet/what-dis.htm

http://www.virtualdrugstore.com

http://www.womens-health.org

α (*var. of* alpha)

A
　　abortus
　　alveolar
　　　A and D Ointment
A₂
　　phospholipase A.
A₄
　　androstenedione
a
　　arterial
A-200 Shampoo
A-a
　　alveolar-arterial
　　　A-a gradient
a-A
　　arterial to alveolar
AAM
　　aggressive angiomyxoma
Aarskog-Scott syndrome (ASS)
Aarskog syndrome
Aase-Smith syndrome
Aase syndrome
AB
　　abdominal
　　abortion
　　abortus
A&B
　　apnea and bradycardia
　　　A&B spell
abacavir sulfate
ABAER
　　automated brainstem auditory evoked
　　response
abandonment
abarelix depot-M
Abbe-McIndoe
　　　A.-M. procedure
　　　A.-M. vaginal reconstruction
Abbe-McIndoe-Williams procedure
Abbe vaginal construction
Abbe-Wharton-McIndoe procedure
Abbokinase
Abbott LifeCare PCA Plus II infusion system
abbreviated atrioventricular conduction
ABC
　　argon beam coagulator
　　aspiration biopsy cytology
Abderhalden-Fanconi syndrome
abdomen
　　pendulous a.
abdominal (AB)
　　a. adhesion
　　a. approach
　　a. auscultation
　　a. ballottement
　　a. binder
　　a. cavity
　　a. circumference (AC)
　　a. colposuspension
　　a. compartment syndrome (ACS)
　　a. delivery
　　a. distention
　　a. dystocia
　　a. enlargement
　　a. epilepsy
　　a. examination
　　a. fetal electrocardiography
　　a. hernia
　　a. hysterectomy
　　a. hysteropexy
　　a. hysterotomy
　　a. irradiation
　　a. metroplasty
　　a. muscle deficiency
　　a. muscle deficiency anomalad
　　a. muscle deficiency syndrome
　　a. musculature aplasia syndrome
　　a. myomectomy
　　a. ostium
　　a. paravaginal repair
　　a. percussion
　　a. peritoneum
　　a. pregnancy
　　a. rescue
　　a. sacral colpoperineopexy
　　a. sacropexy
　　a. sacrospinous ligament colposuspension
　　a. salpingo-oophorectomy
　　a. salpingotomy
　　a. stria
　　a. strip radiotherapy
　　a. tuberculosis
abdominis
　　a. muscle
abdominocyesis
abdominohysterectomy
abdominohysterotomy
abdominopelvic
　　a. irradiation
　　a. scan
abdominoperineal
abdominovaginal hysterectomy
abducens palsy
abducent palsy
abducted thumbs syndrome
Abelcet
abembryonic

Aberdeen knot
Aberfeld syndrome
aberrancy
    pubertal a.
aberrant
    a. systemic feeding artery
aberration
    chromosomal a.
    chromosome a.
    G-banded cytogenetic a.
    heterosomal a.
    newtonian a.
    penta-X chromosomal a.
    steroidogenic a.
    tetra-X chromosomal a.
    triple-X chromosomal a.
abetalipoproteinemia
ABG
    arterial blood gas
ABI
    A. model 373 sequencing gel
      system
    A. model 377 sequencing gel
      system
    A. PRISM Dye Terminator Cycle
      Sequencing Ready Reaction Kit
abiotrophy
    Leber a.
abirritant
Abitrate
ablactation
ablation
    balloon endometrial a.
    endometrial a.
    endometrial resection and a. (ERA)
    hysteroscopic endometrial a.
    laser a.
    laser uterosacral nerve a. (LUNA)
    Nd:YAG laser a.
    ovarian a.
    partial rollerball endometrial a.
    rectoscopic endometrial a.
    rollerball endometrial a.
    rollerbar endometrial a.
    thermal balloon a.
    uterosacral nerve a.
ablatio placentae
ablative therapy with bone marrow
rescue
ablator
    endometrial a.
ablepharon-macrostomia syndrome
abnormal
    a. cortisol secretion
    a. deceleration
    a. embryo
    a. feedback signal
    a. fetal development

    a. karyotype
    a. labor
    a. menstruation
    a. placentation
    a. response
    a. uterine bleeding (AUB)
abnormality
    acral dysostosis with facial and
      genital a.'s
    cervical a.
    chromosomal structural a.
    congenital a.
    cord a.
    cortical gyral a.
    deflexion a.
    electroretinal a.
    epithelial cell a.
    fetal postural a.
    fetal thoracic a.
    fibrinogen a.
    genetic a.
    genitourinary a.
    gestational a.
    gyral a.
    hormonal a.
    hypogonadism, alopecia, diabetes
      mellitus, mental retardation,
      deafness and ECG a.'s
    metabolic a.
    midbrain a.
    müllerian a.
    müllerian, renal, cervicothoracic,
      somite a.'s (MURCS)
    multiple endocrine a.'s (MEA)
    neurobehavioral a.
    neurologic a.
    nonpalpable a.
    oral-facial-digital syndrome with
      retinal a.'s
    ovarian a.
    placentation a.
    possible migrational a.
    reproductive tract a.
    sex chromosomal a.
    situs a.
    skeletal a.
    skeletal abnormalities, cutis laxa,
      craniostenosis, psychomotor
      retardation, facial a.'s (SCARF)
    soft tissue a.
    sporadic chromosome a.
    tuberous breast a.
    umbilical artery waveform
      notching a.
    urinary tract a.
    uterine a.
    vaginal epithelial a.
    X-chromosome a.

ABO
ABO antigen
ABO blood group system
ABO erythroblastosis
ABO hemolytic disease of the newborn
ABO incompatibility
abort
aborted ectopic pregnancy
aborter
habitual a.
abortient
abortifacient
abortigenic
abortion (AB)
Aburel a.
accidental a.
ampullar a.
aneuploid a.
complete a.
criminal a.
Csapo a.
elective a.
euploid a.
habitual a.
imminent a.
incipient a.
incomplete a.
induced a.
inevitable a.
infected a.
justifiable a.
menstrual extraction a.
missed a.
recurrent euploidic a.
recurrent spontaneous a. (RSA)
repeated a.
saline a.
selective a.
septic a.
spontaneous a. (SAB)
surgically-induced a.
therapeutic a. (TAB)
threatened a.
tubal a.
abortionist
abortive
abortus (A, AB)
ABR
auditory brainstem response

abruption
placental a.
sinus a.
abruptio placentae
ABS
arterial blood sample
abscess
amebic a.
Bartholin a.
Bezold a.
Douglas a.
Dubois a.
epidural a.
milk a.
ovarian a.
parametric a.
parametritic a.
pelvic a.
perinephric a.
premammary a.
pyogenic a.
retroesophageal a.
retropharyngeal a.
retrotonsillar a.
stitch a.
subareolar a.
subphrenic a.
tuboovarian a. (TOA)
absence
a. of abdominal muscle syndrome
a. defect
a. seizure
uterine a.
vaginal a.
absent
a. cerebellum
a. splenium
Absolok endoscopic clip applicator
absolute
a. cardiac dullness
a. sterility
a. temperature
absorbable
a. gelatin sponge
a. staple
a. suture
Absorbine
A. Antifungal
A. Antifungal Foot Powder
A. Jock Itch
A. Jr. Antifungal

NOTES

3

absorptiometer
Hologic 1000 QDR dual-energy a.
Lunar DPX dual-energy a.
absorptiometry
dual-energy photon a.
dual-energy x-ray a. (DEXA, DXA)
dual-photon a.
radiographic a.
single-energy photon a.
single-photon a.
x-ray a.
absorption
calcium a.
a. fever
fluorescent treponemal antibody a. (FTA-ABS)
abstinence
a. score
Abt-Letterer-Siwe syndrome
Aburel abortion
abuse
alcohol a.
CAGE test for alcohol a.
child a.
domestic a.
a. dwarfism syndrome
maternal drug a.
sexual a.
AC
abdominal circumference
ACA
anticentromere antibody
acalvaria
acanthocytosis
acantholysis bullosa
acanthosis nigricans
ACAPI
anterior cerebral artery pulsatility index
acarbose
acardia
acardiac
a. fetus
a. twins
acardius
fetus a.
ACAT
automated computerized axial tomography
acatalasemia
acatalasia
accelerated
a. skeletal maturation, Marshall-Smith type
a. starvation
accelerator
betatron electron a.
linear a.

Accelon Combi cervical biosampler
access
fetoplacental a.
transcervical tubal a.
uterine a.
ACCESS immunoassay system
accessory
a. breast
a. chromosome
a. müllerian funnel
a. nipple
a. ovary
a. placenta
a. sex gland
a. spleen
a. tragus
a. yolk
accident
cerebrovascular a. (CVA)
obstetric a.
umbilical cord a.
accidental
a. abortion
a. fetal injury
a. hemorrhage
a. pregnancy
accompanying mood state
accouchement
a. forcé
accoucheur
Accoustix conductivity gel
accretio cordis
accretion
bone a.
Accu-Chek
A.-C. Easy glucose monitor
A.-C. II Freedom blood glucose monitor
A.-C. II glucometer
A.-C. test
AccuPoint hCG Pregnancy Test Disk
accuracy
assay a.
diagnostic a.
Accurbron
Accurette
A. endometrial suction curette
A. microcurettage
Accuscope
A. colposcope
A. II
AccuSite injectable gel
AccuSpan tissue expander
AccuStat
A. hCG pregnancy test
Accustat pulse oximeter
Accutane
A. dysmorphic syndrome

ACD
>alopecia, contracture, dwarfism
>area of cardiac dullness
>>ACD level
>>ACD mental retardation syndrome
>>ACD syndrome

ACE
>angiotensin-converting enzyme
>>ACE genotype

acebutolol
Acel stopcock
acentric
>a. chromosome

acephalia
acephalobrachia
acephalocardia
acephalochiria
acephalogaster
acephalopodia
acephalorrhachia
acephalostomia
acephalothoracia
acephalus, acephalous
Aceta
acetaminophen
Acetasol HC Otic
acetate
>aluminum a.
>calcium a.
>cyproterone a.
>Cytolex pexiganan a.
>depot medroxyprogesterone a. (DMPA)
>desmopressin a.
>estradiol cypionate and medroxyprogesterone a.
>estradiol/norethindrone a.
>flecainide a.
>ganirelix a.
>gonadorelin a.
>goserelin a.
>leuprolide a.
>m-cresyl a.
>medroxyprogesterone a. (MPA)
>megestrol a.
>nafarelin a.
>norethindrone a.
>quingestanol a.
>sermorelin a.

acetazolamide
acetic acid
acetohexamide

acetone
acetonide
>fluocinolone a.
>triamcinolone a.

acetonuria
acetophenazine
acetophenetidin
acetophenide
>dihydroxyprogesterone a.

acetowhite
>a. epithelium
>a. lesion
>a. reaction

acetoxyprogesterone
>a. derivative
>17-α-a. derivative

acetylcholine
>a. chloride

acetylcholinesterase (ACHE, AchE)
acetylcysteine
acetyldigitoxin
acetylsalicylic acid (ASA)
ACF
>asymmetric crying facies

achalasia
>infantile a.

achalasia-microcephaly syndrome
Achard syndrome
Achard-Thiers syndrome
ACHE, AchE
>acetylcholinesterase

acheilia
acheiria, achiria
acheiropody, achiropody
Aches-N-Pain
achievable
>as low as reasonably a. (ALARA)

achiria (*var. of* acheiria)
achiropody (*var. of* acheiropody)
acholic
>a. stool

achondrogenesis
>a. syndrome

achondroplasia
>a. syndrome

achondroplastic dwarfism
achordia
achromasia
achromia
achromiens
>incontinentia pigmenti a.

Achromobacter lwoffi

NOTES

Achromycin
   A. V
achrondoplastic dwarf
acid
   acetic a.
   acetylsalicylic a. (ASA)
   amino a.
   D-amino a.
   L-amino a.
   aminocaproic a.
   arachidonic a.
   arginine-glycine-aspartic a. (arg-gly-asp)
   arginosuccinic a.
   arylalkanoic a.
   arylcarboxylic a.
   arylpropionic a.
   ascorbic a.
   a. aspiration syndrome
   bichloracetic a.
   bile a.
   boric a.
   branched-chain amino a.
   carbonic a.
   9-cis-retinoic a.
   clavulanic a.
   deoxyadenylic a. (dAMP)
   deoxycytidylic a. (dCMP)
   deoxyguanylic a. (dGMP)
   deoxyribonucleic a. (DNA)
   deoxythymidylic a. (dTMP)
   dichloroacetic a.
   diethylenetriamine pentaacetic a. (DPTA)
   diisopropyl iminodiacetic a. (DISIDA)
   dimercaptosuccinic a.
   a. elution test
   epsilon-aminocaproic a. (EACA)
   ethacrynic a.
   ethylenediaminetetraacetic a. (EDTA)
   flufenamic a.
   folic a.
   folinic a.
   formiminoglutamic a. (FIGLU)
   free fatty a.
   fusanic a.
   hydriodic a.
   hydroxyazobenzoic a. (HABA)
   5-hydroxyindoleacetic a. (5-HIAA)
   21-hydroxyindoleacetic a. (21-HIAA)
   iocetamic a.
   iopanoic a.
   linoleic a.
   lipid-associated sialic a.
   lipoic a.

   lysergic a.
   mandelic a.
   mefenamic a.
   messenger ribonucleic a. (mRNA)
   N-acetylneuraminic a. (NANA)
   nalidixic a.
   N-[2-hydroxyethyl]piperazine N′-[2-ethanesulfonic a.] (HEPES)
   nicotinic a.
   noncarbonic a.
   nonvolatile a.
   nucleic a.
   pantothenic a.
   paraaminomethylbenzoic a. (PAMBA)
   paraaminosalicylic a. (PAS, PASA)
   phenylpyruvic a.
   a. phosphatase
   polyglycolic a.
   retinoic a.
   ribonucleic a. (RNA)
   salicylsalicylic a.
   sialic a.
   Slow Fe with folic a.
   sodium citrate with citric a.
   tolfenamic a.
   tranexamic a.
   transfer ribonucleic a. (tRNA)
   trichloroacetic a. (TCA)
   2,4,5-trichlorophenoxyacetic a.
   undecylenic a.
   uric a.
   valproic a.
   volatile a.
acid-base
   a.-b. balance
   a.-b. equilibrium
   a.-b. measurement
   a.-b. status
   a.-b. value
acidemia
   fetal a.
   glutaric a.
   isovaleric a.
   lactic a.
   metabolic a.
   methylmalonic a.
   mixed umbilical arterial a.
   pyroglutamic a.
   trihydroxycoprostanic a.
acidic fibroblast growth factor (FGFa)
acidification
   renal a.
   vaginal a.
acidosis
   diabetic a.
   fetal a.
   hyperchloremic renal a.

A

hyperchromic a.
lactic a.
metabolic a.
perinatal a.
renal tubular a. (RTA)
respiratory a.
transient respiratory a.
acid-Schiff
periodic a.-S. (PAS, PASA)
aciduria
beta-aminoisobutyric a.
hereditary orotic a.
methylmalonic a.
xanthurenic a.
Aci-Jel vaginal jelly
Acinetobacter
A. baumanii
A. lwoffi
ACIP
Advisory Committee on Immunization
Practices
Aciphex
acitretin
aCL
anticardiolipin antibody
aclasis
diaphyseal a.
Aclovate Topical
acne
a. conglobata
halogen a.
neonatal a.
a. neonatorum
a. vulgaris
acorn cannula
Acosta classification
acoustic
a. blink reflex
a. enhancement
a. impedance
a. meningioma
a. neuroma
a. shadow
a. stimulation study
a. stimulation test (AST)
a. trauma
acoustical interference
ACPS
acrocephalopolysyndactyly
acquired
a. agammaglobulinemia
a. hemolytic anemia

a. hypogammaglobulinemia
a. immune deficiency syndrome
(AIDS)
a. immunodeficiency syndrome
(AIDS)
a. immunodeficiency syndrome-
related virus (ARV)
Acrad HS catheter
acral
a. demineralization
a. dysostosis with facial and
genital abnormalities
acral-renal-mandibular syndrome
acrania
acroblast
acrobrachycephaly
acrocallosal syndrome (ACS)
acrocentric chromosome
acrocephalopolysyndactyly (ACPS)
acrocephalosyndactyly, type I–V (ACS)
acrocephaly
acrochordon
acrocraniofacial dysostosis
acrocyanosis
acrodermatitis
a. enteropathica
papular a. of childhood (PAC)
acrodynia
acrodysgenital syndrome
acrodysostosis syndrome
acrodysplasia
acrodysplasia-dysostosis syndrome
acrofacial
a. dysostosis
a. dysostosis with postaxial defects
syndrome
acrofrontofacionasal (AFFN)
a. dysostosis syndrome types 1, 2
acrokeratosis paraneoplastica
acromastitis
acromegaloid
a. facial appearance (AFA)
a. facial appearance syndrome
**acromegaloid-cutis verticis gyrata-
leukoma syndrome**
acromegaly
acromelic frontonasal dysplasia
acromesomelia
acromesomelic
a. dwarfism
a. dysplasia
acromial dimple

NOTES

7

acromicric dysplasia
acromion presentation
acromphalus
acroosteolysis
> cranioskeletal dysplasia with a.
> hereditary osteodysplasia with a.
> a. syndrome
> a. with osteoporosis and changes in skull and mandible

acropectorovertebral dysplasia
acropustulosis
> a. of infancy

acrorenal syndrome
acrorenoocular syndrome
acrosin
acrosomal cap
acrosome
> a. reaction
> a. reaction with ionophore challenge (ARIC)

acrosome-intact sperm
acrosphenosyndactyly
acrosyndactyly
acrotism
ACS
> abdominal compartment syndrome
> acrocallosal syndrome
> acrocephalosyndactyly, type I–V

ACT
> activated clotting time

Actamin
act-FU-Cy
> actinomycin D, 5-fluorouracil, cyclophosphamide

ACTH
> adrenocorticotropic hormone
> ACTH deficiency
> ACTH insufficiency
> ACTH stimulation
> ACTH stimulation test

Acthar
ActHIB *H. influenzae* **type B vaccine**
Acticin Cream
Acticort Topical
Actidil
Actifed Allergy Tablet
Actimmune
actin
actin-myosin interaction
Actinomyces
> *A. israelii*
> *A. naeslundii*
> *A. propionica*

actinomycin D, 5-fluorouracil, cyclophosphamide (act-FU-Cy)
actinomycosis
> pelvic a.

action
> fetal heart a.
> gene a.
> law of mass a.
> luteolytic a.
> mediating a.
> self-priming a.
> uterine a.

Actiprofen
activated
> a. clotting time (ACT)
> a. estrogen receptor
> a. partial thromboplastin time (APTT)
> a. protein C (APC)
> a. protein C resistance

activation
> egg a.
> embryonic genome a.
> endothelial cell a.
> genome a.

activator
> lymphocyte a.
> plasminogen a. (PA)
> Platelin Plus A.
> recombinant tissue type plasminogen a. (rt-PA)
> tissue plasminogen a. (t-PA)
> urokinase plasminogen a. (u-PA)

active
> a. bowel sounds
> Free & A.
> a. phase
> a. phase arrest
> a. phase of labor
> a. range of motion (AROM)
> a. specific immunotherapy (ASI)
> a. third-stage management

Activelle
activin
> a. A

activity
> antigen a.
> colony-stimulating a. (CSA)
> elevated enzyme a.
> endometrial cycling a.
> enzyme a.
> fetal a.
> fetal cardiac a.
> fetal somatic a.
> lupus anticoagulant a.
> lymphotoxin antitumor a.
> Manning score of fetal a.
> mitogenic a.
> opioid a.
> ovarian a.
> peripheral androgen a.
> phospholipase a.

progestational a.
proline aminopeptidase a.
sexual a.
tonicoclonic seizure a.
uterine a.
withdrawal-like a.
actocardiotocograph fetal monitor
Actonel
Actrapid insulin with Ultratard
Actron
actuarial survival
Acular
acuminata
condylomata a.
verruca a.
acuminatum
condyloma a.
giant anorectal condyloma a.
acupuncture
Acuson
A. color Doppler
A. computed sonography
A. 128 Doppler ultrasound
A. Model 128XP machine
A. 128XP ultrasound
A. 128XP-10 ultrasound
acuta
Juliusberg pustulosis
vacciniformis a.
pityriasis lichenoides et
varioliformis a. (PLEVA)
acutainer
acute
a. anaphylaxis
a. atherosis
a. cystitis
a. disseminated histiocytosis X
a. fatty liver
a. fatty liver disease
a. fatty liver of pregnancy (AFLP)
a. intermittent porphyria
a. interstitial pneumonia
a. intrapartum transfusion
a. laryngotracheal bronchitis
a. lymphocytic leukemia (ALL)
a. mountain sickness (AMS)
a. neonatal herpes
a. nonlymphocytic leukemia
a. otitis media (AOM)
a. perinatal transfusion
a. renal failure
a. rheumatic fever (ARF)

a. tocolysis
a. tubular necrosis (ATN)
a. urethral syndrome
a. uterine inversion
acute-on-chronic tissue hypoxemia
acute-phase serum study
AcuTrainer
A. device
Acutrim
A. 16 Hours
A. II
A. Late Day
Acuvel
acyclic pelvic pain
acyclovir
acyesis
acystia
AD
Alzheimer disease
autosomal dominant
ADA
American Diabetes Association
ADA diet
adacrya
adactylia
adactyly
Adair-Dighton syndrome
Adair-Veress needle
Adalat
Adam complex
Adams
A. advancement
A. advancement of round ligaments
A. test for scoliosis
Adams-Oliver syndrome
Adams-Stokes syndrome
Adapin
adaptation
Deyo a.
maternal ocular a.
uterine artery hemodynamic a.
adaptive
a. landscape
a. peak
a. radiation
a. surface
a. value
ADC
AIDS dementia complex
ADC Medicut shears
Adcon-L anti-adhesion barrier

NOTES

ADD
> attention deficit disorder

add-back
> a.-b. therapy
> a.-b. treatment

Adderall

addiction
> alcohol a.
> cocaine a.
> drug a.
> opioid a.

Addison
> A. disease
> A. disease-cerebral sclerosis syndrome
> A. disease-spastic paraplegia syndrome
> A.-Schilder syndrome

addisonian
> a. crisis
> a. pernicious anemia
> a. syndrome

additive genetic variance

additivity
> intralocal a.

address
> chromonemal a.

adducted
> a. thumb-clubfoot syndrome
> a. thumbs-mental retardation syndrome
> a. thumbs syndrome

adductus
> metatarsus a.

adefovir

adelocephaly

adelomorphous

adenine
> a. arabinoside

adenitis
> cervical a.
> mesenteric a.
> sclerosing a.
> vestibular a.

adenoacanthoma
> endometrial a.
> lymph node endometriotic a.

adenocarcinoma
> cervical a.
> cervical clear cell a.
> ciliated cell endometrial a.
> clear cell a.
> endometrial a.
> endometrial clear cell a.
> endometrial secretory a.
> mesonephric a.
> metastatic a.
> microinvasive a.

> ovarian clear cell a.
> papillary a.
> secretory a.
> serous a.
> a. in situ
> vaginal a.
> vaginal clear cell a.
> vulvar adenoid cystic a.

adenofibroma

adenofibromyoma

adenofibrosis

adenohypophysis

adenohypophysitis
> lymphocytic a.

adenoid
> a. cystic carcinoma
> a. facies

adenoidectomy

adenoiditis

adenoleiomyofibroma

adenoma
> apocrine a.
> chromophobic a.
> ductal a.
> growth hormone-secreting a.
> islet cell a.
> lactating a.
> a. malignum
> a. of nipple
> ovarian tubular a.
> parathyroid a.
> pituitary a.
> prolactin-secreting a.
> a. sebaceum
> suspected pituitary a.
> testosterone-secreting adrenal a.
> virilizing a.

adenomatoid
> a. oviduct tumor

adenomatosis
> erosive a. of nipple
> familial multiple endocrine a.
> fibrosing a.

adenomatous
> a. endometrial hyperplasia
> a. hyperplasia (AH)
> a. polyp

adenomegaly

adenomere

adenomyoma

adenomyomatosis

adenomyosis
> stromal a.
> a. uteri

adenopathy
> axillary a.
> cervical a.
> postinflammatory a.

adenosalpingitis
adenosarcoma
    müllerian a.
adenosine
    a. deaminase deficiency
    a. monophosphate (AMP)
    a. phosphate
    a. triphosphate (ATP)
adenosis
    blunt duct a.
    fibrosing a.
    microglandular a.
    sclerosing a.
    vaginal a.
adenosquamous carcinoma
adenoviral pneumonia
adenovirus
adenylate cyclase
adenyl cyclase
adermia
adermogenesis
ADH
    antidiuretic hormone
    atypical ductal hyperplasia
adhalin gene
ADHD
    attention deficit hyperactivity disorder
adherens
    zonula a.
adherent
    a. discharge
    a. placenta
    a. vaginal discharge
adhesiolysis
adhesion
    abdominal a.
    AFS classification of adnexal a.'s
    amniotic a.
    banjo-string a.
    a. barrier
    cell-extracellular matrix a.
    dense a.
    fiddle-string a.
    filmy a.
    intracervical a.
    intrauterine a.
    lysis of a.
    a. lysis
    a. molecule
    omental a.
    periovarian a.
    peritubal a.

    piano-wire a.
    platelet a.
    A. Scoring Group (ASG, ASG
      system, ASG system)
    serosal a.
    sperm-egg a.
    vaginal cuff a.
adhesive
    Biobrane a.
    a. disease
    a. endometriosis
    a. vaginitis
    a. vulvitis
Adie syndrome
Adipex-P
adipocere
adipocyte
adiponecrosis subcutanea neonatorum
adipose tissue
adiposogenital
    a. syndrome
adipsia
adjusted gestational age
adjustment
    psychosocial a.
adjuvant
    a. chemoradiation therapy
    a. chemotherapy
    Freund a.
    a. radiotherapy
    a. therapy
ad lib feeding
administration
    Food and Drug A. (FDA)
    Health Care Financing A. (HCFA)
    oral a.
    parenteral a.
    pulsatile GnRH a.
    sequential a.
    transdermal a.
    transnasal a.
    vaginal a.
adnata
    alopecia a.
adnatum
    filiform a.
adnexa (*pl. of* adnexum)
adnexal
    a. adhesion classification system
    a. cyst
    a. infection
    a. mass

NOTES

adnexal *(continued)*
  a. metastasis
  a. torsion
  a. tumor
adnexectomy
adnexitis
adnexopexy
adnexum, pl. **adnexa**
  transposed adnexa
ADOD
  arthrodentoosteodysplasia
adolescence
adolescent
  a. breast
  a. gynecology
  A. and Pediatric Pain Tool
    (APPT)
  a. pregnancy
  a. sterility
adolescentis
  Bifidobacterium a.
adoption
adoptive immunotherapy
ADOS
  autosomal dominant Opitz syndrome
ADR
  ataxia-deafness-retardation
adrenal
  a. androgen
  a. androgen secretion
  a. cell rest tumor
  a. cortex
  a. crisis
  a. gland
  a. gland morphology
  a. hyperandrogenism
  a. hyperandrogenism marker
  a. hyperplasia
  a. hypofunction
  a. insufficiency
  Marchand a.'s
  a. morphologic consideration
  a. neoplasm
  a. reticularis
  a. steroid
  a. steroidogenesis
  a. virilism
  a. virilizing syndrome
adrenalectomy
Adrenalin
  A. Chloride
adrenaline injection
adrenalitis
  autoimmune a.
adrenarche
  precocious a.
  premature a.

adrenergic
  a. blocker
  a. drug
  a. receptor
  a. stimulator
adrenocortical
  a. atrophy-cerebral sclerosis
    syndrome
  a. function
  a. hormone
  a. insufficiency
  a. steroid
  a. steroidogenesis
adrenocorticotropic, adrenocorticotrophic
  a. hormone (ACTH)
  a. hormone deficiency
adrenocorticotropin
  chorionic a.
adrenogenital
  a. syndrome (AGS)
adrenoleukodystrophy
adrenoleukomyeloneuropathy (ALMN)
adrenomegaly
adrenomyeloneuropathy (AMN)
adrenomyodystrophy
Adriamycin
  cisplatin, Cytoxan, A. (CISCA)
  cisplatin, etoposide, Cytoxan, A.
    (CECA)
  A., fluorouracil, methotrexate
    (AFM)
  A., Oncovin, prednisone, etoposide
    (AOPE)
  A. PFS
  A. RDF
adRP
  autosomal-dominant retinitis pigmentosa
Adrucil
  A. Injection
ADS
  anonymous donor sperm
Adson
  A. forceps
  A. ganglion scissors
  A. pickups
Adsorbocarpine
adult
  a. generalized gangliosidosis
  a. granulosa cell tumor (AGCT)
  a. pseudohypertrophic muscular
    dystrophy
  a. respiratory distress syndrome
    (ARDS)
adult-onset
  a.-o. congenital adrenal hyperplasia
  a.-o. diabetes mellitus (AODM)
  a.-o. polyglandular syndrome

A

advance
    a. directive
    A. formula
Advanced
    A. Collection breast pump
advanced
    a. carcinoma
    a. maternal age
advancement
    Adams a.
    vaginal a.
Advantage
    A. 24
    A. ultrasound
adverse
    a. effect
    a. outcome
Advil Cold & Sinus Caplets
Advisory Committee on Immunization Practices (ACIP)
adynamia episodica hereditaria
adynamic ileus
adysplasia
    hereditary urogenital a.
AEC
    ankyloblepharon, ectodermal defects, cleft lip/palate
    AEC syndrome
AEGIS sonography management system
AENNS
    Albert Einstein Neonatal Developmental Scale
AEP
    auditory evoked potential
Aequitron 9200 apnea monitor
AER
    aldosterone excretion rate
aeration of lung
aerobe
aerobic
    a. metabolism
aerobics
    water a.
Aerochamber
aerocolpos
Aerolate
    A. III
    A. Jr.
    A. SR S
Aerolone
Aeromonas hydrophila
aerophore

Aeroseb-Dex
Aeroseb-HC Topical
aerosol
    Atrisol a.
    Breezee Mist A.
    cromolyn sodium inhalation a.
    Nasalide Nasal A.
    a. therapy
    Virazole A.
aerosolized medication
aeruginosa
    Pseudomonas a.
aestivale
    hydroa a.
AF
    amniotic fluid
    AF WBC
AFA
    acromegaloid facial appearance
    AFA syndrome
AFAFP
    amniotic fluid alpha-fetoprotein
AFE
    amniotic fluid embolism
afetal
affective disorder
Affinity bed
Affirm
    A. VPIII test
    A. VP microbial identification system
AFFN
    acrofrontofacionasal
Affymetrix GeneChip system
AFI
    amniotic fluid index
afibrinogenemia
    congenital a.
Afko-Lube
AFLP
    acute fatty liver of pregnancy
AFM
    Adriamycin, fluorouracil, methotrexate
AFP
    alpha-fetoprotein
    AFP X-tra
AFP-EIA
    alpha-fetoprotein enzyme immunoassay
AFRAX
    autism-fragile X syndrome
African-American variant galactosemia

NOTES

13

Afrin
>A. Children's Nose Drops
>A. Nasal Solution

Afrinol

AFS
>American Fertility Society
>>AFS adhesion scoring system
>>AFS classification of adnexal
>>adhesions

Aftate

afterbirth

afterload
>a. applicator
>a. colpostat
>a. tandem

afterpain

AFV
>amniotic fluid volume

AG1343

AGA
>appropriate for gestational age
>aspartylglucosamide
>>post-term AGA
>>term AGA

agalactia

agalactiae
>*Streptococcus a.*

agalactorrhea

agalactosis

agalactous

agammaglobulinemia
>acquired a.
>Bruton a.
>X-linked a.

aganglionic megacolon

aganglionosis
>colonic a.

agar
>a. gel precipitation technique
>MacConkey II a.

AGCT
>adult granulosa cell tumor

AGCUS, AGUS
>atypical glandular cells of uncertain
>significance
>atypical glandular cells of undetermined
>significance

age
>adjusted gestational a.
>advanced maternal a.
>appropriate for gestational a.
>(AGA)
>bone a.
>childbearing a.
>coital a.
>developmental a. (DA)
>Dubowitz/Ballard Exam for
>Gestational A.

>estimated gestational a. (EGA)
>fertilization a.
>fetal a.
>gestational a. (GA)
>Greulich and Pyle bone a.
>growth-adjusted sonographic a.
>(GASA)
>a. index
>large for gestational a. (LGA)
>maternal a.
>menstrual a.
>ovulatory a.
>paternal a.
>postconceptional a.
>postovulatory a.
>small for gestational a. (SGA)

Agenerase

agenesis
>bilateral facial a.
>callosal a.
>cerebellar vermis a.
>cervical a.
>corpus callosum a.
>a. of corpus callosum
>a. of corpus callosum-mental
>retardation-osseous lesions
>syndrome
>corpus callosum partial a.
>a. of corpus callosum with
>stenogyria
>cortical a.
>diaphragmatic a.
>gonadal a.
>hereditary renal a.
>müllerian a.
>nuclear a.
>ovarian a.
>pancreatic a.
>pulmonary a.
>renal a.
>sacral a.
>septi pellucidi a.
>thymic a.
>vaginal a.

agenitalism

agent
>alkylating a.
>alpha-adrenergic a.
>anthelmintic a.
>anticancer a.
>antifibrinolytic a.
>antifolic a.
>antineoplastic a.
>antiplatelet a.
>antiprostaglandin a.
>beta-adrenergic a.
>beta$_2$-adrenergic a.
>A. Blue

butenafine antifungal a.
chemotherapeutic a.
cycle-nonspecific a.
cycle-specific a.
cytotoxic a.
delta a.
fibrinolytic a.
hyperosmotic a.
infertility a.
myelosuppressive a.
nonalkylating a.
A. Orange
Osteomark a.
pressor a.
progestational a.
satumomab pendetite imaging a.
sclerosing a.
teratogenic a.
tocolytic a.
TWAR a.

age-related risk

agglutination
head-to-head sperm a. (H-H)
head-to-tail sperm a. (H-T)
a. inhibition test
labial a.
latex particle a.

agglutinin
lens culinaris a.
Pisum sativum a. (PSA)
Ricinus communis a.

aggregation
platelet a.

aggressive angiomyxoma (AAM)

aging gamete

agitated depression

aglossia-adactylia syndrome

aglossia congenita

aglossostomia

agnathia

agonadal

agonadism

agonadism, mental retardation, short stature, retarded bone age syndrome

agonal respirations

agonist
beta-adrenergic a.
beta-receptor a.
calcium a.
cholinergic a.
dopamine receptor a.
dopaminergic a.

estrogen a.
GnRH a.
gonadotropin-releasing hormone a.

AGR
aniridia, ambiguous genitalia, mental retardation

agranulocytosis
Kostmann infantile a.

A/G ratio
albumin-globulin ratio

Agris-Dingman submammary dissector

AGS
adrenogenital syndrome

AGT
aminoglutethimide

AGU
aspartylglucosaminuria

AGUS (*var. of* AGCUS)

agyria

agyria-pachygyria
a.-p. band
a.-p. syndrome

AH
adenomatous hyperplasia
assisted hatching

AHD
arteriohepatic dysplasia

AHDS
Allen-Herndon-Dudley syndrome

Ahlfeld sign

AHO
Albright hereditary osteodystrophy

Ahumada-Del Castillo syndrome

AI
artificial insemination
AI 5200 S Open Color Doppler imaging system

Aicardi-Goutieres syndrome

Aicardi syndrome

AID
artificial insemination by donor

aid
Sensability breast self-examination a.

AIDS
acquired immune deficiency syndrome
acquired immunodeficiency syndrome
AIDS dementia complex (ADC)
transfusion-related AIDS (TRAIDS)

AIDS-related
A.-r. complex (ARC)
A.-r. lymphoma

NOTES

15

AIH
> artificial insemination by husband

AIHA
> autoimmune hemolytic anemia

AIM
> area of interest magnification

AIN
> anal intraepithelial neoplasia

air
> a. bronchogram
> a. embolism
> a. leak syndrome

AIRE
> autoimmune regulatory
> AIRE gene

Airlift balloon retractor
Airshields Isolette
airway
> a. conductance
> a. resistance
> a. suction

akathisia
AK-Chlor
Akesson
akinesia algera
Akineton
AK-Mycin
AK-Spore H.C. Otic
AKTob
Ala-Cort Topical
alacrima
> a. congenita

Alagille syndrome
Alagille-Watson syndrome (AWS)
Alajouanine syndrome
alanine aminotransferase
ALARA
> as low as reasonably achievable
> ALARA principle

alar flaring
alarm
> Nite Train'r A.
> Sioux a.

Ala-Scalp Topical
alba
> linea a.
> pityriasis a.
> pneumonia a.

Albers-Schönberg syndrome
Albert Einstein Neonatal Developmental Scale (AENNS)
Albert-Smith pessary
albescens
> retinopathy punctata a.

Albini nodule
albinism
> cutaneous a.
> Forsius-Eriksson type ocular a.
> Nettleship-Falls ocular a.
> ocular a.
> oculocutaneous a.
> tyrosinase-negative oculocutaneous a.
> tyrosinase-positive oculocutaneous a.

albinism-deafness syndrome
albinismus circumscriptus
albipunctatus
> recessive fundus a.

albopapuloid variant
Albright
> A. disease
> A. hereditary osteodystrophy (AHO)
> A. osteodystrophy
> A. syndrome

albuginea
> tunica a.

albumin
> a. gradient
> plasma a.
> serum a.

albumin-globulin ratio (A/G ratio, A/G ratio)
albuminuria
Albunex
albuterol
ALCAPA
> anomalous left coronary artery from pulmonary artery
> ALCAPA syndrome

alclometasone
Alcock canal
alcohol
> a. abuse
> a. addiction
> blood a.
> ethyl a. (EtOH)
> a. ingestion
> nicotinyl a.
> a. related (AR)
> A. Use Disorders Identification Test (AUDIT)

alcoholica
> embryopathy a.

alcoholic embryopathy
alcoholism
> maternal a.

Alconefrin
Aldactone
Aldara
> A. cream

Alder anomaly
Alder-Reilly anomaly
Aldomet
Aldoril

aldosterone
    a. excretion rate (AER)
    a. replacement therapy
aldosteronism
    juvenile a.
aldosteronism-normal blood pressure syndrome
Aldred syndrome
Aldrich syndrome
Aldridge
    A. rectus fascia sling
    A. sling procedure
Aldridge-Studdefort urethral suspension
Ale-Calo syndrome
alendromide
alendronate
    a. sodium
Alesse
aleukia
    congenital a.
Aleve
Alexander
    A. disease
    A. operation
    A. syndrome
    A. unit
Alexander-Adams
    A.-A. hysteropexy
    A.-A. uterine suspension
alexandrite laser
alfa
    dornase a.
    follitropin a.
alfalfa
alfa-n3
alfentanil
Alferon N
algera
    akinesia a.
algesiometer
    vulvar a.
alginate wound dressing
AlgiSite dressing
alglucerase
algomenorrhea
Algo newborn hearing screener
Algosteril dressing
alimentation
    intravenous a.
    parenteral a.
Alimentum
    A. feeding

A-line
    arterial line
Alkaban-AQ
Alka-Butazolidin
alkalemia
alkaline phosphatase
alkaloid
    levorotatory a.
    plant a.
    Veratrum a.
    Vinca a.
alkalosis
    hypokalemic a.
    metabolic a.
    respiratory a.
Alka-Mints
alkaptonuria
Alka-Seltzer
Alkeran
alkylating
    a. agent
    a. chemotherapy
ALL
    acute lymphocytic leukemia
allantoic
    a. artery
    a. cyst
    a. duct
    a. sac
    a. stalk
allantoidoangiopagous
    a. twins
allantoidoangiopagus
allantois
Allegra
Allegra-D
allele
    fixed a.
    a. frequency
    multiple a.'s
    mutant a.
    a. polymerase chain reaction
    premutation a.
    recessive a.
    silent a.
    a. specific associated primer
    a. specific oligo
    wild-type a.
allelic
    a. exclusion
    a. gene
    a. heterogeneity

NOTES

allelism
Allemann syndrome
Allen
    A. and Capute neonatal
       neurodevelopmental examination
    A. fetal stethoscope
    A. laparoscopic stirrups
Allen-Doisy test
Allen-Herndon-Dudley syndrome (AHDS)
Allen-Herndon syndrome
Allen-Masters syndrome
Aller-Chlor Oral
Allerest
    A. 12 Hour Nasal Solution
Allergan Ear Drops
allergen
allergic dermatitis
allergy
    sperm a.
Allevyn dressing
all-fours
    a.-f. maneuver
    a.-f. maneuver for shoulder
      dystocia
alligator
    a. forceps
    a. skin
Allis
    A. clamp
    A. forceps
Allis-Abramson
    A.-A. breast biopsy
    A.-A. breast biopsy forceps
alloantibody
alloantigen
allodiploid
alloenzyme
allogenic, allogeneic
    a. antigen
    a. disease
    a. fetal graft
allogenicity
allograft
    fascia lata a.
    a. membrane
    a. survival
alloimmune
    a. disease
    a. factor
    a. mechanism
    a. neonatal thrombocytopenic
      purpura
    a. thrombocytopenia
    a. thrombocytopenia purpura
alloimmunity
alloimmunization
alloimmunize
alloisoleucine

alloploidy
allopolyploidy
all-or-none phenomenon
allosome
    paired a.
allosyndesis
allotetraploidy
allothreonine
allotropism
allotype
allowance
    recommended dietary a.
allozygote
Allport retractor
allylamine
ALMN
    adrenoleukomyeloneuropathy
alobar holoprosencephaly
Aloka
    A. 650 CL ultrasound
    A. OB/GYN ultrasound
    A. 650 scanner
    A. SD ultrasound system
    A. SSD-720 real-time scanner
alopecia
    a. adnata
    a., anosmia, deafness,
      hypogonadism syndrome
    a. areata
    a. congenitalis
    a., contracture, dwarfism (ACD)
    a., contracture, dwarfism, mental
      retardation syndrome
    a., contracture, dwarfism syndrome
    a., epilepsy, oligophrenia syndrome
    a. hereditaria
    a. mental retardation (AMR)
    a., mental retardation, epilepsy,
      microcephaly syndrome
    a. mental retardation syndrome
    a. universalis with mental
      retardation
Alora
    A. patch
    A. Transdermal
Alpers disease
alpha, α
    a.$_1$-acid glycoprotein levels
    a. antitrypsin level
    Curosurf poractant a.
    a. dimeric protein
    a. error
    estrogen receptor a. (Er alpha)
    follitropin a.
    a. helix
    a. interferon
    a. particle
    Prostin F2 a.

a. thalassemia
transforming growth factor a.
(TGFalpha)
alpha-1
a.-antitrypsin
a. PI
a. protease inhibitor
a. proteinase inhibitor
a. thymosin product
alpha-adrenergic
a.-a. agent
a.-a. blocker
a.-a. receptor
a.-a. stimulator
alpha-chain disorder
Alphaderm
5alpha-dihydroprogesterone
17-alpha-ethinyl testosterone
alpha-fetoprotein (AFP)
amniotic fluid a.-f. (AFAFP)
a.-f. elevation
a.-f. enzyme immunoassay (AFP-
EIA)
a.-f. level
maternal serum a.-f. (MSAFP)
alpha-glucosidase inhibitor
17alpha-hydroxypregnenolone
17alpha-hydroxyprogesterone
17α-h. caproate
alpha-L-fucosidase (FUCA)
a.-L.-f. deficiency
alpha-L-iduronidase (IDA)
a.-L.-i. deficiency
alpha-melanotrophin
alpha-methyl-para-tyrosine (AMPT)
Alphamine
AlphaNine SD
alphaprodine
alpha-recombinant interferon
Alpha-Tamoxifen
alpha-thalassemia
a.-t. intermedia
a.-t. minor
alpha-thalassemia/mental
a.-t. retardation deletion
a.-t. retardation syndrome, deletion
type
a.-t. retardation syndrome,
nondeletion type
alpha-thalassemia mental retardation
(ATR)
a.-t. m. r. syndrome

a.-t. m. r. syndrome deletion type
a.-t. m. r. syndrome nondeletion
type
X-linked a.-t. m. r. (ATRX)
Alphatrex Topical
Alport
A. syndrome
A. syndrome-like nephritis
alprazolam
AL-Rr Oral
Alstrom sign
ALTE
apparent life-threatening event
Altemeier
A. perineal rectosigmoidectomy
A. procedure
alteration
uterine activity a.
alternative
a. hypothesis
a. system
Alteromonas putrefaciens
altretamine
altruism
aludrine
aluminum acetate
Alupent
Alurate
aluteal
alveolar (A)
arterial to a. (a-A)
a. partial pressure (PA)
a. soft part sarcoma
alveolar-arterial (A-a)
a.-a. oxygen diffusing capacity
a.-a. pressure difference $(p(A-a)O_2)$
alveolus, pl. **alveoli**
pulmonary a.
alymphocytosis
alymphoplasia
Alzate catheter
Alzheimer disease (AD)
amalonaticus
Citrobacter a.
amantadine
amastia
amaurosis
a. congenita
a. fugax
Leber congenital a. (LCA)
recessive Leber congenital a.
amaurotic familial idiocy

NOTES

amazia
amazon thorax
ambenonium
Ambien
ambient
    a. oxygen concentration
ambiguity
    frank a.
    genital a.
    sexual a.
ambiguous genitalia
ambiguus
    situs a.
AmBisome
amboceptor
ambosexual area
Ambras syndrome
Ambu
    A. bag
    A. infant resuscitator
    A. respirator
ambulation
ambulatory
    a. antibiotic treatment
    a. care
    a. monitoring
    a. uterine contraction test
AMC
    arthrogryposis multiplex congenita
    ataxia-microcephaly-cataract
Amcill
amcinonide
amebiasis
amebic
    a. abscess
    a. colitis
    a. dysentery
    a. vaginitis
amelia
amelogenesis imperfecta
ameloonychohypohidrotic syndrome
amendment
    Clinical Laboratory Improvement A.
    (CLIA)
amenia
Amen Oral
amenorrhea
    athletic a.
    dietary a.
    emotional a.
    eugonadal a.
    eugonadotropic a.
    exercise-induced a.
    hypergonadotropic a.
    hyperprolactinemic a.
    hypogonadotropic a.
    hypophysial a.
    hypothalamic a.

    jogger's a.
    lactation a.
    ovarian a.
    pathologic a.
    physiologic a.
    postmenopausal a.
    postpartum a.
    postpill a.
    primary a.
    secondary a.
    traumatic a.
amenorrhea-galactorrhea syndrome
amenorrheic, amenorrheal
    a. patient
    a. woman
amentia
    nevoid a.
Americaine
American
    A. College of Nurse Midwives
    A. College of Obstetrics &
        Gynecology network
    A. Diabetes Association (ADA)
    A. Fertility Society (AFS)
    A. Society for Reproductive
        Medicine
    A. Urogynecologic Society
amethopterin
ametria
AMF
    autocrine motility factor
AMH
    antimüllerian hormone
Amicar
Amiel-Tison
    A.-T. score
    A.-T. test
amikacin
    a. sulfate
Amikin
amiloride
amine odor
amino
    a. acid
    a. acid analyzer
    a. acid metabolism
aminoacidemia
aminoacidopathy
aminoaciduria
aminoaciduriasis
aminocaproic acid
Amino-Cerv pH 5.5 cervical Cream
aminoglutethimide (AGT)
aminoglycoside
Amino-Opti-E
aminopenicillin
aminopeptidase
    leucine a.

aminophylline
aminoprine
aminopterin
    aminopterin syndrome sine a.
    (ASSAS)
    aminopterin syndrome without a.
    a. embryopathy syndrome
aminopterin-like embryopathy syndrome
aminopyrine
Aminosyn-PF supplement
amino-terminal peptide
aminotransferase
    alanine a.
    aspartate a. (AST)
    tyrosine a. (TAT)
amiodarone
Amipaque
AMIS
    antibody-mediated immune suppression
Amish brittle hair syndrome
Amitid
Amitone
Amitril
amitriptyline
Amko vaginal speculum
ammonemia, ammoniemia
Ammon fissure
ammoniac
    sal a.
ammoniemia (*var. of* ammonemia)
ammonium
    a. bromide
    a. chloride
AMN
    adrenomyeloneuropathy
amnestic response
Amni-Glove N Gel kit
amniocentesis
    early a. (EA)
    genetic a.
amniochorion
amniocyte
amniogenic cell
amniogram
amniography
    a. in hydatidiform mole
Amniohook
amnioinfusion
    transabdominal a.
amnioma
amnion
    a. nodosum

    a. ring
    a. rupture
amnion-chorion separation
amnionic
    a. caruncle
amnionicity
amnionitis
    silent a.
amniopatch
amniorrhea
amniorrhexis
amnioscope
amnioscopy
Amniostat fetal lung maturity screening
Amniostat-FLM test
amniotic
    a. adhesion
    a. band
    a. band anomalad
    a. banding syndrome
    a. band limb amputation
    a. band sequence
    a. band syndrome
    a. cavity
    a. fluid (AF)
    a. fluid alpha-fetoprotein (AFAFP)
    a. fluid bilirubin
    a. fluid cell culture
    a. fluid embolism (AFE)
    a. fluid embolism syndrome
    a. fluid embolus
    a. fluid embolus syndrome
    a. fluid fluorescence polarization
    a. fluid index (AFI)
    a. fluid level
    a. fluid pocket
    a. fluid quantitation
    a. fluid volume (AFV)
    a. fluid volume disorder
    a. fluid white blood cell count
    a. infection syndrome
    a. infection syndrome of Blane
    a. sac
    a. sheet
amniotome
    Baylor a.
amniotomy
    a. plus oxytocin method
amobarbital
A-mode ultrasound
amorphous
    a. breast calcification

NOTES

amorphous *(continued)*
    a. fetus
    a. inspissation
amorphus
    fetus a.
amotio placentae
amoxapine
amoxicillin
Amoxil
AMP
    adenosine monophosphate
    assisted medical procreation
AmpErase electrocautery
amphetamine
amphidiploidy
amphigonous inheritance
amphimixis
Amphotec
amphotericin B
    a. B
ampicillin
    a. sodium/sulbactam sodium
    a. trihydrate
Ampicin
Amplicor
    A. Chlamydia Assay
    A. HIV-1 test kit
    A. PCR diagnostics
    A. PCR kit
    A. typing kit
amplification
    DNA a.
    gene a.
    Y-specific DNA a.
amplified fragment length polymorphism
AmpliTaq DNA polymerase FS
amplitude
    oscillation a.
amprenavir
AMPT
    alpha-methyl-para-tyrosine
ampulla of oviduct
ampullar
    a. abortion
    a. pregnancy
amputation
    amniotic band limb a.
    birth a.
    cervical a.
    congenital a.
    intrauterine a.
    Jabouley a.
    spontaneous a.
AMR
    alopecia mental retardation
Amreich vaginal extirpation
amrinone lactate

AMS
    acute mountain sickness
    AMS 800 artificial urethral
        sphincter
Amsacrine (mAMSA)
Amsel criteria
amstelodamensis
    status degenerativus a.
    typus a.
Amsterdam
    A. dwarfism
    A. infant ventilator
    A. type
amyelencephalia
amyelencephalic
amyelencephalous
amyelia
amyeloidosis
amyelous
amygdalin
amylase
    serum a.
amylase-creatinine clearance ratio
amyl nitrite
amylobarbitone
amyloidosis
    paraneoplastic a.
amylophagia
amyoplasia
    a. congenita
    oculomelic a.
amyotonia congenita
Amytal
ANA
    antinuclear antibody
anabolic steroid
Anabolin
anacatadidymus *(var. of* anakatadidymus*)*
anadidymus
anadysplasia
    metaphyseal a.
anaerobe
anaerobic
    a. vaginosis
anaerobius
    Peptococcus a.
Anafranil
anagen
anagenesis
anakatadidymus, anacatadidymus
anal
    a. atresia
    a. canal
    a. EMG PerryMeter sensor
    a. incontinence
    a. intraepithelial neoplasia (AIN)
    a. Pap smear
    a. rcflcx

a. sphincter
a. sphincter cinedefecography
a. sphincter disruption
a. sphincter dysplasia
a. sphincter electromyography
a. sphincter laceration
a. squamous intraepithelial lesion (ASIL)
a. wink
anal-ear-renal radial malfunction syndrome
analgesia
  caudal a.
  conduction a.
  congenital a.
  continuous epidural a.
  epidural a.
  narcotic a.
  patient-controlled a. (PCA)
  peridural a.
  perineal a.
  regional a.
  segmental epidural a.
  spinal a.
analgesic
  narcotic a.
analgesic-rebound headache
analogue
  gonadotropin-releasing hormone a.
  luteinizing hormone-releasing hormone a.
  oxytocin a.
  prostaglandin E a.
  tetracycline a.
analysis, pl. **analyses**
  automated multiple a.
  base sequence a.
  blood chromosome a.
  bulked segregant a.
  cell block a.
  chromosomal a.
  chromosome a.
  clinicopathological a.
  computer-assisted semen a. (CASA)
  cytogenetic a.
  endonuclease a.
  fluorescence depolarization a.
  genetic bit a.
  genetic linkage a.
  heteroduplex a.
  karyotype a.
  linkage a.

molecular genetic a.
multivariant a.
postoperative symptom a.
restriction endonuclease a.
saturation a.
semen a.
seminal fluid a. (SFA)
sequential multiple a. (SMA)
analyte
  serum a.
analyzer
  amino acid a.
  AVL 9110 pH a.
  BRACAnalyzer gene a.
  Clinitek 50 Urine Chemistry A.
  computer-assisted semen a. (CASA)
  Coulter Channelyser cell a.
  HemoCue blood glucose a.
  HemoCue blood hemoglobin a.
  Osteomeasure computer-assisted image a.
  Serono SR1 FSH a.
  Sonoclot coagulation a.
  SRI automated immunoassay a.
anamnestic response
Anandron
anaphase
  a. I, II
  a. lag
anaphylactic shock
anaphylactoid
  a. purpura
  a. syndrome of pregnancy
anaphylaxis
  acute a.
  recurrent a.
anaplastic carcinoma
Anaprox
  A.-DS
anasarca
  fetal a.
  fetoplacental a.
anaspadias
Anaspaz
anastomosis, pl. **anastomoses**
  circular end-to-end a. (CEEA)
  Clado a.
  colorectal a.
  cornual a.
  gastrointestinal a. (GIA)
  isthmointerstitial a.
  microsurgical tubocornual a.

NOTES

anastomosis *(continued)*
    onlay patch a.
    placental vascular a.
    tubocornual a.
    ureterotubal a.
    ureteroureteral a.
anastrozole
anatomic conjugate
anatomy
    fetal intracranial a.
    immune system a.
    intracranial a.
Anavar
Ancef
ancestor
    leading a.
anchor
    Mainstay urologic soft tissue a.
    Mitex GII/mini a.
Ancobon
Ancotil
ancylostomiasis
Andermann syndrome
Andernach ossicle
Andersen
    A. disease
    A. syndrome
Anderson
    A. disease
    A. marker
    A. syndrome
Anderson-Fabry disease
Andogsky syndrome
Andrews infant laryngoscope
Andro
androblastoma
Androcur
    A. Depot
Andro-Cyp
Andro/Fem
androgen
    adrenal a.
    a. antagonist
    a. dynamics
    excess a.
    a. excess
    a. insensitivity syndrome
    a. interaction
    a. metabolism
    a. receptor
    a. resistance
    a. resistance syndrome
    a. secretion
androgen-dependent carcinoma
androgenesis
androgenic
androgenicity
androgenized woman

androgenous
androgen-producing tumor
androgynism
androgynous
androgyny
android
    a. obesity
    a. pelvis
Android-F
Andro-L.A.
andrology
Androlone
    A.-D
Andronate
Andropository-200
androstane
    5α-a.-3α,17β-diol glucuronide (3α-diol-G)
3α-androstanediol glucuronide
androstenedione (A₄)
androsterone
    a. glucuronide
Androvite
anechoic
    a. space
    a. tissue
anectasis
anejaculation
anembryonic
    a. gestation
    a. pregnancy
anemia
    acquired hemolytic a.
    addisonian pernicious a.
    angiopathic hemolytic a.
    aplastic a.
    autoimmune acquired hemolytic a.
    autoimmune hemolytic a. (AIHA)
    Benjamin a.
    Blackfan-Diamond a.
    blood-loss a.
    congenital hypoplastic a.
    congenital nonregenerative a.
    congenital nonspherocytic hemolytic a.
    Cooley a.
    crescent cell a.
    Czerny a.
    Diamond-Blackfan congenital hypoplastic a.
    Diamond-Blackfan juvenile pernicious a.
    drug-induced hemolytic a.
    elliptocytic a.
    erythroblastic a.
    familial erythroblastic a.
    Fanconi a.
    fetal a.

globe cell a.
Heinz body a.
Heinz-body hemolytic a.
hemolytic a.
hereditary nonspherocytic a.
Herrick a.
a. hypochromica sideroachrestica
  hereditaria
hypoplastic a.
iron deficiency a. (IDA)
Jaksch a.
juvenile pernicious a.
Larzel a.
macrocytic a. of pregnancy
Mediterranean a.
megaloblastic a.
microangiopathic hemolytic a.
microcytic a.
a. neonatorum
ovalocytary a.
pernicious a. (PA)
PGK hereditary nonspherocytic a.
phosphoglycerate kinase deficiency
  hereditary nonspherocytic a.
physiologic a.
pregnancy-associated hypoplastic a.
a. of prematurity
a. pseudoleukemica infantum
pyridoxine-responsive a.
Runeberg a.
sickle cell a.
sideroblastic a.
Von Jaksch a.

anemic
a. effect

anemicus
nevus a.

anencephalic, anencephalous
anencephaly, anencephalia
anergy
Anestacon
anesthesia
caudal a.
conduction a.
epidural a.
extradural a.
general endotracheal a.
inhalation a.
local a.
lumbar epidural a.
mask inhalation a.
maternal a.

obstetric a.
office laparoscopy under local a.
  (OLULA)
paracervical a.
peridural a.
perineal a.
pudendal a.
regional a.
saddle block a.
spinal a.

anesthesiologist
anesthetic
eutectic mixture of local a.'s
  (EMLA)
gas a.
a. gas exposure
local a.
volatile a.
walking epidural a.

anesthetist
Certified Registered Nurse A.
  (CRNA)

anestrous
aneugamy
aneuploid
a. abortion

aneuploidy
atypical a.
a. infant
mosaic a.
Pallister mosaic a.
recurrent a.
XXXXY a.

aneurysm
aortic a.
arterial a.
dissecting aortic a.
intracranial a.
intrauterine cirsoid a.
ruptured cerebral a.
saccular a.
splenic artery a.

AneuVysion Assay prenatal genetic test
ANF
atrial natriuretic factor

Angelman syndrome
angel-shaped phalangoepiphyseal
  dysplasia (ASPED)
angel's kisses
Anger Expression Scale
angiitis
angina

NOTES

angiocardiogram
Elema a.
angioedema
angiogenesis
placental adaptive a.
angiogenic growth factor
angiographic embolization
angiography
pulmonary a.
angiokeratoma
a. circumscriptum
Mibelli a.
a. of Mibelli
vulvar a.
angiolysis
angioma
a. capillare et venosum calcificans
cerebral a.
cutaneocerebral a.
dural spinal a.
intradural spinal a.
retinal a.
spider a.
spinal a.
AngioMark
angiomatoid tumor
angiomatosis
bacillary a.
cerebrocutaneous a.
cutaneomeningospinal a.
encephalocraniofacial a.
encephalofacial a.
a. encephalofacialis
encephalotrigeminal a.
meningeal capillary a.
meningoculofacialis a.
meningooculofacial a.
a. meningoulofacialis
neurooculocutaneous a.
Sturge-Weber a.
angiomatosis, oculoorbito, thalamo-
encephalic syndrome
angiomyofibroblastoma
angiomyolipoma rupture
angiomyoma of oviduct
angiomyxoma
aggressive a. (AAM)
angioneurotic edema
angioosteohypertrophy
a. syndrome
angiopathic hemolytic anemia
angiopathy
vulvar congenital dysplastic a.
angioplasty
percutaneous transluminal a. (PTA)
angiosarcoma
uterine a.
angiosonography

angiotensin
a. II
angiotensin-converting enzyme (ACE)
angiotensinogen
angle
urethral a.
urethrovesical a.
anhedonia
anhidrotic ectodermal dysplasia
anhydremia
anhydrohydroxyprogesterone
Anhydron
anhydrous
betaine a.
a. magnesium sulfate
ani (*pl. of* anus)
anideus
embryonic a.
anileridine
anion gap
aniridia
a., ambiguous genitalia, mental
retardation (AGR)
a., ambiguous genitalia, mental
retardation triad syndrome
a., cerebellar ataxia-oligophrenia
syndrome
a., genitourinary abnormalities,
mental retardation triad
a., Wilms tumor association
syndrome
a. Wilms tumor, gonadoblastoma
syndrome
anisindione
anisocoria
anisodactyly
anisomastia
anisomelia
anisotropine
ankle clonus
ankyloblepharon
a., ectodermal defects, cleft
lip/palate (AEC)
a., ectodermal dysplasia, clefting
a., ectodermal dysplasia, clefting
syndrome
ankylocheilia
ankylocolpos
ankylodactyly
ankyloglossia
ankyloglosson superius syndrome
ankyloproctia
ankylosing spondylitis
anlage, pl. **anlagen**
anlagen of the auditory ossicle
anneal
annexectomy
annexin V

annexitis
annexopexy
annular-array transducer
annulare
    granuloma a.
annulati
    pili a.
ano
    fissure in a.
anococcygeal raphe
anogenital wart
anomalad
    abdominal muscle deficiency a.
    amniotic band a.
    facioauriculovertebral a.
    holoprosencephaly a.
    Robin a.
    Sturge-Weber a.
anomalous
    a. left coronary artery from
        pulmonary artery (ALCAPA)
    a. left coronary artery from
        pulmonary artery syndrome
    a. pulmonary venous connection
    a. uterus
anomaly
    Alder a.
    Alder-Reilly a.
    aortic arch a.
    arthrogryposis-like hand a.
    Axenfeld-Rieger a.
    birth a.
    body stalk a.
    branchial a.
    cardiac a.
    cervical a.
    Chédiak-Higashi a.
    Chiari a.
    chromosomal a.
    coloboma, heart anomaly, choanal
        atresia, retardation, and genital
        and ear a.'s (CHARGE)
    congenital a.
    DiGeorge a.
    Duane a.
    Ebstein a.
    fetal vascular a.
    genetic a.
    intracranial dural vascular a.
    iridogoniodysgenesis with
        somatic a.'s
    May-Hegglin a.

    microcephaly-cervical spine
        fusion a.'s
    microphthalmia or anophthalmos
        with associated a.'s (MAA)
    Möbius a.
    müllerian duct a.
    multiple congenital a.'s
    Nager a.
    orthopaedic a.
    Pelger-Huet a.
    Peters a.
    Poland a.
    Rieger a.
    sex chromosomal a.
    Shone a.
    spondylar changes-nasal a.-striated-
        metaphyses (SPONASTRIME)
    Sprengel a.
    Uhl a.
    umbilical cord a.
    Undritz a.
    urogenital a.
    uterine a.
    VACTERL a.
    vaginal a.
    vascular a.
    vertebral, anal, cardiac, tracheal,
        esophageal, renal, limb a.
    X-linked mental-retardation-bilateral
        clasp thumb a.
    X-linked mental retardation/multiple
        congenital a. (XLMR/MCA)
anonychia-ectrodactyly
anonychia-onychodystrophy
anonymous donor sperm (ADS)
anophthalmia
    a., hand-foot defects-mental
        retardation syndrome
    a.-Waardenburg syndrome
anophthalmos
    a.-limb anomalies syndrome
    a.-syndactyly syndrome
anorchia
anorchism
anorectal
    a. incontinence
    a. stenosis
anorectic
    a. reaction
Anorex
anorexia
    a. nervosa

NOTES

27

anorgasmia
anorgasmic
anosmia
    congenital a.
anosteogenesis
anotia
anovarianism
anovular
    a. menstruation
anovulation
    hyperandrogenic a. (HA)
    hyperandrogenic chronic a.
    persistent a.
anovulatory
    a. bleeding
    a. infertility
    a. patient
anoxia
    fetal a.
    a. neonatorum
anoxic-ischemic encephalopathy
ANP
    atrial natriuretic peptide
Ansaid
    A. Oral
Ansaldo AU560 ultrasound
Anspor
Answer
    A. Plus
Antabuse
antacid
Antagon
    A. injection
antagonism
antagonist
    androgen a.
    calcium a.
    estrogen a.
    folic acid a.
    follicle-stimulating hormone a.
    gonadotropin-releasing hormone a.
    a.-induced gonadotropin deprivation
    narcotic a.
    opioid receptor a.
    progesterone a.
antecedent
    cerebral palsy a.
    plasma thromboplastin a. (PTA)
anteflex
anteflexed
    anteverted and a. (AV/AF)
anteflexion
antenatal
    a. anti-D immunoglobulin
    a. corticosteroid
    a. corticosteroid therapy
    a. diagnosis
    a. fetofetal transfusion

    a. morbidity
    a. patient
    a. phenobarbital treatment
    a. screening
    a. thyrotropin releasing hormone
    a. treatment
    a. ultrasound
antepartum
    a. asphyxia
    a. bleeding
    a. care
    a. fetal BPP
    a. fetal CST
    a. fetal NST
    a. fetal surveillance
    a. hemorrhage
    a. monitor
    a. pyelonephritis
    a. Rh isoimmunization
anteposition
anterior
    a. apical vault defect
    a. asynclitism
    a. cerebral artery pulsatility index
     (ACAPI)
    a. chamber cleavage syndrome
    a. commissure
    dysgenesis mesostromalis a.
    a. enterocele
    a. fontanel
    a. lie
    a. lip of the cervix
    a. neural tube defect
    occiput a. (OA)
    a. pelvic exenteration
    a. pituitary-like hormone
    a. and posterior (A&P)
    a. and posterior repair
    a. retrosternal hernia of Morgagni
    a. urethritis
    a. vagina
    a. vaginotomy
anterolateral
    a. fontanel
anteroposterior (AP)
    a. diameter of the pelvic inlet
anteversion
anteverted
    a. and anteflexed (AV/AF)
    a. nostril
    a. pinna
anthelix (*var. of* antihelix)
anthelmintic agent
anthrax
    cutaneous a.
    pulmonary a.
anthropoid pelvis
anthropometric measurement

antiandrogen
  a. receptor blocker
antiannexin V antibody
antiarrhythmic
antibacterial therapy
AntibiOtic
  A. Otic
antibiotic
  beta-lactam a.
  beta-lactamase-resistant
    antistaphylococcal a.
  beta-lactamase-resistant
    antistaphylococcal a.
  prophylactic a.
  a. prophylaxis
  a. resistance
antibody
  antiannexin V a.
  anticardiolipin a. (aCL)
  anticardiolipin a.-positive
  anticentromere a. (ACA)
  antiferritin a.
  antigliadin a.
  anti-HIV a.
  antiidiotype a.
  anti-Kell a.
  anti-La a.
  anti-Lewis a.
  anti-M a.
  antimitochondrial a.
  antinuclear a. (ANA)
  antipaternal antileukocytotoxic a.
  antiphospholipid a. (aPL)
  antiplatelet immunoglobulin G a.
  anti-Ro a.
  antisperm a.
  anti-SSA a.
  blood group a.
  circulating platelet a.
  conjugated a.
  cytophilic a.
  fluorescent treponemal a. (FTA)
  Frei a.
  genus-specific monoclonal a.
  hemagglutinating inhibition a. (HIA)
  hepatitis B core a. (HBcAb)
  hepatitis Be a. (HBeAb)
  hepatitis Bs a. (HBsAb)
  HPV type 16 capsid a.
  humoral a.
  IgA a.
  IgD a.

  IgE a.
  IgG a.
  IgM a.
  Kell a.'s
  Ki67 a.
  Kveim a.
  link a.
  lupus anticoagulant a.
  maternal antiplatelet a.
  maternal sperm a.
  monoclonal a.
  natural a.
  neurofilament a.
  OncoScint CR103 monoclonal a.
  ovarian a.
  phospholipid a.
  platelet-associated a.
  polyclonal a.
  polyclonal-monoclonal a.
  a. reaction site
  a. response
  Rh a.
  rhesus a.
  Rh negative a.
  Rh positive a.
  S-100 a.
  a. screening
  serum a.
  species-specific a.
  sperm surface a.
  tissue-specific a.
  Treponema pallidum a.
  xenogeneic a.
antibody-mediated immune suppression
  (AMIS)
anticancer agent
anticardiolipin
  a. antibody (aCL)
  a. antibody-positive
anticentromere antibody (ACA)
anticholinergic
  a. drug
anticholinesterase
anticipation
  evidence of a.
anticoagulant
  lupus a. (LAC)
  a. therapy
anticoagulation
anticodon
anticonvulsant
  a. drug

NOTES

29

anti-D
    a.-D immune globulin
    a.-D immunoglobulin
antide
antidepressant
    a. therapy
    tricyclic a. (TCA)
antidiuretic hormone (ADH)
antiembolism stocking
antiemetic therapy
antiestrogen
antiestrogenic effect
antiferritin antibody
antifibrinolytic agent
antifolic agent
antifungal
    Absorbine A.
    Absorbine Jr. A.
    Breezee Mist A.
    a. drug therapy
antigalactagogue
antigalactic
antigen
    ABO a.
    a. activity
    allogenic a.
    antiproliferating cell nuclear a.
      (anti-PCNA)
    Australia a.
    a. binding site
    CA 125 a.
    CA 19-9 a.
    cancer a. (CA)
    carcinoembryonic a. (CEA)
    carcinoembryonic a. 125 (CEA 125)
    carcinoma a.
    cell surface a.
    a. determinant
    direct fluorescent a. (DFA)
    epithelial membrane a. (EMA)
    fluorescent antibody against
      membrane a. (FAMA)
    Forssman a.
    Goa a.
    hepatitis B a. (HBAg)
    hepatitis B surface a. (HBsAg)
    heterophil a.
    histocompatibility locus a.
    human leukocyte a. (HLA)
    H-Y a.
    incompatible blood group a.
    M a.
    major histocompatibility a.
    melanoma specific a.
    MHC a.
    nuclear a.
    oncofetal a.
    ovarian carcinoma a.

    p24 a.
    pancreatic oncofetal a. (POA)
    platelet a.
    polysaccharide group-specific a.
    prostate-specific a. (PSA)
    red cell a.
    Rh a.
    rhesus a.
    a. screen
    sialyl Tn a.
    sialyted Lewis A a.
    surface a.
    surface a. subtype ayw1
    surface a. subtype ayw2
    surface a. subtype ayw3
    surface a. subtype ayw4
    T6 a.
    thymic lymphocyte a. (TL)
    a. tolerance
    transplantation a.
    Treponema pallidum a.
    tumor a.
    tumor-associated a. (TAA)
    tumor-specific transplantation a.
      (TSTA)
    von Willebrand factor a.
    vWF a.
antigen-antibody complex
antigenic
    a. modulation
    a. paralysis
antigenicity
    tumor a.
antigen-sensitive cell
antigliadin antibody
antiglobulin test
antiglomerular basement membrane
    antibody disease
antigonadotropin
anti-HBc
antihelix, anthelix
antihelminthic therapy
antihemophilic factor
Antihist-1
antihistamine
Antihist-D
anti-HIV antibody
antihypertensive
    a. drug
    a. therapy
antiicteric
antiidiotype antibody
antiincontinence
antiinhibitor coagulant complex
antiinsulin
anti-Kell antibody
anti-La antibody
anti-Lewis antibody

antiluteogenic
antimalarial drug
anti-M antibody
antimesenteric surface
antimetabolite
antimicrobial
    beta-lactam a.
    a. therapy
antimicrosomal
Antiminth
antimitochondrial antibody
antimongolism
antimongoloid
antimüllerian hormone (AMH)
antineoplastic
    a. agent
    a. drug
antinuclear antibody (ANA)
antioxidant
    chain-breaking a.
    preventive a.
antiparasitic drug therapy
antipaternal antileukocytotoxic antibody
anti-PCNA
    antiproliferating cell nuclear antigen
antiphospholipid
    a. antibody (aPL)
    a. syndrome (APS)
antiplatelet
    a. agent
    a. immunoglobulin G antibody
Antipress
antiprogesterone
antiprogestin zk 137316
antiprogestogen
antiproliferating cell nuclear antigen
    (anti-PCNA)
antiprostaglandin
    a. agent
antipsychotic drug
antipyretic
antipyrine
antireceptor
antiretroviral therapy
anti-Rh gamma globulin
anti-RHO-D titer
anti-Ro antibody
antisense
    a. oligonucleotide
    a. strand
antiserum, pl. antisera
    SB-6 a.

antishock trousers
antisialagogue
antisocial personality disorder (ASPD)
antispasmodic
antisperm antibody
anti-SSA antibody
antistreptolysin titer
antithrombin
    a. 3 (AT3)
    a. II
antithymocyte globulin
antithyroglobulin
antithyroid
    a. drug
    a. drug therapy
antitreponemal test
Anti-Tuss
antivesicoureterel reflux surgery
antiviral therapy
Antley-Bixler syndrome
Antocin
antral
    a. follicle
    a. stenosis
antrum, pl. antra
anucleate fragment
annular, anular
    a. band
    a. placenta
    a. tubule
anuria
anuric
anus, pl. ani
    ectopic a.
    imperforate a.
    levator ani
    Paget disease of a.
    patent a.
    pruritus ani
    a. of Rusconi
    vaginal ectopic a.
    vestibular a.
    vulvovaginal a.
anus-hand-ear syndrome
Anusol
    A. HC-1 Topical
    A. HC-2.5% Topical
AO
    arthroophthalmopathy
AODM
    adult-onset diabetes mellitus

NOTES

AOM
>acute otitis media
>arthroophthalmopathy

AOPA
>Ara-C, Oncovin, prednisone,
>asparaginase

AOPE
>Adriamycin, Oncovin, prednisone,
>etoposide

aorta, pl. **aortae**
>coarctation of a.
>descending a.
>fetal a.
>hypoplastic a. (HA)
>overriding a.

aortic
>a. aneurysm
>a. arch anomaly
>a. arch anomaly-peculiar facies
> mental retardation syndrome
>a. blood flow velocity waveform
>a. bruit
>a. laceration
>a. node
>a. node metastasis
>a. regurgitation
>a. root diameter
>a. stenosis
>a. stenosis, corneal clouding,
> growth and mental retardation
> syndrome
>a. valve disease
>a. valve insufficiency

aorticopulmonary (*var. of*
>aortopulmonary)

aortitis

aortocardiotocograph

aortography

aortopulmonary, aorticopulmonary
>a. septum
>a. shunt
>a. window

AP
>anteroposterior

A&P
>anterior and posterior
>A&P repair

Apak syndrome

apareunia

APC
>activated protein C

APECED
>autoimmune polyendocrinopathy,
>candidiasis, ectodermal dystrophy

ape hand

Apert
>A. disease
>A. syndrome

Apert-Crouzon
>A.-C. disease
>A.-C. syndrome

apex, pl. **apices**
>a. of vagina
>vaginal a.

Apgar
>A. rating
>A. scale
>A. score
>A. timer

aphasia
>global a.
>Wernicke a.

aphtha, pl. **aphthae**
>Bednar aphtha

aphthosis
>perianal a.

aphthous
>a. stomatitis
>a. ulcer

APIB
>Assessment of Preterm Infants Behavior

apices (*pl. of* apex)

A.P.L.

aPL
>antiphospholipid antibody

aplasia
>a. axialis extracorticalis congenita
>cerebellar vermis a.
>cutis a.
>a. cutis
>a. cutis congenita
>extracortical axial a.
>gonadal a.
>heminasal a.
>hereditary retinal a.
>Leydig cell a.
>nuclear a.
>ovarian a.
>retinal a.
>thymic a.
>thymic-parathyroid a.
>vas deferens a.

aplastic
>a. abdominal muscle syndrome
>a. anemia
>a. crisis
>a. leukemia
>a. pancytopenia
>a. patella

apnea
>a. alarm mattress
>a. and bradycardia (A&B)
>infantile sleep a.
>initial a.
>late a.
>a. monitor

a. neonatorum
a. of prematurity
reflexic a.
sleep a.
vasovagal reflex a.
apneustic
Apo-
A.-Amoxi
A.-Ampi
A.-Bromocriptine
A.-Cephalex
A.-Cimetidine
A.-Cloxi
A.-Diazepam
A.-Diclo
A.-Diflunisal
A.-Doxy
A.-Doxy Tabs
A.-Erythro E-C
A.Naproxen
A.-Pen VK
A.-Piroxicam
A.-Ranitidine
A.-Sulfamethoxazole
A.-Sulfatrim
A.-Tamox
A.-Terfenadine
A.-Tetra
A.-Zidovudine
apocrine
a. adenoma
a. cyst
a. gland
a. metaplasia
a. miliaria
apodia
apogamy, apomixis
Apogee 800 ultrasound system
apolipoprotein
apomixis (*var. of* apogamy)
apoplectic
apoplexy
parturient a.
uteroplacental a.
apoprotein
a. A
Apoptag Plus kit
apoptosis
neutrophil a.
spontaneous a.
apoptotic body
apparatus, pl. **apparatus**

Barcroft/Haldane a.
Heyns abdominal decompression a.
apparent life-threatening event (ALTE)
appearance
acromegaloid facial a. (AFA)
bull's eye sonographic a.
cushingoid a.
ground-glass a.
meconium ileus a.
peau d'orange a.
snowstorm a.
strawberry a.
Appelt-Gerkin-Lenz syndrome
appendectomy
appendiceal fecalith
appendices (*pl. of* appendix)
appendicitis
suppurative a.
appendicolith
appendicovesicotomy
Mitrofanoff a.
appendix, pl. **appendices**
Apple Medical bipolar forceps
apple-peel atresia
application
bioelectromagnetic a.
nonionizing nonthermal a.
silicone band a.
spring clip a.
topical iodine a.
applicator
Absolok endoscopic clip a.
afterload a.
Bloedorn a.
cesium a.
cotton-tipped a.
Ernst radium a.
Falope ring a.
Filshie clip a.
Fletcher-Suit a.
radioactive a.
ring a.
Ter-Pogossian cervical a.
applier
LDS clip a.
apposition
a. of skull suture
approach
abdominal a.
transrectal a.
appropriate for gestational age (AGA)

NOTES

APPT
Adolescent and Pediatric Pain Tool
apraxia-ataxia-mental deficiency syndrome
apraxia-oculomotor contracture-muscle atrophy syndrome
Apresoline
aprobarbital
aproctia
Apro-Flurbiprofen
apron
Hottentot a.
perineal surgical a.
pudendal a.
aprosencephaly-atelencephaly syndrome
aprosencephaly syndrome
aprosopia
aprotinin
APS
antiphospholipid syndrome
APTT
activated partial thromboplastin time
Apt test
AQ
Nasacort AQ
Aquacel dressing
Aquachloral
A. Supprettes
Aquacort
Aquaflex ultrasound gel pad
Aquagel lubricating gel
AquaMEPHYTON
A. Injection
Aquaphor gauze
Aquaphyllin
AquaSens FMS 1000 Fluid Monitoring System
Aquasol
A. A
A. E
Aquasonic 100 ultrasound transmission gel
Aquasorb dressing
Aquaspirillum itersonii
Aquatensen
aqueductal
a. stenosis
aqueduct of Sylvius
aqueous
a. crystalline penicillin G
a. penicillin G
a. penicillin sodium
a. phase
Aquest
AR
alcohol related
autosomal recessive

arabinoside
adenine a.
cytosine a.
arabinosylcytosine
arachidic bronchitis
arachidonic
a. acid
a. acid level
a. acid metabolite
arachnidism
arachnodactyly
congenital contractural a. (CCA)
arachnoid cyst
arachnoiditis
Ara-C, Oncovin, prednisone, asparaginase (AOPA)
Aralen
Aramine
Aran-Duchenne disease
arbitrarily
a. primed polymerase chain reaction
a. primer
arborization
pulmonary a.
vaginal fluid a.
arbor vitae
arbovirus
ARC
AIDS-related complex
arcade
mitral a.
arch
branchial a.
narrow pubic a.
neural a.
pubic a.
tendinous a.
Zimmermann a.
archencephalon
archenteron
archenteronoma
archiblast
archigastrula
architectural disturbance
architecture
histologic a.
pelvic a.
ARCS
azoospermia, renal anomaly, cervicothoracic spine dysplasia
ARCS syndrome
arcuate
a. artery
a. ligament of pubis
a. nucleus
a. uterus

arcus
    a. tendineus
    a. tendineus fasciae pelvis
ARDS
    adult respiratory distress syndrome
area
    ambosexual a.
    body surface a. (BSA)
    a. of cardiac dullness (ACD)
    a. of interest magnification (AIM)
    skip a.
    subpanicular a.
areata
    alopecia a.
areflexia
Arenavirus
areola, pl. **areolae**
    nevoid hyperkeratosis of the nipple
      and a.
    a. umbilicus
areolar
Arey rule
ARF
    acute rheumatic fever
ArF excimer laser
Arfonad
Argesic-SA
arg-gly-asp
    arginine-glycine-aspartic acid
arginase deficiency
arginine
    a. glutamate
    a. hydrochloride
    a. tolerance test (ATT)
    a. vasopressin (AVP)
    a. vasotocin
arginine-glycine-aspartic acid (arg-gly-asp)
argininemia
argininosuccinicacidemia
argininosuccinicaciduria
arginosuccinic acid
argon
    a. beam coagulation
    a. beam coagulator (ABC)
    a. laser
Argonz-Del Castillo syndrome
Argyle arterial catheter
argyrophilic granule
arhinencephaly, arrhinencephalia, arrhinencephaly
arhinia, arrhinia

Arias-Stella
    A.-S. effect
    A.-S. phenomenon
    A.-S. reaction
ARIC
    acrosome reaction with ionophore
      challenge
Aries-Pitanguy procedure
Arimidex
Aristocort
    A. A Topical
    A. Topical
Arkawa syndrome 1, 2
Arkless-Graham syndrome
Arlidin
ARM
    artificial rupture of membranes
arm
    a. board
    chromosome a.
    a. of chromosome
    nuchal a.
    a. position
    a. presentation
    a. recoil
    sling a.
armamentarium
Army-Navy retractor
Arnold-Chiari
    A.-C. deformity
    A.-C. malformation
    A.-C. syndrome
Arnoux sign
AROM
    active range of motion
    artificial rupture of membranes
Aromasin
aromatase
    a. inhibitor
aromatization
array
    superficial linear a. (SLA)
arrayed library
arrest
    active phase a.
    cardiac a.
    circulatory a.
    deep transverse a.
    a. disorder
    follicular development a.
    a. of labor
    preterm labor a.

NOTES

arrest *(continued)*
    respiratory a.
    sinus a.
    transverse a.
arrested development
Arrestin
arrhenoblastoma
arrhinencephalia *(var. of* arhinencephaly)
arrhinencephaly *(var. of* arhinencephaly)
arrhinia *(var. of* arhinia)
arrhinia, choanal atresia,
  microphthalmia syndrome
arrhythmia
    cardiac a.
    fetal a.
    respiratory sinus a.
    sinus a.
arrival
    born on a. (BOA)
Arruga-Nicetic capsule forceps
ARSB
    arylsulfatase B
    ARSB syndrome
        arylsulfatase B
ART
    assisted reproductive technology
Artane
arterenol
arterial (a)
    a. to alveolar (a-A)
    a. aneurysm
    a. blood gas (ABG)
    a. blood pressure
    a. blood sample (ABS)
    a. calcification
    a. carbon dioxide pressure (tension)
      ($PaCO_2$)
    a. line (A-line)
    a. linear density
    a. occlusive disease
    a. oxygen pressure (tension) (PaO_2)
    a. partial pressure (Pa)
    a. thrombosis
    a. vascular bed
    a. vascular disease
    a. waveform
arteriogram
    pelvic a.
arteriography
arteriohepatic dysplasia (AHD)
arteriolopathy
    decidual a.
arteriosclerosis
    infantile a.
ArterioSonde
arteriosus
    ductus a. (DA)

    patent ductus a. (PDA)
    truncus a.
arteriovenous (AV)
    a. fistula
    a. malformation
    a. oxygen difference
    a. shunt
arteritis
    familial granulomatous a.
    a. umbilicalis
artery
    aberrant systemic feeding a.
    allantoic a.
    anomalous left coronary artery
      from pulmonary a. (ALCAPA)
    arcuate a.
    azygos a.
    basal a.
    central retinal a.
    cervical a.
    circumflex a.
    coiled a.
    colic a.
    deep circumflex iliac a.
    discordant umbilical a.'s
    endometrial spiral a.
    epigastric a.
    femoral circumflex a.
    fetal cranial a.
    a. forceps
    great a.'s
    hemorrhoidal a.
    hypogastric a.
    ileocolic a.
    iliac a.
    inferior mesenteric a.
    lumbar a.
    mesenteric a.
    middle sacral a.
    obturator a.
    omphalomesenteric a.
    ovarian a.
    Parrot a.
    pelvic a.
    posterior inferior cerebellar a.
      (PICA)
    posterior inferior communicating a.
      (PICA)
    pudendal a.
    radial a.
    spiral a.
    superficial external pudendal a.
      (SEPA)
    superior mesenteric a.
    transposition of great a.'s (TGA)
    transposition of great a.'s (TGA)
    umbilical a. (UA)
    uterine a.

vaginal a.
vertebral a.
Artha-G
arthralgia
arthritis, pl. **arthritides**
juvenile rheumatoid a. (JRA)
monarticular a.
polyarticular juvenile rheumatoid a.
psoriatic a.
pyogenic a.
rheumatoid a. (RA)
septic a.
arthrochalasis multiplex congenita
arthrodentoosteodysplasia (ADOD)
arthrogryposis
distal a.
a. multiplex
a. multiplex congenita (AMC)
**arthrogryposis, ectodermal dysplasia,
cleft lip/palate developmental delay
syndrome**
arthrogryposis-like hand anomaly
arthroophthalmopathy (AO, AOM)
hereditary progressive a.
arthroophthalmophathia hereditaria
Arthropan
Arthus reaction
Articu-Lase laser mirror
artifact
cultural a.
deodorant a.
technical a.
artificial
a. chromosome
a. fever
a. insemination (AI)
a. insemination by donor (AID)
a. insemination donor
a. insemination by husband (AIH)
a. intravaginal insemination
a. pacemaker
a. rupture of membranes (ARM,
AROM)
a. spermatocele
a. temperature
a. urethral sphincter (AUS)
a. vagina
art line
Arts syndrome
ARV
acquired immunodeficiency syndrome-
related virus

**Arvee model 2400 infant apnea
monitor**
aryepiglottic fold
arylalkanoic acid
arylcarboxylic acid
arylpropionic acid
**arylsulfatase B (ARSB syndrome, ARSB,
ARSB syndrome)**
a. B. syndrome
AS
Crysticillin AS
Duracillin AS
Pentids-P AS
Pfizerpen AS
as
a. low as reasonably achievable
(ALARA)
a. low as reasonably achievable
principle
ASA
acetylsalicylic acid
asaccharolyticus
Peptococcus a.
Asbee-Hansen disease
ascariasis
ascending
a. colon
a. intrauterine infection
a. venography
ascensus
ascertainment
total a.
Ascher syndrome
Aschheim-Zondek (AZ)
A.-Z. test
aschistodactylia
ascites
chylous a.
fetal a.
lues a.
tumor a.
ascorbic acid
Ascriptin
A. A/D
ASCUS
atypical squamous cells of undetermined
significance
ASCUS smear
ASCUS/AGUS
ASD
atrial septal defect
Asendin

NOTES

aseptic
>a. fever
>a. meningitis
>a. temperature

Asepto syringe
asexual dwarfism
ASG
>Adhesion Scoring Group
>ASG system
>>Adhesion Scoring Group

Asherman syndrome
Ashkenazi
>A. Jew
>A. Jewish heritage

ASI
>active specific immunotherapy

ASIL
>anal squamous intraepithelial lesion

Askanazy cell
Aslan
>A. endoscopic scissors
>A. 2 mm minilaparoscope

Asmalix
asoma
asparaginase
>Ara-C, Oncovin, prednisone, a.
>(AOPA)

aspartate
>a. aminotransferase (AST)

aspartylglucosamide (AGA)
>a. deficiency

aspartylglucosaminuria (AGU)
ASPD
>antisocial personality disorder

ASPED
>angel-shaped phalangoepiphyseal
>dysplasia

Aspen
>A. laparoscopy electrode
>A. ultrasound platform

AspenVAC smoke evacuation system
aspergillosis
>bronchopulmonary a.
>ocular a.
>pulmonary a.

Aspergum
aspermatogenesis
aspermia
asphyctic infant
asphyxia
>antepartum a.
>autoerotic a.
>birth a.
>blue a.
>fetal a.
>a. livida
>neonatal a.
>a. neonatorum

>a. pallida
>perinatal a.
>sexual a.

asphyxiating
>a. thoracic chondrodystrophy
>a. thoracic dysplasia
>a. thoracic dysplasia syndrome
>a. thoracic dystrophy (ATD)
>a. thoracodystrophy syndrome

asphyxiation
>intrapartum a.

aspirate
>nasopharyngeal a.
>surveillance tracheal a.

aspiration
>a. biopsy
>a. biopsy cytology (ABC)
>cyst a.
>epididymal sperm a.
>fine-needle a. (FNA)
>a. of gastric contents
>gastric fluid a.
>a. of mature oocyte
>meconium a.
>menstrual a.
>microsurgical epididymal sperm a.
>(MESA)
>needle a.
>percutaneous epididymal sperm a.
>(PESA)
>a. pneumonia
>a. pneumonitis
>a. prophylaxis
>rete testis a. (RETA)
>testicular sperm a. (TESA)
>vacuum a.

aspiration-tulip device
aspirator
>Aspirette endocervical a.
>blunt a.
>Cavitron ultrasonic surgical a.
>(CUSA)
>Cavitron USA NS100 ultrasonic
>surgical a.
>Cook a.
>Endo-Assist sponge a.
>endocervical a.
>endometrial a.
>GynoSampler endometrial a.
>Nezhat-Dorsey a.
>Sharplan USA ultrasonic surgical a.
>Vabra cervical a.
>vacuum a.

Aspirette endocervical aspirator
asplenia
>a. syndrome

ASS
>Aarskog-Scott syndrome

A

ASSAS
aminopterin syndrome sine aminopterin
assault
sexual a.
assay
a. accuracy
Amplicor Chlamydia A.
automated
immunochemiluminomimetric
insulin a.
BCA protein a.
biologic a.
CA 125 a.
clonogenic a.
Coat-A-Count a.
colongenic a.
cytomegalovirus total
immunoglobulin a.
Detect HIV-1 a.
Digene HPV A.
enzyme a.
enzyme-linked immunosorbent a.
(ELISA)
estradiol a.
FAMA a.
fetal fibronectin a.
FSH MAIAclone
immunoradiometric a.
Gen-Probe amplified CT a.
hamster egg penetration a.
hemizona a.
HemoQuant a.
hermizona a.
hormone a.
Hybrid Capture DNA A.
HZA a.
immunochemiluminomimetric
insulin a.
immunologic a.
immunoradiometric a. (IRMA)
IMx Estradiol A.
limulus amebocyte lysate a.
Lyme enzyme-linked
immunosorbent a.
lysosomal hydrolase enzyme a.
a. marker
microhemagglutination a. (MHA)
Osteomark NTx a.
PCR a.
PIVKA-II a.
prostacyclin a.
Pyrilinks-D a.

radioreceptor a.
RAMP hCG a.
receptor a.
Recombigen a.
recombinant immunosorbent a.
(RIBA)
sandwich a.
a. sensitivity
serum a.
a. specificity
sperm penetration a. (SPA)
stem cell a.
TDxFLM A.
TDxFLx A.
tetrazolium dye a.
thyroid-stimulating hormone a.
tumor-cloning a.
Vidas varicella zoster a.
ViraType HPV DNA typing a.
assessment
Ballard gestational a.
Dubowitz Neurological A.
Erhardt Developmental
Prehension A.
fetal movement a.
four-quadrant a.
gestational age a.
high-risk pregnancy a.
morphologic a.
periodic patient a.
A. of Preterm Infants Behavior
(APIB)
psychometric a.
Scanlon A.
TOVA ADD/ADHD a.
ultrasound a.
ASSI bipolar coagulating forceps
assignment
gender a.
sex a.
assimilation pelvis
assistant
Carter Tubal A.
assisted
a. breech
a. breech delivery
a. cephalic delivery
a. conception
a. fertilization
a. hatching (AH)
a. medical procreation (AMP)
a. reproduction

NOTES

assisted *(continued)*
    a. reproduction technology
    a. reproductive technology (ART)
    a. zonal hatching (AZH)
association
    American Diabetes A. (ADA)
    CHARGE a.
    a. constant
    VACTERL a.
    a. with hydrocephalus syndrome
assortative mating
AST
    acoustic stimulation test
    aspartate aminotransferase
astasia-abasia
Astech meter
astemizole
asteroid body
asthenospermia
asthenozoospermia
asthma
    bronchial a.
    maternal a.
    thymic a.
AsthmaHaler
AsthmaNefrin
asthmaticus
    status a.
astigmatism
Astler-Coller modification of Dukes classification
astomia
Astramorph PF
Astrand 30-beat stopwatch method
astrocyte
    fibrinoid degeneration of a.'s
astrocytoma
AstroGlide personal lubricant
Astrup blood gas value
asymmetric
    a. crying facies (ACF)
    a. short stature syndrome
    a. tonic neck reflex (ATNR)
asymmetrical
    a. conjoined twins
asymmetrically
asymmetrus
    janiceps a.
asymmetry
    a. of face
    facial a.
    nasolabial fold a.
asymptomatic
    a. bacteriuria
    a. dehiscence
    a. infection
    a. infertility
    a. mild endometriosis

    a. myoma
    a. urinary tract infection (AUTI)
    a. viral shedding
asynapsis
asynchronous birth
asynclitic
    a. position
    a. position of fetus
asynclitism
    anterior a.
    posterior a.
AT3
    antithrombin 3
    AT3 deficiency types I, II
    AT3 type II HBS
    AT3 type II PE
    AT3 type II RS
Atabrine
atactica
    heredopathia a.
Atad Ripener Device
Atarax
ataxia
    brachydactyly, nystagmus, cerebellar a.
    cerebellar a.
    a.-deafness syndrome
    Friedreich a.
    a., myoclonic encephalopathy, macular degeneration, recurrent infections syndrome
    spastic a.
    spinocerebellar a.
    a. telangiectasia
    a.-telangiectasia syndrome
    X-linked cerebral a. (CLA)
    X-linked olivopontocerebellar a. (OPCA)
ataxia-deafness-retardation (ADR)
    a.-d.-r. syndrome
    a.-d.-r. with ketoaciduria
ataxia-microcephaly-cataract (AMC)
    a.-m.-c. syndrome
ataxic
ATD
    asphyxiating thoracic dystrophy
atelectasis
    congenital a.
    linear a.
    plate-like a.
    primary a.
    secondary a.
atelectatic
atelencephalia
atelencephalic syndrome
atelia
ateliosis
ateliotic dwarfism

A

atelocardia
atelocephaly-atelocheilia
atelocheiria
ateloglossia
atelognathia
atelomyelia
atelopodia
atelosteogenesis
atelostomia
atenolol
athelia
atheosis
    congenital a.
atherosclerosis
atherosis
    acute a.
    decidual arteriolar a.
athetoid
athlete's foot
athletic amenorrhea
athyroid
athyroidism, athyrea
athyrotic
    a. cretinism
    a. hypothyroidism
Ativan
Atkin-Flaitz-Patil syndrome
Atkin-Flaitz syndrome
Atkins diet
ATL
    A. HDI 3000 ultrasound system
    A. Ultramark 4,8,9 ultrasound
atlantoaxial instability
atlantodidymus
ATN
    acute tubular necrosis
ATnative
ATNR
    asymmetric tonic neck reflex
atomic
    a. absorption spectrophotometry
    a. absorption spectroscopy
    a. milk
atonic
    a. astatic diplegia
    a. seizure
atony
    uterine a.
atopic
    a. dermatitis
    a. erythroderma
atopy

atosiban
atovaquone
ATP
    adenosine triphosphate
ATR
    alpha-thalassemia mental retardation
    ATR syndrome
atracurium
    a. besylate
atraumatic forceps
atresia
    anal a.
    apple-peel a.
    biliary a.
    choanal a.
    duodenal a.
    esophageal a.
    a. folliculi
    a. of the foramen of Luschka and Magendie
    ileal a.
    jejunal a.
    jejunoileal a.
    mitral a.
    mitral valve a.
    oocyte a.
    pulmonary artery a.
    pyloric a.
    tricuspid a.
    vaginal a.
atretic
    a. cervix
    a. follicle
    a. vagina
atretocornus
atretocystia
atretogastria
atria (*pl. of* atrium)
atrial
    a. bigeminy
    a. contraction
    a. fibrillation
    a. flutter
    a. natriuretic factor (ANF)
    a. natriuretic hormone
    a. natriuretic peptide (ANP)
    a. septal defect (ASD)
    a. septostomy
    a. septostomy via balloon
    a. tachycardia
atriodigital dysplasia
atrioventricular (AV)

NOTES

atrioventricular *(continued)*
    a. block
    a. canal defect
    a. conduction delay
    a. discordance
    a. node
    a. reciprocating tachycardia
    a. septal defect
    a. septum
    a. shunt
at-risk pregnancy
Atrisol aerosol
atrium, pl. **atria**
Atromid-S
atrophia bulborum hereditaria
atrophic
    a. change
    a. endometrium
    a. vaginitis
atrophicus
    lichen sclerosus et a. (LS)
atrophy
    cerebral a.
    Dejerine-Sottas a.
    dentatorubral-pallidoluysian a.
      (DRPLA)
    endometrial a.
    epithelial a.
    familial olivopontocerebellar a.
    Fazio-Londe a.
    infantile cerebellooptic a.
    infantile spinal muscular a.
    juvenile spinal muscular a.
    Kjer-type dominant optic a.
    limb-girdle muscular weakness
      and a.
    linear a.
    neonatal olivopontocerebellar a.
      (OPCA)
    Parrot a. of newborn
    peroneal muscle a.
    postmenopausal a.
    progressive encephalopathy, edema,
      hypsarrhythmia, optic a. (PEHO)
    skin a.
    traction a.
    urogenital a. (UGA)
    vaginal a.
    vulvar a.
    Werdnig-Hoffmann muscular a.
atropine
Atrovent
ATRX
    X-linked alpha-thalassemia mental
    retardation
      ATRX syndrome
        X-linked alpha-thalassemia/mental
        retardation syndrome

ATT
    arginine tolerance test
attack
    transient ischemic a. (TIA)
attention
    a. deficit disorder (ADD)
    a. deficit hyperactivity disorder
      (ADHD)
attenuating tissue
Attenuvax
attitude
    fetal a.
    postpartum a.
attorney
    durable power of a.
    power of a.
attrition
    follicular a.
    sperm a.
Attwood staining method
atypia
    bowenoid a.
    cervical a.
    glandular a.
    koilocytic a.
    koilocytotic a.
    vulvar a.
atypical
    a. aneuploidy
    a. cell
    a. chondrodystrophy
    a. ductal hyperplasia (ADH)
    a. endosalpingiosis
    a. epithelium
    a. glandular cells of uncertain
      significance (AGCUS, AGUS)
    a. glandular cells of undetermined
      significance (AGCUS, AGUS)
    a. glandular cells of unknown
      significance
    a. karyotype
    a. lobular hyperplasia
    a. squamous cells of undetermined
      significance (ASCUS)
    a. vasculature
    a. vessel colposcopic pattern
^{198}**Au**
    gold-198
AUB
    abnormal uterine bleeding
Auchincloss modified radical
  mastectomy
Audio Doppler D920
audiogram
audiometer
    Pilot a.
audiometry
    behavioral a.

AUDIT
> Alcohol Use Disorders Identification Test

auditory
> a. brainstem response (ABR)
> a. evoked potential (AEP)
> a. evoked response

audodilation
> Frank nonsurgical perineal a.

Auerbach plexus
Aufricht nasal retractor
augmentation
> labor a.
> a. mammaplasty
> oxytocin a.
> Pitocin a.
> submucosal urethral a.
> transumbilical breast a. (TUBA)

augmented breast
Augmentin
augnathus
aura
Auralgan
aural temperature
Aureomycin
aureus
> *Staphylococcus a.*

auriculobrachiogenital dysplasia
auriculoosteodysplasia
aurocephalosyndactyly
Aurora MR breast imaging system
aurothioglucose
Auroto
AUS
> artificial urethral sphincter

auscultation
> abdominal a.
> chest percussion and a.
> periodic a.

Austin syndrome
Australia antigen
authority
> Human Fertilization and
> Embryology A. (HFEA)

AUTI
> asymptomatic urinary tract infection

autism
> a., dementia, ataxia, loss of
> purposeful hand use syndrome
> a.-fragile X syndrome (AFRAX)

autistic
auto
> A. Suture ABBI system

A. Suture Multifire Endo GIA 30
> stapler
> A. Syringe

autoamputation
autoantibody
> thyroid a.
> typhoid a.

autoantigen
> La (SS-B) a.
> Ro (SS-A) a.

autochthonous tumor
autocrine
> a. communication
> a. motility factor (AMF)

AutoCyte System
autodilation
> Frank nonsurgical perineal a.

autoerotic
> a. asphyxia

autogamy
autogenous vaccine
autograft
AutoGuard catheter
autoimmune
> a. acquired hemolytic anemia
> a. adrenalitis
> a.-associated congenital heart block
> a. disease
> a. factor
> a. hemolytic anemia (AIHA)
> a. mechanism
> a. oophoritis
> a. polyendocrinopathy, candidiasis,
> ectodermal dystrophy (APECED)
> a. polyglandular syndrome
> a. regulatory (AIRE)
> a. regulatory gene
> a. thrombocytopenic purpura

autoimmunity
autoinoculation
autologous
> a. blood
> a. blood donation
> a. bone marrow reinfusion
> a. transfusion

autolysis
automated
> a. brainstem auditory evoked
> response (ABAER)
> a. computerized axial tomography
> (ACAT)

NOTES

automated *(continued)*
    a. immunochemiluminomimetric insulin assay
    a. multiple analysis
automatic
    a. karyotype system database
    a. walking
automaticity
autonomic
    a. nervous system
    a. seizure
    a. walking reflex
autonomous replication sequence
autonomy
AutoPap
    A. 300
    A. automated screening device
    A. 300 QC system
    A. reader
Autoplex T
autopolyploid
autopolyploidy
autoprothrombin I
autopsy
autoradiography
Autoread centrifuge hematology system
autosite
autosomal
    a. chromosome disorder
    a. deletion
    a. dominant (AD)
    a. dominant disorder
    a. dominant inheritance
    a. dominant macrocephaly syndrome
    a. dominant nonsyndromic hearing loss (DFNA3)
    a. dominant Opitz syndrome (ADOS)
    a. dominant trait
    a. gene
    a. heredity
    a. monosomy
    a. recessive (AR)
    a. recessive disorder
    a. recessive inheritance
    a. recessive nonsyndromic hearing loss (DFNB1)
    a. recessive trait
    a. trisomy
autosomal-dominant retinitis pigmentosa (adRP)
autosome
    balanced rearrangement of a.
    group C a.
    a. translocation
autostapling device
autotransfusion
autozygote

Auvard speculum
AV
    arteriovenous
    atrioventricular
    AV malformation
    AV shunt
AV/AF
    anteverted and anteflexed
    AV/AF uterus
AVC
    AVC Cream
    AVC suppository
Aventyl
average
    a. path velocity (VAP)
    a. radiation dose
Avicidin
Avina female urethral plug
Avirax
Avitene
    A. hemostatic material
avium-intracellulare
    Mycobacterium a.-i.
Aviva mammography system
AVL 9110 pH analyzer
AVP
    arginine vasopressin
avuncular relationship
AWS
    Alagille-Watson syndrome
Axenfeld-Rieger
    A.-R. anomaly
    A.-R. syndrome
Axenfeld syndrome
axes *(pl. of* axis)
axetil
    cefuroxime a.
axial
    a. mesodermal dysplasia complex
    a. resolution
axilla, pl. **axillae**
axillary
    a. adenopathy
    a. hematoma
    a. irradiation
    a. irradiation therapy
    a. lymphadenopathy
    a. lymph node
    a. node dissection
    a. skin lesion
    a. tail
    a. tail of Spence
    a. temperature
    a. view
axis, pl. **axes**
    conjugate a.
    embryonic a.
    gonadal a.

HPA a.
HPO a.
hypothalamic-hypophyseal-ovarian-
    endometrial a.
hypothalamic-pituitary a.
hypothalamic-pituitary-adrenal a.
hypothalamic-pituitary-gonadal a.
hypothalamic-pituitary-ovarian a.
pelvic a.
a. of pelvis
pituitary a.
a. traction
axis-traction forceps
axonotmesis
Axotal
Ayercillin
Aygestin
Aylesbury spatula
Ayr
    A. saline drops
    A. saline nasal mist
Ayre
    A. spatula
    A. spatula-Zelsmyr Cytobrush
        technique
AZ
    Aschhcim-Zondek
Azactam
azasteroid
azatadine
azathioprine
AZF
    azoospermia factor
    AZFa region of Yq

AZFb region of Yq
AZFc region of Yq
AZH
    assisted zonal hatching
azidothymidine (AZT)
azithromycin
    a. dihydrate
azlocillin
Azmacort
azo
    a. dye
    A. Gantrisin
Azo-Gamazole
azoospermia
    deleted in a. (DAZ)
    a. factor (AZF)
    obstructive a.
    a., renal anomaly, cervicothoracic
        spine dysplasia (ARCS)
azoospermic man
Azorean disease
Azo-Standard
azotemia
    prerenal a.
Azovan Blue
AZT
    azidothymidine
Aztec
    A. ear
    A. idiocy
A.-Z. test
aztreonam
Azulfidine
azygos artery

NOTES

β (*var. of* beta)
B
    B cell
    B chromosome
    B complex vitamins
    B lymphocyte
B$_{12}$
    vitamin B$_{12}$
Babcock clamp
BABE OB ultrasound reporting system
babesiosis
Babinski
    B. reflex
    B. sign
Babinski-Fröhlich syndrome
Babkin reflex
Babock forceps
Babson chart
baby
    blue b.
    blueberry muffin b.
    boarder b.
    bottle-fed b.
    breast-fed b.
    Clinical Risk Index for B.'s
      (CRIB)
    cocaine b.
    collodion b.
    crack b.
    B. Doe regulations
    B. Dopplex 3000 antepartum fetal
      monitor
    giant b.
    jittery b.
    juice b.
    nipple-fed b.
    test-tube b.
    b. Tischler biopsy punch
    well-hydrated b.
    well-perfused b.
BABYbird
    B. II respirator
    B. II ventilator
Babyflex
    B. heated ventilation system
    B. ventilator
Babytherm IC
BAC
    bacterial artificial chromosome
    blood alcohol concentration
bacampicillin
Bacarate
Baciguent

bacillary
    b. angiomatosis
    b. dysentery
Bacille
    B. bilié de Calmette-Guérin (BCG)
    B. bilié de Calmette-Guérin
      vaccine
bacillus, pl. bacilli
    Calmette-Guérin b.
    b. Calmette-Guérin vaccine
    Döderlein b.
    Ducrey b.
    Gram-negative bacilli
    Gram-positive bacilli
bacitracin
back
    b. board
    b. clamp
backache
backcross
    b. mating
background
    dirty b.
Backhaus clamp
back-up position
baclofen
Bacon-Babcock operation
bacteremia
    clostridial b.
    polymicrobial b.
bacteremic shock
bacteria (*pl. of* bacterium)
bacterial
    b. artificial chromosome (BAC)
    b. artificial chromosome probe
    b. contamination
    b. count
    b. cystitis
    b. endocarditis
    b. enteritis
    b. growth
    b. infection
    b. pneumonia
    b. recovery
    b. toxin
    b. vaginosis (BV)
bacteriology
bacteriophage
bacterium, pl. bacteria
    coccobacillary bacteria
    Gram-negative bacteria
    Gram-positive bacteria
bacteriuria
    asymptomatic b.

B

Bacteroides
    B. capillosus
    B. corrodens
    B. distasonis
    B. fragilis
    B. melaninogenicus
    B. ovatus
    B. thetaiotaomicron
bacteroidosis
Bactine Hydrocortisone
Bactocill
Bactrim
    B. DS
Bactroban
Bact-T-Screen
Badenoch urethroplasty
Baden procedure
BADS syndrome
BAEP
    brainstem auditory evoked potential
BAER
    brainstem auditory evoked response
bag
    Ambu b.
    Barnes b.
    Cardiff resuscitation b.
    Champetier de Ribes b.
    Endopouch Pro specimen-
      retrieval b.
    Hope resuscitation b.
    intestinal b.
    sterile isolation b.
    Vi-Drape bowel b.
    Voorhees b.
    b. of waters (BOW)
bag-and-mask ventilation
bagged mask ventilation
bagging
Baggish hysteroscope
Bagshawe protocol
Bailey Physical Development Index
Baird forceps
Bair Hugger patient warming system
Bakchaus towel forceps
Bakelite cystoscopy sheath
Baker punch
baker's leg
BAL
    blood alcohol level
balance
    acid-base b.
    electrolyte b.
    fetal acid-base b.
    sodium b.
    transcapillary fluid b.
balanced
    b. chromosome rearrangement

    b. rearrangement of autosome
    b. translocation
balanic hypospadias
balanitic hypospadias
balanitis
    b. circinata
    b. circumscripta plasmacellularis
    plasma cell b.
    b. of Zoon
Baldy operation
Baldy-Webster
    B.-W. procedure
    B.-W. uterine suspension
Balfour
    B. bladder blade
    B. retractor
ball
    Bichat fat b.
    birthing b.
    cauterizing b.
    b. electrode
    fungus b.
    B. operation
    B. pelvimetry technique
Ballantine clamp
Ballantyne-Runge syndrome
Ballantyne-Smith syndrome
Ballard
    B. chart
    B. examination
    B. gestational assessment
    B. test
Ball-Burch procedure
Baller-Gerold syndrome (BGS)
Ballinger-Wallace syndrome
balloon
    atrial septostomy via b.
    b. catheter technique
    electrode b.
    b. endometrial ablation
    24-French Foley b.
    b. heating therapy
    Origin b.
    Rashkind b.
    b. septectomy
    b. septostomy
    Soft-Wand atraumatic tissue
      manipulator b.
    ThermaChoice uterine b.
    b. thermoplasty
    b. tuboplasty
    b. valvuloplasty
    b. valvulotomy
ballottable
ballottement
    abdominal b.
    uterine b.

balm
>butt b.

Balmex

Balminil
>B. Decongestant

Baloser hysteroscope

balsa vaginal form

Balthazar Scales of Adaptive Behavior

Baltic myoclonus

Bamberger fluid

Bamforth syndrome

banana
>Kanana B.
>b. sign

bananas, rice cereal, applesauce, and toast (BRAT)

band
>agyria-pachygyria b.
>amniotic b.
>annular b.
>BB b.
>C b.
>chorioamniotic b.
>cytological b.
>G b.
>hymenal b.
>b. keratopathy
>Ladd b.
>limbic b.'s
>MB b.
>MM b.
>oligoclonal b.'s
>Q b.
>R b.
>Silastic b.
>b. stage
>Streeter b.
>T b.

bandage
>Kerlix gauze b.
>Kling b.
>b. scissors

Band-Aid operation

banding
>centromeric b.
>chromosomal b.
>chromosome b.
>Giemsa b.
>high-resolution b.
>low-resolution b.
>b. pattern
>pulmonary artery b.

>quinacrine b.
>reverse b.
>tubal b.

Bandl
>pathologic retraction ring of B.
>B. ring

banjo curette

banjo-string adhesion

bank, banking
>clone b.
>cord blood b.
>sperm b.
>umbilical cord blood b.

banked breast milk

Banki syndrome

Bannayan-Riley-Ruvalcaba syndrome (BRRS)

Bannayan syndrome

Bannayan-Zonna syndrome (BZS)

Banophen Decongestant Capsule

Banthine

Banti syndrome

BAP
>bone alkaline phosphatase

bar
>Bill traction b.
>Denis Browne b.
>Mercier b.

Baraitser-Burn syndrome

Baraitser-Winter syndrome

Barber-Say syndrome

Barbilixir

Barbita

barbiturate
>b. poisoning

barbotage

Barc Liquid

Barcroft/Haldane apparatus

Bard
>B. Biopty cut needle
>B. Biopty gun
>B. cervical cannula

Bardet-Biedl syndrome (BBS)

Bard-Parker blade

bare lymphocyte syndrome

barium
>b. enema
>b. study

barium-impregnated plastic intrauterine device

Barkan infant lens

barking cough

B

NOTES

Barlow
- B. disease
- B. hip dysplasia test
- B. maneuver
- B. syndrome

Barnes
- B. bag
- B. cerclage
- B. curve
- B. zone

baromacrometer
baroreceptor
baroreflex response
barotrauma
Barr body
barrel cervix
barrel-shaped
- b.-s. cervix
- b.-s. lesion

barren
Barrett esophagus
Barrier
- B. gown
- B. laparoscopy drape
- B. pack

barrier
- Adcon-L anti-adhesion b.
- blood-testis b.
- b. contraception
- b. contraceptive
- INTERCEED TC7 absorbable adhesion b.
- b. method
- placental b.
- Sil-K OB b.
- TC7 adhesion b.

Barron pump
Bart
- hemoglobin B.
- B. hemoglobin
- B. syndrome

Bartholin
- B. abscess
- B. cyst
- B. cystectomy
- B. duct
- B. gland carcinoma
- B., urethral, Skene (BUS)
- B., urethral, Skene gland

bartholinitis
Bartholin-Patau syndrome
Bartholomew rule of fourths
Bartonella
- B. henselae
- B. quintana

Barton forceps
Bartsocaas-Papas syndrome
Bartter syndrome (BS)

barymazia
basal
- b. artery
- b. body temperature (BBT)
- b. body temperature chart
- b. body thermometer
- b. cell carcinoma
- b. cell epithelioma
- b. cell hyperplasia
- b. cell nevus syndrome (BCNS)
- b. ganglion
- b. ganglion disorder-mental retardation (BGMR)
- b. ganglion disorder-mental retardation syndrome
- b. lamina
- b. plate

base
- b. deficit
- b. excess
- b. medication
- nitrogenous b.
- b. pair
- b. sequence
- b. sequence analysis

baseball stitch
baseline
- b. fetal heart rate
- b. tonus
- b. value
- b. variability of fetal heart rate

basement
- b. membrane
- b. membrane zone (BMZ)

bases (*pl. of* basis)
bas-fond
basicaryoplastin
basic fibroblast growth factor (bFGF)
basichromatin
basilemma
Basis
- B. breast pump
- B. soap

basis, pl. **bases**
basophilic leukemia
Bassen-Kornzweig
- B.-K. disease
- B.-K. syndrome

Basset radical vulvectomy
bastard
Bastiaanse-Chiricuta procedure
bat ear
bath
- belly b.
- B. respirator
- sitz b.

bathing trunk nevus
bathrocephaly

batrachian position
Batten-Mayou disease
Batten-Turner congenital myopathy
battered
    b. buttock syndrome
    b. child syndrome
    b. fetus syndrome
    b. wife syndrome
    b. woman
battering cycle
battery
    Vulpe Assessment B.
battery-operated breast pump
battledore placenta
Baudelocque
    B. diameter
    B. operation
    B. uterine circle
baumanii
    Acinetobacter b.
Baumberger forceps
Baum bumps
Bayer Timed-Release Arthritic Pain Formula
bayesian hypothesis
Bayley-Pinneau table
Bayley Scale of Infant Development (BSID)
Baylor
    B. amniotic perforator
    B. amniotome
Bayne Pap Brush
bayonet
    b. forceps
    b. leg
Bazex-Dupré-Christol syndrome
Bazex syndrome
BBB, BBBG, BBB/G, BBG
BB band
BBB syndrome
BBG (*var. of* BBB)
BBS
    Bardet-Biedl syndrome
BBT
    basal body temperature
BCA protein assay
BCAVD
    bilateral congenital absence of vas deferens
BCC
    benign cellular changes

BCD
    blepharocheilodontic
    BCD syndrome
BCDDP
    Breast Cancer Detection Demonstration Project
BCDL
    Brachmann-Cornelia de Lange
BCG
    Bacille bilié de Calmette-Guérin
    BCG vaccine
Bcl-2 oncogene
BCNS
    basal cell nevus syndrome
BCP
    birth control pill
BCT
    benign cystic teratoma
    breast-conserving therapy
BD
    BD Sensability breast self-examination
    BD syndrome
BDD
    body dysmorphic disorder
BDLS
    Brachmann-Cornelia de Lange syndrome
    Brachmann-de Lange syndrome
BDNF
    brain-derived neurotrophic factor
BDProbeTec ET system
bead
    Chelex b.
    DEAE b.'s
beading
beaked pelvis
beaking
BEAM
    brain electrical activity map
bean
    castor b.
    jelly b.
Bear
    B. Cub infant ventilator
    B. Hugger warming blanket
    B. NUM-1 tidal volume monitor
    B. respirator
Beare-Stevenson cutis gyrata syndrome
Beare syndrome
bearing down
bearing-down pain

B

NOTES

51

beat
escape b.'s
Beath pin
Beatson ovariotomy
beats per minute (bpm)
beat-to-beat
b.-t.-b. variability
b.-t.-b. variability of fetal heart rate
b.-t.-b. variation of fetal heart rate
Beau line
Beben
Because vaginal foam
Beccaria sign
Beck
B. Depression Inventory score
B. disease
Becker
B. breast prosthesis
B. muscular dystrophy (BMD)
B. pseudohypertrophic muscular dystrophy
B. tissue expander
B. type progressive muscular dystrophy
Becker-Kiener muscular dystrophy
Beckwith syndrome
Beckwith-Wiedemann syndrome
Béclard sign
Becloforte
beclomethasone
b. diproprionate
b. propionate
Beclovent
Beconase
B. AQ Nasal Inhaler
B. Nasal Inhaler
bed
Affinity b.
arterial vascular b.
bumper b.
Ohio b.
oversewing placental b.
placental b.
pulmonary vascular b.
b. rest
Bednar aphtha
bedwetting
beef insulin
Beemer-Langer syndrome
Beemer lethal malformation syndrome
Beepen-VK
Beesix
Begeer syndrome
Béguez César disease

behavior
Assessment of Preterm Infants B. (APIB)
Balthazar Scales of Adaptive B.
behavioral
b. audiometry
b. genetics
Behçet
B. disease
B. syndrome
Behr syndrome
Beighton criteria
Belgian type mental retardation
bell
Gomco b.
B. palsy
B. staging criteria
belladonna
Bell-Buettner hysterectomy
Bellergal
Bellergal-S
Bellucci alligator forceps
belly
b. bath
b. bath therapy
Bel-Phen-Ergot S
belt
Marsupial b.
BEMP
bleomycin, Eldisine, mitomycin, Platinol
Benadryl
B. Decongestant Allergy Tablet
Benahist
Bendectin
Bendopa
bendroflumethiazide
BeneFix
Benelli mastopexy
Benemid
benign
b. breast disease
b. cellular changes (BCC)
b. congenital hypotonia
b. cyst
b. cystic ovarian teratoma
b. cystic teratoma (BCT)
b. familial chronic pemphigus
b. familial macrocephaly (BFM)
b. familial megalencephaly
b. familial neonatal convulsion (BFNC)
b. familial neonatal seizure
b. implant
b. infantile familial convulsion (BIFC)
b. lesion
b. mass
b. mesothelioma of genital tract

b. mucinous cystadenoma
b. neonatal epilepsy
b. nevus
b. nonprolapsed uterus
b. ovarian neoplasm
b. papillomavirus infection
b. tumor
b. X-linked recessive muscular dystrophy
Benjamin
B. anemia
B. syndrome
Bennett
B. PR-2 ventilator
B. respirator
B. small corpuscles
Benoject
Benson baby pylorus separator
Bentyl
benzathine
b. benzylpenicillin
b. penicillin
penicillin G b.
b. penicillin G
Benzedrine
benzocaine
benzodiazepine
benzoin
benzothiophene-derived selective estrogen receptor modulator
benzthiazide
benztropine mesylate
benzylpenicillin
BEP
bleomycin, etoposide, cisplatin
bleomycin, etoposide, Platinol
BEP therapy
beractant
b. surfactant
Berardinelli
B.-Seip-Lawrence syndrome
B.-Seip syndrome
B. syndrome
Berdon syndrome
Berger
B. paresthesia
B. renal disease
Bergia syndrome
beriberi
Shoshin b.
Berkeley
B. suction curette

B. suction machine
B. Vacurette
Berkeley-Bonney retractor
Berkow formula for burns
Berkson-Gage
B.-G. calculation
B.-G. test/assay
Berlin breakage syndrome
Bernard-Soulier syndrome
Bernay uterine packer
Bernoulli trial
Bernstein test
Berry-Kravis and Israel syndrome
Berry syndrome
Berry-Treacher Collins syndrome
Bertini syndrome
Berwick dye
Besnier prurigo of pregnancy
Best disease
bestiality
besylate
atracurium b.
beta, β
b. blocker
b. carotene
b. chain
b. error
estrogen receptor b. (Er beta)
follitropin b.
free b.
b. interferon
b.-lactamase-resistant antistaphylococcal antibiotic
b. ray
b. thalassemia
transforming growth factor b. (TGFbeta)
beta-1 integrin
beta-2 integrin
beta-adrenergic
b.-a. agent
b.-a. agonist
b.-a. drug
b.-a. receptor
beta$_2$-adrenergic agent
beta-aminoisobutyric aciduria
Betachron E-R
Betacort
Betaderm
Betadine
beta-endorphin

NOTES

17beta-estradiol
  17β-e. dehydrogenase
beta-galactosidase-1 (GLB-1)
  b.-g.-1 deficiency
beta-glucuronidase (GUSB gene, GUSB gene, GUSB, GUSB locus, GUSB locus)
beta-glucuronidase deficiency mucopolysaccharidosis
beta-hCG, beta-HCG
  beta-human chorionic gonadotropin
beta-hCG discriminatory zone
beta-hemolytic streptococcus
3betaHSD
  3beta-hydroxysteroid dehydrogenase
beta-human chorionic gonadotropin (beta-hCG, beta-HCG)
betaine anhydrous
beta-lactam
  b.-l. antibiotic
  b.-l. antimicrobial
Betalene Topical
Betalin S
beta-lipotrophin
Betaloc
betamethasone valerate
beta-mimetic
Betapen-VK
beta-receptor agonist
beta-subunit
  human chorionic gonadotropin b.-s.
beta-synthase
  cystathionine b.-s. (CBS)
Betatrex Topical
betatron electron accelerator
Beta-Val Topical
Betaxin
bethanechol
  b. chloride
Bethesda
  B. classification system
  B. II system
  B. Pap smear classification
  B. system guidelines
  B. system Pap classification
Betke-Kleihauer test
Betke stain
Betnelan
Betnesol
Betnovate
Beuren syndrome
Bevan incision
Beverly-Douglas lip-tongue adhesion technique
BeWo cell
Bewon
Bexophene
bezafibrate

Bezalip
bezoar
Bezold abscess
bFGF
  basic fibroblast growth factor
bFGF-stimulated cell proliferation
BFL
  Börjeson-Forssman-Lehmann
BFLS
  Börjeson-Forssman-Lehmann syndrome
BFM
  benign familial macrocephaly
BFNC
  benign familial neonatal convulsion
BF-STS
  biological false-positive serologic test for syphilis
bG
  Chemstrip bG
BGMR
  basal ganglion disorder-mental retardation
  BGMR syndrome
BGS
  Baller-Gerold syndrome
bi-allelic marker
Biamine
Bianchine-Lewis syndrome
bias
  detection b.
biatriatum
  cor triloculare b.
Biaxin
BICAP
  B. cautery
  B. probe
bicarbonate
  sodium b.
bicarbonate-carbonic acid system
bicephalus
Bichat fat ball
bichloracetic acid
bichorial pregnancy
Bicillin
  B. C-R
  B. C-R 900/300
  B. L-A
Bicitra
Bickers-Adams syndrome
BiCNU
BiCoag forceps
bicornuate, bicornate, bicornous
  b. uterus
bicuspid aortic valve
bidet
bidirectional PDA
bidiscoidal placenta

BIDS
> brittle hair, intellectual impairment, decreased fertility, short stature
> BIDS syndrome

Biederman sign
Bielschowsky-Jansky disease
Bielschowsky syndrome
Biemond syndrome 1, 2
Bierer ovum forceps
BIFC
> benign infantile familial convulsion

bifid
> b. cervix
> b. exencephalia
> b. nose
> b. pelvis
> b. scrotum
> b. uterus
> b. uvula
> b. xiphoid

bifida
> spina b.

Bifidobacterium
> *B. adolescentis*
> *B. bifidum*
> *B. infantis*

bifidum
> *Bifidobacterium b.*
> cranium b.

bifidus
> b. factor
> *Lactobacillus b.*

biforate uterus
bigeminal pregnancy
bigeminy
> atrial b.

Biggers medium
Biglieri syndrome
biischial diameter
bikinin (HI-30)
bilaminar blastoderm
BiLAP bipolar laparoscopic probe
bilary colic
bilateral
> b. acoustic neurofibromatosis
> b. breast pump
> b. cerebral ventriculomegaly
> b. choroid plexus cyst
> b. club feet
> b. congenital absence of vas deferens (BCAVD)
> b. ectopic pregnancy

> b. facial agenesis
> b. left-sidedness
> b. mediolateral episiotomies
> b. myocutaneous graft
> b. myringotomy tubes (BMT)
> b. ovarian neoplasm
> b. salpingo-oophorectomy (BSO)
> b. simultaneous tubal pregnancies
> b. tubal ligation (BTL)
> b. ureteral obstruction (BUO)
> b. uropathy

bile
> b. acid
> b. duct

bile-plug syndrome
biliary
> b. atresia
> b. cirrhosis
> b. hypoplasia

BiliBed phototherapy unit
BiliBlanket Plus Phototherapy system
BiliCheck
Bili mask
Bili mask phototherapy eye cover
bilirubin
> amniotic fluid b.
> b. blanket
> b. encephalopathy
> b. infarction

bilirubinometry
Bili-Timer
Bill
> B. maneuver
> B. traction bar
> B. traction handle forceps

Billings method
Billroth tumor forceps
biloba
> placenta b.

biloculare
> cor b.

Bilopaque
Biloptin
bimanual
> b. pelvic examination
> b. version

binary process
binder
> abdominal b.
> breast b.
> Dale abdominal b.

B

NOTES

binder *(continued)*
    obstetric b.
    scultetus b.
Binder syndrome
binding
    breast b.
    fragment antigen b.
    protein b.
    b. protein
    b. site
    sperm-zona pellucida b.
binge drinking
bingeing
binomial distribution
binovular twins
bioactive
    b. hormone
bioassay
BioBands bracelet
bioblast
Biobrane adhesive
Biobrane/HF dressing
Biocef
Biocell RTV saline-filled breast implant
Biocept-5 pregnancy test
Biocept-G pregnancy test
biochemical
    b. genetics
    b. pregnancy
    b. study
biochemistry
Bioclate
biocompatibility
bioelectromagnetic application
biofeedback therapy
biofield therapeutics
bioflavonoid
Biogel
    B. Reveal glove
    B. Reveal puncture indication
      system
bioinformatics
Biojector 2000
biologic
    b. assay
    b. response modifier (BRM)
    b. treatment
biological
    b. false-positive serologic test for
      syphilis (BF-STS)
    b. sampling
Biomerica
biometric profile
biometry
    fetal b.
Biomydrin
Biopatch dressing

biophysical
    b. profile (BPP)
    b. profile score
biopsy
    Allis-Abramson breast b.
    aspiration b.
    cervical cone b.
    chorionic villus b. (CVB)
    coin b.
    cold cup b.
    cold knife cone b.
    cone b.
    cul-de-sac b.
    b. dating
    embryo b.
    endometrial b.
    excisional b.
    fine-needle aspiration b.
    b. forceps
    hemicone b.
    hot b.
    Kevorkian punch b.
    Keyes punch b.
    kidney b.
    lymph node b.
    mirror image breast b.
    needle b.
    negative punch b.
    omental b.
    out-of-phase endometrial b.
    peritoneal b.
    Pipelle b.
    b. probe
    punch b.
    renal b.
    single cell b.
    skinny needle b.
    stereotactic breast b.
    transvaginal fine-needle b.
    trophectoderm b.
    vulvar b.
Biopty cut needle
biosampler
    Accelon Combi cervical b.
Bioself Fertility Indicator
BioStar strep A 1A test
biosynthesis
    prostaglandin b.
    steroid b.
biosynthetic defect
Bio-Tab
biotechnology
biotin
biotinidase
    b. deficiency
biotinylated
Biotirmone

B

BIP
    bleomycin, ifosfamide, Platinol
biparental inheritance
biparietal diameter (BPD)
bipartita
    placenta b.
bipartite
    b. uterus
biperiden
biphenyl
    polychlorinated b.
biplane
    b. cineangiography
    b. intracavitary probe
    b. seriography
bipolar
    b. cautery
    B. Circumactive Probe
    b. coagulation
    b. cutting loop
    b. electrocautery
    b. electrode
    b. laparoscopic forceps
    b. taxis
    b. urological loop
    b. vaporization
    b. version
bipotential
bipotentiality
bipronucleate
Birbeck granule
Bird
    B. Mark 8 respirator
    B. OP cup
    B. vacuum extractor
bird-beak jaw
bird-headed
    b.-h. dwarfism
    b.-h. dwarf of Seckel
    b.-h. dwarf syndrome
birdlike
    b. face syndrome
    b. facies
Birnberg bow
birth
    b. amputation
    b. anomaly
    b. asphyxia
    asynchronous b.
    b. canal
    b. canal laceration
    b. care center

    b. certificate
    b. control
    b. control pill (BCP)
    b. cushion
    date of b. (DOB)
    b. defect
    dry b.
    b. fracture
    gravida, para, multiple births,
      abortions, live b.'s (GPMAL)
    higher-order b.
    home b.
    b. injury
    multiple b.'s
    b. paralysis
    premature b.
    preterm b.
    b. rate
    spontaneous preterm b. (SPTB)
    b. trauma
    twin b.
    b. weight (BW)
    b. weight discordance
    wrongful b.
    year of b. (YOB)
birthing
    b. ball
    b. chair
    b. room
birthmark
bisacodyl
biscoumacetate
    ethyl b.
Bi-Set catheter
Bishop
    B. pelvic scoring system
    B. Prelabor Scoring System
    B. score
    B. score of cervical ripening
Bishop-Harmon forceps
bishydroxycoumarin
bisphosphonate
    b. therapy
Biswas Silastic vaginal pessary
bitartrate
    dihydrocodeine b.
bite
    stork b.
bitemporal
    b. aplasia cutis congenita
    b. diameter
    b. forceps marks syndrome

NOTES

57

bitterling test
bivalent chromosome
bivalve speculum
bivalving of the uterus
biviua
    Prevotella b.
Bixler
    B. hypertelorism
    B. syndrome
black
    b. jaundice
    b. line
    b. lock-albinism-deafness syndrome
    b. tongue
blackened speculum
Blackfan-Diamond
    B.-D. anemia
    B.-D. syndrome
bladder
    b. blade
    b. bubble
    b. catheter
    b. drill
    b. dysfunction
    exstrophy of the b.
    b. flap
    b. function
    b. habit
    b. hypotonia
    hypotonic b.
    b. laceration
    b. muscle stress test
    b. neck
    b. neck elevation test
    b. neck mobility
    b. neck obstruction
    b. neck stenosis
    b. neck suspension
    neurogenic b.
    b. outlet syndrome
    b. pillar
    b. retractor
    b. retraining
    stammering b.
    b. tumor
    urinary b.
**BladderManager portable ultrasound
  scanner**
BladderScan BVI2500
blade
    #15 b.
    Balfour bladder b.
    Bard-Parker b.
    bladder b.
    E-Mac laryngoscope b.
    Endo-Assist retractable b.
    Gott-Balfour b.
    Gott-Harrington b.

    Gott-Seeram b.
    Miller b.
    Orbit b.
Blair-Brown procedure
**Blaivas classification of urinary
  incontinence**
Blake closure of peritoneum
Blalock-Hanlon
    B.-H. operation
    B.-H. procedure
Blalock-Taussig
    B.-T. operation
    B.-T. procedure
    B.-T. shunt
blanching
    laser b.
bland cytology
Blane
    amniotic infection syndrome of B.
blanket
    Bear Hugger warming b.
    bilirubin b.
blastema
    metanephric b.
blastocele
blastocyst
    b. hatching
    b. implantation
    b. splitting
    b. transfer
blastocyte
blastoderm
    bilaminar b.
    embryonic b.
    trilaminar b.
blastodisk
blastogenesis
blastogenic period
blastolysis
blastoma
    nodular renal b.
blastomere
    b. cell
    b. separation
blastomycosis
blastotomy
blastula
blastysis
    trichorrhexis b.
Blaustein classification
bleed
    herald b.
bleeding
    abnormal uterine b. (AUB)
    anovulatory b.
    antepartum b.
    breakthrough b. (BTB)
    b. diathesis

dysfunctional uterine b. (DUB)
estrogen breakthrough b.
estrogen-progesterone withdrawal b.
estrogen withdrawal b.
HydroThermAblator system for
  excessive uterine b.
intermenstrual b.
pelvic b.
placental b.
postcoital b.
postmenarchal b.
postmenopausal b. (PMB)
preadolescent vaginal b.
progesterone breakthrough b.
progesterone withdrawal b.
b. site
b. site ligation
space of Retzius b.
third trimester b.
b. time
uterine b.
vaginal b.
withdrawal b.
Bleier clip
blennorrhagia
blennorrhagic
blennorrhea
blennorrheal
Blenoxane
bleomycin
cisplatin, vinblastine, and b.
b., Eldisine, mitomycin, Platinol
  (BEMP)
b., etoposide, cisplatin (BEP)
b., etoposide, Platinol (BEP)
b., ifosfamide, Platinol (BIP)
b. sulfate
Bleph-10
blepharitis
blepharochalasis
blepharocheilodontic (BCD)
b. syndrome
blepharonasofacial
b. malformation syndrome
blepharophimosis
b., ptosis, epicanthus inversus
  (BPEI)
b., ptosis, epicanthus inversus,
  primary amenorrhea syndrome
b., ptosis, epicanthus inversus
  syndrome (BPEIS)

b., ptosis, epicanthus inversus,
  telecanthus complex
b., ptosis, syndactyly, short stature
  syndrome
b. sequence
blepharophimosis-ptosis syndrome
blepharoptosis
b., blepharophimosis, epicanthus
  inversus, telecanthus syndrome
blepharospasm
blighted ovum
blind-ending vagina
blind loop syndrome
blindness
congenital retinitis b. (CRB)
Episkopi b.
BLIS
breast leakage inhibitor system
BlisterFilm dressing
blistering disease
BLM
borderline malignancy
bloc
en b.
Blocadren
Bloch-Siemens syndrome
Bloch-Sulzberger
B.-S. melanoblastoma
B.-S. syndrome
block
atrioventricular b.
autoimmune-associated congenital
  heart b.
bundle branch b.
Cerrobend b.
complete atrioventricular b.
complete heart b.
congenital complete heart b.
  (CCHB)
congenital heart b.
dorsal penile nerve b.
extradural b.
heart b.
lead b.
nerve b.
paracervical b.
pudendal b.
saddle b.
subarachnoid b.
Wenckebach heart b.

B

NOTES

blockade
>paracervical b.
>spinal b.

blockage
>epiglottal b.
>neuromuscular b.
>proximal tubal b.

blocker
>adrenergic b.
>alpha-adrenergic b.
>antiandrogen receptor b.
>beta b.
>calcium channel b.
>cyproheptadine receptor b.
>ganglionic b.

blocking factor

Bloedorn applicator

blood
>b. alcohol
>b. alcohol concentration (BAC)
>b. alcohol level (BAL)
>autologous b.
>b. chimerism
>b. chromosome analysis
>b. component
>b. component therapy
>cord b.
>b. count
>b. culture
>designated donor b.
>donor-specific b.
>b. dyscrasia
>b. ethanol
>fetal b.
>b. flow ($\dot{Q}$)
>b. gas
>b. gas determination
>b. glucose
>b. group
>b. group antibody
>b. group immunization
>b. grouping
>intervillous b.
>b. level
>b. loss
>maternal peripheral b.
>b. mole
>oxygenated fetal b.
>oxygen concentration in pulmonary capillary b. (CcO_2)
>b. patch
>b. pigment stain
>b. pressure
>b. product
>b. sampling
>b. spot
>b. sugar
>b. sugar monitoring
>b. transfusion
>b. typing
>b. urea nitrogen (BUN)
>b. vessel formation
>b. volume
>whole b.

Bloodgood
>B. disease
>B. syndrome

blood-loss anemia

blood-testis barrier

blood-type test

bloody show

Bloom syndrome

blot, blotting
>Eastern b.
>Northern b.
>Southern b.
>Western b.

Blount disease

blow-by oxygen

blue
>Agent B.
>b. asphyxia
>Azovan B.
>b. baby
>b. cone monochromatism
>b. diaper syndrome
>b. dome cyst
>b. dome syndrome
>methylene b.
>b. navel
>b. nevus
>postpartum b.'s
>b. ring pessary
>b. rubber-bleb nevus syndrome
>b. sclera
>b. spot
>toluidine b.
>Urolene B.

blueberry
>b. muffin baby
>b. muffin nodule
>b. muffin spot

Blumberg sign

Blumer shelf

blunt
>b. aspirator
>b. duct adenosis
>b. probe
>b. and sharp dissection
>b. trauma

Bluntport disposable trocar

BMD
>Becker muscular dystrophy
>bone mineral density

BMI
>body mass index

B-mode ultrasound
BMT
    bilateral myringotomy tubes
BMZ
    basement membrane zone
BNBAS
    Brazelton Neonatal Behavioral
      Assessment Scale
BOA
    born on arrival
board
        arm b.
        back b.
        papoose b.
boarder baby
Boari flap
bobbing
        ocular b.
Bochdalek hernia
BOD
    brachymorphism, onychodysplasia,
      dysphalangism
    BOD syndrome
Bodian-Schwachman syndrome
body
        apoptotic b.
        asteroid b.
        Barr b.
        Call-Exner b.
        b. coils of cord
        dense b.
        Döhle b.
        Donovan b.
        b. dysmorphic disorder (BDD)
        b. fluid
        foreign b.
        Golgi b.
        b. habitus
        Heinz b.
        Howell-Jolly b.
        ketone b.
        Lafora b.
        lamellar b.
        Lostorfer b.
        b. mass index (BMI)
        b. mass index nomogram
        Nissl b.'s
        owl's eye inclusion b.
        perineal b.
        polar b.
        psammoma b.
        refractile b.

        Schaumann b.
        Schiller-Duvall b.
        b. stalk
        b. stalk anomaly
        b. stalk malformation
        b. surface area (BSA)
        b. surface area calculation
        b. temperature
        Wagner-Missner b.
        b. weight
        Winkler b.
BOF
    branchiooculofacial
BOFS
    branchiooculofacial syndrome
boggy uterus
Bogros space
Bohn
        B. epithelial pearl
        B. equation
Bohr effect
Bohring syndrome
Bolt sign
Bombay phenotype
Bonamil formula
Bonamine
bond
        hydrogen b.
bonding
        mother-infant b.
bone
        b. accretion
        b. age
        b. age standard of Greulich and
          Pyle
        b. alkaline phosphatase (BAP)
        b. attenuation coefficient
        brittle b.'s
        b. densitometry
        b. density
        b. density measurement
        dwarfism and cortical thickening of
          tubular b.'s
        b. dysplasia
        b. formation
        innominate b.
        ivory b.'s
        b. loss
        marble b.'s
        b. marrow
        b. marrow failure
        b. marrow puncture

B

NOTES

bone *(continued)*
    b. marrow toxicity
    b. marrow transplantation
    b. mass
    b. mineral
    b. mineral content
    b. mineral density (BMD)
    b. resorption
    b. stippling
    b. tumor
    b. turnover
    weightbearing b.
    wormian b.'s
Bonine
Bonnaire method
Bonnano catheter
Bonneau syndrome
Bonnevie-Ullrich syndrome
Bonney
    B. abdominal hysterectomy
    B. blue stress incontinence test
Bontril
bony metastasis
Bookwalter retractor
boomerang
    b. dysplasia
    b. syndrome
booster
    tetanus toxoid b.
Boost nutritional drink
BOR
    branchiootorenal
    BOR syndrome
borborygmus, pl. **borborygmi**
border
    shaggy heart b.
borderline
    b. amniotic fluid index
    b. diabetes
    b. epithelial ovarian carcinoma
    b. epithelial ovarian neoplasm
    b. epithelial ovarian tumor
    b. malignancy (BLM)
    b. malignant epithelial neoplasm
Bordetella pertussis
Borg
    B. Perceived Exertion Scale
    B. Physical Activity Scale
boric
    b. acid
    b. acid capsule
Börjeson-Forssman-Lehmann (BFL)
    B.-F.-L. syndrome (BFLS)
Börjeson syndrome
born
    b. on arrival (BOA)
borne
Bornholm disease

Borrelia burgdorferi
borreliosis
Borsieri sign
bosselated
bossing
    frontal b.
botryoid
    b. pseudosarcoma
    b. sarcoma
botryoides
    sarcoma b.
bottle
    b. fed
    b. feed
    b. tooth decay
bottle-fed baby
botulinum
    Clostridium b.
botulism
Bouchut respiration
bougie
    Holinger infant b.
Bouin solution
bound
    b. estradiol
    b. testosterone
Bourneville
    B. disease
    B. syndrome
Bourneville-Brissaud disease
Bourneville-Pringle syndrome
Bourns
    B. infant respirator
    B. LS104-150 infant ventilator
boutonnière incision
Bovie
    B. cauterization
    B. cautery
    B. unit
bovina
    facies b.
bovine
    b. dermal collagen
    b. face
    b. facies
    b. mucus penetration test
    pegademase b.
    b. spongiform encephalopathy
    (BSE)
    b. surfactant
BOW
    bag of waters
bow
    Birnberg b.
bowel
    echogenic fetal b.
    b. function
    b. habit

hyperechoic b.
b. infarction
b. obstruction
perforated b.
b. preparation
small b.
b. sounds

Bowen
B. double-bladed scalpel
B. Hutterite syndrome

Bowen-Conradi syndrome
bowenoid
b. atypia
b. papulosis

bowing reflex
bowleg
bowl of pelvis
Bowman layer
box
CCAAT b.
Hogness b.
negative-pressure b.
paired b.
Pribnow h.
TATA b.

Boyle uterine elevator
Bozeman
B. operation
B. position
B. uterine dressing forceps

Bozeman-Fritsch catheter
BPD
biparietal diameter
bronchopulmonary dysplasia

BPEI
blepharophimosis, ptosis, epicanthus
inversus

BPEIS
blepharophimosis, ptosis, epicanthus
inversus syndrome

BPI
brachial plexus injury

bpm
beats per minute
breaths per minute

BPP
biophysical profile
antepartum fetal BPP
fetal BPP
modified BPP
BPP score

BRACA
comprehensive B.
B. gene test
multisite B.
single site B.

BRACAnalysis
BRACAnalyzer gene analyzer
brace
Cruiser hip abduction b.
Rhino Triangle b.

bracelet
BioBands b.

brachial
b. birth palsy
b. plexus
b. plexus injury (BPI)
b. plexus palsy
b. plexus stretching

brachioskeletogenital (BSG)
b. syndrome

Brachmann-Cornelia
B.-C. de Lange (BCDL)
B.-C. de Lange syndrome (BDLS)

Brachmann-de Lange syndrome (BDLS)
Bracht maneuver
brachycamptodactyly
brachycephalosyndactyly
brachycephaly
b., deafness, cataract, microstomia,
mental retardation syndrome

brachydactyly
b., dwarfism, hearing loss,
microcephaly, mental retardation
syndrome
b., mesomelia, mental retardation,
aortic dilation, mitral valve
prolapse, characteristic facies
syndrome
b., nystagmus, cerebellar ataxia
b., nystagmus, cerebellar ataxia
syndrome
Pitt-Williams b.
Sugarman b.

brachydactyly-distal symphalangism
syndrome
brachydactyly-ectrodactyly
brachygnathia
brachymelia
rhizomelic b.

brachymesomelia-renal syndrome
brachymesophalangism, I–V
brachymesophalangy

B

NOTES

brachymetacarpalia, cataract, mesiodens syndrome
brachymetacarpy
brachymetatarsus IV
brachymorphism
    b., onychodysplasia, dysphalangism (BOD)
    b., onychodysplasia, dysphalangism syndrome
brachyolmia
brachypelvic, brachypellic
brachysyndactyly
brachytelephalangy
brachytelomesophalangy
brachytherapy
    interstitial b.
    intracavitary b.
Bradley method of prepared childbirth
bradyarrhythmia
bradycardia
    apnea and b. (A&B)
    fetal b.
    postcordocentesis b.
    sinus b.
bradycardiac
bradygenesis
bradykinin
bradylexia
bradymenorrhea
bradyspermatism
bradytocia
Bragg-Paul respirator
Bragg peak
Brailsford
    B. disease
    B. syndrome
brain
    b. damage
    b. death
    b. disorder
    b. electrical activity map (BEAM)
    fetal b.
    b. peptide
    b. sparing
brain-death syndrome
brain-derived neurotrophic factor (BDNF)
brain-sparing effect
brainstem, brain stem
    b. auditory evoked potential (BAEP)
    b. auditory evoked response (BAER)
branched-chain amino acid
brancher deficiency
branchial
    b. anomaly
    b. arch

    b. arch syndrome
    b. cleft
    b. cleft remnant
    b. clefts-lip pseudocleft syndrome
    b. cyst
    b. ducts
    b. fistula
    b. plexus
branching
    fetal capillary b.
    b. snowflake test
branchiomere
branchiooculofacial (BOF)
    b. syndrome (BOFS)
branchiootic syndrome
branchiootorenal (BOR)
    b. dysplasia
    b. syndrome
branchioskeletogenital (BSG syndrome, BSG syndrome)
Brandt-Andrews maneuver
Brandt syndrome
brash
    weaning b.
BRAT
    bananas, rice cereal, applesauce, and toast
    BRAT diet
Braune canal
Braun episiotomy scissors
Braun-Schroeder single-tooth tenaculum
brawny edema
Braxton
    B. Hicks contraction
    B. Hicks sign
    B. Hicks version
Brazelton Neonatal Behavioral Assessment Scale (BNBAS)
BRCA1
    breast cancer gene 1
        BRCA1 breast cancer gene
        BRCA1 gene mutation
BRCA2
    breast cancer gene 2
        BRCA2 breast cancer gene
        BRCA2 gene mutation
breakage
    chromosome b.
BreakAway dressing
breakdown
    endometrial b.
    germinal vesicle b. (GVBD)
    wound b.
breakpoint
breakthrough bleeding (BTB)
breast
    accessory b.
    adolescent b.

augmented b.
b. binder
b. binding
b. biopsy tissue
b. bud
caked b.
b. cancer
B. Cancer Detection Demonstration Project (BCDDP)
b. cancer gene 1 (BRCA1)
b. cancer gene 2 (BRCA2)
b. carcinoma
carcinoma of the b. (CB)
b. care
b. change
childhood b.
b. conservation
Contour Profile anatomically shaped silicone b.
b. cyst
b. disease
b. embryology
engorged b.
b. engorgement
b. fed
b. feed
b. feeding
fibrocystic b.
b. flush
b. implant
irritable b.
keeled b.
lactating b.
b. leakage inhibitor system (BLIS)
b. malignancy
b. milk
b. milk jaundice
nonlactating b.
Paget disease of b.
peau d'orange appearance of the b.
pigeon b.
b. plate
b. prosthesis
b. pump
b. self-examination (BSE)
b. stimulation contraction test (BSCT)
supernumerary b.
Trilucent b.
BreastAlert differential temperature sensor

BreastCheck
breast-conserving therapy (BCT)
breast-covering therapy
BreastExam
breast-fed baby
breast-preserving therapy
Breathe Right
breath hydrogen excretion test
breathing
  fetal b.
  intermittent positive pressure b. (IPPB)
  periodic b.
  synchronous b.
breaths per minute (bpm)
Brecht feeder
breech
  assisted b.
  b. delivery
  b. extraction
  b. first twin
  frank b.
  b. head
  b. location
  b. location out of pelvis
  nonfrank b.
  b. presentation
  b. singleton
  spontaneous b.
  b. type
breed
breeding
  cross b.
  b. line
breeze
  B. respirator
  B. ventilator
Breezee
  B. Mist Aerosol
  B. Mist Antifungal
bregma
bregmatodymia
bregmocardiac reflex
Breisky-Navratil retractor
Brennen biosynthetic surgical mesh
Brenner tumor
Brentano syndrome
Breonesin
brephic
brephoplastic
brephotrophic
Breslow microstaging system

NOTES

Brethaire
Brethine
bretylium tosylate
Bretylol
Breuer-Hering inflation reflex
Breus mole
Brevibloc
Brevicon
Brevi-Kath epidural catheter
Brevital
Briard-Evans syndrome
Bricanyl
Bricker
>   B. procedure
>   B. ureteroileostomy

bridge
>   membrane b.

bridging
>   b. cross
>   b. flap

Briggance Diagnostic Inventory of Early Development
brim
>   pelvic b.
>   b. sign

brine flotation method
bring-your-own medical record
bris
Brissaud
>   B. dwarfism
>   B. infantilism
>   B. syndrome

brittle
>   b. bones
>   b. diabetes
>   b. hair, intellectual impairment, decreased fertility, short stature (BIDS)
>   b. hair, intellectual impairment, decreased fertility, short stature syndrome
>   b. hair-mental deficit syndrome

BRM
>   biologic response modifier

broad
>   b. ligament
>   b. ligament hernia
>   b. ligament pregnancy
>   b. thumb-hallux syndrome
>   b. thumb-mental retardation syndrome

broad-based gait
broad-spectrum antibiotic therapy
Broca pouch
Brockenbrough technique
Broders index
Brodie-Trendelenburg test
Brofed Elixir

Bromaline Elixir
Bromarest
Bromatapp
Brombay
bromelin method
Bromfed
>   B. Syrup
>   B. Tablet

Bromfenex
>   B. PD

bromide
>   ammonium b.
>   calcium b.
>   distigmine b.
>   ipratropium b.
>   mepenzolate b.
>   methantheline b.
>   pancuronium b.
>   Peacock b.
>   potassium b.
>   pyridostigmine b.
>   sodium b.
>   strontium b.
>   triple b.

bromocriptine
>   Apo-B.
>   injectable b.
>   b. mesylate
>   b. rebound
>   b. resistance
>   b. therapy

bromocriptine-resistant prolactinoma
bromodeoxyuridine (BUdR)
bromodiphenhydramine
bromopheniramine maleate
Bromphen
>   B. Tablet

brompheniramine
Brompton cocktail
bromsulfophthalein (BSP)
Bronalide
bronchi (pl. of bronchus)
bronchial
>   b. asthma
>   b. bud

bronchiectasis
bronchiolectasia
bronchiolitis
>   b. obliterans organizing pneumonia

bronchitis
>   acute laryngotracheal b.
>   arachidic b.
>   chronic obstructive b.
>   epidemic capillary b.

bronchobiliary
>   b. fistula

bronchogenic

bronchogram
air b.
bronchomalacia
bronchopneumonia
bronchopulmonary
b. aspergillosis
b. dysplasia (BPD)
b. lavage
b. malformation
bronchoscope
Holinger infant b.
Storz infant b.
bronchoscopy
bronchospasm
bronchus, pl. **bronchi**
elastic recoil of the b.
Bronitin
B. Mist
Bronkaid
B. Mist
Bronkephrine
Bronkodyl
Bronkometer
Bronkosol
Bronson chewable prenatal vitamins
bronze
b. baby syndrome
b. diabetes
b. Schilder disease
Brooks syndrome
Brooks-Wisniewski-Brown syndrome
Brotane
broth
Lim b.
Todd-Hewitt b.
Brouha test
Broviac catheter
brow
b. position
b. presentation
brow-anterior position
brow-down
b.-d. position
b.-d. presentation
Brown
B. uvula retractor
B. vertical retraction syndrome
brown
b. baby syndrome
b. fat nonshivering thermogenesis
Brown-Adson tissue forceps
Brown-Symmers disease

Brown-Vialetto-Van Laere syndrome
Brown-Wickham technique
brow-posterior position
brow-up position
Broxidine
broxuridine
BRRS
Bannayan-Riley-Ruvalcaba syndrome
Brucella
brucellosis
Bruck-de Lange syndrome
Brudzinski sign
Bruehl-Kjaer transvaginal ultrasound probe
Bruhat
B. laser fimbrioplasty
B. technique
Bruininks-Oseretsky Test of Motor Proficiency
bruit
aortic b.
carotid b.
placental b.
Brunner syndrome
Brunschwig operation
Brusa-Toricelli syndrome
brush
Bayne Pap B.
cytology b.
b. cytology
endocervical sampling b.
Stormby b.
Brushfield spot
Brushfield-Wyatt syndrome
brushing
colposcopically directed b.
Bruton
B. agammaglobulinemia
B. disease
Bryan-Leishman stain
Bryant traction
Bryce-Teacher ovum
BS
Bartter syndrome
BSA
body surface area
B-scanner
real-time B.-s.
static B.-s.
BSCT
breast stimulation contraction test

NOTES

BSE
>bovine spongiform encephalopathy
>breast self-examination

BSG
>brachioskeletogenital
>>BSG syndrome
>>branchioskeletogenital

BSID
>Bayley Scale of Infant Development

BSO
>bilateral salpingo-oophorectomy

BSP
>bromsulfophthalein

BTA stat test

BTB
>breakthrough bleeding

BTL
>bilateral tubal ligation

bubble
>bladder b.
>b. boy disease
>gastric b.
>b. gum cytoplasm
>b. isolation unit
>b. isolette
>b. stability test

bubbly lung syndrome

bubo
>bullet b.
>chancroidal b.
>climatic b.
>primary b.
>tropical b.
>venereal b.
>virulent b.

bubonic

buccal fat pad

Bucladin-S Softabs

buclizine

bucrylate

bud
>breast b.
>bronchial b.
>end b.
>hair b.
>limb b.
>metanephric b.
>syncytial b.
>tail b.
>ureteric b.

Budd-Chiari syndrome

Buddha-like habitus

Buddha stance

budding

budesonide

Budin rule

BUdR
>bromodeoxyuridine

Buenos Aires type mental retardation

buffalo hump

Buffaprin

buffered aspirin

Bufferin

Buffinol

buffy coat component

Bugbee electrode

Buhl disease

Buist method

bulb
>Rouget b.
>sinovaginal b.
>b. suction
>b. suctioning
>b. syringe
>vestibular b.

bulbar polioencephalitis

bulbitis

bulbocavernosus
>b. fat flap
>b. muscle

bulbocavernous reflex

bulbourethral
>b. gland

bulimia
>b. nervosa

bulimorexia

bulked segregant analysis

bulk selection

bulky carcinoma

bulldog syndrome

bullet bubo

bullosa
>acantholysis b.
>concha b.
>epidermolysis b.
>generalized atrophic benign
>>epidermolysis b. (GABEB)
>hereditary macular epidermolysis b.
>varicella b.

bullous
>b. congenital ichthyosiform
>>erythroderma
>b. dermatosis
>b. impetigo
>b. myringitis
>b. pemphigoid

bull's eye sonographic appearance

bumetanide

Bumex

Bumm curette

bump
>Baum b.'s

bumper bed

BUN
>blood urea nitrogen

B

bundle
  b. branch block
  b. of His
  hypertrophic b.
BUO
  bilateral ureteral obstruction
buphthalmia
bupivacaine
buprenorphine
bupropion
Burch
  B. colposuspension
  B. colpourethropexy
  B. modification
  B. procedure
  B. retropubic urethropexy
burden
  genetic b.
  tumor b.
Burger triangle
buried vaginal island procedure
Burkitt lymphoma
burn
  Berkow formula for b.'s
Burnet acquired immunity
burning vulva syndrome
Burn-McKeown syndrome
Burow solution
burp
  wet b.
burping
bursa-dependent system
bursa of Fabricius
BUS
  Bartholin, urethral, Skene
  BUS gland
buserelin
busulfan, busulphan
butabarbital
Butalan
butalbital
Butalgen
Butanefrine
butaperazine

Butazolidin
Butazone
butenafine antifungal agent
butoconazole
  b. nitrate 2%
butorphanol
  b. tartrate
  b. tartrate nasal spray
butoxide
  piperonyl b.
  pyrethrins and piperonyl b.
butriptyline hydrochloride
butt
  b. balm
  b. paste
butterfly
  b. drain
  b. flap
  b. needle
  b. rash
  b. vertebrae
button
  peritoneal b.
buttonhole incision
buttonholing
butyrophenone
Buxton clamp
BV
  bacterial vaginosis
BVI2500
  BladderScan B.
BW
  birth weight
Byers flap
Byler disease
bypass
  cardiopulmonary b.
  gastric b.
  jejunoileal b.
  b. surgery
Byrd-Drew method
BZS
  Bannayan-Zonna syndrome

NOTES

C

C band
C syndrome

3C

craniocerebellocardiac
3C dysplasia
3C syndrome

C-500

Optimox C.

C₁
C₂

c (*var. of* cal)

CA

cancer antigen
carcinoma
cardiac-apnea
CA 125
cancer antigen 125
CA 125 antigen
CA 19-9 antigen
CA 125 assay
CA 15-3 breast cancer marker
CA 72-4 cancer marker
CA 125 endometrial cancer marker
CA 19-9 GI cancer marker
CA 195 GI cancer marker
CA 50 GI cancer marker
CA monitor
CA 549 tumor marker

Ca

calcium
carcinoma

cabbage leaves
cabergoline
Cabot

C. cannula
C. trocar

CAC

cisplatin, Ara-C, caffeine

cachectic

c. infantilism

cachectin
cachexia

cancer c.

cacogenesis
cacomelia
CAD

computer-aided diagnosis

CADD-Prizm pain control system
CAF

cell adhesion factor
cyclophosphamide, doxorubicin, and 5-fluorouracil
Cytoxan, Adriamycin, fluorouracil

café au lait spot

caffeine

cisplatin, Ara-C, c. (CAC)
citrated c.
c. therapy

Caffey

C. disease
C. pseudo-Hurler syndrome

Caffey-Kenny disease
Caffey-Silverman syndrome
Caffey-Smyth-Roske syndrome
CAFTH

Cytoxan, Adriamycin, fluorouracil, tamoxifen, Halotestin

CAGE

cutting, annoyance, guilt, eye-opener
CAGE test
CAGE test for alcohol abuse

CAH

congenital adrenal hyperplasia

CAHMR

cataract, hypertrichosis, mental retardation
CAHMR syndrome

CAIS

complete androgen insensitivity syndrome

cake

omental c.

caked breast
cal, c

calorie

Calabro syndrome
Calan
calcaneovalgus

talipes c.

calcaneovarus

talipes c.

calcaneus

talipes c.

Cal Carb-HD
Calci-Chew
Calciday-667
calcifediol
calciferol
calcificans

angioma capillare et venosum c.
chondrodystrophia fetalis c.

calcification

amorphous breast c.
arterial c.
coarse c.
dystrophic c.
granulomatous c.
malignant c.
popcorn-like c.

C

71

calcification *(continued)*
    skin c.
    sutural c.
    vascular c.
calcified
    c. fetus
    c. myoma
Calcimar
Calci-Mix
calcinosis
    c. cutis, Raynaud phenomenon, esophageal motility disorders, sclerodactyly, telangiectasia (CREST)
    c. cutis, Raynaud phenomenon, sclerodactyly, telangiectasia (CRST)
calciotropic
Calciparine
calcitonin
    c. receptor
    c. salmon
calcitriol
calcitropic hormone
calcium (Ca)
    c. absorption
    c. acetate
    c. agonist
    c. antagonist
    c. bromide
    c. carbonate
    c. channel blocker
    c. citrate
    c. cyclamate
    docusate c.
    c. glubionate
    c. gluconate
    c. heparin
    intracellular c.
    c. ion
    ionized c. (iCa)
    c. pantothenate
    c. polycarbophil
    c. supplement
calculation
    Berkson-Gage c.
    body surface area c.
calculus, pl. **calculi**
    mammary c.
    renal calculi
    urate c.
    urinary c.
    uterine c.
CaldeCort
    C. Anti-Itch Topical Spray
    C. Topical
Calderol

Caldesene Topical
Caldwell-Moloy classification
CALF
    Cytoxan, Adriamycin, leucovorin, calcium, fluorouracil
calfactant
calf compression unit
CALF-E
    Cytoxan, Adriamycin, leucovorin, calcium, fluorouracil, ethinyl estradiol
caliectasis
californium
    c.-252 (^{252}Cf)
calipers
    Harpenden c.
    Tenzel c.
cal/kg/day
    calories per kilogram per day
Calkins sign
Call-Exner body
callosal agenesis
callosum
    agenesis of corpus c.
    congenital thrombocytopenia, Robin sequence, agenesis of corpus c., distinctive facies, developmental delay syndrome
    hereditary agenesis of corpus c.
    X-linked mental retardation-seizures-acquired microcephaly-agenesis of corpus c.
Calmette-Guérin
    Bacille bilié de C.-G. (BCG)
    C.-G. bacillus
calmodulin
calorie (cal, c)
    20-c. formula
    24-c. formula
    c.'s per kilogram per day (cal/kg/day)
    c.'s per ounce (cal/oz)
cal/oz
    calories per ounce
Cal-Plus
Caltrate 600
Caltrate, Jr.
calusterone
calvarial hyperostosis
Calvé-Legg-Perthes syndrome
Calymmatobacterium granulomatis
CAM
    cell adhesion molecule
    child-adult mist
    chorioallantoic membrane
    complementary and alternative medicine
    cystic adenomatous malformation
    CAM tent

CAMAK
    cataract, microcephaly, arthrogryposis,
      kyphosis
      CAMAK syndrome
Camalax
Cameco syringe pistol aspiration device
camera
    ETV8 CCD ColorMicro video c.
Camey
    C. ileocystoplasty
    C. reservoir
CAMFAK
    cataract, microcephaly, failure to thrive,
      kyphoscoliosis
      CAMFAK syndrome
cAMP
    cyclic adenosine monophosphate
Camper fascia
camphor
camphorated oil
CAMP-specific phosphodiesterase
   inhibitor
camptodactyly, camplodactyly
camptomelia
camptomelic
    c. dysplasia
    c. syndrome
Camptosar
Campylobacter
    C. fetus
    C. jejuni
camsylate
    trimethaphan c.
Camurati-Englemann syndrome
canal
    Alcock c.
    anal c.
    birth c.
    Braune c.
    cervical c.
    elastic c.
    endocervical c.
    inguinal c.
    Kovalevsky c.
    c. of Lambert
    Lambert c.'s
    neurenteric c.
    c. of Nuck
    Nuck c.
    parturient c.
    pudendal c.
    Steiner c.

    uterovaginal c.
    vesicourethral c.
canalicular period
canaliculus
Canavan disease
Canavan-van Bogaert-Bertrand disease
Cancell
cancer
    c. antigen (CA)
    c. antigen 125 (CA 125)
    c. antigen 125 test
    breast c.
    c. cachexia
    cervical c.
    cervical stump c.
    c. chemotherapy
    clear cell vaginal c.
    colorectal c.
    C. Committee of College of
      American Pathologists
    endometrial c.
    epithelial c.
    epithelial ovarian c.
    gynecologic c.
    hereditary ovarian c.
    intraepithelial endometrial c.
    invasive c. (IC)
    invasive cervical c.
    lung c.
    microinvasive cervical c.
    c. nest
    occult c.
    ovarian c.
    ovarian epithelial c. (OEC)
    rectal c.
    SGO classification of c.
    stage IV epithelial ovarian c.
    c. and steroid hormone (CASH)
    testicular c.
    c. therapy
    thyroid c.
    vaginal c.
cancericidal dose
Candela laser
candicidin
Candida
    C. albicans
    C. glabrata
    C. krusei
    C. tropicalis
candida
    c. colonization

C

NOTES

candida *(continued)*
  eczematous c.
  recurrent c.
candidal
  c. diaper dermatitis
  c. vulvovaginitis
candidate
  gene c.
  c. gene
candidiasis
  congenital c.
  vaginal c.
  vulvovaginal c. (VVC)
candidosis
  intertriginous c.
  mucocutaneous c.
Candistatin
candle
  cesium c.
  urethral c.
  vaginal c.
candy-cane stirrups
Canesten
  C. Topical
  C. Vaginal
canker
cannabis
cannula
  acorn c.
  Bard cervical c.
  Cabot c.
  cervical c.
  Circon-ACMI c.
  Cohen uterine c.
  Core Dynamics disposable c.
  Dexide disposable c.
  endometrial c.
  Ethicon disposable c.
  Genitor mini-intrauterine
    insemination c.
  Gesco c.
  Hasson c.
  Hunt-Reich c.
  intrauterine balloon-type c.
  intrauterine insemination c.
  IUI disposable c.
  Jarit disposable c.
  Kahn c.
  Karman c.
  KDF-2.3 intrauterine
    insemination c.
  LaparoSAC single-use obturator
    and c.
  Lübke uterine vacuum c.
  Marlow disposable c.
  nasal c.
  Olympus disposable c.
  Rubin c.

  Scott c.
  Semm uterine vacuum c.
  Solos disposable c.
  stable access c.
  step-down c.
  Stortz disposable c.
  trumpet c.
  Vabra c.
  vacuum c.
  Vancaillie uterine c.
  Weck disposable c.
  Wisap disposable c.
  Wolf disposable c.
  Ximed disposable c.
cannulate
cannulation
canopy
  surgical overhead c. (SOC)
cantharidin
cantharis, pl. cantharides
canthomeatal line
Cantil
Cantor tube
Cantrell syndrome
Cantú syndrome
CAP
  cyclophosphamide, Adriamycin, cisplatin
cap
  acrosomal c.
  cervical c.
  cradle c.
  Dumas vault c.
  Dutch c.
  Oves Cervical C.
  ProtectaCap c.
  Universal reducer c.
  vault c.
  Vimule c.
capacitation
  sperm c.
capacitive
capacity
  alveolar-arterial oxygen diffusing c.
  corticosteroid-binding globulin-
    binding c. (CB-GBC)
  cystometric c.
  fetal blood oxygen-carrying c.
  functional residual c. (FRC)
  iron-binding c. (IBC)
  lung c.
  c. of lung (CL)
  maximum breathing c.
  oxygen-diffusing c.
  plasma iron-binding c.
  pulmonary diffusing c.
  total iron-binding c. (TIBC)
  total lung c. (TLC)

urinary concentrating c.
vital c.
Capasee diagnostic ultrasound system
CAPD
    continuous ambulatory peritoneal dialysis
capillary
    c. blood gas (CBG)
    c. blood sampling
    dilated c.
    c. erection
    c. refill time
capillosus
    Bacteroides c.
capita (*pl. of* caput)
capitis
    pediculosis c.
    tinea c.
caplets
    Advil Cold & Sinus C.
    Dimetapp Sinus C.
    Dristan Sinus C.
Capnocytophaga
Capoten
capped uterus
caproate
    17alpha-hydroxyprogesterone c.
    hydroxyprogesterone c.
    17-hydroxyprogesterone c.
Capronor
capsaicin
capsicum
Capsin
capsularis
capsular stripping
capsule
    Banophen Decongestant C.
    boric acid c.
    contraceptive suppository c.
    Crosby-Kugler pediatric c.
    Dimetapp 4-Hour Liqui-Gel C.
    Heyman c.'s
    Kadian sustained-release
      morphine c.'s
    Poly-Histine-D C.
    Redux c.
    ruptured c.
    Uro-Mag c.
    Virilon c.
captopril
capture
    ovum c.
caput, pl. **capita**

c. medusae
c. quadratum
c. succedaneum
Capute scale
Capzasin-P
caramel test
carbachol
carbamazepine
Carbamide
carbamoyltransferase
    ornithine c. (OCT)
carbarsone
carbazole
carbenicillin
carbergoline
carbetocin
Carb-HD
    Cal C.-HD
carbimazole
carbinoxamine maleate
carbogen
carbohydrate
    c. deficient glycoprotein (CDG)
    c. homeostasis
    c. intolerance
    c. metabolism
    c. tolerance
carbohydrate-deficient glycoprotein
    syndrome (CDGS)
Carbolith
carbon
    c. dioxide (CO_2)
    c. dioxide gas
    c. dioxide laser
    c. dioxide tension
    c. monoxide poisoning
carbonate
    calcium c.
carbonic acid
carbonyl iron
carboplatin
carboprost tromethamine
carboxamide
    dimethyl-triazeno-imidazole c.
    (DTIC)
carboxyhemoglobin (COHb)
carboxyl terminal peptide (CTP)
carbuncle
Carcassone perineal ligament
carcinoembryonic
    c. antigen (CEA)
    c. antigen 125 (CEA 125)

NOTES

C

carcinogen
    chemical c.
carcinogenesis
carcinoid
    nonappendiceal c.
    c. syndrome
    c. tumor
carcinoma, pl. carcinomas, carcinomata (CA, Ca)
    adenoid cystic c.
    adenosquamous c.
    advanced c.
    anaplastic c.
    androgen-dependent c.
    c. antigen
    Bartholin gland c.
    basal cell c.
    borderline epithelial ovarian c.
    c. of the breast (CB)
    breast c.
    bulky c.
    cecal c.
    cervical c.
    c. of cervix
    clear cell endometrial c.
    colloid c.
    colon c.
    contralateral synchronous c.
    ductal c.
    embryonal c.
    endometrial c.
    endometrioid c.
    epithelial ovarian c.
    FAB staging of c.
    fallopian tube c.
    focal lobular c.
    gastric c.
    glassy cell c.
    gynecologic c.
    hepatocellular c. (HCC)
    infiltrating ductal c. (IDC)
    infiltrating lobular c. (ILC)
    infiltrating small-cell lobular c.
    inflammatory c.
    intracystic papillary c.
    intraductal papillary c.
    invasive c.
    juvenile c.
    lobular c.
    lung c.
    medullary c.
    Merkel cell c.
    mesometanephric c.
    mesonephric c.
    mesonephroid clear-cell c.
    metaplastic c.
    metastatic c.
    microinvasive c.

    mucinous c.
    multicentric c.
    multiple nevoid-basal cell c. (MNBCC)
    nevoid basal cell c. (NBCC)
    oat cell c.
    ovarian small-cell c.
    papillary endometrial c.
    papillary serous cervical c.
    peritoneal serous papillary c.
    preclinical c.
    primary hepatocellular c. (PHC)
    primary peritoneal c. (PPC)
    recurrent c.
    renal cell c.
    scirrhous c.
    secretory c.
    serous c.
    signet ring cell c.
    c. in situ (CIS)
    small cell c.
    sporadic nonfamilial clear cell c.
    squamous cell c.
    tubular c.
    uterine corpus c.
    uterine papillary serous c.
    vaginal c.
    verrucous c.
    vulvar c.
    vulvovaginal c.
    well-circumscribed c.
    wolffian duct c.
    yolk sac c.
carcinomatosis
carcinosarcoma
    uterine c.
card
    Guthrie c.
    Sono-Gram fetal ultrasound image c.
cardiac
    c. anomaly
    c. arrest
    c. arrhythmia
    c. autonomic modulation
    c. catheterization
    c. defect, abnormal face, thymic hypoplasia, cleft palate, hypocalcemia (CATCH-22)
    c. dysrhythmia
    c. failure
    c. flow
    c. function
    c. glycoside
    c. lesion
    c. massage
    c. output

c. rhabdomyoma
c. tamponade
cardiac-apnea (CA)
c.-a. monitor
cardiac-limb syndrome
Cardiff
C. Count-to-Ten chart
C. resuscitation bag
Cardilate
cardinal
c. ligament
c. movement
c. point
cardinal-uterosacral
c.-u. ligament
c.-u. ligament complex
cardiocranial syndrome
cardiofacial syndrome
cardiofaciocutaneous syndrome (CFC)
cardiogenic shock
cardiogenital syndrome
cardiography
impedance c.
cardiomyopathy
idiopathic dilated c.
peripartum c.
postpartum c.
X-linked c. (XLCM)
X-linked dilated c. (XLCM)
cardioplegia
cold potassium c.
cardioprotective effect
cardiopulmonary
c. bypass
c. collapse
c. resuscitation (CPR)
Cardioquin
cardiorespiratory
c. function
c. syndrome of obesity in child
cardiorespirogram (CR-gram)
cardiospasm
cardiotachometer
cardiothymic shadow
cardiotocogram
terminal c.
cardiotocograph
cardiotocography
intrapartum c.
cardiovascular (CV)
c. complication

c. effect
c. system
cardiovertebral syndrome
carditis
care
ambulatory c.
antepartum c.
breast c.
followup c.
Kangaroo C.
monitored anesthesia c.
obstetric c.
postoperative c.
postpartum c.
preconception c.
prenatal c.
prepregnancy c.
Similac Special C.-24
Carey-Fineman-Ziter syndrome
caries
dental c.
carina, pl. **carinae**
carinatum
pectus c.
Carmault clamp
carmine
indigo c.
Carmi syndrome
Carmol
Carmol-HC Topical
carmustine
Carnation Follow-Up
carneous
c. degeneration
c. mole
Carnevale syndrome
Carney syndrome
carnitine
c. deficiency
c. transferase enzyme disorder
Caroli disease
**Caroline Crachami osteodysplastic
primordial dwarfism**
carotid bruit
carotin
Carpenter syndrome
carphenazine maleate
carp-like mouth
carp mouth
Carrasyn Hydrogel dressing
carrier
embryo c.

C

NOTES

carrier *(continued)*
    Endo-Assist endoscopic ligature c.
    factor V Leiden c.
    fragile X c.
    gene c.
    gestational c.
    heterozygous c.
    latent c.
    linear in-line ligature c.
    Miya hook ligature c.
    obligate c.
    c. protein
    Raz double-prong ligature c.
    c. testing
    translocation c.
cart
    Sensorimedics Horizon
      Metabolic C.
Carter Tubal Assistant
cartilage
    fetal c.
cartridge
    serum pregnancy assay c.
Cartwright blood group
caruncle
    amnionic c.
    myrtiform c.
    urethral c.
caruncula, pl. **carunculae**
    c. hymenalis
    c. myrtiformis
Carus
    C. circle
    C. curve
CASA
    computer-assisted semen analysis
    computer-assisted semen analyzer
casanthranol
cascara sagrada
case
    index c.
caseation
casein hydrolysate
caseosa
    vernix c.
caseous
CASH
    cancer and steroid hormone
    classic abdominal Semm hysterectomy
    CASH study
Casodex
Casser fontanel
casserian fontanel
cast
    decidual c.
    hyaline c.
    uterine c.
Castaneda procedure

casting
    Cerrobend c.
castor bean
Castroviejo
    C. fixation forceps
CAT
    computed axial tomography
cat
    c. cry
    c. eye syndrome (CES)
catadidymus
Cataflam
    C. Oral
catagen phase
catalase
catamenia
catamenial
    c. hemoptysis
catamenogenic
Catania type acrofacial dysostosis
Cataplexy
cataplexy
Catapres
cataract
    c., ataxia, deafness, retardation
      syndrome
    cerulean c.
    congenital c.
    c., hypertrichosis, mental retardation
      (CAHMR)
    c., hypertrichosis, mental retardation
      syndrome
    infantile c.
    c., mental retardation,
      hypogonadism syndrome
    c., microcephaly, arthrogryposis,
      kyphosis (CAMAK)
    c., microcephaly, arthrogryposis,
      kyphosis syndrome
    c., microcephaly, failure to thrive,
      kyphoscoliosis (CAMFAK)
    c., microcephaly, failure to thrive,
      kyphoscoliosis syndrome
    c., motor system disorder, short
      stature, learning difficulty, skeletal
      abnormalities syndrome
cataract-dental syndrome
cataractogenic
cataract-oligophrenia syndrome
catarrhal
    c. jaundice
CATCH-22
    cardiac defect, abnormal face, thymic
      hypoplasia, cleft palate, hypocalcemia
      CATCH-22 syndrome
catecholamine
    endogenous c.
catechol estrogen

Catel-Manzke syndrome
catgut suture
cathartic
   saline c.
cathepsin
   c. D
catheter
   Acrad HS c.
   Alzate c.
   Argyle arterial c.
   AutoGuard c.
   Bi-Set c.
   bladder c.
   Bonnano c.
   Bozeman-Fritsch c.
   Brevi-Kath epidural c.
   Broviac c.
   Caud-A-Kath epidural c.
   central venous c. (CVC)
   central venous pressure c.
   Chemo-Port c.
   ChronoFlex c.
   CliniCath peripherally inserted c.
   Conceptus Soft Torque uterine c.
   Conceptus VS c.
   coudé c.
   Cystocath c.
   Davis bladder c.
   DeLee suction c.
   Dobbhoff c.
   Dorros infusion and probing c.
   double-balloon c.
   double-lumen c.
   Drew-Smythe c.
   Du Pen epidural c.
   EASI c.
   EchoMark salpingography c.
   Ehrlich c.
   Embryon GIFT c.
   Embryon HSG c.
   Evert-O-Cath drug delivery c.
   femoral artery c.
   Flexxican c.
   Foley c.
   French Gesco c.
   Gesco c.
   Groshong c.
   Hickman c.
   Hohn c.
   HUMI c.
   Hurwitz c.
   hysterosalpingography c.

   intrauterine pressure c. (IUPC)
   Johnson transtracheal oxygen c.
   KDF-2.3 intrauterine c.
   Kish urethral illuminating c.
   Koala intrauterine pressure c.
   Labotect c.
   L-Cath peripherally inserted
     neonatal c.
   Leonard c.
   LeRoy ventricular c.
   Malecot c.
   Mentor c.
   microendoscopic optical c.
   Microtip c.
   Micro-Transducer c.
   Millar microtransducer urethral c.
   Neo-Sert umbilical vessel c.
   On-Command c.
   Opti-Flow c.
   percutaneous central venous c.
     (PCVC)
   percutaneous nephrostomy c.
   peripherally inserted central c.
     (PICC)
   peritoneal c.
   PermCath c.
   Per-Q-Cath c.
   Pezzer c.
   Pleur-evac chest c.
   Port-A-Cath c.
   PRO infusion c.
   Quinton dual-lumen c.
   radial arterial c.
   Raimondi c.
   red rubber c.
   Release c.
   Reliance urinary control insert c.
   Soft-Cell c.
   soft seal c.
   Soft Torque uterine c.
   Soules intrauterine insemination c.
   c. specimen
   Spectranetics c.
   split sheath c.
   Stamey c.
   Stamey-Malecot c.
   Stargate falloposcopy c.
   suction c.
   support c.
   suprapubic c.
   Swan-Ganz c.
   Tenckhoff c.

C

NOTES

catheter *(continued)*
TFX c. stylet
c. toes
transcervical tubal access c. (T-TAC)
transurethral c.
triple-lumen c.
T-TAC c.
Tygon c.
umbilical artery c. (UAC)
umbilical vein c. (UVC)
uterine cornual access c. (UCAC)
uterine ostial access c. (UOAC)
Vabra c.
Vas-Cath c.
V-Cath c.
venous c.
Wallace c.
whistle-tip c.
Wholey balloon occlusion c.
c.-within-a-catheter
Word Bartholin gland c.
Word bladder c.
Zynergy Zolution c.
catheterization
cardiac c.
femoral artery c.
fetal bladder c.
pulmonary artery c.
transvaginal tubal c.
umbilical artery c.
umbilical vein c.
urinary c.
cat-scratch
c.-s. disease
c.-s. fever
cat's cry syndrome
CatsEye digital camera system
Cattell Infant Intelligence Scale
caudad
Caud-A-Kath epidural catheter
caudal
c. analgesia
c. anesthesia
c. appendage, short terminal phalanges, deafness, cryptorchidism, mental retardation syndrome
c. duplication
c. dysplasia syndrome
c. pole
c. regression syndrome (CRS)
Cauer forceps
caul
causal
c. embryology
c. independence
c. inference

cauterization
Bovie c.
cauterizing ball
cautery
BICAP c.
bipolar c.
Bovie c.
c. conization
Endoclip c.
laparoscopic c.
monopolar c.
ovarian c.
Oxycel c.
cava
inferior vena c.
superior vena c.
vena c.
CAVD
congenital absence of vas deferens
cavernosum
corpus c.
cavernous
c. hemangioma
c. lymphangioma
c. plexus
c. sinus thrombosis
CAVH
continuous arteriovenous hemofiltration
cavitary white-matter lesion
Cavitron
C. ultrasonic surgical aspirator (CUSA)
C. USA NS100 ultrasonic surgical aspirator
cavity
abdominal c.
amniotic c.
endometrial c.
exocelomic c.
oral c.
pelvic c.
peritoneal c.
pseudomonoamniotic c.
thoracic c.
uterine c.
cavovalgus
talipes c.
cavum
c. septum pellucidum, cavum vergae, macrocephaly, seizures, mental retardation syndrome
cavus
pes c.
talipes c.
Cayler syndrome
CB
carcinoma of the breast

CBCL
Child Behavior Checklist
CBG
capillary blood gas
cord blood gas
corticosteroid-binding globulin
CB-GBC
corticosteroid-binding globulin-binding
capacity
CBS
cystathionine beta-synthase
CC
clomiphene citrate
cc
cubic centimeter
CCA
congenital contractural arachnodactyly
CCAAT box
CCAM
congenital cystic adenomatoid
malformation
CCC
craniocerebellocardiac
CCC dysplasia
CCC syndrome
CCCT
clomiphene citrate challenge test
CCD Spirette
CCHB
congenital complete heart block
CCHD
cyanotic congenital heart disease
cc/hr
cubic centimeter per hour
cc/kg/d
cubic centimeters per kilogram per day
CcO$_2$
oxygen concentration in pulmonary
capillary blood
CCR
cumulative conception rate
CCUP
colpocystourethropexy
CD
Ceclor CD
CD34 hematopoietic progenitor cell
CD4 cell count
CD4$^+$ level
CDAP
continuous distending airway pressure
CDC
Centers for Disease Control

CD4 cell
CD8 cell
CD8 cell count
CDE
color Doppler energy
CDE blood group system
CDG
carbohydrate deficient glycoprotein
CDGS
carbohydrate-deficient glycoprotein
syndrome
CDH
congenital diaphragmatic hernia
congenital dislocated hip
CDI
Children's Depression Inventory
CDIS
continuous distention irrigation system
CDL
Cornelia de Lange
cDNA
complementary DNA
cDNA library
CDS
color Doppler sonography
CEA
carcinoembryonic antigen
CEA 125
carcinoembryonic antigen 125
ceasmic
cebocephalus
cebocephaly
CECA
cisplatin, etoposide, Cytoxan, Adriamycin
cecal carcinoma
Ceclor
C. CD
cecocolic intussusception
Cecon
CED
cranioectodermal dysplasia
Cedax
Cedilanid
C.-D
Cedocard-SR
CEE
conjugated equine estrogen
CEEA
circular end-to-end anastomosis
CeeNU
Ceepryn

C

NOTES

CEF
    Cytoxan, epirubicin, fluorouracil
cefaclor
cefadroxil
Cefadyl
cefamandole
Cefanex
cefazolin
    c. sodium
cefdinir
cefepime
    c. HCl
cefixime
Cefizox
cefmetazole
Cefobid
cefonicid
cefoperazone
    c. sodium
ceforanide
Cefotan
cefotaxime
    c. sodium
cefotetan
    c. disodium
cefoxitin
    c. sodium
cefpodoxime
    c. proxetil
cefprozil
ceftazidime
ceftibuten
Ceftin
    C. Oral
ceftizoxime
ceftriaxone
    c. sodium
cefuroxime
    c. axetil
Cefzil
Celera
celery stalking
Celestoderm-EV/2
Celestoderm-V
Celestone
celiac
    c. disease
    c. infantilism
    c. sprue
celibacy
celibate
celiohysterectomy
celiohysterotomy
celiomyomectomy
celiomyomotomy
celioparacentesis
celiosalpingectomy
celiosalpingotomy

celioscopy
celiotomy
    exploratory c.
    vaginal c.
cell
    c. adhesion factor (CAF)
    c. adhesion molecule (CAM)
    amniogenic c.
    antigen-sensitive c.
    Askanazy c.
    atypical c.
    B c.
    BeWo c.
    blastomere c.
    c. block analysis
    CD4 c.
    CD8 c.
    CD34 hematopoietic progenitor c.
    cell-salvaged packed c.
    ciliated c.
    clue c.
    c. collector
    committed c.
    corona radiata c.
    c. culture
    cumulus c.
    c. cycle
    c. cycling in chemotherapy
    cytotoxic c.
    daughter c.
    decidual c.
    dendritic c.
    c. determination
    diploid c.
    diploid spermatogonial stem c.
    c. division
    dome c.
    double c.
    dysplastic c.
    effector c.
    egg c.
    embryonic stem c.
    endodermal c.
    endothelial c.
    epithelioid c.
    extragonadal germ c.
    fetal c.
    fetouterine c.
    frozen red c.'s
    c. generation time
    germ c.
    glandular c.
    granulosa lutein c.
    c. growth inhibitor
    haploid c.
    HeLa c.
    helper T c.
    hematopoietic stem c. (HSC)

hobnail c.
Hoffbauer c.
human endothelial c. (HEC)
Hurthle c.
immunocompetent c.
c. interaction gene
interstitial c.
Ito c.
K c.
c. kill
killer c.
c. kinetics
koilocytotic c.
Langerhans c.
Langhans c.'s
Leydig c.'s
lipid c.
luteal c.
lutein c.
lymphoblastoid c.
lymphoid c.
lymphokine-activated killer c.
  (LAK)
c. lysate
maturation of c.
MCF-7 breast cancer c.
memory c.
mutant c.
myoepithelial c.
natural killer c.
neuroid c.
NK c.
nucleated red blood c.
osteoblast-like c.
owl's eye c.
packed red blood c.'s
parabasal c.
peripheral blood mononuclear c.
  (PBMC)
pituitary c.
placental giant c.
pregnancy c.
pregranulosa c.
primary embryonic c.
primordial germ c.
primordial pluripotent stem c.
Purkinje c.'s
Raji c.
C. Recovery System (CRS)
red blood c.
C. Saver
Schwann c.

Sertoli c.
Sertoli-Leydig c.
sex c.
sickle c.
silver c.
somatic c.
spindle c.
squamous c.
stem c.
steroid c.
stromal c.
suppressor c.
suppressor T c.
c. surface antigen
syncytial c.
T c.
target c.
theca c.
theca-interstitial c.
thecal interstitial c.
theca lutein c.
totipotent c.
totipotential c.
trophoblastic c.
tuboendometrial c.
vaginal smear intermediate c.
vaginal smear parabasal c.
vaginal smear superficial c.
Vero c.
viable endometrial c.
Vignal c.'s
white blood c.
WI-38 c.
WISH c.
Wistar Institute Susan Hayflick c.
yolk c.
cell-extracellular matrix adhesion
cell-mediated immunity (CMI)
Cellolite patty
cell-salvaged packed cell
cellular
  c. cytotoxic mechanism
  c. division
  c. immunity
  c. migration
  c. viral
cellulicidal
cellulitic phlegmasia
cellulitis
  cuff c.
  pelvic c.
  postoperative cuff c.

C

NOTES

Cell-VU disposable semen analysis
 chamber
celom
celomic
    c. epithelium
    c. metaplasia
Celontin
celosomia
celosomy
CEM
    cytosine arabinoside, etoposide,
    methotrexate
Cemill
cenadelphus
Cenafed
Cena-K
Cenani-Lenz syndactyly
Cenestin
    C. synthetic conjugated estrogens
    C. tablets
center
    birth care c.
    C.'s for Disease Control (CDC)
    electrophilic c.
    epiphyseal ossification c.
    c. for Health Technology (CHT)
    lower limb ossification c.
    Poison Control C. (PCC)
    X inactivation c. (XIC)
centigray (cGy)
centimeter
    cubic c. (cc)
centimorgan
Centocor CA 125 radioimmunoassay kit
central
    c. axis depth dose
    c. defect
    c. dogma
    c. hyperalimentation
    c. jaundice
    c. nervous system (CNS)
    c. nervous system development
    c. nervous system differentiation
    c. nervous system disease
    c. nervous system tumor
    c. placenta previa
    c. Recklinghausen disease type II
    c. retinal artery
    c. type neurofibromatosis
    c. venous catheter (CVC)
    c. venous line (CVL)
    c. venous pressure (CVP)
    c. venous pressure catheter
    c. venous pressure line
central-anterior
centralis
    neurinomatosis c.
    placenta previa c.

centralization
    fetal circulatory c.
centralopathic epilepsy
Centrax
Centre d'Etude de Polymorphism
 Humain
centric fusion translocation
centrifuge
centromere
    c. interference
centromeric
    c. banding
    c. instability-immunodeficiency
    syndrome
    c. region of chromosome
    c. signal
centronuclear myopathy (CNM)
centrotemporal epilepsy
Century urodynamics chair
Ceo-Two
Cepacol
    C. Anesthetic Troche
    C. Troche
cephalad
cephalexin
cephalhematoma
    c. deformans
cephalhydrocele
cephalic
    c. cry
    c. delivery
    c. forceps
    c. pole
    c. presentation
    c. replacement
    c. version
cephalization
cephalocele
cephalocentesis
cephalodactyly
    Vogt c.
cephalodiprosopus
cephalohematoma
cephalomelus
cephalometry
    ultrasonic c.
cephalonia
cephalopagus
cephalopelvic
    c. disproportion (CPD)
cephalopelvimetry
cephalopolysyndactyly syndrome
cephalosporin
cephalothin
    c. sodium
cephalothoracopagus
cephalotome
cephalotomy

cephalotribe
cephapirin
cephazolin
CEPH pedigree
cephradine
Cephulac
Ceporacin
Ceporex
Ceptaz
CeraLyte drink mix
c-*erb* B-2 oncogene
c-*erb* B-2 oncoprotein
c-*erb* B-2 protooncogene
cerclage
    Barnes c.
    cervical c.
    Mann isthmic c.
    McDonald cervical c.
    rescue cervical c.
    Shirodkar cervical c.
    transabdominal cervicoisthmic c.
    web c.
cerebellar
    c. ataxia
    c. degeneration
    c. hemisphere compression
    c. hypoplasia
    c. vermis agenesis
    c. vermis aplasia
    c. vermis hypo/aplasia, oligophrenia, congenital ataxia, ocular coloboma, hepatic fibrosis (COACH)
    c. vermis hypogenesis
cerebelloparenchymal disorder IV (CPD IV)
cerebellotrigeminal
    c. dermal dysplasia
    c. and focal dermal dysplasia
cerebellotrigeminodermal dysplasia
cerebellum
    absent c.
cerebral
    c. angioma
    c. atrophy
    c. blood flow
    c. dysfunction
    c. dysgenesis
    c. edema
    c. embolism
    c. GM1 gangliosidosis
    c. infarction

    c. leukodystrophy
    c. leukomalacia
    c. malformations, seizures, hypertrichosis, overlapping fingers syndrome
    c. palsy (CP)
    c. palsy antecedent
    c. palsy-hypotonic seizures-megalocornea syndrome
    c. thrombosis
    c. vasospasm
cerebrale
    cranium c.
cerebral, ocular, dental, auricular, skeletal syndrome (CODAS)
cerebral-placental ratio
cerebri
    pseudotumor c.
cerebriform
cerebroarthrodigital syndrome
cerebroatrophic hyperammonemia
cerebrocostomandibular syndrome
cerebrocutaneous angiomatosis
cerebrofacioarticular syndrome (CFA)
cerebrofaciothoracic syndrome or dysplasia
cerebrohepatorenal syndrome
cerebromacular degeneration
cerebroocular
    c. dentoauriculoskeletal syndrome (CODAS)
    c. dysgenesis (COD)
    c. dysgenesis-muscular dystrophy (COD-MD)
    c. dysgenesis-muscular dystrophy syndrome
    c. dysplasia-muscular dystrophy
cerebrooculomuscular syndrome (COMS)
cerebrooculonasal syndrome
cerebroosteonephrodysplasia (COND)
    Hutterite c.
cerebroosteonephrosis syndrome
cerebrospinal
    c. fluid (CSF)
cerebrovascular
    c. accident (CVA)
    c. disease
Ceredase
Cerespan
cerevisiae
    Saccharomyces c.
Cerezyme

C

NOTES

85

ceroid lipofuscinosis
Cerose-DM
Cerrobend
    C. block
    C. casting
certificate
    birth c.
certified
    C. Nurse-Midwife (CNM)
    C. Registered Nurse Anesthetist
    (CRNA)
Cerubidine
cerulean cataract
ceruloplasmin
cerumen
Cervagem
Cervex-Brush cervical cell sampler
cervical
    c. abnormality
    c. adenitis
    c. adenocarcinoma
    c. adenopathy
    c. agenesis
    c. amputation
    c. anomaly
    c. artery
    c. atypia
    c. block kit
    c. canal
    c. cancer
    c. cannula
    c. cap
    c. carcinoma
    c. carcinoma stimulation
    c. cerclage
    c. clamp
    c. clear cell adenocarcinoma
    c. cockscomb
    c. combing
    c. condyloma
    c. cone biopsy
    c. conization
    c. culture
    c. cytology
    c. dilation
    c. dysplasia
    c. dystocia
    c. ectopic pregnancy
    c. ectropion
    c. effacement
    c. epithelial neoplasia
    c. epithelium
    c. erosion
    c. eversion
    c. factor
    c. funneling
    c. GIFT
    c. incision

    c. incompetence (CI)
    c. infection
    c. insemination
    c. intraepithelial neoplasia (CIN)
    c. isthmus
    c. laceration
    c. leiomyoma
    c. lesion
    c. motion tenderness (CMT)
    c. mucorrhea
    c. mucosa
    c. mucus
    c. myoma
    c. os
    c. polyp
    c. pregnancy
    c. priming
    c. prolapse
    c. ripening
    c. sarcoma
    c. score
    c. smear
    c. stenosis
    c. stroma
    c. stump cancer
    c. stump tumor
    c. tenaculum
    c. tissue impedance range
    topographic c.
    c. transformation zone
    c. vertebral fusion
cervicectomy
cervices (*pl. of* cervix)
cervicitis
    chlamydial c.
    chronic c.
    gonorrheal c.
    mucopurulent c.
    nongonococcal c.
cervicography
cervicoplasty
cervicotomy
cervicovaginal
    c. fetal fibronectin
    c. fistula
    c. infection
    c. junction
    c. ridge
    c. secretion
Cervidil
    C. C.R.
    C. vaginal insert
Cer-View lateral vaginal retractor
cervigram
Cervilaxin
Cerviprost gel
CerviSoft cytology collection device
cervix, pl. **cervices**

anterior lip of the c.
atretic c.
barrel c.
barrel-shaped c.
bifid c.
carcinoma of c.
cockscomb c.
collared c.
cone biopsy of c.
conization of c.
dilation of c.
effacement of c.
fish-mouth c.
friable c.
incompetent c.
international classification of cancer
   of c.
malignant tumor of c.
multiple c.
c. neoplasm
shortened c.
strawberry c.
CES
cat eye syndrome
cranial electrical stimulation
cesarean, cesarean section
c. delivery
extraperitoneal c.
c. hysterectomy
Kerr c.
Latzko c.
low cervical c.
lower segment c. (LSCS)
low transverse c. (LTC)
c. operation
salvage c.
c. section (C-section)
transperitoneal c.
vaginal birth after c. (VBAC)
cesium
c.-137 (^{137}Cs)
c.-137 level
c. applicator
c. candle
c. cylinder
c. implant
c. irradiation
c. source
cessation
Cetacaine
Cetacort Topical
Cetamide

Cetane
cetirizine
Cetrorelix for injection
Cetrotide
Cetus trial
cetylpyridinium
Cevalin
Cevi-Bid
Ce-Vi-Sol
Cevita
CF
clavicular fracture
cystic fibrosis
^{252}Cf
californium-252
CFA
cerebrofacioarticular syndrome
CFC
cardiofaciocutaneous syndrome
c-fms protooncogene
CFND
craniofrontonasal dysplasia
CFUC
colony-forming unit in culture
CFU-E
colony-forming unit erythroid
CG
chorionic gonadotropin
CGH
comparative genomic hybridization
cGMP
cyclic guanosine monophosphate
cGy
centigray
Chadwick sign
Chagas disease
chain
beta c.
c. cystourethrography
heavy c.
K c.
kappa c.
light c.
chain-breaking antioxidant
chair
birthing c.
Century urodynamics c.
Midmark 413 power female
   procedure c.
chalasia
chalazion

NOTES

challenge
acrosome reaction with
ionophore c. (ARIC)
intravenous glucose c.
progestational c.
chamber
Cell-VU disposable semen
analysis c.
face c.
hyperbaric c.
Makler reusable semen analysis c.
Neubauer c.
Chamberlen forceps
Champetier de Ribes bag
chancre
hunterian c.
chancroid
chancroidal bubo
chandelier sign
change
atrophic c.
benign cellular c.'s (BCC)
breast c.
concomitant c.
failed physiologic c.
fibrocystic breast c.
focal c.
glomerular c.
harlequin color c.
hematological c.
hormone-stimulated endometrial c.
hydatidiform c.
immunohistochemical c.
immunologic c.
libidinal c.
c. of life
ovarian cycle c.
polyneuropathy, organomegaly,
endocrinopathy, M protein,
skin c.'s (POEMS)
retinal c.
Rias-Stella c.
sensorineural c.
structural c.
visual c.
change-point regression
channel
exposed large venous c.
surface epithelium vascular c.
vascular c.
voltage-dependent calcium c.
(VDCC)
Chapple syndrome
character
classifiable c.
denumerable c.
discrete c.
Y-linked c.

characteristic
epidemiological c.
c. face-hypogenitalism-hypotonia-
pachygyria syndrome
c. facies
morphological c.
organoleptic c.
secondary sex c.
characterization
immunohistochemical stromal
leukocyte c.
Charcot-Marie-Tooth (CMT)
C.-M.-T. disease
C. M.-T. neuropathy, X-linked
recessive, type II
C.-M.-T. syndrome (CMTS)
C.-M.-T. syndrome X-linked
recessive type II
C.-M.-T. syndrome, X-linked type
II with deafness and mental
retardation
**Charcot-Marie-Tooth-Hoffmann
syndrome**
CHARGE
coloboma, heart anomaly, choanal atresia,
retardation, and genital and ear
anomalies
CHARGE association
CHARGE syndrome
Charing Cross experience
Charlevoix disease
Charlson
C. comorbidity index
C. score
chart
Babson c.
Ballard c.
basal body temperature c.
Cardiff Count-to-Ten c.
Genentech growth c.
Liley three-zone c.
pedigree c.
POMARD anthropomorphic
measurement reference c.
Ross growth c.
Walker c.
Chassar
C. Moir-Sims procedure
C. Moir sling procedure
chaste
chastity
chat
cri du c.
CHD
congenital hip dislocation
Cheadle
C. disease
C. syndrome

Checklist
 Child Behavior C. (CBCL)
Chédiak-Higashi
 C.-H. anomaly
 C.-H. syndrome
cheek
 chipmunk c.'s
cheese-wiring
cheesy discharge
cheilognathopalatoschisis
cheilognathoprosoposchisis
cheilognathoschisis
cheilognathouranoschisis
cheiloschisis
Chelex
 C. bead
chemical
 c. carcinogen
 c. diabetes
 c. pneumonia
 c. pneumonitis
 c. pregnancy
 c. sampling
chemiluminescence
chemiluminescent
 c. illumination
 c. immunoassay (CIA)
Chemke syndrome
chemoattractant
chemo cinc receptor cx(3) CR1
chemoembolization
chemoimmunotherapy
chemokines
Chemo-Port catheter
chemoprophylaxis
 intrapartum c.
chemoradiation
chemoreceptor
chemotactic factor
chemotaxis
chemotherapeutic
 c. agent
 c. retroconversion
chemotherapy
 adjuvant c.
 alkylating c.
 cancer c.
 cell cycling in c.
 combination c.
 high-dose c.
 intraperitoneal c.
 Karnofsky performance status of c.

 metabolism in intraperitoneal c.
 c. phase trial
 postoperative c.
 prophylactic c.
chemstrip
 C. bG
 Micral C.
 C. 4 The OB
 C. 10 with SG
Cheney syndrome
Cherney incision
cherubism
 c., gingival fibromatosis, epilepsy,
  mental deficiency syndrome
cherub sign
chest
 c. examination
 funnel c.
 keel c.
 c. percussion and auscultation
 c. physical therapy
 c. radiography
 c. x-ray (CXR)
chewable
 E.E.S. C.
Cheyne-Stokes respiration
Chiari anomaly
Chiari-Arnold syndrome
Chiari-Frommel syndrome
chiasma, pl. **chiasmata**
 c. formation
 c. interference
Chiba needle
Chicago classification
Chicco breast pump
chickenpox
 c. pneumonia
Chid breast pump
Chilaiditi syndrome
CHILD
 congenital hemidysplasia with
  ichthyosiform erythroderma and limb
  defects
  CHILD syndrome
child
 c. abuse
 c. abuse dwarfism
 C. Behavior Checklist (CBCL)
 cardiorespiratory syndrome of
  obesity in c.
 C. Protective Services (CPS)

C

NOTES

child (*continued*)
term birth, living c. (TBLC)
unborn c.
child-adult mist (CAM)
childbearing
c. age
childbed fever
childbirth
Bradley method of prepared c.
Kitzinger method of c.
natural c.
physiologic c.
childbirth-related
c.-r. medical condition
c.-r. morbidity
childhood
c. breast
erythroblastic anemia of c.
c. fibromyalgia
c. genital trauma
papular acrodermatitis of c. (PAC)
progressive muscular dystrophy
of c.
c. pseudohypertrophic muscular
dystrophy
childlessness
children
Hospital for Sick C. (HSC)
Kaufman Assessment Battery
for C.
living c. (LC)
puppet c.
term infants, premature infants,
abortions, living c. (TPAL)
Wechsler Intelligence Scale for C.-
Revised (WISC-R)
Children's
C. Advil Suspension
C. Depression Inventory (CDI)
C. Depression Inventory test
C. Motrin Suspension
C. Silfedrine
CHIME
coloboma, heart defects, ichthyosiform
dermatosis, mental retardation, ear
defects syndrome
chimera
chimeric
c. gene
chimerism
blood c.
chin
cleft c.
c. dimple
galoche c.
c. position
CHIP
Coping Health Inventory for Parents

chip
gene c.
chipmunk cheeks
CHL
crown-heel length
Chlamydia
C. pneumoniae
C. psittaci
C. sepsis
C. trachomatis (CT)
C. trachomatis ligase chain
reaction
C. trachomatis tubal infertility
chlamydial
c. cervicitis
c. conjunctivitis
c. infection
c. pneumonia
c. urethritis
c. vaginitis
Chlamydiazyme
C. immunoassay
C. test
Chlo-Amine
C.-A. Oral
chloasma
Chlor-100
chloral hydrate
chlorambucil
chloramphenicol
Chlorate Oral
chlorcyclizine hydrochloride
chlordecone
chlordiazepoxide
chloride
acetylcholine c.
Adrenalin C.
ammonium c.
bethanechol c.
doxacurium c.
methylbenzethonium c.
oxybutynin c.
potassium c.
pralidoxime c.
sodium c.
tubocurarine c.
chloridorrhea
chlormethiazole
Chlor-Niramine
Chlorohist-LA
Chloromycetin
C. Injection
chlorophyllin copper complex
chloroplast
c. DNA
Chloroptic
chloroquine
chlorothiazide

chlorotrianisene
Chlorphed
Chlorphed-LA Nasal Solution
chlorpheniramine
Chlor-Pro
    C.-P. Injection
Chlorpromanyl
chlorpromazine
chlorpropamide
chlorprothixene
Chlortab
chlortetracycline
    c. fluorescence test
chlorthalidone
Chlor-Trimeton
    C.-T. Injection
    C.-T. Oral
Chlor-Tripolon
chlorzoxazone
CHMIS
    community health management
    information system
choanal atresia
chocolate
    c. agar plate
    c. cyst
cholangiogram
    operative c.
cholangiography
    single-film c.
cholangiopancreatography
    endoscopic retrograde c. (ERCP)
Cholebrine
cholecalciferol
cholecystectomy
    laparoscopic c. (LC)
cholecystitis
cholecystokinin
choledochal cyst
choledochojejunostomy
    Roux-en-Y c.
Choledyl
cholelithiasis
cholera
    c. infantum
    c. vaccine
cholestasis
    maternal c.
    c.-peripheral pulmonary stenosis
    c., pigmentary retinopathy, cleft
      palate syndrome

cholestatic
    c. hepatosis of pregnancy
    c. jaundice
cholesterol
    c. 20,22 desmolase
    c. ester storage disease
    c. oxidase
    plasma c.
cholesterolemia
    familial c.
cholestyramine
choline
    c. magnesium trisalicylate
    c. salicylate
    c. theophyllinate
cholinergic
    c. agonist
    c. drug
cholinesterase
    c. inhibitor
Choloxin
chondrodysplasia
    c. ectodermica
    giant cell c.
    Grebe c.
    hereditary c.
    c.-pseudohermaphrodism syndrome
    c. punctata
chondrodystrophia
    c. calcificans congenita
    c. congenita punctata
    c. congenita tarda
    c. fetalis calcificans
    myotonia c.
    c. myotonia
    c. tarda
chondrodystrophic myotonia
chondrodystrophy
    asphyxiating thoracic c.
    atypical c.
    hereditary deforming c.
chondroectodermal
    c. dysplasia
    c. dysplasia-like syndrome
chondromalacia fetalis
chondromere
chondroosteodystrophy
chondroplastic dwarfism
chondrosarcoma
    uterine c.
chondrosome

NOTES

C

Chooz
chorda, pl. chordae
    c. umbilicalis
chordablastoma
chordamesoderm
chordate
chordee
chorea
    c. gravidarum
    hereditary benign c.
    Huntington c.
    Sydenham c.
choreiform movement
choreoathetosis
    familial inverted c.
Chorex
chorioadenoma
    c. destruens
chorioallantoic
    c. membrane (CAM)
    c. placenta
    c. vessel
chorioamnion
    c. infection
chorioamnionic
    c. infection
    c. placenta
chorioamnionitis
    Gardnerella vaginalis c.
    histologic c.
chorioamniotic band
chorioangioma
chorioangiomatosis
chorioangiopagi
chorioangiosis
chorioblastoma
choriocarcinoma
    nongestational c.
choriodecidua
choriodecidual tissue
chorioepithelioma
choriogenesis
choriogonadotropin
choriomeningitis
    lymphocytic c.
chorion
    c. frondosum
    c. laeve
    c. sampling
chorionic
    c. adrenocorticotropin
    c. cyst
    c. gonadotropic hormone
    c. gonadotropin (CG)
    c. growth hormone
    c. plate
    c. sac
    c. somamammotropin

    c. thyrotropin
    c. vascularization
    c. vesicle
    c. villi
    c. villus biopsy (CVB)
    c. villus infarction
    c. villus ischemia
    c. villus sampling (CVS)
chorionicity
chorioretinal anomalies, corpus callosum
    agenesis, infantile spasms syndrome
choriovitelline placenta
choroid
    c. plexus
    c. plexus cyst (CPC)
    c. plexus papilloma
Choron
Chotzen syndrome
Christchurch chromosome
Christian-Andrews-Conneally-Muller
    syndrome
Christian-Opitz syndrome
Christian syndrome 1, 2
Christmas
    C. disease
    C. factor
    C. tree pattern
Christ-Siemens-Touraine syndrome
chromaffinoma
Chromagen
    C. FA
    C. OB
chromatid
    sister c.
chromatin
    sex c.
    X c.
    Y c.
chromatofocusing pH range
chromatography
    denaturing high-performance
      liquid c.
    gas c.
    thin-layer c.
chromatophore nevus of Naegeli
chromic
    #1 c.
    c. gut pelviscopic loop ligature
    c. phosphate
chromogen
chromogene
chromohydrotubation
chromomere
chromonemal address
chromoneme
chromopertubation
chromophobic adenoma

chromosomal

 c. aberration
 c. analysis
 c. anomaly
 c. banding
 c. breakage-immunodeficiency
  syndrome
 c. deletion
 c. inversion
 c. marker
 c. mosaicism
 c. pattern
 c. segment
 c. sex
 c. structural abnormality
 c. translocation

chromosome

 c. 1–23
 c. aberration
 accessory c.
 acentric c.
 acrocentric c.
 c. analysis
 arm of c.
 c. arm
 artificial c.
 B c.
 bacterial artificial c. (BAC)
 c. banding
 bivalent c.
 c. breakage
 centromeric region of c.
 Christchurch c.
 c. complement
 daughter c.
 c. deletion
 deletion of c.
 derivative c.
 dicentric c.
 c. diploid/tetraploid mixoploidy
  syndrome
 c. diploid/triploid mixoploidy
  syndrome
 expansion of c.
 founder c.
 fractured c.
 fragile X c.
 gametic c.
 giant c.
 c. GI deletion syndrome
 heterotropic c.
 heterotypical c.

 homologous c.
 human artificial c. (HAC)
 insertion of c.
 inversion of c.'s
 c. inversion syndrome
 c. inverted duplication
 inverted X c.
 iso-X c.
 c. jumping
 c. knob
 lampbrush c.
 late replicating c.
 long arm of c. (q)
 long arm of Y c.
 c. map
 c. mapping
 marker c.
 marker X c.
 metacentric c.
 metaphase c.
 mitochondrial c.
 mitotic c.
 c. 1–22 monosomy syndrome
 nonhomologous c.
 nucleolar c.
 odd c.
 4p+ c.
 c. paint
 c. pair
 c. pairing
 c. 11p detention syndrome
 c. 9p disorder
 Ph1 c.
  Philadelphia chromosome
 Philadelphia c. (Ph1 chromosome,
  Ph1 chromosome)
 c. 8p mosaic tetrasomy
 polytene c.
 c. 1p–22p deletion syndrome
 c. 1p–22p monosomy
 c. 1p–22p trisomy
 c. 1q–22q deletion syndrome
 c. 1q–22q duplication syndrome
 c. 1q–22q monosomy
 c. 1q–22q tetrasomy syndrome
 c. 1q–22q triplication syndrome
 c. 1q–22q trisomy
 c. 8 recombinant syndrome
 c. reduction
 ring c. 1–22 (r)
 c. 1–22 ring syndrome
 c. sequencing

C

NOTES

chromosome (continued)
sex c.
sex-linked c.
short arm of c. (p)
small c.
somatic c.
submetacentric c.
supernumerary c.
supernumerary marker c. (SMC)
c. 22 supernumerary marker
telocentric c.
c. tetraploidy syndrome
translocation c.
c. triploidy syndrome
c. 1–22 trisomy
c. 1–22 trisomy syndrome
Turner syndrome in female with
  X c.
c. 14 uniparental disomy syndrome
unpaired c.
W c.
c. walking
X c.
c. X autosome translocation
  syndrome
c. X fragility syndrome
c. X inversion syndrome
XO c.
c. XO syndrome
c. Xp21 deletion syndrome
c. Xp22 deletion syndrome
c. X pentasomy
c. Xp21 monosomy
c. Xq deletion syndrome
c. Xq duplication syndrome
c. Xq monosomy
c. Xq trisomy
XX c.
c. XXX syndrome
c. 47,XXX syndrome
c. XXXXX syndrome
c. XXXXY syndrome
c. XXY syndrome
Y c.
yeast artificial c. (YAC)
c. Y-18 translocation syndrome
Z c.
chromospermism
chromotubation
chronic
c. atrophic vulvitis
c. bullous dermatitis
c. cervicitis
c. cystic mastitis
c. hepatitis
c. hypertension
c. hypertrophic vulvitis
c. interstitial salpingitis

c. intertrigo
c. lung disease (CLD)
c. lymphocytic leukemia
c. obstructive bronchitis
c. pelvic pain (CPP)
c. scrotal hypothermia
c. vascular disease
chronicus
lichen simplex c.
ChronoFlex catheter
chronotropic effect
Chronulac
CHT
Center for Health Technology
Chudley-Lowry-Hoar syndrome
Chudley syndrome 1, 2
Chung microstaging system
Chvostek sign
chyle (var. of chylus)
chylomicron
c. retention disease
chylothorax
chylous ascites
chylus, chyle
CI
cervical incompetence
cord insertion
Ci
curie
CIA
chemiluminescent immunoassay
Cianchetti syndrome
ciclopirox
cidal level
Cidex soak
CI Direct Blue #53
cidofovir topical gel
Cidomycin
CIE
counterimmunoelectrophoresis
CI Food Blue #1
ciguatera poisoning
cilastatin
cilia, sing. **cilium**
immotile c.
ciliated
c. cell
c. cell endometrial adenocarcinoma
c. metaplasia
Ciloxan
cimetidine
CIN
cervical intraepithelial neoplasia
cineangiography
biplane c.
cinedefecography
anal sphincter c.

Cineloop
C. image review ultrasound system
C. Ultrasound
cinnarizine
Cinobac
cinoxacin
Cin-Quin
Cipro
C. HC Otic
ciprofloxacin
c. hydrochloride
circadian
c. cycle
c. rhythm
circle
Baudelocque uterine c.
Carus c.
Huguier c.
c. of Willis
Circon-ACMI
C.-A. cannula
C.-A. hysteroscope
C.-A. trocar
CircPlus compression wrap/dressing
circular end-to-end anastomosis (CEEA)
circulating
c. hormone
c. platelet antibody
circulation
ductal-dependent pulmonary c.
extracorporeal c. (ECC)
fetal c.
fetoplacental c.
hypophyseal portal c.
hypothalamic-hypophyseal portal c.
persistent fetal c. (PFC)
pituitary-hypothalamic c.
c. time
umbilical c.
uteroovarian c.
uteroplacental c.
circulatory
c. arrest
c. crossover
circumcise
circumcision
c. clamp
Mogen c.
pharaonic c.
Sunna c.
circumference
abdominal c. (AC)

fetal abdominal c.
head c. (HC)
midarm c.
occipitofrontal c.
circumferential
c. ringed creases of limbs
c. skin crease of limb
c. skin creases-psychomotor
retardation syndrome
circumflexa
ichthyosis linearis c.
circumflex artery
circummarginate placenta
circumscribed
c. mass
well-c.
circumscripta
myositis ossificans c.
circumscriptum
angiokeratoma c.
lymphangioma c.
circumscriptus
albinismus c.
Circumstraint
circumvallata
placenta c.
circumvallate placenta
cirrhonosus
cirrhosis
biliary c.
hypertrophic c.
c. of liver
cirsomphalos
CIS
carcinoma in situ
CISCA
cisplatin, Cytoxan, Adriamycin
cisplatin
c., Ara-C, caffeine (CAC)
bleomycin, etoposide, c. (BEP)
c., Cytoxan, Adriamycin (CISCA)
c., etoposide, Cytoxan, Adriamycin
(CECA)
intraperitoneal c.
c., methotrexate, vinblastine (CMV)
c., vinblastine, and bleomycin
vinblastine, bleomycin, c. (VBP)
9-cis-retinoic acid
cissa, citta, cittosis
cistern
prominent quadrigeminal plate c.
cisternal puncture

C

NOTES

cisternography
    isotope c.
cistron
Citracal
citrate
    calcium c.
    clomiphene c. (CC)
    cyclophenil c.
    fentanyl c.
    magnesium c.
    oral transmucosal fentanyl c.
    potassium c.
    sufentanil c.
    tamoxifen c.
    toremifene c.
citrated caffeine
Citrobacter
    C. amalonaticus
    C. diversus
    C. freundii
citrovorum
    c. factor
citrulline
citrullinemia
citrullinuria
citta (*var. of* cissa)
cittosis (*var. of* cissa)
CK
    creatine kinase
CL
    capacity of lung
    cleft lip
    compliance of lung
CLA
    X-linked cerebral ataxia
clade
Clado anastomosis
cladogenesis
Cladosporium herbarum
Claforan
clamp
    Allis c.
    Babcock c.
    back c.
    Backhaus c.
    Ballantine c.
    Buxton c.
    Carmault c.
    cervical c.
    circumcision c.
    DeBakey c.
    extracutaneous vas fixation c.
    Gomco circumcision c.
    Haney c.
    Heaney c.
    hysterectomy c.
    ICSI Massachusetts c.
    Kelly c.

    Kocher c.
    Lahey c.
    Lem-Blay circumcision c.
    Mogen c.
    Pean c.
    pediatric bulldog c.
    pediatric vascular c.
    pedicle c.
    Pennington c.
    Sztehlo umbilical c.
    thoracic c.
    thyroid Lahey c.
    umbilical c.
    vulsellum c.
    Willett c.
    Winston cervical c.
    Yellen c.
    Zeppelin c.
clamped down
Claripex
clarithromycin
Claritin
    C. RediTab
Claritin-D
    C.-D. 24-Hour
Clark
    C. classification of vulvar
      melanoma
    C. microstaging system
Clarke
    C. Hadfield syndrome
    C. ligator scissor forceps
Clarus model 5169 peristaltic pump
CLAS
    congenital localized absence of skin
clasped thumbs-mental retardation
  syndrome
class
    c. I, II receptor
classic
    c. abdominal Semm hysterectomy
      (CASH)
    c. X-linked recessive muscular
      dystrophy
classical
    c. galactosemia
    c. genetics
    c. incision extension
    c. transverse incision
    c. uterine incision
classifiable character
classification
    Acosta c.
    Astler-Coller modification of
      Dukes c.
    Bethesda Pap smear c.
    Bethesda system Pap c.
    Blaustein c.

Caldwell-Moloy c.
Chicago c.
Cori c.
Denver c.
Dripps–American Society of
   Anesthesiologists c.
Dukes c.
FAB c.
HIV c.
International Federation of
   Gynecology and Obstetrics c.
Jansky c.
Jewett c.
Kajava c.
microinvasive carcinoma c.
Moss c.
Pulec and Freedman c.
Schuknecht c.
TNM c.
tumor, node, metastasis c.
Wassel c.
White c.
Wolfe c. of breast cancer

clastogenic
clathrin
clavicle
congenital pseudoarthrosis of the c.
clavicular fracture (CF)
clavulanate potassium
clavulanic acid
Clavulin
clawfoot
clawhand
CLD
chronic lung disease
clean-catch
c.-c. urinalysis
c.-c. urine specimen
cleaner
Sklar aseptic germicidal c.
clean intermittent self-catheterization
cleanser
Dey-Wash skin wound c.
clear
c. cell adenocarcinoma
c. cell endometrial carcinoma
c. cell sarcoma
c. cell vaginal cancer
clearance
creatinine c.
c. of fetal product
immune c.

ultrafiltration virus c.
urate c.
urea c.
virus c.
Clearblue
C. Easy
C. Improved
ClearCut 2 electrosurgical handpiece
ClearPlan
C. Easy
C. Easy fertility monitor
C. Easy ovulation predictor
ClearSite
C. Hydro Gauze dressing
Clearview
C. hCG pregnancy test
ClearView uterine manipulator
cleavage
embryonic c.
manual c.
c. plan
c. stage
cleaved
c. amplified polymorphic sequence
c. embryo
cleft
branchial c.
c. chin
c. face
c. hand
hyobranchial c.
c. jaw
laryngeal c.
c. lip (CL)
c. lip, cleft palate, lobster claw
   deformity syndrome
c. lip/palate (CLP)
natal c.
orofacial c.
oroorbital c.
c. palate (CP)
c. palate, diaphragmatic hernia,
   coarse facies, acral hypoplasia
   syndrome
c. palate-lateral synechia syndrome
   (CPLS)
c. palate, microcephaly, large ears,
   short stature syndrome
c. spine
Stillman c.
c. vertebrae
visceral c.

NOTES

C

97

clefting
> ankyloblepharon, ectodermal dysplasia, c.
> ectrodactyly, ectodermal dysplasia, c. (EEC)
> hypertelorism, microtia, c. (HMC)
> c., ocular anterior chamber defect, lid anomalies syndrome

cleidocranial
> c. digital dysostosis
> c. dysplasia
> c. dysplasia syndrome

cleidocranialis
> dysostosis c.
> dysplasia c.

cleidocraniodigitalis
> dysostosis c.

cleidocraniopelvina
> dysostosis c.

cleidofacialis
> dysplasia c.

cleidorhizomelic syndrome
cleidorrhexis
cleidotomy
cleidotripsy
clemastine
clenched
> c. fist and pleural effusion
> c. fists

Cleocin
> C. HCl
> C. HCl Oral
> C. Pediatric Oral
> C. Phosphate
> C. T
> C. vaginal cream

Clevedan positive pressure respirator
CLIA
> Clinical Laboratory Improvement Amendment

click
> hip c.
> Ortolani c.
> pulmonary ejection c.

clidinium
clidoic
Clifford syndrome
climacteric
> grand c.
> c. psychosis
> c. syndrome

climacterium
Climara
> C. estradiol transdermal system
> C. estradiol transdermal system patch
> C. Transdermal

climatic bubo

clinch knot
clindamycin
> c. phosphate
> c. phosphate topical solution

clinical
> c. crib
> c. finding
> C. Laboratory Improvement Amendment (CLIA)
> c. pregnancy
> c. prognosis
> C. Risk Index for Babies (CRIB)
> c. staging

CliniCath peripherally inserted catheter
clinicopathologic
clinicopathological analysis
Clinistix
Clinitek 50 Urine Chemistry Analyzer
Clinitest
clinocephaly
clinodactyly
> fifth finger c.

clinometacarpy
Clinoril
clip
> Bleier c.
> Colotzmark C.
> FETENDO tracheal c.
> Filshie c.
> Hulka c.
> Hulka-Clemens c.
> Liga c.
> c. technique
> towel c.

Clistin
clitoral
> c. hood
> c. hypertrophy
> c. neurofibroma
> c. therapy device (CTD)

clitoridectomy
clitorides (*pl. of* clitoris)
clitoridis
> phimosis c.

clitoriditis, clitoritis
clitoris, pl. **clitorides**
> c. crisis
> c. enlargement
> c. tourniquet syndrome (CTS)

clitorism
clitoritis (*var. of* clitoriditis)
clitoromegaly
clitoroplasty
CLN
> neuronal ceroid lipofuscinosis

cloaca
> congenital c.

cloacal
    c. duct
    c. exstrophy
    c. malformation
    c. membrane
clobetasol
    c. propionate
Clocort Maximum Strength
clocortolone
Cloderm Topical
clodronate
clofibrate
Clomid
clomiphene
    c. citrate (CC)
    c. citrate challenge test (CCCT, C3T)
    c. fetal malformation
clomiphene-resistant polycystic ovary syndrome
clomipramine
clomocycline
clonality
clonal selection theory
clonazepam
clone
    c. bank
    DNA c.
    molecular c.
    overlapping c.'s
    recombinant c.
clonidine
    c. hydrochloride
cloning
    DNA c.
    embryo c.
    gene c.
    positional c.
    c. vector
clonogenic
    c. assay
    c. technique
clonus
    ankle c.
Clopra
Cloquet node
clorazepate
Clorpactin
closed
    c. chest massage
    c. drainage

closing
    c. coagulum
    c. ring of Winkler-Waldeyer
    c. volume
clostridia (*pl. of* clostridium)
clostridial bacteremia
Clostridium
    C. botulinum
    C. difficile
    C. perfringens
clostridium, pl. **clostridia**
closure
    incision c.
    nonlocking c.
    premature airway c.
    premature ductus arteriosus c.
    Smead-Jones c.
    Steri-Strip skin c.
    Tom Jones c.
clot
Clotrimaderm
clotrimazole
    c. vaginal cream 2%
Clouston syndrome
cloven spine
cloverleaf
    c. skull
    c. skull deformity
    c. skull syndrome
cloxacillin
Cloxapen
CLP
    cleft lip/palate
CLS
    Cornelia de Lange syndrome
club
    c. foot
    c. hand
clubbing
    hereditary c.
clubfoot
    Turco posteromedial release of c.
clubhand, club hand
    radial c.
clue cell
cluster
    DAZ gene c.
Clutton joints
Clyman endometrial curette
CMF
    cyclophosphamide, methotrexate, 5-fluorouracil

NOTES

C

CMFP
cyclophosphamide, methotrexate, 5-fluorouracil, and prednisone
Cytoxan, methotrexate, fluorouracil, prednisone
CMFPT
Cytoxan, methotrexate, fluorouracil, prednisone, tamoxifen
CMFVP
cyclophosphamide, methotrexate, 5-fluorouracil, vincristine, and prednisone
Cytoxan, methotrexate, fluorouracil, vincristine, prednisone
CMI
cell-mediated immunity
CMI-Mityvac cup
CMI-O'Neil cup
CMP
cow's milk protein
CMT
cervical motion tenderness
Charcot-Marie-Tooth
CMTS
Charcot-Marie-Tooth syndrome
CMV
cisplatin, methotrexate, vinblastine
congenital cytomegalovirus infection
controlled mechanical ventilation
cytomegalovirus
disseminated CMV
c-*myc*
c. oncogene
CNAP
continuous negative airway pressure
CNFS
craniofrontonasal syndrome
CNM
centronuclear myopathy
Certified Nurse-Midwife
CNS
central nervous system
CNS development
CO$_2$
carbon dioxide
CO$_2$ laser
^{60}Co
cobalt-60
COACH
cerebellar vermis hypo/aplasia, oligophrenia, congenital ataxia, ocular coloboma, hepatic fibrosis
Coactin
coagulase-positive staphylococcus
coagulation
argon beam c.
bipolar c.
cold c.
c. defect

diffuse intravascular c. (DIC)
c. disorder
disseminated intravascular c. (DIC)
c. factor
c. factor zymogen
c. profile
c. test
coagulator
argon beam c. (ABC)
cold c.
Elmed BC 50 M/M digital bipolar c.
Malis CMC-II bipolar c.
coagulopathy
consumption c.
heritable c.
incipient c.
inherited c.
maternal c.
coagulum
closing c.
coaptation
c. bipolar forceps
urethral c.
coarctation
c. of aorta
juxtaductal c.
juxtaductal aortic c.
coarse
c. calcification
c. facies
c. rale
Coat-A-Count assay
coated Vicryl Rapide suture
Coats disease
coaxial
c. position
c. sheath cut-biopsy needle
cobalt
c.-60 (^{60}Co)
c.-60 moving strip technique
c. megavoltage machine
radioactive c.
Coban dressing
Cobantril
Cobb-Ragde needle
Cobe gun
Coblation-Channeling surgical procedure
cocaine
c. addiction
c. baby
crack c.
cocci (*pl. of* coccus)
coccidioidomycosis
coccobacillary
c. bacteria
c. form
coccus, pl. **cocci**

Gram-negative cocci
Gram-positive cocci
coccygeus muscle
coccygodynia
coccyx
cochleopalpebral reflex
cochleosaccular degeneration
cochleovestibular paresis
Cochrane Pregnancy and Childbirth Database
Cockayne syndrome
cockscomb
cervical c.
c. cervix
cocktail
Brompton c.
GI c.
c. party syndrome
pediatric c.
COD
cerebroocular dysgenesis
CODAS
cerebral, ocular, dental, auricular, skeletal syndrome
cerebroocular dentoauriculoskeletal syndrome
code
genetic c.
codeine
cod liver oil
Codman Accu-Flow shunt
COD-MD
cerebroocular dysgenesis-muscular dystrophy
COD-MD syndrome
codominance
codominant
c. gene
c. inheritance
codon
stop c.
termination c.
coefficient
bone attenuation c.
c. of inbreeding
inbreeding c.
c. of parentage
Spearman correlation c.
coexistent fetus
cofactor
ristocetin c.

COF/COM
Cytoxan, Oncovin, fluorouracil plus Cytoxan, Oncovin, methotrexate
Coffey suspension
Coffin-Lowry syndrome
Coffin-Siris syndrome
Coffin-Siris-Wegienka syndrome
Coffin syndromes 1, 2
Cogentin
CO₂Guard
COHb
carboxyhemoglobin
Cohen
C. syndrome
C. uterine cannula
Cohnheim theory
cohort study
coil
Margulies c.
metal c.
coiled artery
coiling
coin biopsy
coincident
coital
c. age
c. factor
coitarche
coition
coitus
c. incompletus
c. interruptus
c. la vache
c. reservatus
c. Saxonius
Colace
cola-colored neonate
colarium
Colax
Colcher-Sussman technique
colchicine
cold
C. & Allergy Elixir
c. biopsy forceps
c. coagulation
c. coagulator
c. cup biopsy
c. knife cone
c. knife cone biopsy
c. knife conization
c. knife method
c. potassium cardioplegia

NOTES

C

Cold-Eeze
Cole
    C. endotracheal tube
    C. intubation procedure
    C. orotracheal tube
    C. syndrome
Cole-Carpenter syndrome
coleitis
coleocele
coleotomy
Cole-Rauschkolb-Toomey syndrome
colfosceril palmitate
coli
    Escherichia c.
Colibri forceps
colic
    c. artery
    bilary c.
    infantile c.
    c. intussusception
    meconial c.
    menstrual c.
    ovarian c.
    renal c.
    tubal c.
    uterine c.
colicky
Colifoam
coliform
colistimethate sodium
colistin
colitis
    amebic c.
    granulomatous c.
    infectious c.
    pseudomembranous c.
    tuberculous c.
    ulcerative c.
CollaCote dressing
collagen
    bovine dermal c.
    GAX c.
    microfibrillar c.
    c. vascular disease
    c. weakness
collagenase
collagenization
collagenosis
    mediastinal c.
collapse
    cardiopulmonary c.
collapsed subpectoral implant
collar
    c. of pearls
    venereal c.
    c. of Venus
collared cervix
Collastat

collateral
    venous c.
collection
    oocyte c.
collector
    cell c.
    Cytobrush cell c.
    Cytobrush-Plus cell c.
    Cytopick endocervical and
      uterovaginal cell c.
    Endocell endometrial cell c.
    Flexi-Seal fecal c.
    Leukotrap red cell c.
    Papette cervical c.
    Uterobrush endometrial sample c.
    Wallach-Papette disposable cervical
      cell c.
Colles
    C. fascia
    C. fracture
colli
    pterygium c.
collimator
Collins test
collodion
    c. baby
    c. skin
colloid
    c. carcinoma
    c. cyst
    c. infusion
    c. oncotic pressure
    c. osmotic pressure (COP)
    radioactive c.
    c. solution
Collyrium Fresh Ophthalmic
coloboma
    c.-anal atresia syndrome
    c., clefting, mental retardation
      syndrome
    c. cleft lip/palate-mental retardation
      syndrome
    c., heart anomaly, choanal atresia,
      retardation, and genital and ear
      anomalies (CHARGE)
    c., heart defects, ichthyosiform
      dermatosis, mental retardation, ear
      defects syndrome (CHIME)
    c.-hepatic fibrosis
    c., mental retardation,
      hypogonadism, obesity syndrome
    c., microphthalmos, hearing loss,
      hematuria, cleft lip/palate
      syndrome
    c.-microphthalmos syndrome
    c., obesity, hypogenitalism, mental
      retardation syndrome
colocolic intussusception

colocolponeopoiesis
colocolpopoiesis
    Kun c.
colon
    ascending c.
    c. carcinoma
    descending c.
    rectosigmoid c.
    sigmoid c.
    transverse c.
colongenic assay
colonic
    c. aganglionosis
    c. obstruction
    c. polyp
    c. polyposis
colonization
    candida c.
    oropharyngeal c.
    stool c.
    vaginal c.
colonoscopic release
colonoscopy
colony-forming
    c.-f. unit in culture (CFUC)
    c.-f. unit erythroid (CFU-E)
colony-stimulating
    c.-s. activity (CSA)
    c.-s. factor (CSF)
color
    c. Doppler energy (CDE)
    c. Doppler sonography (CDS)
    c. Doppler ultrasonography
    c. echocardiogram
    C. Power Angio imaging
Colorado tick fever
colorectal
    c. anastomosis
    c. cancer
    c. tumor
color-flow Doppler
colorimetric reverse-dot-blot-hybridization
ColorpHast Indicator Strips
coloscope
    ZM-1 c.
colostomy
    diverting c.
    fecal diversion c.
    temporary diverting c.
colostration
colostric
colostrorrhea

colostrous
colostrum
Colotzmark Clip
colovaginal fistula
colpatresia
colpectasis, colpectasia
colpectomy
    skinning c.
colpitis
    c. mycotica
colpocele
colpocleisis
    Latzko c.
    LeFort partial c.
colpocystitis
colpocystocele
colpocystoplasty
colpocystotomy
colpocystoureterotomy
colpocystourethropexy (CCUP)
colpodynia
colpohyperplasia
    c. cystica
    c. emphysematosa
colpohysterectomy
colpohysteropexy
colpohysterotomy
colpomicrohysteroscope
    Hamou c.
colpomicroscopy
colpomycosis
colpomyomectomy
colpopathy
colpoperineopexy
    abdominal sacral c.
colpoperineoplasty
colpoperineorrhaphy
colpopexy
    sacral c.
    sacrospinous c.
    transvaginal sacrospinous c.
colpoplasty
colpopoiesis
colpoptosis, colpoptosia
colporectopexy
colporrhagia
colporrhaphy
    Goffe c.
colporrhexis
colposcope
    Accuscope c.
    CooperSurgical overhead c.

NOTES

colposcope *(continued)*
 Leisegang c.
 OMPI c.
 Zeiss c.
 Zoomscope c.
colposcopic
 c. diagnosis
 c. grading of cervical dysplasia
 c. screening
colposcopically directed brushing
colposcopist
colposcopy
 digital imaging c.
 endocervical canal c.
 estrogen-assisted c.
 vulvar c.
colposcopy-trained
colpospasm
colpostat
 afterload c.
 Henschke c.
colpostenosis
colpostenotomy
colposuspension
 abdominal c.
 abdominal sacrospinous ligament c.
 Burch c.
 laparoscopic retropubic c.
colpotomy
 c. incision
colpoureterotomy
colpourethrocystopexy
 retropubic c.
colpourethropexy
 Burch c.
 modified Burch c.
colpoxerosis
COL 2301(terbutaline gel)
columnar
 c. epithelium
 c. epithelium papilla
Coly-Mycin
 C.-M. M
 C.-M. S
 C.-M. S Otic Drops
Colyte
coma
 diabetic c.
 hyperosmolar c.
combination
 c. chemotherapy
 estrogen-progestogen c.
 norgestrel/ethinyl estradiol c.
 c. oral contraceptive
combined
 c. birth control pill
 c. immunodeficiency

 c. pregnancy
 c. version
combing
 cervical c.
CombiPatch
Combivir
Comby sign
comedocarcinoma
Comfort personal lubricant gel
Comhist
 C. LA
commensal flora
commission
 physician payment review c.
 (PPRC)
commissioning couple
commissural lip pit
commissure
 anterior c.
 posterior c.
commissurotomy
 mitral c.
committed cell
communicable disease
communicating uterus
communication
 autocrine c.
 fetal-maternal c.
 paracrine c.
 vascular c.
communis
 truncus arteriosus c.
community
 c. health management information
 system (CHMIS)
 c. medicine
compacta
Companion 318 Nasal CPAP System
comparative
 c. embryology
 c. genomic hybridization (CGH)
 c. mapping
compatibility
 maternal-fetal HLA c.
Compazine
complement
 chromosome c.
 c. fixation
 c. fixation test
 c. value
complemental inheritance
complementary
 c. and alternative medicine (CAM)
 c. DNA (cDNA)
 c. gene
 c. RNA
 c. sequence

complementation
  c. test
complete
  c. abortion
  c. androgen insensitivity syndrome (CAIS)
  c. androgen resistance syndrome
  c. atrioventricular block
  c. breech presentation
  c. feminizing testes syndrome
  c. heart block
  c. hydatidiform mole
  c. linkage
  c. placenta previa
  c. precocious puberty
  c. remission
  c. testicular feminization
complete/complete/+ station
complex
  Adam c.
  c. adnexal endometrioma
  AIDS dementia c. (ADC)
  AIDS-related c. (ARC)
  amphotericin B cholesteryl sulfate c.
  antigen-antibody c.
  antiinhibitor coagulant c.
  axial mesodermal dysplasia c.
  blepharophimosis, ptosis, epicanthus inversus, telecanthus c.
  cardinal-uterosacral ligament c.
  chlorophyllin copper c.
  c. congenital heart disease
  del 11/aniridia c.
  Diana c.
  early amnion vascular disruption c.
  Eisenmenger c.
  Electra c.
  facioauriculovertebral malformation c.
  Fallot c.
  GALT-UDP-galactose c.
  gene c.
  Ghon c.
  Gollop-Wolfgang c.
  human factor IX c.
  Jocasta c.
  Lear c.
  major histocompatibility c. (MHC)
  c. mass
  nipple-areola c.
  Oedipus c.

Phaedra c.
  QRS c.
  spike and wave c.
  synaptonemal c.
  Thomson c.
  tuberous sclerosis c.
  tuboovarian c. (TOC)
  uterosacral c.
  VATER c.
  vitamin B c.
complexion
  florid c.
  pallid c.
compliance
  lung c.
  c. of lung (CL)
complication
  cardiovascular c.
  intraoperative c.
  neurologic c.
  obstetric c.
  operative site c.
  postoperative c.
  pregnancy c.
  pulmonary c.
  respiratory c.
  vascular c.
component
  blood c.
  buffy coat c.
  extensive intraductal c. (EIC)
  extracellular matrix c.
compound
  c. heterozygote
  nitroimidazole c.
  c. pregnancy
  c. presentation
Comprecin
comprehensive BRACA
compressibility
compression
  cerebellar hemisphere c.
  cord c.
  external pneumatic calf c. (EPC)
  c. force
  head c.
  intermittent pneumatic c.
  pneumatic c.
  spot c.
  thorax c.
  tracheal c.
  c. ultrasonography

NOTES

C

compression *(continued)*
    uterine c.
    vein c.
compressive dressing
compressus
    fetus c.
compromise
    fetal c.
Compton effect
computed
    c. axial tomography (CAT)
    c. tomographic pelvimetry
    c. tomography (CT)
    c. tomography laser mammography
      (CTLM)
computer-aided diagnosis (CAD)
computer-assisted
    c.-a. semen analysis (CASA)
    c.-a. semen analyzer (CASA)
COMS
    cerebrooculomuscular syndrome
concealed hemorrhage
conceive
    C. Ovulation Predictor
    PreCare C.
concentrate
    factor c.
    platelet c.
concentration
    ambient oxygen c.
    blood alcohol c. (BAC)
    fetal steroid c.
    inhibin c.
    lipoprotein c.
    maternal steroid c.
    mean hemoglobin c.
    mean plasma iron c.
    minimum inhibitory c. (MIC)
    plasma iron c.
    serum lithium c.
    steroid c.
concepti (*pl. of* conceptus)
conception
    assisted c.
    estimated date of c. (EDC)
    evacuation of retained products
      of c. (ERPC)
    natural c.
    products of c. (POC)
    retained products of c. (RPC)
    wrongful c.
Conceptrol
conceptus, pl. **concepti**
    C. fallopian tube catheterization
      system
    C. Soft Torque uterine catheter
    C. VS catheter
concha bullosa

Concise Plus hCG urine test
concomitant change
concordant twins
concrete pelvis
concussion
COND
    cerebroosteonephrodysplasia
condition
    childbirth-related medical c.
    fetal c.
    local ovarian c.
    maternal c.
    neonatal c.
    orthopaedic c.
    X-linked dominant c.
    X-linked recessive c.
conditional probability
condom
    female c.
    intravaginal c.
    vaginal c.
conductance
    airway c.
conduction
    abbreviated atrioventricular c.
    c. analgesia
    c. anesthesia
conduit
    ileal c.
    intestinal c.
    urinary c.
conduplicato corpore
condyloma, pl. **condylomata**
    condylomata acuminata
    c. acuminatum
    cervical c.
    c. latum
    recalcitrant c.
    resistant c.
condylomatous
Condylox
cone
    c. biopsy
    c. biopsy of cervix
    cold knife c.
    transvaginal c.
    vaginal c.
Confide HIV test kit
configuration
    villoglandular c.
confinement
    estimated date of c. (EDC)
    expected date of c. (EDC)
Conformant dressing
congenita
    aglossia c.
    alacrima c.
    amaurosis c.

amyoplasia c.
amyotonia c.
aplasia axialis extracorticalis c.
aplasia cutis c.
arthrochalasis multiplex c.
arthrogryposis multiplex c. (AMC)
bitemporal aplasia cutis c.
chondrodystrophia calcificans c.
cutis marmorata telangiectatica c.
dyskeratosis c.
hypertrichosis universalis c.
ichthyosis c.
macrosomia adiposa c.
pachyonychia c.
pseudoglioma c.
pterygo arthromyodysplasia c.
syngnathia c.
Thomsen myotonia c.

congenital

c. abducens facial paralysis
c. abnormality
c. absence of vas deferens
   (CAVD)
c. acromicria syndrome
c. adrenal hyperplasia (CAH)
c. adrenal hypoplasia
c. adrenal lipoid hyperplasia
c. afibrinogenemia
c. aleukia
c. alveolar dysplasia
c. amaurosis of retinal origin
c. amputation
c. analgesia
c. anemia of newborn
c. anomaly
c. anosmia
c. anosmia-hypogonadotropic
   hypogonadism syndrome
c. arthromyodysplastic syndrome
c. articular rigidity
c. atelectasis
c. atheosis
c. bullous urticaria pigmentosa
c. candidiasis
c. cataract
c. cataracts, sensorineural deafness,
   Down syndrome facial appearance,
   short stature, mental retardation
   syndrome
c. central hypoventilation syndrome
c. clasped thumbs

c. clasped thumbs-mental retardation
   syndrome
c. cloaca
c. complete heart block (CCHB)
c. contractural arachnodactyly
   (CCA)
c. contracture of extremity
c. cystic adenomatoid malformation
   (CCAM)
c. cytomegalovirus infection (CMV)
c. diaphragmatic hernia (CDH)
c. dislocated hip (CDH)
c. dyskeratosis
c. ectodermal dysplasia of face
c. ectodermic scalp defect
c. ectropion
c. elephantiasis
c. elevation of the scapula
c. emphysema, cryptorchidism,
   penoscrotal web, deafness, mental
   retardation syndrome
c. encephalo-ophthalmic dysplasia
c. endothelial corneal dystrophy
c. epulis
c. epulis of the newborn
c. erythroderma
c. erythropoietic porphyria
c. facial diplegia
c. folate malabsorption
c. glaucoma
c. heart block
c. heart defect
c. heart disease
c. hemidysplasia with ichthyosiform
   erythroderma and limb defects
   (CHILD)
c. hemolytic jaundice
c. hereditary hematuria
c. hip dislocation (CHD)
c. hydantoin syndrome
c. hypertrichosis-
   osteochondrodysplasia-cardiomegaly
   syndrome
c. hypoaldosteronism
c. hypocupremia syndrome
c. hypofibrinogenemia
c. hypogammaglobulinemia
c. hypomyelination
c. hypoplastic anemia
c. hypothalamic hamartoblastoma
c. hypothyroidism
c. hypothyroidism syndrome

NOTES

congenital *(continued)*
  c. hypotonia
  c. ichthyosis
  c. ichthyosis-mental retardation-spasticity syndrome
  c. ichthyosis-trichodystrophy syndrome
  c. injury
  c. lipoatrophic diabetes
  c. lipodystrophy
  c. listeriosis
  c. lobar emphysema
  c. localized absence of skin (CLAS)
  c. macular degeneration
  c. malformation
  c. megacolon
  c. microcephaly-hiatus hernia-nephrotic syndrome
  c. miosis
  c. muscular dystrophy
  c. muscular dystrophy with central nervous system involvement
  c. muscular hypertrophy-cerebral syndrome
  c. myotonic dystrophy
  c. nephrosis
  c. neutropenia
  c. nonhemolytic jaundice
  c. nonregenerative anemia
  c. nonspherocytic hemolytic anemia
  c. nystagmus
  c. obliterative jaundice
  c. oculofacial paralysis
  c. photosensitive porphyria
  c. pneumonia
  c. postural deformity
  c. progressive muscular dystrophy with mental retardation
  c. progressive oculoacousticocerebral degeneration
  c. pseudoarthrosis of the clavicle
  c. pseudohydrocephalic progeroid syndrome
  c. pterygium
  c. ptosis
  c. pulmonary lymphangiectasia
  c. retinitis blindness (CRB)
  c. retinitis pigmentosa
  c. rubella
  c. rubella syndrome (CRS)
  c. sensory neuropathy
  c. spastic paraplegia
  c. suprabulbar paresis
  c. supraspinous fossa
  c. syphilis
  c. thoracic scoliosis

  c. thrombocytopenia, Robin sequence, agenesis of corpus callosum, distinctive facies, developmental delay syndrome
  c. toxoplasmosis
  c. tubular stenosis
  c. universal muscular hypoplasia
  c. warfarin syndrome
congenitalis
  alopecia c.
  heredoretinopathia c.
Congest
congestion
  pelvic vein c.
congestive
  c. cardiomyopathy-hypergonadotropic hypogonadism syndrome
  c. heart failure
conglobata
  acne c.
conglutination
conidium
conization
  cautery c.
  cervical c.
  c. of cervix
  cold knife c.
  hot knife c.
  laser cervical c.
  loop diathermy cervical c.
conjoined twins
conjugata diagonalis
conjugate
  anatomic c.
  c. axis
  diagonal c.
  c. diameter of the pelvic inlet
  effective c.
  external c.
  false c.
  c. of inlet
  internal c.
  obstetric c.
  obstetric c. of outlet
  c. of pelvic outlet
  true c.
  urinary steroid c.
conjugated
  c. antibody
  c. equine estrogen (CEE)
  c. estrogen and meprobamate
  c. estrogens
conjugation
conjunctiva, pl. conjunctivae
  xerosis c.
conjunctivitis
  chlamydial c.
  gonococcal c.

infantile purulent c.
neonatal c.
silver nitrate c.
trachoma inclusion c. (TRIC)
ConMed electrosurgical pencil
connection
anomalous pulmonary venous c.
decidua-macrophage c.
total cavopulmonary c.
connective
c. tissue
c. tissue disorder
Conners Scale
connexin
Conn syndrome
conotruncal
c. anomaly face syndrome (CTAF)
c. heart defects
conotruncus
Conradi
C. disease
C. syndrome
Conradi-Hünermann syndrome
consanguineous
c. mating
consanguinity
consensus sequence
consent
informed c.
written c.
conservation
breast c.
conservative
c. drug use
c. surgery
conserved
evolutionarily c.
c. sequence
consideration
adrenal morphologic c.
constant
association c.
dissociation c.
pulmonary time c.
constipation
constitutional
c. delay
c. dwarfism
c. hirsutism
c. precocious puberty
c. short stature

constriction
fetal ductus arteriosus c.
c. ring
constrictive pericarditis-dwarfism syndrome
construction
Abbe vaginal c.
consumption
c. coagulopathy
maternal alcohol c.
oxygen c.
contact
sperm-cervical mucus c. (SCMC)
contagiosa
impetigo c.
contagiosum
molluscum c.
contaminant
measurable undesirable
respiratory c.'s (MURCS)
contaminate
contamination
bacterial c.
content
aspiration of gastric c.'s
bone mineral c.
evacuation of uterine c.'s
contig
c. map
c. map cosmid
Contigen
C. Bard collagen implant
C. glutaraldehyde cross-linked
collagen implant
contiguous gene syndrome
Contin
MS C.
continence
c. ring
continent
c. ileostomy
c. supravesical bowel urinary
diversion
c. urinary pouch
contingency table
continua
epilepsia partialis c.
continuous
c. ambulatory peritoneal dialysis
(CAPD)
c. arteriovenous hemofiltration
(CAVH)

NOTES

continuous *(continued)*
    c. distending airway pressure (CDAP)
    c. distention irrigation system (CDIS)
    c. epidural analgesia
    c. intravenous oxytocin drip
    c. murmur
    c. negative airway pressure (CNAP)
    c. positive airway pressure (CPAP)
    c. random variable
    c. running monofilament suture
    c. subcutaneous infusion (CSQI)
    c. subcutaneous insulin infusion
    c. subcutaneous insulin injection
    c. venovenous hemodialysis (CVVHD)
continuous/combined treatment
continuous-wave
    c.-w. Doppler
    c.-w. ultrasound imaging
contour
    cranial c.
    C. Profile anatomically shaped silicone breast
contoured tilting compression mammography
contraception
    barrier c.
    emergency c.
    hormonal c.
    long-acting c.
    morning-after c.
    oral c.
    postcoital c. (PCC)
    vaginal ring c.
contraceptive
    barrier c.
    combination oral c.
    c. device
    c. diaphragm
    c. effectiveness
    estrogen-progestin c.
    estrophasic oral c.
    Estrostep oral c.
    c. failure
    c. failure rate
    c. film
    c. foam
    c. implant
    injectable c.
    intrauterine c. device (IUCD)
    intravaginal c.
    c. jelly
    long-acting c.
    low-dose oral c.

    low steroid content combined oral c.
    c. method
    monophasic oral c.
    oral c. (OC)
    oral steroid c.
    ParaGard T380A intrauterine copper c.
    progestin oral c.
    c. ring
    sequential oral c.
    c. sponge
    steroid c.
    c. suppository capsule
    c. technique
    triphasic oral c.
    vaginal c.
contracted pelvis
contractility
contraction
    atrial c.
    Braxton Hicks c.
    myometrial c.
    pelvic c.
    premature ventricular c. (PVC)
    smooth muscle c.
    c. stress test (CST)
    tetanic c.
    tetanic uterine c.'s
    uterine c.
    Z degree of c.'s
    Z' degree of c.'s
    Z" degree of c.'s
contractural
    c. arachnodactyly disease
    c. arachnodactyly syndrome
contracture
    multiple articular c.
    c., muscle atrophy, oculomotor apraxia syndrome
contralateral
    c. ovary
    c. ovulation
    c. synchronous carcinoma
contrast
    Echovist c.
Control
control
    birth c.
    Centers for Disease C. (CDC)
    glycemic c.
    seizure c.
controlled
    c. mechanical ventilation (CMV)
    c. ovarian hyperstimulation
    c. vaginal delivery
Controller
    Pepcid AC Acid C.

controversy
 transfusion c.
conundrum
convergent sidewalls
conversion
 extraglandular c.
convex probe
convulsion
 benign familial neonatal c. (BFNC)
 benign infantile familial c.'s
  (BIFC)
 febrile c.
 neonatal c.
 puerperal c.
 salaam c.
Cook aspirator
Cooke-Medley Hostility Scale
Cooks syndrome
Cooley anemia
Coombs test
Cooper
 C. fascia
 C. Surgical Monopolor ELSG
  LEEP System
 suspensory ligaments of C.
 C. syndrome
cooperativity
 theca-granulosa cell c.
Coopernail sign
CooperSurgical
 C. LEEP system 1000
 C. overhead colposcope
COP
 colloid osmotic pressure
Copeland fetal scalp electrode
Cophene-B
Coping
 C. Health Inventory for Parents
  (CHIP)
copious
 c. antibiotic irrigation
 c. discharge
Copper-7 (Cu-7)
 C. intrauterine device
copper
 C. T intrauterine device
 c. transport disease
copulating pouch
copulation
 c. plug
copy
 DNA c.

large single c.
small single c.
cor
 c. biloculare
 c. pulmonale
 c. triatriatum
 c. triatriatum dexter
 c. triloculare biatriatum
cord
 c. abnormality
 c. blood
 c. blood bank
 c. blood erythropoietin level
 c. blood gas (CBG)
 c. blood transplantation
 body coils of c.
 c. compression
 c. entanglement
 furcate insertion of c.
 genital c.
 c. insertion (CI)
 kinked c.
 medullary c.
 nephrogenic c.
 nuchal c.
 omphalomesenteric c.
 palpable c.
 c. of Pflüger
 presentation of c.
 prolapse of umbilical c.
 rete c.'s
 c. serum level
 sex c.
 spermatic c.
 spinal c.
 tethered c.
 three-vessel c.
 c. torsion
 two-vessel c.
 umbilical c.
 vitelline c.
Cordarone
cordate pelvis
Cordguard umbilical cord sampler
cordiform
 c. pelvis
 c. uterus
cordis
 accretio c.
 ectopia c.
cordocentesis
cordotomy

C

NOTES

Cordran
> C. SP

core
> C. Dynamics disposable cannula
> C. Dynamics disposable trocar
> c. temperature

Corey ovum forceps
Corgard
Corgonject
Cori
> C. classification
> C. disease
> C. enzyme deficiency

Coricidin
coring
> c. biopsy gun
> intramyometrial c.
> myometrial c.
> uterine c.

corkscrew
> c. maneuver
> c. vessel

Cormax Ointment
cornea
> xerosis c.

corneal
> c. leukoma
> c. opacity

Cornelia
> C. de Lange (CDL)
> C. de Lange syndrome (CLS)

Corner-Allen test
Corning method
Cornoy solution
cornu, pl. **cornua**
> c. uteri
> uterine c.

cornual
> c. anastomosis
> c. gestation
> c. pregnancy

Corometrics
> C. fetal monitor
> C. Gold Quik Connect Spiral electrode tip
> C. 118 maternal/fetal monitor
> C. Model 900SC in-office mammography
> C. Quik Connect Spiral electrode tip

corona
> c. of penis
> c. radiata
> c. radiata cell

coronal
> c. craniosynostosis
> c. suture line of skull

coronary
> c. artery disease
> c. heart disease
> c. thrombosis
> c. vasospasm

Coronex
corpus, pl. **corpora**
> c. albicans cyst
> c. callosum agenesis
> c. callosum agenesis, chorioretinal abnormality syndrome
> c. callosum agenesis, chorioretinopathy, infantile spasms syndrome
> c. callosum agenesis, facial anomalies, salaam seizures syndrome
> c. callosum hypoplasia, retardation, adjusted thumbs, spastic paraparesis, hydrocephalus syndrome (CRASH)
> c. callosum partial agenesis
> c. cavernosum
> c. fibrosum
> c. hemorrhagicum
> corpora lutea
> corpora lutea cysts
> c. luteum
> c. luteum cyst
> c. luteum deficiency syndrome
> c. luteum dysfunction
> c. luteum function
> c. luteum insufficiency
> c. luteum size
> c. luteum spurium
> c. luteum verum
> pediculosis c.
> tinea c.
> uterine c.
> c. of uterus

corpuscle
> Bennett small c.'s
> genital c.
> meconium c.
> Nunn engorged c.'s

corrodens
> *Bacteroides c.*
> *Eikenella c.*

CorrTest method
Corson myoma forceps
CortaGel Topical
Cortaid
> C. Maximum Strength Topical
> C. with Aloe Topical

Cortamed
Cortate
Cortatrigen Otic
Cort-Dome Topical

Cortef Feminine Itch Topical
Cortenema enema
cortex, pl. cortices
    adrenal c.
    double c.
    ovarian c.
    renal c.
cortical
    c. agenesis
    c. granule
    c. granule exocytosis
    c. gyral abnormality
    c. hemorrhage
    c. implantation
    c. mantle
    c. mass
    c. necrosis
    c. nephron
    c. reaction
    c. supremacy stage
    c. tuber
corticalis
cortices (*pl. of* cortex)
corticoid
corticosteroid
    antenatal c.
    fluorinated c.
corticosteroid-binding
    c.-b. globulin (CBG)
    c.-b. globulin-binding capacity (CB-GBC)
corticosterone
corticotrope
corticotropin-like intermediate lobe peptide
corticotropin-releasing
    c.-r. factor
    c.-r. hormone (CRH)
    c.-r. inhibitor
Corticreme
Cortiment
cortisol
    24-hour urinary free c.
    c. level
    c. response
cortisone
Cortisporin Otic
Cortisporin-TC
    C.-TC Otic
    C.-TC Otic Suspension
Cortizone-10 Topical
Cortizone-5 Topical

Cortoderm
cortol
Cortone
Cortrosyn
    C. stimulation test
Corynebacterium
    C. parvum
    C. vaginitis
coryza
co-segregation
Cosmegen
cosmid
    contig map c.
Costello syndrome
costovertebral
    c. angle tenderness (CVAT)
    c. dysplasia
cosyntropin
    c. stimulation test
cot
    finger c.
Cotazym
Cotrim
    C. DS
co-trimoxazole
Cotte operation
Cottle-Neivert retractor
cotton swab test
cotton-tipped applicator
co-twin
    death of c.
cotyledon
    fetal c.
    maternal c.
    c. perfusion system
    placental c.
cotyledonary placenta
coudé catheter
cough
    barking c.
    croupy c.
    Pedituss C.
    staccato c.
    c. test
    uterine c.
    whooping c.
cough-pressure transmission ratio
Cough-X lozenge
Coulter
    C. Channelyser cell analyzer
    C. counter
Coumadin

NOTES

C

coumarin
    c. derivative
    c. syndrome
counseling
    genetic c.
Counsellor-Davis artificial vagina operation
Counsellor-Flor modification of McIndoe technique
Counsellor vaginal mold
count
    amniotic fluid white blood cell c.
    bacterial c.
    blood c.
    CD4 cell c.
    CD8 cell c.
    differential cell c.
    erythrocyte c.
    hemolysis, elevated liver enzymes, and low platelet c. (HELLP)
    kick c.
    lamellar body c. (LBC)
    lap c.
    leukocyte c.
    platelet c.
    reticulocyte c.
    sperm c.
counter
    Coulter c.
    c. stab wound incision
counterimmunoelectrophoresis (CIE)
couple
    commissioning c.
    c. testing
coupling
    receptor c.
course
    long c.
    short c.
court-ordered obstetrical intervention
couvade
Couvelaire uterus
cover
    Bili mask phototherapy eye c.
    Sheathes ultrasound probe c.
    Ultra Cover transducer c.
Coverlet dressing
Cover-Roll gauze
Cover-Strip wound closure strip
Cowchock-Fischbeck syndrome
Cowchock syndrome
Cowper gland
cowperian duct
cow's milk protein (CMP)
COX
    cyclooxygenase
        COX pathway

COX-1
    cyclooxygenase-1
COX-2
    cyclooxygenase-2
coxa
    c. valga
    c. vara
coxarthrosis
coxoauricular syndrome
Coxsackievirus
    C. B
    C. infection
CP
    cerebral palsy
    cleft palate
CPAP
    continuous positive airway pressure
        nasal CPAP
        CPAP ventilator
CPC
    choroid plexus cyst
CPD
    cephalopelvic disproportion
CPD IV
    cerebelloparenchymal disorder IV
C-peptide
CPK
    creatine phosphokinase
CPLS
    cleft palate-lateral synechia syndrome
CPP
    chronic pelvic pain
CPR
    cardiopulmonary resuscitation
C_{21} progestin
C_{21} progestogen
CPS
    Child Protective Services
        CPS ID chromogenic medium
cps
    cycles per second
CR1
    chemo cinc receptor cx(3) C.
C.R.
    Cervidil C.R.
crab louse
crack
    c. baby
    c. cocaine
cracked pot sound
crackling rale
cradle
    c. cap
    Criss Cross C.
cramp
    leg c.
cramping
    uterine c.

Crane-Heise syndrome
cranial
- c. contour
- c. duplication
- c. electrical stimulation (CES)
- c. meningocele
- c. sclerosis, osteopathia striata, macrocephaly syndrome
- c. sclerosis with striated bone disease
- c. sign
- c. suture
- c. synostosis
- c. ultrasound

cranioacrofacial syndrome
craniocarpotarsal
- c. dystrophy
- c. syndrome

craniocaudal view
craniocerebellocardiac (3C, CCC)
- c. dysplasia
- c. syndrome

cranioclasia, cranioclasis
cranioclast
Zweifel-DeLee c.
craniocleidodysostosis
craniodiaphysial dysplasia
craniodidymus
cranioectodermal dysplasia (CED)
craniofacial
- c. anomalies, polysyndactyly syndrome
- c. dysmorphism, absent corpus callosum, iris colobomas, connective tissue dysplasia syndrome
- c. dysmorphism-polysyndactyly syndrome
- c. dysmorphology
- c. dysostosis

craniofacial-deafness-hand syndrome
craniofenestria
craniofrontal dysplasia
craniofrontonasal
- c. dysostosis
- c. dysplasia (CFND)
- c. syndrome (CNFS)

craniolacunia
craniomalacia
craniomeningocele
craniometaphysial dysplasia
craniooculofrontonasal malformation

crani oorodigital syndrome
craniopagus
craniopathy
craniopharyngioma
craniorhiny
craniorrachischisis
cranioschisis
cranioskeletal dysplasia with acroosteolysis
craniostenosis
craniostosis
craniosynostosis
- c. arachnodactyly, abdominal hernia syndrome
- c. arthrogryposis, cleft palate syndrome
- c. ataxia, trigeminal anesthesia, parietal anesthesia and pons, vermis fusion syndrome
- coronal c.
- c.-lid anomalies syndrome
- c.-radial aplasia syndrome

craniotabes
craniothoracopagus syncephalus
craniotome
craniotomy
craniotubular
- c. dysplasia
- c. dysplasia, growth retardation, mental retardation, ectodermal dysplasia, loose skein

cranium
- c. bifidum
- c. bifidum occultum
- c. cerebrale
- c. viscerale

Cranley Maternal-Fetal Attachment Scale
CRASH
corpus callosum hypoplasia, retardation, adjusted thumbs, spastic paraparesis, hydrocephalus syndrome
craving
pica c.
CRB
congenital retinitis blindness
C-reactive protein (CRP)
cream, creme
Acticin C.
Aldara c.
Amino-Cerv pH 5.5 cervical C.
AVC C.

NOTES

115

cream *(continued)*
 Cleocin vaginal c.
 clotrimazole vaginal c. 2%
 dihydrotestosterone c.
 DV Vaginal C.
 Elimite C.
 EMLA c.
 Estrace vaginal c.
 estradiol vaginal c.
 Fungoid C.
 imiquimod c.
 intravaginal c.
 Lamisil C.
 lidocaine-prilocaine c.
 Masse Breast C.
 Maximum Strength Desenex Antifungal C.
 miconazole nitrate vaginal c.
 Nupercainal c.
 nystatin and triamcinolone c.
 Ogden vaginal c.
 penciclovir c.
 PEN-Kera moisturizing c.
 Preparation-H hydrocortisone c.
 Sklar c.
 Sween C.
 Terazol 3 vaginal c.
 Terazol 7 vaginal c.
 triple sulfa c.
 vaginal c.
 Vite E C.
creamy vulvitis
crease
 palmar c.
 plantar c.
 simian c.
 sole c.
 Sydney c.
creatine
 c. kinase (CK)
 c. phosphokinase (CPK)
creatinine
 c. clearance
Credé
 C. maneuver
 C. maneuver of eyes
 C. maneuver of uterus
creeping
cre-loxP system
cremasteric fascia
creme *(var. of* cream)
cremnocele
crenation sign
Creon
crepitant rale
crepitation
crepitus

crescent
 c. cell anemia
 C. pillow
Crescormon
CREST
 calcinosis cutis, Raynaud phenomenon, esophageal motility disorders, sclerodactyly, telangiectasia
 CREST syndrome
crest
 neural c.
Cresylate
cretin
 c. dwarfism
cretinism
 athyrotic c.
cretinism-muscular hypertrophy syndrome
cretinoid dysplasia
Creutzfeldt-Jakob syndrome
CR-gram
 cardiorespirogram
CRH
 corticotropin-releasing hormone
cri
 c. du chat
 c. du chat syndrome
CRIB
 Clinical Risk Index for Babies
crib
 clinical c.
 c. death
 open c.
 tongue c.
Crib-O-Gram
cribriform hymen
cricoid
 c. pressure
 c. split
Crigler-Najjar syndrome
Crile
 C. forceps
 C. hemostat
Crile-Wood needle holder
criminal abortion
Crinone
 C. bioadhesive progesterone gel
 C. progesterone gel
crisis, pl. **crises**
 addisonian c.
 adrenal c.
 aplastic c.
 clitoris c.
 hypercalcemic c.
 lupus c.
 sickle cell c.
 thyroid c.
Crisponi syndrome

Criss Cross Cradle
crisscross heart
crista dividens
criteria, sing. **criterion**
    Amsel c.
    Beighton c.
    Bell staging c.
    Friedrich c.
    Jones c.
    Kass c.
    modified Beighton c.
    Norris-Carrol c.
    Nugent c.
    Spiegel c.
    Spiegelberg c.
critical
    c. care monitoring
    c. temperature
    c. weight hypothesis
Crixivan
CRL
    crown-rump length
CRNA
    Certified Registered Nurse Anesthetist
crocodile
    c. skin
    c. tongue
Crohn disease
Crolom
Crome syndrome
Cromoglycate
    PMS-Sodium C.
cromolyn
    c. sodium
    c. sodium inhalation aerosol
crooked fingers syndrome
Crosby-Kugler pediatric capsule
cross
    c. breeding
    bridging c.
    dihybrid c.
    monohybrid c.
    reciprocal c.
    C. syndrome
    c. table
crossed
    c. adductor reflex
    c. extension reflex
    c. polysyndactyly
crossing
    c. over
cross-linking

Cross-McKusick-Breen syndrome
crossover
    circulatory c.
cross-reaction
cross-sectional
crotamiton
croup
croupette
croupy cough
Crouzon
    C. craniofacial dysostosis
    C. disease
    C. syndrome
crowding
    fetal c.
crown-heel length (CHL)
crowning
crown-rump length (CRL)
CRP
    C-reactive protein
CRS
    caudal regression syndrome
    Cell Recovery System
    congenital rubella syndrome
CRST
    calcinosis cutis, Raynaud phenomenon,
        sclerodactyly, telangiectasia
    CRST syndrome
cruciferous vegetable
crude risk ratio
Cruiser hip abduction brace
cruris
    tinea c.
crusta lactea
cry
    cat c.
    cephalic c.
    high-pitched c.
    uterine c.
crying cat syndrome
CRYOcare
    C. cryoablation system
cryocauterization
cryocautery
cryoconization
CryoGenetics CryoPrism
cryoglobulinemia
cryogun
    Wallach LL100 cryosurgical c.
Cryomedics
    C. disposable LLETZ electrode
    C. electrosurgery system

C

NOTES

cryomyolysis
cryoprecipitate
cryopreservation
    oocyte c.
cryopreserved
    c. embryo
CryoPrism
    CryoGenetics C.
cryoprotectant
cryosurgery
cryotherapy
cryovial
cryptic hyperandrogenism
cryptoazoospermia
cryptocephalus
cryptococcal meningitis
cryptococcosis
cryptodidymus
cryptomenorrhea
cryptomerorachischisis
cryptomicrotia-brachydactyly syndrome
cryptophthalmia
    c.-syndactyly syndrome
    c. syndrome
cryptophthalmos
    c.-syndactyly syndrome
    c. syndrome
cryptorchid
cryptorchidism
cryptotia
CrystalEyes endoscopic video system
crystallina
    miliaria c.
crystallization
    fern leaf c.
crystalloid
    Reinke c.'s
    c.'s of Reinke
    c. solution
crystal violet
Crystapen
Crysticillin AS
Crystodigin
CS
    cyclic sedentary
        Poly-Histine CS
C&S
    culture and sensitivity
^{137}Cs
    cesium-137
CS-5 cryosurgical system
CSA
    colony-stimulating activity
Csapo abortion
C-section
    cesarean section
        LUST C-section

CSF
    cerebrospinal fluid
    colony-stimulating factor
Csillag disease
CSQI
    continuous subcutaneous infusion
CST
    contraction stress test
        antepartum fetal CST
        fetal CST
C$_{19}$-steroid
CT
    Chlamydia trachomatis
    computed tomography
        CT guidance
        CT laser mammography
        CT pelvimetry
C3T
    clomiphene citrate challenge test
CTAF
    conotruncal anomaly face syndrome
CTD
    clitoral therapy device
C-telopeptide
    type I collagen C.-t.
C-thalassemia
CTL
    cytotoxic T lymphocyte
CTLM
    computed tomography laser
      mammography
CTP
    carboxyl terminal peptide
CTS
    clitoris tourniquet syndrome
Cu-7
    Copper-7
      Cu-7 intrauterine device
cube pessary
cubic
    c. centimeter (cc)
    c. centimeter per hour (cc/hr)
    c. centimeters per kilogram per
      day (cc/kg/d)
cubitus valgus
Cue Fertility Monitor
cuff
    c. cellulitis
    Ethox c.
    posterior vaginal c.
    vaginal c.
cuirass respirator
Culcher-Sussman technique
culdeplasty
cul-de-sac
    c.-d.-s. biopsy
    Douglas c.-d.-s.
    c.-d.-s. of Douglas

posterior c.-d.-s.
rectouterine c.-d.-s.
culdocele
culdocentesis
nondiagnostic c.
Wiegernick c.
culdoplasty
Halban c.
high McCall c.
Marion-Moschcowitz c.
Mayo c.
McCall c.
prophylactic c.
Torpin-Waters-McCall c.
culdoscope
culdoscopy
culdotomy
Cullen sign
Culpolase laser
cultivar
cultural artifact
culture
amniotic fluid cell c.
blood c.
cell c.
cervical c.
colony-forming unit in c. (CFUC)
extended c.
fibroblast c.
gonorrhea c.
c.-negative cytomegalovirus infection
oocyte c.
c. and sensitivity (C&S)
sputum c.
tracheal-aspirate c.
Ureaplasma c.
URICULT c.
urine c.
cumulative
c. conception rate (CCR)
c. gene
c. sum test
cumulus
c. cell
c. oophorus
c. ovaricus
cup
Bird OP c.
CMI-Mityvac c.
CMI-O'Neil c.
c. ear
c. insemination

Instead feminine protection c.
Malström c.
Milex cervical c.
Mityvac obstetric vacuum
extractor c.
Mityvac Super M c.
O'Neil c.
Tender-Touch vacuum birthing c.
Cuprimine
Curaderm dressing
Curafil dressing
curage
Curagel Hydrogel dressing
Curasorb calcium alginate dressing
curdy discharge
Curet
curetment (*var. of* curettement)
curettage
dilation and c. (D&C)
endocervical c. (ECC)
endometrial c. (EMC)
fractional dilation and c.
radial c.
repeat c.
sharp c.
suction c.
vaginal interruption of pregnancy
with dilatation and c. (VIP-DAC)
curette, curet
Accurette endometrial suction c.
banjo c.
Berkeley suction c.
Bumm c.
Clyman endometrial c.
Duncan c.
Green uterine c.
Heaney c.
Helix endocervical c.
Helix uterine biopsy c.
Kelly-Gray c.
Kevorkian c.
Kevorkian-Younge c.
Mi-Mark disposable endocervical c.
Novak c.
Pipelle endometrial suction c.
Pipet C.
Randall suction c.
Ridpath c.
Shapleigh c.
Sims c.
St. Clair-Thompson c.
Tabb c.

NOTES

curette *(continued)*
    Thomas c.
    Townsend endocervical biopsy c.
    uterine c.
    Vabra suction c.
    Yankauer c.
    Z-Sampler endometrial suction c.
curettement, curetment
curetting/curettage
    endocervical c./c.
curie (Ci)
Curling ulcer
Curosurf poractant alpha
Curran syndrome
currant jelly stool
Currarino triad
currens
    larva c.
current
    diathermy c.
Curretab
    C. Oral
Curry-Jones syndrome
curse
    Ondine c.
    c. of Ondine
Curtis syndrome
curve
    Barnes c.
    Carus c.
    Friedman c.
    isodose c.
    Kaplan-Meier survival c.
    Liley c.
    oxygen dissociation c.
    sexual response c.
    standard c.
curved
    c. hemostat
    c. Mayo scissors
curvilinear velocity (VCL)
CUSA
    Cavitron ultrasonic surgical aspirator
CUSALap
    C. accessory needle
Cushing
    C. disease
    C. forceps
    C. syndrome
cushingoid
    c. appearance
    c. facies
    c. syndrome
Cushing-Rokitansky ulcer
cushion
    birth c.
    c. defect of heart

    scintimammography prone breast c.
    sucking c.
cutaneocerebral angioma
cutaneomastocystosis
cutaneomeningospinal angiomatosis
cutaneous
    c. albinism
    c. anthrax
    c. blood flow
    c. hepatic porphyria
    c. larva migrans
    c. leishmaniasis
    c. lesion
    c. melanoma
    c. mucormycosis
    c. nevi
    c. pressure pulse
    c. vasculitis
cutis
    aplasia c.
    c. aplasia
    c. aplasia of scalp
    c. elastica
    c. hyperelastica
    c. laxa
    c. marmorata
    c. marmorata telangiectatica
      congenita
    c. navel
    c. verticis gyrata
    c. verticis gyrata, thyroid aplasia,
      mental retardation syndrome
    xanthosis c.
Cutivate
cut section
cutter
    Endopath ETS-FLEX endoscopic
      articulating linear c.
    Endopath EZ35 endoscopic
      linear c.
    endoscopic linear c.
    Polaris reusable c.'s
cutting
    c., annoyance, guilt, eye-opener
      (CAGE)
    c. loop
CV
    cardiovascular
CVA
    cerebrovascular accident
CVAT
    costovertebral angle tenderness
CVB
    chorionic villus biopsy
CVC
    central venous catheter
CVL
    central venous line

CVP
central venous pressure
CVP line
CVS
chorionic villus sampling
CVVHD
continuous venovenous hemodialysis
CXR
chest x-ray
CY 2010
cyanocobalamin
cyanosis
perioral c.
cyanotic
c. congenital heart disease (CCHD)
cybernin
cyclacillin
cyclamate
calcium c.
sodium c.
cyclandelate
cyclase
adenylate c.
cyclazocine
cycle
battering c.
cell c.
circadian c.
endometrial c.
estrogen-progestin artificial c.
fertility c.
futile c.'s
genesial c.
glutamate c.
initiated c.
intrauterine pressure c.
itch-scratch c.
menstrual c.
menstrual-ovarian c.
c.-monitoring detail
c.-nonspecific agent
ovarian c.
ovulatory menstrual c.
c.'s per second (cps)
reproductive c.
sexual response c.
spontaneous menstrual c.
cyclencephalus
cycle-specific agent
cyclic
c. adenosine monophosphate
(cAMP)

c. adenosine 3′,5′-monophosphate
c. guanosine monophosphate
(cGMP)
c. guanosine 3′,5′-monophosphate
c. mastalgia
c. proliferative endometrium
c. sedentary (CS)
c. vulvitis
c. vulvodynia
c. vulvovaginitis
cyclicity
postmenarchal c.
cyclizine lactate
Cyclocort Topical
Cyclofem
cycloheximide
Cyclomen
cyclooxygenase (COX)
c.-1 (COX-1)
c.-2 (COX-2)
platelet c.
cyclooxygenase-2-derived prostanoid
cyclopentanoperhydrophenanthrene
cyclopenthiazide
cyclopentolate hydrochloride
cyclophenil citrate
cyclophosphamide
c., Adriamycin, cisplatin (CAP)
c., doxorubicin, cisplatin
c., doxorubicin, and 5-fluorouracil
(CAF)
c., methotrexate, 5-fluorouracil
(CMF)
c., methotrexate, 5-fluorouracil, and
prednisone (CMFP)
vincristine, actinomycin D, c.
(VAC)
cyclophosphamide, methotrexate, 5-
fluorouracil, vincristine, and
prednisone (CMFVP)
cyclopia
cyclopism
cyclopropane
Cyclo-Provera
cyclops
c. hypognathus
C. procedure
cycloserine
Cyclospasmol
cyclosporin a
cyclosporine
cyclothiazide

NOTES

121

cycrimine hydrochloride
Cycrin Oral
cyesis
Cylert
cylinder
    cesium c.
    Delclos c.
cylindrical embryo
cyllosoma
cymbocephaly
CYP21A
CYP11B1
CYP11B2
CYP21B
cypionate
    estradiol c.
    testosterone c.
Cypress facial neuromusculoskeletal
  syndrome
cyproheptadine
    c. hydrochloride
    c. receptor blocker
cyproterone
    c. acetate
cyst
    adnexal c.
    allantoic c.
    apocrine c.
    arachnoid c.
    c. aspiration
    Bartholin c.
    benign c.
    bilateral choroid plexus c.
    blue dome c.
    branchial c.
    breast c.
    chocolate c.
    choledochal c.
    chorionic c.
    choroid plexus c. (CPC)
    colloid c.
    corpora lutea c.'s
    corpus albicans c.
    corpus luteum c.
    Dandy-Walker c.
    dermoid c.
    dermoid c. of ovary
    dysontogenetic c.
    echinococcal c.
    endometrial c.
    epidermal inclusion c.
    epithelial c.
    fetal ovarian c.
    follicular c.
    functional ovarian c.
    Gartner c.
    gartnerian c.
    glioependymal c.

    hydatid c.
    inclusion c.
    involution c.
    keratin c.
    lacteal c.
    luteal ovarian c.
    massive ovarian c.
    milk c.
    müllerian c.
    multilocular c.
    multiloculated c.
    multiple c.
    nabothian c.
    neoplastic c.
    oil c.
    omental c.
    omphalomesenteric c.
    oophoritic c.
    ovarian c.
    paraovarian c.
    paratubal c.
    perimesonephric c.
    pilonidal c.
    popliteal c.
    porencephalic c.
    posterior fossa c.
    rete c. of ovary
    Sampson c.
    sebaceous c.
    siderophagic c.
    simple c.
    subcapsular c.
    tarry c.
    tension c.
    theca lutein c.
    umbilical c.
    unilocular ovarian c.
    urachal c.
    vaginal dysontogenetic c.
    vaginal embryonic c.
    vaginal inclusion c.
    vermina c.
    vestibular c.
    vitelline duct c.
    vitellointestinal c.
    vulvar inclusion c.
    wolffian-remnant c.
Cystadane
cystadenocarcinoma
    mucinous c.
    ovarian c.
    papillary serous c.
cystadenofibroma
cystadenoma
    benign mucinous c.
    ovarian c.
    ovarian proliferative c.
    serous c.

cystathionine
    c. beta-synthase (CBS)
    c.-beta-synthase deficiency
cystathioninemia
cystathioninuria
cystectomy
    Bartholin c.
    ovarian c.
    vulvovaginal c.
cystencephalus
cystic
    c. adenomatous malformation
    (CAM)
    c. adnexal mass
    c. disease of the breast
    c. fibrosis (CF)
    c. fibrosis transmembrane
    conductance regulator mutation
    c. glandular hyperplasia
    c. hygroma
    c. hyperplasia of the breast
    c. kidney
    c. lesion
    c. lymphangioma
    c. mole
    c. myoma degeneration
    c. Walthard rest
cystica
    osteogenesis imperfecta c.
    spina bifida c.
cysticum
    epithelioma adenoides c.
    lymphangioma c.
cystinosis
cystinuria
cystitis
    acute c.
    bacterial c.
    hemorrhagic c.
    honeymoon c.
    Hunner interstitial c.
    interstitial c.
    irradiation c.
    postoperative c.
    postradiation c.
    radiation c.
cystoblast
Cystocath catheter
cystocele
    paravaginal c.
    c. repair
cystography

cystological
cystometer
    Lewis recording c.
cystometric capacity
cystometrics
    office c.
cystometrogram
    eyeball c.
    multichannel c.
cystometry
cystopexy
cystosarcoma
    c. phyllodes
cystoscopic transurethral tumor resection
cystoscopy
    rigid c.
Cystospaz
cystostomy
cystotomy
    suprapubic c.
cystourethrocele
cystourethrogram
    voiding c. (VCU, VCUG)
cystourethrography
    chain c.
    metallic bead-chain c.
cystourethropexy
    needle c.
    retropubic c.
    vaginal c.
cystourethroscopy
Cytadren
cytarabine
cytidine
    c. diphosphate-choline
    c. diphosphate-diacylglycerol
    c. monophosphate
    c. triphosphate-phosphocholine
    cytidyltransferase
cytidyltransferase
    cytidine triphosphate-
    phosphocholine c.
Cytobrush
    C. cell collector
    C.-Plus cell collector
    C. Plus endocervical cell sampler
    C. spatula
    Zelsmyr C.
cytochrome
    c. *c* oxidase deficiency

C

NOTES

cytochrome *(continued)*
 c. oxidase test
 c. P450scc
cytogenetic, cytogenic
 c. analysis
 c. line
 c. map
 c. study
cytogenetics
cytokine
 pleiotrophic c.
Cytolex pexiganan acetate
cytologic
 c. screening
 c. smear
 c. study
 c. washing
cytological
 c. band
 c. map
cytology
 aspiration biopsy c. (ABC)
 bland c.
 brush c.
 c. brush
 cervical c.
 peritoneal c.
 scrape c.
 sputum c.
 urine c.
 vaginal c.
cytolysis
cytomegalic inclusion disease
cytomegalovirus (CMV)
 c. disease
 c. immunoglobulin
 c. infection
 c. seropositive
 c.-specific immunoglobulin
 c. total immunoglobulin assay
Cytomel
cytometry
 flow c.
cytopathogenesis
cytophagocytosis
cytophilic antibody
**Cytopick endocervical and uterovaginal
 cell collector**
cytopipette
cytoplasm
 bubble gum c.

cytoplasmic
 c. inheritance
 c. trait
cytoreduction
cytoreductive surgery
CytoRich process
Cytosar-U
cytosine
 c. arabinoside
 c. arabinoside, etoposide,
  methotrexate (CEM)
cytoskeleton
**cytosolic tyrosine transaminase
 deficiency**
Cytospray
Cytotec
 C. induction
cytotechnician
cytotoxic
 c. agent
 c. cell
 c. effect
 c. factor
 c. T lymphocyte (CTL)
cytotoxicity
cytotrophoblast
 malignant c.
Cytovene
Cytoxan
 C., Adriamycin, fluorouracil (CAF)
 C., Adriamycin, fluorouracil,
  tamoxifen, Halotestin (CAFTH)
 C., Adriamycin, leucovorin,
  calcium, fluorouracil (CALF)
 C., Adriamycin, leucovorin,
  calcium, fluorouracil, ethinyl
  estradiol (CALF-E)
 C., epirubicin, fluorouracil (CEF)
 fluorouracil, epirubicin, C. (FEC)
 C., methotrexate, fluorouracil,
  prednisone (CMFP)
 C., methotrexate, fluorouracil,
  prednisone, tamoxifen (CMFPT)
 C., methotrexate, fluorouracil,
  vincristine, prednisone (CMFVP)
 C., Oncovin, fluorouracil plus
  Cytoxan, Oncovin, methotrexate
  (COF/COM)
Czerny anemia

δ (*var. of* delta)
2D
2D Doppler
2D echocardiogram
D920
Audio Doppler D920
DA
developmental age
ductus arteriosus
dacarbazine
mesna, Adriamycin, Ifosfamide, D.
(MAID)
Dacron fiber
dacryoadenitis
dacryocystitis
dacryocystostenosis
dacryostenosis
dactinomycin
dactylitis
sickle cell d.
tuberculous d.
dactylomegaly
DA-DAPI stain
DAG
diacylglycerol
Dagenan
daily fetal movement record
Dakin antibacterial solution
Dalacin C
Dale
D. abdominal binder
D. Foley catheter holder
Dalkon shield
Dall-Miles cable grip procedure
Dalmane
Dalrymple sign
damage
brain d.
hepatocellular .d.
obstetric d.
obturator nerve d.
tubal inflammatory d. (TID)
D-amino acid
dAMP
deoxyadenylic acid
danazol
low-dose d.
dance
St. Vitus d.
Dandy-Walker
D.-W. cyst
D.-W. deformity
D.-W. malformation

D.-W. malformation-basal ganglia
disease-seizures syndrome
D.-W. syndrome (DWS)
Dandy-Walker-like syndrome
Dane particle
Danforth sign
danger
radiation d.
Danlos
D. disease
D. syndrome
Dann respirator
Danocrine
danthron
D-antigen isoimmunization
dantrolene
Danus-Fontan procedure
Danus-Stanzel repair
DAP
direct agglutination pregnancy
DAP test
Dapa
Daranide
Darbid
Daricon
Darier
D. disease
D. sign
dark-field
d.-f. examination
d.-f. microscopy
Darrow-Gamble syndrome
Dartal
Darvocet-N
Darvon
D.-N
Darwin
D. ear
D. theory of evolution
darwinian
d. evolution
d. fitness
d. reflex
dashboard perineum
DAT
dementia of the Alzheimer type
data
mortality d.
database
automatic karyotype system d.
Cochrane Pregnancy and
Childbirth D.
OSSUM d.
POSSUM d.

D

date
> d. of birth (DOB)
> post d.'s

dating
> biopsy d.
> endometrial d.
> pregnancy d.

Datril

daughter
> d. cell
> d. chromosome
> DES d.

daunomycin

daunorubicin

DAV
> dibromodulcitol with Adriamycin and vincristine

David-O'Callaghan syndrome

Davis
> D. bladder catheter
> D. & Geck (DG)

Davydov procedure

dawn phenomenon

Dawson
> D. disease
> D. encephalitis

day
> Acutrim Late D.
> calories per kilogram per d. (cal/kg/day)
> cubic centimeters per kilogram per d. (cc/kg/d)
> d. of life (DOL)

DAZ
> deleted in azoospermia
> DAZ gene cluster

DAZL1 **autosomal homologue**

dazzle reflex

dB
> decibel

DBCP
> dibromochloropropane

D&C
> dilation and curettage

DCIS
> ductal carcinoma in situ

dCMP
> deoxycytidylic acid

DCR
> digitorenocerebral syndrome

DDAVP
> DDAVP nasal spray

ddC
> zalcitabine

ddI
> didanosine

DDP
> *cis*-diamminedichloroplatinum

DDST
> Denver Developmental Screening Test

D&E
> dilation and evacuation

de
> D. Crecchio syndrome
> d. Grouchy syndrome 1, 2
> d. la Chapelle dysplasia
> d. Lange syndrome 1, 2
> d. Morsier-Gauthier syndrome
> d. Morsier syndrome
> d. novo balanced chromosome rearrangement
> d. novo balanced translocation
> d. novo deletion
> D. Sanctis-Cacchione syndrome
> D. Toni-Fanconi-Debré syndrome
> D. Toni-Fanconi syndrome
> D. Vaal syndrome

dead
> d. fetus syndrome
> full term, born d. (FTBD)

DEAE beads

deafness
> d.-craniofacial syndrome
> diabetes insipidus and mellitus with optic atrophy and d. (DIDMOAD)
> d., femoral epiphyseal dysplasia, short stature, developmental delay syndrome
> hereditary progressive bulbar paralysis with d.
> d., hypogonadism, hypertrichosis, short stature syndrome
> d., imperforate anus, hypoplastic thumbs syndrome
> lentigines (multiple), electrocardiographic abnormalities, ocular hypertelorism, pulmonary stenosis, abnormalities of genitalia, retardation of growth, and d. (sensorineural) (LEOPARD)
> d.-nephritis syndrome
> d., onychodystrophy, osteodystrophy, retardation
> d., onychoosteodystrophy, mental retardation (DOOR)
> pontobulbar palsy with d.
> pontobulbar palsy with neurosensory d.
> progressive bulbar palsy with perceptive d.

death
> brain d.
> d. of co-twin
> crib d.

early embryonic d.
false-negative d.
fetal d.
fetal brain d.
infant d.
injury-related maternal d.
intrapartum d.
intrauterine d. (IUD)
intrauterine fetal d.
maternal d.
neonatal d.
nonmaternal d.
perinatal d.
single intrauterine d.
sudden infant d. (SID)
Deaver retractor
DeBakey
D. clamp
D. tissue forceps
debendox
Debetrol
debrancher enzyme deficiency
Debré-Fibiger syndrome
Debré-Sémélaigne syndrome
debris
fetal d.
debulking
ovarian carcinoma d.
surgical d.
d. of tumor
debut
sexual d.
Decadron
Deca-Durabolin
decalvans
keratosis follicularis spinulosa d.
decamethonium bromide
decannulate
Decanoate
Hybolin D.
decapitate
decapitation
decarboxylase
ornithine d.
pyruvate d.
Decaspray
decay
bottle tooth d.
deceleration
abnormal d.
early d.'s
fetal heart rate d.

late d.
variable d.
decibel (dB)
decidua
d. basalis
d. capsularis
d. compacta
ectopic d.
d.-macrophage connection
d. menstrualis
d. parietalis
d. polyposa
tuberous subchorial hematoma of the d.
d. vera
decidual
d. arteriolar atherosis
d. arteriolopathy
d. cast
d. cell
d. endometritis
d. fibrinoid necrosis
d. fibrin thrombosis
d. floor
d. lumen
d. mural thickening
d. prolactin synthesis
d. reaction
decidualized endometrium
deciduate placenta
deciduation
deciduitis
deciduoma
Loeb d.
decline
recurrent premenstrual lithium serum concentration d.
Declomycin
Decofed Syrup
decompensation
decomposition
decompression
small intestine d.
uterine d.
vaginal d.
Decon
Par D.
decondensation
sperm chromatin d.
decongestant
Balminil D.
New D.

D

NOTES

127

decreased
  d. biparietal diameter
  d. libido
decubitus position
deep
  d. circumflex iliac artery
  d. dyspareunia
  d. tendon reflex (DTR)
  d. transverse arrest
  d. vein thrombophlebitis
  d. vein thrombosis (DVT)
  d. venous thrombosis (DVT)
defecography
defect
  absence d.
  anterior apical vault d.
  anterior neural tube d.
  atrial septal d. (ASD)
  atrioventricular canal d.
  atrioventricular septal d.
  biosynthetic d.
  birth d.
  central d.
  coagulation d.
  congenital ectodermic scalp d.
  congenital heart d.
  congenital hemidysplasia with
    ichthyosiform erythroderma and
    limb d.'s (CHILD)
  conotruncal heart d.'s
  diaphragmatic d.
  distal sacral d.
  endocardial cushion d. (ECD)
  fetal structural d.
  genetic d.
  genitourinary d.
  intrauterine filling d.
  iodide trap d.
  luteal phase d. (LPD)
  microphthalmia with linear
    skin d.'s (MLS)
  müllerian fusion d.
  neural tube d. (NTD)
  ostium primum d.
  ostium secundum d.
  ovulatory d.
  paravaginal d.
  pelvic support d.
  perineal d.
  pleiotrophic functional d. (PE)
  sacral neural tube d.
  secundum atrial septal d.
  single gene d.
  sinus venosus d.
  structural heart d.
  strutural brain d.
  terminal transverse acheiria d.
  terminal transverse limb d.

  uterine lateral fusion d.
  ventricular septal d. (VSD)
  ventriculoseptal d. (VSD)
  vertebral defects, imperforate anus,
    tracheoesophageal fistula,
    renal d.'s (VATER)
  X-linked uric aciduria enzyme d.
defective abdominal wall syndrome
defect-specific repair
defeminization
defensive obstetrics
deferens
  bilateral congenital absence of
    vas d. (BCAVD)
  congenital absence of vas d.
    (CAVD)
  vas d.
deferoxamine
  d. mesylate
defervesced
defervescence
defibrillator
  LifePak d.
deficiency
  abdominal muscle d.
  ACTH d.
  adenosine deaminase d.
  adrenocorticotropic hormone d.
  alpha-L-fucosidase d.
  alpha-L-iduronidase d.
  antithrombin III d.
  arginase d.
  aspartylglucosamide d.
  AT3 d. types I, II
  3beta-dehydrogenase d.
  beta-galactosidase-1 d.
  biotinidase d.
  brancher d.
  carnitine d.
  Cori enzyme d.
  cystathionine-beta-synthase d.
  cytochrome *c* oxidase d.
  cytosolic tyrosine transaminase d.
  debrancher enzyme d.
  developmental d.
  dihydrotestosterone receptor d.
    (DHTR)
  disaccharidase d.
  enterokinase d.
  enzyme d.
  erythrocyte enzyme d.
  erythrocyte glutathione
    peroxidase d.
  erythrocyte phosphoglycerate
    kinase d.
  factor I–XIII d.
  factor D d.
  factor H d.

fibrinogen d.
folic acid d.
fructose galactokinase d.
galactocerebrosidase d. (GALC)
galactosylceramide beta-
 galactosidase d.
glucosamine-6-sulfate d.
glucose-6-phosphatase d. (G6PD)
glucose-6-phosphate
 dehydrogenase d.
glutamate formiminotransferase d.
glycerol kinase d. (GKD)
GnRH d.
gonadotropic d.
growth hormone d.
GUSB d.
hexosaminidase A d.
homozygous glucose-6-phosphate
 dehydrogenase d.
17-hydroxylase d.
21-hydroxylase d.
hyperammonemia due to ornithine
 transcarbamoylase d.
hypoxanthine-guanine
 phosphoribosyltransferase d.
 (HGPRT)
hypoxanthine-
 phosphoribosyltransferase d.
iduronate sulfatase d. (IDS)
IgA d.
IgE d.
IgM d.
immune d.
immunoglobulin d.
intrinsic sphincter d. (IDS, ISD)
LCHAD d.
long-chain-3-hydroxyacyl-CoA
 dehydrogenase d.
methionine synthase d.
monoamine oxidase A d. (MAOA)
multiple sulfatase d. (MSD)
N-acetylgalactosamine-4-sulfatase d.
nonclassic 21-hydroxylase d.
OCT d.
ornithine carbamoyltransferase d.
ornithine transcarbamylase d.
 (OTCD)
PGK d.
phosphorylase kinase d.
placental sulfatase d.
prolactin d.
prostacyclin d.

protein C d.
protein S d.
proximal femoral focal d.
pyruvate dehydrogenase complex d.
 (PDHC)
pyruvate kinase d.
red cell enzyme d.
red cell phosphoglycerate kinase d.
riboflavin d.
skeletal calcium d.
sphincter d.
steroid sulfatase d.
sulfoiduronate sulfatase d. (SIDS)
thymic-dependent d.
thyroid d.
thyrotropin d.
transcobalamin d.
type I, II AT3 d.
tyrosine aminotransferase d.
 (TATD)
tyrosine transaminase d.
UDP-galactose-4-epimerase d.
 (GALE)
X-linked congenital glycerol
 kinase d.
X-linked monoamine oxidase d.
xylulose dehydrogenase d.
y-cystathionase d.

deficit

base d.
focal neurological d.
sensory d.

deflazacortum
deflexion

d. abnormality

defloration
deformans

cephalhematoma d.
dystonia musculorum d.
myodysplasia fetalis d.
myodystrophia fetalis d.
osteochondrodystrophia d.

deformation
deformity

Arnold-Chiari d.
cloverleaf skull d.
congenital postural d.
Dandy-Walker d.
gibbous d.
gooseneck d.
hip d.
joint d.

D

NOTES

deformity *(continued)*
    limb reduction d. (LRD)
    lobster-claw d.
    Madelung d.
    Michel d.
    postural d.
    sabre shin d.
    Sandle gap foot d.
    spinning-top d.
    split-foot d.
    Sprengel d.
    Volkmann d.
degenerate
    d. consensus primer
    d. oligonucleotide primer (DOP)
degenerated fibroadenolipoma
degenerating myoma
degeneration
    carneous d.
    cerebellar d.
    cerebromacular d.
    cochleosaccular d.
    congenital macular d.
    congenital progressive
      oculoacousticocerebral d.
    cystic myoma d.
    dominant Dovne honeycomb
      retinal d.
    familial striatal d.
    hepatolenticular d.
    hereditary oligophrenic
      cerebellolental d.
    hyaline myoma d.
    infantile striatonigral d.
    Leber congenital tapetoretinal d.
    malignant d.
    molar d.
    mucoid myoma d.
    red d.
    sarcomatous myoma d.
    vitelliform d.
degenerativa
    melanosis corli d.
degradable starch microsphere (DSM)
degradation
degree
    d. of kindred
    30-d. lens
    70-d. lens
dehiscence
    asymptomatic d.
    episiotomy d.
    wound d.
dehydration
    d. fever
    hypernatremia d.
dehydroepiandrosterone (DHEA)

    d. sulfate (DHEAS)
    d. sulfate loading test
dehydrogenase
    20alpha-dihydroprogestin d.
    3beta-d. deficiency
    17beta-estradiol d.
    3beta-hydroxysteroid d. (3betaHSD)
    glucose-6-phosphate d. (G6PD)
    15-hydroxyprostaglandin d.
    18-hydroxysteroid d.
    lactate d. (LDH)
    lactic acid d.
    long-chain-3-hydroxyacyl-CoA d.
      (LCHAD)
dehydroisoandrosterone
    d. sulfate
Dejerine disease
Dejerine-Klumpke syndrome
Dejerine-Sottas
    D.-S. atrophy
    D.-S. disease
DeJuan forceps
del
    deletion
    del 11/aniridia complex
    del Castillo syndrome
    del (1p)–(22p) syndrome
    del (1q)–(22q) syndrome
    del (1)–(22) syndrome
    del (Xp21) syndrome
    del (Xq) syndrome
Deladroxate
Deladumone
Delalutin
Delaprem
Delatest
Delatestryl
delavirdine
delay
    atrioventricular conduction d.
    constitutional d.
    developmental d.
    short stature, hyperextensibility of
      joints and/or inguinal hernia,
      ocular depression, Rieger
      anomaly, teething d. (SHORT)
    d. syndrome
delayed
    d. fertilization
    d. first stage
    d. hypersensitivity
    d. implantation
    d. menarche
    d. menstruation
    d. puberty
Delclos
    D. cylinder
    D. ovoid

Delcort Topical
DeLee
    D. forceps
    D. instrumentation
    D. maneuver
    D. suction catheter
    D. suctioning
    D. Universal retractor
Delestrogen
deleted in azoospermia (DAZ)
deletion (del)
    alpha-thalassemia/mental
      retardation d.
    autosomal d.
    chromosomal d.
    chromosome d.
    d. of chromosome
    de novo d.
    gene d.
    interstitial d.
    d. 1p–22p syndrome
    d. 1q–22q syndrome
    13q-d. syndrome
    d. 1–22 syndrome
    terminal d.
    d. Xp21 syndrome
    d. Xp22 syndrome
    d. Xq syndrome
Delfen
delinquency
delirium
    d., infection, atrophic
      urethritis/vaginitis, pharmaceuticals,
      psychological, excess urine
delivered
    pregnancy, uterine, not d. (PUND)
delivery
    abdominal d.
    assisted breech d.
    assisted cephalic d.
    breech d.
    cephalic d.
    cesarean d.
    controlled vaginal d.
    Duncan mechanism of placental d.
    en caul d.
    expected day of d. (EDD)
    failed forceps d.
    fear of d.
    fear-causing d.
    fearless first d.
    forceps d. (FD)

    high forceps d.
    instrumental d.
    labor and d. (L&D)
    low forceps d.
    midforceps d.
    d. mode
    natural d.
    operative vaginal d.
    outlet forceps d.
    oxygen d.
    perimortem d.
    postmortem d.
    precipitate labor and d.
    precipitous d.
    premature d.
    preterm d.
    rotational d.
    sequential d.
    spontaneous cephalic d.
    spontaneous preterm d.
    spontaneous vaginal d. (SVD)
    sterile, spontaneous, controlled
      vaginal d. (SSCVD)
    sterile, spontaneous vaginal d.
      (SSVD)
    sunny-side up d.
    term d.
    twin d.
    underwater d.
    vacuum-assisted d.
    vacuum extractor d.
    vaginal d.
    vertex d.
Delleman syndrome
Dellepiane hysterectomy
delta, δ
    d. agent
    d. hepatitis
    d. OD_{450}
delta-F508 cystic fibrosis
Deltalin
Delta-Tritex Topical
Del-Vi-A
demand
    d. effect
    d. feeding
demarcated
demecarium
demeclocycline
dementia of the Alzheimer type (DAT)
Demerol
4-demethyl-epipodophyllotoxin

NOTES

D

demineralization
> acral d.

demise
> fetal d.
> intrapartum d.
> spontaneous preterm labor with
>   intrapartum d.
> twin d.

Demons-Meigs syndrome
Demser
Demulen
> D. 1/35
> D. 1/50

demyelinating encephalopathy
demyelinogenic leukodystrophy
denature
denaturing
> d. gradient gel electrophoresis
> d. high-performance liquid
>   chromatography

Denavir
dendritic cell
dengue
Denhardt solution
denidation
Denis
> D. Browne bar
> D. Browne clubfoot splint
> D. Browne pouch

Denman spontaneous evolution
Dennen forceps
Dennett diet
Dennie-Marfan syndrome
Denonvilliers fascia
de novo
dense
> d. adhesion
> d. body

densitometer
> DXA d.
> Expert bone d.
> Hologic 1000 QDR d.
> OsteoAnalyzer d.
> PIXI bone d.

densitometry
> bone d.

density
> arterial linear d.
> bone d.
> bone mineral d. (BMD)
> fat d.
> d. gradient
> hip bone d.
> hypoechoic d.
> optical d.

dental caries
**dentatorubral-pallidoluysian atrophy
(DRPLA)**

dentia praecox
denticulate hymen
dentin dysplasia
dentinogenesis imperfecta
dentooculoosseous dysplasia
denuded surface
denumerable character
Denver
> D. classification
> D. Developmental Screening Test
>   (DDST)
> D. hydrocephalus shunt

Denys-Drash syndrome
deodorant artifact
deoxyadenosine
deoxyadenylic acid (dAMP)
deoxycorticosterone (DOC)
deoxycytidylic acid (dCMP)
deoxy-D-glucose
deoxyguanylic acid (dGMP)
deoxypyridinoline (Dpd)
deoxyribonuclease (DNase)
deoxyribonucleic
> d. acid (DNA)
> d. acid index

deoxyribonucleotide
deoxythymidylic acid (dTMP)
Depakene
depAndrogyn
**Department of Children and Youth
Services**
Depen
dependent edema
depGynogen
> d. Injection

depigmentation
depigmentosus
> nevus d.

depletion
> juvenile spermatogonial d.

DepoCyst
Depo-Estradiol
> D.-E. Injection

Depogen
> D. Injection

depolarization
> ectopic ventricular d.

Depo-Medrol
depo-MPA
Deponit
Depo-Provera
> D.-P. Injection

deposit
> macrocephaly with feeblemindedness
>   and encephalopathy with
>   peculiar d.'s

deposition
> diffuse perivillous fibrinoid d.

depot
    Androcur D.
    fat d.
    Lupron D.
    Lupron D.-3 Month
    Lupron D.-4 Month
    d. medroxyprogesterone
    d. medroxyprogesterone acetate
      (DMPA)
Depotest
Depo-Testadiol
Depotestogen
Depo-Testosterone
Depot-Ped
    Lupron D.-P.
depression
    agitated d.
    fetal skull d.
    inbreeding d.
    intimate partner d.
    narcotic d.
    paternal postnatal d.
    postpartum d.
    respiratory d.
depressor anguli oris muscle hypoplasia syndrome
deprivation
    antagonist-induced gonadotropin d.
    estrogen d.
depth dose
depth-dose distribution
Dercum disease
derepressed gene
derivation
    sexual d.
derivative
    acetoxyprogesterone d.
    17-α-acetoxyprogesterone d.
    d. chromosome
    coumarin d.
    19-nortestosterone d.
Dermacort Topical
Dermarest Dricort Topical
Derma-Smoothe/FS Topical
Dermasone
dermatitis, pl. **dermatitides**
    allergic d.
    atopic d.
    candidal diaper d.
    chronic bullous d.
    diaper d.
    d. excoriativa infantum

    d. exfoliativa infantum
    d. gangrenosa infantum
    d. herpetiformis
    irradiation d.
    Jacquet d.
    juvenile plantar d.
    nickel d.
    papular d. of pregnancy (PDP)
    periorificial d.
    progesterone d.
    rebound d.
    seborrheic d.
    d. venenata
    vulvar d.
    vulvar d.
dermatofibroma
dermatofibrosarcoma protuberans
dermatoglyphic
    d. finding
    d. pattern
dermatoleukodystrophy
dermatomyositis
    primary idiopathic d.
dermatoosteolysis
    Kirghizian d.
Dermatop
dermatophytosis
dermatosis, pl. **dermatoses**
    bullous d.
    dermolytic bullous d.
    plantar d.
    Siemens-Bloch pigmented d.
    vulvar d.
dermis
    papillary d.
    reticular d.
dermocyma
dermoid
    d. cyst
    d. cyst of ovary
    epibulbar d.
Dermolate Topical
dermolytic bullous dermatosis
dermopathy
    restrictive d.
Dermoplast
dermotrichic syndrome
Dermovate
Dermtex HC with Aloe Topical
Derogatis Brief Symptom Inventory
Deronil
Derry syndrome

D

NOTES

DES
  diethylstilbestrol
  dysequilibrium syndrome
  DES daughter
  DES exposure
DESAD
  National Collaborative Diethylstilbestrol
  Adenosis Project
desaturation
Desbuquois syndrome
Descemet membrane
descending
  d. aorta
  d. colon
descensus
  d. uteri
descent
  fetal d.
  rapid d.
  rotation and d.
Deschamps ligature
descriptive embryology
Desenex
  Prescription Strength D.
desensitization
  pituitary d.
Desferal
desferrioxamine mesylate
desiccate
design
  family-based d.
designated donor blood
desipramine
desire
  inhibited sexual d.
Desitin
deslanoside
deslorelin
Desmarres forceps
desmins
desmiognathus
desmocranium
desmoid tumor
desmolase
  cholesterol 20,22 d.
Desmons syndrome
desmopressin
  desmopressin acetate
desmosome
Desocort
Desogen
desogestrel
  ethinyl estradiol and d.
  d. and ethinyl estradiol
desonide
DeSouza exercises
DesOwen Topical

desoximetasone
desoxycorticosterone (DOC)
Desoxyn
desquamation
desquamative inflammatory vaginitis
desquamativum
  erythroderma d.
desultory labor
Desyrel
detachment
  retinal d.
detail
  cycle-monitoring d.
Detect HIV-1 assay
detection
  d. bias
  mammographic d.
  prenatal d.
deterioration
  fetal d.
determinant
  antigen d.
  d. group
determination
  blood gas d.
  cell d.
  emesis pH d.
  gender d.
  hemoglobin A_{Ic} d.
  parentage d.
  prenatal sex d.
  scalp pH d.
  sex d.
  sweat chloride d.
  testis d.
detrusor
  d. instability
  d. muscle
  d. sphincter dyssynergia
development
  abnormal fetal d.
  arrested d.
  Bayley Scale of Infant D. (BSID)
  Briggance Diagnostic Inventory of
    Early D.
  central nervous system d.
  CNS d.
  early follicular d.
  endometrial d.
  excretory system d.
  fetal d.
  genital d.
  goiter d.
  male genital duct d.
  mental d.
  National Institute of Child Health
    and Human D. (NICHD)
  neurologic d.

reproductive system d.
stromal d.
Tanner stages of d.
developmental
    d. age (DA)
    d. deficiency
    d. delay
    d. delay-multiple strawberry nevi
      syndrome
    d. diapause
    d. level
    d. stage
Deventer pelvis
deviation
    sexual d.
    standard d. (SD)
device
    AcuTrainer d.
    aspiration-tulip d.
    Atad Ripener D.
    AutoPap automated screening d.
    autostapling d.
    barium-impregnated plastic
      intrauterine d.
    Cameco syringe pistol aspiration d.
    CerviSoft cytology collection d.
    clitoral therapy d. (CTD)
    contraceptive d.
    Copper-7 intrauterine d.
    Copper T intrauterine d.
    Cu-7 intrauterine d.
    Diva laparoscopic morcellator d.
    Donnez d.
    double-balloon d.
    double-umbrella d.
    Eder cord blood collection d.
    Endosample endometrial
      sampling d.
    Endo Stitch laparoscopic
      suturing d.
    external urethral barrier d.
    Femcept d.
    GynoSampler endometrial
      sampling d.
    Hollister collecting d.
    intrauterine d. (IUD)
    intrauterine contraceptive device
      (IUCD)
    Lippes-type intrauterine d.
    Macroplastique implantable d.
    Makler insemination d.
    M-cup vacuum extraction d.

Medilog 9000 polysomnography d.
Model 5315 sequential
    compression d.
Mucat cervical sampling d.
Multiload Cu-375 intrauterine d.
Multispatula cervical sampling d.
Niplette d.
Nite Train-R enuresis
    conditioning d.
OraSure d.
OsteoView d.
Poly CS d.
Progestasert intrauterine d.
progesterone-releasing T-shaped d.
Protocult stool sampling d.
Scopette d.
Seitzinger d.
Shug male contraceptive d.
Uterine Explora Curette endometrial
    sampling d.
Venodyne pneumatic compressive d.
Wallach Endocell collection d.
Z sampler endometrial sampling d.
DeWeese axis traction forceps
Dewey obstetrical forceps
DEXA
    dual-energy x-ray absorptiometry
Dexacort Phosphate Turbinaire
Dex-A-Diet
    Maximum Strength D.-A.-D.
dexamethasone
    d. suppression test
    d. therapy
Dexasone
Dexatrim
    Maximum Strength D.
    D. Pre-Meal
dexbrompheniramine
Dexchlor
dexchlorpheniramine
Dexedrine
Dexide
    D. disposable cannula
    D. disposable trocar
Dexon
    D. II suture
    D. mesh
    D. Plus suture
Dexone
    D. LA
dexpanthenol

D

NOTES

dexter
> cor triatriatum d.

dextran
> d. 40
> d.-70 barrier material
> high molecular weight d.
> iron d.
> low molecular weight d.
> d. sulfate

dextrin
> d. sulfate
> d. sulfate gel

dextroamphetamine
dextrocardia
dextrocardia/situs inversus syndrome
dextro formula
dextromethorphan
dextroposition of heart
dextrose
> d. in water

Dextrostix (D-stix)
dextrosuria
dextrothyroxine
dextrotransposition (D-transposition)
dextroversion
Dey-Drop Ophthalmic Solution
Deyo adaptation
Dey-Wash skin wound cleanser
DFA
> direct fluorescent antigen
> DFA test

D-fenfluramine
DFFRY **gene**
DFNA3
> autosomal dominant nonsyndromic
> hearing loss

DFNB1
> autosomal recessive nonsyndromic
> hearing loss

DG
> Davis & Geck
> DG Softgut suture

D-galactose
dGMP
> deoxyguanylic acid

DGSX
> X-linked dysplasia-gigantism syndrome

DHE
> dihydroergotamine

DHEA
> dehydroepiandrosterone
> DHEA sulfate

DHEAS
> dehydroepiandrosterone sulfate

DHT
> dihydrotestosterone

DHTR
> dihydrotestosterone receptor deficiency

DI
> donor insemination

DiaBeta
diabetes
> borderline d.
> brittle d.
> bronze d.
> chemical d.
> congenital lipoatrophic d.
> drug-induced d.
> gestational d.
> d. insipidus
> d. insipidus, diabetes mellitus, optic
> atrophy syndrome (DIDMO)
> d. insipidus and mellitus with
> optic atrophy and deafness
> (DIDMOAD)
> insulin-dependent d. mellitus
> (IDDM)
> juvenile d.
> juvenile-onset d.
> ketosis-prone d.
> ketosis-resistant d.
> latent d.
> lipoatrophic d.
> maternal d.
> d. mellitus
> d. mellitus, mental retardation,
> lipodystrophy, dysmorphic traits
> syndrome
> d. neonatorum
> non-insulin-dependent d. mellitus
> (NIDDM)
> overt insulin-dependent d. mellitus
> streptozocin-induced d.

diabetes-deafness syndrome
diabetic
> d. acidosis
> d. coma
> d. embryopathy
> d. fetopathy
> insulin-dependent d.
> d. ketoacidosis (DKA)
> d. mother
> d. nephropathy
> non-insulin-dependent d.
> pregnant d.
> d. retinopathy

Diabinese
diacetate
> ethinyl estradiol and ethynodiol d.
> ethynodiol d.
> propylene glycol d.

diacylglycerol (DAG)
diadochokinesis
diagnosis, pl. **diagnoses**
> antenatal d.
> colposcopic d.

computer-aided d. (CAD)
fetal d.
genetic d.
histologic d.
histopathological d.
neonatal d.
preimplantation d.
preimplantation genetic d. (PGD)
prenatal genetic d.
ultrasound d.
diagnostic
d. accuracy
d. hysteroscope
d. imaging
d. procedure
d. radiation
diagnostics
Amplicor PCR d.
DNA d.
Quest D.
Roche D.
Vivigen d.
diagonal conjugate
diakinesis
d. stage
d. stage of oocyte meiosis
Dialose
dialysate
dialysis, pl. **dialyses**
continuous ambulatory peritoneal d.
(CAPD)
kidney d.
peritoneal d.
dialytic parabiosis
diameter
anteroposterior d. of the pelvic
inlet
aortic root d.
Baudelocque d.
biischial d.
biparietal d. (BPD)
bitemporal d.
conjugate d. of the pelvic inlet
decreased biparietal d.
fetal biparietal d.
gestational sac mean d.
internal d.
intertuberous d.
left atrial d. (LAD)
Loehlein d.
d. mediana
d. obliqua

oblique d.
occipitofrontal d.
occipitomental d.
2-d. pocket technique
posterior sagittal d.
suboccipitobregmatic d.
trachelobregmatic d.
d. transversa
transverse d.
diamine
d. oxidase
D. T.D.
diaminobenzidine
3,3′ d.
cis-**diamminedichloroplatinum (DDP)**
c.-d. II
diamniotic
d. dichorionic placenta
d. twins
Diamond-Blackfan
D.-B. congenital hypoplastic anemia
D.-B. juvenile pernicious anemia
D.-B. syndrome
diamond-shaped murmur
Diamox
Diana
D. complex
D. Project
Dianabol
Dianeal dialysis solution
Diaparene
diapause
developmental d.
embryonic d.
diaper
d. dermatitis
double d.'s
d. rash
triple d.'s
diaphanography
diaphragm
contraceptive d.
everted d.
intrauterine d.
Ortho All-Flex d.
pelvic d.
d. pessary
urogenital d.
diaphragmatic
d. agenesis
d. defect
d. hernia

NOTES

D

diaphragmatic *(continued)*
    d. hernia, abnormal face, distal
      limb anomalies syndrome
    d. hernia-distal digital hypoplasia
      syndrome
    d. hernia-exophthalmos-hypertelorism
      syndrome
    d. hernia-myopia-deafness syndrome
diaphyseal
    d. aclasis
    d. dysplasia
Diapid
diarrhea
diary
    food d.
    retraining d.
    voiding d.
diastasis
    d. recti
diastematomyelia
diastolic murmur
diastrophic
    d. dwarfism
    d. dysplasia
diathermy
    d. current
    electrocoagulation d.
    excision d.
    laparoscopic ovarian d.
    d. loop excision
diathesis, pl. **diatheses**
    bleeding d.
    familial d.
    gouty d.
diatrizoate
Diazemuls Injection
diazepam
diazoxide
Dibbell
    D. cleft lip-nasal reconstruction
    D. cleft lip-nasal revision
Dibenzyline
dibrachia
dibromochloropropane (DBCP)
dibromodulcitol
    d. with Adriamycin and vincristine
      (DAV)
dibucaine
DIC
    diffuse intravascular coagulation
    disseminated intravascular coagulation
Dicarbosil
dicentric
    d. chromosome
dicephalic twins
dicephalus
dicheilia
dicheiria

dichloroacetic acid
dichlorphenamide
dichogamy
dichorionic, dichorial
    d. placenta
    d. twins
dichorionic-diamniotic (di-di)
    d.-d. placenta
    d.-d. twins
dichotomous
Dickinson syndrome
diclofenac
    d. potassium
    d. sodium
dicloxacillin
    d. sodium
dictyate
    d. stage
    d. stage of oocyte meiosis
dicumarol
    d. resistance
dicyclomine
didanosine (ddI)
Diday law
didehydrondideoxythymidine
didelphic
di-di
    dichorionic-diamniotic
DIDMO
    diabetes insipidus, diabetes mellitus,
      optic atrophy syndrome
DIDMOAD
    diabetes insipidus and mellitus with optic
      atrophy and deafness
didymus
Dieckmann intraosseous needle
diembryony
diencephalic syndrome
dienestrol
Dienestrol Vaginal
Dienst test
diet
    ADA d.
    Atkins d.
    BRAT d.
    Dennett d.
    galatose-free d.
    gluten-free d.
    high-calorie d.
    high-fiber d.
    ketogenic d.
    lactose-free d.
    low-cholesterol d.
    low-fat d.
    low-protein d.
    low-residue d.
    low-salt d.
    low-sodium d.

Moro-Heisler d.
d. and nutrition
Pulmocare d.
Sippy d.
vegetarian d.
dietary
d. amenorrhea
d. fat
d. fiber
d. problem
Dieter forceps
diethylamide
lysergic acid d. (LSD)
diethylaminoethyl
diethyldithiocarbamate
diethylenetriamine pentaacetic acid (DPTA)
diethylpropion
diethylstilbestrol (DES)
d.-exposed
DiFerrante syndrome
difference
alveolar-arterial pressure d. (p(A-a)O$_2$)
arteriovenous oxygen d.
intrapair birth weight d.
racial d.
differential
d. cell count
d. effect
d. temperature sensor (DTS)
d. vascular resistance
differentiation
central nervous system d.
embryonic d.
extraembryonic d.
fetal sexual d.
genital d.
male sex d.
sexual d.
somatic d.
testicular d.
difficile
Clostridium d.
diffusa
neurospongioblastosis d.
diffuse
d. fasciitis
d. fibrocystic disease
d. globoid body sclerosis
d. globoid cell cerebral sclerosis

d. intravascular coagulation (DIC)
d. mesangial sclerosis-ocular abnormalities syndrome
d. perivillous fibrinoid deposition
diflorasone
Diflucan
diflunisal
Digene
D. HPV Assay
D. Hybrid Capture II HPV Test
DiGeorge
D. anomaly
D. microdeletion syndrome
D. sequence
digestive system
Dighton-Adair syndrome
Digilab FTS 40A spectrometer
digit
supernumerary d.
digital
d. anomalies, short palpebral fissures, atresia of esophagus or duodenum syndrome
d. imaging colposcopy
d. mammography
d. mammography system
d. radiography
d. ultrasound
digitalis
digitata
Laminaria d.
digitoorofacial syndrome I–V
digitooropalatal syndrome
digitorenocerebral syndrome (DCR)
digitotalar dysmorphism
digitoxin
dignathus
digoxigenin-labeled deoxyuridine triphosphate
digoxin
digoxin-like immunoreactive factor (DLIF)
digyny
dihybrid cross
dihydralazine
dihydrate
azithromycin d.
dihydrocodeine bitartrate
dihydrocodeinone
dihydroergotamine (DHE)
d. mesylate

NOTES

139

dihydrofolate
    d. reductase
    d. reductase inhibitor
dihydromorphinone hydrochloride
20alpha-dihydroprogestin dehydrogenase
dihydrotachysterol
dihydrotestosterone (DHT)
    d. cream
    d. receptor deficiency (DHTR)
dihydroxyphenylalanine (dopa, DOPA, Dopa)
dihydroxyprogesterone acetophenide
1,25-dihydroxyvitamin D_3
diiodohydroxyquin
diiodohydroxyquinoline
diiodothyronine (T_2)
    d. test
diiodotyrosine (DIT)
diisopropyl
    d. fluorophosphate
    d. iminodiacetic acid (DISIDA)
Dilantin
    D. syndrome
Dilapan
    D. hygroscopic cervical dilator
    D. laminaria
dilatation
dilated
    d. capillary
    d. collateral vein
    fingertip d.
dilation
    cervical d.
    d. of cervix
    d. and curettage (D&C)
    esophageal d.
    d. and evacuation (D&E)
    Frank technique of d.
    premature cervical d.
    d. of ventricle
dilator
    #10 d.
    Dilapan hygroscopic cervical d.
    Goodell d.
    Hanks d.
    Hegar d.
    Hurtig d.
    hygroscopic d.
    laminaria cervical d.
    Lucite d.
    mechanical cervical d.
    osmotic d.
    Pharmaseal disposable cervical d.
    Pratt d.
    retained hygroscopic cervical d.
    rocket d.
    Soehendra d.

    vaginal d.
    Walther d.
Dilaudid
dildo, dildoe
dilemma
Dimaphen
    D. Elixir
    D. Tablets
dimelia
Dimelor
dimenhydrinate
dimension
dimercaptosuccinic acid
dimeric
    d. inhibin A level
    d. protein
Dimetabs
Dimetane
    D. Decongestant Elixir
    D. Extentabs
Dimetapp
    D. Extentabs
    D. 4-Hour Liqui-Gel Capsule
    D. Sinus Caplets
    D. Tablet
dimethindene maleate
dimethothiazine mesylate
dimethylsulfoxide (DMSO)
dimethyl-triazeno-imidazole carboxamide (DTIC)
dimethylxanthine
dimetria
dimidiate
Dimitri disease
dimorphism
    sexual d.
dimple
    acromial d.
    chin d.
    pilonidal d.
    sacral d.
    vaginal d.
Dinamap blood pressure monitor
dinitrate
    isosorbide d.
dinitrochlorobenzene (DNCB)
    d. therapy
dinitrofluorobenzene
dinoprost
    d. tromethamine
dinoprostone
    d. cervical gel
Diochloram
Diocto
Diodoquin
dioecious
Dioeze

3α-diol-G
> 5α-androstane-3α,17β-diol glucuronide

Diomycin
diosgenin
Diosuccin
Dio-Sul
Dioval
> D. Injection

Diovan
diovular
> d. twins

dioxide
> carbon d. (CO_2)
> fraction of alveolar carbon d. ($FACO_2$)
> fraction in expired gas of carbon d. ($FECO_2$)
> partial pressure of arterial carbon d. ($PaCO_2$)
> partial pressure of carbon d. (PCO_2)

dioxin
dioxyline
dipalmitoyl phosphatidylcholine (DPPC)
diphallus
diphemanil methylsulfate
diphenadione
Diphenadryl
Diphenatol
diphenhydramine
Diphenylan sodium
diphenylhydantoin
diphenylthiourea
diphosphate
> menadiol sodium d.
> uridine d. (UDP)

diphosphate-choline
> cytidine d.-c.

diphosphate-diacylglycerol
> cytidine d.-d.

diphosphoglycerate
diphtheria
> d., pertussis, tetanus (DPT)

diphyllobothriasis
Diphyllobothrium latum
DIPI
> direct intraperitoneal injection
> direct intraperitoneal insemination

diplegia
> atonic astatic d.
> congenital facial d.
> facial d.
> infantile d.
> spastic d.

Diplococcus pneumoniae
diplogenesis
diploid
> d. cell
> d. distribution
> d. merogony
> d. spermatogonial stem cell

diploid/tetraploid mixoploidy
diploid/triploid mixoploidy
diploidy
diplomyelia
diplopagus
diplopia
diplosomatia
diplosome
diplotene
> d. phase of meiosis

diploteratology
dipodia
Diprolene
> D. AF Topical
> D. Glycol
> D. Topical

diproprionate
> beclomethasone d.

dipstick
dipus
dipygus parasiticus
dipyridamole
direct
> d. agglutination pregnancy (DAP)
> d. agglutination pregnancy test
> d. egg injection
> d. fluorescent antigen (DFA)
> d. fluorescent antigen test
> d. intraperitoneal injection (DIPI)
> d. intraperitoneal insemination (DIPI)
> d. oocyte sperm transfer (DOST)
> d. oocyte transfer (DOT)
> d. vision internal urethrotomy (DVIU)

directed amplification of minisatellite region DNA
direction
> pelvic d.

directive
> advance d.

dirithromycin
dirty background

D

NOTES

disability
> mobility d.
> neurodevelopmental d.
> neurologic d.

disaccharidase deficiency
DiSala syndrome
Disalcid
disappearance
> fetal d.

disaturated
> d. lecithin
> d. phosphatidylcholine

disc (*var. of* disk)
discharge
> adherent d.
> adherent vaginal d.
> cheesy d.
> copious d.
> curdy d.
> frothy d.
> gleety d.
> mucous d.
> nipple d.
> pulsatile d.
> urethral d.
> vaginal d.

DisCide disinfecting towel
discoplacenta
discordance
> atrioventricular d.
> birth weight d.
> twin birth weight d.

discordancy
> growth d.

discordant
> d. artery flow velocity waveform
> d. twins
> d. umbilical arteries

discrepancy
> size-date d.

discrete
> d. character
> d. random variable

discus proligerus
disease
> ABO hemolytic d. of the newborn
> acute fatty liver d.
> Addison d.
> adhesive d.
> Albright d.
> Alexander d.
> allogenic d.
> alloimmune d.
> Alpers d.
> Alzheimer d. (AD)
> Andersen d.
> Anderson d.
> Anderson-Fabry d.

antiglomerular basement membrane antibody d.
aortic valve d.
Apert d.
Apert-Crouzon d.
Aran-Duchenne d.
arterial occlusive d.
arterial vascular d.
Asbee-Hansen d.
autoimmune d.
Azorean d.
Barlow d.
Bassen-Kornzweig d.
Batten-Mayou d.
Beck d.
Béguez César d.
Behçet d.
benign breast d.
Berger renal d.
Best d.
Bielschowsky-Jansky d.
blistering d.
Bloodgood d.
Blount d.
Bornholm d.
Bourneville d.
Bourneville-Brissaud d.
Brailsford d.
breast d.
bronze Schilder d.
Brown-Symmers d.
Bruton d.
bubble boy d.
Buhl d.
Byler d.
Caffey d.
Caffey-Kenny d.
Canavan d.
Canavan-van Bogaert-Bertrand d.
Caroli d.
cat-scratch d.
celiac d.
central nervous system d.
central Recklinghausen d. type II
cerebrovascular d.
Chagas d.
Charcot-Marie-Tooth d.
Charlevoix d.
Cheadle d.
cholesterol ester storage d.
Christmas d.
chronic lung d. (CLD)
chronic vascular d.
chylomicron retention d.
Coats d.
collagen vascular d.
communicable d.
complex congenital heart d.

congenital heart d.
Conradi d.
contractural arachnodactyly d.
copper transport d.
Cori d.
coronary artery d.
coronary heart d.
cranial sclerosis with striated
  bone d.
Crohn d.
Crouzon d.
Csillag d.
Cushing d.
cyanotic congenital heart d.
  (CCHD)
cystic d. of the breast
cytomegalic inclusion d.
cytomegalovirus d.
Danlos d.
Darier d.
Dawson d.
Dejerine d.
Dejerine-Sottas d.
Dercum d.
diffuse fibrocystic d.
Dimitri d.
distal tubal d.
Dukes d.
Duroziez d.
end-stage kidney d.
Erb-Goldflam d.
ethanolaminosis in glycogen-
  storage d.
extramammary Paget d.
Fabry d.
Fairbank d.
familial Alzheimer d. (FAD)
Farber d.
Feer d.
fetal heart d.
fibrocystic d.
fifth venereal d.
FIGO stage I d.
Filatov-Dukes d.
Folling d.
Fong d.
Forbes d.
Fordyce d.
fourth venereal d.
Fox-Fordyce d.
free neuraminic acid storage d.
free sialic acid storage d.

Freiberg d.
gallbladder d.
ganglioside storage d.
gastroesophageal reflux d. (GERD)
Gaucher d.
Gee d.
Gee-Herter d.
Gee-Herter-Heubner d.
genetic d.
genetotrophic d.
genital ulcer d. (GUD)
gestational trophoblastic d. (GTD)
Gierke d.
Gilbert d.
Glanzmann d.
Glenard d.
glycogen-storage d. type I–VII
Goldstein d.
graft-versus-host d. (GVHD)
Graves d.
Greenfield d.
Günther d.
Hailey-Hailey d.
Hallervorden-Spatz d.
Hand-Schüller-Christian d.
Hansen d.
Hartnup d.
heart d.
Heller-Döhle d.
helminthic d.
hemoglobin C d.
hemoglobin H d.
hemoglobin M d.
hemolytic d. of the newborn
  (HDN)
hemorrhagic d.
Henoch d.
hereditary d.
heredoconstitutional d.
heredodegenerative d.
Hers d.
Hirschsprung d.
Hodgkin d.
homologous d.
Hünermann d.
Huntington d.
Hurler d.
Hutinel d.
hyaline membrane d. (HMD)
hydatid d.
hydrocephaloid d.
hypothalamic d.

D

NOTES

disease *(continued)*

I-cell d.
infantile celiac d.
infantile polycystic kidney d. (IPKD)
infectious d.
inflammatory bowel d.
intraperitoneal endometrial metastatic d.
ischemic heart d.
Jaksch d.
Jeune d.
Joseph d.
Kashin-Bek d.
Kawasaki d. (KD)
kidney d.
Kikuchi d.
Kimmelstiel-Wilson d. (KW)
kinky-hair d.
Kohdschnetter d.
Köhler bone d.
Kok d.
Kozlowski d.
Krabbe d. (KD)
Kramer d.
Krause d.
Kufs d.
Kugelberg-Welander d.
kyphoscoliotic heart d.
Lafora body d.
Langdon Down d.
Leber d.
Legg-Calvé-Perthes d.
Legionnaire d.
Leigh d.
Leiner d.
Lesch-Nyhan d.
Letterer-Siwe d.
Little d.
liver d.
Lobstein d.
Lou Gehrig d.
Luft d.
Lyell d.
Lyme d.
lysosomal storage d.
Machado-Joseph d.
mad cow d.
Maher d.
maple syrup urine d. (MSUD)
Marion d.
Maroteaux-Lamy d.
maternal cyanotic heart d.
McArdle d.
MEB d.
medullary cystic d.
Melnick-Needles d.
Menkes d.

Merzbacher-Pelizaeus d.
Milroy d.
Minamata d.
Minot d.
mixed connective tissue d. (MCTD)
Moeller-Barlow d.
Mondor d.
Morquio d.
Morquio-Brailsford d.
Morquio-Ullrich d.
moyamoya d.
mucopolysaccharide storage d. I–VIII
muscle-eye-brain d.
N-acetylneuraminic acid storage d. (NSD)
neonatal hyaline membrane d.
neoplastic trophoblastic d.
New York Heart Association classification of heart d.
Nicolas-Favre d.
Niemann-Pick d.
nonmetastatic gestational trophoblastic d. (NMGTD)
Norrie d.
Oguchi d.
Oppenheim d.
Osgood-Schlatter d.
Osler-Weber-Rendu d.
Owren d.
oxygen toxicity lung d.
Paas d.
Paget d.
Panner d.
parathyroid d.
Parkinson d.
Pelizaeus-Merzbacher d. (PMD)
pelvic adhesive d. (PAD)
pelvic inflammatory d. (PID)
peptic ulcer d.
Peyronie d.
Phocas d.
phytanic acid storage d.
pink d.
placental site gestational trophoblastic d.
polycystic kidney d. (PKD)
polycystic ovarian d. (PCOD)
polycystic renal d.
Pompe d.
Portuguese d.
postpartum pleuropulmonary and cardiac d.
Potter d.
preinvasive cervical d.
premalignant d.
Pringle d.

pseudo-Crouzon d.
psychosomatic d.
pulmonary parenchymal d.
pulseless d.
Raussly d.
reactive airway d. (RAD)
Recklinghausen d. type I, II
Reclus d.
Refsum d.
renal d.
retroviral d.
Rh d.
rhesus D d.
rhesus hemolytic d.
rheumatic d.
rheumatic heart d.
Rh hemolytic d.
Riga-Fede d.
Ritter d.
runt d.
Salla d.
Sandhoff d.
Sanfilippo d. A, B, C, D
Saunders d.
S-C d.
Scheuermann d.
Schilder d.
Schimmelbusch d.
Scholz d.
sclerocystic d. of the ovary
S-D d.
secretory d.
Seitelberger d.
severe combined
  immunodeficiency d. (SCID)
sexually transmitted d. (STD)
sialic acid storage d.
sickle cell d.
sickle cell-beta-thalassemia d.
sickle cell-hemoglobin C d. (S-C
  disease)
sickle cell-hemoglobin D d.
sickle cell-hemoglobin S d.
sickle cell-thalassemia d.
silent pelvic inflammatory d.
Simmonds d.
sixth d.
skin d.
Sly d.
Spielmeyer-Vogt d.
spinocerebellar degenerative d.
Stargardt d.

startle d.
Steinert d.
Sticker d.
Still d.
storage d.
Strumpell-Lorrain d.
Sturge d.
Swift d.
Tangier d.
Taussig-Bing d.
Tay-Sachs d.
Thiemann d.
third d.
Thomsen d.
Thomson d.
thromboembolic d. (TED)
thyroid d.
Tillaux d.
Trevor d.
trophoblastic d.
trophoblastic neoplastic d.
tubulointerstitial d.
type I glycogen storage d.
underlying d.
Underwood d.
Unverricht d.
valvular heart d.
van Bogaert d.
vascular d.
venereal d. (VD)
Vogt-Spielmeyer d.
Volkmann d.
Voltolini d.
von Gierke d.
von Hippel-Lindau d.
von Recklinghausen d.
von Willebrand d.
von Willebrand d. type IIB, III
Vrolik d.
Wegner d.
Weil d.
Werdnig-Hoffmann d.
Werlhof d.
Whipple d.
Wilkins d.
Williams d.
Wilson d.
Winckel d.
Wolman d.
wooly-hair d.
X-linked d.
X-linked dominant d.

D

NOTES

disease *(continued)*
    X-linked recessive d.
    Zuska d.
disengagement
disequilibrium
    linkage d.
    d. syndrome
    test of linkage d.
disgerminoma
dish
    d. face
    insemination d.
    Uri-Two petri d.
DISIDA
    diisopropyl iminodiacetic acid
disiens
        Prevotella d.
disinfectant
    Sklar aseptic germicidal d.
    Sklarsoak d.
disintegrin
Disipal
disk, disc
    AccuPoint hCG Pregnancy Test D.
    embryonic d.
    EMLA anesthetic d.
    placental d.
dislocated elbow, bowed tibiae, scoliosis, deafness, cataract, microcephaly, mental retardation syndrome
dislocation
    congenital hip d. (CHD)
disodium
    cefotetan d.
    etidronate d.
    intermittent cyclical etidronate d.
    moxalactam d.
    ticarcillin d.
D isoimmunization
disomus
disomy
    uniparental d.
    uniparental d. of chromosome 14 (UPD)
Disonate
disopyramide
disorder
    affective d.
    alpha-chain d.
    amniotic fluid volume d.
    antisocial personality d. (ASPD)
    arrest d.
    attention deficit d. (ADD)
    attention deficit hyperactivity d. (ADHD)
    autosomal chromosome d.
    autosomal dominant d.
    autosomal recessive d.

body dysmorphic d. (BDD)
brain d.
carnitine transferase enzyme d.
cerebelloparenchymal d. IV (CPD IV)
chromosome 9p d.
coagulation d.
connective tissue d.
dominant d.
eating d.
endocrine d.
familial bipolar mood d.
fetal iodine deficiency d. (FIDD)
gamma-chain d.
gamma-loop d.
gastrointestinal d.
genetic d.
hematologic d.
hematopoietic d.
heme metabolism d.
hepatic d.
hypothalamic-pituitary d.
infantile sialic acid storage d. (ISSD)
inherited d.
kifafa seizure d.
lactation d.
late luteal phase dysphoric d. (LLPDD)
lipid metabolism d.
lysosomal enzyme d.
mendelian genetic d.
metabolic d.
metal metabolism d.
migrational d.
multifactorial d.
multigenic d.
musculoskeletal d.
myeloproliferative d.
neurologic d.
ovarian d.
panic d.
parathyroid d.
pervasive developmental d. (PDD)
pituitary d.
platelet d.
polygenic d.
porphyrin metabolism d.
premenstrual dysphoric d. (PMDD)
Primary Care Evaluation of Mental D.'s (PRIME-MD)
pulmonary d.
purine metabolism d.
recessive d.
reproductive d.
rheumatologic d.
seizure d.

severe combined
immunodeficiency d. (SCID)
single-gene d.
transport d.
urinary tract d.
vulvovaginal d.
X-linked d.

dispermy
disperse placenta
displacement
d. implantation
lateral head d. (LHD)
lateral sperm head d.

display
M-mode d.

disproportion
cephalopelvic d. (CPD)
fetopelvic d.

disputed maternity
disruption
anal sphincter d.
perineal sphincter d.

disruptive proboscis
dissecans
osteochondritis d.

dissecting aortic aneurysm
dissection
axillary node d.
blunt and sharp d.
gauze d.
groin d.
inguinal-femoral node d.
partial zona d. (PZD)
pelvic node d.
selective inguinal node d.

dissector
Agris-Dingman submammary d.
Endo-Assist cutting d.
Kitner d.
Maryland d.
Polaris reusable d.
spud d.

disseminated
d. CMV
d. gonococcal infection
d. gonorrhea
d. intravascular coagulation (DIC)

dissemination
iatrogenic d.
peritoneal d.

disseminatus
lupus erythematosus d.

dissimilar twins
dissociation
d. constant
hemoglobin-oxygen d.

distal
d. arthrogryposis
d. arthrogryposis, hypopituitarism, mental retardation, facial anomalies syndrome
d. arthrogryposis, mental retardation, characteristic facies syndrome
d. limb deficiency-mental retardation syndrome
d. occlusion
d. sacral defect
d. transverse limb defects-mental retardation-spasticity syndrome
d. tubal disease
d. tubal microsurgery
d. tubal obstruction

distance
genetic d.
intercapillary d.
source-to-axis d. (SAD)
source-to-skin d. (SSD)

distasonis
Bacteroides d.

distention
abdominal d.

distichiasis
distigmine bromide
distinctive facies
distortion
segregation d.

distress
fetal d.
iatrogenic fetal d.
neonatal d.
respiratory d.
transient fetal d.

distribution
binomial d.
depth-dose d.
diploid d.
tetraploid d.

disturbance
architectural d.
mental d.
ovulatory d.
visual d.

disturbed equilibrium syndrome

D

NOTES

disulfiduria
    mercaptolactate-cysteine d. (MCDU)
disulfiram
DIT
    diiodotyrosine
Ditropan
    D. XL
Diucardin
Diuchlor H
Diulo
Diurese-R
diuresis
diuretic
    thiazide d.
Diuril
diurnal
    d. micturition
    d. rhythm
diurnus
    pavor d.
Diva
    D. laparoscopic morcellator
    D. laparoscopic morcellator device
divergent fetal growth
diversion
    continent supravesical bowel
     urinary d.
    ileal d.
    urinary d.
diversity
    genetic d.
diversus
    Citrobacter d.
diverticula (*pl. of* diverticulum)
diverticulectomy
    urethral d.
diverticulitis
diverticulosis
diverticulum, pl. **diverticula**
    Hutch d.
    Meckel d.
    Nuck d.
    pharyngeal d.
    tubal d.
    urethral d.
diverting colostomy
diving reflex
division
    cell d.
    cellular d.
    equatorial d.
    meiotic d.
    miotic d.
    premature centromere d.
    transcervical d.
dizygotic twins
dizygous

DKA
    diabetic ketoacidosis
DLIF
    digoxin-like immunoreactive factor
DM
    Poly-Histine DM
6-DMAP
DMD
    Duchenne de Boulogne muscular
     dystrophy
    Duchenne muscular dystrophy
DMD/BMD
    Duchenne de Boulogne muscular
     dystrophy/Becker muscular dystrophy
    Duchenne muscular dystrophy/Becker
     muscular dystrophy
D-mosaic blood type
DMPA
    depot medroxyprogesterone acetate
DMSO
    dimethylsulfoxide
DNA
    deoxyribonucleic acid
    DNA amplification
    DNA amplification fingerprinting
    chloroplast DNA
    DNA clone
    DNA cloning
    complementary DNA (cDNA)
    DNA copy
    DNA diagnostics
    directed amplification of
     minisatellite region DNA
    DNA dual color probe
    exogenous DNA
    DNA fingerprinting
    genomic DNA
    heterochromatic DNA
    heteroduplex DNA
    homoduplex DNA
    DNA hybridization
    DNA index
    DNA library
    DNA ligase
    DNA marker
    mitochondrial DNA
    DNA nucleotidylexotransferase
    DNA nucleotidyltransferase
    plasmid DNA
    DNA ploidy
    DNA polymerase
    randomly amplified polymorphic
     DNA
    recombinant DNA
    repetitive DNA
    DNA replication
    satellite DNA
    DNA sequence

DNA typing
Y-specific DNA
DNA-directed
DNA-d. RNA polymerase
DNase
deoxyribonuclease
DNCB
dinitrochlorobenzene
DNCB therapy
Doan's
Extra Strength D.
Original D.
DOB
date of birth
Dobbhoff
D. catheter
D. nasogastric feeding tube
DOBI
dynamic optical breast imaging system
dobutamine
d. hydrochloride
Dobutrex
DOC
deoxycorticosterone
desoxycorticosterone
docetaxel
docusate
d. calcium
d. sodium
Döderlein
D. bacillus
D. method
D. method of vaginal hysterectomy
D. roll-flap operation
D. technique
dog-like facies
dogma
central d.
Döhle body
DOL
day of life
Dolacet
Dolanex
Dolene
dolichocephalic
d. head
dolichopellic, dolichopelvic
doll's
d. eye reflex
d. head maneuver
Dolobid
Dolophine

Dolorac
Dolphin
D. hysteroscopic fluid management system
D. instrument
domain
shedding d.
Domeboro
Otic D.
dome cell
domestic
d. abuse
d. violence
domesticum
Pronema d.
dominance
incomplete d.
thromboxane d.
d. variance
dominant
autosomal d. (AD)
d. choroidal sclerosis
d. disorder
d. Dovne honeycomb retinal degeneration
d. follicle
d. gene
d. inheritance
d. lethal trait
domperidone
Donahue-Uchida syndrome
Donald-Fothergill operation
Donald procedure
donation
autologous blood d.
egg d.
embryo d.
oocyte d. (OD)
ovum d.
sperm d.
Donnez device
Donohue syndrome
donor
artificial insemination d.
artificial insemination by d. (AID)
d. egg
egg d.
embryo d.
d. insemination (DI)
oocyte d.
d. oocyte transfer

D

NOTES

donor *(continued)*
    sperm d.
    d. sperm
donor-specific blood
Donovan body
donovanosis
DOOR
    deafness, onychoosteodystrophy, mental
      retardation
      DOOR syndrome
DOP
    degenerate oligonucleotide primer
dopa, DOPA, Dopa
    dihydroxyphenylalanine
L**-dopa**
    levodopa
Dopamet
dopamine
    d. hydrochloride
    d. receptor agonist
dopaminergic agonist
Dopar
Dopcord recorder
Doppler
    Acuson color D.
    color-flow D.
    continuous-wave D.
    2D D.
    D. effect
    D. evaluation
    FetalPulse Plus fetal D.
    D. flow
    D. flow-velocity wave form
    Imex Pocket-Dop OB D.
    D. principle
    pulsed D.
    range-gated D. (RGD)
    D. shift
    D. shift spectra
    spectral D.
    Ultrascope obstetrical D.
    D. ultrasonography
    D. ultrasound
    D. velocimetry
    D. waveform
Dopplex
Dopram
Doptone fetal stethoscope
dornase alfa
Dorros infusion and probing catheter
dorsal
    d. birthing position
    d. lithotomy position
    d. penile nerve block
    d. supine position
Doryx

DOS
    dysosteosclerosis
dosage-sensitive sex reversal (DSS)
dose, dosage
    average radiation d.
    cancericidal d.
    central axis depth d.
    depth d.
    gene d.
    maximal permissible d.
    minimum lethal d. (MLD)
    physiologic replacement d.
    radiation d.
    radiation-absorbed d. (rad)
DOST
    direct oocyte sperm transfer
Dostinex
DOT
    direct oocyte transfer
dot
    Schüffner d.
dot-blot
    d.-b. HPV hybridization test
    d.-b. hybridization
    d.-b. procedure
    d.-b. technique
dothiepin
double
    d. cell
    d. cortex
    d. diapers
    d. gloving
    d. helix
    d. hump sign of an enterocele
    d. intussusception
    d. lip
    d. ring
    d. ureter
double-balloon
    d.-b. catheter
    d.-b. device
double-bank bili lights
double-blanket phototherapy
double-bleb sign
double-breech presentation
double-bubble
    d.-b. flushing reservoir
    d.-b. isolette
    d.-b. sign
double-focus tube
double-footling presentation
double-freeze technique
double-inlet
    d.-i. left ventricle
    d.-i. right ventricle
double-J stent
double-lumen catheter
double-mouthed uterus

double-tooth tenaculum
double-umbrella device
double-volume exchange transfusion
double-walled
    d.-w. bubble isolette
    d.-w. incubator
doubling time
douche, douching
    fan d.
    Fritsch d.
    iodine d.
    Massengill d.
    maternal d.
    povidone-iodine d.
    vaginal d.
    yogurt d.
doughnut pessary
Douglas
    D. abscess
    cul-de-sac of D.
    D. cul-de-sac
    D. mechanism
    D. method
    D. pouch
    D. spontaneous evolution
douglascele
doula
dovetail sign
dowager's hump
Down
    morbus D.
    D. syndrome
down
    clamped d.
    testes d.
downregulation
    d. regimen
downslanting palpebral fissure
downsloping palpebral fissure
downstream
doxacurium chloride
doxapram hydrochloride
doxazosin
doxepin
Doxil
doxorubicin
    d. hydrochloride
Doxy-200
Doxy-Caps
Doxychel
Doxycin

doxycycline
    d. hyclate
    d. monohydrate
doxylamine
Doxy Oral
Doxy-Tabs
Doxytec
Doyen vaginal hysterectomy
Doyle operation
Dpd
    deoxypyridinoline
D^u-positive mother
DPPC
    dipalmitoyl phosphatidylcholine
DPT
    diphtheria, pertussis, tetanus
    DPT vaccine
DPTA
    diethylenetriamine pentaacetic acid
DR
    human lymphocyte antigen-locus
     DR
dragon
    d. pyelogram
    d. sign
drain
    butterfly d.
    fluted d.
    Freyer suprapubic d.
    Hysto-vac d.
    Jackson-Pratt suction d.
    Penrose d.
    retroperitoneal d.
    Shirley wound d.
drainage
    closed d.
    hematoma d.
    lymphatic d.
    lymph node d.
    percutaneous d.
    pulmonary venous d.
    suction d.
    waterseal d.
drain-pipe urethra
Dramamine
Dramilin
Dramocen
Dramoject
drape
    Barrier laparoscopy d.
    Ioban d.

D

NOTES

drape *(continued)*
    iodophor-impregnated adhesive d.
    OPMI d.
Drash syndrome
DRD
    dystrophia retinae pigmentosa-dysostosis syndrome
Drenison
dressing
    alginate wound d.
    AlgiSite d.
    Algosteril d.
    Allevyn d.
    Aquacel d.
    Aquasorb d.
    Biobrane/HF d.
    Biopatch d.
    BlisterFilm d.
    BreakAway d.
    Carrasyn Hydrogel d.
    ClearSite Hydro Gauze d.
    Coban d.
    CollaCote d.
    compressive d.
    Conformant d.
    Coverlet d.
    Curaderm d.
    Curafil d.
    Curagel Hydrogel d.
    Curasorb calcium alginate d.
    ExuDerm d.
    d. forceps
    Fuller shield rectal d.
    FyBron calcium alginate d.
    hydrogel d.
    Iodosorb absorptive d.
    Kalginate d.
    Nu Gauze d.
    Nu-Gel d.
    PanoPlex d.
    PolyMem d.
    SignaDress d.
    Silon wound d.
    StrataSorb d.
    SurePress d.
    SureSite d.
    Thinsite d.
    Transorbent d.
    Veingard d.
    Viasorb d.
Drews forceps
Drew-Smythe catheter
Dri-Dot
    Prognosticon D.-D.
drift
    genetic d.
    pure random d.

drill
    bladder d.
drilling
    zona d. (ZD)
drink
    Boost nutritional d.
Drinker respirator
drinking
    binge d.
    social d.
drip
    continuous intravenous oxytocin d.
Dripps–American Society of Anesthesiologists classification
Drisdol
Dristan
    D. Long Lasting Nasal Solution
    D. Sinus Caplets
driver
    Laurus ND-260 needle d.
Drixoral
    D. Nasal
    D. Non-Drowsy
    D. Syrup
Drize
droloxifene
dromedary hump
drooling
droperidol
drops
    Afrin Children's Nose D.
    Allergan Ear D.
    Ayr saline d.
    Coly-Mycin S Otic D.
    Mallazine Eye D.
    Mylicon d.
    Phazyme infant d.
    Polytrim eye d.
    Triaminic Oral Infant D.
dropsy
    d. of the fetus
    iatrogenic preterm birth maternal d.
Droxia
DRPLA
    dentatorubral-pallidoluysian atrophy
drug
    d. addiction
    adrenergic d.
    anticholinergic d.
    anticonvulsant d.
    antihypertensive d.
    antimalarial d.
    antineoplastic d.
    antipsychotic d.
    antithyroid d.
    beta-adrenergic d.
    cholinergic d.
    d. fever

d. holiday
illicit d.
intramuscular d.
intranasal d.
intravenous d.
nonestrogen d.
nonsteroidal antiinflammatory d. (NSAID)
over-the-counter d.
prescription d.
psychiatric d.
psychoactive d.
psychotropic d.
recreational d.
social d.
drug-depressed infant
drug-induced
    d.-i. diabetes
    d.-i. hemolytic anemia
dry
    d. birth
    d. labor
    d. vagina
DS
    Bactrim DS
    Cotrim DS
    Septra DS
    Sulfatrim DS
    Uroplus DS
    WinRho DS
DSM
    degradable starch microsphere
DSS
    dosage-sensitive sex reversal
D-stix
    Dextrostix
d4T
    stavudine
D-Tach removal needle
DTIC
    dimethyl-triazeno-imidazole carboxamide
    DTIC-Dome
dTMP
    deoxythymidylic acid
DTR
    deep tendon reflex
D-transposition
    dextrotransposition
DTS
    differential temperature sensor
dual-energy
    d.-e. photon absorptiometry

    d.-e. x-ray absorptiometry (DEXA, DXA)
dual nucleoside analog reverse transcriptase inhibitor
Dual-Pak
    Monistat D.-P.
dual-photon absorptiometry
Duane
    D. anomaly
    D. syndrome
Duarte variant galactosemia
DUB
    dysfunctional uterine bleeding
Dubin-Johnson syndrome
Dubois abscess
Dubowitz
    D. evaluation
    D. examination
    D. Neurological Assessment
    D. scale for infant maturity
    D. score
    D. syndrome
Dubowitz/Ballard Exam for Gestational Age
Duchenne
    D. de Boulogne muscular dystrophy (DMD)
    D. de Boulogne muscular dystrophy/Becker muscular dystrophy (DMD/BMD)
    D. muscular dystrophy (DMD)
    D. muscular dystrophy/Becker muscular dystrophy (DMD/BMD)
    D. myodystrophy
    D. palsy
    D. paralysis
    D. pseudohypertrophic muscular dystrophy
    D. syndrome
    D. type progressive muscular dystrophy
    D. type pseudohypertrophic progressive muscular dystrophy
Duchenne-Griesinger syndrome
duckbill speculum
Duckett procedure
Duckey nipple
Ducrey bacillus
duct
    allantoic d.
    Bartholin d.
    bile d.

D

NOTES

duct *(continued)*
    branchial d.'s
    cloacal d.
    cowperian d.
    ejaculatory d.
    focally dilated d.
    Gartner d.
    gartnerian d.
    genital d.
    imperforate nasolacrimal d.
    lacrimal d.
    lactiferous d.
    male genital d.
    mesonephric d.
    metanephric d.
    müllerian d.
    nasolacrimal d.
    omphalomesenteric d.
    paramesonephric d.
    paraurethral d.
    paucity of interlobular bile d.
     (PILBD)
    Reichel cloacal d.
    Skene d.
    solitary dilated d.
    Stensen d.
    vestibular d.
    vitelline d.
    vitellointestinal d.
    wolffian d.
ductal
    d. adenoma
    d. carcinoma
    d. carcinoma in situ (DCIS)
    d. ectasia
    d. hyperplasia
    d. papilloma
ductal-dependent
    d.-d. lesion
    d.-d. pulmonary circulation
ductule
ductuli efferentia
ductus
    d. arteriosus (DA)
    d. arteriosus persistens
    d. venosus
Duffy system
Dufourmentel technique
Dührssen incision
Dukes
    D. classification
    D. disease
Dulbecco phosphate buffered saline
Dulcolax
dullness
    absolute cardiac d.
    area of cardiac d. (ACD)
3-D ultrasound

Dumas vault cap
dummy
    d. source
    d. spacer
Dumontpallier pessary
Duncan
    D. curette
    D. folds
    D. knot
    D. mechanism
    D. mechanism of placental delivery
    D. placenta
    D. position
    D. presentation
    D. slipknot
Duo-Cyp
duodenal
    d. atresia
    d. duplication
    d. ileus
    d. obstruction
    d. ulcer
Duosol
Duotrate
dup
    duplication
    dup (10p)/del (10q) syndrome
    dup (1p)–(22p) syndrome
    dup (9q/del (9p) syndrome
    dup (1q)–(22q) syndrome
    dup (Xq) syndrome
Du Pen epidural catheter
Duphalac
duplex
    d. scanning
    d. ultrasound
    d. uterus
duplication (dup)
    caudal d.
    chromosome 15 inverted d.
    cranial d.
    duodenal d.
    fetal d.
    gene d.
    d. of ileum
    intrachromosomal d.
    mirror d.
    d. 1p–22p syndrome
    d. 1q–22q syndrome
    renal d.
    ureteral d.
    d. Xq syndrome
duplication-deficiency syndrome
duplicitas
    d. anterior
    d. symmetros
durable power of attorney
Durabolin

Duracillin AS
Duraclon
Dura-Estrin
Duragen
dural
    d. arteriovenous malformation
    d. spinal angioma
Duralon-UV nylon membrane
Duralutin
Duramist Plus
Duramorph
Dura Neb portable nebulizer
DuraPrep surgical solution
Duraquin
Duratest
Duratestrin
Durathate
Duration Nasal Solution
Duretic
Duricef
Duroziez disease
duskiness
dusky skin
Dutch
    D. cap
    D. pessary
Duval forceps
Duverney gland
Duvoid
D^U variant blood type
DVIU
    direct vision internal urethrotomy
DVT
    deep vein thrombosis
    deep venous thrombosis
DV Vaginal Cream
dwarf
    achrondoplastic d.
    d. pelvis
    Russell d.
    d. syndrome
dwarfism
    achondroplastic d.
    acromesomelic d.
    alopecia, contracture, d. (ACD)
    Amsterdam d.
    asexual d.
    ateliotic d.
    bird-headed d.
    Brissaud d.
    Caroline Crachami osteodysplastic primordial d.

d., cerebral atrophy, keratosis follicularis syndrome
child abuse d.
chondroplastic d.
d., congenital medullary stenosis syndrome
constitutional d.
d. and cortical thickening of tubular bones
d., cortical thickening of tubular bones, transient hypocalcemia
cretin d.
diastrophic d.
d., eczema, peculiar facies syndrome
familial bird-headed d.
geleophysic d.
genital d.
hypophysial d.
hypothyroid d.
d., ichthyosiform erythroderma, mental deficiency syndrome
infantile d.
Kniest d.
Langer mesomelic d.
Laron d.
lean spastic d.
Lenz-Majewski hyperostotic d.
lethal d.
Levi-Lorain d.
microcephalic d.
microcephalic primordial d. 1
micromelic d.
mulibrey nanism or d.
nanocephalic d.
d., onychodysplasia syndrome
osteodysplastic primordial d.
osteoglophonic d.
ovarian d.
parastremmatic d.
d., pericarditis syndrome
pituitary d.
d., polydactyly, dysplastic nails syndrome
polydystrophic d.
primordial d.
rhizomelic d.
Robinow d.
Russell-Silver d.
Saldino-Noonan d.
Seckel d.
sexual d.

NOTES

dwarfism *(continued)*
    short-limb d.
    short-rib d.
    Silver d.
    Silver-Russell d.
    six-fingered d.
    Smith-McCort d.
    thanatophoric d.
    Walt Disney d.
DWS
    Dandy-Walker syndrome
DXA
    dual-energy x-ray absorptiometry
    DXA densitometer
Dyadic Adjustment Scale
Dyazide
Dyban technique
Dycill
Dyclone
dyclonine
dydrogesterone
dye
    azo d.
    Berwick d.
    Evans blue d.
    indigo carmine d.
    iodinated d.
    methylene blue d.
    triple d.
Dyflos
Dyggve-Melchior-Clausen syndrome
Dyke-Davidoff syndrome
Dymelor
Dymenate
Dynabac
Dynacin
    D. Oral
dynamic
    androgen d.'s
    d. graciloplasty
    d. image on ultrasound
    d. optical breast imaging system (DOBI)
    d. posturography
Dynamite mattress system
Dynapen
dynorphin
Dynothel
Dyonic DyoBride 300 illuminator
dyphylline
Dyrenium
dysadrenalism
dysautonomia
    familial d.
dysbetalipoproteinemia
dyscephalia, dyscephaly
    Francois d.
    Hallermann-Streiff d.

    mandibulooculofacial d.
    d. mandibulooculofacialis
    oculomandibular d.
dyscephaly-congenital cataract-hypotrichosis syndrome
dyschezia
dyschondroplasia
    Voorhoeve d.
dyschondrosteosis
dyscrasia
    blood d.
dysdiadochokinesis
dysembryoma
dysembryoplasia
dysencephalia splanchnocystica
dysentery
    amebic d.
    bacillary d.
dysequilibrium syndrome (DES)
dysesthetic vulvodynia
dysfibrinogenemia
dysfolliculogenesis
dysfunction
    bladder d.
    cerebral d.
    corpus luteum d.
    endothelial d.
    familial vocal cord d.
    family d.
    hypertonic uterine d.
    hypothalamic d.
    hypothalamic-pituitary axis d.
    hypotonic uterine d.
    lymphocyte d.
    placental d.
    postoperative bladder d.
    postoperative voiding d.
    sexual d.
    thyroid d.
    urinary bladder d.
    uterine d.
    voiding d.
dysfunctional
    d. family
    d. labor pattern
    d. uterine bleeding (DUB)
dysgammaglobulinemia
dysgenesia
dysgenesis
    cerebral d.
    cerebroocular d. (COD)
    gonadal d.
    iridocorneal mesodermal d.
    d. mesodermalis cormeae et irides
    d. mesostromalis anterior
    mixed gonadal d.
    d. neuroepithelialis retinae
    ovarian d.

pure gonadal d.
renal d.
seminiferous tubule d.
X-linked recessive d.
XX-type gonadal d.
XY gonadal d.
dysgenetic gonads
dysgenic
dysgenitalism
dysgerminoma
ovarian d.
dyshepatia
lipogenic d.
dyshidrosis
dyshidrotic eczema
dyshormonogenesis
dyskaryosis
dyskeratosis
d. congenita
congenital d.
dyskinesia
dyslexia
dysmature
dysmaturity
d. syndrome
dysmenorrhea
essential d.
functional d.
intrinsic d.
mechanical d.
membranous d.
obstructive d.
ovarian d.
primary d.
secondary d.
spasmodic d.
tubal d.
ureteral d.
uterine d.
vaginal d.
dysmenorrheal membrane
dysmorphic
d. facial feature
d. facies
d. syndrome
dysmorphism
digitotalar d.
major d.
mandibulooculofacial d.
minor d.
dysmorphogenesis
otomandibular facial d.

dysmorphologist
dysmorphology
craniofacial d.
dysmotility
dysmyelinisatus
status d.
dysmyelinogenic
dysnomia
dysontogenesis
dysontogenetic cyst
dysorganoplasia
dysosteogenesis
dysosteosclerosis (DOS)
dysostosis, pl. **dysostoses**
acrocraniofacial d.
acrofacial d.
Catania type acrofacial d.
cleidocranial digital d.
d. cleidocranialis
d. cleidocraniodigitalis
d. cleidocraniopelvina
craniofacial d.
d. craniofacialis with hypertelorism
craniofrontonasal d.
Crouzon craniofacial d.
d. enchondralis metaepiphysaria
epiphyseal d.
frontofacionasal d.
Genée-Wiedemann acrofacial d.
(GWAFD)
d. generalisata
hereditary polytopic enchondral d.
hypomandibular faciocranial d.
mandibulofacial d. (MFD)
d. mandibulofacials
maxillofacial d.
metaphyseal d.
d. multiplex
mutational d.
Nager acrofacial d.
orodigitofacial d.
otofacial d.
otomandibular d.
pelvicocleidocranial d.
preaxial acrofacial d.
preaxial mandibulofacial d.
Rodriquez lethal acrofacial d.
Treacher Collins mandibulofacial d.
unilateral mandibulofacial d.
dysostotic-idiocy-gargoylism-
lipochondrodystrophy syndrome

D

NOTES

157

dyspareunia
    deep d.
    Friedrich criteria for d.
    secondary vestibular d.
    vestibular d.
dysphoria
    premenstrual d.
dysplasia
    acromelic frontonasal d.
    acromesomelic d.
    acromicric d.
    acropectorovertebral d.
    anal sphincter d.
    angel shaped phalangoepiphyseal d.
      (ASPED)
    anhidrotic ectodermal d.
    arteriohepatic d. (AHD)
    asphyxiating thoracic d.
    atriodigital d.
    auriculobrachiogenital d.
    azoospermia, renal anomaly,
      cervicothoracic spine d. (ARCS)
    bone d.
    boomerang d.
    branchiootorenal d.
    bronchopulmonary d. (BPD)
    3C d.
    camptomelic d.
    CCC d.
    cerebellotrigeminal and focal
      dermal d.
    cerebrofaciothoracic syndrome or d.
    cervical d.
    chondroectodermal d.
    cleidocranial d.
    d. cleidocranialis
    d. cleidofacialis
    colposcopic grading of cervical d.
    congenital alveolar d.
    congenital encephalo-ophthalmic d.
    costovertebral d.
    craniocerebellocardiac d.
    craniodiaphysial d.
    cranioectodermal d. (CED)
    craniofrontal d.
    craniofrontonasal d. (CFND)
    craniometaphysial d.
    craniotubular d.
    cretinoid d.
    de la Chapelle d.
    dentin d.
    dentooculoosseous d.
    diaphyseal d.
    diastrophic d.
    dyssegmental d.
    ectodermal d.
    endocervical glandular d.
    epiphyseal d.

    d. epiphysealis hemimelia
    d. epiphysealis hemimelica
    d. epiphysealis punctata
    facial ectodermal d.
    faciocardiomelic d.
    faciogenital d.
    familial focal facial dermal d.
    Fanconi d.
    fetal skeletal d.
    focal facial dermal d.
    focal facial dermal d. II (FFDD II)
    frontofacionasal d.
    frontometaphyseal d.
    frontonasal d. (FND)
    geleophysic d.
    gracile bone d.
    Grebe d.
    hereditary bone d.
    hereditary expansile polyostotic
      osteolytic d.
    hereditary retinal d.
    hidrotic ectodermal d.
    high-grade cervical d.
    hip d.
    Holt-Oram atriodigital d.
    hydrocephalus, agyria, retinal d.
      (HARD)
    hypohidrotic ectodermal d.
    immunoosseous d.
    intraepithelial cervical d.
    iridodental d.
    ischiopatellar d.
    Kniest d.
    Kniest-like d.
    kyphomelic d.
    Langer d.
    lateral facial d. (LFD)
    Lenz d.
    lethal d.
    lethal bone d.
    d. linguofacialis
    low-grade d. (LGD)
    mammary d.
    mandibuloacral d.
    Margarita Island type ectodermal d.
    d. marginalis posterior
    maxillonasal d.
    mesectodermal d.
    mesomelic d.
    metaphoric d.
    metaphyseal d.
    metatropic d.
    Mexican cardiomelic d.
    microglandular d.
    multicystic renal d.
    multiple epiphyseal d.
    necrotic facial d.
    neonatal osseous d.

oculoauricular d.
d. oculoauricularis
oculoauriculovertebral d.
oculodentodigital d.
d. oculodentodigitalis
oculodentoosseous d. (ODOD)
odontoonychodermal d.
olfacto-ethmoidohypothalamic d.
olfactogenital d.
ophthalmomandibulomelic d.
orofaciodigital syndrome with
  tibial d.
osseous-oculodental d.
OSSUM database of skeletal d.'s
osteodental d. (ODD)
osteoglophonic d.
otospondylomegaepiphyseal d.
pelvis capsular d.
pelvis-shoulder d.
polycystic fibrous d.
polyostotic fibrous d.
pseudoachondroplastic d.
pseudodiastrophic d.
ptosis of eyelids-diastasis recti-
  hip d.
punctate epiphyseal d.
radiation d.
Rapp-Hodgkin ectodermal d.
renal d.
Robinow mesomelic d.
Schneckenbecken d.
septooptic d.
skeletal d.
split hand-cleft lip/palate and
  ectodermal d. (SCE)
SPONASTRIME d.
spondyloepimetaphyseal d. (SEMD)
spondyloepiphyseal d.
spondylometaphyseal d.
spondyloperipheral d.
spondylothoracic d.
squamous d.
Streeter d.
thanatophoric d.
thymic d.
trichorhinoauriculophalangeal
  multiple exostoses d.
trichorhinophalangeal d.
trichorhinophalangeal multiple
  exostosis d.
dysplastic
d. cell

dyspnea
paroxysmal nocturnal d. (PND)
dyspraxia
dysraphia
olfactoethmoidohypothalamic d.
d. of spine
dysraphism
dysrhythmia
cardiac d.
gastric d.
dyssegmental dysplasia
dysspermia
dyssynergia
detrusor sphincter d.
dystaxia cerebralis infantilis
dysthymic
dysthyroidal infantilism
dystocia
abdominal d.
all-fours maneuver for shoulder d.
cervical d.
fetal d.
maternal d.
placental d.
shoulder d.
dystocia-dystrophia syndrome
dystonia
d. musculorum deformans
torsion d.
dystonia-deafness syndrome
dystrophia
d. bullosa hereditaria, typus
  maculosus
d. retinae-dysacousis syndrome
d. retinae pigmentosa-dysostosis
  syndrome (DRD)
dystrophica
epidermis bullosa d.
dystrophic calcification
dystrophin gene
dystrophy
adult pseudohypertrophic
  muscular d.
asphyxiating thoracic d. (ATD)
autoimmune polyendocrinopathy,
  candidiasis, ectodermal d.
  (APECED)
Becker-Kiener muscular d.
Becker muscular d. (BMD)
Becker pseudohypertrophic
  muscular d.

D

NOTES

dystrophy *(continued)*
Becker type progressive
muscular d.
benign X-linked recessive
muscular d.
cerebroocular dysgenesis-muscular d.
(COD-MD)
cerebroocular dysplasia-muscular d.
childhood pseudohypertrophic
muscular d.
classic X-linked recessive
muscular d.
congenital endothelial corneal d.
congenital muscular d.
congenital myotonic d.
craniocarpotarsal d.
Duchenne de Boulogne muscular d.
(DMD)
Duchenne de Boulogne muscular
dystrophy/Becker muscular d.
(DMD/BMD)
Duchenne muscular d. (DMD)
Duchenne muscular
dystrophy/Becker muscular d.
(DMD/BMD)
Duchenne pseudohypertrophic
muscular d.
Duchenne type pseudohypertrophic
progressive muscular d.
Erb juvenile muscular d.
facioscapulohumeral muscular d.
familial osseous d.
Fukuyama congenital muscular d.
(FCMD)
gingival fibromatosis-corneal d.
hereditary bullous d.
infantile neuroaxonal d.

infantile thoracic d.
juvenile epithelial corneal d.
Landouzy-Dejerine muscular d.
Leyden-Möbius muscular d.
limb-girdle muscular d.
macular d.
Meesmann corneal d.
micropolygyria with muscular d.
mild X-linked recessive
muscular d.
muscular d. (MD)
myotonic d. (MD)
myotonic muscular d.
oculocerebrorenal d.
oculopharyngeal muscular d.
osteochondromuscular d.
pseudohypertrophic adult
muscular d.
pseudohypertrophic progressive
muscular d.
reflex sympathetic d. (RSD)
Schnyder crystalline corneal d.
(SCCD)
scleroatonic muscular d.
severe childhood autosomal
recessive muscular d. (SCARMD)
short-limb d.
Steinert myotonic d.
thoracic asphyxiant d.
thoracic-pelvic-phalangeal d.
twenty-nail d.
vulvar d.
X-linked recessive muscular d.
dysuria
dysuria-sterile pyuria syndrome
dyszoospermia

E3
    unconjugated estriol
E$_1$
    estrone
    synthetic prostaglandin E$_1$, E$_2$
E$_2$
    estradiol
    eutopic endometrium prostaglandin
      E$_2$
E$_3$
    estriol
E6-E7 gene
EA
    early amniocentesis
EABT
    estrogen add back therapy
EACA
    epsilon-aminocaproic acid
Eagle-Barrett syndrome
Eagle test
ear
    Aztec e.
    bat e.
    cup e.
    Darwin e.
    lop e.
    low-set e.'s
    Morel e.
    Mozart e.
    Otocalm E.
    e., patella, short stature syndrome
      (EPS)
    satyr e.
    scroll e.
    Wildermuth e.
Earle
    E. balanced salt solution
    E. culture medium
    E. medium
early
    e. amniocentesis (EA)
    e. amnion vascular disruption
      complex
    e. cardiac motion
    e. deceleration
    e. embryonic death
    e. embryonic loss
    e. follicular development
    E. Neonatal Neurobehavioural Scale
    e. neonate
    e. onset Parkinsonism-mental
      retardation syndrome
    E. and Periodic Screening,
      Diagnosis, and Treatment program
      (EPSDT)

    e. pregnancy factor
    e. pregnancy loss
    e. pregnancy test (EPT)
    e. pregnancy wastage
    e. proliferative phase
    e. stromal invasion
early-onset
    e.-o. diabetes mellitus-epiphyseal
      dysplasia syndrome
    e.-o. preeclampsia
EASI
    extraamniotic saline infusion
    EASI catheter
Eastern
    E. blot
    E. blot test
Eastman-Bixler syndrome
Easy
    Clearblue E.
    ClearPlan E.
eating
    e. disorder
    e. habit
EBM
    epidermolysis bullosa, macular type
EBNS
    endoscopic bladder neck suspension
Ebola hemorrhagic fever
Ebstein anomaly
EBV
    Epstein-Barr virus
E-C
E-cadherin protein
ECBI
    Eyberg Child Behavior Inventory
ecbolic
ECC
    endocervical curettage
    extracorporeal circulation
Eccentric Y retractor
eccentrochondrodysplasia
eccentroosteochondrodysplasia
ecchymosis
    trocar site e.
ECCL
    encephalocraniocutaneous lipomatosis
Eccocee ultrasound system
eccouchement forcé
eccrine gland
eccyesis
ECD
    endocardial cushion defect
ECG, EKG
    electrocardiogram
echinococcal cyst

E

echinocyte
Echistatin
ECHO
  enteric cytopathogenic human orphan
  enterocytopathogenic human orphan
    ECHO virus
echo
  echocardiogram
    echo formation
    scattered echo
    single shot fast spin echo (SSFSE)
    specular echo
echocardiogram (echo)
  color e.
  2D e.
  fetal e.
  M-mode e.
  two-dimensional e.
echocardiography
echodensity
echoencephalogram
echoencephalography
echogenic
  e. cardiac focus
  e. fetal bowel
  e. tissue
echogenicity
echogram
  M-mode e.
echolucency
echolucent
EchoMark salpingography catheter
echoplanar magnetic resonance imaging
echothiophate iodide
Echovist
  E. contrast
Echowarm gel warmer
eclampsia
  e. nutans
  puerperal e.
  superimposed e.
eclamptic
  e. retinopathy
  e. seizure
eclamptogenic, eclamptogenous
ECM
  erythema chronicum migrans
    ECM rash
ECMO
  extracorporeal membrane oxygenation
E-Complex-600
econazole
ECOR
  extracorporeal CO_2 removal
Ecostatin
Ecotrin
ECP
  emergency contraceptive pill

ECR
  endocervical resection
ectasia, ectasis
  ductal e.
  familial aortic e.
  mammary duct e.
  scoliosis with dural e.
ecthyma
ectocervical
  e. lesion
ectocervix
ectoderm
ectodermal
  e. dysplasia
  e. dysplasia, cleft lip and palate,
    hand and foot deformity, mental
    retardation syndrome
  e. dysplasia, cleft lip and palate,
    mental retardation, syndactyly
    syndrome I, II
  e. dysplasia, mental retardation,
    syndactyly syndrome
  e. and mesodermal dysplasia with
    osseous involvement ectodermal
    dysplasia-central
ectodermica
  chondrodysplasia e.
ectoendocervical junction
ectolecithal
ectomere
ectomesoblast
ectoneurodermal hamartoma
ectopagus
ectoparasite
ectopia
  e. cordis
  e. lentis
  renal e.
  ureteral e.
ectopic
  e. anus
  e. decidua
  e. endometrial tissue
  e. endometrium
  e. implant
  e. pregnancy (EP)
  e. ureter
  e. ventricular depolarization
Ectosone
ectrodactyly
  e.-cleft lip/palate syndrome
  e., ectodermal dysplasia, clefting
    (EEC)
  e., ectodermal dysplasia, clefting
    syndrome
  e., ectodermal dysplasia and cleft
    lip/palate syndrome

e.-mandibulo facial dysostosis syndrome
e.-spastic paraplegia-mental retardation syndrome
ectrodactyly-ectodermal
    e.-e. dysplasia, cleft palate syndrome
ectromelia
ectrometacarpia
ectrometatarsia
ectrophalangia
ectropion
    cervical e.
    congenital e.
ectrosyndactyly
ECV
    external cephalic version
eczema
    dyshidrotic e.
    e. herpeticum
    infantile e.
    e. marginatum
    e. neonatorum
    nummular e.
    e. vaccinatum
eczematoid
eczematous candida
EDC
    estimated date of conception
    estimated date of confinement
    expected date of confinement
EDD
    expected day of delivery
Eddowes syndrome
Edebohl position
Edecrin
edema
    angioneurotic e.
    brawny e.
    cerebral e.
    dependent e.
    gestational e.
    hereditary e.
    high-altitude cerebral e. (HACE)
    high-altitude pulmonary e. (HAPE)
    labial e.
    leg e.
    lung e.
    menstrual e.
    e. neonatorum
    premenstrual e.

    pulmonary e.
    vulvar e.
Eden-Lawson hysterectomy
EdenTec 2000W in-home cardiorespiratory monitor
Eder cord blood collection device
edge effect
ediburgensis
    typus e.
Edinburgh malformation syndrome
EDRF
    endothelium-derived relaxing factor
edrophonium chloride
EDS
    Ehlers-Danlos syndrome
EDTA
    ethylenediaminetetraacetic acid
EDTA-Vacutainer
Edwardsiella tarda
Edwards syndrome
EE
    electrosurgical excision
    ethinyl estradiol
EEA stapler
EEC
    ectrodactyly, ectodermal dysplasia, clefting
        EEC syndrome
EECS
    extraembryonic celomic space
EEG
    electroencephalogram
eelworm
EEP
    end-expiratory phase
    end-expiratory pressure
E.E.S.
    E.E.S. Chewable
    E.E.S. Granules
EFAS
    embryofetal alcohol syndrome
efavirenz
EFE
    endocardial fibroelastosis
efface
effacement
    cervical e.
    e. of cervix
effect
    adverse e.
    anemic e.
    antiestrogenic e.

E

NOTES

effect *(continued)*
    Arias-Stella e.
    Bohr e.
    brain-sparing e.
    cardioprotective e.
    cardiovascular e.
    chronotropic e.
    Compton e.
    cytotoxic e.
    demand e.
    differential e.
    Doppler e.
    edge e.
    estrogen e.
    estrogen-agonist uterine e.
    fetal alcohol e. (FAE)
    founder e.
    Hawthorne e.
    hormonal e.
    hypnotic e.
    iatrogenic e.
    jet cooling e.
    Mach band e.
    marijuana e.
    mass e.
    neurologic adverse e.
    perinatal e.
    pituitary gonadotropin e.
    Posiero e.
    psychological e.
    returning-soldier e.
    salutary e.
    sedative e.
    side e.
    social factor e.
    star e.
    teratogenic e.
    Yom Kippur e.
effective
    e. conjugate
    e. refractory period
effectiveness
    contraceptive e.
    use e.
effector cell
efferentia
    ductuli e.
efficacious
efficacy
    oral contraceptive e.
efficiency
    female fertility e.
effluvium
    telogen e.
efflux
Effortil
effusion
    clenched fist and pleural e.

    malignant e.
    malignant pleural e.
    pleural e.
Efidac/24
EFM
    electronic fetal monitoring
    external fetal monitoring
Efroxine
Efudex
    E. Topical
EFW
    estimated fetal weight
EGA
    estimated gestational age
Egan mammography
EG/BUS
    external genitalia/Bartholin, urethral, and
      Skene glands
EGF
    epidermal growth factor
egg
    e. activation
    e. cell
    e. donation
    donor e.
    e. donor
    frozen e.
    e. membrane
    e. retrieval
    e. transport
    e. yolk extender
Egnell
    E. breast pump
    E. vacuum
egophony
EH
    endometrial hyperplasia
Ehlers-Danlos
    E.-D. syndrome (EDS)
    E.-D. syndrome type VII
    E.-D. syndrome type VIIA1
    E.-D. syndrome type VIIA2
    E.-D. syndrome type VIIB
    E.-D. syndrome type VIII–XI
EHM
    embryonic heart motion
Ehrlich catheter
EIA
    enzyme immunoassay
EIC
    extensive intraductal component
eicosanoid
EIFT
    embryo intrafallopian transfer
eight-cell embryo
Eikenella corrodens
EIN
    endometrial intraepithelial neoplasia

Einthoven triangle
Eisenmenger
    E. complex
    E. syndrome
ejaculate
    sperm-free e.
ejaculation
    e. failure
    premature e.
    retrograde e.
ejaculatory duct
ejecta
    laser e.
ejection
    milk e.
EKG (*var. of* ECG)
    electrocardiogram
Eklund
    E. positioning system
    E. technique
elastic
    e. canal
    e. recoil of the bronchus
    e. stocking
elastica
    cutis e.
elastin
elastosis performans serpiginosa
Elavil
ELBW
    extremely low birth weight
ELBWI
    extremely low birth weight infant
Eldecort Topical
elderly
    e. primigravida
elective
    e. abortion
    e. abortion material
Electra complex
electric breast pump
electrocardiogram (ECG, EKG)
electrocardiography
    abdominal fetal e.
    fetal e.
electrocautery
    AmpErase e.
    bipolar e.
    Endoclip monopolar e.
    laparoscopic e.
    transurethral e.

electrocoagulation
    e. diathermy
electrode
    Aspen laparoscopy e.
    ball e.
    e. balloon
    bipolar e.
    Bugbee e.
    Copeland fetal scalp e.
    Cryomedics disposable LLETZ e.
    fetal scalp e. (FSE)
    Kontron e.
    Littmann ECG e.
    LLETZ-LEEP active loop e.
    loop e.
    Medi-Trace e.
    Megadyne/Fann E-Z clean
      laparoscopic e.
    REM PolyHesive II patient
      return e.
    rollerball e.
    rollerbar e.
    roller-barrel e.
    Soderstrom-Corson e.
    spiral e.
    unipolar e.
    Valleylab ball e.
    Valleylab loop e.
electrodesiccation
electroejaculation
electroencephalogram (EEG)
electroencephalography
Electro-Gel conductivity gel
electrohysterograph
electrolysis
electrolyte
    e. balance
    e. imbalance
electromagnetic
    e. radiation
    e. spectrum
electromechanical morcellation
electrometrogram
electrometrography
electromyelogram (EMG)
electromyogram (EMG)
electromyography
    anal sphincter e.
    surface e. (SEMG)
electron
electronic fetal monitoring (EFM)
electron-volt (eV, ev)

E

NOTES

electrophilic center
electrophoresis
    denaturing gradient gel e.
    e. gel
    thermal gel gradient e.
electroporation
electroretinal abnormality
Electroscope disposable scissors
electroshield monitoring system
electroshock therapy
electrosurgery
electrosurgical
    e. excision (EE)
    e. loop excision
    e. plume
    e. wire
elegance
    sartorial e.
Elejalde syndrome
Elema angiocardiogram
element
    immature neural e.
    IS e.
    transposable e.
elemental iron
elephantiasis
    congenital e.
    genital e.
    e. vulvae
elephant pelvis
elevated
    e. cardiac output
    e. enzyme activity
    e. liver enzymes
    e. obstetric risk
elevation
    alpha-fetoprotein e.
    fetus growth e.
elevator
    Boyle uterine e.
    lemon-squeezer obstetrical e.
    Somer uterine e.
    uterine e.
    Wadia e.
elfin
    e. facies
    e. facies hypercalcemia syndrome
    e.facies hypercalcemia syndrome
elimination
    Hofmann e.
Elimite
    E. Cream
Elipten
ELISA
    enzyme-linked immunosorbent assay
Elispot test
Elixir
    Brofed E.

    Bromaline E.
    Cold & Allergy E.
    Dimaphen E.
    Dimetane Decongestant E.
    Genatap E.
Elixomin
Elixophyllin
Elliot
    E. forceps
    E. sign
elliptical uterine incision
elliptocytic anemia
elliptocytosis
Ellis-Sheldon syndrome
Ellis-van Creveld syndrome
Elmed
    E. BC 50 M/M digital bipolar
      coagulator
    E. peristaltic irrigation pump
Elmiron
Elocon
elongation
ElSahy-Waters syndrome
Elscint ESI-3000 ultrasound
Eltor
Eltroxin
elucidation
EMA
    epithelial membrane antigen
E-Mac
    English MacIntosh
    E-Mac laryngoscope blade
Embden-Meyerhof pathway
Embolex
emboli (pl. of embolus)
embolism
    air e.
    amniotic fluid e. (AFE)
    cerebral e.
    pulmonary e.
embolization
    angiographic e.
    fibroid e.
    pelvic arterial e.
    e. therapy
    transcatheter uterine artery e.
    uterine artery e. (UAE)
embolus, pl. emboli
    amniotic fluid e.
    gas e.
    Gianturco spring e.
    pulmonary emboli
    trophoblastic e.
Embosphere microsphere
embrace reflex
embroscopy
embryatrics
embryectomy

embryo
    abnormal e.
    e. biopsy
    e. carrier
    cleaved e.
    e. cloning
    cryopreserved e.
    cylindrical e.
    e. donation
    e. donor
    eight-cell e.
    e. encapsulation
    endometrial e.
    four-cell e.
    frozen e.
    e. intrafallopian transfer (EIFT)
    Janosik e.
    multicelled e.
    nodular e.
    preimplantation e. (PIE)
    presomite e.
    previllous e.
    pronuclear e.
    e. reduction
    somite e.
    Spee e.
    e. splitting
    stunted e.
    tetraploid e.
    e. thawing
    e. thawing with transfer
    e. transfer (ET)
    triploid e.
    two-cell e.
embryoblast
embryocardia
embryocide
embryoctony
embryofetal alcohol syndrome (EFAS)
embryofetoscopy
    transabdominal thin-gauge e.
    (TGEF)
embryogenesis
embryography
embryoid
embryology
    breast e.
    causal e.
    comparative e.
    descriptive e.
    experimental e.
    genital tract e.

    Leydig cell e.
    reproductive tract e.
embryoma
embryomorphous
Embryon
    E. GIFT catheter
    E. GIFT transfer catheter set
    E. HSG catheter
embryonal
    e. carcinoma
    e. rhabdomyosarcoma
    e. sarcoma
embryonate
embryonic
    e. anideus
    e. axis
    e. blastoderm
    e. cleavage
    e. diapause
    e. differentiation
    e. disk
    e. genome activation
    e. heart motion (EHM)
    e. loss
    e. neural tube
    e. period
    e. sac
    e. stem cell
    e. testicular regression syndrome
embryoniform
embryonization
embryonum
    smegma e.
embryony
embryopathology
embryopathy
    alcoholic e.
    e. alcoholica
    diabetic e.
    fetal cytabarine/thioguanine e.
    heparin e.
    isotretinoin e.
    retinoic acid e.
    rubella e.
    trimethadione e.
    valproic acid e.
    warfarin e.
embryoplastic
embryoscope
embryoscopy
embryotome
embryotomy

E

NOTES

embryotoxic
embryotrophy
EMC
  endometrial curettage
Emcyt
emergency
  e. contraception
  e. contraceptive pill (ECP)
  hypertensive e.
Emerson respirator
emesis
  e. pH determination
  salivation, lacrimation, urination,
    defecation, gastrointestinal distress
    and e. (SLUDGE)
emetic
EMG
  electromyelogram
  electromyogram
  exomphalos, macroglossia, and gigantism
  MyoTrac EMG
  EMG syndrome
eminence
  median e.
EMIT
  enzyme-multiplication immunoassay
    technique
Emko
EMLA
  eutectic mixture of local anesthetics
  EMLA anesthetic disk
  EMLA cream
emmenagogic
emmenagogue
emmenia
emmenic
emmeniopathy
emmenology
Emmet operation
Emmet-Studdiford perineorrhaphy
Emmett cervical tenaculum
Emmett-Gelhorn pessary
Emo-Cort
emotional amenorrhea
emphysema
  congenital lobar e.
  perivascular e.
  pulmonary interstitial e. (PIE)
empirically
Empirin
Empirynol Plus contraceptive gel
emprosthotonos
empty
  e. scrotum syndrome
  e. sella syndrome
emptying
  gastric e.

empyema
  epidural e.
EMR
  endomyometrial resection
emulsion
  Pusey e.
  Soyacal IV fat e.
  Travamulsion IV fat e.
E-Mycin
  E.-M.-E
en
  e. bloc
  e. bloc resection
  e. caul delivery
  e. face position
EnAbl thermal ablation system
enalapril
enanthate
  estradiol e.
  norethindrone e. (NET-EN)
  testosterone e.
Enbrel
encainide
Encap
  Novo-Rythro E.
encapsulation
  embryo e.
Encare
enceinte
encelitis, enceliitis
encephalitis
  Dawson e.
  herpes e.
  neonatal HSV-1, -2 e.
  Rasmussen e.
  Schilder e.
  St. Louis e.
  viral e.
encephalocele
encephaloclasis
encephalocraniocutaneous lipomatosis
  (ECCL)
encephalocraniofacial
  e. angiomatosis
  e. neuroangiomatosis
encephaloduroarterial synangiosis
encephalofacial angiomatosis
encephalofacialis
  angiomatosis e.
  neuroangiomatosis e.
encephalomalacia
encephalomeningocele
encephalomere
encephalomyelitis
  postinfectious e. (PIE)
encephalomyelocele
encephalomyoarterial synagiosis
encephalomyopathy of Leigh

encephalopathy
anoxic-ischemic e.
bilirubin e.
bovine spongiform e. (BSE)
demyelinating e.
hypertensive e.
hypoxic-ischemic e. (HIE)
neonatal e.
spongiform e.
e. with prolinemia
encephalotomy
encephalotrigeminal
e. angiomatosis
e. syndrome
enchondroma
enchondromatosis
encode
encopresis
encranius
encu method
encyesis
end
e. bud
E. Lice Liquid
endadelphos
Endcaps
Glucolet E.
Endep
end-expiratory
e.-e. phase (EEP)
e.-e. pressure (EEP)
End-Flo laparoscopic irrigating system
endoanal
e. ultrasonography
e. ultrasound
Endo-Assist
E.-A. cutting dissector
E.-A. endoscopic forceps
E.-A. endoscopic knot pusher
E.-A. endoscopic ligature carrier
E.-A. endoscopic needle holder
E.-A. retractable blade
E.-A. retractable scalpel
E.-A. sponge aspirator
Endo-Avitene
E.-A. hemostatic material
E.-A. microfibrillar collagen
hemostat
endocarcinoma
endocardial
e. cushion defect (ECD)
e. fibroelastosis (EFE)

endocarditis
bacterial e.
infective e.
subacute bacterial e. (SBE)
endocavitary radiation therapy
Endocell endometrial cell collector
endocervical
e. aspirator
e. canal
e. canal colposcopy
e. curettage (ECC)
e. curetting/curettage
e. glandular dysplasia
e. mucosa
e. polyp
e. resection (ECR)
e. sampling
e. sampling brush
e. sinus tumor
endocervicitis
endocervix
endochondroma
endochorion
Endoclip
E. cautery
E. monopolar electrocautery
endocolpitis
endocrine
e. disorder
e. factor
e. gland
e. system
endocrinologic sex
endocrinology
reproductive e.
endocrinopathy
endocytosis
receptor-mediated e.
endoderm
endodermal
e. cell
e. sinus tumor (EST)
endodermatosis
Endodermologie LPG system
endogenous
e. catecholamine
e. estrogen
e. estrogenic stimulation
e. gonadotropin
e. gonadotropin activity suppression
e. hormone
e. opiate

E

NOTES

endogenous *(continued)*
  e. opiate receptor
  e. opioid peptide
  e. pyrogen
  e. steroid
Endo-GIA30 suture stapler
Endo-GIA stapler
endograsper
  roticulating e.
Endolive 3-D stereo video endoscope
Endoloop suture
EndoMax advanced laparoscopic instrument
EndoMed LSS laparoscopy system
endometria (*pl. of* endometrium)
endometrial
  e. ablation
  e. ablator
  e. adenoacanthoma
  e. adenocarcinoma
  e. aspirator
  e. atrophy
  e. biopsy
  e. breakdown
  e. cancer
  e. cannula
  e. carcinoma
  e. cavity
  e. clear cell adenocarcinoma
  e. curettage (EMC)
  e. cycle
  e. cycling activity
  e. cyst
  e. dating
  e. development
  e. embryo
  e. histology
  e. hyperplasia (EH)
  e. implant
  e. intraepithelial neoplasia (EIN)
  e. metastasizing leiomyoma
  e. milieu
  e. morphology
  e. neoplasia
  e. neoplasm
  e. polyp
  e. proliferation
  e. protein
  e. receptor
  e. resection
  e. resection and ablation (ERA)
  e. sampling
  e. sarcoma
  e. secretory adenocarcinoma
  e. shedding
  e. spiral artery
  e. stimulation
  e. stripe

  e. stromal sarcoma (ESS)
  e. thickness
  e. tuberculosis
  e. vaporization
endometrioid
  e. carcinoma
  e. tumor
endometrioma
  complex adnexal e.
  large e.
endometriosis
  adhesive e.
  e.-associated infertility
  asymptomatic mild e.
  extrapelvic e.
  e. interna
  internal e.
  nonpigmented e.
  ovarian e.
  pelvic e.
  pulmonary e.
  rectosigmoid colon e.
  retroperitoneal e.
  small bowel e.
  stromal e.
  tubal e.
  urinary tract e.
  vulvar e.
endometriotic
  e. focus
  e. implant
  e. lesion
endometritis
  decidual e.
  e. dissecans
  postpartum e.
  puerperal e.
endometrium, pl. endometria
  atrophic e.
  cyclic proliferative e.
  decidualized e.
  ectopic e.
  eutopic e.
  hyperechoic e.
  inactive e.
  menstrual e.
  nonpregnant e.
  Swiss cheese e.
  tubal e.
endometropic
endomyometrial resection (EMR)
endomyometritis
  postpartum e.
endomyometrium
endonuclease
  e. analysis
  restriction e.
endoparametritis

Endopath
  E. endoscopic articulating stapler
  E. ETS-FLEX endoscopic
    articulating linear cutter
  E. EZ35 endoscopic linear cutter
  E. needle tip electrosurgery probe
  E. Optiview optical surgical
    obturator
  E. TriStar trocar
  E. Ultra Veress needle
endopelvic fascia
endoperoxide
endophytic
endoplasmic
  e. reticulum (ER)
Endopouch
  E. Pro specimen-retrieval bag
endorectal flap
endorphin
  e., dopamine, and prostaglandin
    theory
endosalpingiosis
  atypical e.
endosalpingitis
endosalpingoblastosis
endosalpinx
**Endosample endometrial sampling
  device**
endoscope
  Endolive 3-D stereo video e.
  MicroLap e.
  mother and baby e.
  Storz e.
endoscopic
  e. bladder neck suspension (EBNS)
  fetal e. (FETENDO)
  e. linear cutter
  e. microsurgery
  e. retrograde
    cholangiopancreatography (ERCP)
endoscopy
Endoshears
Endosound endoscopic ultrasound
endosperm
**Endo Stitch laparoscopic suturing
  device**
Endotek
  E. OM-3 Urodata monitor
  E. UDS-1000 monitor
  E. urodynamics system
endothelial
  e. cell

  e. cell activation
  e. dysfunction
endothelin
  e. plasma level
  e. receptor-B gene
endothelin-1 (ET-1)
endothelin-2 (ET-2)
endothelin-3 (ET-3)
  e. gene
endotheliochorial placenta
endothelio-endothelial placenta
endotheliosis
  glomerular capillary e.
**endothelium-derived relaxing factor
  (EDRF)**
endotoxic shock
endotoxin
  lipopolysaccharide e.
endotracheal (ET)
  e. intubation
  e. tube
endotrachelitis
endovaginal
  e. finding
  e. imaging
  e. probe
  e. transducer
  e. ultrasonography
  e. ultrasound (EVUS)
endovasculitis
  hemorrhagic e.
end-stage kidney disease
end-tidal CO2 monitoring
Enduron
enema
  barium e.
  Cortenema e.
  Fleet e.
  lactulose e.
  oil e.
  retention e.
  Rowasa e.
  soapsuds e.
energy
  color Doppler e. (CDE)
  e. expenditure
Enfamil
  E. human milk fortifier
  E. Next Step toddler formula
  E. Premature 20 Formula
enflurane
engaged head

E

NOTES

engagement
Engerix-B hepatitis B vaccine
engineering
    genetic e.
Englert forceps
English
    E. lock
    E. MacIntosh (E-Mac)
    E. position
    E. yew
Engman syndrome
engorged breast
engorgement
    breast e.
    vascular e.
engraftment
    fetal e.
Engstrom respirator
enhancement
    acoustic e.
    e. factor
    immunologic e.
enhancer
    XP Xcelerator ultrasound e.
enigmatic fever
Enkaid
enkephalin
enlarged
    e. cavum pellucidum
    e. cavum septum
    e. posterior fossa
enlargement
    abdominal e.
    clitoris e.
    parotid e.
    pelvis e.
    sellar e.
enolase
    neuron-specific e. (NSE)
Enovid
enoxacin
Enseals
    potassium iodide E.
ensu method
Ensure
Entamoeba histolytica
entanglement
    cord e.
    fetal e.
enteral feeding
enteric
    e. cytopathogenic human orphan
      (ECHO)
    e. feeding
enteritidis
    Salmonella e.

enteritis
    bacterial e.
    regional e.
Enterobacter
enterobiasis
Enterobius vermicularis
enterocele
    anterior e.
    double hump sign of an e.
    pulsion e.
    traction e.
enterococcus, pl. enterococci
    E. faecalis
enterocolic
enterocolitica
    Yersinia e.
enterocolitis
    necrotizing e. (NEC)
    pseudomembranous e.
enterocytopathogenic human orphan
    (ECHO)
enterokinase deficiency
enteromenia
enteropathica
    acrodermatitis e.
enteropathy
enterostomal therapy
enterostomy
enterotoxin F
enterovaginal fistula
enteroviral
enterovirus
Entex LA
Entocort
entrapment
    intrapartum head e.
entrapped temporal horn
Entree
    E. II trocar and cannula system
    E. Plus trocar and cannula system
enuresis
envelope
    peritoneal e.
environment
    Home Observation for the
      Measurement of the E. (HOME)
    hormonal e.
environmental
    e. toxin
    e. trigger
    e. variance
Envision endocavity probe
enzmatic
Enzone
enzygotic
    e. twin
enzyme
    e. activity

angiotensin-converting e. (ACE)
e. assay
e. deficiency
elevated liver e.'s
e. immunoassay (EIA)
inosinate pyrophoshorylase e.
major detoxificating e.
proteolytic e.
rare-cutter e.
restriction e.
TNF-alpha converting e.
zinc-dependent e.

enzyme-linked
e.-l. antiglobulin test
e.-l. immunosorbent assay (ELISA)

enzyme-multiplication immunoassay technique (EMIT)

enzymopathy

eosin

eosin-4 stain

eosinophilia
pulmonary e.
pulmonary infiltrate with e. (PIE)

eosinophilic
e. fasciitis
e. gastroenteritis
e. leukemia
e. pneumonia

EP
ectopic pregnancy
etoposide

E Pam

EPC
external pneumatic calf compression

ependymoblastoma

ependymoma

ephebiatrics

ephebic

ephebogenesis

ephebogenic

ephebology

ephedrine
racemic e.
e. sulfate

ephelis, pl. **ephelides**

ephemeral
e. fever
e. temperature

epibulbar dermoid

epicanthal
e. fold

epicanthus
e. inversus

epicondyle

epidemic
e. capillary bronchitis

epidemiologic

epidemiological
e. characteristic
e. feature
e. genetics

epidemiology

epidermal
e. growth factor (EGF)
e. inclusion cyst
e. melanocyte
e. nevus syndrome

epidermidis
Staphylococcus e.

epidermis, pl. **epidermides**
e. bullosa dystrophica

epidermodysplasia verruciformis

epidermolysis
e. bullosa
e. bullosa, macular type (EBM)
e. bullosa simplex

epidermolytic hyperkeratosis

epidermophytosis

epididymal
e. sperm
e. sperm aspiration

epididymis, pl. **epididymides**

epididymitis
postoperative congestive e.

epididymoorchitis

epididymovasostomy

epidural
e. abscess
e. analgesia
e. anesthesia
e. empyema
e. hemangioma
e. hemorrhage
e. morphine
e. opioid

Epifoam

Epifrin

epigastric
e. artery
e. hernia
e. pain

epigastrius

epigenesis

E

NOTES

epigenetic
epiglottal blockage
epignathus
epilepsia partialis continua
epilepsy
    abdominal e.
    benign neonatal e.
    centralopathic e.
    centrotemporal e.
    focal e.
    grand mal e.
    jacksonian e.
    e. management
    myoclonic e. with ragged red
      fibers (MERRF)
    myoclonus e.
    nocturnal e.
    petit mal e.
    photosensitive e.
    psychomotor e.
    rolandic e.
    temporal lobe e.
epileptic
    e. seizure
epilepticus
    status e.
epileptiform
epiloia
epimenorrhagia
epimenorrhea
Epinal
epinephrine
    Marcaine with e.
    racemic e.
EpiPen
    E. Jr.
epiphyseal
    e. dysostosis
    e. dysplasia
    e. dysplasia-microcephaly-nystagmus
      syndrome
    e. dysplasia, short stature,
      microcephaly, nystagmus syndrome
    e. growth
    e. ossification center
epiphysis, pl. epiphyses
epiploia
epipodophyllotoxin
epipygus
epirubicin
episioperineoplasty
episioperineorrhaphy
episioplasty
episiorrhaphy
episiostenosis
episiotomy
    bilateral mediolateral e.'s
    e. dehiscence

    median e.
    mediolateral e.
    midline e.
    e. repair
    ruptured e.
Episkopi blindness
episode
    nonprimary first e.
episome
epispadias
epistasis
epistatic
epistaxis
epistemology
episthotonos
epitarsus
17-epitestosterone
epithelial
    e. atrophy
    e. autoantibody localization
    e. cancer
    e. cell abnormality
    e. cyst
    e. hyperplasia
    e. membrane antigen (EMA)
    e. ovarian cancer
    e. ovarian carcinoma
    e. plug
    e. stromal ovarian neoplasm
    e. stromal tumor
epithelialization
epitheliochorial placenta
epithelioid
    e. cell
epithelioma
    e. adenoides cysticum
    basal cell e.
epithelium
    acetowhite e.
    atypical e.
    celomic e.
    cervical e.
    columnar e.
    germinal e.
    luminal e.
    native squamous e.
    ovarian germinal e.
    papilla of columnar e.
    squamous e.
    stratified squamous e.
    surface e.
    urogenital e.
    uterine e.
    white e.
epitope
epituberculosis
Epivir
    E.-HBV

EPO
    evening primrose oil
Epo
    erythropoietin
epoophorectomy
epoophoron
Eppendorfer
    E. biopsy forceps
    E. biopsy punch
Eppy/N
Eprolin
EPS
    ear, patella, short stature syndrome
    expressed prostatic secretion
EPSDT
    Early and Periodic Screening, Diagnosis,
      and Treatment program
epsilon-aminocaproic acid (EACA)
Epstein-Barr virus (EBV)
Epstein pearls
EPT
    early pregnancy test
epulis
    congenital e.
    e. gravidarum
    e. of newborn
    e. of pregnancy
Equagesic
equal
    e. conjoined twins
Equanil
equation
    Bohn e.
    Hadlock e.
    Henderson-Hasselbalch e.
    Kučera e.
    Rose e.
    Shepard e.
    Starling e.
equatorial
    e. division
    e. plane
Equilet
equilibrium
    acid-base e.
    genetic e.
    Hardy-Weinberg e.
    linkage e.
    e. phase
    Starling e.
equinovalgus
    talipes e.

equinovarus
    fetal talipes e.
    pes e.
    talipes e.
equinus
    talipes e.
Equitest
equivalent
    lethal e.
E-R
    Betachron E-R
ER
    endoplasmic reticulum
    estrogen receptor
ERA
    endometrial resection and ablation
    ERA resectoscope sheath
Er alpha
    estrogen receptor alpha
Erb
    E. juvenile muscular dystrophy
    E. palsy
    E. syndrome
Erb-Charcot syndrome
Erb-Duchenne paralysis
Er beta
    estrogen receptor beta
Erb-Goldflam
    E.-G. disease
    E.-G. syndrome
ERCP
    endoscopic retrograde
      cholangiopancreatography
erection
    capillary e.
ergocalciferol
2-Br-alpha-ergocryptine mesylate
Ergomar
ergonovine
    e. maleate
ergot
ergotamine
    e. tartrate
Ergotrate
Erhardt
    E. Developmental Prehension
      Assessment
    E. forceps
erigentes
    nervi e.
Erlacher-Blount syndrome
Ernst radium applicator

E

NOTES

Eronen syndrome
Eros-CTD
E-rosette receptor
erosion
>cervical e.
>mesh e.

erosive
>e. adenomatosis of nipple
>e. vulvitis

ERPC
>evacuation of retained products of conception

error
>alpha e.
>beta e.
>e. of the first kind
>metabolic e.
>e. of the second kind

ERT
>estrogen replacement therapy

eruption
>Kaposi varicelliform e.

eruptive
>e. fever
>e. temperature

Er:YAG laser
Erybid
Eryc
EryPed
erysipelas
>e. internum

Ery-Tab
erythema
>e. chronicum migrans (ECM)
>e. infectiosum
>Jacquet e.
>e. marginatum
>e. multiforme exudativum
>e. neonatorum toxicum
>e. nodosum
>e. nodosum leprosum
>palmar e.
>e. streptogenes
>e. toxicum neonatorum

erythematosus
>lupus e. (LE)
>neonatal lupus e. (NLE)
>systemic lupus e. (SLE)

erythredema polyneuropathy
erythrityl tetranitrate
Erythro-Base
erythroblastic
>e. anemia
>e. anemia of childhood

erythroblastopenia
erythroblastosis
>ABO e.

>e. fetalis
>e. neonatorum

Erythrocin
erythrocyte
>e. count
>e. enzyme deficiency
>e. glutathione peroxidase deficiency
>e. membrane
>e. mosaicism
>packed e.
>e. phosphoglycerate kinase deficiency
>e. sedimentation rate (ESR)
>e. transfusion

erythrocytosis
erythroderma
>atopic e.
>bullous congenital ichthyosiform e.
>congenital e.
>e. desquamativum

erythrohepatic porphyria
erythroid
>colony-forming unit e. (CFU-E)
>e. hypoplasia
>mature burst-forming unit e. (M-BFU-E)

erythrokeratodermia
>e. variabilis

erythroleukoblastosis
Erythromid
erythromycin
erythron
erythrophagocytic lymphohistiocytosis
erythroplasia
>Queyrat e.
>Zoon e.

erythropoiesis
erythropoietic porphyria
erythropoietin (Epo)
Eryzole
Escalante syndrome
escape beats
Escherichia
>*E. coli*
>*E. coli* sepsis
>*E. coli* vaccine

Esclim
>E. estradiol transdermal system
>E. Transdermal
>E. transderm system

Escobar syndrome
escutcheon
Eserine
Esidrix
Eskalith
esmolol hydrochloride
esophageal
>e. atresia

e. dilation
e. pressure
e. reflux
e. sphincter
e. stenosis
e. varix
esophagitis
infectious e.
monilial e.
reflux e.
esophagoscope
Holinger infant e.
esophagus
Barrett e.
esotropia
ESR
erythrocyte sedimentation rate
ESS
endometrial stromal sarcoma
essential
e. dysmenorrhea
e. vulvodynia
Essiac
EST
endodermal sinus tumor
Estalis
ester
L-valine e.
phorbol e.
esterified
e. estrogen
Estes
E. operation
E. procedure
estetrol
esthesioneuroblastoma
esthiomene
estimated
e. date of conception (EDC)
e. date of confinement (EDC)
e. fetal weight (EFW)
e. gestational age (EGA)
estimation
gestational age e.
estimator
maximum likelihood e.
Estinyl
Estrace
E. Oral
E. vaginal cream
Estra-D

Estraderm
E. Transdermal
E. transdermal system
E. TTS
estradiol (E$_2$)
17beta-e.
e. assay
17beta-e. dehydrogenase
bound e.
e. cypionate
e. cypionate and
medroxyprogesterone acetate
Cytoxan, Adriamycin, leucovorin,
calcium, fluorouracil, ethinyl e.
(CALF-E)
desogestrel and ethinyl e.
e. enanthate
ethinyl e. (EE)
ethynodiol diacetate and e.
exogenous e.
free e.
e. level
levonorgestrel and ethinyl e.
micronized e.
micronized 17beta e.
norethindrone acetate and ethinyl e.
norethindrone and ethinyl e.
norgestimate and ethinyl e.
pre-hCG e.
e. receptor
e.-releasing silicone vaginal ring
e. suppression
e. transdermal system
e. vaginal cream
e. valerate
estradiol-17 beta
estradiol/levonorgestrel
estradiol/norethindrone acetate
estradiol/norgestimate
Estradurin
Estra-L
E.-L. Injection
estramustine
e. phosphate sodium
estrane
Estrasorb
Estratab
Estratest
E. H.S.
E. Oral
Estraval

E

NOTES

Estring
  E. estradiol vaginal ring
estriol (E$_3$)
  e. level
  salivary e.
  unconjugated e. (E3)
Estro-Cyp
  E.-C. Injection
estrogen
  e. add back therapy (EABT)
  e. agonist
  e. antagonist
  e. breakthrough bleeding
  catechol e.
  Cenestin synthetic conjugated e.'s
  conjugated e.'s
  conjugated equine e. (CEE)
  e. deprivation
  e. effect
  endogenous e.
  esterified e.
  e. excess
  e. level
  e. loss
  e. metabolism
  oral conjugated e.
  orally administered e.
  postcoital e.
  e. receptor (ER)
  e. receptor alpha (Er alpha)
  e. receptor beta (Er beta)
  e. receptor localization
  e. replacement
  e. replacement therapy (ERT)
  serum e.
  e. sulfotransferase
  e. surge
  e. synthesis
  synthetic conjugated e.
  transdermal e.
  uninterrupted e.
  unopposed e.
  e. window etiologic hypothesis
  e. withdrawal
  e. withdrawal bleeding
  e.'s with methyltestosterone
estrogen-agonist uterine effect
estrogen-assisted colposcopy
estrogen-dependent neoplasia
estrogenic
estrogen-induced prolactinoma
estrogen-only HRT
estrogen/progesterone ratio
**estrogen-progesterone withdrawal
  bleeding**
estrogen-progestin
  e.-p. artificial cycle
  e.-p. contraceptive

  e.-p. replacement therapy
  e.-p. test
estrogen-progestogen combination
Estroject
  E.-L.A.
estrone (E$_1$)
  oral e.
  e. sulfate
estrophasic oral contraceptive
estropipate
Estrostep
  E. 21
  E. Fe
  E. oral contraceptive
Estrovis
estrus
ET
  embryo transfer
  endotracheal
ET-1
  endothelin-1
ET-2
  endothelin-2
ET-3
  endothelin-3
ethacrynic acid
ethambutol
Ethamide
ethanol (EtOH)
  blood e.
**ethanolaminosis in glycogen-storage
  disease**
ethchlorvynol
ether
Ethibond
  E. polybutilate-coated polyester
    suture
Ethicon
  E. disposable cannula
  E. disposable trocar
Ethiguard needle
ethinamate
ethinyl
  e. estradiol (EE)
  e. estradiol and desogestrel
  e. estradiol and ethynodiol
    diacetate
  e. estradiol and norethindrone
  e. estradiol and norgestimate
  e. estradiol and norgestrel
  e. testosterone
ethiodized oil
Ethiodol
ethionamide
ethisterone
ethmocephalus
ethmocephaly syndrome
ethnic heritage

ethnicity
ethoheptazine citrate
ethopropazine hydrochloride
ethosuximide
ethotoin
Ethox cuff
ethoxzolamide
Ethrane
ethyl
    e. alcohol (EtOH)
    e. biscoumacetate
ethylenediaminetetraacetic acid (EDTA)
ethylene glycol
ethylnorepinephrine
ethylphenylephrine
ethynodiol
    e. diacetate
    e. diacetate and estradiol
Ethyol
etidronate
    e. disodium
    sodium e.
etiocholanolone
etiologic fraction of tubal infertility
etiology
    fever of unknown e. (FUE)
etiopathogeny
etoglucid
EtOH
    ethanol
    ethyl alcohol
etomidate
Etopophos
etoposide (EP)
    Adriamycin, Oncovin,
      prednisone, e. (AOPE)
    e. phosphate
etretinate
ETV8 CCD ColorMicro video camera
EUB-405
    E. ultrasound scanner
    E. ultrasound system
Eubacterium
euchromatin
eugenics
eugonadal
    e. amenorrhea
eugonadism
eugonadotropic
    e. amenorrhea
    e. hypogonadism
    e. oligospermia

eukaryote
Eulipos
eumenorrheic
    e. woman
eunuch
eunuchoid gigantism
eunuchoidism
    female e.
    hypergonadotropic e.
    hypogonadotropic e.
euploid
    e. abortion
    e. pregnancy
euploidy
eupnea
eupneic
euprolactinemic woman
Eurax Topical
Euro-Med FNA-21 aspiration needle
europium
eustachian tube
eutectic mixture of local anesthetics
    (EMLA)
Euthroid
euthyroid
Eutonyl
eutopic
    e. endometrium
    e. endometrium prostaglandin E_2
eV, ev
    electron-volt
evacuation
    dilation and e. (D&E)
    fimbrial e.
    molar e.
    e. proctography
    e. of retained products of
      conception (ERPC)
    e. of uterine contents
evacuator
    smoke e.
    uterine e.
evaluation
    Doppler e.
    Dubowitz e.
    fluoro-urodynamic e.
    infertility e.
    mental status e.
    physical capacity e. (PCE)
    urological e.
    uterine e.
Evans blue dye

E

NOTES

Evans-Steptoe procedure
evaporated
    e. milk
    e. milk formula
evaporation
evening primrose oil (EPO)
event
    apparent life-threatening e. (ALTE)
    iatrogenic e.
    life e.
eventration
Everard Williams procedure
Everone
Evershears
    E. bipolar laparoscopic forceps
    E. bipolar laparoscopic scissors
eversion
    cervical e.
    vaginal e.
everted diaphragm
Evert-O-Cath drug delivery catheter
evidence
    e. of anticipation
    rape e.
    sexual assault forensic e. (SAFE)
evisceration
    vaginal e.
Evista
E-Vitamin
evoked potential technique
evolution
    darwinian e.
    Darwin theory of e.
    Denman spontaneous e.
    Douglas spontaneous e.
    spontaneous e.
evolutionarily conserved
EVUS
    endovaginal ultrasound
Ewing
    E. sarcoma
    E. tumor
ex
    e. lap
    e. utero intrapartum tracheloplasty
    (EXIT)
    e. vivo
    e. vivo fertilization
    e. vivo liver-directed gene therapy
Exact-Touch Saccomanno Pap smear
  collection system
exaggerated craniocaudal view
Exami-Gown gown
examination, exam
    abdominal e.
    Allen and Capute neonatal
      neurodevelopmental e.
    Ballard e.

    bimanual pelvic e.
    chest e.
    dark-field e.
    Dubowitz e.
    eye e.
    neonate e.
    neurologic e.
    newborn e.
    Pediatric Early Elemental E.
     (PEEX)
    pelvic e.
    radiological e.
    rectal e.
    rectovaginal e.
    self-breast e. (SBE)
    speculum e.
    sterile vaginal e. (SVE)
    stool e.
    swab e.
    vaginal e.
excavator
    Merlis obstetrical e.
excavatum
    pectus e.
Excedrin IB
eXcel-DR Glasser laparoscopic needle
excess
    androgen e.
    e. androgen
    base e.
    estrogen e.
    glucocorticoid e.
    soluble antigen e.
excessive
    e. blood loss
    e. villous maturation
exchange
    fetal-maternal e.
    sister chromatid e.
    Starling law of transcapillary e.
    e. transfusion
excision
    e. diathermy
    diathermy loop e.
    electrosurgical e. (EE)
    electrosurgical loop e.
    local e.
    radical e.
    repeat-loop e.
excisional biopsy
excitation
exclusion
    allelic e.
excrescence
    internal e.
excretion
    sodium e.
    water e.

excretory system development
Exelderm Topical
exemestane
exencephalia
    bifid e.
exencephaly
exenteration
    anterior pelvic e.
    pelvic e.
    posterior pelvic e.
    pyelonephritis in e.
    stress reaction in e.
    total pelvic e.
exercise
    DeSouza e.'s
    Kegel e.'s
    maternal e.
    pelvic e.
exercise-induced
    e.-i. amenorrhea
    e.-i. incontinence
exfoliation
    lamellar e.
exhaustion
    ovarian follicle e.
EXIT
    ex utero intrapartum tracheloplasty
exit plan
Exna
exocelomic
    e. cavity
    e. membrane
exocervix
exocytosis
    cortical granule e.
exogenous
    e. DNA
    e. estradiol
    e. estrogenic stimulation
    e. gonadotropin
    e. gonadotropin stimulation
    e. hormone
    e. obesity
exomphalos
    e., macroglossia, and gigantism
      (EMG)
    e., macroglossia, and gigantism
      syndrome
exon
exonuclease
exophthalmos

exophytic
exostosis, pl. exostoses
    multiple exostoses
    pelvic e.
    trichorhinophalangeal multiple e.
      (TRAMPE)
Exosurf
    E. Neonatal
exotoxin
    e. C
    Pseudomonas e.
exotropia
expander
    AccuSpan tissue e.
    Becker tissue e.
    Meshgraft Skin E.
expansion
    e. of chromosome
    plasma volume e.
    volume e.
expectancy
    life e.
expectant management
expected
    e. date of confinement (EDC)
    e. day of delivery (EDD)
expenditure
    energy e.
experience
    Charing Cross e.
    Positive Reinforcement in the
      Menopausal E. (PRIME)
experimental
    e. embryology
    e. obesity
Expert bone densitometer
exploratory
    e. celiotomy
    e. laparotomy
exposed large venous channel
eXpose retractor
exposure
    anesthetic gas e.
    DES e.
    maternal mercury e.
    methamphetamine e.
    prenatal diethylstilbestrol e.
    in utero e.
expressed
    e. prostatic secretion (EPS)
    e. sequence tag

E

NOTES

expression
: gene e.
: Human Gene E. (HuGe index)
: oncogene e.

expressivity

expulsion

expulsive
: e. force
: e. pains

exsanguination
: fetal e.

exstrophy
: e. of the bladder
: cloacal e.

extended
: e. culture
: e. field irradiation therapy
: e. radical mastectomy
: e. rubella syndrome

extended-spectrum penicillin

extender
: egg yolk e.

extension
: classical incision e.
: midline vertical uterine e.

extensive intraductal component (EIC)

Extentabs
: Dimetane E.
: Dimetapp E.

exteriorization
: uterine e.

externa
: otitis e.

external
: e. anal sphincter
: e. beam irradiation
: e. cephalic version (ECV)
: e. cephalic version and spontaneous vertex
: e. conjugate
: e. fetal monitoring (EFM)
: e. genitalia
: e. genitalia/Bartholin, urethral, and Skene glands (EG/BUS)
: e. hemorrhage
: e. os
: e. pneumatic calf compression (EPC)
: e. radiation therapy
: e. rotation
: e. surface
: e. urethral barrier device
: e. version
: e. x-ray therapy

exterogestate

extirpation
: Amreich vaginal e.

extraamniotic
: e. pregnancy
: e. saline infusion (EASI)

extracapsular growth

extracellular
: e. fluid
: e. matrix
: e. matrix component

extrachorial pregnancy

extrachromosomal inheritance

extracorporeal
: e. circulation (ECC)
: e. CO_2 removal (ECOR)
: e. membrane oxygenation (ECMO)

extracortical axial aplasia

extract

extraction
: breech e.
: internal podalic version and breech e.
: Marshall-Taylor vacuum e.
: menstrual e.
: partial breech e.
: podalic e.
: spontaneous breech e.
: testicular sperm e. (TESE)
: total breech e.
: vacuum e.

extractor
: Bird vacuum e.
: Kobayashi vacuum e.
: Malström vacuum e.
: Mityvac vacuum e.
: M-type e.
: mucus e.
: Murless head e.
: plastic cup vacuum e.
: Silastic cup e.
: Silc e.
: Tender-Touch e.
: Trizol RNA e.
: vacuum e.
: Vantos vacuum e.

extracutaneous vas fixation clamp

extradural
: e. anesthesia
: e. block
: e. hematoma

extraembryonic
: e. celomic space (EECS)
: e. differentiation
: e. fetal membrane
: e. location
: e. mesoderm

extrafascial hysterectomy

extraglandular conversion

extragonadal
: e. germ cell

extrahepatic
extralobar sequestration
extramammary Paget disease
extramedullary hematopoiesis
extramembranous pregnancy
extramural upper airway obstruction
extranumerary nipple
extraovular
    e. placement
extrapelvic
    e. endometriosis
    e. malignancy
extraperitoneal cesarean
extraplacental membrane
extrapulmonary
extrapyramidal
extrarenal
Extra Strength Doan's
extrasystole
extrathoracic tuberculosis
extrauterine
    e. pregnancy
extravillous trophoblast
extremely
    e. low birth weight (ELBW)
    e. low birth weight infant
    (ELBWI)
extremity
    congenital contracture of e.
    flaccid e.

extrinisic pathway inhibitor
extrinsic
extrusion
    oocyte e.
    placental e.
extubated
extubation
exudativum
    erythema multiforme e.
ExuDerm dressing
Eyberg Child Behavior Inventory
  (ECBI)
eye
    Credé maneuver of e.'s
    e. defects-diffuse renal mesangial
    sclerosis syndrome
    e. examination
eyeball cystometrogram
eyelids
    fusion of e.
eye-opener
    cutting, annoyance, guilt, e.-o.
    (CAGE)
Eyesine Ophthalmic
E-Z-EM
    E-Z-EM BioGun automated biopsy
    system
    E-Z-EM PercuSet amniocentesis
    tray

NOTES

E

183

$F_{2\alpha}$
prostaglandin F. ($PGF_{2\alpha}$)

FA
Chromagen FA
Nestabs FA
Pramet FA
Pramilet FA

F.A.
Mission Prenatal F.A.

FAB
French-American-British
FAB classification
FAB staging of carcinoma

Fabricius
bursa of F.

Fabry disease

face
asymmetry of f.
bovine f.
f. chamber
cleft f.
congenital ectodermal dysplasia
of f.
dish f.
fetal f.
f. presentation

face/chin presentation

FACE-IT protective shield

FACES-III
Family Adaptability and Cohesion Scale-
III

face-to-pubes position

facial
f. asymmetry
f. clefting syndrome, Gypsy type
f. diplegia
f. dysplasia, hyperextensibility of
joints, clinodactyly, growth
retardation, mental retardation
syndrome
f. ectodermal dysplasia
f. nerve
f. nerve injury (FNI)
f. nerve palsy
f. nerve paralysis

facial-digital-genital syndrome

facies
f.
adenoid f.
asymmetric crying f. (ACF)
birdlike f.
f. bovina
bovine f.
characteristic f.
coarse f.

cushingoid f.
distinctive f.
dog-like f.
dysmorphic f.
elfin f.
familial asymmetric crying f.
femoral hypoplasia unusual f.
(FHUF)
flat f.
gargyloid f.
leonine f.
leprechaun f.
Marshall Hall f.
peculiar f.
Potter f.
smiling f.
unusual f.

facioauriculovertebral (FAV)
f. anomalad
f. malformation complex

faciocardiomelic dysplasia

faciocardiorenal syndrome

faciocerebroskeletocardiac
f. syndrome

faciocutaneoskeletal (FCS)
f. syndrome

faciodigitogenital syndrome

faciogenital
f. dysplasia
f. syndrome

faciooculoacousticorenal (FOAR)
f. syndrome

faciopalatoosseous (FPO)
f. syndrome

facioscapulohumeral
f. muscular dystrophy
f. syndrome of Landouzy-Dejerine

faciotelencephalic malformation

faciotelencephalopathy

Facit polyp forceps

FACO$_2$
fraction of alveolar carbon dioxide

FACS
fluorescence-activated cell sorter

FACT
Functional Assessment of Cancer
Therapy

factitious precocious puberty

factor
f. I
f. II
f. III
f. IV
f. V
f. VI

F

factor *(continued)*
f. VII
f. VIII
f. VIII:C
f. IX
f. X
f. XI
f. XII
f. XIII
f. Xa
acidic fibroblast growth f. (FGFa)
alloimmune f.
angiogenic growth f.
antihemophilic f.
atrial natriuretic f. (ANF)
autocrine motility f. (AMF)
autoimmune f.
azoospermia f. (AZF)
basic fibroblast growth f. (bFGF)
bifidus f.
blocking f.
brain-derived neurotrophic f.
  (BDNF)
cell adhesion f. (CAF)
cervical f.
chemotactic f.
Christmas f.
citrovorum f.
coagulation f.
coital f.
colony-stimulating f. (CSF)
f. concentrate
corticotropin-releasing f.
cytotoxic f.
f. D deficiency
digoxin-like immunoreactive f.
  (DLIF)
early pregnancy f.
endocrine f.
endothelium-derived relaxing f.
  (EDRF)
enhancement f.
epidermal growth f. (EGF)
fecundity f.
fetal f.
fibrin stabilizing f.
fibroblast growth f.
fibroblast pneumocyte f.
follicular f.
glass f.
granulocyte colony-stimulating f.
  (G-CSF)
granulocyte-macrophage colony-
  stimulating f. (GM-CSF)
growth f. (GF)
growth-like f.
Hageman f.
f. H deficiency

helper f.
hepatocyte growth f. (HGF)
HG f.
home f.
human antihemophilic f.
human sperm cytosolic f.
humoral f.
immune f.
infertility f.
insulin-like growth f. (IGF)
insulin-like growth f.-1
intrauterine f.'s
f. I–XIII deficiency
keratinocyte growth f. (KGF)
kidney-derived growth f.
leukemia inhibitory f. (LIG)
lifestyle f.
lymphocyte-activating f.
macrophage-activating f. (MAF)
macrophage colony-stimulating f.
macrophage-inhibition f.
male f.
maternal f.
maturation-promoting f. (MPF)
menarche f.
microbial f.
migration-inhibitory f. (MIF)
mitosis-promoting f. (MPF)
mixed lymphocyte reaction
  blocking f.
mortality risk f.
M-phase-promoting f.
müllerian duct inhibitory f.
müllerian inhibiting f. (MIF)
nerve growth f. (NGF)
obstetric risk f.
ovarian f.
perinatal risk f.
placental growth f. (PGF)
plasma thromboplastin antecedent f.
platelet-activating f.
platelet-derived growth f.
precipitatory f.
prenatal risk f.
prognostic f.
prolactin inhibiting f. (PIF)
prolactin releasing f. (PRF)
purified human f. IX
recombinant antihemophilic f.
recombinant f. VIIa
Rh f.
rheumatoid f.
risk f.
somatic cell-derived growth f.
  (SCF)
Stuart f.
Stuart-Prower f.
testis-determining f. (TDF)

tissue f.
transfer f.
transforming growth f. (TGF)
tubal f.
tumor angiogenesis f. (TAF)
tumor-limiting f.
tumor necrosis f. (TNF)
tumor necrosis f.-alpha
uterine f.
f. V^{Leiden}
vascular endothelial growth f. (VEGF)
f. VII, VIII inhibitor
f. V Leiden carrier
f. V Leiden mutation
f. V Leiden mutation test
f. V Leiden thrombophilia
von Willebrand f. (vWF)
f. V polypeptide
work f.

Fact Plus
Factrel
FAD
familial Alzheimer disease
Fader Tip ureteral stent
Fadhil syndrome
fadir sign
FADS
fetal akinesia deformation sequence
FAE
fetal alcohol effect
faecalis
Enterococcus f.
Fagan Test of Infant Intelligence
failed
f. forceps delivery
f. physiologic change
failure
acute renal f.
bone marrow f.
cardiac f.
congestive heart f.
contraceptive f.
ejaculation f.
fertility f.
fimbrial f.
gonadal f.
heart f.
hypothalamic f.
implantation f.
kidney f.
multiple organ f.

ovarian f.
premature ovarian f. (POF)
f. to progress
renal f.
reproductive f.
f. to thrive (FTT)
failure-to-thrive syndrome
Fairbank disease
Fairbank-Keats syndrome
falciform hymen
Falk-Shukuris operation
falling of the womb
fallopian
f. pregnancy
f. tube
f. tube carcinoma
f. tube mass
f. tube metastasis
f. tube prolapse
f. tube sperm perfusion (FTSP)
falloposcope
falloposcopy
Fallot
F. complex
pentalogy of F.
pink tetralogy of F.
F. syndrome
tetralogy of F. (TF, TOF)
trilogy of F.
Falope
F. ring
F. ring applicator
false
f. conjugate
f. heteroovular twins
f. knot
f. knot of umbilical cord
f. labor
f. mole
f. pains
f. pregnancy
f. twins
f. waters
false-negative
f.-n. death
false-positive
falx laceration
FAMA
fluorescent antibody against membrane antigen
fluorescent antimembrane antibody test
FAMA assay

F

NOTES

famciclovir
familial
    f. alobar holoprosencephaly
    f. Alzheimer disease (FAD)
    f. aortic ectasia
    f. aortic ectasia syndrome
    f. asymmetric crying facies
    f. ataxia-hypogonadism syndrome
    f. atypical multiple mole melanoma
      syndrome
    f. bipolar mood disorder
    f. bird-headed dwarfism
    f. cardiac myxoma
    f. cardiac myxoma syndrome
    f. centrolobal sclerosis
    f. cholesterolemia
    f. congenital alopecia, mental
      retardation, epilepsy, unusual EEG
      syndrome
    f. congenital fourth cranial nerve
      palsy
    f. congenital superior oblique
      oculomotor palsy
    f. congenital trochlear nerve palsy
    f. diathesis
    f. dysautonomia
    f. endocrine-neuroectodermal
      abnormalities syndrome
    f. erythroblastic anemia
    f. erythrophagocytic
      lymphohistiocytosis (FEL)
    f. exudative vitreoretinopathy
    f. focal facial dermal dysplasia
    f. granulomatous arteritis
    f. hirsutism
    f. hypercholesterolemia
    f. hyperglycerolemia
    f. hyperprolinemia
    f. hypofibrinogenemia
    f. infertility
    f. inverted choreoathetosis
    f. juvenile nephrophthisis (FJN)
    f. lipodystrophy
    f. lipoid adrenal hyperplasia
    f. macroglossia-omphalocele
      syndrome
    f. Mediterranean fever
    f. multiple endocrine adenomatosis
    f. neurovisceral lipidosis
    f. olivopontocerebellar atrophy
    f. osseous dystrophy
    f. osteochondrodystrophy
    f. polysyndactyly-craniofacial
      anomalies syndrome
    f. pterygium syndrome
    f. striatal degeneration
    f. tendency

    f. third and fourth pharyngeal
      pouch syndrome
    f. thrombophilia
    f. trait
    f. Turner syndrome
    f. visceral neuropathy
    f. vocal cord dysfunction
familiaris
    osteopathia dysplastica f.
family
    F. Adaptability and Cohesion
      Scale-III (FACES-III)
    F. Apgar Questionnaire
    f. dysfunction
    dysfunctional f.
    F. Environment Scale (FES)
    gene f.
    f. history
    f. planning
    f. selection
family-based
    f.-b. design
    f.-b. test
famotidine
Famvir
Fanconi
    F. anemia
    F. dysplasia
    F. syndrome
Fanconi-Albertini Zellweger syndrome
Fanconi-Petrassi syndrome
Fanconi-Prader syndrome
Fanconi-Schlesinger syndrome
fan douche
Fansidar
faradism
Farber
    F. disease
    F. test
Fareston
Farre white line
Farris test
FAS
    fetal alcohol syndrome
fascia
    Camper f.
    Colles f.
    Cooper f.
    cremasteric f.
    Denonvilliers f.
    endopelvic f.
    Gerota f.
    f. lata allograft
    f. lata suburethral sling
    obturator f.
    perirectal f.
    pubocervical f.
    Scarpa f.

Smead-Jones closure of peritoneum and f.
subserous f.
fascial
f. flap
f. necrosis
f. sling procedure
f. strip
fasciculation
fasciitis
diffuse f.
eosinophilic f.
necrotizing f.
fascioscapulohumeral
fashion
forward roll f.
Halban f.
Moschowitz f.
FASIAR
follicle aspiration, sperm injection, and assisted rupture
fastener
ROC XS suture f.
Fastin
fasting
f. blood sugar
f. plasma glucose (FPG)
fast neutron
fat
f. density
f. depot
dietary f.
f. flap
f. metabolism
f. necrosis
preperitoneal f.
unsaturated f.
fate mapping
fatty
f. acid-coenzyme A
f. halo
f. infiltration
f. liver of pregnancy
faun tail nevus
FAV
facioauriculovertebral
Fazio-Londe atrophy
FBM
fetal breathing movement
5-FC
5-fluorocytosine

FCMD
Fukuyama congenital muscular dystrophy
FCS
faciocutaneoskeletal
FCS syndrome
FD
forceps delivery
FDA
Food and Drug Administration
FDIU
fetal death in utero
Fe
iron
Estrostep Fe
Loestrin Fe
Norlestrin Fe
Slow Fe
fear
f. of delivery
pregnancy f.
fear-causing delivery
fearless first delivery
feature
dysmorphic facial f.
epidemiological f.
grotesque f.'s
mongoloid f.'s
febrile
f. convulsion
f. morbidity
f. seizure
FEC
fluorouracil, epirubicin, Cytoxan
fecal
f. diversion colostomy
f. impaction
f. incontinence
f. occult blood testing (FOBT)
f. streptococci
fecalith
appendiceal f.
feces
FECO$_2$
fraction in expired gas of carbon dioxide
fecund
fecundability
fecundate
fecundation
fecundity
f. factor
natural f.
f. selection

F

NOTES

fed
> bottle f.
> breast f.
> nipple f.

Federici sign

feed
> bottle f.
> breast f.

feeder
> Brecht f.
> f. layer

feeding
> ad lib f.
> Alimentum f.
> breast f.
> demand f.
> enteral f.
> enteric f.
> Finkelstein f.
> gastric enteral f.
> gavage f.
> nasoduodenal f.
> nipple f.
> f. regimen
> scheduled f.
> syringe f.
> transpyloric enteral f.

Feelings About Yourself instrument

Feer disease

feet (*pl. of* foot)

Feiba VH Immuno

Feilchenfeld forceps

Feingold syndrome

Feinmesser-Zelig syndrome

FEL
> familial erythrophagocytic lymphohistiocytosis

Feldene

Felty syndrome

female
> f. condom
> f. eunuchoidism
> f. fertility efficiency
> f. fertility inefficiency
> f. genital mutilation (FMG)
> f. genital tract mutilation (FGTM)
> f. karyotype
> f. pseudohermaphroditism
> f. pseudo-Turner syndrome
> super f.
> triple-X f.

Femara

Femcaps

Femcept device

Femguard

Femhrt

feminization
> complete testicular f.

> f. syndrome
> testicular f.

feminizing
> f. adrenal tumor
> f. surgery
> f. testes syndrome

Feminone

Femizol-M

Femogen
> F. Forte

Femogex

femoral
> f. artery catheter
> f. artery catheterization
> f. circumflex artery
> f. hernia
> f. hypoplasia unusual facies (FHUF)
> f. lymph node
> f. nerve
> f. neuropathy

femoral-facial syndrome

FemPatch

Femsoft continence insert

Femstat
> F. 3
> F. Prefill

femur
> f. length (FL)

femur-fibula-ulna syndrome

FEN
> fluids, electrolytes, nutrition

fenamate

fencing reflex

fenestrata
> placenta f.

fenestration
> laparoscopic f.

fenfluramine

fenoterol

fentanyl
> f. citrate
> f. group
> f. lollipop

fentomoles/milligram (fm/mg)

Fenton vaginoplasty

Fe_3O_4
> magnetite

Feosol

FER
> frozen embryo replacement

Ferguson reflex

Fer-In-Sol
> F.-I.-S. supplement
> F.-I.-S. vitamins

fern
> f. leaf crystallization
> f. leaf pattern

f. leaf tongue
f. test
ferning
f. technique
vaginal fluid f.
fern-positive Nitrazine
Ferralet
ferric sulfate
Ferriman-Gallwey
F.-G. hirsutism score
F.-G. hirsutism scoring system
Ferris Smith-Sewall retractor
ferritin
Ferrold-Hisaw unit
ferrous
f. fumarate
f. gluconate
f. sulfate (FeSO$_4$)
Fertil-A-Chron
fertile
f. period
f. phase
fertility
f. cycle
f. failure
f. rate
subsequent f.
fertilizable life span
fertilization
f. age
assisted f.
delayed f.
ex vivo f.
micro assisted f.
in vitro f. (IVF)
in vivo f.
fertilized ovum
Fertinex
FES
Family Environment Scale
FeSO$_4$
ferrous sulfate
fes **protooncogene**
fetal
f. abdominal circumference
f. Accutane syndrome
f. acid-base balance
f. acidemia
f. acidosis
f. acoustic stimulation test
f. activity
f. activity test

f. adrenal gland
f. age
f. akinesia deformation sequence
(FADS)
f. akinesia syndrome
f. alcohol effect (FAE)
f. alcohol syndrome (FAS)
f. aminopterin like syndrome
f. aminopterin syndrome
f. anasarca
f. anemia
f. anoxia
f. anticoagulant syndrome
f. aorta
f. aortic blood flow
f. aortic Doppler velocimetry
f. arrhythmia
f. arterial oxygen saturation
(FS$_p$O$_2$)
f. ascites
f. asphyxia
f. aspiration syndrome
f. attitude
f. biometry
f. biophysical profile
f. biparietal diameter
f. bladder catheterization
f. blood
f. blood flow
f. blood gases
f. blood glucose
f. blood oxygen-carrying capacity
f. blood pH
f. blood study
f. blood value
f. blood volume
f. body movement
f. BPP
f. bradycardia
f. brain
f. brain death
f. brain disruption sequence
f. brain stem auditory evoked
potential
f. breathing
f. breathing movement (FBM)
f. capillary branching
f. cardiac activity
f. cardiac function
f. cardiac motion
f. cartilage
f. cell

F

NOTES

fetal *(continued)*
- f. cellular growth
- f. circulation
- f. circulatory centralization
- f. cocaine syndrome
- f. compromise
- f. condition
- f. congenital hyperplasia
- f. cortisol infusion
- f. cotyledon
- f. cranial artery
- f. crowding
- f. CST
- f. cystic hygroma
- f. cytabarine/thioguanine embryopathy
- f. cytomegalovirus infection
- f. death
- f. death rate
- f. death in utero (FDIU)
- f. debris
- f. demise
- f. descent
- f. deterioration
- f. development
- f. diabetes insipidus
- f. diagnosis
- f. Dilantin syndrome
- f. disappearance
- f. distress
- f. distress syndrome
- F. Dopplex monitor
- f. dose limit
- f. drug therapy
- f. ductus arteriosus constriction
- f. duplication
- f. dystocia
- f. echocardiogram
- f. electrocardiography
- f. endoscopic (FETENDO)
- f. engraftment
- f. entanglement
- f. erythroid progenitor
- f. exsanguination
- f. face
- f. face syndrome
- f. facies syndrome
- f. factor
- f. fibronectin (FFN, fFN)
- f. fibronectin assay
- f. fibronectin test
- f. foot length (FFL)
- f. fracture
- f. gigantism
- f. gigantism-renal hamartoma-nephroblastomatosis syndrome
- f. goiter
- f. grasping
- f. growth
- f. growth measurement
- f. growth parameters
- f. growth restriction (FGR)
- f. growth retardation
- f. habitus
- f. head
- f. head:abdominal circumference ratio
- f. head position
- f. heart
- f. heart action
- f. heart disease
- f. heart rate (FHR)
- f. heart rate deceleration
- f. heart rate monitor
- f. heart rate monitoring
- f. heart rate pattern
- f. heart rate reactivity
- f. heart rate variability
- f. heart sounds
- f. hemoglobin (HbF)
- f. hemorrhage
- f. hiccup
- f. hormone
- f. hydantoin syndrome (FHS)
- f. hydrops
- f. hypercarbia
- f. hypotrophy
- f. hypoxia
- f. imaging
- f. intracranial anatomy
- f. iodine deficiency disorder (FIDD)
- f. isotretinoin syndrome
- f. jeopardy
- f. karyotype
- f. LDL level
- f. lie
- f. limb
- f. liver
- f. LOU
- f. lower obstructive uropathy
- f. lung liquid
- f. lung maturation
- f. lung maturity (FLM)
- F. Lung Maturity test
- f. maceration
- f. macrosomia
- f. malformation
- f. malpresentation
- f. maturity
- f. membrane
- f. metabolism
- f. methotrexate syndrome
- f. monitoring strip
- f. morbidity
- f. mortality

f. movement assessment
f. NST
f. nuchal translucency thickness
f. nutrition
f. oculocerebrorenal syndrome of Lowe
f. outcome
f. outline
f. ovarian cyst
f. oximetry monitoring
f. oxygenation
f. oxygen saturation
f. paramethadione-trimethadione syndrome
f. phenytoin syndrome
f. physiologist
f. pole
f. position
f. postural abnormality
f. posture
f. pulmonary hypoplasia
f. pulmonary sequestration
f. pyelectasis
f. reduction
f. rejection
f. resorption
f. retention
f. risk
f. scalp blood sampling
f. scalp electrode (FSE)
f. scalp oxygenation
f. scalp platelet sampling
f. seizure
f. serum
f. sexual differentiation
f. shoulder extraction force
f. skeletal dysplasia
f. skin sampling
f. skull depression
f. small parts
f. somatic activity
f. souffle
f. spine position
f. startle response
f. station
f. steroid concentration
f. structural defect
f. surgery
f. surveillance
f. surveillance technique
f. surveillance test
f. swallowing

f. talipes equinovarus
f. thalidomide
f. thoracic abnormality
f. thrombotic vasculopathy (FTV)
f. tissue implant
f. tissue transplant
f. toxoplasmosis
f. transfusion
f. trauma
f. trimethadione syndrome
f. ultrasound
f. urination
f. uropathy
f. valproate syndrome (FVS)
f. varicella syndrome (FVS)
f. vascular anomaly
f. ventriculomegaly
f. version in utero
f. viability
f. warfarin syndrome
f. wastage
f. weight
f. well-being
Fetalert fetal heart rate monitor
fetalis
    chondromalacia f.
    erythroblastosis f.
    hydrops f.
    ichthyosis f.
    maternal hydrops f.
    maternal parvovirus f.
    nonimmune hydrops f. (NIHF)
    opisthotonos f.
    rachitis f.
fetalism
fetal-maternal
    f.-m. communication
    f.-m. exchange
    f.-m. hemorrhage
    f.-m. medicine
fetal-neonatal transition
fetal-pelvic index
fetal-placental
    f.-p. steroidogenesis
    f.-p. unit
FetalPulse
    F. Plus fetal Doppler
    F. Plus monitor
Fetasonde fetal monitor
fetation
FETENDO
    fetal endoscopic

NOTES

F

FETENDO (continued)
    FETENDO method
    FETENDO PLUG
    FETENDO surgery
    FETENDO tracheal clip
feticide
    selective f.
fetoamniotic
    f. shunt
fetofetal
    f. transfusion
    f. transfusion syndrome
fetography
fetology
fetomaternal
    f. hemorrhage (FMH)
    f. transfusion
fetometry
    ultrasonic f.
    ultrasound f.
feto-neonatal estrogen-binding protein
fetopathy
    diabetic f.
fetopelvic disproportion
fetoplacental
    f. access
    f. anasarca
    f. blood volume
    f. circulation
    f. function
    f. transfusion
fetoprotein
fetoscope
    Hillis-DeLee f.
fetoscopy
fetotoxic
fetouterine cell
fetu
    fetus in f.
fetus, pl. **fetuses**
    acardiac f.
    f. acardius
    amorphous f.
    f. amorphus
    asynclitic position of f.
    calcified f.
    Campylobacter f.
    coexistent f.
    f. compressus
    dropsy of the f.
    f. in fetu
    f. growth elevation
    growth-retarded f.
    habitus of f.
    harlequin f.
    HLA-compatible f.
    ichthyosis f.
    impacted f.

    just-viable f.
    macerated f.
    mummified f.
    near-viable f.
    nonstressed f.
    nonvertex f.
    nonviable f.
    nuchal translucency in a f.
    paper-doll f.
    papyraceous f.
    f. papyraceus
    parasitic f.
    presentation of f.
    previable f.
    retroperitoneal f.
    Rh-sensitized f.
    f. sanguinolentis
    singleton f.
    sireniform f.
    sirenomelic f.
    stressed f.
    stunted f.
    triploid f.
    trisomic f.
    vanishing f.
    viable f.
    Vibrio f.
fetus-to-fetus transplant
Feuerstein-Mims syndrome
Feulgen stain
FEV
    forced expiratory volume
fever
    absorption f.
    acute rheumatic f. (ARF)
    artificial f.
    aseptic f.
    cat-scratch f.
    childbed f.
    Colorado tick f.
    dehydration f.
    drug f.
    Ebola hemorrhagic f.
    enigmatic f.
    ephemeral f.
    eruptive f.
    familial Mediterranean f.
    Haverhill f.
    hay f.
    hemorrhagic f.
    inanition f.
    intermittent f.
    milk f.
    paratyphoid f.
    parturient f.
    periodic f.
    puerperal f.
    Q f.

Queensland f.
quotidian f.
rat-bite f.
relapsing f.
rheumatic f.
Rocky Mountain spotted f.
scarlet f.
South African tick f.
spotted f.
tactile f.
tick f.
typhoid f.
typhus f.
undulant f.
unexplained f.
f. of unknown etiology (FUE)
f. of unknown origin (FUO)
Valley f.
West Nile f.
yellow f.
Fèvre-Languepin syndrome
fexofenadine
FFDD II
focal facial dermal dysplasia II
FFL
fetal foot length
FFN, fFN
fetal fibronectin
FFP
fresh frozen plasma
FFU syndrome
FGFa
acidic fibroblast growth factor
FGF-R
fibroblast growth factor receptor
FGR
fetal growth restriction
fgr protooncogene
FG syndrome
FGTM
female genital tract mutilation
FHR
fetal heart rate
nonreassuring FHR
FHS
fetal hydantoin syndrome
floating harbor syndrome
FHUF
femoral hypoplasia unusual facies
FHUF syndrome
FIA
fluorescent immunoassay

fiber
Dacron f.
dietary f.
Purkinje f.'s
FiberCon
fibrillation
atrial f.
fibrillin
fibrin
f. degradation product
f. sheath
f. stabilizing factor
Fibrindex test
fibrinogen
f. abnormality
f. deficiency
plasma f.
fibrinogen-fibrin conversion syndrome
fibrinoid
f. degeneration of astrocytes
f. leukodystrophy
fibrinolysis
tissue activator-induced f.
fibrinolytic
f. agent
f. and clotting system
fibrinous polyp
fibroadenolipoma
degenerated f.
fibroadenoma
giant f.
intracanalicular f.
pericanalicular f.
fibroblast
f. culture
f. growth factor
f. growth factor receptor (FGF-R)
human embryo f. (HEF)
f. pneumocyte factor
fibrochondrogenesis
fibrocystic
f. breast
f. breast change
f. disease
fibrodysplasia ossificans progressiva (FOP)
fibroelastosis
endocardial f. (EFE)
fibroepithelial polyp
fibroid
f. embolization
intramural f.

F

NOTES

fibroid *(continued)*
    submucous f.
    subserous f.
    uterine f.
fibroidectomy
fibroma
    histiocytic f.
    f. molle gravidarum
    ovarian f.
    vulvar f.
fibromatosis
    gingival f.
    ovarian f.
    pelvic f.
fibromectomy
fibromuscular cervical stroma
fibromyalgia
    childhood f.
fibromyoma, pl. **fibromyomata**
    uterine f.
fibronectin
    cervicovaginal fetal f.
    fetal f. (FFN, fFN)
    oncofetal f.
fibroplasia
    retrolental f.
fibrosa
    osteitis f.
fibrosarcoma
fibrosing
    f. adenomatosis
    f. adenosis
fibrosis
    cerebellar vermis hypo/aplasia,
      oligophrenia, congenital ataxia,
      ocular coloboma, hepatic f.
      (COACH)
    coloboma-hepatic f.
    cystic f. (CF)
    delta-F508 cystic f.
    focal f.
    hepatic f.
    horseshoe f.
    postradiation periureteral f.
    pulmonary f.
    retropubic f.
fibrosum
    molluscum f.
fibrothecoma
fibrotic ophthalmoplegia
fibrous
    f. connective tissue
    f. dysplasia of jaw
fibroxanthoma
Fick
    F. method
    F. principle

FIDD
    fetal iodine deficiency disorder
fiddle-string adhesion
field
    high-powered f. (hpf)
    low-powered f. (lpf)
    visual f.
fifth
    f. digit syndrome
    f. finger clinodactyly
    f. phacomatosis
    f. venereal disease
FIGLU
    formiminoglutamic acid
FIGLU-uria
    formiminoglutamicaciduria
FIGO
    International Federation of Gynecology
      and Obstetrics
      FIGO nomenclature
      FIGO stage I disease
      FIGO staging
figure
    mitotic f.
filament
    myosin f.
Filatov-Dukes disease
filial
    f. generation
Filibon
filiform adnatum
Filippi syndrome
filled and spilled
fillet
film
    contraceptive f.
    vaginal contraceptive f. (VCF)
    VCF vaginal contraceptive f.
filmy adhesion
Filshie
    F. clip
    F. clip applicator
filter
    Greenfield f.
    vena caval f.
filtered specimen trap
filtrate
    glomerular f.
filtration
    glass-wool f.
    glomerular f.
    rate of fluid f. (Qf)
fimbria, pl. **fimbriae**
    flowering-out of f.
    mushrooming of f.
    f. ovarica
fimbrial
    f. ectopic pregnancy

f. evacuation
f. failure
f. obstruction
f. prolapse
fimbriated end of fallopian tube
fimbriectomy
Uchida f.
fimbriocele
fimbrioplasty
Bruhat laser f.
finasteride
finding
clinical f.
dermatoglyphic f.
endovaginal f.
histopathological f.
laboratory f.
pathologic f.
sonographic f.
ultrasonic endovaginal f.
Findley folding pessary
Fine-Lubinsky syndrome
fine Metzenbaum scissors
fine-needle
f.-n. aspiration (FNA)
f.-n. aspiration biopsy
finger
f. cot
f. grasp
Madonna f.
seal f.'s
spider f.
webbed f.
fingerbreadth
fingerprinting
DNA f.
DNA amplification f.
plasmid f.
fingerstick
fingertip dilated
Finkelstein feeding
first
f. arch syndrome
f. degree prolapse
f. parallel pelvic plane
F. Response ovulation predictor
f. and second branchial arch
syndrome
f. stage of labor
f. trimester
first-cycle clinical pregnancy
first-degree laceration

first-generation progesterone
FISH
fluorescence in situ hybridization
Fisher
infantile choreoathetosis of F.
F. and Paykel RD1000 resuscitator
fisherman's knot
Fisher-Yates test
Fishman syndrome
fish-mouth cervix
fish skin
fishy odor
FISS
Flint Infant Security Scale
fissure
Ammon f.
f. in ano
downslanting palpebral f.
downsloping palpebral f.
palpebral f.
upslanting palpebral f.
fisting of hands
fists
clenched f.
fistula, pl. **fistulae, fistulas**
arteriovenous f.
branchial f.
bronchobiliary f.
cervicovaginal f.
colovaginal f.
enterovaginal f.
gastrointestinal f.
genitourinary f.
H-type f.
intestinal f.
lacteal f.
mammary f.
metroperitoneal f.
perineovaginal f.
postradiation f.
rectolabial f.
rectovaginal f.
rectovestibular f.
rectovulvar f.
renal f.
systemic arteriovenous f.
tracheoesophageal f. (TEF)
tubovaginal f.
umbilical f.
urachal f.
ureter f.
ureterovaginal f.

F

NOTES

fistula *(continued)*
    urethrovaginal f.
    urinary f.
    urogenital f.
    uteroperitoneal f.
    vesicouterine f.
    vesicovaginal f.
    vesicovaginorectal f.
    vitelline f.
fitness
    darwinian f.
Fitz-Hugh and Curtis syndrome
Fitzsimmons syndrome
five-hour glucose tolerance test
fixation
    complement f.
    iliococcygeal f.
    Nichols sacrospinous f.
fixator
    Ilizarov external f.
fixed allele
FJN
    familial juvenile nephrophthisis
FL
    femur length
flaccid
    f. extremity
flag sign
Flagyl
    F. 375
flail
    f. foot
FLAIR sequence
flammeus
    nevus f.
flanking region
flank pain
flap
    bladder f.
    Boari f.
    bridging f.
    bulbocavernosus fat f.
    butterfly f.
    Byers f.
    endorectal f.
    fascial f.
    fat f.
    gluteal free f.
    lateral transverse thigh f.
    latissimus dorsi f.
    maple leaf f.
    Martius bulbocavernosus fat f.
    McCraw gracilis myocutaneous f.
    myocutaneous f.
    Pontén fasciocutaneous f.
    Rubens f.
    saddlebag f.
    TRAM f.

    TRAMP f.
    transverse rectus abdominis
      myocutaneous f.
    transversus and rectus
      musculoperitoneal f.
    vaginal wall f.
    Warren f.
flare
flaring
    f. of ala nasi
    alar f.
    f. and grunting
    grunting and f.
    nasal f.
flash
    hot f.
flat
    f. facies
    f. pelvis
    f. wart
flatfoot
flatus vaginalis
flecainide acetate
Fleet
    F. enema
    F. Pain Relief
Fleming
    F. afterloading tandem
    F. ovoid
Flents breast comfort pack
fleshy mole
Fletcher-Suit
    F.-S. afterloading tandem
    F.-S. applicator
**fleur de lis breast reconstruction
pattern**
flexible hysteroscope
flexion
    f. spasm
Flexi-Seal fecal collector
flexure
    splenic f.
Flexxican catheter
Fliess treatment
Flint Infant Security Scale (FISS)
FLM
    fetal lung maturity
    FLM test
floating
    f. harbor syndrome (FHS)
    f. membrane
Flonase
flood
flooding
floor
    decidual f.
    pelvic f.
floppy infant syndrome

flora
>commensal f.
>microbial f.
>uterine f.
>vaginal f.

Florical

florid
>f. complexion
>f. toxemia

Florida pouch

Florone
>F. E

flow
>blood f. (Q̇)
>cardiac f.
>cerebral blood f.
>cutaneous blood f.
>f. cytometry
>Doppler f.
>fetal aortic blood f.
>fetal blood f.
>gene f.
>f. karyotyping
>menstrual f.
>myocardial blood f.
>pulmonary blood f. (PBF)
>f. rate
>renal plasma f.
>retrograde menstrual f.
>umbilical blood f.
>urine f.
>uterine blood f.
>uteroplacental blood f.
>f. velocity waveform
>volume f.

flowering-out of fimbria

FlowGel
>F. barrier material

flowmeter
>laser-Doppler f.

Flowstat

Flowtron DVT prophylaxis unit

flow-volume loop

Floxin

floxuridine
>f. in hepatic metastasis

Fluanxol

fluconazole

flucytosine

Fludara

flufenamic acid

Fluhmann test

fluid
>amniotic f. (AF)
>Bamberger f.
>body f.
>cerebrospinal f. (CSF)
>extracellular f.
>follicular f.
>foul-smelling amniotic f.
>human oviduct f. (HOF)
>hysteroscopy f.
>meconium-stained amniotic f. (MSAF)
>f. overload
>peritoneal f.
>f. replacement
>seminal f.
>serosanguineous f.
>sperm-counting f.
>f. therapy
>transcapillary f.
>f. wave

fluids, electrolytes, nutrition (FEN)

fluke

flumazenil

flumecinol

flunisolide

fluocinolone
>f. acetonide

fluocinonide

Fluoderm

Fluogen

Fluonid Topical

fluorescein
>f. isothiocyanate
>f. sodium
>f. treponema antibody test

fluorescein-conjugated monoclonal antibody test

fluorescein-labeled milk

fluorescence
>f. depolarization analysis
>f. in situ hybridization (FISH)

fluorescence-activated cell sorter (FACS)

fluorescent
>f. antibody against membrane antigen (FAMA)
>f. antimembrane antibody test (FAMA)
>f. immunoassay (FIA)
>f. polarization
>f. treponemal antibody (FTA)

F

NOTES

fluorescent *(continued)*
    f. treponemal antibody absorption
    (FTA-ABS)
    f. treponemal antibody-absorption
    test
fluoridation
fluoride
    slow-release sodium f.
    sodium f.
fluorinated corticosteroid
fluorine
fluorocortisone
5-fluorocytosine (5-FC)
fluorodeoxyuridine (FUDR)
9-fluorohydrocortisone
fluorophosphate
    diisopropyl f.
Fluoroplex
    F. Topical
fluoroscopic guidance
fluoroscopy
fluorouracil
    Cytoxan, Adriamycin, f. (CAF)
    Cytoxan, Adriamycin, leucovorin,
    calcium, f. (CALF)
    Cytoxan, epirubicin, f. (CEF)
    f., epirubicin, Cytoxan (FEC)
5-fluorouracil (5-FU)
    intraperitoneal 5-f.
fluoro-urodynamic evaluation
Fluothane
fluoxetine HCl
fluoxymesterone
flupenthixol
fluphenazine
flurandrenolide
flurazepam
flurbiprofen
Flurosyn Topical
flush
    breast f.
    hot f.
    f. method
    orgasmic f.
    vasomotor f.
flutamide
fluted drain
Flutex Topical
fluticasone
flutter
    atrial f.
Fluvoxame maleate
fluvoxamine
Fluzone
fly
    Spanish f.
flying-T pelvis

FMG
    female genital mutilation
FMH
    fetomaternal hemorrhage
fm/mg
    fentomoles/milligram
FMR1
    fragile site mental retardation 1
        FMR1 gene
FMR2
    fragile site mental retardation 2
FNA
    fine-needle aspiration
FNA-21
    F. needle
    F. syringe
FND
    frontonasal dysplasia
FNI
    facial nerve injury
foam
    Because vaginal f.
    contraceptive f.
    intravaginal f.
    f. rubber vaginal form
    Sklar f.
    F. Stability Index (FSI)
    f. stability test
FOAR
    faciooculoacousticorenal
        FOAR syndrome
FOBT
    fecal occult blood testing
focal
    f. change
    f. dermal hypoplasia syndrome
    f. epilepsy
    f. facial dermal dysplasia
    f. facial dermal dysplasia II
    (FFDD II)
    f. fibrosis
    f. lobular carcinoma
    f. mucopolysaccharidosis
    f. neurological deficit
    f. spot size
    f. villitis
    f. vulvitis
focally dilated duct
focus, pl. **foci**
    echogenic cardiac f.
    endometriotic f.
    hyperechogenic foci
    occult f.
Foerster
    F. sign
    F. sponge-holding forceps
Foille
folate

fold
> aryepiglottic f.
> Duncan f.'s
> epicanthal f.
> genitocrural f.
> head and tail f.
> f. of Hoboken
> Juvara f.
> nuchal f.
> Pawlik f.
> rectouterine f.
> rugal f.
> skin f.
> splanchnic f.
> urogenital f.

Foldan
folding frequency
Folex PFS
Foley catheter
folic
> f. acid
> f. acid antagonist
> f. acid deficiency

folinic acid
folk medicine
follicle
> antral f.
> f. aspiration, sperm injection, and assisted rupture (FASIAR)
> f. aspiration tube
> atretic f.
> dominant f.
> graafian f.
> luteinized unruptured f.
> f. maturation stimulation
> nabothian f.
> preantral f.
> preovulatory f.
> primary f.
> primordial f.
> f. regulatory protein (FRP)
> f. steroidogenesis
> f. stimulating hormone (FSH)

follicle-stimulating
> f.-s. hormone antagonist
> f.-s. hormone binding inhibitor
> f.-s. hormone inhibition
> f.-s. hormone secretion

follicular
> f. atrophoderma-basal cell carcinoma syndrome
> f. atrophoderma-basocellular proliferation-hypotrichosis syndrome
> f. attrition
> f. cyst
> f. development arrest
> f. factor
> f. fluid
> f. function
> f. hematoma
> f. maturation
> f. phase
> f. phase gonadotropin secretion
> f. scoring
> f. urethritis
> f. vulvitis

follicularis
> keratosis f.

folliculitis
folliculogenesis
folliculostatin
Folling disease
follistatin
Follistim
follitropin
> f. alfa
> f. alpha
> f. beta

followup, follow-up
> f. care
> Carnation F.'s
> long-term f.

Follutein
Folvite
fomite
Fong disease
Fontaine syndrome
fontanel, fontanelle
> anterior f.
> anterolateral f.
> Casser f.
> casserian f.
> Gerdy f.
> mastoid f.
> posterior f.
> posterolateral f.
> sagittal f.
> scaphoid f.
> f. sign
> sphenoid f.
> sunken f.

Fontan procedure
fonticulus, pl. **fonticuli**

F

NOTES

food
    f. diary
    F. and Drug Administration (FDA)
foot, pl. **feet**
    athlete's f.
    bilateral club feet
    club f.
    flail f.
    Freidreich f.
    Morand f.
    f. presentation
    reel f.
    rockerbottom f.
    spatula f.
foot-drop
footling breech presentation
footprinting
FOP
    fibrodysplasia ossificans progressiva
foramen, pl. **foramina**
    f. magnum
    f. of Monro
    f. ovale
    pleuroperitoneal f.
    f. primum
    f. secundum
Forane
Forbes-Albright syndrome
Forbes disease
force
    compression f.
    expulsive f.
    fetal shoulder extraction f.
    F. GSU argon-enhanced
      electrosurgery system
    physician-applied f.
forcé
    eccouchement f.
forced
    f. expiratory volume (FEV)
    f. grasping reflex
forceps
    Adson f.
    alligator f.
    Allis f.
    Allis-Abramson breast biopsy f.
    Apple Medical bipolar f.
    Arruga-Nicetic capsule f.
    artery f.
    ASSI bipolar coagulating f.
    atraumatic f.
    axis-traction f.
    Babock f.
    Baird f.
    Bakchaus towel f.
    Barton f.
    Baumberger f.
    bayonet f.

    Bellucci alligator f.
    BiCoag f.
    Bierer ovum f.
    Billroth tumor f.
    Bill traction handle f.
    biopsy f.
    bipolar laparoscopic f.
    Bishop-Harmon f.
    Bozeman uterine dressing f.
    Brown-Adson tissue f.
    Castroviejo fixation f.
    Cauer f.
    cephalic f.
    Chamberlen f.
    Clarke ligator scissor f.
    coaptation bipolar f.
    cold biopsy f.
    Colibri f.
    Corey ovum f.
    Corson myoma f.
    Crile f.
    Cushing f.
    DeBakey tissue f.
    DeJuan f.
    DeLee f.
    f. delivery (FD)
    Dennen f.
    Desmarres f.
    DeWeese axis traction f.
    Dewey obstetrical f.
    Dieter f.
    dressing f.
    Drews f.
    Duval f.
    Elliot f.
    Endo-Assist endoscopic f.
    Englert f.
    Eppendorfer biopsy f.
    Erhardt f.
    Evershears bipolar laparoscopic f.
    Facit polyp f.
    Feilchenfeld f.
    Foerster sponge-holding f.
    Francis f.
    Fujinon biopsy f.
    Gelhorn f.
    Gerald f.
    Glassman f.
    Graddle f.
    Greven f.
    Haig-Fergusson f.
    Halsted mosquito f.
    Haugh f.
    Hawk-Dennen f.
    Heaney f.
    Heaney-Ballantine f.
    Heaney-Hyst f.
    Hildebrandt uterine hemostatic f.

Hirst placental f.
Hodge f.
hot biopsy f.
Hunt bipolar f.
Iselin f.
Jacobson f.
Jaws f.
jeweler's f.
Juers f.
Keiland f.
Kevorkian-Younge biopsy f.
Kjelland f.
Kjelland-Barton f.
Kjelland-Luikart f.
Kleppinger bipolar f.
Lahey f.
Lalonde delicate hook f.
Laufe-Piper f.
Laufe polyp f.
Laurer f.
Leff f.
Levret f.
Livernois-McDonald f.
Llorente dissecting f.
low f.
Luikart f.
Luikart-Simpson f.
f. maneuver
Mazzariello-Caprini f.
McGee f.
McGill f.
McKerman-Adson f.
McKerman-Potts f.
McLane f.
McPherson f.
Nadler f.
Nägele f.
Neville-Barnes f.
nonfenestrated f.
obstetric f.
Ochsner f.
ovum f.
paddle f.
Palmer ovarian biopsy f.
Pean f.
pelvic f.
Pennington f.
Perez-Castro f.
Phaneuf uterine artery f.
Piper f.
Pistofidis cervical biopsy f.
pituitary f.

Polaris reusable f.
Pollock f.
premie Simpson f.
punch biopsy f.
Quinones-Neubüser uterine-grasping f.
Quinones uterine-grasping f.
radial jaw biopsy f.
Randall stone f.
Reddick-Saye f.
Reiner-Knight f.
ring f.
Rochester-Ochsner f.
Rochester-Pean f.
Roger f.
f. rotation
Russian tissue f.
Saenger ovum f.
Scanzoni f.
Schroeder tenaculum f.
Schroeder vulsellum f.
Schubert uterine biopsy f.
Seitzinger tripolar cutting f.
Semken f.
Shea f.
Shearer f.
Shute f.
Simpson f.
Singley f.
Sopher ovum f.
sponge f.
sponge-holding f.
spoon f.
Stolte f.
suture grasper f.
Tarnier axis-traction f.
tendon f.
Therma Jaw hot urologic f.
Thomas-Gaylor biopsy f.
Tischler cervical biopsy f.
Tischler-Morgan uterine biopsy f.
tissue f.
trial f.
Tucker-McLane f.
Tucker-McLane-Luikart f.
uterine tenaculum f.
Utrata f.
Willett f.
Winter placental f.
Yeoman f.
Z-Clamp hysterectomy f.

F

NOTES

Fordyce
- F. disease
- F. granule
- F. spot

forebag

foregut malformation

foreign
- f. body
- f. body salpingitis

forekidney

foremilk

foreplay

foreskin

forewaters

fork
- replication f.

form
- balsa vaginal f.
- coccobacillary f.
- Doppler flow-velocity wave f.
- foam rubber vaginal f.
- Ware Short F.-35

formaldehyde

formation
- blood vessel f.
- bone f.
- chiasma f.
- echo f.
- somite f.

formative yolk

Forma water-jacketed incubator

forme fruste

formiminoglutamic acid (FIGLU)

formiminoglutamicaciduria (FIGLU-uria)

formiminotransferase deficiency syndrome

formula
- Advance f.
- Bonamil f.
- 20-calorie f.
- 24-calorie f.
- dextro f.
- Enfamil Next Step toddler f.
- Enfamil Premature 20 F.
- evaporated milk f.
- fortified f.
- full-strength f.
- glucose f.
- goat's milk f.
- Good Nature f.
- Good Start HA f.
- half-strength f.
- Hardy-Weinberg f.
- high-calorie f.
- high-fructose f.
- high-glucose f.
- hypercaloric f.
- Infalyte f.
- I-Soyalac f.
- Lactofree f.
- lactose-free f.
- Lytren f.
- Mall f.
- MCT oil f.
- Mead-Johnson f.
- menstrual f.
- Mollison f.
- Neocate f.
- Nursoy f.
- Nutramigen f.
- Osmolite f.
- Pedialyte f.
- PediaSure f.
- PEF-24 f.
- PM-60/40 f.
- Polycose f.
- Portagen f.
- Pregestimil f.
- Premature Special Care f.
- ProSobee f.
- Pulmocare f.
- quarter-strength f.
- Rehydralyte f.
- Sabbagha f.
- Similac Special Care f.
- Sim SC-20 f.
- Sim SC-24 f.
- Sim SC-40 f.
- S-M-A f.
- sodium-free f.
- Soyalac f.
- Spearman-Brown prediction f.
- Special Care f.
- SSC-20 f.
- SSC-24 f.
- sucrose-free f.
- Sustacal f.
- Sustagen f.
- three-quarter strength f.
- Triaminic AM Decongestant F.

formula-feeding woman

fornix, pl. **fornices**
- posterior f.
- vaginal f.

Forsius-Eriksson type ocular albinism

Forssman antigen

Fortaz

Forte
- Niferex-PN F.
- Zone-A F.

Fortel ovulation test

fortified formula

fortifier
- Enfamil human milk f.
- human milk f.

Fortovase

Forvade
forward
    f. roll fashion
    f. roll method
Fosamax
foscarnet
Foscavir
fosfomycin tromethamine
fosinopril
fossa, pl. **fossae**
    congenital supraspinous f.
    enlarged posterior f.
    ischiorectal f.
    f. navicularis
    posterior f.
    shallow acetabular fossae
Fothergill-Donald operation
Fothergill-Hunter operation
Fothergill operation
foul-smelling amniotic fluid
founder
    f. chromosome
    f. effect
Fountain syndrome
four-cell embryo
four-chamber view
fourchette
four-day syndrome
four-flap Z-plasty
four-hour rule
Fourier transform infrared
  microspectroscopy
four-leaf clover pattern
Fournier teeth
four-quadrant assessment
fourth
    Bartholomew rule of f.'s
    f. parallel pelvic plane
    f. phacomatosis
    f. venereal disease
fourth-degree laceration
foveal hypoplasia
Fowler position
Fowler-Stephens orchiopexy
Fox-Fordyce disease
FPG
    fasting plasma glucose
FPO
    faciopalatoosseous
    FPO syndrome
FRA, fra
    fragile

fragile chromosome site
fragile gene
fraction
    f. of alveolar carbon dioxide
    (FACO$_2$)
    f. in expired gas of carbon
    dioxide (FECO$_2$)
    plasma f.
    recombination f.
    S phase f.
fractional dilation and curettage
fractionated radiation therapy
fractionation
fracture
    birth f.
    clavicular f. (CF)
    Colles f.
    fetal f.
    hip f.
    intrauterine f.
    maternal f.
    osteoporotic f.
    pelvic f.
    skull f.
    spinal compression f.
    vertebral compression f.
fractured chromosome
fragile (FRA, fra)
    f. chromosome site (FRA, fra)
    f. gene (FRA, fra)
    f. site mental retardation 1
    (FMR1)
    f. site mental retardation 2
    (FMR2)
    f. X (FRAX, fra(x))
    f. X carrier
    f. X chromosome
    f. X chromosome testing
    f. X mental retardation
    f. X-mental retardation syndrome
    f. Xq syndrome
    f. X syndrome
fragile X gene (FRAX, fra(x))
fragility
    X-linked first site of f. (FRAXA)
    X-linked second site of f.
    (FRAXE)
fragment
    f. antigen binding
    anucleate f.
    Okazaki f.
    placental f.

NOTES

F

fragment *(continued)*
    restriction f.
    urinary beta-core f. (UbetaCF)
fragmentation
    uterine f.
Fragmin
frame
    open reading f.
    reading f.
frameshift
    f. mutation
Franceschetti-Goldenhar syndrome
Franceschetti-Jadassohn syndrome
Franceschetti-Klein syndrome
Franceschetti syndrome
Franceschetti-Zwahlen-Klien syndrome
Franceschetti-Zwahlen syndrome
Francis forceps
Francois
    F. dyscephalia
    F. dyscephalic syndrome
Frangenheim-Goebell-Stoeckel operation
Frank
    F. nonsurgical perineal audodilation
    F. nonsurgical perineal autodilation
    F. procedure
    F. technique
    F. technique of dilation
frank
    f. ambiguity
    f. breech
    f. breech presentation
Frankenhäuser
    F. ganglion
    F. plexus
Franklin-Dukes test
Frank-Starling
    F.-S. principle
    F.-S. relation
frappage therapy
Fraser-Francois syndrome
Fraser-like syndrome
Fraser syndrome
fraternal
    f. twins
FRAX, fra(x)
    fragile X
    fragile X gene
fra(X)
    fra(X)(q27) syndrome
    fra(X)(28) syndrome
    fra(X) syndrome
FRAXA
    X-linked first site of fragility
FRAXE
    X-linked second site of fragility
    FRAXE-associated mental
      retardation

FRAXE1 X-linked mental
    retardation fragile site 1
FRAXE2 X-linked mental
    retardation fragile site 2
FRAXE1
    X-linked mental retardation-fragile site 1
FRAXE2
    X-linked mental retardation-fragile site 2
FRC
    functional residual capacity
FreAmine
Fredet-Ramstedt
    F.-R. operation
    F.-R. procedure
free
    F. & Active
    f. androgen index
    f. beta
    f. beta hCG
    f. beta test
    f. estradiol
    f. fatty acid
    f. hydroxyl radical
    f. neuraminic acid storage disease
    f. sialic acid storage disease
    f. testosterone
    f. testosterone index
    f. thyroxine index
free-floating loop
Freeman cookie cutter areola marker
Freeman-Sheldon syndrome
freeze-thaw-freeze
freezing process
Frei
    F. antibody
    F. test
Freiberg
    F. disease
    F. infraction
Freidman splint
Freidreich foot
frena (*pl. of* frenum)
French
    24-F. Foley balloon
    F. Gesco catheter
    F. hysteroscope
    F. lock
French-American-British (FAB)
frenectomy
frenoplasty
frenotomy
frenulum, pl. **frenula**
    f. labiorum pudendi
frenum, pl. **frena**
frequency
    allele f.
    folding f.
    Nyquist f.

pulse f.
recombination f.
urinary f.
fresh frozen plasma (FFP)
freudian
Freud theory
Freund
    F. adjuvant
    F. operation
freundii
    Citrobacter f.
Freyer suprapubic drain
friable cervix
FRIA muscle training device & program
Friberg microsurgical agglutination test
Friedman
    F. curve
    F. rabbit test
Friedman-Lapham test
Friedreich ataxia
Friedrich
    F. criteria
    F. criteria for dyspareunia
Friend syndrome
frigid
frigidity
Fritsch
    F. douche
    F. syndrome
Fritsch-Asherman syndrome
Froben
    F.-SR
frogleg
    f. position
    f. view
froglegged
Fröhlich syndrome
frondlike
frontal
    f. bones scalloping
    f. bossing
    f. horn
    f. suture
frontoanterior
    left f. position (LFA)
    f. position
    right f. position (RFA)
frontodigital syndrome
frontofacionasal
    f. dysostosis
    f. dysplasia

frontometaphyseal dysplasia
frontonasal dysplasia (FND)
frontoposterior
    left f. position (LFP)
    f. position
    right f. position (RFP)
frontotransverse
    left f. position (LFT)
    f. position
    right f. position (RFT)
frostbite
frothy discharge
frozen
    f. egg
    f. embryo
    f. embryo replacement (FER)
    f. ovum
    f. pelvis
    f. plasma
    f. red cells
    f. section
    f. semen
    f. smile puckered lips
    f. sperm
    f. zygote
frozen-thawed embryo transfer
FRP
    follicle regulatory protein
fructokinase
fructose
    f. galactokinase deficiency
    f. intolerance
fructosuria
frusemide
fruste
    forme f.
Fryns
    F. syndrome 1–3
Fryns-Moerman syndrome
Fryns-van den Berghe syndrome
FS
    AmpliTaq DNA polymerase FS
    FS Shampoo Topical
FSE
    fetal scalp electrode
FSH
    follicle stimulating hormone
    highly purified FSH
    FSH MAIAclone immunoradiometric assay
    purified urinary FSH

NOTES

FSI
Foam Stability Index
FS$_p$O$_2$
fetal arterial oxygen saturation
FTA
fluorescent treponemal antibody
FTA test
FTA-ABS
fluorescent treponemal antibody
absorption
FTBD
full term, born dead
FTSP
fallopian tube sperm perfusion
FTT
failure to thrive
FTV
fetal thrombotic vasculopathy
5-FU
5-fluorouracil
FUCA
alpha-L-fucosidase
fucosidosis
FUDR
fluorodeoxyuridine
FUE
fever of unknown etiology
fugax
amaurosis f.
fugue state
Fuhrmann syndrome
Fujinon
F. biopsy forceps
F. flexible hysteroscope
Fukuyama
F. congenital muscular dystrophy
(FCMD)
F. syndrome
FUL
functional urethral length
fulguration
full-breech presentation
Fuller
F. Albright syndrome 1
F. shield
F. shield rectal dressing
full-strength formula
full term, born dead (FTBD)
full-term pregnancy
full-thickness skin graft
fulminans
purpura f.
fulminant
f. hepatitis
Fulvicin
fumarate
ferrous f.

function
adrenocortical f.
bladder f.
bowel f.
cardiac f.
cardiorespiratory f.
corpus luteum f.
fetal cardiac f.
fetoplacental f.
follicular f.
hypothalamic-pituitary f.
kidney f.
leukocyte f.
luteal f.
myocardial f.
oromotor f.
ovarian f.
pituitary gland f.
pituitary-ovarian f.
pulmonary f.
renal f.
renal tubular f.
reproductive f.
sexual f.
thyroid f.
urethral f.
ventricular f.
functional
F. Assessment of Cancer Therapy
(FACT)
f. dysmenorrhea
f. murmur
f. ovarian cyst
f. prepubertal castrate syndrome
f. residual capacity (FRC)
f. urethral length (FUL)
function-enhancing mutation
fundal
f. height
f. placentation
fundectomy
fundipuncure
fundus, pl. fundi
ghost vessel f.
incarcerated f.
optic f.
salt and pepper f.
fundusectomy
fungal
f. infection
fungating mass
fungi (*pl. of* fungus)
Fungizone
Fungoid
F. AF Topical Solution
F. Cream
F. Tincture
fungus, pl. fungi

f. ball
umbilical f.
funic
f. presentation
f. reduction
f. souffle
funicular souffle
funiculus, pl. **funiculi**
f. umbilicalis
funipuncture
funis
funisitis
necrotizing f.
funnel
accessory müllerian f.
f. chest
f. length
f. width
funneling
cervical f.
funnel-shaped pelvis
Funston syndrome
FUO
fever of unknown origin
Furadantin
Furalan
furan
furazolidone
furcate insertion of cord

furoate
mometasone f.
furosemide
Furoxone
furrowlike umbilicus
furuncle
furunculosis
fusanic acid
fused frontal horn
fusion
cervical vertebral f.
f. of eyelids
f. implantation
incomplete müllerian f.
labial f.
müllerian duct f.
robertsonian f.
Fusobacterium
F. gonidiaformans
F. mortiferum
F. necrophorum
F. nucleatum
futile cycle
FVS
fetal valproate syndrome
fetal varicella syndrome
FyBron
F. calcium alginate dressing
fyn protooncogene

NOTES

F

G

gravida
G band
G protein
G syndrome

G$_0$

gap$_0$

G$_1$

gap$_1$

G$_2$

gap$_2$

GA

gestational age

Gabastou hydraulic method

GABEB

generalized atrophic benign
epidermolysis bullosa

GAG

glycosaminoglycan

gag reflex

Gailliard syndrome

gain

weight g.

gait

broad-based g.
toe-in g.
waddling g.

galactacrasia

galactagogue

galactic

galactobolic

galactocele

galactocerebrosidase deficiency (GALC)

galactography

galactokinase

galactokinesis

galactopoiesis

galactorrhea

g.-amenorrhea syndrome

galactose

g. 1-phosphate uridyltransferase
(GALT)

galactosemia

African-American variant g.
classical g.
Duarte variant g.
GALT deficiency g.
Los Angeles variant g.

galactosis

galactosylceramide

g. beta-galactosidase deficiency

galactosylsphinogosine lipidosis

galactotherapy

galatose-free diet

Galbiati bilateral fetal ischiopubiotomy

GALC

galactocerebrosidase deficiency

GALE

UDP-galactose-4-epimerase deficiency

galea

Galeazzi sign

Galen

vein of G.

Galileo rigid hysteroscope

Gallant reflex

gallbladder

g. disease

Galli-Mainini test

gallinatum

pectus g.

gallium-67 citrate contrast medium

gallium citrate Ga-67

gallium scan

Galloway-Mowat syndrome

Galloway syndrome

gallstone

g. pancreatitis

galoche chin

GALT

galactose 1-phosphate uridyltransferase
GALT deficiency galactosemia

galtonian-Fisher genetics

galtonian trait

Galton law of regression

GALT-UDP-galactose complex

Gambee suture

Gamble-Darrow syndrome

gametangium

gamete

aging g.
g. intrafallopian transfer (GIFT)
g. manipulation
g. micromanipulation

gametic

g. chromosome
g. selection

gametogenesis

ovarian g.

Gamimune N

gamma

g. globulin
g. glutamyl transferase (GGT)
g. interferon
g. ray

gamma-1b

gamma-benzene hexachloride

Gamma BHC

Gammabulin Immuno

gamma-chain disorder

G

Gammagard
  G. S/D
gamma-loop disorder
Gamper bowing reflex
gampsodactyly
Gamulin Rh
ganciclovir
gangliocytoma
ganglion, pl. **ganglia**
  basal g.
  Frankenhäuser g.
  g. trigger theory
ganglioneuroblastoma
ganglioneuroma
ganglioneuromatosis
ganglionic blocker
ganglioside storage disease
gangliosidosis
  GM2 g.
gangliosidosis
  adult generalized g.
  cerebral GM1 g.
  generalized g. GM1 type I
  generalized infantile g.
  generalized juvenile g.
  generalized g. juvenile type
  GM3 g.
  g. GM3
  g. GM1 juvenile type
  g. GM1 late onset without bony
    involvement
  g. GM1 type II
  juvenile GM1 g.
  neuronal GM1 g.
gangrene
  gas g.
  pulmonary g.
  spontaneous g. of newborn
gangrenosa
  varicella g.
ganirelix
  g. acetate
  g. acetate injection
Gantanol
Gantrisin
  Azo G.
GAP
  gonadotropin-releasing hormone-
    associated peptide
gap$_0$ (G$_0$)
gap$_1$ (G$_1$)
gap$_2$ (G$_2$)
gap
  anion g.
  g. junction
GAPS
  Guidelines for Adolescent Preventive
    Services

Garamycin
Garatec
Garcia-Lurie syndrome
Gardnerella
  G. vaginalis
  G. vaginalis chorioamnionitis
  G. vaginitis
Gardner-Silengo-Wachtel syndrome
Gardner syndrome
Gareis-Mason syndrome
gargantuan mastitis
gargoylism
gargyloid facies
Gariel pessary
Gartner
  G. cyst
  G. duct
gartnerian
  g. cyst
  g. duct
gas, pl. **gases**
  g. anesthetic
  arterial blood g. (ABG)
  blood g.
  capillary blood g. (CBG)
  carbon dioxide g.
  g. chromatography
  cord blood g. (CBG)
  g. embolus
  fetal blood gases
  g. gangrene
  inspired g. (I)
  intervillous blood g.
  sweep g.
  g. transfer
  venous blood g. (VBG)
  volume of expired g. (VE)
GASA
  growth-adjusted sonographic age
Gaskin maneuver
gasless laparoscopy
Gas-Pak jar
Gasser syndrome
GAST
  gonadotropin agonist stimulation test
gastric
  g. bubble
  g. bypass
  g. carcinoma
  g. dysrhythmia
  g. emptying
  g. emptying time
  g. enteral feeding
  g. fluid aspiration
  g. lavage
  g. reduction surgery
  g. volvulus
gastritis

gastroacephalus
gastroamorphus
gastrocolic reflex
Gastrocrom
gastrodidymus
gastroenteritis
    eosinophilic g.
gastroesophageal
    g. incompetence
    g. reflux (GER)
    g. reflux disease (GERD)
Gastrografin
gastrointestinal
    g. anastomosis (GIA)
    g. disorder
    g. fistula
    g. infection
    g. malignancy
    g. obstruction
    g. series
    g. tract
    g. tuberculosis
gastromelus
gastropagus
gastroplasty
gastroschisis
    Silastic silo reduction of g.
gastrostomy
gastrothoracopagus
gastrula
gate control theory
Gaucher disease
Gauss sign
gauze
    Aquaphor g.
    Cover-Roll g.
    g. dissection
    g. wick
gavage
    g. feeding
GAX collagen
Gazayerlie endoscopic retractor
G-banded cytogenetic aberration
G-banding
GBBB syndrome
GBS
    group B streptococcus
    GBS types Ia, Ib, Ic, II, III
GCI
    gestational carbohydrate intolerance
GCPS
    Greig cephalopolysyndactyly syndrome

G-CSF
    granulocyte colony-stimulating factor
GDM
    gestational diabetes mellitus
GDP
    guanyldiphosphate
GE
    GE RT 3200 Advantage II
      ultrasound
    GE Senographe 2000D
Geck
    Davis & G. (DG)
Gee disease
Gee-Herter disease
Gee-Herter-Heubner
    G.-H.-H. disease
    G.-H.-H. syndrome
Gehrung pessary
geitonogamy
gel
    Accoustix conductivity g.
    AccuSite injectable g.
    Aquagel lubricating g.
    Aquasonic 100 ultrasound
      transmission g.
    Cerviprost g.
    cidofovir topical g.
    COL 2301(terbutaline g.)
    Comfort personal lubricant g.
    Crinone bioadhesive progesterone g.
    Crinone progesterone g.
    dextrin sulfate g.
    dinoprostone cervical g.
    1D sodium dodecyl sulfate g.
    Electro-Gel conductivity g.
    electrophoresis g.
    Empirynol Plus contraceptive g.
    Itch-X g.
    keratolytic g.
    MetroGel-Vaginal g.
    metronidazole vaginal g.
    Monsel g.
    Multidex g.
    Panretin topical g.
    povidone-iodine g.
    Prepidil Vaginal G.
    PRO 2000 G.
    PRO/Gel ultrasound transmission g.
    prostaglandin E_2 g.
    Protectaid contraceptive sponge
      with F-5 g.
    Scan ultrasound g.

G

NOTES

gel *(continued)*
    SonoMix ultrasound g.
    Tisit Blue G.
gelastic seizure
gelatin agglutination test
gelatinous
    g. skin
    g. varix
geleophysic
    g. dwarfism
    g. dysplasia
Gelfoam
Gelhorn
    G. forceps
    G. pessary
Gelpi-Lowrie
    G.-L. hysterectomy
    G.-L. retractor
Gelpi perineal retractor
gel-transfer
gemcitabine
gemellary pregnancy
gemellipara
gemellology
gemeprost
geminus, pl. **gemini**
gemmule
    Hoboken g.
Gemonil
Gemzar
Genapax
Genaspor
Genatap Elixir
Gencalc 600
Gen-Clobetasol
gender
    g. assignment
    g. determination
    g. dysphoria syndrome
    g. reversal
    g. role
gene
    g. action
    adhalin g.
    AIRE g.
    allelic g.
    g. amplification
    autoimmune regulatory g.
    autosomal g.
    BRCA1 breast cancer g.
    BRCA2 breast cancer g.
    breast cancer g. 1 (BRCA1)
    breast cancer g. 2 (BRCA2)
    g. candidate
    candidate g.
    g. carrier
    cell interaction g.
    chimeric g.

    g. chip
    g. cloning
    codominant g.
    complementary g.
    g. complex
    cumulative g.
    g. deletion
    derepressed g.
    DFFRY g.
    dominant g.
    g. dose
    g. duplication
    dystrophin g.
    E6-E7 g.
    endothelin-3 g.
    endothelin receptor-B g.
    g. expression
    g. family
    g. flow
    FMR1 g.
    fragile g. (FRA, fra)
    fragile X g. (FRAX, fra(x))
    GUSB g.
        beta-glucuronidase
    H g.
    hair cortex keratin g. type II
    herpes simplex virus thymidine
        kinase g.
    histocompatibility g.
    holandric g.
    homeobox 2 g.
    homeotic g.'s
    housekeeping g.
    HOX A g.
    HPV E7 g.
    immune response g.
    immune suppressor g.
    immunoglobulin g.
    Ir g.
    Is g.
    jumping g.
    KAL g.
    leaky g.
    lethal g.
    g. library
    g. location
    g. locus
    major g.
    major capsid protein g.
    g. map
    g. mapping
    marker g.
    methylenetetrahydrofolate
      reductase g.
    MOMP g.
    g. mosaic
    mutant g.
    g. mutation

NF1 g.
nomp C g.
nonstructural g.
od g.
operator g.
p53 g.
partner g.
penetrant g.
pleiotropic g.
g. pool
pRb tumor suppressor g.
g. probe
g. product
p53 tumor suppressor g.
RBM g.
recessive g.
reciprocal g.
regulator g.
repressed g.
repressor g.
RhCE g.
    rhesus CE gene
RhD g.
    rhesus D gene
rhesus CE g. (RhCE gene, RhCE gene)
rhesus D g. (RhD gene)
g. sequencing
sex-conditioned g.
sex-influenced g.
sex-limited g.
sex-linked g.
silent g.
g. splicing
structural g.
sublethal g.
suicide g.
supplementary g.
suppressor g.
syntenic g.
g. targeting
tdy g.
g. therapy
g. transcription
g. transfer
tumor suppression g.
tumor suppressor g.
wild-type g.
Wilms tumor suppression g.
X-linked g.
Y-linked g.
Z-linked g.

Genée-Wiedemann
G.-W. acrofacial dysostosis (GWAFD)
G.-W. syndrome
Genentech
G. biosynthetic human growth hormone
G. growth chart
general
G. Electric Model RT-3200 ultrasound
g. endotracheal anesthesia
generalisata
dysostosis g.
osteitis condensans g.
generalized
g. atrophic benign epidermolysis bullosa (GABEB)
g. gangliosidosis GM1 adult type
g. gangliosidosis GM1 type I
g. gangliosidosis juvenile type
g. hypertrichosis terminals-gingival hyperplasia syndrome
g. hypotonia, congenital hydronephrosis, characteristic face syndrome
g. infantile gangliosidosis
g. infantile gangliosidosis with bony involvement
g. juvenile gangliosidosis
g. linear interactive modeling (GLIM)
g. lipodystrophy
generation
filial g.
generational
generator
Valleylab Force IC electrosurgical g.
genesial cycle
genetic
g. abnormality
g. amniocentesis
g. anomaly
g. bit analysis
g. burden
g. code
g. counseling
g. defect
g. diagnosis
g. disease
g. disorder

NOTES

G

genetic *(continued)*
  g. distance
  g. diversity
  g. drift
  g. engineering
  g. engineering technology
  g. equilibrium
  g. fine structure
  g. interference
  g. line
  g. linkage
  g. linkage analysis
  g. locus
  g. map
  g. marker
  g. material
  g. model
  g. mutation
  g. polymorphism
  g. predisposition
  g. screening
  g. sex
  g. susceptibility
  g. switch
  g. test
  g. variance
geneticist
  medical g.
genetics
  behavioral g.
  biochemical g.
  classical g.
  epidemiological g.
  galtonian-Fisher g.
  human g.
  medical g.
  modern g.
  multilocal g.
  prenatal g.
  reproductive g.
  reverse g.
genetotrophic
  g. disease
genetous
Geneye Ophthalmic
genital
  g. ambiguity
  g. anomaly-cardiomyopathy
    syndrome
  g. cord
  g. corpuscle
  g. crisis of newborn
  g. development
  g. differentiation
  g. duct
  g. dwarfism
  g. elephantiasis
  g. herpes

  g. infection
  g. lesion
  g. mycoplasma
  g. prolapse
  g. prolapse staging
  g. ridge
  g. tract
  g. tract embryology
  g. tract malignancy
  g. tract trauma
  g. tract tumor
  g. tubercle
  g. tuberculosis
  g. ulcer
  g. ulcer disease (GUD)
  g. ulcer syndrome
  g. wart
genitalia
  ambiguous g.
  external g.
  indifferent g.
  internal g.
  lymphatic drainage of g.
genitalis
  herpes g.
genitocrural fold
genitofemoral nerve
genitogram with or without IVP
genitopalatocardiac syndrome
Genitor mini-intrauterine insemination
  cannula
genitourinary
  g. abnormality
  g. defect
  g. fistula
Genoa syndrome
genoblast
genocopy
genome
  g. activation
  Human G. (HuGe)
  g. project
genomic
  g. DNA
  g. imprinting
  g. library
  g. in situ hybridization
Genoptic
  G. S.O.P.
Genora
  G. 0.5/35
  G. 1/35
  G. 1/50
Genotropia
genotype
  ACE g.
genotypic
Genpril

Gen-Probe
    G.-P. amplified CT assay
    G.-P. amplified CT test
Gentacidin
Gent-AK
gentamicin
    g. sulfate
gentian violet
Gentrasul
genu
    g. recurvatum
    g. valgum
genucubital position
genuine stress incontinence (GSI)
Genupak tampon
genupectoral position
genus
genus-specific monoclonal antibody
Geocillin
geographic tongue
Geopen
geophagia
geotaxis
Gepfert procedure
GER
    gastroesophageal reflux
Gerace reflex
Gerald forceps
GERD
    gastroesophageal reflux disease
Gerdy fontanel
Geref
Gerhardt syndrome
Gerimed
germ
    g. cell
    g. cell ovarian neoplasm
    g. cell teratoma
    g. cell testicular tumor
    g. layer
    g. line
    g. line mosaicism
    g. ridge
    g. tube test
German
    G. lock
    G. measles
    G. syndrome
germinal
    g. epithelium
    g. epithelium of Waldeyer
    g. membrane

    g. pole
    g. vesicle
    g. vesicle breakdown (GVBD)
    g. vesicle stage
germinoma
germ-line mutation
germplasm
Gerota fascia
Gerstmann syndrome
Gesco
    G. cannula
    G. catheter
Gesell
    G. Developmental Scale
    G. test
    G. test with Knobloch modification
gestagen
gestagenic
gestation
    anembryonic g.
    cornual g.
    higher-order g.
    multifetal g.
    multiple g.
    prolonged g.
    term g.
    tubal g.
    twin g.
    unruptured tubal g.
gestational
    g. abnormality
    g. age (GA)
    g. age assessment
    g. age estimation
    g. carbohydrate intolerance (GCI)
    g. carrier
    g. diabetes
    g. diabetes mellitus (GDM)
    g. edema
    g. hypertension
    g. lupus
    g. mother
    g. proteinuria
    g. psychosis
    g. ring
    g. sac (GS)
    g. sac mean diameter
    g. sac size (GSS)
    g. surrogacy
    g. surrogate
    g. thyrotoxicosis
    g. trophoblastic disease (GTD)

G

NOTES

gestational *(continued)*
    g. trophoblastic neoplasia (GTN)
    g. trophoblastic tumor (GTT)
gestationis
    herpes g. (HG)
    pemphigoid g.
gestation-specific nomogram
gestator
Gesterol
Gestodene
gestodene
Gestogen
gestosis, pl. **gestoses**
    second trimester acute g.
Gestrinone
GF
    growth factor
GFR
    glomerular filtration rate
GGT
    gamma glutamyl transferase
GH
    growth hormone
Ghon
    G. complex
    G. tubercle
ghost vessel fundus
GH-RH
    growth hormone-releasing hormone
GHST
    growth hormone stimulation test
GIA
    gastrointestinal anastomosis
    GIA 60 stapler
    GIA 80 stapler
Gianotti-Crosti syndrome
giant
    g. anorectal condyloma acuminatum
    g. baby
    g. cell chondrodysplasia
    g. cell hepatitis
    g. cell pneumonia
    g. chromosome
    g. fibroadenoma
    g. platelet syndrome
Gianturco spring embolus
Giardia
giardiasis
gibbous deformity
gibbus
    thoracolumbar g.
Gibco BRL sperm preparation medium
GI cocktail
Giemsa
    G. banding
    G. stain
Gierke disease

GIFT
    gamete intrafallopian transfer
    cervical GIFT
    intrauterine GIFT
    vaginal GIFT
gigantism
    cerebral g.
    eunuchoid g.
    exomphalos, macroglossia, and g.
      (EMG)
    fetal g.
    hyperpituitary g.
    pituitary g.
gigantoblast
gigantomastia
giggle incontinence
Gigli
    G. operation
    G. saw
Gilbert disease
Gilbert-Dreyfus syndrome
Gilbert-Lereboullet syndrome
gill
    g. arch skeleton
Gilles de la Tourette syndrome
Gillespie syndrome 1, 2
Gilliam
    G. operation
    G. round ligament
Gilliam-Doleris
    G.-D. operation
    G.-D. uterine suspension
Gill respirator
Gimbernat reflex ligament
gingival
    g. fibromatosis
    g. fibromatosis-corneal dystrophy
    g. fibromatosis, hypertrichosis,
      cherubism, mental retardation,
      epilepsy syndrome
    g. fibromatosis, hypertrichosis,
      mental retardation, epilepsy
      syndrome
    g. hyperplasia, hirsutism,
      convulsions syndrome
    g. hypertrophy-corneal dystrophy
      syndrome
gingivitis
    herpetic g.
gingivostomatitis
    herpetic g.
Giordano operation
girdle
    Hitzig g.
gitalin
Gittes
    G. operation
    G. procedure

G. technique
G. urethral suspension

GK
glycerol kinase

GKD
glycerol kinase deficiency

glabella

glabrata
Torueopsis g.

gland
accessory sex g.
adrenal g.
apocrine g.
Bartholin, urethral, Skene g.
bulbourethral g.
BUS g.
Cowper g.
Duverney g.
eccrine g.
endocrine g.
external genitalia/Bartholin, urethral, and Skene g.'s (EG/BUS)
fetal adrenal g.
Littre g.
mammary g.
Mery g.
Montgomery g.
nabothian g.
parathyroid g.
paraurethral g.
parotid g.
periurethral g.
Philip g.
pineal g.
pituitary g.
salivary g.
Skene g.
sublingual g.
submaxillary g.
thymus g.
thyroid g.
urethral g.
uterine g.
vestibular g.

glandular
g. atypia
g. atypia lesion
g. cell
g. hyperplasia
g. mastitis
g. neoplasia
g. tissue

glans

Glanzmann
G. disease
G. syndrome
G. thrombasthenia

Glanzmann-Riniker syndrome

glass factor

Glassman forceps

glass-wool filtration

glassy cell carcinoma

glaucoma
congenital g.
primary congenital g.

Glaucon

Glazunov tumor

GLB-1
beta-galactosidase-1

glcyoprotein-producing tumor

Gleeson FloVAC Hi-Flo laparoscopic suction/irrigation system

gleet
vent g.

gleety discharge

Glen Anderson ureteroneocystostomy

Glenard disease

Glenn operation

GLIM
generalized linear interactive modeling

glimepiride

glioependymal cyst

glioma
optic g.
pontine g.

gliomatosis peritonei

gliosis
g. uteri

glipizide

global aphasia

globe cell anemia

globin
α-g.
β-g.

globoid
g. cell cerebral sclerosis
g. cell leukodystrophy

globozoospermia

globulin
anti-D immune g.
anti-Rh gamma g.
antithymocyte g.
corticosteroid-binding g. (CBG)
gamma g.

G

globulin *(continued)*
    hepatitis B immune g. (HBIG, HBIg)
    hepatitis immune g.
    human g.
    hyperimmune serum g.
    immune g.
    immune serum g. (ISG)
    pregnancy-associated g.
    rabies immune g.
    Rh immune g. (RhIg)
    rho(D) human immune g.
    sex hormone-binding g. (SHBG)
    testosterone-estrogen-binding g.
    tetanus immune g.
    thyroid-binding g. (TBG)
    thyroxine-binding g.
    varicella-zoster immune g. (VZIG)
    zoster immune g. (ZIG)
globus hystericus
glomerular
    g. capillary endotheliosis
    g. change
    g. filtrate
    g. filtration
    g. filtration rate (GFR)
    g. insufficiency
    g. proteinuria
    g. sclerosis
glomeruli (*pl. of* glomerulus)
glomerulonephritis
    membranous g.
glomerulopathy
glomerulosclerosis
glomerulotubular
glomerulus, pl. **glomeruli**
glossitis
glossopalatine ankylosis syndrome
glossoptosis
glove
    Biogel Reveal g.
    latex g.
    Micro-Touch Platex medical g.
    powder-free g.
gloving
    double g.
glubionate
    calcium g.
glucagon
Glucamide
glucocorticoid
    g. excess
glucocorticosteroid therapy
glucoglycinuria
Glucola
    G. screen
Glucolet Endcaps

glucometer
    Accu-Chek II g.
    G. Elite diabetes care system
    Glucostar II g.
    G. II
gluconate
    calcium g.
    ferrous g.
    potassium g.
gluconeogenesis
Glucophage
glucosamine-6-sulfate deficiency
glucose
    blood g.
    g. challenge test
    fasting plasma g. (FPG)
    fetal blood g.
    g. formula
    hypertonic g.
    g. intolerance
    g. meter
    postprandial g.
    g. tolerance
    g. tolerance test (GTT)
    urinary g.
glucose-galactose intolerance
glucose-intolerant gravida
glucose-6-phosphatase deficiency (G6PD)
glucose-6-phosphate
    g.-p. dehydrogenase (G6PD)
    g.-p. dehydrogenase deficiency
glucosiduronate
Glucostar II glucometer
Glucostix
glucosuria
Glucotrol
glucuronate pregnanediol
glucuronide
    3α-androstanediol g.
    5α-androstane-3α,17β-diol g. (3α-diol-G)
    androsterone g.
    virilizing 3α-androstanediol g.
Glukor
glutamate
    arginine g.
    g. cycle
    g. formiminotransferase deficiency
glutamyl transpeptidase (GTP)
glutaraldehyde
    g. cross-linked collagen injection
glutaric acidemia
glutathione
    g. peroxidase
    g. S-transferase
    g. S-transferase isoform
glutathionemia

gluteal
  g. free flap
  g. lymph node
gluten-free diet
glyburide
glycemia
glycemic control
glycerin
  g., lanolin and peanut oil
  g. suppository
glycerol
  iodinated g.
  g. kinase (GK)
  g. kinase deficiency (GKD)
glycerophospholipid
glyceryl
  g. guaiacolate
  g. trinitrate (GTN)
glycine
glycinuria
glycocalyx
glycodelin-A
glycogen
glycogenesis
glycogenolysis
  von Gierke g.
glycogenosis
  type 7 g.
glycogen-storage
  g.-s. disease type I–VII
glycohemoglobin
Glycol
  Diprolene G.
glycol
  ethylene g.
glycoprotein
  carbohydrate deficient g. (CDG)
  heterodimeric integral membrane g.
  g. IB/IX
glycopyrrolate
glycorrhachia
glycosaminoglycan (GAG)
glycoside
  cardiac g.
glycosuria
glycosylated hemoglobin
Glyrol
Glyset
GM3
  gangliosidosis G.
  G. gangliosidosis
  G. hematoside sphingolipodystrophy

gm
  gram
GM-CSF
  granulocyte-macrophage colony-
    stimulating factor
GM2 gangliosidosis
GMS
  goniodysgenesis, mental retardation, short
    stature
  GMS syndrome
G-myticin
gnathocephalus
GnRH
  gonadotropin-releasing hormone
  GnRH agonist
  GnRH deficiency
GnRH-facilitated
  G.-f. FSH release
  G.-f. LH release
Goa antigen
goat's milk formula
Goebell-Frangenheim-Stoeckel technique
Goebell procedure
Goebell-Stoeckel-Frangenheim procedure
Goffe colporrhaphy
goiter
  g. development
  fetal g.
  simple colloid g.
goiter-deafness syndrome
Golabi-Ito-Hall syndrome
Golabi-Rosen syndrome (GRS)
gold
  g.-198 (^{198}Au)
  radioactive g.
  g. salt
  g. seed
  g. sodium thiomalate
  g. therapy
Goldblatt-Vilijoen radial ray hypoplasia
Goldenhar
  G. sequence
  G. syndrome
Goldenhar-Gorlin syndrome
Golden sign
Goldie-Coldman hypothesis
Goldstein
  G. disease
  G. sign
Goldston syndrome
Golgi body
Gollop-Wolfgang complex

G

NOTES

Goltz-Gorlin syndrome
Goltz-Peterson-Gorlin-Ravitz syndrome
Goltz syndrome
GoLYTELY
GOMBO
    growth retardation, ocular abnormalities, microcephaly, brachydactyly, oligophrenia
    GOMBO syndrome
Gomco
    G. bell
    G. circumcision clamp
    G. technique
Gomez and López-Hernández syndrome
gonad
    dysgenetic g.'s
    indifferent g.'s
    maternal g.'s
    streak g.'s
    undifferentiated g.
gonadal
    g. agenesis
    g. agenesis syndrome
    g. aplasia
    g. axis
    g. dysgenesis
    g. dysgenesis syndrome
    g. failure
    g. failure, short stature, mitral valve prolapse, mental retardation syndrome
    g. mosaicism
    g. ridge
    g. sex
    g. steroid
    g. steroid suppression
    g. streak
    g. stroma
    g. stromal ovarian tumor
gonadarche
gonadectomy
gonadoblastoma
    ovarian g.
gonadocrinin
gonadoliberin
gonadorelin
    g. acetate
gonadostat
gonadotoxin
gonadotrope
gonadotrophin (*var. of* gonadotropin)
gonadotrophin-releasing hormonelike protein
gonadotrophin-resistant ovary syndrome
gonadotropic deficiency
gonadotropin, gonadotrophin
    g. agonist stimulation test (GAST)

beta-human chorionic g. (beta-hCG, beta-HCG)
chorionic g. (CG)
endogenous g.
exogenous g.
human chorionic g. (hCG, HCG)
human menopausal g. (hMG, HMG)
g. level
nicked free beta subunit of human chorionic g.
pituitary g.
pulsatile human menopausal g.
g. pulsatile release
g. regulation
g.-releasing hormone-associated peptide (GAP)
g. secretion
g. secretion inhibitor
urinary chorionic g. (UCG)
urinary menopausal g.
gonadotropin-induced ovarian hyperstimulation
gonadotropin-releasing
    g.-r. hormone (GnRH)
    g.-r. hormone agonist
    g.-r. hormone analogue
    g.-r. hormone antagonist
    g.-r. hormone-like peptide
gonadotropin-resistant testis
Gonal-F
gonane
Gonic
gonidiaformans
    Fusobacterium g.
goniodysgenesis
    g., mental retardation, short stature (GMS)
    g., mental retardation, short stature syndrome
gonoblennorrhea
gonococcal
    g. conjunctivitis
    g. infection
    g. ophthalmia
gonococcus
Gono Kwik test
gonorrhea
    g. culture
    disseminated g.
gonorrheal
    g. cervicitis
    g. ophthalmia
    g. salpingitis
gonorrhoeae
    Neisseria g.
Gonozyme test
Gonzales blood group

good
    G. Nature formula
    G. Start HA formula
Goodell
    G. dilator
    G. sign
Goodman syndrome
Goodpasture syndrome
gooseneck deformity
Gordan-Overstreet syndrome
Gordon syndrome
Gore-Tex
    G.-T. Soft Tissue Patch
Gore-Tex surgical membrane
Gorlin-Goltz syndrome
Gorlin-Psaume syndrome
Gorlin syndrome 1, 2
goserelin
    g. acetate
    g. acetate implant
gossypiboma
gossypol
Gott-Balfour blade
Gott-Harrington blade
Gott malleable retractor
Gottron sign
Gott-Seeram blade
Gougerot-Carteaud syndrome
Goulet retractor
gouty diathesis
Gower
    hemoglobin G.-1, -2
Gowers sign
gown
    Barrier g.
    Exami-Gown g.
G6PD
    glucose-6-phosphatase deficiency
    glucose-6-phosphate dehydrogenase
GPMAL
    gravida, para, multiple births, abortions, live births
G-protein-coupled receptor
graafian
    g. follicle
    g. vesicle
gracile bone dysplasia
gracilis
    g. flap neovagina
    g. flap technique
    g. muscle

graciloplasty
    dynamic g.
Graddle forceps
grade
    g. 1 germinal matrix hemorrhage
    placental g.
Gradenigo syndrome
gradient
    A-a g.
    albumin g.
    density g.
    g. recalled acquisition in the steady state (GRASS)
grading
    histopathological g.
    placental g.
    tumor g.
Graefenberg ring
Graefe-Usher syndrome
Grafco breast pump
graft
    allogenic fetal g.
    bilateral myocutaneous g.
    full-thickness skin g.
    heterologous g.
    isogeneic g.
    Martius g.
    polytetrafluoroethylene g.
    sacrocolpopexy g.
    seromuscular intestinal patch g.
    split-thickness g.
graft-versus-host
    g.-v.-h. disease (GVHD)
    g.-v.-h. reaction
Graham
    G. Steell murmur
    G. syndrome
Graham-Rosenblith scale
gram (gm)
    oxy-CR g.
    G. stain
Gram-negative, gram-negative
    G.-n. bacilli
    G.-n. bacteria
    G.-n. cocci
    G.-n. organism
    G.-n. pneumonia
    G.-n. rods
Gram-positive, gram-positive
    G.-p. bacilli
    G.-p. bacteria
    G.-p. cocci

G

NOTES

Gram-positive *(continued)*
  G.-p. organism
  G.-p. rods
Gram-Weigert stain
grand
  g. climacteric
  g. mal
  g. mal epilepsy
  g. multipara
  g. multiparity
  g. pregnancy
granddad syndrome
grandmother theory
GraNee needle
Granger sign
granisetron
granny knot
Grant syndrome
Grant-Ward operation
granular
  g. cell myoblastoma
  g. lung
  g. urethritis
granulation tissue
granule
  argyrophilic g.
  Birbeck g.
  cortical g.
  E.E.S. G.'s
  Fordyce g.
  sulfur g.
granulocyte colony-stimulating factor (G-CSF)
granulocyte-macrophage colony-stimulating factor (GM-CSF)
granulocytic leukemia
granuloma, pl. **granulomata**
  g. annulare
  g. gravidarum
  g. inguinale
  Langhans giant cell g.
  pyogenic g.
  sperm g.
  telangiectatic g.
granulomatosis
  Wegener g.
granulomatous
  g. calcification
  g. colitis
  g. infection
  g. mastitis
  g. salpingitis
granulosa
  g. cell tumor
  g. lutein cell
  membrane g.
granulosa-stromal cell tumor
granulosa-theca cell tumor

grape mole
graphesthesia
grasp
  finger g.
  pincer g.
  g. reflex
grasper
  Lion's Claw g.
  MetraGrasp ligament g.
  Polaris reusable g.
grasping
  fetal g.
GRASS
  gradient recalled acquisition in the steady state
  GRASS MRI
  GRASS MRI technique
grating sound
Graves
  G. bivalve speculum
  G. disease
gravid
  g. uterus
gravida (G)
  glucose-intolerant g.
  g., para, multiple births, abortions, live births (GPMAL)
gravidarum
  chorea g.
  epulis g.
  hydrops g.
  hyperemesis g.
gravidic
  g. retinitis
  g. retinopathy
gravidism
graviditas
  g. examnialis
  g. exochorialis
gravidity
Gravindex test
gravis
  icterus g.
  myasthenia g.
  neonatal myasthenia g.
Gravlee jet washer
Gravol
gray
  g. baby syndrome
gray-scale ultrasonography
great
  g. arteries
  g. vessel
great-grand multipara
Grebe
  G. chondrodysplasia
  G. dysplasia

green
> g. hemoglobin

Greenfield
> G. disease
> G. filter

Green uterine curette
greeting spasm
Gregersen U-elevator
Greig
> G. cephalopolysyndactyly syndrome (GCPS)
> G. syndrome

Greulich and Pyle bone age
Greven forceps
grid
gridiron incision
Grifulvin V
grimace
Grimelius stain
Grisactin
Griscelli syndrome
griseofulvin
Grisolle sign
Grisovin
Gris-PEG
groin
> g. dissection
> g. hernia

grommet
groove
> Harrison g.
> neural g.
> primitive g.
> g. sign

Groshong catheter
grotesque features
ground-glass appearance
group
> Adhesion Scoring G. (ASG, ASG system, ASG system)
> g. A streptococcus
> blood g.
> g. B streptococcus (GBS)
> Cartwright blood g.
> g. C autosome
> determinant g.
> g. D streptococcus
> fentanyl g.
> Gonzales blood g.
> Kidd blood g.
> Lewis blood g.
> linkage g.

> Lutheran blood g.
> private blood g.
> Rh blood g.
> streptococcal g.
> support g.

grouping
> blood g.

growth
> bacterial g.
> g. discordancy
> divergent fetal g.
> epiphyseal g.
> extracapsular g.
> g. factor (GF)
> g. failure-pericardial constriction syndrome
> fetal g.
> fetal cellular g.
> g. hormone (GH)
> g. hormone deficiency
> g. hormone release
> g. hormone-releasing hormone (GH-RH)
> g. hormone-secreting adenoma
> g. hormone stimulation test (GHST)
> longitudinal g.
> nonestrogen-regulated g.
> placental g.
> g. problem
> pubertal g.
> g. retardation
> g. retardation, ocular abnormalities, microcephaly, brachydactyly, oligophrenia (GOMBO)
> g. retardation, ocular abnormalities, microcephaly, brachydactyly, oligophrenia syndrome
> g. retardation, small and puffy hands, eczema syndrome
> retarded fetal g.
> skeletal g.

growth-adjusted sonographic age (GASA)
growth-discordant twins
growth-like factor
growth-retarded fetus
GRS
> Golabi-Rosen syndrome

Grubben syndrome
Gruber syndrome
Grünfelder reflex

G

NOTES

grunting
> flaring and g.
> g. and flaring
> g. respirations

GS
> gestational sac

GSI
> genuine stress incontinence

G-spot

GSS
> gestational sac size

GTD
> gestational trophoblastic disease

GTN
> gestational trophoblastic neoplasia
> glyceryl trinitrate

G-to-T transversion mutation

GTP
> glutamyl transpeptidase
> guanosine triphosphate
> guanosine 5'-triphosphate
> guanyltriphosphate

GTT
> gestational trophoblastic tumor
> glucose tolerance test

guaiacolate
> glyceryl g.

guaiac test

guaifenesin

guanabenz

guanadrel

guanethidine

guanfacine

guanine

guanosine
> g. triphosphate (GTP)
> g. 5'-triphosphate (GTP)

guanyldiphosphate (GDP)

guanyltriphosphate (GTP)

Guardian DNA system

gubernaculum

GUD
> genital ulcer disease

Guerin-Stein syndrome

guidance
> CT g.
> fluoroscopic g.
> ultrasound g.

guide
> Pilot suturing g.
> trocar g.

guidelines
> Bethesda system g.

Guidelines for Adolescent Preventive Services (GAPS)

Guillain-Barré syndrome

gulf
> Lecat g.

gumma

gun
> Bard Biopty g.
> Cobe g.
> coring biopsy g.

Günther disease

Gurrieri syndrome

GUSB
> beta-glucuronidase
> > GUSB deficiency
> > GUSB gene
> > > beta-glucuronidase
> > GUSB locus
> > > beta-glucuronidase

gustatory lacrimation

Gustavson syndrome

Guthrie
> G. card
> G. muscle
> G. test

gut suture

guttata
> parapsoriasis g.
> g. parapsoriasis
> g. psoriasis

gutter
> paracolic g.
> pelvic g.

GVBD
> germinal vesicle breakdown

GVHD
> graft-versus-host disease

GV oocytes

GWAFD
> Genée-Wiedemann acrofacial dysostosis

G-well

gymnast's wrist

GYN
> gynecologic
> gynecologist
> gynecology

gynadromorph

gynandroblastoma

gynatresia

Gynecare
> G. Thermachoice uterine balloon therapy system
> G. Verascope Hysteroscopy System

gynecic

gynecogenic

gynecography

gynecoid
> g. obesity
> g. pelvis

gynecologic, gynecological (GYN)
> g. cancer
> g. cancer patient
> g. carcinoma

g. history
g. oncology
gynecologist (GYN)
gynecology (GYN)
adolescent g.
pediatric g.
gynecomastia
Gynecort Topical
Gyne-Lotrimin
G.-L. Vaginal
Gyne-Sulf
gyniatrics
gyniatry
Gynogen
G. L.A.
G. L.A. Injection

gynogenetic
Gynol II
Gynol II contraceptive jelly
gynopathy
gynoplasty, gynoplastics
GynoSampler
G. endometrial aspirator
G. endometrial sampling device
Gynoscann
Gynos perineometer
gyral
g. abnormality
g. malformation
gyrata
cutis verticis g.

NOTES

G

H2O syndrome
HA
 hyperandrogenic anovulation
 hypoplastic aorta
Haase rule
HABA
 hydroxyazobenzoic acid
 HABA binding test
habit
 bladder h.
 bowel h.
 eating h.
 sexual h.
 sleeping h.
HabitEX smoking cessation system
habitual
 h. aborter
 h. abortion
habituation
habitus
 body h.
 Buddha-like h.
 fetal h.
 h. of fetus
HAC
 hexamethylmelamine with doxorubicin
  and cyclophosphamide
 human artificial chromosome
HACE
 high-altitude cerebral edema
Hacker hypospadias
Hadlock equation
Haemophilus
 H. ducreyi
 H. influenzae
 H. influenzae type B (HIB, Hib)
 H. pertussis
 H. pertussis vaccine (HPV)
 H. vaginalis
 H. vaginitis
Hagedorn
 neutral protamine H. (NPH)
Hageman factor
Hahn sign
Haig-Fergusson forceps
Haight baby retractor
Hailey-Hailey disease
hair
 h. bud
 h. cortex keratin gene type II
 h. growth phase
 h. loss
 mental retardation, polydactyly,
  phalangeal hypoplasia, syndactyly,
  unusual face, uncombable h.

 moniliform h.
 h. monster
 pubic h.
 sexual h.
 terminal h.
 vellus h.
HAIR-AN
 hyperandrogenism, insulin resistance,
  acanthosis nigricans
 HAIR-AN syndrome
hairball
hair-brain syndrome
hairless pseudofemale
hairpin vessel
hairy tongue
Hajdu-Cheney syndrome
Hakim-Adams syndrome
Hakim syndrome
Halban
 H. culdoplasty
 H. fashion
 H. syndrome
Halbrecht syndrome
halcinonide
Haldol
half-Fourier acquisition single-shot turbo
 spin-echo sequence
half-life
half-strength formula
half-value layer (HVL)
Halle
 H. infant nasal speculum
 H. point
Hallermann-Streiff
 H.-S. dyscephalia
 H.-S. syndrome
Hallermann-Streiff-François syndrome
Hallermann syndrome
Hallervorden-Spatz
 H.-S. disease
 H.-S. syndrome
Hallgen syndrome
Hallopeau-Siemens syndrome
Hall-Pallister syndrome
Hall-Riggs syndrome
Hall syndrome 1, 2
hallux valgus
halo
 fatty h.
 h. nevus
 osteopenic h.
 h. sign
 h. sign of hydrops
halobetasol

H

Halog
  H.-E
halogen acne
haloperidol
Halotestin
  Cytoxan, Adriamycin, fluorouracil,
  tamoxifen, H. (CAFTH)
halothane
Halpern syndrome
Halsted
  H. mastectomy
  H. mosquito forceps
  H. operation
halstedian concept of tumor spread
Haltran
hamartoblastoma
  congenital hypothalamic h.
  hypothalamic h.
  renal, anus, lung, polydactyly, h.
  (RALPH syndrome, RALPH,
  RALPH syndrome)
hamartoma
  ectoneurodermal h.
  mesenchymal h.
  neuroectodermal h.
hamartomatosis
  hereditary multiple system h.
hamartoneoplastic syndrome
hamartopolydactyly syndrome
Hamel syndrome
Ham F10 medium
Hamilton method
Hamman-Rich syndrome
hammer
  Quisling h.
  h. toe
hammock
  Mersilene gauze h.
Hamou
  H. colpomicrohysteroscope
  H. contact microhysteroscope
  H. hysteroscope
  H. microcolpohysteroflator
  H. Micro-Hysteroflator
  H. technique
hamster
  h. egg penetration assay
  h. ovum
  h. test
hand
  ape h.
  cleft h.
  fisting of h.'s
  h. and head presentation
  lobster-claw h.
  mitten h.
  vaginal h.
  h. ventilation

hand-foot-genital syndrome
hand-foot-mouth syndrome
hand-foot syndrome
hand-foot-uterus syndrome
handpiece
  ClearCut 2 electrosurgical h.
  PhotoDerm PL h.
Hand-Schüller-Christian
  H.-S.-C. disease
  H.-S.-C. syndrome
Hanely-McDermitt pelvimeter
Haney clamp
hanging-drop test
Hanhart syndrome
Hanks dilator
Hansen disease
H$_2$-antagonist
HAODM
  hypoplasia of anguli oris depressor
  muscle
HAPE
  high-altitude pulmonary edema
haploid
  h. cell
  h. set
  h. sperm
  h. spermatozoon
haploinsufficiency
haplotype
  maternal HLA h.
happy puppet syndrome
hapten
haptoglobin
HARD
  hydrocephalus, agyria, retinal dysplasia
  HARD syndrome
HARD+/-E
  hydrocephalus, agyria, retinal dysplasia
  with or without encephalocele
  syndrome
Hardikar syndrome
Hardy-Weinberg
  H.-W. equilibrium
  H.-W. formula
  H.-W. law
harelip
harlequin
  h. color change
  h. fetus
  h. ichthyosis
  h. reaction
  h. sign
Harmonic scalpel
harness
  Pavlik h.
  Wheaton Pavlik h.

Harpenden
- H. calipers
- H. stadiometer

Harrington retractor

Harris
- H. growth arrest line
- H. uterine injector (HUI)

Harris-Kronner uterine manipulator-injector (HUMI)

Harrison
- H. groove
- H. method

Harrod syndrome

Hart
- H. line
- H. syndrome

Hartmann solution

Hartman sign

Hartnup disease

Hashimoto thyroiditis

hashish

Hasson cannula

hat
- measuring h.

Hata phenomenon

hatching
- assisted h. (AH)
- assisted zonal h. (AZH)
- blastocyst h.

HATT
- hemagglutination treponemal test

Haugh forceps

Haultain operation

Hautain uterine inversion

Haverhill fever

Hawk-Dennen forceps

Hawkins breast localization needle

Hawthorne effect

hay fever

Hayflick
- Wistar Institute Susan H. (WISH)

Hay-Wells syndrome

hazy lung

HB
- hepatitis B
  - Recombivax HB

Hb
- hemoglobin

HBAg
- hepatitis B antigen

HBcAb
- hepatitis B core antibody

HBeAb
- hepatitis Be antibody

HbF
- fetal hemoglobin

HbH disease-mental retardation syndrome

HBIG, HBIg
- hepatitis B immune globulin
- hepatitis B immunoglobulin

HbP
- primitive (fetal) hemoglobin

HBS
- AT3 type II HBS
- HBS mutation

HbS
- hemoglobin S
- sickle cell hemoglobin

HBsAb
- hepatitis Bs antibody

HBsAg
- hepatitis B surface antigen

HbsC
- sickle cell hemoglobin C

HbSS
- homozygosity for hemoglobin S

HbS-Thal
- hemoglobin S-thalassemia
- sickle thalassemia

HBV
- hepatitis B vaccine
- hepatitis B virus

HC
- head circumference
  - Prevex HC

HCC
- hepatocellular carcinoma

HCFA
- Health Care Financing Administration

hCG, HCG
- human chorionic gonadotropin
  - free beta hCG
  - OvuDate hCG
  - Pro-Step hCG
  - Tandem Icon II hCG
  - Test Pack hCG

hck protooncogene

HCl
- hydrochloride
  - cefepime HCl
  - Cleocin HCl
  - fluoxetine HCl
  - ondansetron HCl

NOTES

H

HCl (*continued*)
    oxytetracycline HCl
    phenazopyridine HCl
    ropivacaine HCl
    sertraline HCl
    valacyclovir HCl
    venlafaxine HCl
Hct
    hematocrit
    pretransfusion Hct
HCTZ
    hydrochlorothiazide
HCV
    hepatitis C virus
HDC-ABMT
    high-dose chemotherapy with autologous
    bone marrow transplantation
HDI 3000 ultrasound
HDL
    high-density lipoprotein
    HDL-cholesterol
HDN
    hemolytic disease of the newborn
head
    breech h.
    h. circumference (HC)
    h. circumference/abdominal
      circumference ratio
    h. compression
    dolichocephalic h.
    engaged h.
    fetal h.
    hourglass h.
    molding of h.
    h. pole
    h. presentation
    H. reflex
    h. and tail fold
    transillumination of h.
    h. ultrasound
headache
    analgesic-rebound h.
    migraine h.
    preeclampsia h.
    spinal h.
    vascular h.
head:body ratio
headlight
    Keeler fiberoptic h.
head-to-head sperm agglutination (H-H)
head-to-tail sperm agglutination (H-T)
health
    H. Care Financing Administration
      (HCFA)
    h. maintenance organization (HMO)
**HealthCheck One-Step One Minute
pregnancy test**

Healthdyne
    H. apnea monitor
    H. oximeter
    H. ventilator
Heaney
    H. clamp
    H. curette
    H. forceps
    H. needle holder
    H. operation
    H. technique
Heaney-Ballantine forceps
Heaney-Hyst
    H.-H. forceps
    H.-H. retractor
Heaney-Simon retractor
**hearing loss, mental deficiency, growth
retardation, clubbed digits, EEG
abnormalities syndrome**
hearing-loss nephritis syndrome
heart
    h. block
    crisscross h.
    cushion defect of h.
    dextroposition of h.
    h. disease
    H. and Estrogen/Progestin
      Replacement Study (HERS)
    h. failure
    fetal h.
    hypoplasia of left h. (HLL)
    midline h.
    h. rate
    h. rate monitoring
    h. sounds
    h. valve replacement
heartburn
heart-hand syndrome
heart-shaped
    h.-s. pelvis
    h.-s. uterus
heat
    h. loss
    prickly h.
    h. transfer mechanism
heavy chain
heavy-ion
    h.-i. irradiation
    h.-i. mammography
hebetic
HEC
    human endothelial cell
Hecht pneumonia
heelstick
heel-to-ear maneuver
HEF
    human embryo fibroblast

Hegar
> H. dilator
> H. sign

height
> fundal h.
> midparental h.
> symphysis-fundus h.
> h. table

Heineke-Mikulicz pyloroplasty
Heinz
> H. body
> H. body anemia

Heinz-body hemolytic anemia
HeLa
> Henrietta Lacks
> HeLa cell

helcomenia
Helicobacter pylori
> PYtest for H. p.

helix, pl. **helices**
> alpha h.
> double h.
> H. endocervical curette
> h. termination peptide
> H. uterine biopsy curette
> Watson-Crick h.

heliX
> h. knot pusher

Helixate
helix-loop-helix transcription
Hellendall sign
Heller
> H. myotomy
> H. test

Heller-Belsey operation
Heller-Döhle disease
Heller-Nissen operation
Hellin law
Hellin-Zeleny law
HELLP
> hemolysis, elevated liver enzymes, and
> low platelet count
> HELLP syndrome

Helmex
helminth
helminthic disease
helper
> h. factor
> h. T cell

Hemabate

Hemaflex sheath
hemagglutinating inhibition antibody (HIA)
hemagglutination
> h. inhibition (HI)
> h. treponemal test (HATT)
> *Treponema pallidum* h. (TPHA)

hemagglutinin
hemagogue
hemangiectasia hypertrophica
hemangiectatic hypertrophy of Parkes-Weber
hemangioblastoma
hemangioma
> cavernous h.
> epidural h.
> infantile hepatic h.
> macular h.
> port-wine h.
> strawberry h.
> vulvar h.

hemangioma-thrombocytopenia syndrome
hemangiomatosis
hemangiomatous branchial clefts-lip pseudocleft syndrome
hemarthrosis
hematemesis
Hematest
> H. positive
> H. test

hematocele
> pelvic h.
> pudendal h.

hematocephalus
hematochezia
hematocolpometra
hematocolpos
hematocrit (Hct)
> mean menstrual cycle h.
> posttransfusion h.
> spun h.

hematogenously
hematologic
> h. disorder
> h. neoplasia

hematological change
hematology
hematoma
> axillary h.
> h. drainage
> extradural h.
> follicular h.

NOTES

H

hematoma *(continued)*
 infected cuff h.
 interstitial and loculated h.
 intrauterine h.
 puerperal h.
 retroperitoneal h.
 subchorionic h.
 subdural h.
 sublingual h.
 submental h.
 umbilical cord h.
 uterine h.
 vaginal h.
 vulvar h.
hematometra
hematometrocolpos
hematomphalocele
hematopoiesis
 extramedullary h.
hematopoietic
 h. disorder
 h. stem cell (HSC)
 h. stem cell transplantation
 h. system stimulator
hematosalpinx
hematotrachelos
hematoxylin
hematuria
 congenital hereditary h.
 macroscopic h.
 transient h.
hematuria-nephropathy-deafness syndrome
heme
 h. metabolism disorder
 h. positive
hemelytrometra
hemiacardius
hemiacephalus
hemiagnathia
hemianencephaly
hemiatrophy
hemibody irradiation
hemicardia
hemicephalia
hemicervix
hemicolectomy
hemicone biopsy
hemicrania
hemifacial microsomia (HM)
hemi-Fontan procedure
hemignathia and microtia syndrome
hemihypertrophy
hemihysterectomy
hemimelia
 dysplasia epiphysealis h.
 transverse h.
hemimelus

hemin
heminasal
 h. aplasia
 h. hypoplasia
hemipagus
hemiparesis
hemiplegia
hemisection uterine morcellation technique
hemiuterus
hemivertebra
hemivulvectomy
hemizona assay
hemizygous
hemoblastic leukemia
Hemoccult II test
hemochorial
 h. placenta
hemochorioendothelial placentation
hemochromatosis
 HFE gene for h.
hemoconcentration
HemoCue
 H. blood glucose analyzer
 H. blood glucose system
 H. blood hemoglobin analyzer
 H. blood hemoglobin system
 H. glucose test
 H. hemoglobin photometer
 H. hemoglobin test
 H. microcurette
hemocytometer
 Neubauer h.
hemodialysis
 continuous venovenous h. (CVVHD)
hemodilution
hemodynamic
 h. monitoring
hemodynamics
 maternal central h.
 uterine h.
Hemofil M
hemofilter
hemofiltration
 continuous arteriovenous h. (CAVH)
hemoglobin (Hb)
 h. A
 h. A_{Ic} determination
 h. Bart
 h. C disease
 h. E
 h. F
 fetal h. (HbF)
 glycosylated h.
 h. Gower-1, -2
 green h.

h. H
h. H disease
h. H disease-mental retardation
   syndrome
h. H related mental retardation
h. level
h. M disease
mean corpuscular h. (MCH)
h. Portland
primitive (fetal) h. (HbP)
reduced h.
h. S (HbS)
h. SC
serum free h.
sickle cell h. (HbS)
h. SS
h. S-thalassemia (HbS-Thal)
unstable h.
variant h.
hemoglobinemia
hemoglobinopathy
hemoglobin-oxygen dissociation
hemoglobinuria
paroxysmal nocturnal h.
hemolysin
hemolysis
h., elevated liver enzymes, and
   low platelet count (HELLP)
microangiopathic h.
hemolytic
  ABO h. disease of the newborn
h. anemia
h. disease of the newborn (HDN)
hemolytic-uremic syndrome
hemolyzed specimen
hemometra
Hemonyne
Hemopad
hemopathy
maternal h.
hemoperitoneum
ovulation-associated h.
hemophilia
h. A, B
hemophiliac
hemopneumothorax
hemoptysis
catamenial h.
HemoQuant assay
hemorrhage
accidental h.
antepartum h.

concealed h.
cortical h.
epidural h.
external h.
fetal h.
fetal-maternal h.
fetomaternal h. (FMH)
grade 1 germinal matrix h.
intracranial h. (ICH)
intrapartum h.
intraventricular h. Grade I–IV
  (IVH)
periventricular-intraventricular h.
placental h.
postcesarean h.
postpartum h. (PPH)
pulmonary h.
scleral h.
sternocleidomastoid h.
subarachnoid h.
subchorionic h.
subdural h.
subependymal h. (SEH)
subgaleal h.
transplacental h.
trauma-related acute pelvic h.
unavoidable h.
hemorrhagic
h. cystitis
h. disease
h. disease of the newborn
h. endovasculitis
h. familial nephritis
h. fever
h. hereditary nephritis
h. scurvy
h. shock
h. telangiectasia
hemorrhagica
purpura h.
hemorrhagicum
corpus h.
hemorrhoid
hemorrhoidal
h. artery
h. nerve
hemosalpinx
hemosiderin laden macrophage
hemosiderosis
pulmonary h.
hemospermia
hemostasis

NOTES

H

hemostat
> Crile h.
> curved h.
> Endo-Avitene microfibrillar
>   collagen h.

hemostatic staple line
Hemotene
Henderson-Hasselbalch equation
Henle
> loop of H.

Hennekam
> H. lymphangiectasia-lymphedema
>   syndrome

Henoch disease
Henoch-Schönlein purpura
Henrietta Lacks (HeLa)
Henschke colpostat
henselae
> *Bartonella h.*

Hepalean
heparin
> calcium h.
> h. challenge test
> h. embryopathy
> h. lock (hep lock)
> low-dose h.
> minidose h.
> prophylactic h.
> h. sodium
> unfractionated h.

heparin-binding site mutation
heparinization
heparinized lactated Ringer
heparitinuria
hepatic
> h. disorder
> h. ductular hypoplasia
> h. ductular hypoplasia-multiple
>   malformations syndrome
> h. fibrosis
> h. focal nodular hyperplasia
> h. infantilism
> h. metastasis
> h. necrosis
> h. neoplasm
> h. porphyria
> h. pregnancy
> h. rupture

hepatitis
> h. A
> h. A immunization
> h. A virus
> h. B (HB)
> h. B antigen (HBAg)
> h. B core antibody (HBcAb)
> h. Be antibody (HBeAb)
> h. B immune globulin (HBIG,
>   HBIg)

> h. B immunoglobulin (HBIG,
>   HBIg)
> h. Bs antibody (HBsAb)
> h. B surface antigen (HBsAg)
> h. B vaccine (HBV)
> h. B virus (HBV)
> h. C
> chronic h.
> h. C virus (HCV)
> h. D
> delta h.
> h. E
> fulminant h.
> h. G
> giant cell h.
> herpes h.
> h. immune globulin
> infectious h.
> maternal h.
> neonatal cholestatic h.
> non-A, non-B h. (NANBH)
> h. screening
> serum h.
> viral h.

hepatobiliary
hepatoblastoma
hepatocellular
> h. carcinoma (HCC)
> h. damage

hepatocyte growth factor (HGF)
hepatoerythropoietic porphyria
hepatofacioneurocardiovertebral
>   **syndrome**
hepatolenticular
> h. degeneration

hepatoma
hepatomegaly
hepatorenal
hepatosplenomegaly
Hep-B-Gammagee
HEPES
> N-[2-hydroxyethyl]piperazine N′-[2-
>   ethanesulfonic acid]

hep lock
> heparin lock

Heprofile ELISA test
Heptavax-B
HER-2/neu protooncogene
Heraeus LaserSonics InfraGuide
herald bleed
herbal medicine
herbarum
> *Cladosporium h.*

herbicide
Herbst registry
Herceptin
Hercules
> infant H.

hereditaria
    adynamia episodica h.
    alopecia h.
    anemia hypochromica
      sideroachrestica h.
    arthroophthalmophathia h.
    atrophia bulborum h.
    keratitis fugax h.
    porphyria cutanea tarda h.

hereditary
    h. abductor vocal cord paralysis
    h. agenesis of corpus callosum
    h. benign chorea
    h. benign intraepithelial dyskeratosis
      syndrome
    h. blepharophimosis, ptosis,
      epicanthus inversus syndrome
    h. bone dysplasia
    h. bullous dystrophy
    h. bullous skin dystrophy, macular
      type
    h. chondrodysplasia
    h. clubbing
    h. cutaneomandibular polyoncosis
    h. deforming chondrodystrophy
    h. disease
    h. dysplastic nevus syndrome
    h. ectodermal polydysplasia
    h. edema
    h. epithelial dysplasia of the
      retinae
    h. expansile polyostotic osteolytic
      dysplasia
    h. familial congenital nephritis
    h. hematuria syndrome
    h. interstitial pyelonephritis
    h. macular epidermolysis bullosa
    h. motor and sensory neuropathy
      II (HMSN II)
    h. motor sensory neuropathy II-
      deafness-mental retardation (MSN)
    h. motor sensory neuropathy II-
      deafness-mental retardation
      syndrome
    h. multiple system hamartomatosis
    h. nephritis deafness-abnormal
      thrombogenesis syndrome
    h. nephritis-deafness syndrome
    h. nonspherocytic anemia
    h. oligophrenic cerebellolental
      degeneration
    h. orotic aciduria

    h. osteochondrodysplasia
    h. osteodysplasia with acroosteolysis
    h. ovarian cancer
    h. polytopic enchondral dysostosis
    h. progressive arthroophthalmopathy
    h. progressive bulbar paralysis with
      deafness
    h. renal agenesis
    h. retinal aplasia
    h. retinal dysplasia
    h. spherocytosis
    h. syphilis
    h. trait
    h. urogenital adysplasia

heredity
    autosomal h.
    sex-linked h.
    X-linked h.

heredoataxia
heredobiologic
heredoconstitutional disease
heredodegeneration
heredodegenerative disease
heredodiathesis
heredofamilial
heredoimmunity
heredolues
heredopathia atactica
heredoretinopathia congenitalis
heredosyphilis
Hering-Breuer reflex
heritability
heritable
    h. coagulopathy

heritage
    Ashkenazi Jewish h.
    ethnic h.
    H. Panel genetic screening test

herkogamy
Hermansky-Pudlak syndrome
hermaphrodite
hermaphroditism
    true h.
    XX h.

hermaphroditismus
hermizona assay
Hernandez syndrome
hernia
    abdominal h.
    Bochdalek h.
    broad ligament h.
    congenital diaphragmatic h. (CDH)

NOTES

H

hernia *(continued)*
    diaphragmatic h.
    epigastric h.
    femoral h.
    groin h.
    hiatal h.
    incarcerated h.
    incisional h.
    inguinal h.
    h. inguinale
    labial h.
    linea alba h.
    Morgagni h.
    paraduodenal h.
    peritoneal h.
    pleuroperitoneal h.
    port site h.
    reducible h.
    retrocecal h.
    retrosternal h.
    Richter h.
    sliding h.
    spigelian h.
    strangulated h.
    transmesenteric h.
    umbilical h.
    h. uteri inguinale
    ventral h.
herpangina
Herp-Check test
herpes
    acute neonatal h.
    h. encephalitis
    genital h.
    h. genitalis
    h. gestationis (HG)
    h. hepatitis
    intrauterine h.
    h. labialis
    neonatal h.
    h. neonatorum
    h. simplex
    h. simplex genitalis (HSG)
    h. simplex virus (HSV)
    h. simplex virus thymidine kinase
     gene
    h. simplex virus type 1 (HSV-1)
    h. simplex virus type 2 (HSV-2)
    syphilis, toxoplasmosis, rubella,
     cytomegalovirus, and h.
     (STORCH)
    h. whitlow infection
    h. zoster
    h. zoster virus (HZV)
Herpesvirus
    H. hominis
    H. suis
    H. varicellae

herpesvirus
    human h. 6 (HHV 6)
herpetic
    h. gingivitis
    h. gingivostomatitis
    h. stomatitis
herpeticum
    eczema h.
herpetiformis
    dermatitis h.
Herplex Liquifilm
Herrick anemia
HERS
    Heart and Estrogen/Progestin
    Replacement Study
Hers disease
Herter infantilism
Hertig-Rock ovum
hertz (Hz)
Hesselbach triangle
hetacillin
hetastarch
    h. solution
heteradelphus
heteralius
heterocephalus
heterochromatic DNA
heterochromatin
heterochromia
heterochromosome
heterodimer
heterodimeric integral membrane
    glycoprotein
heterodisomy
    maternal uniparental h.
heteroduplex
    h. analysis
    h. DNA
heterodymus
heterogamete
heterogeneity
    allelic h.
    locus h.
    loss of h. (LOH)
heterogeneous
heterogenicity
heterograft
heterokaryon
heterologous
    h. graft
    h. insemination
    h. surfactant
    h. twins
    h. uterine sarcoma
heteromorphous
heteroovular
    h. twins
heteropagus

heterophil
    h. antigen
heterophilic
heteroploid
heteroprosopus
heterosexual
    h. precocious puberty
heterosomal aberration
heterosome
heterostyly
heterotaxia syndrome
heterotaxy
    visceral h.
heterotopia
    neuronal h.
    periventricular nodular gray
      matter h.
    subcortical band h.
heterotopic
    h. pregnancy
heterotropic
    h. chromosome
heterotypic
heterotypical chromosome
heterozygosity
heterozygote
    compound h.
    obligate h.
heterozygous
    h. carrier
Heuser membrane
hew mutation
Hexa-Betalin
Hexa-CAF
    hexamethylmelamine, cyclophosphamide,
    doxorubicin, and 5-fluorouracil
hexachloride
    gamma-benzene h.
hexachlorocyclohexane
hexadactyly
    preaxial h.
Hexadrol
Hexalen
hexamethonium chloride
hexamethylmelamine
    h., cyclophosphamide, doxorubicin,
    and 5-fluorouracil (Hexa-CAF)
    h. with doxorubicin and
    cyclophosphamide (HAC)
hexamine
hexaploidy
Hexastat

hexenmilch
hexestrol
Hexit
hexocyclium methylsulfate
hexoprenaline sulfate
hexosaminidase
    h. A deficiency
hexylresorcinol
Heyman capsules
Heyman-Herndon clubfoot procedure
Heyns abdominal decompression
    apparatus
HFEA
    Human Fertilization and Embryology
    Authority
HFE gene for hemochromatosis
HFJ
    high-frequency jet
      HFJ ventilation
      HFJ ventilator
HFJV
    high-frequency jet ventilation
HFO
    high-frequency oscillatory
      HFO ventilation
      HFO ventilator
HFOV
    high-frequency oscillatory ventilation
HFPP
    high-frequency positive pressure
      HFPP ventilation
      HFPP ventilator
HFPPV
    high-frequency positive pressure
    ventilation
HG
    herpes gestationis
      HG factor
H gene
HGF
    hepatocyte growth factor
hGH
    human growth hormone
HGPRT
    hypoxanthine-guanine
    phosphoribosyltransferase deficiency
HGSIL
    high-grade squamous intraepithelial
    lesion
H-H
    head-to-head sperm agglutination
      H-H neonatal shunt

NOTES

H

HHA
> hypogonadotropic hypogonadism-anosmia syndrome

HHHO
> hypotonia, hyperphagia, hypogonadism, obesity
> HHHO syndrome

HHV 6
> human herpesvirus 6

HI
> hemagglutination inhibition
> HI titer

HI-30
> bikinin

HIA
> hemagglutinating inhibition antibody

5-HIAA
> 5-hydroxyindoleacetic acid

21-HIAA
> 21-hydroxyindoleacetic acid

hiatal hernia

hiatus
> h. hernia, microcephaly, nephrosis syndrome
> urogenital h.

HIB, Hib
> *Haemophilus influenzae* type B
> HIB polysaccharide vaccine

hiccup, hiccough
> fetal h.

Hickman catheter

Hicks version

Hi-Cor-1.0 Topical

Hi-Cor-2.5 Topical

hidradenitis suppurativa

hidradenoma
> papillary h.

hidrotic ectodermal dysplasia

HIE
> hypoxic-ischemic encephalopathy

high
> h. altitude perinatal mortality
> h. fetal order
> h. forceps delivery
> h. intrauterine insemination
> h. McCall culdoplasty
> h. molecular weight dextran
> h. stirrups

high-altitude
> h.-a. cerebral edema (HACE)
> h.-a. pulmonary edema (HAPE)

high-calorie
> h.-c. diet
> h.-c. formula

high-contrast Bucky imaging

high-density lipoprotein (HDL)

high-dose
> h.-d. chemotherapy

> h.-d. chemotherapy with autologous bone marrow transplantation (HDC-ABMT)

higher-order
> h.-o. birth
> h.-o. gestation

high-fiber diet

high-frequency
> h.-f. jet (HFJ)
> h.-f. jet ventilation (HFJV)
> h.-f. oscillatory (HFO)
> h.-f. oscillatory ventilation (HFOV)
> h.-f. positive pressure (HFPP)
> h.-f. positive pressure ventilation (HFPPV)
> h.-f. ventilator

high-fructose formula

high-glucose formula

high-grade
> h.-g. cervical dysplasia
> h.-g. squamous intraepithelial lesion (HGSIL, HSIL)

highly purified FSH

high-pitched cry

high-powered field (hpf)

high-resolution
> h.-r. banding
> h.-r. ultrasonography
> h.-r. ultrasound

high-risk
> h.-r. infant
> h.-r. mother
> h.-r. obstetrician
> h.-r. patient
> h.-r. pregnancy
> h.-r. pregnancy assessment

Hi-Gonavis test

Higoumenakia sign

hilar
> h. cell hyperplasia
> h. cell pathology
> h. cell tumor

Hildebrandt uterine hemostatic forceps

Hillis-DeLee fetoscope

Hillis-Müller maneuver

hindbrain

hindgut

hindwater

Hinton test

hip
> h. bone density
> h. click
> congenital dislocated h. (CDH)
> h. deformity
> h. dysplasia
> h. fracture

Hiprex

Hirschberg test

Hirschsprung
  H. disease
  H. disease, microcephaly, mental
    retardation, characteristic facies
    syndrome
Hirst placental forceps
hirsute woman
hirsutism
  constitutional h.
  familial h.
  hormonal h.
  idiopathic h.
  male-pattern h.
  postmenopausal h.
  h., skeletal dysplasia, mental
    retardation syndrome
His
  bundle of H.
  H. rule
Hismanal
Histalet Forte Tablet
histaminase
histamine
Histantil
Histatan
Hista-Vadrin Tablet
Histerone
histidinemia
histidinuria
histiocyte
histiocytic fibroma
histiocytoma
histiocytosis
  acute disseminated h. X
  Langerhans cell h.
  malignant h.
  sinus h.
  h. X
histocompatibility
  h. gene
  h. locus antigen
Histofreezer cryosurgical system
histogenesis
histoimmunological origin
histoincompatibility
  maternal-fetal h.
histologic
  h. architecture
  h. chorioamnionitis
  h. diagnosis
  h. placental inflammation

histology
  endometrial h.
  proliferative h.
histomorphometry
histone
  h. H1 kinase
histopathological
  h. diagnosis
  h. finding
  h. grading
  h. study
histopathology
histoplasmosis
  pulmonary h.
history
  family h.
  gynecologic h.
  menstrual h.
  obstetric h.
  occupational h.
  reproductive h.
  sexual h.
  urologic h.
histrelin
Histussin NC
Hitachi
  H. EUB 420 digital ultrasound
  H. EUB 405 imaging system
  H. UB 420 digital ultrasound
    system
hitchhiker thumb
Hitzig girdle
HIV
  human immunodeficiency virus
  HIV classification
  HIV test
HIVAGEN test
Hivid
HIV-seropositive
HL
  humerus length
HLA
  human leukocyte antigen
  HLA-A3
  HLA-B
  HLA-B14,DR1
  HLA-compatible fetus
  HLA-D
  HLA-DR
  HLA-matched platelet transfusion
HLAB 5701
HLA-D

NOTES

H

HLHS
      hypoplastic left heart syndrome
      HLHS syndrome
HLI
      human leukocyte interferon
HLL
      hypoplasia of left heart
HM
      hemifacial microsomia
HMC
      hypertelorism, microtia, clefting
      HMC syndrome
HMD
      hyaline membrane disease
      neonatal HMD
hMG, HMG
      human menopausal gonadotropin
      hMG/IUI
HMO
      health maintenance organization
HMSN II
      hereditary motor and sensory neuropathy
      II
HN
      Two-Cal HN
HNU
      human *neu* unit
hobnail cell
Hoboken
      fold of H.
      H. gemmule
      H. nodule
hockey stick incision
Hodge
      H. forceps
      H. maneuver
      H. pessary
Hodgkin
      H. disease
      H. lymphoma
Hoehne sign
HOF
      human oviduct fluid
Hoffbauer cell
Hofmann elimination
Hogben test
Hogness box
Hohn catheter
holandric
      h. gene
      h. inheritance
holder
      Dale Foley catheter h.
hold technique
hole
      Murphy h.

holiday
      drug h.
      weekend drug h.
Holinger
      H. infant bougie
      H. infant bronchoscope
      H. infant esophageal speculum
      H. infant esophagoscope
      H. infant laryngoscope
Hollister collecting device
hollow of sacrum
holmium:yttrium-aluminum-garnet
holoacardius
holoblastic
      h. ovum
hologastroschisis
Hologic
      H. 1000 QDR densitometer
      H. 1000 QDR dual-energy
      absorptiometer
hologynic inheritance
holoprosencephalic proboscis
holoprosencephaly
      alobar h.
      h. anomalad
      familial alobar h.
      lobar h.
holorachischisis
holosystolic murmur
Holter monitor
Holt-Oram
      H.-O. atriodigital dysplasia
      H.-O. syndrome
Holzgreve syndrome
Homans sign
homatropine
HOME
      Home Observation for the Measurement
      of the Environment
home
      h. birth
      h. factor
      H. Observation for the
      Measurement of the Environment
      (HOME)
      h. oxygen
      h. pregnancy test
      h. uterine activity monitor
      (HUAM)
      h. uterine activity monitoring
      (HUAM)
      h. uterine monitoring (HUM)
homeobox (HOX)
      h. 2 gene
homeostasis
      carbohydrate h.
homeostatic lag
homeotic genes

homicide
  intimate partner h.
hominis
  Herpesvirus h.
  Mycoplasma h.
  Poliovirus h.
homochronous inheritance
homocysteine
homocystinemia
homocystinuria
homoduplex DNA
homogamete
homogeneity
  tissue h.
homograft
homokaryon
homolog
HomoloGene
homologous
  h. chromosome
  h. disease
  h. insemination
  h. recombination
  h. uterine sarcoma
homologue
  DAZL1 autosomal h.
homology
homophilic
homosexuality
homotropic inheritance
homotypic
homozygosity for hemoglobin S (HbSS)
homozygote
homozygous
  h. glucose-6-phosphate
    dehydrogenase deficiency
honeymoon cystitis
Honvol
hood
  clitoral h.
  h. mist
  h. O_2
  h. oxygen
  Oxy-Hood oxygen h.
  H. procedure
  Rock-Mulligan h.
  vaginal h.
hook
  Mayo h.
  Miya h.
  tenaculum h.
  h. traction technique

Hooker-Farbes test
hookworm
Hootnick-Holmes syndrome
HOP
  hypothalamic-pituitary-ovarian
Hope resuscitation bag
hordeolum
horizon
  Streeter h.
horizontal transmission of virus
hormonal
  h. abnormality
  h. antineoplastic therapy
  h. contraception
  h. effect
  h. environment
  h. hirsutism
  h. implant
  h. level
  h. pregnancy test tablet
  h. treatment
hormone
  adrenocortical h.
  adrenocorticotropic h. (ACTH)
  antenatal thyrotropin releasing h.
  anterior pituitary-like h.
  antidiuretic h. (ADH)
  antimüllerian h. (AMH)
  h. assay
  atrial natriuretic h.
  bioactive h.
  calcitropic h.
  cancer and steroid h. (CASH)
  chorionic gonadotropic h.
  chorionic growth h.
  circulating h.
  h. complex receptor
  corticotropin-releasing h. (CRH)
  endogenous h.
  exogenous h.
  fetal h.
  follicle stimulating h. (FSH)
  Genentech biosynthetic human
    growth h.
  gonadotropin-releasing h. (GnRH)
  growth h. (GH)
  growth h.-releasing hormone (GH-
    RH)
  human chorionic
    adrenocorticotropic h.
  human growth h. (hGH)

NOTES

H

243

hormone *(continued)*
>human urinary follicle-stimulating h. (hu-FSH)
>Humatrope growth h.
>hypothalamic luteinizing hormone-releasing h.
>inappropriate antidiuretic h. (IADH)
>LATS h.
>luteinizing h. (LH)
>luteinizing hormone-releasing h. (LH-RH)
>lutein-stimulating h. (LSH)
>luteotropic h.
>melanocyte-stimulating h.
>müllerian inhibiting h. (MIH)
>ovarian h.
>parathyroid h. (PTH)
>pituitary h.
>placental h.
>placental growth h. (PGH)
>pregnancy h.
>purified h.
>recombinant follicle-stimulating h. (rFSH)
>recombinant human growth h.
>h. replacement therapy (HRT)
>serum parathyroid h.
>sex h.
>somatrem growth h.
>somatropin growth h.
>steroid h.
>syndrome of inappropriate secretion of antidiuretic h. (SIADH)
>h. therapy
>thyroid h.
>thyroid-stimulating h. (TSH)
>thyrotropic h.
>thyrotropin-releasing h. (TRH)
>tropic h.
>urinary-derived human follicle-stimulating h. (u-hFSH)
>urinary luteinizing h. (uLH)

hormone-receptor complex internalization
hormone-stimulated endometrial change
hormonogenesis
hormonotherapy
horn
>entrapped temporal h.
>frontal h.
>fused frontal h.
>noncommunicating uterine h.
>rudimentary uterine h.
>uterine h.
>h. of uterus

Horner syndrome
horseshoe
>h. fibrosis
>h. placenta

HOS
>hypoosmotic swelling

hospital
>H. Recliner seat
>H. for Sick Children (HSC)

hospitalization
>prolonged delivery h.

host
>h. defense mechanism
>h. response mechanism

hot
>h. biopsy
>h. biopsy forceps
>h. cross bun skull
>h. flash
>h. flush
>h. knife conization

Hottentot apron
24-Hour
>Claritin-D 24-Hour

hour
>Acutrim 16 H.'s
>cubic centimeter per h. (cc/hr)
>h. of sleep (hs)
>24-h. urinary free cortisol

hourglass
>h. head
>h. uterus

housekeeping gene
Howell biopsy aspiration needle
Howell-Jolly body
HOX
>homeobox
>HOX A gene

Ho:YAG laser
Hoyeraal-Hreidarsson syndrome
HP
>Profasi HP

H.P.
>Mission Prenatal H.P.

HPA
>hypothalamic-pituitary-adrenal
>HPA axis

hpf
>high-powered field

hPL, HPL
>human placental lactogen

HPO
>hypothalamic-pituitary-ovarian
>HPO axis

HPV
>*Haemophilus pertussis* vaccine
>human papilloma virus
>HPV-associated lesion
>HPV E7 gene
>HPV type 16 capsid antibody

H-*ras*
    H-*ras* oncogene
    H-*ras* p21 protein
HRHS
    hypoplastic right heart syndrome
HRT
    hormone replacement therapy
      estrogen-only HRT
HS
    hysterosalpingography
      HS mucopolysaccharidosis
H.S.
      Estratest H.S.
hs
    hour of sleep
HSAS
    hydrocephalus due to congenital stenosis
      of aqueduct of Sylvius
HSC
    hematopoietic stem cell
    Hospital for Sick Children
      HSC Scale
HSG
    herpes simplex genitalis
    hysterosalpingography
      HSG tray
HSI
    human seminal (plasma) inhibitor
HSIL
    high-grade squamous intraepithelial
      lesion
HSV
    herpes simplex virus
HSV-1
    herpes simplex virus type 1
HSV-2
    herpes simplex virus type 2
H-T
    head-to-tail sperm agglutination
HTLV
    human T-cell leukemia virus
      HTLV-I
        human T-cell leukemia virus type I
      HTLV-II
        human T-cell leukemia virus type
          II
      HTLV-III
        human T-cell leukemia virus type
          III
H-type fistula

HUAM
    home uterine activity monitor
    home uterine activity monitoring
Hudson T Up-Draft II disposable
   nebulizer
Huffman
    H. infant vaginal speculum
    H. infant vaginoscope
Huffman-Huber
    H.-H. infant urethrotome
    H.-H. infant vaginoscope
hu-FSH
    human urinary follicle-stimulating
      hormone
HuGe
    Human Genome
      HuGe index
HuGE Net
    Human Genome Epidemiology Network
Hughes syndrome
Huguier circle
Huhner test
HUI
    Harris uterine injector
      HUI Mini-Flex
Hulka-Clemens clip
Hulka clip
HUM
    home uterine monitoring
Humain
      Centre d'Etude de
        Polymorphism H.
Humalog
    H. insulin lispro injection
    H. Pen
human
    h. antihemophilic factor
    h. artificial chromosome (HAC)
    h. chorionic adrenocorticotropic
      hormone
    h. chorionic gonadotropin (hCG,
      HCG)
    h. chorionic gonadotropin beta-
      subunit
    h. chorionic gonadotropin level
    h. embryo fibroblast (HEF)
    h. endothelial cell (HEC)
    h. epidermal growth factor-2
      oncogene
    h. EP1 receptor
    h. factor IX complex

NOTES

H

human *(continued)*
  H. Fertilization and Embryology Authority (HFEA)
  H. Gene Expression (HuGe index)
  H. Gene Expression index
  h. gene therapy
  h. genetics
  H. Genome (HuGe)
  H. Genome Epidemiology Network (HuGE Net)
  H. Genome Initiative
  H. Genome Project
  h. globulin
  h. growth hormone (hGH)
  h. herpesvirus 6 (HHV 6)
  h. immunodeficiency virus (HIV)
  h. immunodeficiency virus test
  Insulatard NPH h.
  h. insulin
  h. leukocyte antigen (HLA)
  h. leukocyte interferon (HLI)
  h. lymphocyte antigen-locus DR
  h. menopausal gonadotropin (hMG, HMG)
  h. milk fortifier
  Nabi-HB hepatitis B immune globulin h.
  h. *neu* unit (HNU)
  h. oviduct fluid (HOF)
  h. ovum fertilization test
  h. papilloma virus (HPV)
  h. placental lactogen (hPL, HPL)
  h. seminal (plasma) inhibitor (HSI)
  h. sperm cytosolic factor
  H. Surf surfactant
  h. T-cell leukemia virus (HTLV)
  h. T-cell leukemia virus type I (HTLV-I)
  h. T-cell leukemia virus type II (HTLV-II)
  h. T-cell leukemia virus type III (HTLV-III)
  h. urinary follicle-stimulating hormone (hu-FSH)
  Velosulin H.
Humate-P
Humatin
Humatrope
  H. growth hormone
HumatroPEN
Humegon
humeroradial synostosis
humerus length (HL)
HUMI
  Harris-Kronner uterine manipulator-injector
  HUMI catheter
humidification ventilator

humidifier
  Ohio h.
humoral
  h. antibody
  h. factor
  h. immunity
Humorsol
hump
  buffalo h.
  dowager's h.
  dromedary h.
Humulin
  H. 50/50
  H. 70/30
  H. L
  H. N
  H. Pen
  H. R
  H. U Ultralente
Hünermann disease
Hunner
  H. interstitial cystitis
  H. ulcer
Hunt bipolar forceps
Hunter-Fraser syndrome
hunterian chancre
Hunter-MacMurray syndrome
Hunter-McAlpine
  H.-M. craniosynostosis syndrome
  H.-M. syndrome
Hunter syndrome
Huntington
  H. chorca
  H. disease
Hunt-Reich cannula
Hurler
  H. disease
  H. syndrome
  H. variant
Hurler-like syndrome
Hurler-Pfaundler syndrome
Hurler-Scheie syndrome
hurry
  intestinal h.
Hurst syndrome
Hurthle cell
Hurtig dilator
Hurwitz catheter
husband
  artificial insemination by h. (AIH)
Hutch diverticulum
Hutchinson
  H. sign
  H. syndrome
  H. teeth
  H. triad
Hutchinson-Gilford syndrome
Hutchison triad

Hutinel disease
Hutterite
 H. cerebroosteonephrodysplasia
Huxley respirator
HVF ventilator
HVL
 half-value layer
hyaline
 h. cast
 h. membrane
 h. membrane disease (HMD)
 h. membrane syndrome
 h. myoma degeneration
hyalinosis
 infantile system h.
hyaluronidase
H-Y antigen
Hyate:C
Hybolin
 H. Decanoate
 H. Improved
hybrid
 H. Capture DNA Assay
 h. capture system
 somatic h.
hybridization
 comparative genomic h. (CGH)
 DNA h.
 dot-blot h.
 fluorescence in situ h. (FISH)
 genomic in situ h.
 papillomavirus h.
 polar body in-situ h.
 in situ h.
 in situ nucleic acid h.
hybridoma technique
Hybritech
Hycamtin
hyclate
 doxycycline h.
Hycodan
Hycort Topical
HYCX
 hydrocephalus due to congenital stenosis
  of aqueduct of Sylvius
hydantoin syndrome
hydatid
 h. cyst
 h. cyst of Morgagni
 h. disease
 h. mole

 h. polyp
 h. pregnancy
hydatidiform
 h. change
 h. mole
Hyde-Forster syndrome
Hyderm
hydralazine
hydramnion, hydramnios
 idiopathic h.
hydranencephaly
Hydrate
hydrate
 chloral h.
hydration
 maternal h.
Hydrazide
Hydrea
hydremica
 plethora h.
hydrencephalocele
hydrencephalomeningocele
hydriodic acid
hydroa
 h. aestivale
 h. gestationis
 h. puerorum
 h. vacciniforme
hydroalcoholic
hydrocele
 h. feminae
 Maunoir h.
 h. muliebris
 Nuck h.
hydrocelectomy
hydrocephalic
 h. lissencephaly
hydrocephalocele
hydrocephaloid
 h. disease
hydrocephalus
 h. agyria, retinal dysplasia (HARD)
 h. agyria, retinal dysplasia with or
  without encephalocele syndrome
  (HARD+/-E)
 h. due to congenital stenosis of
  aqueduct of Sylvius (HSAS,
  HYCX)
 LICAM gene for X-linked h.
 normal pressure h. (NPH)

NOTES

H

hydrocephalus *(continued)*
    h. skeletal anomalies, mental disturbances syndrome
    X-linked h.
hydrocephalus-cerebellar agenesis syndrome
hydrocephaly
    h. with features of VATER
Hydro-chlor
hydrochloride (HCl)
    arginine h.
    butriptyline h.
    chlorcyclizine h.
    ciprofloxacin h.
    clonidine h.
    cyclopentolate h.
    cycrimine h.
    cyproheptadine h.
    dihydromorphinone h.
    dobutamine h.
    dopamine h.
    doxapram h.
    doxorubicin h.
    esmolol h.
    ethopropazine h.
    isoprenaline h.
    mechlorethamine h.
    meclizine h.
    meperidine h.
    methacycline h.
    methadone h.
    methamphetamine h.
    methdilazine h.
    methixene h.
    methoxamine h.
    methylphenidate h.
    metoclopramide h.
    mitoxantrone h.
    Mustargen H.
    naftifine h.
    naloxone h.
    nylidrin h.
    opipramol h.
    paroxetine h.
    phenazopyridine h.
    piperidolate h.
    propranolol h.
    pseudoephedrine h.
    quinacrine h.
    ranitidine h.
    ritodrine h.
    ropivacaine h.
    sertraline h.
    sulfamethoxazole/phenazopyridine h.
    sulfisoxazole/phenazopyridine h.
    tetracycline h.
    thiphenamil h.
    tolazoline h.

    topotecan h.
    trifluoperazine h.
    triflupromazine h.
    trimethobenzamide h.
    tripelennamine h.
    triprolidine h.
    tritodrine h.
    valacyclovir h.
    vancomycin h.
hydrochlorothiazide (HCTZ)
hydrocodone
hydrocolpocele, hydrocolpos
hydrocortisone
    Bactine H.
Hydrocort Topical
hydrodensitometry
hydrodissection
HydroDIURIL
hydroepiandrosterone
hydroflotation
hydroflumethiazide
hydrogel dressing
hydrogen
    h. bond
    h. breath test
    h. peroxide
    h. peroxide-producing lactobacillus
hydrolase
    microsound epoxide h.
hydrolethalis syndrome
hydrolysate
    casein h.
hydrolysis
    steroid conjugate h.
hydrolysis-resistant
hydroma *(var. of* hygroma)
hydromeningocele
hydrometra
hydrometrocolpos
hydromicrocephaly
hydromorphone
Hydromox
hydromphalus
hydromyelocele
hydromyelomeningocele
hydronephrocolpos, postaxialpolydactyly, congenital heart disease syndrome
hydronephrosis
hydroparasalpinx
hydropertubation
hydrophila
    Aeromonas h.
hydrophobia
hydrophthalmos
hydropic chorionic villus
hydrops
    fetal h.
    h. fetalis

h. folliculi
h. gravidarum
immune fetal h.
Kell h.
maternal h.
nonimmune fetal h.
h. ovarii
placental h.
h. tubae profluens
hydroquinone
3% h.
hydrorrhea
h. gravidae
h. gravidarum
hydrosalpinx
intermittent h.
hydrostatic
Hydro-Tex Topical
Hydro TherAblator
HydroThermAblator system for excessive uterine bleeding
hydrothorax
tension h.
hydrotubation
hydroureter
hydroureteronephrosis
hydrovarium
hydroxide
magnesium h.
potassium h. (KOH)
11β-hydroxyandrosterone
hydroxyazobenzoic acid (HABA)
17-hydroxycorticosteroid
11-hydroxyetiocholanolone
hydroxyindoleacetic
5-h. acid (5-HIAA)
21-h. acid (21-HIAA)
hydroxyindole-*o*-methyltransferase
hydroxylase
17 alpha-h.
21-h.
phenylalanine h.
17-hydroxylase
-h. deficiency
-h. deficiency syndrome
21-hydroxylase
-h. deficiency
-h. deficiency syndrome
hydroxylation
hydroxylysine
3-hydroxy-3-methylglutaryl coenzyme A
hydroxyphenyluria

17-hydroxypregnenolone
hydroxyprogesterone
h. caproate
h. and estradiol valerate
17-hydroxyprogesterone (17-OHP)
17-h. caproate
hydroxyproline
hydroxyprolinemia
15-hydroxyprostaglandin dehydrogenase
hydroxysteroid
3beta-h. dehydrogenase (3betaHSD)
18-h. dehydrogenase
hydroxyurea
1,25-hydroxyvitamin D
25-hydroxyvitamin-D$_3$
hydroxyzine
hyfrecation
hyfrecator
Hy-Gene seminal fluid collection kit
Hy-Gestrone
hygiene
perineal h.
hygroma, hydroma
cystic h.
fetal cystic h.
nuchal cystic h.
hygroscopic dilator
Hygroton
Hylorel
Hylutin
hymen
h. bifenestratus
h. biforis
cribriform h.
denticulate h.
falciform h.
imperforate h.
infundibuliform h.
microperforate h.
redundant h.
h. sculptatus
septate h.
h. subseptus
vertical h.
virginal h.
hymenal
h. band
h. membrane
h. ring
h. tag
hymenectomy
hymenitis

NOTES

H

hymenorrhaphy
hymenotomy
hyobranchial cleft
hyoscine
    h. methylbromide
hyoscyamine sulfate
hypamnion, hypamnios
Hypan tent
Hypaque
    H. Meglumine
Hyperab
hyperacidity
hyperactive bowel sounds
hyperactivity
hyperadrenalism
hyperalaninemia
hyperaldosteronism
hyperalimentation
    central h.
    Intralipid h.
    intravenous h. (IVH)
    Pedtrace-4 h.
    peripheral h.
    TrophAmine h.
hyperalphalipoproteinemia
hyperammonemia, hyperammoniemia
    cerebroatrophic h.
    h. due to ornithine
      transcarbamoylase deficiency
hyperammonemic syndrome
hyperandrogenemia
hyperandrogenic
    h. anovulation (HA)
    h. chronic anovulation
hyperandrogenism
    adrenal h.
    cryptic h.
    h. insulin resistance, acanthosis
      nigricans (HAIR-AN)
    h. insulin resistance, acanthosis
      nigricans syndrome
    ovarian h.
    h. reversal
hyperargininemia
hyperbaric
    h. chamber
    h. oxygen
    h. oxygen therapy
    h. oxygen treatment
hyperbilirubinemia
hyperbilirubinemic
hypercalcemia
    idiopathic h.
    h., peculiar facies, supravalvular
      aortic stenosis syndrome
hypercalcemia/Williams-Beuren syndrome
hypercalcemic crisis
hypercalciuria

hypercaloric formula
hypercapnia
hypercarbia
    fetal h.
hypercellular uterine leiomyoma
hyperchloremic renal acidosis
hypercholesterolemia
    familial h.
hyperchromic acidosis
hypercoagulability
hypercortisolism
hypercyesis, hypercyesia
hyperdactyly
hyperdibasicaminoaciduria
hyperdiploid
hyperdynamia
    h. uteri
hyperechogenic
    h. foci
hyperechogenicity
    renal h.
hyperechoic
    h. bowel
    h. endometrium
hyperelastica
    cutis h.
hyperemesis
    h. gravidarum
    h. lactentium
hyperemia
hyperencephalus
hyperestrogenism
hyperexpansion
hyperexplexia
hyperfolliculoidism
hypergalactosis
hypergenitalism
hyperglycemia
    ketotic h.
    nonketotic h.
hyperglycerolemia
    familial h.
hyperglycinemia
    nonketotic h.
hypergonadism
hypergonadotropic, hypergonadotrophic
    h. amenorrhea
    h. eunuchoidism
    h. hypogonadism
hypergynecosmia
hyperhaploidy
HyperHep
hyperhidrosis
hyperhomocystinemia
hyper-IgE syndrome
hyperimmune serum globulin
hyperimmunoglobulin E
hyperimmunoglobulinemia A

hyperinsulinemia
hyperinsulinemic-euglycemic clamp
  technique
hyperinsulinism
hyperinvolution
hyperkalemia
hyperkeratosis
  epidermolytic h.
hyperlactation
hyperlacticacidemia
hyperlaxity
  joint h.
hyperlipidemia
hyperlipoproteinemia
hyperlucency
hyperlucent lung syndrome
hyperluteinization
hyperlysinemia
hypermagnesemia
hypermastia
hypermenorrhea
hypermetropia
hypermobility
  joint h.
  urethral h.
hypernatremia
  h. dehydration
hyperopia
hyperornithinemia
hyperosmolar coma
hyperosmotic agent
hyperostosis
  calvarial h.
  h. generalisata with striation
  infantile cortical h.
hyperovarianism
hyperoxia
hyperparathyroidism
  maternal h.
  neonatal h.
hyperphenylalaninemia
hyperphosphatemia
hyperpigmentation
hyperpigmented lesion
hyperpipecolic acidemia-hepatomegaly-
  mental retardation-optic dysplasia-
  progressive
hyperpituitary gigantism
hyperplasia
  adenomatous h. (AH)
  adenomatous endometrial h.
  adrenal h.

adult-onset congenital adrenal h.
atypical ductal h. (ADH)
atypical lobular h.
basal cell h.
congenital adrenal h. (CAH)
congenital adrenal lipoid h.
cystic h. of the breast
cystic glandular h.
ductal h.
endometrial h. (EH)
epithelial h.
familial lipoid adrenal h.
fetal congenital h.
glandular h.
hepatic focal nodular h.
hilar cell h.
late-onset h.
Leydig cell h.
lipoid adrenal h.
lymphoid h.
microglandular cervical h.
nonclassical adrenal h. (NCAH)
nonclassic congenital adrenal h.
  (NC-CAH)
21-OH nonclassical adrenal h.
polypoid h.
salt-wasting congenital adrenal h.
  (SW-CAH)
sebaceous h.
simple virilizing congenital
  adrenal h. (SV-CAH)
stromal h.
Swiss cheese h.
vulvar squamous h.
hyperplastic
  h. polyp
hyperploidy
hyperprogesteronemia
hyperprolactinemia
  tumorous h.
hyperprolactinemia-associated luteal
  phase
hyperprolactinemic amenorrhea
hyperprolinemia
  familial h.
hyperpyrexia
hyperreactio luteinalis
hyperreflexia
hyperreflexic
hyperrelaxinemia
hypersecretion
hypersegmentation

NOTES

H

hypersensitivity
  delayed h.
hypersensitization
hypersplenism
Hyperstat
hyperstimulation
  controlled ovarian h.
  gonadotropin-induced ovarian h.
  ovarian h.
  uterine h.
hypertelorism
  Bixler h.
  dysostosis craniofacialis with h.
  h.-hypospadias syndrome
  h., microtia, clefting (HMC)
  h., microtia, clefting syndrome
  ocular h.
hypertension
  chronic h.
  gestational h.
  intracranial h.
  malignant h.
  maternal h.
  portal h.
  postpartum h.
  pregnancy associated h.
  pregnancy-induced h. (PIH)
  pulmonary h.
  renal h.
  transient h.
  white coat h.
hypertension-preeclampsia
hypertensive
  h. emergency
  h. encephalopathy
Hyper-Tet
hyperthecosis
  ovarian stromal h.
  h. ovarii
  stromal h.
hyperthermia
hyperthyroid
hyperthyroidism
hypertonia
hypertonic
  h. glucose
  h. saline
  h. uterine dysfunction
hypertonus
  uterine h.
hypertransaminasemia
hypertrichosis
  h., coarse face, brachydactyly,
    obesity, mental retardation
    syndrome
  h. universalis congenita
  vellus h.
hypertrichotic osteochondrodysplasia

hypertriglyceridemia
hypertrophic
  h. bundle
  h. bundle of smooth muscle
  h. cirrhosis
  h. stenosis
hypertrophica
  hemangiectasia h.
hypertrophy
  clitoral h.
  infantile myxedema-muscular h.
  left ventricular h. (LVH)
  massive breast h.
  myocardial h.
  ovarian h.
  right ventricular h. (RVH)
  septal h.
  virginal breast h.
hypertympany
hypertyrosinemia II
hyperuricemia
  X-linked primary h.
hyperuricosuria
hypervalinemia
hyperventilation
hyperviscosity
  h. syndrome
hypervitaminosis
hypervolemia
hypha, pl. **hyphae**
hyphema
hypnotic effect
hypoactive bowel sounds
hypoadrenalism
hypoalbuminemia
hypoaldosteronism
  congenital h.
hypoallergenic
hypobetalipoproteinemia
hypocalcemia
  cardiac defect, abnormal face,
    thymic hypoplasia, cleft palate, h.
    (CATCH-22)
  h., dwarfism, cortical thickening
    syndrome
  dwarfism, cortical thickening of
    tubular bones, transient h.
  neonatal h.
hypocalvaria
hypocapnia
hypochloremia
hypochondriasis
hypochondrogenesis
hypochondroplasia
  h. syndrome
hypocycloidal tomography
hypodactyly
hypodense

hypodermoclysis
hypodiploid
hypodontia
hypoechogenic
hypoechoic density
hypoestrogenemia
hypoestrogenic woman
hypoestrogenism
hypofertility
hypofibrinogenemia
    congenital h.
    familial h.
hypofolliculogenesis
hypofunction
    adrenal h.
hypogalactia
hypogalactous
hypogammaglobulinemia
    acquired h.
    congenital h.
    physiologic h.
    transient h.
    X-linked h.
hypogastric
    h. artery
    h. artery ligation
    h. lymph node
    h. pain
    h. plexus
hypogastropagus
hypogastroschisis
hypogenesis
    cerebellar vermis h.
hypogenital dystrophy with diabetic tendency syndrome
hypogenitalism
hypoglossia-hypodactyly syndrome
hypoglycemia
    neonatal h.
hypoglycemic
    oral h.
hypognathus
    cyclops h.
hypogonadal woman
hypogonadism
    h., alopecia, diabetes mellitus, mental retardation, deafness and ECG abnormalities
    eugonadotropic h.
    hypergonadotropic h.
    hypogonadotropic h.

    idiopathic hypothalamic h. (IHH)
    primary h.
hypogonadism-anosmia syndrome
hypogonadotropic, hypogonadotrophic
    h. amenorrhea
    h. eunuchoidism
    h. hypogonadism
    h. hypogonadism-anosmia syndrome (HHA)
    h. hypogonadism, mental retardation, microphthalmia syndrome
hypohaploidy
hypohidrotic
    h. ectodermal dysplasia
    h. ectodermal dysplasia-hypothyroidism-agenesis of corpus callosum syndrome
hypokalemia
hypokalemic alkalosis
hypoleptinemia
hypomagnesemia
hypomandibular faciocranial dysostosis
hypomastia, hypomazia
hypomaturation-hypoplasia
hypomelanosis
    Ito h.
    h. of Ito
hypomelia, hypotrichosis, facial hemangioma syndrome
hypomenorrhea
hypomenorrheic woman
hypomyelination
    congenital h.
hyponatremia
hyponatremic
hypoosmotic
    h. swelling (HOS)
    h. swelling test
hypoovarianism, hypovarianism
hypoparathyroidism
    h., stature, mental retardation, seizures syndrome
hypoperistalsis
    megacystis, microcolon, intestinal h. (MMIH)
hypopharynx
hypophosphatasia
hypophosphatemia
hypophosphatemic rickets
hypophyseal portal circulation
hypophysectomy

NOTES

H

hypophysial
- h. amenorrhea
- h. dwarfism
- h. infantilism

hypophysitis
- lymphocytic h.

hypopituitarism

hypoplacentosis

hypoplasia
- h. of anguli oris depressor muscle (HAODM)
- biliary h.
- cerebellar h.
- congenital adrenal h.
- congenital universal muscular h.
- h., endocrine disturbances, tracheostenosis syndrome
- erythroid h.
- fetal pulmonary h.
- foveal h.
- Goldblatt-Vilijoen radial ray h.
- heminasal h.
- hepatic ductular h.
- iris h.
- h. of left heart (HLL)
- lipoid adrenal gland h.
- malar h.
- müllerian h.
- oromandibular limb h.
- pancreatic h.
- pulmonary h.
- spondylohumerofemoral h.
- transient erythroid h.
- velofacial h.

hypoplasia/hydrocephalus
- X-linked cerebral h.

hypoplastic
- h. anemia
- h. aorta (HA)
- h. congenital anemia syndrome
- h. labia
- h. left heart syndrome (HLHS)
- h. lung
- h. nails
- h. patella
- h. penis
- h. philtrum
- h. right heart syndrome (HRHS)
- h. sacrum
- h. uterus

hypopotassemia

hypoproteinemia

hypoprothrombinemia

hyposegmentation

hyposensitization

hyposmia
- h.-hypogonadotropic hypogonadism syndrome

hypospadias
- balanic h.
- balanitic h.
- h.-dysphagia syndrome
- Hacker h.
- h.-mental retardation syndrome
- penoscrotal h.

hyposplenism

hypostatic
- h. pneumonia

hypotelorism
- ocular h.

hypotension
- maternal h.
- orthostatic h.

hypothalamic
- h. amenorrhea
- h. disease
- h. dysfunction
- h. failure
- h. hamartoblastoma
- h. hamartoblastoma, hypopituitarism, imperforate anus, postaxial polydactyly syndrome
- h. hamartoblastoma syndrome
- h.-hypophyseal-ovarian-endometrial axis
- h.-hypophyseal portal circulation
- h. hypothyroidism
- h. luteinizing hormone-releasing hormone

hypothalamic-pituitary
- h.-p. axis
- h.-p. axis dysfunction
- h.-p. disorder
- h.-p. function
- h.-p. system

hypothalamic-pituitary-adrenal (HPA)
- h.-p.-a. axis

hypothalamic-pituitary-gonadal axis

hypothalamic-pituitary-ovarian (HOP, HPO)
- h.-p.-o. axis

hypothalamus

hypothermia
- chronic scrotal h.

hypothesis, pl. **hypotheses**
- alternative h.
- bayesian h.
- critical weight h.
- estrogen window etiologic h.
- Goldie-Coldman h.
- Korenman estrogen window h.
- log kill h.
- Lyon h.
- Neyman-Pearson statistical h.
- two-hit h.
- Wramsby h.

hypothyroid
    h. dwarfism
    h.-large muscle syndrome
    h. myopathy
hypothyroidism
    athyrotic h.
    congenital h.
    hypothalamic h.
    subclinical h.
    transient congenital h.
hypotonia
    benign congenital h.
    bladder h.
    congenital h.
    h., hyperphagia, hypogonadism,
      obesity (HHHO)
    h., hypopigmentia, hypogonadism,
      obesity syndrome
    infantile muscular h.
    muscle h.
    h., obesity, hypogonadism, mental
      retardation syndrome
    h., obesity, prominent incisors
      syndrome
    Oppenheim congenital h.
    uterine h.
hypotonic
    h. bladder
    h. myometrium
    h. uterine dysfunction
hypotony
    ocular h.
hypotrichosis
hypotrophy
    fetal h.
hypouricemia
hypovarianism (*var. of* hypoovarianism)
hypoventilation
hypovitaminemia
    thiamine h.
hypovitaminosis
hypovolemia
hypovolemic
    h. shock
hypoxanthine
    h.-guanine phosphoribosyltransferase
      deficiency (HGPRT)
    h.-phosphoribosyltransferase
      deficiency
hypoxemia
    acute-on-chronic tissue h.
hypoxemic

hypoxia
    fetal h.
    perinatal h.
hypoxic
    h. cell sensitizer
hypoxic-ischemic
    h.-i. encephalopathy (HIE)
    h.-i. injury
HypRho-D Mini-Dose
Hyprogest
    H. 250
Hyproval
hypsarhythmia, hypsarrhythmia
hypsicephaly
Hyrexin
Hyskon
hysteralgia
hysteratresia
hysterectomized
hysterectomy
    abdominal h.
    abdominovaginal h.
    Bell-Buettner h.
    Bonney abdominal h.
    cesarean h.
    h. clamp
    classic abdominal Semm h.
      (CASH)
    Dellepiane h.
    Döderlein method of vaginal h.
    Doyen vaginal h.
    Eden-Lawson h.
    extrafascial h.
    Gelpi-Lowrie h.
    laparoscopic-assisted abdominal h.
      (LAAH)
    laparoscopic-assisted vaginal h.
      (LAVH)
    laparoscopic Döderlein h.
    laparoscopic supracervical h. (LSH)
    Mayo h.
    Meigs-Werthein h.
    modified radical h.
    Munro and Parker classification for
      laparoscopic h.
    obstetric h.
    paravaginal h.
    Pelosi vaginal h.
    pelviscopic intrafascial h.
    Porro h.
    radical h.
    Reis-Wertheim vaginal h.

NOTES

H

hysterectomy *(continued)*
 Rutledge classification of
  extended h.
 Semm h.
 subtotal h.
 supracervical h.
 total abdominal h. (TAH)
 vaginal h.
 Ward-Mayo vaginal h.
hysteresis
hystereurysis
hysteria
hysterical
 h. mother
 h. paralysis
 h. seizure
hystericus
 globus h.
hysterocele
hysterocleisis
hysterocolposcope
hysterocystopexy
hysterodynia
hysterofiberscope
 Olympus flexible h.
hysterogram
hysterograph
hysterography
hysterolith
hysterolysis
hysterometer
hysteromyoma
hysteromyomectomy
hysteromyotomy
hystero-oophorectomy
hysteropathy
hysteropexy
 abdominal h.
 Alexander-Adams h.
hysterophore
hysteroplasty
hysterorrhaphy
hysterorrhexis
hysterosalpingectomy
hysterosalpingogram
hysterosalpingography (HS, HSG)
 h. catheter
hysterosalpingo-oophorectomy
hysterosalpingosonography
hysterosalpingostomy
hysteroscope
 Baggish h.

 Baloser h.
 Circon-ACMI h.
 diagnostic h.
 flexible h.
 French h.
 Fujinon flexible h.
 Galileo rigid h.
 Hamou h.
 Karl Storz 15 French flexible h.
 Liesegang LM-FLEX 7 flexible h.
 Olympus h.
 OPERA Star SL h.
 Scopemaster contact h.
 Valle h.
hysteroscopic
 h. endometrial ablation
 h. insufflator
 h. metroplasty
 h. myomectomy
 h. surgery
hysteroscopy
 h. fluid
 laparoscopic-assisted vaginal h.
  (LAVH)
Hysteroser system
hysterosonography
hysterospasm
hysterothermometry
hysterotomy
 abdominal h.
 low transverse h.
 Pelosi h.
 vaginal h.
hysterotonin
hysterotrachelectomy
hysterotracheloplasty
hysterotrachelorrhaphy
hysterotrachelotomy
hysterotubography
Hysto-vac drain
hystrix
 ichthyosis h.
Hytakerol
Hytone
 H. Topical
Hytuss
Hz
 hertz
HZA assay
HZV
 herpes zoster virus

I
  inspired gas
    I IFG-binding protein
^{123}**I**
  iodine-123
^{125}**I**
  iodine-125
^{127}**I**
  iodine-127
^{131}**I**
  iodine-131
^{132}**I**
  iodine-132
IADH
  inappropriate antidiuretic hormone
    IADH syndrome
IAHS
  infection-associated hemophagocytic
    syndrome
iatrogenic
  i. dissemination
  i. effect
  i. event
  i. fetal distress
  i. infertility
  i. menopause
  i. multiple pregnancy (IMP)
  i. precocious puberty
  i. preterm birth maternal dropsy
  i. ureteral injury
  i. urethral obstruction
IB
  Excedrin IB
  Midol IB
  Motrin IB
  Pamprin IB
  Sine-Aid IB
IBC
  iron-binding capacity
IBIDS
  ichthyosis, brittle hair, impaired
    intelligence, decreased fertility, short
    stature syndrome
IB/IX
  glycoprotein I.
  membrane glycoprotein I.
IBR
  Infant Behavior Record
IBS
  irritable bowel syndrome
IBSN
  infantile bilateral striatal necrosis
    syndrome
IBT
  immunobead test

Ibuprin
ibuprofen
Ibuprohm
Ibu-Tab
IC
  invasive cancer
    Babytherm IC
iCa
  ionized calcium
ICAM-1
  intercellular adhesion molecule-1
ICCR
  International Committee for
    Contraceptive Research
I-cell disease
ICF
  immunodeficiency, centromeric
  instability, facial anomalies syndrome
    ICF syndrome
ICH
  intracranial hemorrhage
ichthyosiform
    i. erythroderma, corneal
      involvement, deafness syndrome
    i. erythroderma, hair abnormality,
      mental and growth retarding
      syndrome
ichthyosis
    i., alopecia, ectropion, mental
      retardation syndrome
    i., brittle hair, impaired
      intelligence, decreased fertility,
      short stature syndrome (IBIDS)
    i., characteristic appearance, mental
      retardation syndrome
    i., cheek, eyebrow syndrome
    i. congenita
    congenital i.
    i. fetalis
    i. fetus
    i., follicularis, atrichia (or
      alopecia), photophobia syndrome
      (IFAP)
    harlequin i.
    i., hypogonadism, mental
      retardation, epilepsy syndrome
    i. hystrix
    lamellar i.
    i. linearis circumflexa
    i., male hypogonadism syndrome
    i., mental retardation, dwarfism,
      renal impairment syndrome
    i., mental retardation, epilepsy,
      hypogonadism syndrome
    i., oligophrenia, epilepsy syndrome

ichthyosis *(continued)*
> i., spastic neurologic disorder, oligophrenia syndrome
> i. spinosa
> i., split hair, aminoaciduria syndrome
> i. uteri
> i. vulgaris
> X-linked i.

ichthyotic idiocy
Icon
> I. serum pregnancy test
> I. strep B test
> I. urine pregnancy test

ICRF 159
ICSHI
> intracytoplasmic sperm head injection

ICSI
> intracytoplasmic sperm injection
> ICSI Massachusetts clamp

icteric
icterus
> i. gravis
> i. gravis neonatorum
> Liouville i.
> i. neonatorum
> physiologic i.
> i. praecox

IDA
> alpha-L-iduronidase
> iron deficiency anemia

Idaho syndrome
Idamycin
> I. PFS

idarubicin
IDC
> infiltrating ductal carcinoma

IDDM
> insulin-dependent diabetes mellitus

identical
> i. twins

identity matrix
ideogram
idiocy
> amaurotic familial i.
> Aztec i.
> ichthyotic i.
> Kalmuk i.
> xerodermic i.

idiopathic
> i. cholestasis of pregnancy
> i. dilated cardiomyopathy
> i. hirsutism
> i. hydramnion
> i. hypercalcemia
> i. hypercalcemia-supravalvular aortic stenosis syndrome
> i. hypertrophic subaortic stenosis

> i. hypothalamic hypogonadism (IHH)
> i. infantile hypercalcemia syndrome
> i. infertility
> i. polyserositis
> i. precocious puberty
> i. respiratory distress syndrome (IRDS)
> i. short stature
> i. steatorrhea
> i. thrombocytopenic purpura (ITP)
> i. venous thromboembolism
> i. vulvodynia

idiot
> mongolian i.

IDM
> infant of diabetic mother

idoxifene
idoxuridine
IDS
> iduronate sulfatase deficiency
> intrinsic sphincter deficiency

iduronate sulfatase deficiency (IDS)
I/E
> inspiratory/expiratory
> I/E ratio

IES
> Impact of Events Scale

IFAP
> ichthyosis, follicularis, atrichia (or alopecia), photophobia syndrome
> IFAP syndrome

Ifex
IFI
> intrafollicular insemination

IFN
> interferon

ifosfamide
Ig
> immunoglobulin

IgA
> immunoglobulin A
> IgA antibody
> IgA deficiency
> IgA HIV antibody test
> secretory IgA

IgD antibody
IgE
> immunoglobulin E
> IgE antibody
> IgE deficiency

IGF
> insulin-like growth factor
> IGF-1, -2

IgF
> immunoglobulin F

IgG
> immunoglobulin G

IgG antibody
serovar-specific immunoglobulin IgG

IgM
immunoglobulin M
IgM antibody
IgM deficiency
serovar-specified immunoglobulin IgM

IGT
impaired glucose tolerance

IHH
idiopathic hypothalamic hypogonadism

IIQ-R
Incontinence Impact Questionnaire-Revised

IL
interleukin
IL-1, -2, -8

ILC
infiltrating lobular carcinoma

ileal
i. atresia
i. conduit
i. diversion
i. intussusception
i. perforation

ileitis
terminal i.

ileocecal intussusception
ileocolic
i. artery
i. intussusception

ileocystoplasty
Camey i.

ileoentectropy
ileoileal intussusception
ileostomy
continent i.

Iletin
ileum
duplication of i.

ileus
adynamic i.
duodenal i.
meconium i.
paralytic i.
i. subparta

Ilfeldt splint
iliac
i. artery
i. node
i. vein

iliococcygeal
i. fixation
i. muscle

ilioneoureterocystotomy
iliopagus
iliopectinate line
iliopectineal line
iliothoracopagus
ilioxiphopagus
Ilizarov
I. external fixator
I. procedure

ill-defined mass
illegitimacy
illegitimate
illicit
i. drug
i. drug use
i. sex

illness
psychiatric i.
systemic i.

Illumina
I. Pro Series CO2 surgical laser system
I. Pro Series laparoscopic laser

illuminated vaginal speculum
illumination
chemiluminescent i.

illuminator
Dyonic DyoBride 300 i.

Illum syndrome
Ilosone
I. Pulvules

Ilotycin
IL-1R
maternal endometrial I.

IL-1ra
ILS17 gene for isolated lissencephaly
ILSX gene for X-linked lissencephaly
IM
intramuscular

image recording system
imaging
Color Power Angio i.
continuous-wave ultrasound i.
diagnostic i.
echoplanar magnetic resonance i.
endovaginal i.
fetal i.
high-contrast Bucky i.
magnetic resonance i. (MRI)

NOTES

imaging *(continued)*
>magnetic source i. (MSI)
>M-mode i.
>prenatal magnetic resonance i.
>Rho i.
>SieScape i.
>Tissue Specific i.
>ultrafast magnetic resonance i.

imbalance
>electrolyte i.

Imerslund-Graesback syndrome
Imerslund syndrome
Imex
>I. antepartum monitor
>I. Pocket-Dop OB Doppler

IMEXLAB vascular diagnostic system
imidazole
imidazopyridine
imiglucerase
iminoglycinuria
imipemide
imipenem-cilastaten sodium
imipramine
imiquimod
>i. cream

Imitrex
Imlach ring
immature
>i. infant
>i. neural element
>i. ovarian teratoma

immersion
>static i.

imminent abortion
immobilization
>*Treponema pallidum* i. (TPI)

immobilizer
>Olympic Neostraint i.

immotile
>i. cilia
>i. cilia syndrome

immune
>i. clearance
>i. deficiency
>i. factor
>i. fetal hydrops
>i. globulin
>i. monitoring technique
>i. process
>i. response
>i. response gene
>rubella i.
>i. separation technique
>i. serum globulin (ISG)
>i. suppressor gene
>i. surveillance
>i. system
>i. system anatomy

>i. thrombocytopenia
>i. thrombocytopenic purpura (ITP)

immunity
>Burnet acquired i.
>cell-mediated i. (CMI)
>cellular i.
>humoral i.
>passive i.
>previous maternal i.

immunization
>blood group i.
>hepatitis A i.
>prophylactic i.
>Rh i.

Immuno
>Feiba VH I.
>Gammabulin I.

immunoassay
>alpha-fetoprotein enzyme i. (AFP-EIA)
>chemiluminescent i. (CIA)
>Chlamydiazyme i.
>enzyme i. (EIA)
>fluorescent i. (FIA)
>nonradioactive i.
>nonreactive i.
>Quantikine human IL-6 I.
>radioactive i.
>SalEst i.
>solid-phase enzyme i.

immunobead test (IBT)
immunochemiluminomimetric insulin assay
immunochemistry
immunochemotherapy
immunocompetent
>i. cell

immunocompromise
immunocytochemical
immunodeficiency
>i., centromeric heterochromatin instability, facial anomalies syndrome
>i., centromeric instability, facial anomalies syndrome (ICF)
>combined i.

immunodiagnosis
immunoelectrophoresis
immunofluorescence
immunofluorescent
>i. antibody test
>i. *Chlamydia* test

immunogen
immunogenetics
immunoglobulin (Ig)
>i. A (IgA)
>antenatal anti-D i.
>anti-D i.

cytomegalovirus i.
cytomegalovirus-specific i.
i. deficiency
i. E (IgE)
i. F (IgF)
i. G (IgG)
i. gene
hepatitis B i. (HBIG, HBIg)
intravenous i.
intravenous anti-D i.
i. M (IgM)
Rh i.
Rh$_o$(D) i.
surface i.
thyroid-stimulating i.
immunohistochemical
   i. change
   i. stromal leukocyte characterization
immunologic
   i. assay
   i. change
   i. enhancement
   i. maladaptation
   i. paralysis
   i. pregnancy test
   i. suppression
   i. surveillance
   i. tolerance
   i. unresponsiveness
immunological infertility
immunology
   maternal i.
   placental i.
   transplantation i.
   tumor i.
immunoosseous dysplasia
immunoperoxidase
   i. technique
immunoprophylaxis
immunoprotein
immunoradiometric assay (IRMA)
immunoreaction
immunoreactivity
immunoresistance
immunosuppression
immunosuppressive
   i. therapy
immunotherapy
   active specific i. (ASI)
   adoptive i.
   nonspecific i.
   specific i.

systemic-active nonspecific i.
tumor i.
Imodium
Imogam
Imovax
IMP
   iatrogenic multiple pregnancy
impacted
   i. fetus
   i. twins
   i. uterus
Impact of Events Scale (IES)
impaction
   fecal i.
   psychological causes, excessive
   urine production, restricted
   mobility, stool i.
   stool i.
impaired
   i. glucose tolerance (IGT)
   i. secretion
impairment
   inherited androgen uptake i.
   intrauterine growth i.
   opioidergic control i.
impedance
   acoustic i.
   i. cardiography
   i. plethysmography
   i. pneumography
   transcephalic i.
imperfecta
   amelogenesis i.
   dentinogenesis i.
   osteogenesis i. type I–IV (OI)
imperforate
   i. anus
   i. anus-hands and foot anomalies
   syndrome
   i. anus-polydactyly syndrome
   i. hymen
   i. nasolacrimal duct
   i. urethra
impervious
   i. sheet
   i. stockinette
impetigo
   bullous i.
   i. contagiosa
   i. herpetiformis
   i. neonatorum
Implanon

I

NOTES

implant
>benign i.
>Biocell RTV saline-filled breast i.
>breast i.
>cesium i.
>collapsed subpectoral i.
>Contigen Bard collagen i.
>Contigen glutaraldehyde cross-linked collagen i.
>contraceptive i.
>ectopic i.
>endometrial i.
>endometriotic i.
>fetal tissue i.
>goserelin acetate i.
>hormonal i.
>iridium i.
>levonorgestrel i.
>metastatic i.
>Norplant i.
>Organon percutaneous E2 i.
>peritoneal i.
>radioactive i.
>radium i.
>saline i.
>silicone i.
>subdermal i.
>subdermal levonorgestrel i. (SLI)
>subpectoral i.
>transperineal i.
>transvaginal i.
>Zoladex I.

implantation
>blastocyst i.
>cortical i.
>delayed i.
>displacement i.
>i. failure
>fusion i.
>intrusive i.
>i. phase
>placental i.
>radioactive seed i.
>i. theory
>tubouterine i.

impotence
>psychogenic i.

impotent
impregnate
Impress Softpatch
Impril
imprint
>touch i.

imprinting
>genomic i.

improved
>Clearblue I.
>Hybolin I.

Imuran
Imuthiol
IMV
>intermittent mandatory ventilation
>intermittent mechanical ventilation

IMx Estradiol Assay
^{111}In
>indium-111

in
>i. situ hybridization
>i. situ nucleic acid hybridization
>toeing i.
>i. utero
>i. utero exposure
>i. vitro
>i. vitro fertilization (IVF)
>i. vitro fertilization-embryo transfer (IVF-ET)
>i. vitro maturation
>i. vivo
>i. vivo fertilization
>i. vivo gene therapy

inactivated poliovirus vaccine (IPV)
inactivation
>i. pattern
>X i.

inactive endometrium
inadequacy
>luteal phase i.

inadequate luteal phase
inanition fever
inappropriate
>i. antidiuretic hormone (IADH)
>i. lactation

Inapsine
inborn error of metabolism
inbreeding
>coefficient of i.
>i. coefficient
>i. depression

incarcerated
>i. fundus
>i. gravid uterus
>i. hernia
>i. placenta

incarceration
>uterine i.

Incert bioabsorbable sponge
incessant ovulation
incest
incestuous
incidentaloma
incipient
>i. abortion
>i. coagulopathy

incision
>Bevan i.
>boutonnière i.

buttonhole i.
cervical i.
Cherney i.
classical transverse i.
classical uterine i.
i. closure
colpotomy i.
counter stab wound i.
Dührssen i.
elliptical uterine i.
gridiron i.
hockey stick i.
infraumbilical i.
inverted T uterine i.
Joel-Cohen i.
Kehr i.
laparotomy i.
lazy-S i.
low-segment transverse i.
low transverse uterine i.
low vertical uterine i.
Maylard i.
midline i.
paramedian i.
periumbilical i.
Pfannenstiel i.
prior low transverse uterine i.
prior low vertical uterine i.
Rockey-Davis i.
Sanger i.
Schuchardt i.
Sellheim i.
smiling i.
supraumbilical i.
transverse i.
uterine i.
incisional hernia
inclination
pelvic i.
inclusion cyst
incompatibility
ABO i.
Rh i.
incompatible blood group antigen
incompetence
cervical i. (CI)
gastroesophageal i.
palatopharyngeal i.
incompetent cervix
incomplete
i. abortion
i. breech presentation

i. conjoined twins
i. dominance
i. foot presentation
i. knee presentation
i. müllerian fusion
i. precocious puberty
incompletus
coitus i.
incontinence
anal i.
anorectal i.
Blaivas classification of urinary i.
exercise-induced i.
fecal i.
genuine stress i. (GSI)
giggle i.
I. Impact Questionnaire-Revised (IIQ-R)
key-in-lock i.
i. of milk
Miyazaki-Bonney test for stress i.
overflow i.
paradoxical i.
passive i.
postpartum i.
stress i.
stress urinary i. (SUI)
true i.
type III i.
urge i.
urinary exertional i.
urinary stress i.
incontinentia
i. pigmenti
i. pigmenti achromiens
i. pigmenti type I, II
increase
plasma prorenin i.
increta
placenta i.
incubation
incubator
double-walled i.
Forma water-jacketed i.
Ohmeda Care-Plus i.
incudiform uterus
indapamide
independence
causal i.
Inderal
I. LA
index, pl. **indices, indexes**

NOTES

extremely low birth weight i.
(ELBWI)
i. Hercules
high-risk i.
immature i.
jittery i.
large-for-dates i.
liveborn i.
low birth weight i. (LBWI)
mature i.
i. morbidity
i. mortality
i. mortality rate
Neurodevelopmental Assessment
Procedure for Preterm I.'s (NAPI)
postmature i.
post-term i.
premature i. (PI)
preterm i.
i. respiratory distress syndrome
(IRDS)
Rh-positive i.
singleton i.
small-for-gestational-age i.
i. Star high-frequency ventilator
stillborn i.
term i.
very-low-birth-weight i.
viable i.
vigorous i.
well-oxygenated i.

infanticide

infantile
i. achalasia
i. arteriosclerosis
i. bilateral striatal necrosis
syndrome (IBSN)
i. breath-holding response
i. cataract
i. celiac disease
i. cerebellooptic atrophy
i. choreoathetosis of Fisher
i. colic
i. cortical hyperostosis
i. diplegia
i. dwarfism
i. eczema
i. hepatic hemangioma
i. muscular hypotonia
i. myoclonic seizure
i. myxedema
i. myxedema-muscular hypertrophy

i. neuroaxonal dystrophy
i. optic atrophy-ataxia syndrome
i. osteopetrosis
i. polycystic kidney disease (IPKD)
i. polyneuritis
i. purulent conjunctivitis
i. respiratory distress syndrome
i. salaam
i. scurvy
i. sialic acid storage disorder
(ISSD)
i. sleep apnea
i. spasm
i. spasms, hypsarrhythmia, mental
retardation syndrome
i. spasms with mental retardation
i. spastic paraplegia
i. spinal muscular atrophy
i. striatonigral degeneration
i. system hyalinosis
i. thoracic dystrophy

infantilis
dystaxia cerebralis i.
poliodystrophia cerebri
progressiva i.

infantilism
Brissaud i.
cachectic i.
celiac i.
dysthyroidal i.
hepatic i.
Herter i.
hypophysial i.
Levi-Lorain i.
Lorain i.
muscular i.
myxedematous i.
pseudonuchal i.
regressive i.
sexual i.

infantis
Bifidobacterium i.

infantum
anemia pseudoleukemica i.
cholera i.
dermatitis excoriativa i.
dermatitis exfoliativa i.
dermatitis gangrenosa i.
lichen i.
roseola i.
tabes i.

NOTES

infarction, infarct
- bilirubin i.
- bowel i.
- cerebral i.
- chorionic villus i.
- limb i.
- maternal floor i.
- myocardial i.
- placental i.
- pulmonary i.
- renal i.
- uric acid i.
- white i.

In-Fast bone screw system

Infasurf

infected
- i. abortion
- i. cuff hematoma

infection
- adnexal i.
- ascending intrauterine i.
- asymptomatic i.
- asymptomatic urinary tract i. (AUTI)
- bacterial i.
- benign papillomavirus i.
- cervical i.
- cervicovaginal i.
- chlamydial i.
- chorioamnion i.
- chorioamnionic i.
- congenital cytomegalovirus i. (CMV)
- Coxsackievirus i.
- culture-negative cytomegalovirus i.
- cytomegalovirus i.
- disseminated gonococcal i.
- fetal cytomegalovirus i.
- fungal i.
- gastrointestinal i.
- genital i.
- gonococcal i.
- granulomatous i.
- herpes whitlow i.
- intraabdominal i.
- intraamniotic i.
- intrauterine i.
- IUD-related i.
- latent herpes simplex virus i.
- lower genital tract i.
- maternal i.
- neisserial i.
- neonatal i.
- nonprimary i.
- nosocomial i.
- papillomavirus i.
- paronychial i.
- pelvic i.

- pharyngeal gonococcal i.
- polymicrobial pelvic i.
- postpartum i.
- primary i.
- puerperal i.
- recurrent i.
- respiratory tract i.
- retroperitoneal i.
- streptococcal i.
- surgical i.
- symptomatic i.
- transplacental i.
- *Trichomonas* i.
- upper genital tract i.
- upper respiratory i. (URI)
- urinary tract i. (UTI)
- uterine i.
- vaginal i.
- varicella i.
- varicella-zoster virus i.
- vertically-acquired i.
- viral i.
- vulvar i.
- vulvovaginal premenarchal i.
- wound i.
- yeast i.

infection-associated hemophagocytic syndrome (IAHS)

infectiosum
- erythema i.

infectious
- i. colitis
- i. disease
- i. esophagitis
- i. hepatitis
- i. mononucleosis

infective endocarditis

infecundity

inference
- causal i.

inferior
- inferior i.
- i. mesenteric artery
- i. straight
- i. vena cava

infertile
- i. patient
- i. woman

infertility
- i. agent
- anovulatory i.
- asymptomatic i.
- *Chlamydia trachomatis* tubal i.
- endometriosis-associated i.
- etiologic fraction of tubal i.
- i. evaluation
- i. factor
- familial i.

iatrogenic i.
idiopathic i.
immunological i.
inherited i.
i. investigation
male factor i.
i. management
i. perceptions inventory (IPI)
primary i.
secondary i.
i. treatment
unexplained i.
infestation
threadworm i.
infibulation
infiltrate
interstitial i.
patchy i.
streaky i.
infiltrating
i. ductal carcinoma (IDC)
i. lobular carcinoma (ILC)
i. small-cell lobular carcinoma
infiltration
fatty i.
inflammation
histologic placental i.
intraamniotic i.
pelvic i.
inflammatory
i. bowel disease
i. carcinoma
i. molecule
influenza
i. A,B,C
i. vaccine
i. virus
influenzae
Haemophilus i.
informatics
informed
i. consent
i. consent disclosure rules
informosome
infraction
Freiberg i.
InfraGuide
I. delivery system
Heraeus LaserSonics I.
inframammary
infrapubic ramus
infrared spectroscopy

Infrasurf surfactant
infraumbilical incision
infundibular
i. stalk
i. stenosis
infundibuliform hymen
infundibulopelvic
i. ligament
i. vessel
infundibulum
infusion
colloid i.
continuous subcutaneous i. (CSQI)
continuous subcutaneous insulin i.
extraamniotic saline i. (EASI)
fetal cortisol i.
intraarterial i.
intralymphatic i.
intravenous i.
laminaria i.
vasopressin i.
Ingelman-Sundberg gracilis muscle procedure
ingestion
alcohol i.
Ingram bicycle seat
inguinal
i. canal
i. hernia
i. lymphadenectomy
i. lymph node
i. lymph node metastasis
inguinale
hernia i.
hernia uteri i.
lymphogranuloma i.
inguinal-femoral node dissection
INH
isoniazid
inhalation
i. anesthesia
i. of nitrous oxide
smoke i.
inhaler
Beconase AQ Nasal I.
Beconase Nasal I.
metered dose i. (MDI)
Vancenase AQ I.
Vancenase Nasal I.
inheritance
amphigonous i.
autosomal dominant i.

NOTES

inheritance *(continued)*
    autosomal recessive i.
    biparental i.
    codominant i.
    complemental i.
    cytoplasmic i.
    dominant i.
    extrachromosomal i.
    holandric i.
    hologynic i.
    homochronous i.
    homotropic i.
    maternal i.
    matroclinous i.
    mendelian i.
    monofactorial i.
    multifactorial i.
    oligiogenic i.
    polygenic i.
    quantitative i.
    quasicontinuous i.
    recessive i.
    sex-linked i.
    unit i.
    X-linked dominant i.
    X-linked recessive i.
inherited
    i. androgen uptake impairment
    i. coagulopathy
    i. disorder
    i. infertility
    i. thrombophilia in pregnancy
inhibin
    i. A, B
    i. concentration
    i. subunit
    i. test
inhibin-A subunit
inhibited sexual desire
inhibiting
    serum müllerian i.
inhibition
    follicle-stimulating hormone i.
    hemagglutination i. (HI)
    labor i.
    luteinization i.
    pituitary gonadotropin i.
    premature uterine contraction i.
    prostaglandin synthesis i.
    steroid secretion i.
inhibitor
    alpha-glucosidase i.
    alpha-1 protease i.
    alpha-1 proteinase i.
    aromatase i.
    CAMP-specific phosphodiesterase i.
    cell growth i.
    cholinesterase i.

    corticotropin-releasing i.
    dihydrofolate reductase i.
    dual nucleoside analog reverse
      transcriptase i.
    extrinisic pathway i.
    factor VII, VIII i.
    follicle-stimulating hormone
      binding i.
    gonadotropin secretion i.
    human seminal (plasma) i. (HSI)
    luteinization i.
    luteinizing hormone receptor-
      binding i.
    monoamine oxidase i. (MAOI)
    nonnucleoside analog reverse
      transcriptase i.
    oocyte maturation i. (OMI)
    ovum-capture i.
    plasminogen activator i. (PAI)
    prostaglandin synthetase i. (PGSI)
    protease i. (PI)
    serine protease i. (SERPIN)
    serotonin reuptake i.
    specific phosphodiesterase i.
    tissue factor pathway i. (TFPI)
    topoisomerase I i.
    urinary trypsin i.
inhibitor-1
    plasminogen-activator i. (PAI-1)
inhibitor-2
    plasminogen-activator i. (PAI-2)
iniencephaly
iniodymus
iniopagus
iniops
initial apnea
initiated cycle
initiation
    labor i.
    lactation i.
    puberty i.
    sexual i.
initiative
    Human Genome I.
injectable
    i. bromocriptine
    i. contraceptive
injection
    adrenaline i.
    Adrucil I.
    Antagon i.
    AquaMEPHYTON I.
    Cetrorelix for i.
    Chloromycetin I.
    Chlor-Pro I.
    Chlor-Trimeton I.
    continuous subcutaneous insulin i.
    depGynogen I.

Depo-Estradiol I.
Depogen I.
Depo-Provera I.
Diazemuls I.
Dioval I.
direct egg i.
direct intraperitoneal i. (DIPI)
Estra-L I.
Estro-Cyp I.
ganirelix acetate i.
glutaraldehyde cross-linked
  collagen i.
Gynogen L.A. I.
Humalog insulin lispro i.
intracytoplasmic sperm i. (ICSI)
intracytoplasmic sperm head i.
  (ICSHI)
Kefurox I.
Konakion I.
local methotrexate i.
Minocin IV I.
Monistat i.v. I.
monotropins for i.
Nebcin I.
Osmitrol I.
paracervical i.
periurethral collagen i.
pessary i.
Phenazine I.
Phenergan I.
predecondensed spermhead i.
Prometh I.
Prorex I.
round spermatid nuclei i. (ROSNI)
silicone i.
Teflon periurethral i.
Terramycin I.M. I.
Toposar I.
transurethral collagen i.
Valium I.
Zinacef I.

injector

Harris-Kronner uterine manipulator-i.
  (HUMI)
Harris uterine i. (HUI)
MadaJet XL needle-free i.
Mini-Flex flexible Harris uterine i.
Rowden uterine manipulator i.
  (RUMI)

injury

accidental fetal i.
birth i.

brachial plexus i. (BPI)
congenital i.
facial nerve i. (FNI)
hypoxic-ischemic i.
iatrogenic ureteral i.
intraoperative gastrointestinal i.
irradiation i.
lumbosacral plexus i.
mechanical birth i.
muscular i.
nerve i.
neurological i.
obstetric traction i.
pelvic nerve i.
pulmonary i.
rectal i.
i.-related maternal death
spinal cord i.
straddle i.
thermal i.
traumatic i.
traumatic birth i.
urological i.
vaginal i.
vascular i.

inlet

pelvic i.

inner cell mass

innominate bone

Innova

I. electrotherapy system
I. feminine incontinence treatment
  system
I. pelvic floor stimulator

Innovar

Inocor

inoculata

varicella i.

inosinate pyrophoshorylase enzyme

inositol

i. trisphosphate
i. 1,4,5-trisphosphate

inotropic

InPouch TV subculture kit

insemination

artificial i. (AI)
artificial intravaginal i.
cervical i.
cup i.
direct intraperitoneal i. (DIPI)
i. dish
donor i. (DI)

NOTES

insemination *(continued)*
    heterologous i.
    high intrauterine i.
    homologous i.
    intrafollicular i. (IFI)
    intraperitoneal i. (IPI)
    intratubal i. (ITI)
    intrauterine i. (IUI)
    intravaginal i. (IVI)
    Makler i.
    subzonal i. (SUZI)
    i. swim-up technique
    therapeutic i.
    therapeutic donor i. (TDI)
    therapeutic husband i. (THI)
    washed intrauterine i.
insensible fluid loss
insensitive ovary syndrome
insert
    Cervidil vaginal i.
    Femsoft continence i.
    Reliance urinary control i.
insertion
    i. of chromosome
    cord i. (CI)
    interchromosomal i.
    marginal i.
    i. sequence
    subzonal i. (SUZI)
    velamentous i.
insertional mutagenesis
InSight prenatal test
insipidus
    diabetes i.
    fetal diabetes i.
    nephrogenic diabetes i.
insomnia
inspiration time (I-time)
inspiratory/expiratory (I/E)
inspiratory stridor
inspired gas (I)
inspissated
    i. bile syndrome
    i. milk syndrome
inspissation
    amorphous i.
INSS
    International Neuroblastoma Staging
    System
instability
    atlantoaxial i.
    detrusor i.
Instead feminine protection cup
instillation
institute
    National Cancer I. (NCI)
instrument
    Dolphin i.

    EndoMax advanced laparoscopic i.
    Feelings About Yourself i.
    Kevorkian-Younge cervical
    biopsy i.
    LDS i.
    myoma fixation i.
    Newport medical i.
    Polaris reusable i.
instrumental
    i. delivery
    i. vertex
instrumentation
    DeLee i.
insufficiency
    ACTH i.
    adrenal i.
    adrenocortical i.
    aortic valve i.
    corpus luteum i.
    glomerular i.
    mitral i.
    mitral valve i.
    placental i.
    primary ovarian i.
    renal i.
    respiratory i.
    tricuspid i.
    uterine i.
    uteroplacental i.
insufflation
    i. needle
    peritoneal i.
    tubal i.
insufflator
    hysteroscopic i.
    Kidde tubal i.
    laparoscopic i.
    Neal i.
    Semm Pelvi-Pneu i.
Insulatard NPH human
insulin
    beef i.
    human i.
    i.-resistant diabetes, acanthosis
    nigricans, hypogonadism
    pigmentary retinopathy, deafness,
    mental retardation syndrome
    maternal i.
    neutral protamine Hagedorn i.
    Novolin i.
    NPH i.
    i. pen
    pork i.
    i. pump
    regular purified pork i.
    i. resistance
    I. RIA 100
    i. sensitivity

i. shock
i. tolerance test
Ultralente i.
insulinase
insulin-dependent
i.-d. diabetes mellitus (IDDM)
i.-d. diabetic
insulinemia
insulin-like
i.-l. growth factor (IGF)
i.-l. growth factor-1
insulinoma
insulopenia
Insul-Sheath vaginal speculum sheath
inSync miniform
intact membrane
intake and output (I&O)
Intal
integration
visuomotor i.
integrin
beta-1 i.
beta-2 i.
integrin-binding
intelligence
Fagan Test of Infant I.
Wechsler Preschool and Primary
Scale of I.-Revised (WPPSI-R)
intensity/duration (I/T)
interaction
actin-myosin i.
androgen i.
sperm-mucus i.
sperm-oocyte i.
stromal-epithelial i.
interassay
intercapillary distance
INTERCEED
I. barrier material
I. TC7 absorbable adhesion barrier
intercellular adhesion molecule-1
(ICAM-1)
Intercept
interchange
interchromosomal insertion
intercostal retractions
intercourse
sexual i.
timed i.
intercross
interdigitation

interference
acoustical i.
centromere i.
chiasma i.
genetic i.
negative i.
positive i.
interferon (IFN)
i. alfa-2a
i. alfa-2b
i. alfa-N1
i. alfa-n2
i. alfa-n3
i. alfa-NL
alpha i.
alpha-recombinant i.
i. beta
beta i.
i. beta-1a
i. beta-1b
i. beta-recombinant
gamma i.
i. gamma-1b
human leukocyte i. (HLI)
leukocyte i.
lymphoblastoid i.
interfetal membrane
interkinesis
interleukin (IL)
i.-1
i.-1β
i.-10
i.-12
i.-2 through -6
intermedia
alpha-thalassemia i.
thalassemia i.
intermenstrual
i. bleeding
i. pain
intermittent
i. cyclical etidronate disodium
i. fever
i. hydrosalpinx
i. mandatory ventilation (IMV)
i. mechanical ventilation (IMV)
i. pneumatic compression
i. porphyria
i. positive pressure breathing
(IPPB)

NOTES

intermittent *(continued)*
    i. positive pressure ventilation (IPPV)
    i. sterilization
interna
    endometriosis i.
internal
    i. anal sphincter
    i. conjugate
    i. diameter
    i. endometriosis
    i. excrescence
    i. generative organ
    i. genitalia
    i. os
    i. podalic version
    i. podalic version and breech extraction
    i. radiation therapy
    i. rotation
    i. septation
internalization
    hormone-receptor complex i.
    receptor i.
international
    i. classification of cancer of cervix
    I. Committee for Contraceptive Research (ICCR)
    I. Continence Society prolapse stage I, II
    I. Federation of Gynecology and Obstetrics (FIGO)
    I. Federation of Gynecology and Obstetrics classification
    I. Neuroblastoma Staging System (INSS)
    I. Reference Preparation (IRP)
    I. Society for Gynecologic Pathology (ISGP)
    I. Staging System
    I. Union Against Cancer (UICC)
interobserver variation
Interpersonal Support Evaluation List
interphase
interplant
interpretation
    mirror image i.
    i. variability
interrupted suture
interruption
    vena caval i.
intersex
    i. problem
intersexuality
inter-simple sequence repeat
InterStim therapy
interstitial
    i. brachytherapy

    i. cell
    i. cystitis
    i. deletion
    i. infiltrate
    i. irradiation
    i. keratitis
    i. and loculated hematoma
    i. mastitis
    i. nephritis
    i. plasma cell pneumonia
    i. pregnancy
    i. therapy
interthreshold zone
intertriginous candidosis
intertrigo
    chronic i.
intertuberous diameter
interval
    induction-to-delivery i.
    Q-T i.
intervention
    court-ordered obstetrical i.
    legal i.
    mind-body i.
    postmenopausal estrogen and progestin i. (PEPI)
    salvage i.
interventional procedure
intervillous
    i. blood
    i. blood gas
    i. space
intestinal
    i. bag
    i. bypass procedure
    i. conduit
    i. fistula
    i. hurry
    i. lymphangiectasia, lymphedema, mental retardation syndrome
    i. mesentery
    i. obstruction
    i. parasite
    i. peristalsis
    i. tract
intestinalis
    pneumatosis cystoides i.
intimate
    i. partner
    i. partner depression
    i. partner homicide
intoeing
intolerance
    carbohydrate i.
    fructose i.
    gestational carbohydrate i. (GCI)
    glucose i.
    glucose-galactose i.

lactose i.
pigmentary retinopathy,
  hypogonadism, mental retardation,
  nerve deafness, glucose i.
intoxication
  intrarenal androgenic i.
  water i.
intraabdominal
  i. infection
  i. pressure
  i. streak
  i. surgery
intraamniotic
  i. infection
  i. inflammation
intraarterial infusion
intraassay
Intrabutazone
intracanalicular fibroadenoma
intracavitary
  i. brachytherapy
  i. irradiation
  i. myoma
  i. radium
intracellular
  i. calcium
  i. mediator
intracervical
  i. adhesion
  i. purified porcine relaxin
  i. tent
intrachromosomal duplication
intracranial
  i. anatomy
  i. aneurysm
  i. dural vascular anomaly
  i. hemorrhage (ICH)
  i. hypertension
  i. pathology
  i. tumor
intracystic papillary carcinoma
intracytoplasmic
  i. sperm head injection (ICSHI)
  i. sperm injection (ICSI)
IntraDop probe
intraductal
  i. papillary carcinoma
  i. papilloma
intradural spinal angioma
intraepithelial
  i. cervical dysplasia
  i. disease progression

  i. endometrial cancer
  i. lesion
  i. neoplasia
intrafallopian transfer
intrafollicular insemination (IFI)
intrahepatic cholestasis of pregnancy
intraligamentary pregnancy
intraligamentous myoma
Intralipid
  I. hyperalimentation
intralobular connective tissue
intralocal additivity
intraluminal upper airway obstruction
intralymphatic infusion
intramammary
  i. lymph node
intramural
  i. fibroid
  i. myoma
  i. pregnancy
  i. thrombus
  i. upper airway obstruction
intramuscular (IM)
  i. drug
  i. injection of pethidine
intramyometrial
  i. coring
  i. mass
intranasal drug
intranatal
**Intran disposable intrauterine pressure
  measurement**
intranuclear virion
intraoperative
  i. complication
  i. gastrointestinal injury
  i. lymphatic mapping
  i. radiation
intrapair birth weight difference
intrapartum
  i. antibiotic prophylaxis
  i. asphyxiation
  i. cardiotocography
  i. chemoprophylaxis
  i. cord prolapse
  i. death
  i. demise
  i. fetal monitoring
  i. fetoplacental transfusion
  i. head entrapment
  i. hemorrhage

NOTES

intrapartum *(continued)*
    i. monitor
    i. period
intrapelvic
intraperitoneal
    i. blood transfusion
    i. chemotherapy
    i. cisplatin
    i. endometrial metastatic disease
    i. fetal transfusion
    i. 5-fluorouracil
    i. insemination (IPI)
    i. pregnancy
    i. radiation therapy
intrapulmonary
    i. shunt
    i. shunt ratio (Qs/Qt)
intrarenal
    i. androgenic intoxication
    i. reflux
intratesticularly
intrathecal
    i. narcotic
intrathoracic
    i. airway obstruction
    i. tuberculosis
intratubal insemination (ITI)
intrauterine
    i. adhesion
    i. amputation
    i. balloon-type cannula
    i. cirsoid aneurysm
    i. contraceptive device (IUCD)
    i. death (IUD)
    i. device (IUD)
    i. diaphragm
    i. facial necrosis
    i. factors
    i. fetal death
    i. fetal monitoring
    i. filling defect
    i. fracture
    i. GIFT
    i. growth impairment
    i. growth restriction (IUGR)
    i. growth retardation (IUGR)
    i. growth retardation-microcephaly-mental retardation syndrome
    i. hematoma
    i. herpes
    i. infection
    i. insemination (IUI)
    i. insemination cannula
    i. insemination cannula with mandrel
    i. intraperitoneal fetal transfusion
    i. parabiotic syndrome
    i. pneumonia

    i. pregnancy (IUP)
    i. pressure catheter (IUPC)
    i. pressure cycle
    i. pressure measurement
    i. resuscitation
    i. synechia
    i. volume
intravaginal
    i. condom
    i. contraceptive
    i. cream
    i. foam
    i. insemination (IVI)
    i. pouch
    i. prostaglandin
    i. sponge
intravasation
intravascular transfusion
intravenous (I.V.)
    i. alimentation
    i. anti-D immunoglobulin
    i. drug
    i. glucose challenge
    i. hyperalimentation (IVH)
    i. immunoglobulin
    i. infusion
    i. leiomyomatosis
    i. line
    i. medication
    i. pyelogram (IVP)
    i. pyelography (IVP)
    i. urogram (IVU)
intraventricular hemorrhage Grade I–IV (IVH)
intravesical pressure
InTray CCD test
intrinsic
    i. dysmenorrhea
    i. pulsatility
    i. sphincter deficiency (IDS, ISD)
introducer
    P.D. Access with Peel-Away needle i.
introital
    i. papillosis
introitus
    marital i.
    parous i.
    virginal i.
Introl bladder neck support prosthesis
intromission
intron
Intropin
intrusive
    i. implantation
    i. stress reaction
intubation
    endotracheal i.

neonatal i.
tracheal i.
intussusception
cecocolic i.
colic i.
colocolic i.
double i.
ileal i.
ileocecal i.
ileocolic i.
ileoileal i.
jejunogastric i.
retrograde i.
invaginata
trichorrhexis i.
invagination
invasion
early stromal i.
lymphovascular space i. (LVSI)
trophoblastic i.
invasive
i. cancer (IC)
i. carcinoma
i. cervical cancer
i. hydatidiform mole
i. neoplasia
inventory
Children's Depression I. (CDI)
Derogatis Brief Symptom I.
Eyberg Child Behavior I. (ECBI)
infertility perceptions i. (IPI)
Marital Dyadic I.
Minnesota Multiphasic
Personality I. (MMPI)
Spielberger State Anxiety I.
Urogenital Distress I. (UDI)
West Haven-Yale Multidimensional
Pain I.
inverse
i. polymerase chain reaction
i. square law
Inversine
inversion
acute uterine i.
chromosomal i.
i. of chromosomes
i. duplication (15) chromosome
syndrome
i. duplication (8p) syndrome
Hautain uterine i.
paracentric i.
pericentric i.

puerperal i.
i. 9 syndrome
uterine i.
i. of the uterus
i. of viscera
inversus
blepharophimosis, ptosis,
epicanthus i. (BPEI)
epicanthus i.
situs i.
inverted
i. duplication of chromosome 15
i. nipple
i. pelvis
i. repeat
i. subcuticular suture
i. T uterine incision
i. X chromosome
Investa suture
investigation
infertility i.
Invirase
involuntary sterilization
involute
involution
i. cyst
spontaneous i.
i. of the uterus
involutional
i. melancholia
i. psychosis
involvement
congenital muscular dystrophy with
central nervous system i.
gangliosidosis GM1 late onset
without bony i.
generalized infantile gangliosidosis
with bony i.
metastatic axillary i.
multisite lower genital tract i.
neurologic i.
nodal i.
ocular i.
pelvic i.
postponing sexual i. (PSI)
Tay-Sachs disease with visceral i.
INVOS
I. 2100
I. 3100 cerebral oximeter
inv (X)
I&O
intake and output

NOTES

Ioban
  I. drape
  I. 2 iodophor cesarean sheet
iocetamic acid
iodamide
iodide
  i.-containing medication
  potassium i.
  propidium i.
  saturated solution of potassium i.
    (SSKI)
  sodium i.
  i. therapy
  i. trap defect
iodinated
  i. dye
  i. glycerol
iodine
  i.-123 (^{123}I)
  i.-125 (^{125}I)
  i.-127 (^{127}I)
  i.-131 (^{131}I)
  i.-132 (^{132}I)
  i. douche
  i. 125-labeled fibrinogen scan
  i. stain
  i. supply
iodipamide
iodomethyl-norcholesterol scanning
iodophor
  i.-impregnated adhesive drape
iodoquinol
Iodosorb absorptive dressing
iodothyronine
  i. level
Iofed
  I. PD
iohexol
ion
  calcium i.
Ionamin
Ionasescu syndrome
ionization
ionized calcium (iCa)
ionizing radiation
iopanoic acid
iothalamate sodium
Iowa trumpet
Ipecac
IPI
  infertility perceptions inventory
  intraperitoneal insemination
IPKD
  infantile polycystic kidney disease
ipodate sodium
IPPB
  intermittent positive pressure breathing

IPPV
  intermittent positive pressure ventilation
ipratropium
  i. bromide
  i. bromide nasal spray
iprindole
iproniazid
ipsilateral
ipsilon zone
IPV
  inactivated poliovirus vaccine
Ir
  iridium
    Ir gene
^{192}Ir
  iridium-192
IRDS
  idiopathic respiratory distress syndrome
  infant respiratory distress syndrome
irides (*pl. of* iris)
iridium (Ir)
  i.-192 (^{192}Ir)
  i. implant
  i. wire
iridocorneal mesodermal dysgenesis
iridocyclitis
iridodental dysplasia
iridogoniodysgenesis with somatic
  anomalies
irinotecan
iris, pl. **irides**
  i., coloboma, ptosis, hypertelorism,
    mental retardation syndrome
  dysgenesis mesodermalis cormeae et
    irides
  i. hypoplasia
  speckled irides
iritis
  i. catamenialis
IRMA
  immunoradiometric assay
Iromin-G
iron (Fe)
  carbonyl i.
  i. deficiency anemia (IDA)
  i. dextran
  elemental i.
  I. Intern retractor
  MicroIron II carbonyl i.
  Pedicran with I.
  i. requirement
  serum i.
  Similac-24-LBW with whey and i.
  Similac-24 with i.
  i. turnover
iron-binding capacity (IBC)
Irospan

IRP
International Reference Preparation
irradiation
abdominal i.
abdominopelvic i.
axillary i.
cesium i.
i. cystitis
i. dermatitis
external beam i.
heavy-ion i.
hemibody i.
i. injury
interstitial i.
intracavitary i.
local i.
ovarian-sparing i.
paraaortic node i.
pelvic i.
surface i.
whole-abdomen i.
whole-body i.
whole-pelvis i.
irregularity
menstrual i.
irrigant
Neosporin G.U. I.
irrigation
copious antibiotic i.
irritability
reflex i.
irritable
i. bowel syndrome (IBS)
i. breast
Irving
I. method
I. tubal ligation
Isaac syndrome
ischemia
chorionic villus i.
myocardial i.
ischemic heart disease
ischiadelphus
ischiocavernosus muscle
ischiocavernous
ischiodidymus
ischiomelus
ischiopagus
i. tripus separation
i. tripus twins
ischiopatellar dysplasia

ischiopubica
osteochondritis i.
ischiopubic ramus
ischiopubiotomy
Galbiati bilateral fetal i.
ischiorectal fossa
ischiothoracopagus
ISD
intrinsic sphincter deficiency
IS element
Iselin forceps
isethionate
pentamidine i.
ISG
immune serum globulin
Is gene
ISGP
International Society for Gynecologic
Pathology
Ishihara
POU theory of I.
I. theory
island
Langerhans i.'s
Pander i.'s
islet
i. cell adenoma
i. cell tumor
i.'s of Langerhans
Ismelin
isoamylase
isoantibody
isoantigen
Iso-Bid
isocarboxazid
isochromosome
i. 10p syndrome
i. 12p syndrome
isodose curve
isoenzyme
isoetharine
isofluorphate
isoflurane
isoform
glutathione S-transferase i.
isogeneic graft
isograft
isohemagglutinin
isoimmune
i. fetal thrombocytopenia
isoimmunization
antepartum Rh i.

NOTES

isoimmunization (*continued*)
    D i.
    D-antigen i.
    Kell i.
    i. in pregnancy
    Rh i.
    rhesus i.
isointense
Isojima-Koyama test
Isojima test
isolated
    i. autosomal dominant syndrome
    i. double outlet right ventricle
isolette
    Airshields I.
    bubble i.
    double-bubble i.
    double-walled bubble i.
Isolin
isologous neoplasm
isomer
isomerase
    triose phosphate i.
isomerism
    left atrial i. (LAI)
Isomil
isoniazid (INH)
    prophylactic i.
Isopaque
Isophrin
isoprenaline hydrochloride
isopropamide iodide
isoproterenol
Isoptin
Isopto
    I. Carbachol
    I. Carpine
    I. Cetamide
    I. Eserine
    I. Frin
sordil
sosexual idiopathic precocious puberty
isosorbide dinitrate
Isotamine
isothenuria
isothiocyanate
    fluorescein i.
isotonic PBS
isotope
    i. cisternography
    radioactive i.
    i. scanning
Isotrate
isotretinoin
    i. dysmorphic syndrome
    i. embryopathy
    i. teratogenic syndrome
isovaleric acidemia, isovalericacidemia

iso-X chromosome
I-Soyalac formula
isozyme
ISSD
    infantile sialic acid storage disorder
issue
    legal i.
    social i.
isthmi (*pl. of* isthmus)
isthmica nodosa
isthmic occlusion
isthmointerstitial anastomosis
isthmorrhaphy
isthmus, pl. **isthmuses, isthmi**
    cervical i.
Isuprel
I/T
    intensity/duration
    I/T ratio
itch
    Absorbine Jock I.
    i.-scratch cycle
Itch-X
    I.-X gel
    I.-X spray
itersonii
    Aquaspirillum i.
ITI
    intratubal insemination
I-time
    inspiration time
Ito
    I. cell
    hypomelanosis of I.
    I. hypomelanosis
    I. method
    nevus of I.
    I. nevus
    I. syndrome
ITP
    idiopathic thrombocytopenic purpura
    immune thrombocytopenic purpura
itraconazole
IUCD
    intrauterine contraceptive device
IUD
    intrauterine death
    intrauterine device
    ParaGard T380 copper IUD
IUD-related infection
IUGR
    intrauterine growth restriction
    intrauterine growth retardation
IUI
    intrauterine insemination
    IUI disposable cannula
IUP
    intrauterine pregnancy

IUPC
    intrauterine pressure catheter
I.V.
    intravenous
        Indocin I.V.
        Merrem I.V.
        Metro I.V.
        Monistat I.V.
Ivemark syndrome
IVF
    in vitro fertilization
IVF-ET
    in vitro fertilization-embryo transfer
IVF-induced abdominal pregnancy

IVH
    intravenous hyperalimentation
    intraventricular hemorrhage Grade I–IV
IVI
    intravaginal insemination
ivory bones
IVP
    intravenous pyelogram
    intravenous pyelography
        genitogram with or without IVP
IVU
    intravenous urogram
Ivy bleeding time

NOTES

J

joule
   J needle
   J pulmonary receptor
Jabouley amputation
Jabs syndrome
jackknife
   j. position
   j. seizure
   j. spasm
Jackson
   J. membrane
   J. right-angle retractor
jacksonian epilepsy
Jackson-Pratt suction drain
Jackson-Weiss syndrome (JWS)
Jacobsen-Brodwall syndrome
Jacobsen syndrome
Jacobson forceps
Jacobs tenaculum
Jacob syndrome
Jacquemier sign
Jacquet
   J. dermatitis
   J. erythema
Jadassohn
   J. nevus phakomatosis (JNP)
   nevus sebaceous of J. (SNJ)
   J. test
Jadassohn-Lewandowski syndrome
Jadassohn-Tieche nevus
Jaeken syndrome
Jaffe-Campanacci syndrome
Jaffe-Gottfried-Bradley syndrome
Jaffe-Lichtenstein syndrome
Jahnke syndrome
Jaksch
   J. anemia
   J. disease
   J. syndrome
jamais vu
Janacek reimplantation set
Jancar syndrome
Janeway lesion
janiceps
   j. asymmetrus
   j. parasiticus
   j. twins
Janimine
Janosik embryo
Jansen syndrome
Jansky-Bielschowsky syndrome
Jansky classification
Janus report

japonicum
   Laminaria j.
jar
   Gas-Pak j.
Jarcho-Levin syndrome
Jarisch-Herxheimer reaction
Jarit
   J. disposable cannula
   J. disposable trocar
Jatene procedure
jaundice
   black j.
   breast milk j.
   catarrhal j.
   central j.
   cholestatic j.
   congenital hemolytic j.
   congenital nonhemolytic j.
   congenital obliterative j.
   neonatal cholestatic j.
   j. of newborn
   nuclear j.
   obstructive j.
   peripheral j.
   physiologic j.
   Schmorl j.
jaw
   bird-beak j.
   cleft j.
   j. cysts, basal cell tumors, skeletal
      anomalies syndrome
   fibrous dysplasia of j.
   J.'s forceps
   j. myoclonus
   parrot j.
JDM
   juvenile diabetes mellitus
Jehovah's Witness
jejunal
   j. atresia
jejuni
   Campylobacter j.
jejunogastric intussusception
jejunoileal
   j. atresia
   j. bypass
jello sign
jelly
   Aci-Jel vaginal j.
   j. bean
   contraceptive j.
   Gynol II contraceptive j.
   K-Y lubricating j.
   Wharton j.
Jenamicin

J

Jenest-28
Jensen syndrome
jeopardy
    fetal j.
Jervell-Lange-Nielson syndrome
Jervis syndrome
Jessner-Cole syndrome
jet
    j. cooling effect
    high-frequency j. (HFJ)
    ureteral j.
Jeune
    J. disease
    J. syndrome
Jew
    Ashkenazi J.
jeweler's forceps
Jewett classification
Jirasek stage
jitteriness
jittery
    j. baby
    j. infant
JMS
    Juberg-Marsidi syndrome
JNP
    Jadassohn nevus phakomatosis
Job syndrome
Jocasta complex
JODM
    juvenile-onset diabetes mellitus
Joel-Cohen incision
jogger's amenorrhea
Johanson-Blizzard syndrome
Johnie Mel syndrome
Johnson
    J. method
    J. neuroectodermal syndrome
    J. score 1–10
    J. transtracheal oxygen catheter
Johnson-McMillin syndrome
joint
    Clutton j.'s
    j. deformity
    j. hyperlaxity
    j. hypermobility
    pelvic j.
    j. probability
    sacroiliac j.
Jones
    J. criteria
    J. and Jones wedge technique
Jorgenson scissors
Joseph
    J. disease
    J. syndrome
Josephs-Blackfan-Diamond syndrome
Joubert-Boltshauser syndrome

Joubert syndrome
joule (J)
Jr.
    Aerolate Jr.
    Caltrate, Jr.
    EpiPen Jr.
JRA
    juvenile rheumatoid arthritis
Juberg-Hayward syndrome
Juberg-Holt syndrome
Juberg-Marsidi syndrome (JMS)
Juers forceps
jugular
    j. occlusion plethysmography
    j. venous pressure (JVP)
juice
    j. baby
    white grape j.
Juliusberg
    J. pustulosis
    J. pustulosis vacciniformis acuta
jumping
    chromosome j.
    j. gene
junction
    cervicovaginal j.
    ectoendocervical j.
    gap j.
    squamocolumnar j.
    ureterovesical j. (UVJ)
    uterotubal j. (UTJ)
junctional rhythm
Junior Strength Motrin
Junius-Kuhnt syndrome
junky lung
justifiable abortion
just-viable fetus
JustVision diagnostic ultrasound system
Juvara fold
juvenile
    j. aldosteronism
    j. carcinoma
    j. cataract, cerebellar atrophy, mental retardation, myopathy syndrome
    j. diabetes
    j. diabetes mellitus (JDM)
    j. epithelial corneal dystrophy
    j. GM1 gangliosidosis
    j. hyperuricemia syndrome
    j. kyphosis
    j. osteomalacia
    j. papillomatosis
    j. pelvis
    j. pernicious anemia
    j. plantar dermatitis
    j. polyarthritis
    j. rheumatoid arthritis (JRA)

j. spermatogonial depletion
j. spinal muscular atrophy
j. sulfatidosis
j. xanthogranuloma (JXG)
juvenile-onset
j.-o. diabetes
j.-o. diabetes mellitus (JODM)
juvenilis
kyphoscoliosis dorsalis j.
kyphosis dorsalis j.
osteochondritis deformans j.
osteodystrophia j.
verruca plana j.
juxtaductal
j. aortic coarctation
j. coarctation

juxtaglomerular
j. hyperplasia syndrome
juxtamedullary
juxtaposition
Juzo-Hostess two-way stretch compression stocking
JVP
jugular venous pressure
JWS
Jackson-Weiss syndrome
JXG
juvenile xanthogranuloma

J

NOTES

K

K cell
K chain
Kabikinase
Kabuki
   K. makeup syndrome (KMS)
   K. syndrome (KS)
Kadian sustained-release morphine capsules
Kagan staging system
Kahn
   K. cannula
   K. test
Kajava classification
KAL
   KAL gene
   KAL protein
kala-azar
Kalcinate
Kalginate dressing
Kalischer syndrome
Kaliscinski ureteral procedure
kallikrein
Kallmann-de Morsier syndrome
Kallmann syndrome
Kalmuk idiocy
Kaltostat packing
kanamycin
Kanana Banana
kangaroo
   K. Care
   K. infusion pump
   k. pouch
kangarooing
Kanner syndrome
Kanter sign
Kantrex
Kantu sign
Kaochlor
kaolin clotting time
Kaon
   K.-Cl
Kapeller-Adler test
Kaplan-Meier
   K.-M. method
   K.-M. survival curve
Kaplan syndrome
Kaposi
   K. sarcoma
   K. varicelliform eruption
   K. varicelliform sarcoma
kappa chain
Kapur-Toriello syndrome
karaya

Karl
   K. Storz flexible ureteropyeloscope
   K. Storz 15 French flexible hysteroscope
Karman cannula
Karnofsky
   K. performance status
   K. performance status of chemotherapy
Karo syrup
Kartagener syndrome
karyogenesis
karyokinesis
karyopyknosis
karyopyknotic index
karyosome
karyothecakatadidymus
karyotype
   abnormal k.
   k. analysis
   atypical k.
   female k.
   fetal k.
   male k.
   maternal k.
   parental k.
   paternal k.
   spectral k. (SKY)
   Turner phenotype with normal k.
   45,X k.
   XX k.
   46,XX k.
   47,XX k.
   XXX k.
   47,XXY k.
   XY k.
   46,XY k.
   47,XY k.
   46,XY/47,XY k.
karyotyping
   flow k.
Kasabach-Merritt syndrome
Kasai
   K. peritoneal venous shunt
   K. procedure
Kashin-Bek disease
Kasnelson syndrome
Kasof
Kass criteria
Kato
Kaufman
   K. Assessment Battery for Children
   K. oculocerebrofacial syndrome
   K. pneumonia
   K. syndrome 3

K

Kaufman-McKusick syndrome
Kaveggia syndrome
Kawasaki disease (KD)
Kay Ciel
Kayexalate
Kaylixir
Kayser-Fleischer ring
kb
  kilobase
    kb pair
KBG syndrome
kc
  kilocycle
kcal
  kilocalorie
KD
  Kawasaki disease
  Krabbe disease
KDC-Healthdyne nonfluorescent spotlight
KDF-2.3
    K. intrauterine catheter
    K. intrauterine insemination cannula
Kearns-Sayre syndrome
keel chest
keeled breast
Keeler
    K. fiberoptic headlight
    K. loupe
to keep open (TKO)
keep vein open (KVO)
Keflex
Keflin
Keftab
Kefurox
    K. Injection
Kefzol
Kegel exercises
Kehr
    K. incision
    K. technique
Keiland forceps
Keipert syndrome
Kell
    K. antibodies
    K. hydrops
    K. isoimmunization
    K. sensitization
    K. test
Keller syndrome
Kelly
    K. clamp
    K. operation
    K. plication
    K. plication procedure
    K. retractor
    K. syndrome
Kelly-Gray curette
Kelly-Kennedy plication

keloid
Kemadrin
Kenalog
    K. in Orabase
    K. Topical
Kendall McGaw Intelligent pump
Kennedy-Pacey operation
Kennedy procedure
Kenny-Caffey syndrome
Kenny-Linarelli-Caffey syndrome
Kenny-Linarelli syndrome
Kenny syndrome
Kent Infant Development Scale (KIDS)
Keofeed tube
Keplerian parabola
keratan sulfaturia
keratin cyst
keratinization
keratinocyte
    k. growth factor (KGF)
keratitis
    k. fugax hereditaria
    interstitial k.
    palmar and plantar keratosis
    and k.
keratitis-ichthyosis-deafness (KID)
keratitis, ichthyosis, deafness syndrome
keratoconjunctivitis
    tuberculous k.
keratoconus
keratoderma
    mutilating k.
keratolysis neonatorum
keratolytic gel
keratoma hereditarium mutilans
keratopathy
    band k.
keratosis
    k. follicularis
    k. follicularis spinulosa decalvans
    k. palmaris et plantaris
    k. palmaris et plantaris-corneal
      dystrophy syndrome
    k. palmoplantaris
    k. palmoplantaris-corneal dystrophy
      syndrome
    k. pilaris
    seborrheic k.
Kergaradec sign
kerion
Kerlix gauze bandage
kernicterus
Kernig sign
Kerr cesarean
Kesaree-Wooley syndrome
Kessner Index
Kestrone
Ketalar

ketamine
ketoacidosis
    diabetic k. (DKA)
    starvation k.
ketoaciduria
    ataxia-deafness-retardation with k.
ketoaciduria-mental deficiency syndrome
ketoconazole
3-ketodesogestrel
11-ketoetiocholanolone
ketogenic
    k. diet
    17-k. steroid
ketone body
ketonemia
ketonuria
ketoprofen
ketorolac tromethamine
ketosis
    k.-prone diabetes
    k.-resistant diabetes
    starvation k.
17-ketosteroid reductase deficiency
17-ketosteroids (17-KS)
ketotic
    k. hyperglycemia
Kety-Schmidt technique
Keutel syndrome 1 & 2
kev
    kilo-electronvolt
Kevorkian
    K. curette
    K. punch biopsy
Kevorkian-Younge
    K.-Y. biopsy forceps
    K.-Y. cervical biopsy instrument
    K.-Y. curette
Keyes
    K. dermatologic punch
    K. punch biopsy
    K. vulvar punch
key-in-lock
    k.-i.-l. incontinence
    k.-i.-l. maneuver
kg
    kilogram
KGF
    keratinocyte growth factor
kHz
    kilohertz
Ki67 antibody
Kibrick-Isojima infertility test

Kibrick test
kick count
KID
    keratitis-ichthyosis-deafness
    KID syndrome
Kidd blood group
Kidde
    K. cannula technique
    K. tubal insufflator
kidney
    k. biopsy
    cystic k.
    k. dialysis
    k. disease
    k. failure
    k. function
    k. morphology
    pelvic k.
    single k.
    solitary k.
    k. stone
    k. transplantation
kidney-derived growth factor
KIDS
    Kent Infant Development Scale
Kielland rotation
kifafa seizure disorder
Kikuchi disease
Kilian pelvis
kill
    cell k.
    log cell k.
killed virus vaccine
killer
    k. cell
    natural k. cell
Killian syndrome
kilobase (kb)
    k. pair
kilocalorie (kcal)
kilocycle (kc)
kilo-electronvolt (kev)
kilogram (kg)
kilohertz (kHz)
Kimmelstiel-Wilson
    K.-W. disease (KW)
    K.-W. syndrome
kinase
    creatine k. (CK)
    glycerol k. (GK)
    histone H1 k.
    myosin light-chain k.

K

NOTES

287

kinase *(continued)*
    phosphoglycerate k. (PGK)
    pyruvate k. (PK)
    serine-threonine k.
    tyrosine k.
kind
    error of the first k.
    error of the second k.
kindred
    degree of k.
kinesiologic
kinetics
    cell k.
kinetocardiotocograph
kinetochore
kinin
kinked
    k. cord
    k. midbrain
kinky-hair
    k.-h. disease
    k.-h. syndrome
Kinsbourne syndrome
Kinyoun stain
KIO syndrome
Kirghizian dermatoosteolysis
Kirmisson respirator
Kirsten murine sarcoma virus
Kish urethral illuminating catheter
kisses
    angel's k.
kissing ulcer
kit
    ABI PRISM Dye Terminator Cycle
      Sequencing Ready Reaction K.
    Amni-Glove N Gel k.
    Amplicor HIV-1 test k.
    Amplicor PCR k.
    Amplicor typing k.
    Apoptag Plus k.
    Centocor CA 125
      radioimmunoassay k.
    cervical block k.
    Confide HIV test k.
    Hy-Gene seminal fluid collection k.
    InPouch TV subculture k.
    Male FactorPak seminal fluid
      collection k.
    Metra PS procedure k.
    newborn screening k.
    Ortho diaphragm k.
    Otovent autoinflation k.
    OvuGen test k.
    OvuKIT Self-Test k.
    OvuQuick One-Step ovulation k.
    OvuQuick Self-Test k.
    PregnaGen test k.

    Preven emergency contraception k.
    Progesterone Radioimmunoassay K.
    QIAamp Tissue k.
    rape evidence k.
    SAFE k.
    SureCell Chlamydia Test k.
    SureCell rapid test k.
    TAGO diagnostic k.
    urinary ovulation predictor k.
    Uri-Three urine culture k.
    Vesica sling k.
    VIDAS Estradiol II assay k.
Kitano knot
Kitchen postpartum gauze packer
Kitner dissector
Kitzinger method of childbirth
Kjelland-Barton forceps
Kjelland forceps
Kjelland-Luikart forceps
Kjer-type dominant optic atrophy
Klavikordal
Klebsiella
    K. oxytoca
    K. ozaenae
    K. pneumoniae
kleeblattschädel syndrome
Kleihauer
    K. technique
    K. test
Kleihauer-Betke
    K.-B. stain
    K.-B. test
Klein-Waardenburg syndrome
Kleppinger
    K. bipolar forceps
    K. envelope sign
Klinefelter
    K. syndrome
    K. variant
Klinefelter-Reifenstein-Albright syndrome
Klinefelter-Reifenstein syndrome
Kline test
Kling bandage
Klippel-Feil syndrome
Klippel-Trenaunay-Parkes-Weber
    syndrome
Klippel-Trenaunay syndrome
Klippel-Trenaunay-Weber syndrome
Kloepfer syndrome
Klonopin
K-Lor
Klor-Con
Klorominr Oral
Klorvess
Klotrix
Klotz syndrome
Kluge method

Klumpke
    K. palsy
    K. paralysis
Klumpke-Dejerine paralysis
Klüver-Bucy syndrome
K-Lyte
KM-1 breast pump
K-Medic
KMS
    Kabuki makeup syndrome
knee
    knock k.'s
    k. presentation
knee-chest position
knee-elbow position
Kniest
    K. dwarfism
    K. dysplasia
    K.-like dysplasia
    K. syndrome
knob
    chromosome k.
Knobloch-Gesell test
knock knees
knockout
knot
    Aberdeen k.
    clinch k.
    Duncan k.
    false k.
    false k. of umbilical cord
    fisherman's k.
    granny k.
    Kitano k.
    laparoscopic slip k.
    modified Roeder k.
    primitive k.
    k. pusher
    Roeder k.
    4S k.
    square k.
    surgeon's k.
    syncytial k.
    true k. of umbilical cord
    Weston k.
knuckle of tube
Koala intrauterine pressure catheter
Koate-HP
Kobayashi vacuum extractor
Kobberling-Dunnigan syndrome
Köbner phenomenon
Koby syndrome

Kocher clamp
Kocher-Debré-Sémélaigne syndrome
kocherize
Koch postulates
Kock pouch
Kodak
    K. hCG serum test
    K. SureCell Chlamydia Test
    K. SureCell hCG-Urine Test
    K. SureCell Herpes (HSV) Test
    K. SureCell LCH in-office
        pregnancy test
    K. SureCell Strep A test
Koerber-Salus-Elschnig syndrome
Kogan endocervical speculum
KoGENate
KOH
    potassium hydroxide
    KOH colpotomizer system
    KOH prep
    KOH stain
    KOH test
Kohdschnetter disease
Köhler bone disease
koilocyte
koilocytic atypia
koilocytosis
koilocytotic
    k. atypia
    k. cell
koilonychia
Kok disease
Kolmer-Kline-Kahn test
Kolmer test
Konakion
    K. Injection
Kondon's Nasal
Kono procedure
Kontron electrode
Konyne 80
Kopan needle
Koplik spot
Korenman estrogen window hypothesis
Koromex
Korotkoff
    K. sound
    K. test
Kosenow-Sinios syndrome
Kostmann
    K. infantile agranulocytosis
    K. neutropenia
    K. syndrome

K

NOTES

Kovalevsky canal
Kowarski syndrome
Kozlowski disease
Krabbe
    K. disease (KD)
    K. leukodystrophy
Kramer
    K. disease
    K. syndrome
Kraske position
K-*ras* oncogene
kraurosis vulvae
Krause
    K. disease
    K. syndrome
Krause-Kivlin syndrome
Krause-van Schooneveld-Kivlin syndrome
Kreiselman infant warmer
Kremer
    K. penetration test
Krisovski sign
Kristeller
    K. maneuver
    K. method
Kroner tubal ligation
Krönig technique
Kronner
    K. Manipujector
    K. Manipujector uterine
      manipulator-injector
    K. manipulator
Kruger index
Krukenberg tumor
krusei
    Candida k.
Kruskal-Wallis test
KS
    Kabuki syndrome
    KS mucopolysaccharidosis
17-KS
    17-ketosteroids
KT
    Orudis K.

K-Tab
KTP laser
Kučera equation
Kufs disease
Kugelberg-Welander disease
Kun colocolpopoiesis
Kupperman
    K. index
    K. test
Kurzrok-Miller test
Kurzrok-Ratner test
Kussmaul respiration
Küstner
    K. law
    K. sign
Kveim
    K. antibody
    K. test
KVO
    keep vein open
KW
    Kimmelstiel-Wilson disease
kwashiorkor
Kwell
Kwellada
K-Y lubricating jelly
kynocephalus
Kyotest
kyphomelic dysplasia
kyphoscoliosis
    cataract, microcephaly, failure to
      thrive, k. (CAMFAK)
    k. dorsalis juvenilis
kyphoscoliotic heart disease
kyphosis
    cataract, microcephaly,
      arthrogryposis, k. (CAMAK)
    k. dorsalis juvenilis
    juvenile k.
kyphotic pelvis
Kytril

L
   liter
LA
      Comhist LA
      Dexone LA
      Entex LA
      Inderal LA
      Zephrex LA
L.A.
      Gynogen L.A.
      Theoclear L.A.
La
      La Leche League
      La (SS-B) autoantigen
LA:A
   left atrial to aortic ratio
LAAH
   laparoscopic-assisted abdominal
   hysterectomy
Laband syndrome
labeling
      primed in situ l. (PRINS)
labetalol
labia (*pl. of* labium)
labial
      l. agglutination
      l. edema
      l. fusion
      l. hernia
      l. reflex
labialis
      herpes l.
      micropapillomatosis l.
labioperineal pouch
labioscrotal
      l. swelling
labium, pl. labia
      hypoplastic labia
      labia majora
      l. majus pudendi
      labia minora
labor
      abnormal l.
      active phase of l.
      arrest of l.
      l. augmentation
      l. augmentation induction
      l. and delivery (L&D)
      l., delivery, and recovery (LDR)
      desultory l.
      dry l.
      false l.
      first stage of l.
      induced l.
      l. inhibition

      l. initiation
      latent phase of l.
      mimetic l.
      missed l.
      oxytocin stimulation of l.
      l. pains
      placental stage of l.
      precipitate l.
      precipitous l.
      premature l.
      preterm l. (PTL)
      primary dysfunctional l.
      prodromal l.
      prolonged l.
      second stage of l.
      spontaneous l.
      stages of l.
      third stage of l.
      l. trial
      true l.
      vaginal birth after cesarean—trial
      of l. (VBAC-TOL)
laboratory
      l. finding
      l. test
Labotect catheter
labyrinthine
      l. placenta
      l. reflex
labyrinthitis
LAC
   lupus anticoagulant
laceration
      anal sphincter l.
      aortic l.
      birth canal l.
      bladder l.
      cervical l.
      falx l.
      first-degree l.
      fourth-degree l.
      perineal l.
      scalp l.
      second-degree l.
      tentorial l.
      third-degree l.
      vaginal l.
Lacks
      Henrietta L. (HeLa)
lacmoid staining solution
lacrimal
      l. duct
lacrimation
      gustatory l.
lacrimoauriculodentodigital syndrome

L

Lact-Aid STARTrainer Nursing System & trade
lactate
    amrinone l.
    cyclizine l.
    l. dehydrogenase (LDH)
lactated
    l. Ringer
    l. Ringer solution
lactating
    l. adenoma
    l. breast
    l. woman
lactation
    l. amenorrhea
    l. disorder
    inappropriate l.
    l. initiation
    l. letdown response
lactational
    l. amenorrhea method (LAM)
    l. mastitis
lactea
    crusta l.
lacteal
    l. cyst
    l. fistula
lactic
    l. acid dehydrogenase
    l. acidemia
    l. acidosis
LactiCare-HC Topical
lactiferous
    l. duct
lactifugal
lactifuge
lactigenous
lactigerous
Lactina breast pump
Lactobacillus
    L. acidophilus
    L. bifidus
lactobacillus, pl. **lactobacilli**
    hydrogen peroxide-producing l.
lactobezoar
lactocele
lactoferrin
lactoflavin
Lactofree formula
lactogen
    human placental l. (hPL, HPL)
    placental l.
lactogenesis
lactogenic
lactorrhea
lactose
    l. intolerance

lactose-free
    l.-f. diet
    l.-f. formula
lactosuria
lactotropin
lactulose
    l. enema
lacuna, pl. **lacunae**
lacunar skull
LAD
    left atrial diameter
Ladd
    L. band
    L. procedure
    L. syndrome
Ladin sign
Laerdal resuscitator
laetrile
Lafora
    L. body
    L. body disease
lag
    anaphase l.
    homeostatic l.
lagophthalmia
lagophthalmos
Lahey
    L. clamp
    L. forceps
LAI
    left atrial isomerism
LAIT
    latex agglutination inhibition test
LAK
    lymphokine-activated killer cell
lake
    subchorial l.
Lalonde delicate hook forceps
LAM
    lactational amenorrhea method
Lamaze method
lambda suture line
Lambert
    canal of L.
    L. canals
    L. syndrome
Lambotte syndrome
lamellar
    l. body
    l. body count (LBC)
    l. desquamation of newborn
    l. exfoliation
    l. ichthyosis
lamina, pl. **laminae**
    basal l.
    l. propria

Laminaria
    L. digitata
    L. japonicum
laminaria
    l. cervical dilator
    Dilapan l.
    l. infusion
    l. tent
lamination
laminin
L-amino acid
Lamisil Cream
lamivudine
lamp
    Nightingale examining l.
    Wood l.
lampbrush chromosome
Lanacane
Lanacort Topical
Landau
    L. reflex
    L. test
Landing syndrome
Landouzy-Dejerine
    facioscapulohumeral syndrome
      of L.-D.
    L.-D. muscular dystrophy
Landovski nucleoid
Landry-Guillain-Barré syndrome
Landry palsy
landscape
    adaptive l.
Langdon
    L. Down disease
    L. Down syndrome
Lange
    Brachmann-Cornelia de L. (BCDL)
    Cornelia de L. (CDL)
    L. test
Lange-Akeroyd syndrome
Langer
    L. dysplasia
    L. mesomelic dwarfism
    L. syndrome
Langer-Giedion syndrome
Langerhans
    L. cell
    L. cell histiocytosis
    L. granule
    L. islands
    islets of L.
Langer-Petersen-Spranger syndrome

Langer-Saldino syndrome
Langhans
    L. cell
    L. giant cell granuloma
    L. layer
    L. stria
Laniazid
lanolin
Lanophyllin
Lanoxin
lansoprazole
lanuginous
lanugo
Lanvis
lap
    laparotomy
    lap count
    ex lap
    Lap Sac
    lap tape
laparoelytrotomy
laparohysterectomy
laparohystero-oophorectomy
laparohysteropexy
laparohysterosalpingo-oophorectomy
laparohysterotomy
Laparolift system
laparomyomectomy
LaparoSAC single-use obturator and
   cannula
laparosalpingectomy
laparosalpingo-oophorectomy
laparosalpingotomy
laparoscope
    Lent l.
    MiniSite l.
    Storz l.
    Surgiview l.
    Weerda l.
    Wolf l.
laparoscopic
    l. cautery
    l. cholecystectomy (LC)
    l. Döderlein hysterectomy
    l. electrocautery
    l. fenestration
    l. insufflator
    l. leash
    l. lymphadenectomy
    l. management
    l. microsurgery
    l. multiple-punch resection

L

NOTES

laparoscopic *(continued)*
   l. oophoropexy
   l. ovarian diathermy
   l. resection of the ureterosacral
     ligament (LUSLR)
   l. retropubic colposuspension
   l. sacrocolpopexy
   l. slip knot
   l. sonography
   l. supracervical hysterectomy (LSH)
   l. trocar
   l. tubal ligation (LTL)
   l. unipolar coagulation procedure
laparoscopic-assisted
   l.-a. abdominal hysterectomy
     (LAAH)
   l.-a. vaginal hysterectomy (LAVH)
   l.-a. vaginal hysteroscopy (LAVH)
laparoscopy
   gasless l.
   laser l.
   operative l.
   pelvic l.
   l. port
   second-look l.
LaparoSonic coagulating shears (LCS)
laparotomy (lap)
   exploratory l.
   l. incision
   second-look l.
laparotrachelotomy
laparouterotomy
Lapides technique
Lapro-Clip ligating clip system
LapTie
Lapwall sponge
Largactil
large
   l. cisterna magna
   l. endometrioma
   l. for gestational age (LGA)
   l. intestine neoplasm
   l. loop excision of transformation
     zone (LLETZ)
   l. single copy
large-for-dates
   l.-f.-d. infant
   l.-f.-d. uterus
Larmarck theory
Larodopa
Laron
   L. dwarfism
   L. syndrome
Larsen syndrome
larva
   l. currens
   l. migrans

laryngeal
   l. abductor paralysis
   l. adductor paralysis
   l. cleft
   l. papillomatosis
   l. stridor
   l. web
laryngitis
laryngomalacia
laryngoscope
   Andrews infant l.
   Holinger infant l.
   pencil-handled l.
   Siker l.
laryngoscopy
laryngospasm
laryngotracheobronchitis
laryngotracheoesophageal
laryngotracheomalacia
larynx
Larzel anemia
laser
   l. ablation
   alexandrite l.
   ArF excimer l.
   argon l.
   l. blanching
   Candela l.
   carbon dioxide l.
   l. cervical conization
   CO_2 l.
   Culpolase l.
   l. ejecta
   Er:YAG l.
   Ho:YAG l.
   Illumina Pro Series laparoscopic l.
   KTP l.
   l. laparoscopy
   Merimack 1040 CO2 l.
   l. method
   Nd:YAG l.
   OPMILAS CO2 l.
   l. photocoagulation
   l. photocoagulation of the
     communicating vessel (LPCV)
   l. photovaporization
   l. plume
   l. reaction
   SPTL vascular lesion l.
   l. surgery
   Surgicenter 40 CO2 l.
   Surgilase 55W l.
   l. therapy
   l. treatment
   l. uterosacral nerve ablation
     (LUNA)
   l. vaporization

YAG l.
 yttrium-aluminum-garnet l.
laser-Doppler flowmeter
lasered
Lash
 L. operation
 L. procedure
Lasix
last
 l. menstrual period (LMP)
 l. normal menstrual period (LNMP)
lata (*pl. of* latus)
latae
 tensor fasciae l. (TFL)
late
 l. apnea
 l. deceleration
 l. embryonic testicular regression
  syndrome
 l. infantile systemic lipidosis
 l. luteal phase dysphoric disorder
  (LLPDD)
 l. pregnancy
 l. replicating chromosome
 l. uterine wedge resection
latency
 pudendal nerve terminal motor l.
 terminal motor l.
latent
 l. carrier
 l. diabetes
 l. herpes simplex virus infection
 l. phase
 l. phase of labor
 l. syphilis
late-onset
 l.-o. hyperplasia
 l.-o. local junctional epidermolysis
  bullosa-mental retardation
  syndrome
latera (*pl. of* latus)
lateral
 l. facial dysplasia (LFD)
 l. head displacement (LHD)
 l. nasal proboscis
 l. oblique view
 l. ovarian transposition
 l. plate mesoderm
 l. recumbent position
 l. resolution
 l. sperm head displacement

 l. transverse thigh flap
 l. wall retractor
lateralis
 proboscis l.
laterality sequence
lateralization
laterally extended endopelvic resection
lateris
lateromedial oblique view
lateroversion
latex
 l. agglutination inhibition test
  (LAIT)
 l. fixation test
 l. glove
 l. particle agglutination
 l. sensitization
latissimus dorsi flap
Latrobe retractor
LATS
 long-acting thyroid stimulator
 LATS hormone
LATS-protector
latum
 condyloma l.
 Diphyllobothrium l.
latus, pl. **lata**
 nevus unius lateris
latus, pl. **latera**
Latzko
 L. cesarean
 L. colpocleisis
Laufe-Piper forceps
Laufe polyp forceps
Launois-Cléret syndrome
Launois syndrome
Laurence-Moon-Biedl-Bardet (LMBB)
 L.-M.-B.-B. syndrome (LMBBS)
Laurence-Moon-Biedl syndrome
Laurence-Moon syndrome
Laurer forceps
Laurus ND-260 needle driver
LAV
 lymphadenopathy-associated virus
lavage
 bronchopulmonary l.
 gastric l.
 peritoneal l.
LAVH
 laparoscopic-assisted vaginal
  hysterectomy

L

NOTES

LAVH *(continued)*
    laparoscopic-assisted vaginal
      hysteroscopy
law
    Diday l.
    Hardy-Weinberg l.
    Hellin l.
    Hellin-Zeleny l.
    inverse square l.
    Küstner l.
    Leopold l.
    l. of mass action
    Mendel first l.
    Mendel second l.
    Poiseville l.
Läwen-Roth syndrome
Lawford syndrome
Lawrence syndrome
laxa
    cutis l.
laxity
    mental retardation, overgrowth,
      craniosynostosis, distal
      arthrogryposis, sacral dimple,
      joint l.
layer
    Bowman l.
    feeder l.
    germ l.
    half-value l. (HVL)
    Langhans l.
    Nitabuch l.
    Rauber l.
lazaroid
Lazarus-Nelson technique
lazy leukocyte syndrome
lazy-S incision
LBC
    lamellar body count
LBW
    low birth weight
LBWI
    low birth weight infant
LC
    laparoscopic cholecystectomy
    living children
LCA
    Leber congenital amaurosis
L-Caine
**L-Cath peripherally inserted neonatal
catheter**
LCHAD
    long-chain-3-hydroxyacyl-CoA
      dehydrogenase
      LCHAD deficiency
lck protooncogene
LCS
    LaparoSonic coagulating shears

LCx Probe System test
L&D
    labor and delivery
LDH
    lactate dehydrogenase
LDL
    low-density lipoprotein
LDL-cholesterol
l-dopa
    levodopamine
LDR
    labor, delivery, and recovery
      LDR room
LDS
    ligate-divide-staple
      LDS clip applier
      LDS instrument
LE
    lupus erythematosus
lead
    l. block
    l. pipe urethra
    l. poisoning
Leadbetter-Politano ureterovesicoplasty
leading ancestor
leaf, pl. **leaves**
    l. of broad ligament
    cabbage leaves
league
    La Leche L.
leakage
    placental l.
    silicone implant l.
leak-point
    l.-p. pressure
    l.-p. pressure test
leaky gene
lean spastic dwarfism
Lear complex
***Learning to Eat* manual**
leash
    laparoscopic l.
Lea Shield
leather-bottle stomach
leaves (*pl. of* leaf)
Leber
    L. abiotrophy
    L. congenital amaurosis (LCA)
    L. congenital tapetoretinal
      degeneration
    L. disease
    L. disease 2
    L. hereditary optic neuropathy
      (LHON)
Leboyer
    L. method
    L. technique
lecanopagus

Lecat gulf
lecithin
>disaturated l.
>l./sphingomyelin (l/s)
>l./sphingomyelin ratio
>l. and sphingomyelin ratio

Lecompte maneuver
lectin
Lectromed urinary investigation system
Ledbetter-Politano procedure
Ledercillin VK
Leder stain
LEEP
>loop electrosurgical excision procedure
>LEEP Redi-kit

Lee-White clotting time
Leff forceps
LeFort
>L. operation
>L. partial colpocleisis

LeFort-Neugebauer operation
LeFort-Wehrbein-Duplay hypospadias repair
left
>l. atrial to aortic ratio (LA:A)
>l. atrial diameter (LAD)
>l. atrial isomerism (LAI)
>l. frontoanterior position (LFA)
>l. frontoposterior position (LFP)
>l. frontotransverse position (LFT)
>l. lateral position
>l. mentoanterior position (LMA)
>l. mentoposterior position (LMP)
>l. mentotransverse position (LMT)
>l. occipitoanterior position (LOA)
>l. occipitoposterior position (LOP)
>l. occipitotransverse position (LOT)
>l. to right (L-R)
>l. sacroanterior
>l. sacroanterior position (LSA)
>l. sacroposterior position (LSP)
>l. sacrotransverse position (LST)
>l. scapuloanterior position (LScA)
>l. scapuloposterior position (LScP)
>l. ventricle (LV)
>l. ventricular hypertrophy (LVH)

left-sidedness
>bilateral l.-s.

left-to-right shunt
leg
>baker's l.
>bayonet l.

>l. cramp
>l. edema
>milk l.
>white l.

legal
>l. intervention
>l. issue

Legat point
leg-compression stocking
Legg-Calvé-Perthes disease
leggings
Legionnaire disease
Leiden mutation
Leigh
>L. disease
>encephalomyopathy of L.
>L. syndrome

Leiner disease
leiomyoma, pl. **leiomyomata**
>cervical l.
>endometrial metastasizing l.
>hypercellular uterine l.
>ovarian l.
>parasitic l.
>submucous l.
>leiomyomatata uteri
>uterine leiomyomata
>vascular l.

leiomyomatosis
>intravenous l.
>l. peritonealis disseminata (LPD)

leiomyosarcoma (LMS)
Leisegang colposcope
leishmaniasis
>cutaneous l.
>mucocutaneous l.

Leiter test
Lejeune syndrome
Lejour-type modified breast reduction
Lem-Blay circumcision clamp
lemon sign
lemon-squeezer obstetrical elevator
Lendersloot version test
length
>crown-heel l. (CHL)
>crown-rump l. (CRL)
>femur l. (FL)
>fetal foot l. (FFL)
>functional urethral l. (FUL)
>funnel l.
>humerus l. (HL)
>long bone l.

L

NOTES

Lennox-Gastaut syndrome
Lennox syndrome
lens
>Barkan infant l.
>l. culinaris agglutinin
>30-degree l.
>70-degree l.

Lente
>L. Iletin I, II
>L. L

lentigo, pl. **lentigines**
>lentigines (multiple),
>electrocardiographic abnormalities,
>ocular hypertelorism, pulmonary
>stenosis, abnormalities of
>genitalia, retardation of growth,
>and deafness (sensorineural)
>(LEOPARD)
>lentigines (multiple),
>electrocardiographic abnormalities,
>ocular hypertelorism, pulmonary
>stenosis, abnormalities of
>genitalia, retardation of growth,
>and deafness (sensorineural)
>syndrome

lentis
>ectopia l.

lentivirus
Lent laparoscope
Lenz
>L. dysmorphogenic syndrome
>L. dysplasia
>L. microphthalmia syndrome

Lenz-Majewski
>L.-M. hyperostotic dwarfism
>L.-M. syndrome

Lenz-Majewski-like syndrome
Leonard catheter
leonine facies
LEOPARD
>lentigines (multiple), electrocardiographic
>abnormalities, ocular hypertelorism,
>pulmonary stenosis, abnormalities of
>genitalia, retardation of growth, and
>deafness (sensorineural)
>LEOPARD syndrome

Leopold
>L. law
>L. maneuver

LePad breast exam training pad
leperous salpingitis
Lepore thalassemia
leprechaun facies
leprechaunism
leprosum
>erythema nodosum l.

leprosy

leptin
>maternal plasma l.
>umbilical cord l.

leptomeningitis
leptometacarpy
leptospirosis
leptotene
>l. phase of meiosis
>l. stage

leptotrichosis
Leri
>L. pleonosteosis
>L. syndrome

Leri-Weill syndrome
Leroy syndrome
LeRoy ventricular catheter
LES
>lower esophageal sphincter

lesbian
lesbianism
Leschke syndrome
Lesch-Nyhan
>L.-N. disease
>L.-N. syndrome

lesion
>acetowhite l.
>anal squamous intraepithelial l.
>(ASIL)
>axillary skin l.
>barrel-shaped l.
>benign l.
>cardiac l.
>cavitary white-matter l.
>cervical l.
>cutaneous l.
>cystic l.
>ductal-dependent l.
>ectocervical l.
>endometriotic l.
>genital l.
>glandular atypia l.
>high-grade squamous
>intraepithelial l. (HGSIL, HSIL)
>HPV-associated l.
>hyperpigmented l.
>intraepithelial l.
>Janeway l.
>low-grade squamous
>intraepithelial l. (LGSIL, LSIL)
>lumbosacral plexus l.
>lumbosacral root l.
>Lynch and Crues type 2 l.
>lytic l.
>metastatic l.
>Noonan-like giant cell l.
>pigmented l.
>plexus l.
>powder-burn endometrial l.

precancerous l.
preinvasive l.
premalignant l.
radial sclerosing l.
satellite l.
sclerosing l.
SIL/ASCUS l.
skin l.
spiculated l.
squamous intraepithelial l. (SIL)
vermiform l.
violaceous l.
vulvar pigmented l.
vulvovaginal l.
LET
linear energy transfer
letdown
milk l.
l. reflex
lethal
l. bone dysplasia
l. dwarfism
l. dysplasia
l. equivalent
l. gene
l. multiple pterygium syndrome
lethargic
lethargy
letrozole
Letterer-Siwe disease
LETZ
loop excision of the transformation zone
LETZ procedure
leucine
l. aminopeptidase
l. tolerance test
leucocyte detection strip
leucovorin
leukanakmesis
leukapheresis
leukemia
acute lymphocytic l. (ALL)
acute nonlymphocytic l.
aplastic l.
basophilic l.
chronic lymphocytic l.
eosinophilic l.
granulocytic l.
hemoblastic l.
l. inhibitory factor (LIG)
leukopenic l.
lymphocytic l.

lymphosarcoma cell l.
mast cell l.
megakaryocytic l.
micromyeloblastic l.
myeloblastic l.
myelogenous l.
Leukeran
leukocoria
leukocyte
l. count
l. function
l. integrin lymphocyte function-associated antigen 1
l. interferon
l. transfusion
leukocytosis
leukocytospermia
leukocyturia
leukodystrophy
cerebral l.
demyelinogenic l.
fibrinoid l.
globoid cell l.
Krabbe l.
melanodermic l.
metachromatic l.
sudanophilic l.
leukoencephalopathy
leukoerythroblastic syndrome
leukokraurosis
leukoma
corneal l.
leukomalacia
cerebral l.
periventricular l.
leukopenic leukemia
leukophlegmasia
l. dolens
leukoplakia
l. vulvae
leukoplakic vulvitis
leukorrhagia
leukorrhea
menstrual l.
leukorrheal
leukospermia
Leukotrap red cell collector
leukotriene
leuprolide
l. acetate
levallorphan tartrate
levamisole

L

NOTES

Levaquin
levarterenol
Levate
levator
  l. ani
  l. ani muscle
  l. ani spasm
  l. sling
LeVeen shunt
level
  ACD l.
  alpha₁-acid glycoprotein l.'s
  alpha antitrypsin l.
  alpha-fetoprotein l.
  amniotic fluid l.
  arachidonic acid l.
  blood l.
  blood alcohol l. (BAL)
  CD4⁺ l.
  cesium-137 l.
  cidal l.
  cord blood erythropoietin l.
  cord serum l.
  cortisol l.
  developmental l.
  dimeric inhibin A l.
  endothelin plasma l.
  estradiol l.
  estriol l.
  estrogen l.
  fetal LDL l.
  gonadotropin l.
  hemoglobin l.
  hormonal l.
  human chorionic gonadotropin l.
  l. III ultrasonography
  iodothyronine l.
  maternal estriol l.
  maternal serum l.
  methemoglobin l.
  peak-and-trough l.'s
  peripheral hormone l.
  plasma l.
  postmenopausal l.
  progesterone myometrial l.
  prolactin l.
  relaxin serum l.
  serum l.
  somatomedin l.
  sweat chloride l.
  theophylline l.
  uterine lysosome l.
Levi-Lorain
  L.-L. dwarfism
  L.-L. infantilism
Levin syndrome
Levlen

Levlite
  L. tablet
levocardia
levodopa, levodopamine (L-dopa)
Levo-Dromoran
levofloxacin
Levoid
Levonogesterel
levonorgestrel
  ethinyl estradiol and l.
  l. and ethinyl estradiol
  l. implant
Levophed
levoposition
Levoprome
Levora
levorotatory alkaloid
levorphan
levorphanol tartrate
levoscoliosis
Levothroid
levothyroxine
  l. test
levotransposition (L-transposition)
levoversion
Levret
  L. forceps
  L. maneuver
Levsin
Levsinex
Levy-Hollister syndrome
Lewandowsky
  nevus elasticus of L.
Lewis
  L. blood group
  L. recording cystometer
Leyden-Möbius muscular dystrophy
Leydig
  L. cell aplasia
  L. cell embryology
  L. cell hyperplasia
  L. cells
  L. cell tumor
LFA
  left frontoanterior position
LFD
  lateral facial dysplasia
LFP
  left frontoposterior position
LFT
  left frontotransverse position
LGA
  large for gestational age
  post-term LGA
  term LGA
LGD
  low-grade dysplasia

LGS
    limb-girdle syndrome
LGSIL
    low-grade squamous intraepithelial lesion
LGV
    lymphogranuloma venereum
LH
    luteinizing hormone
        LH Color test
        LH surge
LHD
    lateral head displacement
LHON
    Leber hereditary optic neuropathy
LH-releasing hormone agonist therapy
LH-RH
    luteinizing hormone-releasing hormone
liberty
    reproductive l.
libidinal change
libido
    decreased l.
library
    arrayed l.
    cDNA l.
    DNA l.
    gene l.
    genomic l.
    l. ligation
Librax
Librium
LICAM gene for X-linked hydrocephalus
lice (*pl. of* louse)
Lice-Enz Shampoo
Licentiate in Midwifery (LM)
lichen
    l. infantum
    l. nitidus
    l. planus
    l. ruber planus (LRP)
    l. sclerosus
    l. sclerosus et atrophicus (LS)
    l. scrofulosorum
    l. simplex
    l. simplex chronicus
    l. spinulosus
    l. striatus
lichenification
lichenoides
    pityriasis l.
Lich technique

Liddle test
Lidemol
Lidex
    L.-E
lidocaine
    l. toxicity
lidocaine-prilocaine cream
Lidoject
lie
    anterior l.
    fetal l.
    longitudinal l.
    oblique l.
    posterior l.
    transverse l.
    transverse fetal l.
    unstable l.
Liesegang LM-FLEX 7 flexible hysteroscope
life
    day of l. (DOL)
    l. event
    l. expectancy
    l. stress
    l. table method
    l. table survival
    wrongful birth and l.
LifePak defibrillator
lifestyle factor
Li-Fraumeni cancer syndrome
LIG
    leukemia inhibitory factor
Liga clip
liga-clipped
ligament
    Adams advancement of round l.'s
    broad l.
    Carcassone perineal l.
    cardinal l.
    cardinal-uterosacral l.
    Gilliam round l.
    Gimbernat reflex l.
    infundibulopelvic l.
    laparoscopic resection of the ureterosacral l. (LUSLR)
    leaf of broad l.
    Mackenrodt l.
    ovarian l.
    Petit l.
    posterior uterosacral l.
    Poupart l.
    round l.

L

NOTES

ligament *(continued)*
    sacrospinous l.
    transverse cervical l.
    l. of Treitz
    triangular l.
    umbilical l.
    uteroovarian l.
    uterosacral l.
    Waldeyer preurethral l.
ligamentary ectopic pregnancy
ligamentopexis, ligamentopexy
ligamentous ectopic pregnancy
ligamentum
    l. teres
    l. venosum
ligand-binding
ligand-receptor
ligase
    DNA l.
ligate-divide-staple (LDS)
ligation
    bilateral tubal l. (BTL)
    bleeding site l.
    hypogastric artery l.
    Irving tubal l.
    Kroner tubal l.
    laparoscopic tubal l. (LTL)
    library l.
    modified Irving-type tubal l.
    Parkland tubal l.
    Pomeroy tubal l.
    tubal l. (TL)
    Uchida tubal l.
ligature
    chromic gut pelviscopic loop l.
    Deschamps l.
light
    bili l.
    l. chain
    double-bank bili l.'s
    MultArray l.
    Right Light examination l.
    Solar Beam medical examination l.
    Speculite chemiluminescent l.
    Wood l.
lightening
lightning seizure
LighTouch Neonate thermometer
Lightwood-Albright syndrome
lignocaine
Liley
    L. curve
    L. three-zone chart
Lilliput neonatal oxygenator
limb
    l. bud
    circumferential ringed creases
      of l.'s

    circumferential skin crease of l.'s
    fetal l.
    l. infarction
    l. motion
    multiple benign circumferential skin
      creases on l.
    l. reduction
    l. reduction deformity (LRD)
    short l.
    tripus l.
    vertebral, anal, cardiac, tracheal,
      esophageal, renal, l. (VACTERL)
Limberg technique
limb-girdle
    l.-g. muscular dystrophy
    l.-g. muscular weakness and
      atrophy
    l.-g. syndrome (LGS)
limbic bands
Lim broth
limit
    fetal dose l.
    radiation dose l.
limp infant syndrome
limulus amebocyte lysate assay
Lincocin
lincomycin
lindane
line
    art l.
    arterial l. (A-line)
    Beau l.
    black l.
    breeding l.
    canthomeatal l.
    central venous l. (CVL)
    central venous pressure l.
    CVP l.
    cytogenetic l.
    Farre white l.
    genetic l.
    germ l.
    Harris growth arrest l.
    Hart l.
    hemostatic staple l.
    iliopectinate l.
    iliopectineal l.
    intravenous l.
    lambda suture l.
    long l.
    multiple resistant cell l.'s
    neonatal l.
    percutaneous l.
    peripheral arterial l.
    radial arterial l.
    recombinant substitution l.
    sagittal suture l.
    Shenton l.

simian l.'s
Sydney l.
l. of Toldt
umbilical artery l.
V l.
venous l.

linea, pl. **lineae**
l. alba
l. alba hernia
lineae albicantes
lineae atrophicae
l. nigra
l. terminalis

lineage
neural crest-derived cell l.

linear
l. accelerator
l. atelectasis
l. atrophy
l. energy transfer (LET)
l. IgM disease of pregnancy
l. in-line ligature carrier
l. nevus sebaceous syndrome
l. salpingostomy
l. sebaceous nevus syndrome
l. verrucous epidermal nevus

linearis
nevus sebaceus l.

Lin-Gettig syndrome

lingua, pl. **linguae**
l. nigra
short frenulum linguae

linguofacialis
dysplasia l.

linitis plastica

linkage
l. analysis
complete l.
l. disequilibrium
l. equilibrium
genetic l.
l. group
l. map
partial l.

link antibody

linoleic acid

Lion's Claw grasper

Lioresal

liothyronine

liotrix

Liouville icterus

lip
cleft l. (CL)
cleft l./palate (CLP)
double l.
frozen smile puckered l.'s
nodular blueberry l.'s
l. phenomenon
l. pseudocleft-hemangiomatous branchial cyst syndrome
l. reflex

lipase

lipid
l. cell
l. cell neoplasm
l. cell ovarian tumor
l. metabolism
l. metabolism disorder
l. peroxide

lipid-associated sialic acid

lipidosis
familial neurovisceral l.
galactosylsphinogosine l.
late infantile systemic l.
neurovisceral l.
psychosine l.

Lipidox

Lipiodol

lipoatrophic diabetes

lipoatrophy

lipochondrodystrophy

lipodystrophy
congenital l.
familial l.
generalized l.

lipodystrophy-acromegaloid gigantism syndrome

lipofuscinosis
ceroid l.
neuronal ceroid l. (CLN)

lipogenic dyshepatia

lipogranulomatosis subcutanea

lipoic acid

lipoid
l. adrenal gland hypoplasia
l. adrenal hyperplasia
l. ovarian neoplasm
l. ovarian tumor
l. pneumonia
l. proteinosis

lipolysis

lipoma, pl. **lipomata**
vulvar l.

L

NOTES

lipomatosis
  encephalocraniocutaneous l. (ECCL)
lipomeningocele
lipomyelomeningocele
liponecrosis microcystica calcificans
lipoplasty
  suction-assisted l. (SAL)
lipopolysaccharide (LPS)
  l. endotoxin
lipoprotein
  l. concentration
  high-density l. (HDL)
  low-density l. (LDL)
  l. receptor-related protein (LRP)
  very low-density l. (VLDL)
lipoprotein(a) (Lp(a))
lipoprotein-cholesterol metabolism
Liposyn
  L. II
lipotropin
  beta-l.
  mu-l.
lip-palate syndrome
Lippes loop
Lippes-type intrauterine device
Lipschütz ulcer
Liquaemin
liquefaction
  semen l.
liquid
  Barc L.
  End Lice L.
  fetal lung l.
  Lotrimin AF Spray L.
  Pyrinyl II L.
  Sklar Kleen l.
  Tisit L.
  Titralac Plus L.
  Triple X L.
Liquiprin
liquor folliculi
LIS1 gene for lissencephaly
Lisch nodule
lisinopril
Lisolipin
Lison syndrome
Lispro
lissencephaly
  hydrocephalic l.
  ILS17 gene for isolated l.
  ILSX gene for X-linked l.
  LIS1 gene for l.
list
  Interpersonal Support Evaluation L.
Listeria
  L. meningitis
  L. monocytogenes

listeriosis
  congenital l.
  neonatal l.
Lister scissors
liter (L)
  millequivalent per l. (mEq/L)
  l.'s per minute (L/min)
Lithane
lithiasis
lithium
Lithobid
lithokelyphopedion, lithokelyphopedium
Lithonate
lithopedion, lithopedium
Lithotabs
lithotomy
  marian l.
  l. position
  vaginal l.
Little disease
Littmann ECG electrode
Littre gland
Litzmann obliquity
Livadatis circular myotomy
livebirth, live birth
liveborn infant
live poliovirus vaccine
liver
  acute fatty l.
  cirrhosis of l.
  l. disease
  fetal l.
  l. function tests
  l. metastasis
  l. transplant
  l. transplantation
  l. tumor
Livernois-McDonald forceps
live-virus vaccine
livial
living children (LC)
LLETZ
  large loop excision of transformation
  zone
LLETZ-LEEP active loop electrode
Llorente dissecting forceps
Lloyd-Davies stirrups
LLPDD
  late luteal phase dysphoric disorder
LM
  Licentiate in Midwifery
LMA
  left mentoanterior position
LMBB
  Laurence-Moon-Biedl-Bardet
LMBBS
  Laurence-Moon-Biedl-Bardet syndrome

L/min
> liters per minute

LMP
> last menstrual period
> left mentoposterior position
> low malignant potential

LMS
> leiomyosarcoma

LMT
> left mentotransverse position

LNMP
> last normal menstrual period

LOA
> left occipitoanterior position

lobar
> l. holoprosencephaly
> l. pneumonia

Lobstein
> L. disease
> L. syndrome

lobster-claw
> l.-c. deformity
> l.-c. hand
> l.-c. with ectodermal defects
> syndrome

lobular
> l. carcinoma
> l. neoplasia

lobulation-polydactyly
> l.-p. syndrome

lobule
> placental l.

local
> l. anesthesia
> l. anesthetic
> l. excision
> l. irradiation
> l. methotrexate injection
> l. ovarian condition

localization
> epithelial autoantibody l.
> estrogen receptor l.
> needle l.
> placental l.

localize
location
> breech l.
> extraembryonic l.
> gene l.
> noncornual placental l.

lochia
> l. alba

> l. cruenta
> l. purulenta
> l. rubra
> l. sanguinolenta
> l. serosa

lochial
lochiometra
lochiometritis
lochioperitonitis
lochiorrhagia
lochiorrhea
loci (*pl. of* locus)
lock
> English l.
> French l.
> German l.
> heparin l. (hep lock)
> pivot l.
> sliding l.

locked
> l. twins

Locke solution
Locke-Wallace Marital Adjustment test
locking twins
Locoid Topical
locus, pl. **loci**
> gene l.
> genetic l.
> GUSB l.
> beta-glucuronidase
> l. heterogeneity
> operator l.
> quantitative trait l.

locus-specific probe
LOD score
Loeb deciduoma
Loehlein diameter
Loestrin
> L. 1.5/30
> L. 1/20
> L. 21 1/20
> L. Fe

Lofene
Löffler syndrome
log
> l. cell kill
> l. kill hypothesis

LOH
> loss of heterogeneity

lollipop
> fentanyl l.

Lomanate

NOTES

L

lomefloxacin
Lomotil
lomustine
Lonalac
long
  l. arm of chromosome (q)
  l. arm of Y chromosome
  l. atraumatic retractor
  l. bone length
  l. course
  l. line
  l. weighted speculum
long-acting
  l.-a. contraception
  l.-a. contraceptive
  l.-a. contraceptive steroid
  Sinex L.-a.
  l.-a. thyroid stimulator (LATS)
long-axis view
long-chain-3-hydroxyacyl-CoA
  l.-c.-h.-C. dehydrogenase (LCHAD)
  l.-c.-h.-C. dehydrogenase deficiency
longitudinal
  l. growth
  l. lie
  l. oval pelvis
  l. presentation
  l. scan
long-term
  l.-t. followup
  l.-t. sequelae
  l.-t. survival
Loniten
Lonox
loop
  bipolar cutting l.
  bipolar urological l.
  cutting l.
  l. diathermy cervical conization
  l. electrode
  l. electrosurgical excision procedure (LEEP)
  l. excision of the transformation zone (LETZ)
  flow-volume l.
  free-floating l.
  l. of Henle
  Lippes l.
  low-voltage diathermy l.
  Medevice surgical l.
  physiologic endometrial ablation/resection l. (PEARL)
  polysomnogram with flow-volume l.
  Schroeder tenaculum l.
  somatic nervous system feedback l.
  tenaculum hook l.
  vaginal speculum l.
Loosett maneuver

Lo/Ovral
LOP
  left occipitoposterior position
lop ear
loperamide
lophosphamide
Lopressor
Loprox
Lorabid
loracarbef
Lorain infantilism
loratadine
lorazepam
Lorcet
lordosis
Los Angeles variant galactosemia
Losec
losoxantrone
loss
  autosomal dominant nonsyndromic hearing l. (DFNA3)
  autosomal recessive nonsyndromic hearing l. (DFNB1)
  blood l.
  bone l.
  early embryonic l.
  early pregnancy l.
  embryonic l.
  estrogen l.
  excessive blood l.
  hair l.
  heat l.
  l. of heterogeneity (LOH)
  insensible fluid l.
  menstrual blood l.
  normal blood l.
  l.-of-resistance technique
  postmenopausal bone l.
  pregnancy l.
  rapid bone l.
  recurrent pregnancy l.
  repetitive pregnancy l.
  RP with progressive sensorineural hearing l.
  sensory l.
  spinal bone l.
  surgical weight l.
  vertebral bone l.
  water l.
  weight l.
Lossen rule
Lostorfer body
lost surgical specimen
LOT
  left occipitotransverse position
lotion
  Polysonic ultrasound l.

Lotrimin
    L. AF
    L. AF Powder
    L. AF Spray Liquid
    L. AF Spray Powder
    L. AF Topical
    L. Topical
LOU
    lower obstructive uropathy
      fetal LOU
Lou Gehrig disease
Louis-Bar syndrome
loupe
    Keeler l.
    l. magnification
louse, pl. **lice**
    crab l.
Lovas training
Lovset maneuver
low
    l. birth weight (LBW)
    l. birth weight infant (LBWI)
    l. cervical cesarean
    l. forceps
    l. forceps delivery
    l. malignant potential (LMP)
    l. molecular weight dextran
    l. rectal resection
    l. steroid content combined oral
      contraceptive
    l. transverse cesarean (LTC)
    l. transverse hysterotomy
    l. transverse uterine incision
    l. vertical uterine incision
low-back pain
low-cholesterol diet
low-density lipoprotein (LDL)
low-dose
    l.-d. danazol
    l.-d. heparin
    l.-d. oral contraceptive
    l.-d. steroids
Lowe
    fetal oculocerebrorenal syndrome
      of L.
    L. oculocerebrorenal syndrome
    L. syndrome (LS)
lower
    l. abdominal pain
    l. abdominal tenderness
    l. esophageal sphincter (LES)

    l. genital tract infection
    l. limb ossification center
    l. obstructive uropathy (LOU)
    l. segment cesarean (LSCS)
    l. uterine segment
    l. uterine segment transverse
      (LUST)
    l. uterine segment transverse
      cesarean section
lower-segment scar
Lowe-Terry-MacLachlan syndrome
low-fat diet
low-grade
    l.-g. dysplasia (LGD)
    l.-g. mosaicism
    l.-g. positive smear
    l.-g. squamous intraepithelial lesion
      (LGSIL, LSIL)
Low-Ogestrel-21
Low-Ogestrel-28
low-powered field (lpf)
low-pressure urethra
low-protein diet
low-residue diet
low-resolution banding
Lowry-Maclean
    L.-M. syndrome
Lowry syndrome
Lowry-Wood syndrome (LWS)
low-salt diet
low-segment transverse incision
low-set ears
low-sodium diet
low-voltage diathermy loop
loxapine
Loxitane
lozenge
    Cough-X l.
Lozol
LP
    lumbar puncture
Lp(a)
    lipoprotein(a)
L-PAM
    L-phenylalanine mustard
LPCV
    laser photocoagulation of the
      communicating vessel
LPD
    leiomyomatosis peritonealis disseminata
    luteal phase defect

L

NOTES

lpf
low-powered field
L-phenylalanine mustard (L-PAM)
LPS
lipopolysaccharide
L-R
left to right
LRD
limb reduction deformity
LRP
lichen ruber planus
lipoprotein receptor-related protein
LS
lichen sclerosus et atrophicus
Lowe syndrome
l/s
lecithin/sphingomyelin
LSA
left sacroanterior position
LScA
left scapuloanterior position
LScP
left scapuloposterior position
LSCS
lower segment cesarean
LSD
lysergic acid diethylamide
LSH
laparoscopic supracervical hysterectomy
lutein-stimulating hormone
LSIL
low-grade squamous intraepithelial lesion
LSIL Pap smear
LSP
left sacroposterior position
L/S ratio
LST
left sacrotransverse position
LTC
low transverse cesarean
LTL
laparoscopic tubal ligation
L-transposition
levotransposition
Lubchenco nomogram
lube
Sklar l.
Lübke uterine vacuum cannula
lubricant
AstroGlide personal l.
Maxilube personal l.
Replens l.
vaginal l.
Lubri-Flex stent
Lub syndrome
Lucey-Driscoll syndrome
lucinactant
Lucite dilator

Ludiomil
Luer-Lok syringe
Luer retractor
lues
l. ascites
luetic
LUFS
luteinized unruptured follicle syndrome
Luft disease
Lugol iodine solution
Luikart forceps
Luikart-Simpson forceps
Lujan-Fryns syndrome
luliberin
Lumadex-FSI test
lumbar
l. artery
l. epidural anesthesia
l. puncture (LP)
lumbosacral
l. meningomyelocele
l. plexus injury
l. plexus lesion
l. root lesion
lumen, pl. lumina
decidual l.
urethral l.
Luminal
luminal epithelium
Lumopaque
lump
lumpectomy
LUNA
laser uterosacral nerve ablation
Lunar DPX dual-energy absorptiometer
Lunelle
lung
aeration of l.
l. cancer
capacity of l. (CL)
l. capacity
l. carcinoma
l. compliance
compliance of l. (CL)
l. edema
granular l.
hazy l.
hypoplastic l.
junky l.
microphallus, imperforate anus, syndactyly, hamartoblastoma, abnormal l. (MISHAP)
l. profile
SciMed-Kolobow membrane l.
sequestered l.
l. strip
l. volume

Lupron
  L. Depot
  L. Depot-3 Month
  L. Depot-4 Month
  L. Depot-Ped
lupus
  l. anticoagulant (LAC)
  l. anticoagulant activity
  l. anticoagulant antibody
  l. crisis
  l. erythematosus (LE)
  l. erythematosus disseminatus
  gestational l.
  l. nephritis
  l. obstetric syndrome
  l. vulgaris
Lurline PMS
LUSLR
  laparoscopic resection of the
  ureterosacral ligament
LUST
  lower uterine segment transverse
  LUST C-section
lutea
  corpora l.
luteal
  l. cell
  l. function
  l. ovarian cyst
  l. phase
  l. phase defect (LPD)
  l. phase inadequacy
  l. phase support
luteectomy
luteinalis
  hyperreactio l.
lutein cell
luteinization
  l. inhibition
  l. inhibitor
  l. stimulator
luteinized
  l. thecoma
  l. unruptured follicle
  l. unruptured follicle syndrome
  (LUFS)
luteinizing
  l. hormone (LH)
  l. hormone receptor-binding
  inhibitor
  l. hormone-releasing hormone (LH-
  RH)

  l. hormone-releasing hormone
  analogue
  l. hormone secretion
luteinoma
lutein-stimulating hormone (LSH)
Lutembacher syndrome
luteolysis
luteolytic
  l. action
luteoma
  pregnancy l.
  l. of pregnancy
  stromal l.
luteoplacental shift
luteotropic hormone
luteum
  corpus l.
Lutheran blood group
Lutrepulse
Lutz-Jeanselme nodule
LV
  left ventricle
L-valine ester
LVH
  left ventricular hypertrophy
LVSI
  lymphovascular space invasion
lwoffi
  Achromobacter l.
  Acinetobacter l.
LWS
  Lowry-Wood syndrome
Lyderm
Lyell
  L. disease
  L. syndrome
Lyme
  L. disease
  L. enzyme-linked immunosorbent
  assay
Lymerix
lymph
  l. node
  l. node biopsy
  l. node drainage
  l. node endometriotic
  adenoacanthoma
  l. node metastasis
  l. node positivity
lymphadenectomy
  inguinal l.
  laparoscopic l.

L

NOTES

lymphadenectomy *(continued)*
    Meigs pelvic l.
    paraaortic l.
    pelvic l.
    retroperitoneal l.
lymphadenitis
    mediastinal l.
    mesenteric l.
lymphadenopathy
    l.-associated virus (LAV)
    axillary l.
lymphangiectasia, lymphangiectasis
    congenital pulmonary l.
lymphangiography
lymphangioma
    cavernous l.
    l. circumscriptum
    cystic l.
    l. cysticum
lymphangiosarcoma
lymphangitis
lymphatic
    l. drainage
    l. drainage of genitalia
    paracervical l.'s
    l. spread
    l. system
lymphedema
lymphoblastoid
    l. cell
    l. interferon
lymphocyst
    l. omentum
lymphocyte
    l.-activating factor
    l. activator
    B l.
    cytotoxic T l. (CTL)
    l. dysfunction
    l. function-associated antigen 1
    natural killer l.
    peripheral blood l. (PBL)
    T, T_3, T_4 l.
lymphocytic
    l. adenohypophysitis
    l. choriomeningitis
    l. hypophysitis
    l. leukemia
lymphocytosis
lymphogranuloma
    l. inguinale
    venereal l.
    l. venereum (LGV)
lymphography
lymphohistiocytosis
    erythrophagocytic l.
    familial erythrophagocytic l. (FEL)

lymphoid
    l. cell
    l. hyperplasia
    l. tissue
lymphokine
lymphokine-activated killer cell (LAK)
lymphoma
    AIDS-related l.
    Burkitt l.
    Hodgkin l.
    malignant l.
    metastatic l.
    non-Hodgkin l.
    ovarian l.
    recurrent l.
    retroorbital l.
    true histiocytic l. (THL)
lymphonodular pharyngitis
lymphopenia
lymphoproliferative syndrome
lymphoreticulosis
lymphosarcoma
    l. cell leukemia
lymphotoxin antitumor activity
lymphovascular
    l. space
    l. space invasion (LVSI)
Lynch
    L. and Crues type 2 lesion
    L. syndrome
lynestrenol
lyn protooncogene
lyodura sling procedure
Lyon hypothesis
lyonization
Lyphocin
lypressin
Lyrelle patch
lysate
    cell l.
lysergic
    l. acid
    l. acid diethylamide (LSD)
lysine
    l. malabsorption syndrome
lysis
    adhesion l.
    l. of adhesion
Lysodren
lysoPC diagnostic ovarian cancer test
lysosomal
    l. enzyme disorder
    l. hydrolase enzyme assay
    l. storage disease
lysosome
lytic lesion
Lytren
    L. formula

MAA
microphthalmia or anophthalmos with associated anomalies
Maalox
MAb
OvaRex M.
MacConkey II agar
MacDermot-Winter syndrome
MacDonald sign
MACDP
Metropolitan Atlanta Congenital Defects Program
Macer abdominal cystocele repair
macerated fetus
maceration
fetal m.
Macewen sign
Machado-Joseph disease
Mach band effect
machine
Acuson Model 128XP m.
Berkeley suction m.
cobalt megavoltage m.
Mayo-Gibbon heart-lung m.
megavoltage m.
SDU-400 EchoView ultrasound m.
machinery murmur
MacIntosh
English M. (E-Mac)
Mackenrodt ligament
Macleod syndrome
macrencephaly
macroadenoma
prolactin-secreting m.
macrobead
methyltestosterone m.
Macrobid
macroblepharon
macrocardius
macrocephaly
benign familial m. (BFM)
m., cutis marmorata, telangiectatica congenita syndrome
m., facial abnormalities, disproportionate tall stature mental retardation syndrome
m., hypertelorism, short limbs, hearing loss, developmental delay syndrome
m., multiple lipomas, hemangiomata syndrome
m., pseudoepithelioma, multiple hemangiomas syndrome

m. with feeblemindedness and encephalopathy with peculiar deposits
macrocephaly-hamartomas syndrome
macrocrania
macrocrystal
monohydrate m.
nitrofurantoin monohydrate m.
macrocytic anemia of pregnancy
macrodactyly
Macrodantin
macroevolution
macrogamete
macrogenitosomia
m. praecox
macroglobinemia
Waldenstrom m.
macroglobulin
α_2-m.
macroglossia
m.-omphalocele syndrome
m.-omphalocele-visceromegaly syndrome
macrogyria
macrolecithal
macromastia, macromazia
macromelus
macroorchidism
m. marker X syndrome
macroorchidism marker X (MOMX)
macrophage
m.-activating factor (MAF)
m. colony-stimulating factor
hemosiderin laden m.
m.-inhibition factor
peritoneal m.
macrophallus
Macroplastique implantable device
macroprolactinoma
macrorestriction map
macroscopic hematuria
macrosomia
m. adiposa congenita
fetal m.
m.-mental retardation syndrome
m., obesity, macrocephaly, ocular abnormality syndrome (MOMO)
macrostomia
MACS
magnetically activated cell sorter
macular
m. dystrophy
m. hemangioma

M

maculosus
> dystrophia bullosa hereditaria,
> typus m.

MadaJet XL needle-free injector
mad cow disease
Madden technique
Madelung deformity
Madlener operation
Madonna finger
Maestre de San Juan-Kallmann-de Morsier syndrome
Maestre-Kallmann-de Morsier syndrome
MAF
> macrophage-activating factor

mafenide
Maffucci syndrome
mag
> magnesium

magaldrate
Magan
Magendie
> atresia of the foramen of Luschka
> and M.

MAGGI disposable biopsy needle guide for ultrasound
magma reticulare
magna
> large cisterna m.
> megacysterna m.

magnesia
> milk of m. (MOM)
> Phillips' Milk of M.

magnesium (mag)
> m. citrate
> m. hydroxide
> m. oxide
> m. salicylate
> m. sulfate ($MgSO_4$)

magnetic
> m. resonance imaging (MRI)
> m. resonance mammography (MRM)
> m. resonance urography (MRU)
> m. source imaging (MSI)

magnetically
> m. activated cell sorter (MACS)
> m. responsive microsphere

magnetite (Fe_3O_4)
Magnetrode cervical unit
magnification
> area of interest m. (AIM)
> loupe m.
> m. mammography
> spot m.

magnum
> foramen m.

Magnus and de Kleijn tonic neck reflex

Mag-Ox 400
Magpi hypospadius repair
Magrina-Bookwalter vaginal retractor
Magsal
Maher disease
MAID
> mesna, Adriamycin, Ifosfamide,
> Dacarbazine

maidenhead
Mainstay urologic soft tissue anchor
maintenance medication
Mainz
> M. pouch
> M. pouch urinary reservoir

Majewski syndrome
major
> beta-thalassemia m.
> m. capsid protein gene
> m. detoxificating enzyme
> m. dysmorphism
> m. gene
> m. histocompatibility antigen
> m. histocompatibility complex (MHC)
> thalassemia m.

majora
> labia m.

Makler
> M. insemination
> M. insemination device
> M. reusable semen analysis chamber

mal
> grand m.
> petit m.

malabsorption
> congenital folate m.
> m. syndrome
> tryptophan m.

malacoplakia
maladaptation
> immunologic m.

maladie
> m. de Roger
> m. des tics

malaise
malar
> m. fat pad
> m. hypoplasia
> m. rash

malaria
Malassezia furfur pustulosis
malathion
maldescensus testis
male
> m. factor
> m. factor infertility
> M.-FactorPak

M. FactorPak seminal fluid collection kit
m. genital duct
m. genital duct development
m. karyotype
m. pseudohermaphroditism
m. pseudohermaphroditism-persistent müllerian structures-mental retardation syndrome
m. reproductive system
m. sex differentiation
m. Turner syndrome
46,XX m.
XXY m.
XYY m.
ZZ m.

maleate
ergonovine m.
Fluvoxame m.
methylergonovine m.
methysergide m.

Malecot catheter
male-pattern hirsutism
malformation
Arnold-Chiari m.
arteriovenous m.
AV m.
body stalk m.
bronchopulmonary m.
cloacal m.
clomiphene fetal m.
congenital m.
congenital cystic adenomatoid m. (CCAM)
craniooculofrontonasal m.
cystic adenomatous m. (CAM)
Dandy-Walker m.
dural arteriovenous m.
faciotelencephalic m.
fetal m.
foregut m.
gyral m.
pelvic arteriovenous m.
Rieger m.
teratogen-induced m.
uterine arteriovenous m.
vascular m.
Walker-Warburg m.

malignancy
borderline m. (BLM)
breast m.
extrapelvic m.

gastrointestinal m.
genital tract m.
metastatic m.
ovarian m.
vulvar m.

malignant
m. calcification
m. cytotrophoblast
m. degeneration
m. effusion
m. histiocytosis
m. hypertension
m. lymphoma
m. melanoma
m. mesothelioma (MM)
m. mixed müllerian tumor (MMMT)
m. neoplasm
m. nephrosclerosis
m. ovarian germ cell tumor
m. ovarian neoplasm
m. ovarian teratoma
m. pleural effusion
m. syncytiotrophoblast
m. tumor of cervix

malignum
adenoma m.

Malis CMC-II bipolar coagulator
Mallamint
Mallazine Eye Drops
malleable retractor
Mall formula
Mallory-Weiss syndrome
malnutrition
malodorous
Malotuss
Malouf syndrome
malplacement
malposition
uterine m.

malpresentation
fetal m.

Malpuech
M. facial clefting syndrome

malrotation
renal m.

Malström
M. cup
M. vacuum extractor

mamma, pl. **mammae**
m. accessoria

M

NOTES

mamma *(continued)*
    m. erratica
    supernumerary m.
mammalgia
mammalian transgenesis
mammaplasty, mammoplasty
    augmentation m.
    postreduction m.
    reconstructive m.
    reduction m.
mammary
    m. calculus
    m. duct ectasia
    m. dysplasia
    m. fistula
    m. gland
    m. neuralgia
    m. souffle
mammectomy
mammillaplasty
mammillitis
mammitis
mammoglobin
mammogram
    x-ray m.
mammographic
    m. detection
    m. screening
mammography
    computed tomography laser m.
     (CTLM)
    contoured tilting compression m.
    Corometrics Model 900SC in-
     office m.
    CT laser m.
    digital m.
    Egan m.
    heavy-ion m.
    magnetic resonance m. (MRM)
    magnification m.
    x-ray m.
Mammomat C3 mammography system
mammoplasty *(var. of* mammaplasty)
Mammoscan digital imaging system
mammose
Mammotest breast biopsy system
mammotomy
mammotropic, mammotrophic
mAMSA
    Amsacrine
man
    azoospermic m.
    Mendelian Inheritance in M.
     (MIM)
    oligo m.
managed care organization (MCO)
management
    active third-stage m.

    epilepsy m.
    expectant m.
    infertility m.
    laparoscopic m.
    metabolic m.
    noninvasive m.
    physiologic third-stage m.
    pregnancy m.
    risk m.
    surgical m.
    total quality m. (TQM)
Manchester
    M. operation
    M. ovoid
Manchester-Fothergill operation
Mancini plate
Mandelamine
mandelic acid
Mandelurine
mandible
    acroosteolysis with osteoporosis and
     changes in skull and m.
mandibuloacral dysplasia
mandibulofacial
    m. dysostosis (MFD)
    m. dysostosis with epibulbar
     dermoids syndrome
    m. dysostosis with limb
     malformations syndrome
mandibulofacials
    dysostosis m.
mandibulooculofacial
    m. dyscephalia
    m. dysmorphism
mandibulooculofacialis
    dyscephalia m.
Mandol
mandrel, mandril
    intrauterine insemination cannula
     with m.
maneuver
    all-fours m.
    Barlow m.
    Bill m.
    Bracht m.
    Brandt-Andrews m.
    corkscrew m.
    Credé m.
    DeLee m.
    doll's head m.
    forceps m.
    Gaskin m.
    heel-to-ear m.
    Hillis-Müller m.
    Hodge m.
    key-in-lock m.
    Kristeller m.
    Lecompte m.

Leopold m.
Levret m.
Loosett m.
Lovset m.
Massini m.
Mauriceau m.
Mauriceau-Levret m.
Mauriceau-Smellie-Veit m.
McDonald m.
McRoberts m.
midforceps m.
modified Ritgen m.
Müller-Hillis m.
Munro-Kerr m.
Ortolani m.
ostrich m.
Pajot m.
Pinard m.
Prague m.
reverse form McRoberts m.
Ritgen m.
Rubin m.
Saxtorph m.
Scanzoni m.
Scanzoni-Smellie m.
scarf m.
Schatz m.
Sellick m.
Thorn m.
Valsalva m.
Van Hoorn m.
Wigand m.
Woods screw m.
Zavanelli m.

manifestation
ocular m.
renal m.

manifold

Manipujector
Kronner M.

manipulation
gamete m.

manipulator
ClearView uterine m.
Kronner m.
RUMI uterine m.
uterine m.
Valchev uterine m.

manipulator-injector
Harris-Kronner uterine m.-i.
(HUMI)

Kronner Manipujector uterine m.-i.
Zinnanti uterine m.-i. (ZUMI)

Manning
M. score
M. score of fetal activity

Mann isthmic cerclage

mannitol

mannosidosis

Mantel-Cox test

Mantel-Haenszel
Mantel-Haenszel Q
Mantel-Haenszel test

M antigen

mantle
cortical m.
visible cortical m.

manual
m. breast pump
m. cleavage
m. healing method
Learning to Eat m.
m. pelvimetry
m. rotation

MAOA
monoamine oxidase A deficiency

MAOI
monoamine oxidase inhibitor

Maox

MAP
mean arterial pressure

map
brain electrical activity m. (BEAM)
chromosome m.
contig m.
cytogenetic m.
cytological m.
gene m.
genetic m.
linkage m.
macrorestriction m.
physical m.
recombination m.
restriction m.

maple
m. leaf flap
m. syrup urine
m. syrup urine disease (MSUD)

maplike skull

MapMarkers fluorescent DNA sizing standard

mapping
chromosome m.

M

NOTES

mapping *(continued)*
    comparative m.
    fate m.
    gene m.
    intraoperative lymphatic m.
maprotiline
MAR
    mixed agglutination reaction
    MAR test
mar22
    marker 22
Marañón syndrome
marasmus
marble bones
Marcaine with epinephrine
Marchand
    M. adrenals
    M. rest
Marchetti test
March technique
Marcillin
Marckwald operation
Marcus Gunn phenomenon
Marden-Walker syndrome
Marezine
marfanoid
    m. craniosynostosis syndrome
    m. habitus-mental retardation
        syndrome
    m. habitus-microcephaly-
        glomerulonephritis syndrome
Marfan syndrome
Margarita Island type ectodermal dysplasia
Margesic A-C
marginal
    m. insertion
    m. sinus rupture
marginatum
    eczema m.
    erythema m.
Margulies coil
marian lithotomy
Marie-Sainton syndrome
Marie syndrome
marijuana
    m. effect
Marinesco-Garland syndrome
Marinesco-Sjögren-Garland syndrome
Marinesco-Sjögren-like syndrome
Marinesco-Sjögren syndrome
Marinol
Marion disease
Marion-Moschcowitz culdoplasty
Marital
    M. Dyadic Inventory
marital introitus

mark
    port-wine m.
    strawberry m.
    Unna m.
marker
    m. 22 (mar22)
    adrenal hyperandrogenism m.
    Anderson m.
    assay m.
    bi-allelic m.
    CA 15-3 breast cancer m.
    CA 72-4 cancer m.
    CA 125 endometrial cancer m.
    CA 19-9 GI cancer m.
    CA 195 GI cancer m.
    CA 50 GI cancer m.
    CA 549 tumor m.
    chromosomal m.
    m. chromosome
    chromosome 22 supernumerary m.
    DNA m.
    Freeman cookie cutter areola m.
    m. gene
    genetic m.
    pericentromeric m.
    peripheral androgen activity m.
    protein m.
    sigma tumor m.
    tumor m.
    m. X (marX)
    m. X chromosome
    m. X syndrome
Marlex
Marlow
    M. disposable cannula
    M. disposable trocar
    M. Primus handle, shaft, and TIP
Marmine
Marmo method
marmorata
    cutis m.
marmoratus
    status m.
Maroteaux-Lamy
    M.-L. disease
    M.-L. syndrome
Maroteaux-Malamut syndrome
Marplan
marrow
    bone m.
Marshall
    M. Hall facies
    M. syndrome
    M. test
Marshall-Marchetti-Krantz (MMK)
    M.-M.-K. operation
    M.-M.-K. procedure

Marshall-Marchetti procedure
Marshall-Smith syndrome (MSS)
Marshall-Tanner pubertal staging
Marshall-Taylor vacuum extraction
Marsupial
    M. belt
    M. Pouch
marsupialization
    Spence and Duckett m.
    m. technique
    transurethral m.
Marthritic
Martin-Bell-Renpenning syndrome
Martin-Bell syndrome (MBS)
Martius
    M. bulbocavernosus fat flap
    M. flap and fascial sling
    M. graft
    M. procedure
Martsolf syndrome
marX
    marker X
    marX syndrome
Mary Jane breast pump
Maryland dissector
MAS
    Maternal Attitude Scale
    meconium aspiration syndrome
MASA syndrome
masculine pelvis
masculinization
masculinovoblastoma
mask
    Bili m.
    m. inhalation anesthesia
    m. of pregnancy
    ventilation by m.
    Venturi m.
mass
    adnexal m.
    benign m.
    bone m.
    circumscribed m.
    complex m.
    cortical m.
    cystic adnexal m.
    m. effect
    fallopian tube m.
    fungating m.
    ill-defined m.
    inner cell m.
    intramyometrial m.

    mixed-density m.
    noncalcified nodular m.
    ovarian m.
    pelvic m.
    persistent ovarian m.
    poorly marginated m.
    postmenopausal body m.
    potato-like m.
    m. spectrometry
    stellate m.
    tubal m.
    tumor m.
    umbilical cord m.
    uterine m.
    vertebral bone m.
    well-defined m.
massage
    cardiac m.
    closed chest m.
    perineal m.
    Shiatsu therapeutic m.
    uterine m.
Masse Breast Cream
Massengill douche
Massini maneuver
massive
    m. breast hypertrophy
    m. genital prolapse
    m. ovarian cyst
    m. transfusion
Masson-Fontana stain
MAST
    military antishock trousers
    MAST suit
mast
    m. cell leukemia
mastadenitis
mastadenoma
mastalgia
    cyclic m.
mastatrophy, mastatrophia
mastectomy
    Auchincloss modified radical m.
    extended radical m.
    Halsted m.
    McKissick m.
    McWhirter m.
    modified radical m.
    prophylactic m.
    radical m.
    simple m.
    subcutaneous m.

M

NOTES

mastectomy *(continued)*
    total m.
    Willy Meyer m.
Masters-Allen syndrome
mastitis
    chronic cystic m.
    gargantuan m.
    glandular m.
    granulomatous m.
    interstitial m.
    lactational m.
    m. neonatorum
    parenchymatous m.
    phlegmonous m.
    plasma cell m.
    puerperal m.
    retromammary m.
    stagnation m.
    submammary m.
    suppurative m.
mastocytosis
mastodynia
Mastodynon
mastoid fontanel
mastoiditis
mastology
mastoncus
mastopathy
mastopexy
    Benelli m.
mastoplasia
mastoplasty
mastoptosis
mastorrhagia
mastoscirrhus
mastotomy
masturbation
Mateer-Streeter ovum
material
    Avitene hemostatic m.
    dextran-70 barrier m.
    elective abortion m.
    Endo-Avitene hemostatic m.
    FlowGel barrier m.
    genetic m.
    INTERCEED barrier m.
    metal suture m.
    nylon suture m.
    Poloxamer 407 barrier m.
    polyester suture m.
    polyethylene suture m.
    polypropylene suture m.
    radiocontrast m.
    spontaneous abortion m.
    suture m.
    synthetic suture m.

Materna
    M. prenatal vitamin
    M. Tablet
maternal
    m. abdominal pressure
    m. age
    m. age-related risk
    m. alcohol consumption
    m. alcoholism
    m. anesthesia
    m. antiplatelet antibody
    m. asthma
    M. Attitude Scale (MAS)
    m. Bernard-Soulier syndrome
    m. birthing position
    m. blood clot patch therapy
    m. central hemodynamics
    m. cholestasis
    m. coagulopathy
    m. cocaine use
    m. condition
    m. cortical vein
    m. cortical vein thrombosis
    m. cotyledon
    m. cyanotic heart disease
    m. death
    m. death rate
    m. deprivation syndrome
    m. diabetes
    m. douche
    m. drug abuse
    m. dystocia
    m. endometrial IL-1R
    m. estriol level
    m. exercise
    m. factor
    m. febrile morbidity
    m. floor infarction
    m. fracture
    m. gonads
    m. hemopathy
    m. hepatitis
    m. HLA haplotype
    m. hydration
    m. hydrops
    m. hydrops fetalis
    m. hydrops syndrome
    m. hyperparathyroidism
    m. hypertension
    m. hypotension
    m. immune response
    m. immunology
    m. indications
    m. infection
    m. inflammatory response
    m. inheritance
    m. insulin
    m. karyotype

m. meiosis I (MMI)
m. meiosis II (MMII)
m. mercury exposure
m. mortality
m. mortality rate
m. nutrition
m. ocular adaptation
m. outcome
m. parvovirus fetalis
m. peripheral blood
m. phenylketonuria
m. physiology
m. plasma leptin
m. pulse
m. pyrexia
m. screening
m. serum
m. serum alpha-fetoprotein
  (MSAFP)
m. serum level
m. size
m. sperm antibody
m. stature
m. steroid concentration
m. stress
m. surveillance
m. tissue
m. titer
m. trauma
m. undernourishment
m. uniparental heterodisomy
m. vascular response
m. weight
maternal-fetal
  m.-f. histoincompatibility
  m.-f. HLA compatibility
  m.-f. medicine
  m.-f. transmission
maternal-placental-fetal unit
maternal-placental unit
maternity
  disputed m.
mating
  assortative m.
  backcross m.
  consanguineous m.
  nonrandom m.
  random m.
  m. type
Matritech NMP22 bladder cancer test
matrix, pl. **matrices**
  extracellular m.

identity m.
myxoid m.
square m.
matroclinous inheritance
matter
  particulate m.
mattress
  apnea alarm m.
maturation
  m. of cell
  excessive villous m.
  fetal lung m.
  follicular m.
  ovum m.
  premature accelerated lung m.
  (PALM)
  pulmonary m.
  skeletal m.
  m. value
  in vitro m.
maturation-promoting factor (MPF)
mature
  m. burst-forming unit erythroid (M-
  BFU-E)
  m. cystic ovarian teratoma
  m. infant
  m. neutrophil
maturity
  Dubowitz scale for infant m.
  fetal m.
  fetal lung m. (FLM)
  neuromuscular m.
  physical m.
maturity-onset diabetes of the young
  (MODY)
Maturna bra system
Maunoir hydrocele
Mauriac syndrome
Mauriceau-Levret maneuver
Mauriceau maneuver
Mauriceau-Smellie-Veit maneuver
Maxaquin
Maxeran
Maxiflor
maxillofacial dysostosis
maxillonasal dysplasia
Maxilube personal lubricant
maximal permissible dose
maximum
  m. breathing capacity
  m. likelihood estimator
  m. oxygen uptake (VO_2max)

M

NOTES

maximum *(continued)*
    M. Strength Desenex Antifungal
       Cream
    M. Strength Dex-A-Diet
    M. Strength Dexatrim
    m. temperature (T-max)
    m. urethral closure pressure
       (MUCP)
Maxipime
Maxivate Topical
Maxolon
Maxon
    M. delayed-absorbable suture
Mayer pessary
Mayer-Rokitansky-Küster-Hauser
  syndrome
May-Hegglin anomaly
Maylard incision
Mayo
    M. culdoplasty
    M. hook
    M. hysterectomy
    M. scissors
Mayo-Fueth inversion procedure
Mayo-Gibbon heart-lung machine
Mayo-Hegar needle holder
Mayor sign
Mazanor
mazindol
mazodynia
mazolysis
mazopathy, mazopathia
mazopexy
mazoplasia
Mazzariello-Caprini forceps
Mazzini test
MB band
M-BFU-E
    mature burst-forming unit erythroid
MBM
    mother's breast milk
MBP
    modified Bagshawe protocol
MBS
    Martin-Bell syndrome
McArdle disease
McBurney point
McCall
    M. culdoplasty
    M. stitch
McCall-Schumann procedure
McCaman-Robins test
McCraw gracilis myocutaneous flap
McCune-Albright syndrome
McDonald
    M. cervical cerclage
    M. maneuver

    M. measurement
    M. procedure
McDonough syndrome
MCDU
    mercaptolactate-cysteine disulfiduria
MCF-7 breast cancer cell
mcg
    norethindrone acetate 1 mg/ethinyl
      estradiol 5 mcg
    Vancenase AQ 84 mcg
McGee forceps
McGill forceps
mcg/kg/min
    micrograms per kilogram per minute
McGovern nipple
MCH
    mean corpuscular hemoglobin
MCi
    megacurie
McIndoe-Hayes procedure
McIndoe operation
McKerman-Adson forceps
McKerman-Potts forceps
McKissick mastectomy
McKusick-Kaufman syndrome
McLane forceps
MCO
    managed care organization
McPherson forceps
MCR
    metabolic clearance rate
m-cresyl acetate
McRoberts maneuver
MCS
    Miles-Carpenter syndrome
MCT
    medium chain triglyceride
      MCT Oil
      MCT oil formula
MCTD
    mixed connective tissue disease
M-cup
    Mityvac M-c.
    M-c. vacuum extraction device
MCV
    mean corpuscular volume
    methotrexate, cisplatin, vinblastine
    molluscum contagiosum virus
McWhirter mastectomy
MD
    muscular dystrophy
    myotonic dystrophy
MDI
    mental development index
    metered dose inhaler
MDLS
    Miller-Dieker lissencephaly syndrome

M/E
    myeloid/erythrocyte
      M/E ratio
MEA
    multiple endocrine abnormalities
Mead-Johnson formula
Meadows syndrome
mean
    m. arterial pressure (MAP)
    m. corpuscular hemoglobin (MCH)
    m. corpuscular volume (MCV)
    m. hemoglobin concentration
    m. menstrual cycle hematocrit
    m. plasma iron concentration
    regression of the m.
    standard error of the m. (SEM)
measles
    German m.
    m., mumps, rubella (MMR)
    three-day m.
    m. vaccine
measurable undesirable respiratory contaminants (MURCS)
measurement
    acid-base m.
    anthropometric m.
    bone density m.
    fetal growth m.
    Intran disposable intrauterine pressure m.
    intrauterine pressure m.
    McDonald m.
    midluteal progesterone m.
    nuchal translucency m.
    optical density m.
    pascal unit of pressure m. (Pa)
    transcutaneous m.
Measurin
measuring hat
meatus
    urethral m.
MEB
    muscle-eye-brain
      MEB disease
mebanazine
Mebaral
Mebendacin
mebendazole
MEBS
    muscle-eye-brain syndrome
Mebutar
mecamylamine

mechanical
    m. birth injury
    m. cervical dilator
    m. dysmenorrhea
    m. respirator
mechanism
    alloimmune m.
    autoimmune m.
    cellular cytotoxic m.
    Douglas m.
    Duncan m.
    heat transfer m.
    host defense m.
    host response m.
    ovum pickup m.
    pathophysiologic m.
    Schultze m.
    two-cell m.
mechlorethamine
    m. hydrochloride
Meckel
    M. diverticulum
    M. syndrome
Meckel-Gruber syndrome
meclofenamate
Meclomen
mecometer
meconial colic
meconiorrhea
meconium
    m. aspiration
    m. aspiration syndrome (MAS)
    m. blockage syndrome
    m. corpuscle
    m. ileus
    m. ileus appearance
    m. peritonitis
    m. plug
    m. plug syndrome
    m. stain
meconium-stained
    m.-s. amniotic fluid (MSAF)
    m.-s. skin
MED-1/InfoChart paperless medical record system
Medasonic first beat ultrasound stethoscope
Medela
    M. Dominant vacuum delivery pump
    M. manual breast pump
    M. membrane regulator

M

NOTES

Medevice
- M. surgical loop
- M. surgical paws

Medfusion 1001 syringe infusion pump

media (*pl. of* medium)

medial oblique view

median
- m. cleft upper lip, mental retardation, pugilistic facies syndrome
- m. eminence
- m. episiotomy
- m. facial cleft syndrome
- multiples of the m. (mom)
- multiples of the appropriate gestational m. (MoM)
- m. raphe

mediana
- diameter m.

mediastinal
- m. collagenosis
- m. lymphadenitis

mediastinitis

mediastinum

mediating action

mediator
- intracellular m.

medical
- m. geneticist
- m. genetics
- M. Manager software

medicamentosa
- rhinitis m.

medication
- aerosolized m.
- base m.
- intravenous m.
- iodide-containing m.
- maintenance m.
- over-the-counter m.
- parenteral m.
- pressor m.
- prophylactic m.
- teratogenic m.

medicine
- American Society for Reproductive M.
- community m.
- complementary and alternative m. (CAM)
- fetal-maternal m.
- folk m.
- herbal m.
- maternal-fetal m.
- neonatal m.
- perinatal m.

medicolegal

MED-IDDM
- multiple epiphyseal dysplasia-early onset diabetes mellitus syndrome

Medihaler-Epi

Medihaler-Iso

Medilium

Medilog 9000 polysomnography device

mediolateral
- m. episiotomy
- m. view

Medipore H soft cloth surgical tape

Medipren

Mediterranean anemia

Medi-Trace electrode

Meditran

medium, pl. **media**
- acute otitis media (AOM)
- Biggers m.
- m. chain triglyceride (MCT)
- CPS ID chromogenic m.
- Earle m.
- Earle culture m.
- gallium-67 citrate contrast m.
- Gibco BRL sperm preparation m.
- Ham F10 m.
- mucoid otitis media (MOM)
- Nickerson m.
- OncoScint CR/OV contrast m.
- Sabouraud m.
- selective broth m. (SBM)
- sperm capacitation m.
- Thayer-Martin m.
- thioglycollate broth m.
- transfer m.
- transmission m.
- Whitten m.
- Whittingham m.

MEDLINE search

Med-Neb respirator

medorrhea

medroxyprogesterone
- m. acetate (MPA)
- depot m.

MEDS
- microsurgical extraction of ductal sperm

Medscan

medulla oblongata

medullary
- m. carcinoma
- m. cord
- m. cystic disease

medulloblastoma

medusae
- caput m.

Meesmann corneal dystrophy

mefenamic acid

Mefoxin

megabase

megabladder
megacalicosis
megacardia
Megace
megacephaly, megalocephalia, megalocephaly
Megacillin Susp
megacolon
    aganglionic m.
    congenital m.
    toxic m.
megacurie (MCi)
megacysterna magna
megacystis
    m.-megaureter syndrome
    m., microcolon, intestinal hypoperistalsis (MMIH)
    m., microcolon, intestinal hypoperistalsis syndrome
megadosing
Megadyne/Fann E-Z clean laparoscopic electrode
megaelectron volt (MeV)
megahertz (MHz)
megakaryocytic leukemia
megalencephaly
    benign familial m.
    m., cranial sclerosis, osteopathia striata syndrome
    m. with hyaline panneuropathy
megaloblastic anemia
megalocardia
megalocephalia (*var. of* megacephaly)
megalocephaly (*var. of* megacephaly)
megaloclitoris
megalocornea
    m., developmental retardation, dysmorphic syndrome
    m.-macrocephaly-mental and motor retardation syndrome (MMMM)
    m.-mental retardation syndrome (MMR)
megalodactyly
megalomelia
Megalone
megalopenis
megalophthalmos
megalosyndactyly
megaloureter, megaureter
megavolt (MeV)
megavoltage machine
megestrol acetate

meglumine
    m. diatrizoate
    Hypaque M.
Meier-Gorlin syndrome
Meigs
    M. pelvic lymphadenectomy
    M. syndrome
Meigs-Kass syndrome
Meigs-Okabayashi procedure
Meigs-Werthein hysterectomy
Meinecke-Peper syndrome
Meinicke test
meiosis
    diakinesis stage of oocyte m.
    dictyate stage of oocyte m.
    diplotene phase of m.
    m. I (MI)
    m. II (MII)
    leptotene phase of m.
    maternal m. I (MMI)
    maternal m. II (MMII)
    oocyte m.
    pachytene phase of m.
    paternal m. I (PMI)
    paternal m. II (PMII)
    zygotene phase of m.
meiotic division
Meissner plexus
melancholia
    involutional m.
Melanex topical solution
melaninogenicus
    Bacteroides m.
melanoblastoma
    Bloch-Sulzberger m.
melanoblastosis cutis linearis sive systematisata
melanocyte
    epidermal m.
melanocyte-stimulating hormone
melanocytic nevus
melanodermic leukodystrophy
melanoma
    Clark classification of vulvar m.
    cutaneous m.
    malignant m.
    metastatic m.
    nodular m.
    m. specific antigen
    superficial spreading m.
    vulvar m.

M

NOTES

melanosis
   m. corli degenerativa
   neonatal pustular m.
   transient neonatal pustular m.
melasma
   m. gravidarum
melatonin
   m. secretion
melena
   m. neonatorum
   m. spuria
Melfiat
Melinck-Needles syndrome
Melkersson-Rosenthal syndrome
Mellaril
mellitus
   adult-onset diabetes m. (AODM)
   diabetes m.
   gestational diabetes m. (GDM)
   insulin-dependent diabetes m.
    (IDDM)
   juvenile diabetes m. (JDM)
   juvenile-onset diabetes m. (JODM)
   non-insulin-dependent diabetes m.
    (NIDDM)
   overt insulin-dependent diabetes m.
   transient neonatal diabetes m.
    (TNDM)
   type 1, 2 diabetes m.
Melnick-Fraser syndrome
Melnick-Needles
   M.-N. disease
   M.-N. syndrome
melorheostosis
melphalan
membrana granulosa
membrane
   allograft m.
   artificial rupture of m.'s (ARM,
    AROM)
   basement m.
   m. bridge
   chorioallantoic m. (CAM)
   cloacal m.
   Descemet m.
   Duralon-UV nylon m.
   dysmenorrheal m.
   egg m.
   erythrocyte m.
   exocelomic m.
   extraembryonic fetal m.
   extraplacental m.
   fetal m.
   floating m.
   germinal m.
   m. glycoprotein IB/IX
   Gore-Tex surgical m.
   m. granulosa

   Heuser m.
   hyaline m.
   hymenal m.
   intact m.
   interfetal m.
   Jackson m.
   mucous m.
   placental m.
   Preclude peritoneal m.
   prelabor rupture of the m.'s
    (PROM)
   premature rupture of m.'s (PROM)
   preterm premature rupture of m.'s
    (PPROM)
   preterm rupture of m.'s
   prolonged rupture of m.'s (PROM)
   m. rupture
   Seprafilm bioresorbable m.
   Slavianski m.
   spontaneous rupture of m.'s
    (SROM)
   m. stripping
   ultrafiltration m.
   vernix m.
   Viresolve ultrafiltration m.
   vitelline m.
   Wachendorf m.
   yolk m.
membranous
   m. dysmenorrhea
   m. glomerulonephritis
   m. twins
memory
   m. cell
   m. phenomenon
   semantic m.
MEMR
   multiple exostosis mental retardation
MEN
   multiple endocrine neoplasia, types I, II,
    III
menacme
menadiol
   m. sodium diphosphate
menadione
Menadol
menarche
   delayed m.
   m. factor
menarcheal, menarchial
Mendel
   M. first law
   M. second law
mendelian
   m. genetic disorder
   m. inheritance
   M. Inheritance in Man (MIM)
   m. trait

mendelizing
Mendelson syndrome
Mendenhall syndrome
Menest
Menge pessary
Mengert
    M. index
    M. shock syndrome
meningeal capillary angiomatosis
meningioma
    acoustic m.
meningismus
meningitis, pl. meningitides
    aseptic m.
    cryptococcal m.
    Listeria m.
    tuberculous m.
meningocele
    cranial m.
    spinal m.
meningococcal polysaccharide vaccine
meningococcemia
Meningococcus vaccine
meningoculofacialis angiomatosis
meningoencephalitis
meningoencephalocele
meningoencephalomyelitis
meningomyelocele
    lumbosacral m.
meningooculofacial angiomatosis
meningoulofacialis
    angiomatosis m.
Menkes
    M. disease
    M. kinky hair syndrome (MKHS)
    M. syndrome
Menkes-Kaplan syndrome
menocelis
menometrorrhagia
menopausal
    m. estrogen replacement therapy
    m. syndrome
menopause
    iatrogenic m.
    premature m.
menophania
Menorest hormone replacement patch
menorrhagia
menorrhalgia
menoschesis
menostasis, menostasia
menostaxis

menotropin
menotropins
menouria
menoxenia
menses
    m. phase
menstrual
    m. age
    m. aspiration
    m. blood loss
    m. colic
    m. cycle
    m. cycle hemodynamic response
    m. cycle induction
    m. cycle regulation
    m. cycle resumption
    m. edema
    m. endometrium
    m. extraction
    m. extraction abortion
    m. flow
    m. formula
    m. history
    m. irregularity
    m. leukorrhea
    m. molimina
    m. period (MP)
    m. phase
    m. reflux
    m. sclerosis
    m. state
menstrual-ovarian cycle
menstruant
menstruate
menstruation
    abnormal m.
    anovular m.
    delayed m.
    retained m.
    retrograde m.
    supplementary m.
    suppressed m.
    vicarious m.
mental
    m. deficiency, spasticity, congenital
    ichthyosis syndrome
    m. development
    m. development index (MDI)
    m. disturbance
    m. and growth retardation-
    amblyopia syndrome

M

NOTES

mental *(continued)*

m. and physical retardation, speech disorders, peculiar facies syndrome

m. retardation

m. retardation-absent nails of hallux and pollex syndrome

m. retardation-adducted thumbs syndrome

m. retardation, ataxia, hypotonia, hypogonadism, retinal dystrophy syndrome

m. retardation, blepharonasofacial abnormalities, hand malformations syndrome

m. retardation-clasped thumb syndrome

m. retardation, coarse face, microcephaly, epilepsy, skeletal abnormalities syndrome

m. retardation, coarse facies, epilepsy, joint contracture syndrome

m. retardation, congenital contracture, low fingertip arches syndrome

m. retardation, congenital heart disease, blepharophimosis, blepharoptosis, hypoplastic teeth

m. retardation-distal arthrogryposis syndrome

m. retardation, dysmorphism, cerebral atrophy syndrome

m. retardation, dystonic movements, ataxia, seizures syndrome

m. retardation, epilepsy, short stature, skeletal dysplasia syndrome

m. retardation, facial anomalies, hypopituitarism, distal arthrogryposis syndrome

m. retardation, gynecomastia, obesity syndrome

m. retardation, hearing impairment, distinct facies, skeletal anomalies syndrome

m. retardation, hip luxation, G6PD variant syndrome

m. retardation, macroorchidism syndrome

m. retardation, microcephaly, blepharochalasis syndrome

m. retardation, mitral valve prolapse, characteristic face syndrome

m. retardation, optic atrophy, deafness, seizures syndrome

m. retardation, overgrowth, craniosynostosis, distal arthrogryposis, sacral dimple, joint laxity

m. retardation-overgrowth sequence

m. retardation-overgrowth syndrome

m. retardation, polydactyly, phalangeal hypoplasia, syndactyly, unusual face, uncombable hair

m. retardation, pre-and postnatal overgrowth, remarkable face, acanthosis nigricans syndrome

m. retardation-psoriasis syndrome

m. retardation, retinopathy, microcephaly syndrome

m. retardation, scapuloperoneal muscular dystrophy, lethal cardiomyopathy syndrome

m. retardation, short stature, hypertelorism syndrome

m. retardation, short stature, obesity, hypogonadism syndrome

m. retardation, skeletal dysplasia, abducens palsy syndrome (MRSD)

m. retardation-sparse hair syndrome

m. retardation, spasticity, distal transverse limb defects syndrome

m. retardation, spastic paraplegia, palmoplantar hyperkeratosis syndrome

m. retardation-spastic paraplegia syndrome

m. retardation, typical facies, aortic stenosis syndrome

m. status evaluation

m. subnormality

mentoanterior

left m. position (LMA)

m. position

m. presentation

right m. position (RMA)

mentoposterior

left m. position (LMP)

m. position

m. presentation

right m. position (RMP)

mentor

M. catheter

M. female self-catheter

mentotransverse

left m. position (LMT)

m. position

right m. position (RMT)

mentum

m. anterior position

m. posterior position

m. transverse position

mepenzolate bromide

meperidine
    m. hydrochloride
mephentermine
mephenytoin
mephobarbital
Mephyton
    M. Oral
mepindolol
mepivacaine
meprobamate
    conjugated estrogen and m.
Meprospan
mepyramine
mEq
    millequivalent
mEq/L
    millequivalent per liter
MER
    methanol extraction residue
meralgia paresthetica
2-mercaptoethane sulfonate (MESNA)
mercaptolactate-cysteine disulfiduria (MCDU)
mercaptopurinc
6-mercaptopurine
Mercier bar
Merck respirator
Merimack 1040 CO2 laser
Merkel cell carcinoma
Merlis obstetrical excavator
meroacrania
meroanencephaly
merocyte
merogastrula
merogenesis
merogony
    diploid m.
meromelia
meromicrosomia
meromorphysis
meropenem
merorachischisis, merorrhachischisis
merozygote
Merrem I.V.
MERRF
    myoclonic epilepsy with ragged red fibers
Mersilene
    M. fascial strip
    M. gauze hammock
    M. mesh

    M. mesh sling
    M. suture
Meruvax II
merycism
Mery gland
Merzbacher-Pelizaeus disease
MESA
    microsurgical epididymal sperm aspiration
Mesantoin
mesaraica
mesatipellic pelvis
mesectodermal dysplasia
mesenchymal
    m. hamartoma
    m. neoplasm
mesenchyme
    nonspecific m.
mesenchymoma
mesenteric
    m. adenitis
    m. artery
    m. lymphadenitis
mesenterica
    tabes m.
mesentery
    intestinal m.
mesh
    Brennen biosynthetic surgical m.
    Dexon m.
    m. erosion
    Mersilene m.
Meshgraft Skin Expander
Mesigyna
mesiodens-cataracts syndrome
MESNA
    2-mercaptoethane sulfonate
mesna, Adriamycin, Ifosfamide, Dacarbazine (MAID)
Mesnex
mesoaxial hexadactyly-cardiac malformation syndrome
mesoblast
mesoblastic
    m. nephroma
mesoblastoma
    m. ovarii
    m. vitellinum
mesocephalic
mesoderm
    extraembryonic m.

M

NOTES

mesoderm *(continued)*
    lateral plate m.
    paraxial m.
mesodermal
    m. dysgenesis of anterior segment
    m. sarcoma
mesomelic
    m. dwarfism-small genitalia
      syndrome
    m. dysplasia
mesometanephric carcinoma
mesometric pregnancy
mesometritis
meson
    negative π m.
mesonephric
    m. adenocarcinoma
    m. carcinoma
    m. duct
    m. rest
    m. ridge
    m. tubule
mesonephroi (*pl. of* mesonephros)
mesonephroid
    m. clear-cell carcinoma
    m. tumor
mesonephroma
mesonephros, pl. **mesonephroi**
mesorchium
mesoridazine besylate
mesosalpingeal
mesosalpinx
mesothelioma
    benign m. of genital tract
    malignant m. (MM)
mesovarium, pl. **mesovaria**
messenger
    m. ribonucleic acid (mRNA)
    m. RNA
    second m.
Mestatin
Mestinon
mestranol
    m. and norethindrone
    m. and norethynodrel
mesylate
    benztropine m.
    2-Br-alpha-ergocryptine m.
    bromocriptine m.
    deferoxamine m.
    desferrioxamine m.
    dihydroergotamine m.
    dimethothiazine m.
meta-analysis
metabolic
    m. abnormality
    m. acidemia

    m. acidosis
    m. acidosis syndrome
    m. alkalosis
    m. clearance rate (MCR)
    m. disorder
    m. error
    m. management
metabolism
    aerobic m.
    amino acid m.
    androgen m.
    carbohydrate m.
    estrogen m.
    fat m.
    fetal m.
    inborn error of m.
    m. in intraperitoneal chemotherapy
    lipid m.
    lipoprotein-cholesterol m.
    mineral m.
    neonatal m.
    organic acid m.
    progesterone m.
    prostaglandin m.
    steroid m.
    vitamin m.
    water m.
metabolite
    arachidonic acid m.
    steroid m.
metacentric
    m. chromosome
    m. metaphase
metachromatic leukodystrophy
metachromosome
metacyesis
metaepiphysaria
    dysostosis enchondralis m.
metafemale
metagaster
Metahydrin
metal
    m. coil
    m. metabolism disorder
    m. suture material
    trace m.
metallic
    m. bead-chain cystourethrography
    m. skin staple
metalloproteinase
    tissue inhibitors of m.
Metandren
metanephric
    m. blastema
    m. bud
    m. duct
metanephros, pl. **metanephroi**

metaphase
    m. chromosome
    metacentric m.
metaphoric dysplasia
metaphyseal
    m. anadysplasia
    m. dysostosis
    m. dysplasia
metaphysis
    rachitic m.
metaplasia
    apocrine m.
    celomic m.
    ciliated m.
    squamous m.
    squamous m. of amnion
    tubal m.
    vaginal squamous m.
metaplastic carcinoma
Metaprel
metaproterenol sulfate
metaraminol bitartrate
metastasis, pl. **metastases**
    adnexal m.
    aortic node m.
    bony m.
    fallopian tube m.
    floxuridine in hepatic m.
    hepatic m.
    inguinal lymph node m.
    liver m.
    lymph node m.
    ovarian cancer m.
    placental m.
    pulmonary m.
    stomach cancer m.
    trocar implantation m.
    tumor, node, m. (TNM)
    uterine sarcoma m.
    vascular m.
metastatic
    m. adenocarcinoma
    m. axillary involvement
    m. carcinoma
    m. implant
    m. lesion
    m. lymphoma
    m. malignancy
    m. melanoma
Metastron

metatarsus
    m. adductus
    m. varus
metatropic
    m. dysplasia
Metenier sign
metenkephalin
meter
    Astech m.
    glucose m.
    MiniWright peak flow m.
    Pocketpeak peak flow m.
    US 1005 uroflow m.
    Wright peak flow m.
metered dose inhaler (MDI)
metergoline
metformin
methacycline hydrochloride
methadone hydrochloride
methamphetamine
    m. exposure
methandrostenolone
methanol extraction residue (MER)
methantheline bromide
methaqualone
metharbital
methazolamide
methdilazine hydrochloride
methemoglobinemia
methemoglobin level
methemoglobinuria
methenamine
Methergine
methergoline
methicillin sodium
methimazole
methionine
    m. malabsorption syndrome
    m. synthase
    m. synthase deficiency
methixene hydrochloride
method
    amniotomy plus oxytocin m.
    Astrand 30-beat stopwatch m.
    Attwood staining m.
    barrier m.
    Billings m.
    Bonnaire m.
    brine flotation m.
    bromelin m.
    Buist m.
    Byrd-Drew m.

M

NOTES

method *(continued)*
    cold knife m.
    contraceptive m.
    Corning m.
    CorrTest m.
    Döderlein m.
    Douglas m.
    encu m.
    ensu m.
    FETENDO m.
    Fick m.
    flush m.
    forward roll m.
    Gabastou hydraulic m.
    Hamilton m.
    Harrison m.
    Irving m.
    Ito m.
    Johnson m.
    Kaplan-Meier m.
    Kluge m.
    Kristeller m.
    lactational amenorrhea m. (LAM)
    Lamaze m.
    laser m.
    Leboyer m.
    life table m.
    manual healing m.
    Marmo m.
    pilocarpine iontophoresis m.
    Pomeroy m.
    Prochownik m.
    Puzo m.
    reverse dot blot sequence-specific oligonucleotide m.
    rhythm m.
    Rodeck m.
    shotgun m.
    Smellie m.
    Smellie-Veit m.
    sperm washing insemination m. (SWIM)
    Spiegel m.
    Stroganoff m.
    symptothermal m.
    Tarkowski m.
    thermodilution m.
    Towako m.
    twin m.
    Uchida m.
    u-score m.
    Vecchietti m.
    Victor Gomel m.
    Video Overlay M.
    Wardill four-flap m.
    Wardill-Kilner advancement flap m.
    Watson m.
    Watson-Crick m.

methodology
methotrexate (MTX)
    Adriamycin, fluorouracil, m. (AFM)
    m., cisplatin, vinblastine (MCV)
    cytosine arabinoside, etoposide, m. (CEM)
    Cytoxan, Oncovin, fluorouracil plus Cytoxan, Oncovin, m. (COF/COM)
methotrimeprazine
methoxamine hydrochloride
methoxyflurane
methscopolamine
methsuximide
methyclothiazide
methyl
    Oreton M.
    15-m. prostaglandin
methylation
methylbenzethonium chloride
methylbromide
    hyoscine m.
    scopolamine m.
methyl-CCNU
methyldopa
methylene
    m. blue
    m. blue dye
    m. tetrahydrofolate reductase (MTHFR)
    m. tetrahydrofolate reductase thermolability
5,10-methylenetetrahydrofolate
methylenetetrahydrofolate reductase gene
methylergometrine maleate
methylergonovine
    m. maleate
methylmalonic
    m. acidemia
    m. aciduria
methylmalonicaciduria
methylmercury
methylphenidate
    m. hydrochloride
methylprednisolone
15-methylprostaglandin $F_{2\alpha}$
α-methyl-p-tyrosine
methyltestosterone
    estrogens with m.
    m. macrobead
    Premarin With M.
    Premarin With M. Oral
methylxanthine
methysergide maleate
metoclopramide hydrochloride
metolazone
metopic ridging

Metopirone
    M. test
metopopagus
metoprolol tartrate
Metra
    M. PS procedure kit
MetraGrasp ligament grasper
MetraPass suture passer
MetraTie knot pusher
metratonia
metratrophy, metratrophia
metrectomy
metria
metritis
    postpartum m.
metrizamide
metrizoate sodium
Metrodin
metrodynamometer
metrodynia
MetroGel-Vaginal
metrography
Metro I.V.
metrolymphangitis
metromalacia
metromalacoma, metromalacosis
metronidazole
    m. vaginal gel
metroparalysis
metropathia
    m. hemorrhagica
metropathic
metropathy
metroperitoneal fistula
metroperitonitis
metrophlebitis
metroplasty
    abdominal m.
    hysteroscopic m.
    Strassman m.
Metropolitan Atlanta Congenital Defects Program (MACDP)
metrorrhagia
    m. myopathica
metrorrhea
metrorrhexis
metrosalpingitis
metrosalpingography
metroscope
metrostaxis
metrostenosis
metrotomy

metyrapone
    m. test
metyrosine
Metzenbaum scissors
MeV
    megaelectron volt
    megavolt
MEVA
    MEVA probe
    MEVA Probe for endovaginal scanning
Meval
Mexican cardiomelic dysplasia
mexiletine
Mexitil
Meyer-Schwickerath and Weyers syndrome
Meyer-Schwinkerath syndrome
Mezlin
mezlocillin
    m. sodium
MFD
    mandibulofacial dysostosis
M-FISH cytogenetic technique
mg
    milligram
mg%
    milligrams percent
MgSO$_4$
    magnesium sulfate
MHA
    microhemagglutination assay
MHA-TP
    microhemagglutination assay for antibodies to Treponema pallidum
MHC
    major histocompatibility complex
    MHC antigen
MHz
    megahertz
MI
    meiosis I
Miacalcin
Mibelli
    angiokeratoma of M.
    M. angiokeratoma
MIC
    minimum inhibitory concentration
Micardis
Micatin
    M. Topical
Michel deformity

M

NOTES

Michelin tire baby syndrome
Michels syndrome
miconazole
    m. nitrate
    m. nitrate vaginal cream
Micral
    M. Chemstrip
    M. urine dipstick test
micrencephaly
MICRhoGAM
microadenoma
microangiopathic
    m. hemolysis
    m. hemolytic anemia
microangiopathy
    thrombotic m.
microaspiration
micro assisted fertilization
microatelectasis
microbial
    m. factor
    m. flora
    m. sensitivity
microbiology
microbrachia
microcalcification
microcephalic
    m. dwarfism
    m. primordial dwarfism 1
    m. primordial dwarfism-cataracts
    syndrome
microcephaly
    m.-calcification of cerebral white
    matter syndrome
    m.-cardiomyopathy syndrome
    m.-cervical spine fusion anomalies
    m.-chorioretinopathy syndrome
    m.-deafness syndrome
    m.-digital anomalies syndrome
    m., hiatus hernia, nephrotic
    syndrome
    m. hypergonadotropic
    hypogonadism, short stature
    syndrome
    m. infantile spasm, psychomotor
    retardation, nephrotic syndrome
    m. mental retardation, cataract,
    hypogonadism syndrome
    m. mental retardation, retinopathy
    syndrome
    m. mesobrachyphalangy,
    tracheoesophageal fistula syndrome
    (MMT)
    m. microphthalmia, ectrodactyly,
    prognathism syndrome (MMEP)
    m. mild developmental delay, short
    stature, distinctive face syndrome

    m. mild mental retardation, short
    stature, skeletal anomalies
    syndrome
    m. muscular build, rhizomelia-
    cataracts syndrome
    m.-oculo-digito-esophageal-duodenal
    syndrome (MODED)
    m. sparse hair, mental retardation,
    seizures syndrome
    m.-spastic diplegia syndrome
microcolpohysteroflator
    Hamou m.
microcolpohysteroscopy
microcrania
microcurettage
    Accurette m.
microcurette
    HemoCue m.
microcyst
    milk of calcium m.
microcystica
microcytic anemia
microcytosis
microdactyly
microdeletion syndrome
microdontia
    m.-microcephaly-short stature
    syndrome
microencephaly
microendoscopic optical catheter
microendoscopy
microenvironment
microfibrillar collagen
microflora
    vaginal m.
microgastria
microgenitalism
microglandular
    m. adenosis
    m. cervical hyperplasia
    m. dysplasia
microglobulin
    β_2 m.
β_2 microglobulin
micrognathia
    m.-glossoptosis syndrome
micrograms per kilogram per minute
    (mcg/kg/min)
microhemagglutination
    m. assay (MHA)
    m. assay for antibodies to
    Treponema pallidum (MHA-TP)
microhematometra
Micro-Hysteroflator
    Hamou M.-H.
microhysteroscope
    Hamou contact m.

microimplant
  silicone m.
microinjection
microinvasion
  stromal m.
microinvasive
  m. adenocarcinoma
  m. carcinoma
  m. carcinoma classification
  m. cervical cancer
MicroIron II carbonyl iron
MicroLap
  M. endoscope
  M. Gold system
microlaparoscopic sterilization
microlaparoscopy
microlithiasis
micromanipulation
  gamete m.
micromanipulator
micromazia
micromelia
micromelic dwarfism
Micro-Mist disposable nebulizer
micromyeloblastic leukemia
Micronase
microneedle
microNefrin
micronized
  m. 17beta estradiol
  m. estradiol
  m. progesterone
Micronor
microorchidism
micropapillomatosis labialis
micropenis
microperforate hymen
microphallus
  m., imperforate anus, syndactyly, hamartoblastoma, abnormal lung (MISHAP)
microphthalmia
  m. or anophthalmos with associated anomalies (MAA)
  m.-arhinia
  m., dermal aplasia, sclerocornea syndrome (MIDAS)
  m.-mental deficiency syndrome
  m. with linear skin defects (MLS)
microphthalmos
micropodia

micropolygyria with muscular dystrophy
microprosopus
microretrognathia
microsatellite
  sequence tagged m.
microscope
  Optiphot-2UD m.
microscopy
  dark-field m.
  scanning electron m. (SEM)
microsomia
  hemifacial m. (HM)
  unilateral facial m.
microsound epoxide hydrolase
MicroSpan
  M. microhysterescopy system
  M. minihysteroscopy system
  M. sheath
microspectroscopy
  Fourier transform infrared m.
microsphere
  degradable starch m. (DSM)
  Embosphere m.
  magnetically responsive m.
  radioactive m.
microstomia
Microsulfon
microsurgery
  distal tubal m.
  endoscopic m.
  laparoscopic m.
  transanal endoscopic m. (TEM)
  tubal m.
  tubocornual m.
microsurgical
  m. epididymal sperm aspiration (MESA)
  m. extraction of ductal sperm (MEDS)
  m. tubocornual anastomosis
microthelia
microtia
  unilateral m.
microtia-absent patellae-micrognathia syndrome
Microtip catheter
Micro-Touch Platex medical glove
MicroTrak test
Micro-Transducer catheter
microvillus, pl. **microvilli**
microviscometry

M

NOTES

micturition
    diurnal m.
    nocturnal m.
mid
    midposition
Midamor
midarm circumference
MIDAS
    microphthalmia, dermal aplasia,
      sclerocornea syndrome
midazolam
midbrain
    m. abnormality
    kinked m.
midcycle
    m. cervical mucus
    m. surge
middle sacral artery
midfetal testicular regression syndrome
midforceps
    m. delivery
    m. maneuver
midgut volvulus
midline
    m. cleft syndrome
    m. episiotomy
    m. heart
    m. incision
    m. vertical uterine extension
midluteal progesterone measurement
**Midmark 413 power female procedure
  chair**
midmenstrual
midodrine
Midol IB
midpain
midparental height
midpelvis
midplane
midposition (mid)
midsecretory
midstream urine specimen
midurethra
midwife, pl. **midwives**
    American College of Nurse
      Midwives
midwifery
    Licentiate in M. (LM)
Miescher syndrome
Mietens syndrome
Mietens-Weber syndrome
MIF
    migration-inhibitory factor
    müllerian inhibiting factor
Mifeprex
mifepristone
miglitol

migraine
    m. headache
migrans
    cutaneous larva m.
    erythema chronicum m. (ECM)
    larva m.
migration
    cellular m.
    m.-inhibitory factor (MIF)
    placental m.
migrational disorder
MIH
    müllerian inhibiting hormone
MII
    meiosis II
Mikity-Wilson syndrome
**mild X-linked recessive muscular
  dystrophy**
Miles-Carpenter syndrome (MCS)
Miles syndrome
Milex
    M. cervical cup
    M. spatula
milia (*pl. of* milium)
miliaria
    apocrine m.
    m. crystallina
    m. profunda
    m. pustulosa
    m. rubra
    sebaceous m.
miliary
    m. calcified necrosis
    m. tuberculosis
milieu
    endometrial m.
military antishock trousers (MAST)
milium, pl. **milia**
    milia neonatorum
milk
    m. abscess
    atomic m.
    banked breast m.
    breast m.
    m. of calcium microcyst
    m. cyst
    m. ejection
    m. ejection reflex
    evaporated m.
    m. fever
    fluorescein-labeled m.
    m. leg
    m. letdown
    m. of magnesia (MOM)
    mother's m.
    mother's breast m. (MBM)
    nuclear m.
    m. teeth

uterine m.
witch's m.
milk-plasma ratio
Millar microtransducer urethral catheter
Millen-Read modification
Millen technique
millequivalent (mEq)
    m. per liter (mEq/L)
Miller
    M. blade
    M. ovum
    M. syndrome
Miller-Abbott tube
Miller-Dieker
    M.-D. lissencephaly syndrome
    (MDLS)
    M.-D. syndrome
milleri
    Streptococcus m.
milligram (mg)
milligrams percent (mg%)
millijoule (mJ)
millirad (mrad)
milliroentgen (mr)
millivolt (mV, mv)
mill wheel murmur
Milontin
Milophene
Milroy disease
Miltex disposable biopsy punch
Miltown
MIM
    Mendelian Inheritance in Man
Mi-Mark
    M.-M. disposable endocervical
    curette
    M.-M. endocervical curette set
    M.-M. endometrial curette set
mimetic labor
Mims
    nevus sebaceous of Feuerstein
    and M.
Minamata disease
mind-body intervention
mineral
    bone m.
    m. metabolism
    m. oil
    m. requirement
mineralocorticoid
mineralocorticosteroid
Minesse

Mini-Dose
    HypRho-D M.-D.
minidose heparin
Mini-Flex
    M.-F. flexible Harris uterine
    injector
    HUI M.-F.
miniform
    inSync m.
Mini-Gamulin Rh
miniguard
minilap
    minilaparotomy
minilaparoscope
    Aslan 2 mm m.
    Pixie m.
minimal-incision pubovaginal suspension
minimally invasive surgical technique
    (MIST)
Minims
minimum
    m. inhibitory concentration (MIC)
    m. lethal dose (MLD)
MiniOX I, II, III, 100-IV oxygen
    monitor
Minipress
minisatellite
MiniSite laparoscope
mini Vidas automated immunoassay
    system
MiniWright peak flow meter
Minizide
Minkowski-Chauffard syndrome
Minnesota Multiphasic Personality
    Inventory (MMPI)
Minocin
    M. IV Injection
    M. Oral
minocycline
minor
    alpha-thalassemia m.
    beta-thalassemia m.
    m. dysmorphism
    thalassemia m.
minora
    labia m.
Minot disease
Minot-von Willebrand syndrome
minoxidil
Mintezol
minute
    beats per m. (bpm)

M

NOTES

minute *(continued)*
    breaths per m. (bpm)
    liters per m. (L/min)
    micrograms per kilogram per m.
      (mcg/kg/min)
    m. oxygen uptake
    m. ventilatory volume
Minzolum
Miocarpine
miodidymus
miopus
miosis
    congenital m.
Miostat
miotic division
mirabilis
    Proteus m.
Miradon
Miraluma test
Mircette tablet
Mirchamp sign
Mirena
Mirhosseini-Holmes-Walton syndrome
mirror
    Articu-Lase laser m.
    m. duplication
    m. image breast biopsy
    m. image interpretation
MIS
    müllerian inhibiting substance
miscarriage
    recurrent m.
    spontaneous m.
    threatened m.
    unexplained recurrent m.
miscarry
MISHAP
    microphallus, imperforate anus,
      syndactyly, hamartoblastoma, abnormal
      lung
mismatch
    ventilation/perfusion m.
misonidazole
misoprostol
    oral m.
    vaginal m.
mispairing
missed
    m. abortion
    m. labor
    m. period
missense mutation
mission
    M. Prenatal F.A.
    M. Prenatal H.P.
    M. Prenatal Rx
missionary position

MIST
    minimally invasive surgical technique
mist
    Ayr saline nasal m.
    Bronitin M.
    Bronkaid M.
    child-adult m. (CAM)
    hood m.
    Primatene M.
    tent m.
    m. tent
Mitex GII/mini anchor
Mithracin
mithramycin
mitis
    Streptococcus m.
mitochondrial
    m. chromosome
    m. DNA
    m. myopathy
mitochondrion, pl. **mitochondria**
mitogen
    pokeweed m. (PWM)
mitogenic activity
mitomycin
    m. C
mitoplasm
mitosis
    m. phase
    m.-promoting factor (MPF)
mitotane
mitotic
    m. chromosome
    m. figure
mitoxantrone
    m. hydrochloride
mitral
    m. arcade
    m. atresia
    m. commissurotomy
    m. insufficiency
    m. regurgitation
    m. stenosis
    m. valve
    m. valve atresia
    m. valve insufficiency
    m. valve prolapse
Mitrofanoff
    M. appendicovesicotomy
    M. neourethral procedure
    M. principle
mittelschmerz
Mittendorf-Williams rule
mitten hand
Mityvac
    M. M-cup
    M. obstetric vacuum extractor cup
    M. reusable vacuum pump

M. Super M cup
M. vacuum delivery system
M. vacuum extractor

mix

CeraLyte drink m.
preMA prenatal drink m.

mixed

m. agglutination reaction (MAR)
m. connective tissue disease (MCTD)
m.-density mass
m. discrete-continuous random variable
m. germ cell tumor
m. gonadal dysgenesis
m. lymphocyte reaction blocking factor
m. mesodermal sarcoma (MMS)
m. mesodermal tumor
m. müllerian sarcoma
m. müllerian tumor (MMT)
m. ovarian mesodermal sarcoma
m. pattern
m. porphyria
m. sclerosing bone dysplasia, small stature, seizures, mental retardation syndrome
m. umbilical arterial acidemia
m. uterine tumor

mixoploid
mixoploidy

diploid/tetraploid m.
diploid/triploid m.

Mixtard
Miya

M. hook
M. hook ligature carrier

Miyazaki-Bonney

M.-B. test
M.-B. test for stress incontinence

Miyazaki technique
mJ

millijoule

MKHS

Menkes kinky hair syndrome

MLD

minimum lethal dose

MLNS

mucocutaneous lymph node syndrome

MLS

microphthalmia with linear skin defects

MM

malignant mesothelioma
MM band

MMEP

microcephaly, microphthalmia, ectrodactyly, prognathism syndrome

MMI

maternal meiosis I

MMIH

megacystis, microcolon, intestinal hypoperistalsis
MMIH syndrome

MMII

maternal meiosis II

MMK

Marshall-Marchetti-Krantz
MMK procedure

MMMM

megalocornea-macrocephaly-mental and motor retardation syndrome

MMMT

malignant mixed müllerian tumor

M-mode

M-m. display
M-m. echocardiogram
M-m. echogram
M-m. imaging
M-m. ultrasound

MMPI

Minnesota Multiphasic Personality Inventory

MMR

measles, mumps, rubella
megalocornea-mental retardation syndrome
MMR vaccine

MMS

mixed mesodermal sarcoma

MMT

microcephaly, mesobrachyphalangy, tracheoesophageal fistula syndrome
mixed müllerian tumor

10-mm trocar
MMTV

mouse mammary tumor virus

MNBCC

multiple nevoid-basal cell carcinoma
MNBCC syndrome

MOA

monoamine oxidase

Moban
Mobenol

M

NOTES

Mobidin
mobility
>bladder neck m.
>m. disability

Mobiluncus
>*M. vaginitis*

Möbius
>M. anomaly
>M. sequence
>M. syndrome

modality
Modane
mode
>delivery m.

Modecate
MODED
>microcephaly-oculo-digito-esophageal-
>duodenal syndrome

model
>genetic m.
>pathological m.
>Rossavik growth m.
>M. 5315 sequential compression
>device
>statistical m.

modeling
>generalized linear interactive m.
>(GLIM)

modern genetics
Modicon
modification
>Burch m.
>Gesell test with Knobloch m.
>Millen-Read m.

modified
>m. Bagshawe protocol (MBP)
>m. Beighton criteria
>m. BPP
>m. Burch colpourethropexy
>m. Ham F-10 solution
>m. Irving-type tubal ligation
>m. Pomeroy technique
>m. radical hysterectomy
>m. radical mastectomy
>m. Ritgen maneuver
>m. Roeder
>m. Roeder knot
>m. sling

modifier
>biologic response m. (BRM)

Moditen
MODS
>multiple organ dysfunction syndrome

modulation
>antigenic m.
>cardiac autonomic m.
>sex steroid m.

modulator
>benzothiophene-derived selective
>estrogen receptor m.
>selective estrogen receptor m.
>(SERM)
>triphenylethylene selective estrogen
>receptor m.

Modumate
MODY
>maturity-onset diabetes of the young

Moeller-Barlow disease
Mogen
>M. circumcision
>M. clamp

Mohr-Claussen syndrome
Mohr syndrome
Mohr-Tranebjaerg syndrome (MTS)
moiety, pl. **moieties**
moisturizer
>Vagisil intimate m.

molar
>m. degeneration
>m. evacuation
>Moon m.'s
>mulberry m.
>m. pregnancy

Molatoc
mold
>Counsellor vaginal m.

molding of head
mole
>amniography in hydatidiform m.
>blood m.
>Breus m.
>carneous m.
>complete hydatidiform m.
>cystic m.
>false m.
>fleshy m.
>grape m.
>hydatid m.
>hydatidiform m.
>invasive hydatidiform m.
>partial hydatidiform m.
>prior complete m.
>prior partial m.
>repeated complete m.
>tuberous m.
>vesicular m.
>vulvar hydatidiform m.

molecular
>m. clone
>m. genetic analysis
>m. genetic technique

molecule
>adhesion m.
>cell adhesion m. (CAM)
>inflammatory m.

intercellular adhesion m.-1 (ICAM-1)
recombinant DNA m.
vascular cell adhesion m.
molestation
sexual m.
molimen, pl. **molimina**
menstrual molimina
molindone
Mollica-Pavone-Anterer syndrome
Mollica syndrome
Mollison formula
molluscum
m. contagiosum
m. contagiosum virus (MCV)
m. fibrosum
m. fibrosum gravidarum
molybdenum
m. rotating anode x-ray tube
m. target
MOM
milk of magnesia
mucoid otitis media
MoM
multiples of the appropriate gestational median
mom
multiples of the median
mometasone furoate
MOMO
macrosomia, obesity, macrocephaly, ocular abnormality syndrome
MOMP gene
MOMX
macroorchidism marker X
Monaghan respirator
monarticular arthritis
Mondor disease
mongolian
m. idiot
m. spot
mongolism
mongoloid
m. features
monilethrix
sex-linked neurodegenerative disease with m.
Monilia
monilial
m. esophagitis
m. rash
m. vaginitis

moniliasis
moniliform hair
Monistat
M. Dual-Pak
M. I.V.
M. i.v. Injection
M. Vaginal
Monistat-3 vaginal suppository
Monistat-Derm Topical
monitor
Accu-Chek Easy glucose m.
Accu-Chek II Freedom blood glucose m.
actocardiotocograph fetal m.
Aequitron 9200 apnea m.
antepartum m.
apnea m.
Arvee model 2400 infant apnea m.
Baby Dopplex 3000 antepartum fetal m.
Bear NUM-1 tidal volume m.
CA m.
cardiac-apnea m.
ClearPlan Easy fertility m.
Corometrics fetal m.
Corometrics 118 maternal/fetal m.
Cue Fertility M.
Dinamap blood pressure m.
EdenTec 2000W in-home cardiorespiratory m.
Endotek OM-3 Urodata m.
Endotek UDS-1000 m.
Fetal Dopplex m.
Fetalert fetal heart rate m.
fetal heart rate m.
FetalPulse Plus m.
Fetasonde fetal m.
Healthdyne apnea m.
Holter m.
home uterine activity m. (HUAM)
Imex antepartum m.
intrapartum m.
MiniOX I, II, III, 100-IV oxygen m.
Nellcor N-499 fetal oxygen saturation m.
Nellcor-Puritan-Bennett oxygen saturation m.
neonatal m.
Neo-trak 515A neonatal m.
OMRON m.

M

NOTES

monitor (*continued*)
>Oxisensor fetal oxygen saturation m.
>Pocket-Dop 3 m.
>Press-Mate model 8800T blood pressure m.
>ProDynamic m.
>Profilomat m.
>QuietTrak m.
>Quik Connect fetal m.
>Sonicaid Axis m.
>Sonicaid SYSTEM 8000 fetal m.
>Toitu MT-810 cardiographic m.
>Tokos m.
>transcutaneous m.
>uterine activity m.
>virtual labor m. (VLM)

monitored anesthesia care
monitoring
>ambulatory m.
>blood sugar m.
>critical care m.
>electronic fetal m. (EFM)
>end-tidal CO2 m.
>external fetal m. (EFM)
>fetal heart rate m.
>fetal oximetry m.
>heart rate m.
>hemodynamic m.
>home uterine m. (HUM)
>home uterine activity m. (HUAM)
>intrapartum fetal m.
>intrauterine fetal m.
>tactile sensory m.
>telefetal m.
>tissue pH m.
>transcutaneous oxygen tension m.

monitrice
monoamine
>m. oxidase (MOA)
>m. oxidase A deficiency (MAOA)
>m. oxidase inhibitor (MAOI)

monoamnionicity
monoamniotic twins
monobactam
monobrachius
monocephalus
monochorial twins
monochorionic
>m. diamniotic placenta
>m. monoamniotic placenta
>m. placenta
>m. placentation
>m. twin pregnancy
>m. twins

monochromatism
>blue cone m.
>pi cone m.

Monocid
Monoclate-P
monoclonal
>m. antibody
>m. antibody coagglutination test
>m. antibody therapy

monocranius
Monocryl
>M. suture

monocyte
monocytogenes
>*Listeria* m.

monodactyly
monodermal tumor
Monodox
monoecious
monofactorial inheritance
monofluorophosphate
monogamous
monogamy
monogenic
Mono-Gesic
monohybrid
>m. cross

monohydrate
>doxycycline m.
>m. macrocrystal
>nitrofurantoin m.

monokine
monomelic
mononeuropathy
>peripheral m.

Mononine
mononucleosis
>infectious m.

4-monooxygenase
>phenylalanine -m.

monophasic
>m. oral contraceptive
>m. regimen

monophosphate
>adenosine m. (AMP)
>cyclic adenosine m. (cAMP)
>cyclic adenosine 3′,5′-m.
>cyclic guanosine m. (cGMP)
>cyclic guanosine 3′,5′-m.
>cytidine m.

monoplegia
monoploid
monopodia
monopolar cautery
monops
monopus
monorchidic
Monoscopy locking trocar with Woodford spike
monosome

monosomy
    m. 1–22
    autosomal m.
    chromosome 1p–22p m.
    chromosome 1q–22q m.
    chromosome Xp21 m.
    chromosome Xq m.
    m. G
    m. G syndrome
    partial m. 1p–22p
    partial m. XP21, XP22
    partial m. Xq
    m. 1p–22p
    m. 7 syndrome
    m. X
    m. Xp21
    m. Xp22
    m. Xq
Monospot test
Monosticon Dri-Dot test
Mono-Sure
monosymptomatic delusional pseudocyesis
Mono-Test (FTB)
monotherapy
    zidovudine m.
monotonically
monotropins for injection
Mono-VaccTest (O.T.)
monovular twins
monozygosity
monozygotic twins
monozygous
Monro
    foramen of M.
Monsel
    M. gel
    M. paste
    M. solution
mons pubis
monster
    hair m.
monstrosity
Montefiore syndrome
Montevideo unit
Montgomery
    M. gland
    M. strap
    M. tubercle
Month
Monurol

mood state
Moon
    M. molars
    M. teeth
Moore-Federman syndrome
Morand foot
morbidity
    antenatal m.
    childbirth-related m.
    febrile m.
    fetal m.
    infant m.
    maternal febrile m.
    neonatal m.
    perinatal m.
    postpartum febrile m.
    m. predictor
    puerperal m.
morbilliform rash
morbus Down
morcellation
    electromechanical m.
    m. operation
    uterine m.
    vaginal m.
morcellator
    Diva laparoscopic m.
    motorized m.
    OPERA Star m.
    Semm RX m.
    Steiner electromechanical m.
Morch respirator
Morel ear
Morgagni
    anterior retrosternal hernia of M.
    M. hernia
    hydatid cyst of M.
    M. tubercle
morgagnian
Morgagni-Turner-Albright syndrome
Morgagni-Turner syndrome
Morganella
    M. morganii
morganii
    Morganella m.
    Proteus m.
moribund
Morison pouch
morning
    m. glory syndrome
    m. sickness

M

NOTES

morning-after
> m.-a. contraception
> m.-a. pill

Moro-Heisler diet
Moro reflex
morphea
morphine
> epidural m.
> m. sulfate

morphogen
morphogenesis
morphologic
> m. assessment

morphological
> m. characteristic
> m. sex

morphology
> adrenal gland m.
> endometrial m.
> kidney m.
> QRS m.

morphometric
Morquio
> M. disease
> M. syndrome

Morquio-Brailsford
> M.-B. disease
> M.-B. syndrome

Morquio-Ullrich
> M.-U. disease
> M.-U. syndrome

Morsch-Retec respirator
morselize
mortality
> m. data
> fetal m.
> high altitude perinatal m.
> infant m.
> maternal m.
> neonatal m.
> perinatal m.
> m. predictor
> prenatal m.
> m. rate
> reproductive m.
> m. risk factor

mortiferum
> *Fusobacterium m.*

morula
mosaic
> m. aneuploidy
> gene m.
> m. pattern
> m. tetrasomy 8p syndrome
> m. translocation
> m. trisomy 14
> m. Turner syndrome

mosaicism
> chromosomal m.
> erythrocyte m.
> germ line m.
> gonadal m.
> low-grade m.
> trisomy 8 m.
> Turner m.

Moschowitz
> M. fashion
> M. procedure

mos **protooncogene**
Moss
> M. classification
> M. tube

mother
> m. and baby endoscope
> diabetic m.
> D^u-positive m.
> gestational m.
> high-risk m.
> hysterical m.
> m.-infant bonding
> infant of diabetic m. (IDM)
> rubella-immune m.
> rubella-negative m.
> serology-negative m.
> surrogate m.
> m. wort

motherhood
> surrogate gestational m.

mother's
> m. breast milk (MBM)
> m. milk

moth patch
motif
motile sperm
motilin
motility
> sperm m.

motion
> active range of m. (AROM)
> early cardiac m.
> embryonic heart m. (EHM)
> fetal cardiac m.
> limb m.

motorized morcellator
motor-sensory neuropathy, X-linked Type II, with deafness and mental retardation
Motrin
> M. IB
> M. IB Sinus
> Junior Strength M.

mottling
> m. of skin

mount
> wet m.

mouse
>   m. mammary tumor virus (MMTV)
>   peritoneal m.
>   m. uterine unit

mouth
>   m.-and-hand synkinesia
>   carp m.
>   carp-like m.
>   purse-string m.
>   tapir m.
>   m.-to-mouth resuscitation

Movat stain

movement
>   cardinal m.
>   choreiform m.
>   fetal body m.
>   fetal breathing m. (FBM)
>   rapid eye m. (REM)
>   sound-stimulated fetal m.
>   tonic-clonic m.'s
>   vibroacoustic-induced fetal m.

moxalactam
>   m. disodium

Moxam

moxibustion

moyamoya disease

Moynahan
>   M. alopecia syndrome

Moynihan respirator

Mozart ear

MP
>   menstrual period

MPA
>   medroxyprogesterone acetate

MPF
>   maturation-promoting factor
>   mitosis-promoting factor

M-phase-promoting factor

mr
>   milliroentgen

mrad
>   millirad

MRI
>   magnetic resonance imaging
>   GRASS MRI
>   Siemens Vision MRI
>   ultrafast MRI

MRM
>   magnetic resonance mammography

mRNA
>   messenger ribonucleic acid

MRSD
>   mental retardation, skeletal dysplasia,
>   abducens palsy syndrome

MRU
>   magnetic resonance urography

MRX1
>   X-linked mental retardation

MRXA
>   X-linked mental retardation-aphasia
>   syndrome

MRXS1–6
>   X-linked mental retardation syndrome
>   1–6

MSAF
>   meconium-stained amniotic fluid

MSAFP
>   maternal serum alpha-fetoprotein

MSBP
>   Münchausen syndrome by proxy

MS Contin

MSD
>   multiple sulfatase deficiency

MSI
>   magnetic source imaging

MSIR Oral

MSN
>   hereditary motor sensory neuropathy II-
>   deafness-mental retardation
>   MSN syndrome

MSS
>   Marshall-Smith syndrome

MSUD
>   maple syrup urine disease

MTHFR
>   methylene tetrahydrofolate reductase

MTMX
>   X-linked myotubular myopathy

MTS
>   Mohr-Tranebjaerg syndrome

MTX
>   methotrexate

M-type extractor

Mucat
>   M. cervical sampling
>   M. cervical sampling device

mucinous
>   m. carcinoma
>   m. cystadenocarcinoma
>   m. ovarian neoplasm
>   m. tumor

mucocolpos

M

NOTES

mucocutaneous
m. candidosis
m. leishmaniasis
m. lymph node syndrome (MLNS)
mucoid
m. myoma degeneration
m. otitis media (MOM)
mucolipidosis
Mucomyst
mucopolysaccharide
m. storage disease I–VIII
mucopolysaccharidosis
beta-glucuronidase deficiency m.
m. F
focal m.
HS m.
m. I, II
m. IV A, B
KS m.
m. (MPS) I, II
m. (MPS) III A, B, C, D
m. (MPS) VI–VIII
type I H/S m.
type II, III, VI–VIII m.
type IS m.
type IVA, B m.
mucoprotein
Tamm-Horsfall m.
mucopurulent cervicitis
mucormycosis
cutaneous m.
pulmonary m.
mucorrhea
cervical m.
mucosa
cervical m.
endocervical m.
vaginal m.
mucosal
m. neuroma syndrome
m. transudate
Mucosil
mucosotropic
mucosulfatidosis
mucous
m. discharge
m. membrane
m. plug
mucoviscidosis
MUCP
maximum urethral closure pressure
mucus
cervical m.
m. extractor
midcycle cervical m.
ovulatory m.
mu dimeric protein
Mueller (*var. of* Müller)

MUGA
multiple gated acquisition scan
MUGA scan
mulberry
m. molar
m. ovary
mulibrey
muscle, liver, brain, eye
mulibrey nanism
mulibrey nanism or dwarfism
Müller, Mueller
M. syndrome
M. tubercle
Müller-Hillis maneuver
müllerian
m. abnormality
m. adenosarcoma
m. agenesis
m. cyst
m. duct
m. duct anomaly
m. duct fusion
m. duct inhibitory factor
m. fusion defect
m. hypoplasia
m. inhibiting factor (MIF)
m. inhibiting hormone (MIH)
m. inhibiting substance (MIS)
m., renal, cervicothoracic, somite
    abnormalities (MURCS)
m., renal, cervicothoracic, somite
    abnormalities syndrome
Mullin system
MultArray light
multicelled embryo
multicentric
m. carcinoma
m. lower genital tract neoplasia
multichannel
m. cystometrogram
m. recorder
multicolor FISH cytogenetic technique
multicystic renal dysplasia
Multidex gel
multifactorial
m. disorder
m. inheritance
m. trait
multifetal
m. gestation
m. pregnancy
m. pregnancy reduction
multifetation
multifocal clonic seizure
multifollicular ovary
multigenic disorder
multigravida
Multiload Cu-375 intrauterine device

multiloba
placenta m.
multilocal genetics
multilocular cyst
multiloculated cyst
multimammae
multinucleate
multinucleation parakeratosis
multipara
grand m.
great-grand m.
multiparity
grand m.
multiparous
multiplane intracavitary probe
multiple
m. alleles
m. arbitrary amplicon profiling
m. articular contracture
m. articular rigidity
m. basal cell carcinoma syndrome
multiple basal cell nevus
syndrome
m. benign circumferential skin
creases on limb
m. births
m. cervix
m. congenital anomalies
m. cyst
m. endocrine abnormalities (MEA)
m. endocrine neoplasia, types I, II,
III (MEN)
m. epiphyseal dysplasia
m. epiphyseal dysplasia-early onset
diabetes mellitus syndrome (MED-
IDDM)
m. epiphyseal dysplasia tarda
syndrome
m. exostoses
m. exostosis mental retardation
(MEMR)
m. exostosis mental retardation
syndrome
m. gated acquisition scan (MUGA)
m. gestation
m. hereditary cutaneomandibular
polyoncosis
m. lentigines syndrome
m. marker screening
m. myeloma
m. neuroma syndrome

m. nevoid-basal cell carcinoma
(MNBCC)
m. nevoid-basal cell carcinoma
syndrome
m. nevoid, basal cell epithelioma,
jaw cysts, bifid rib syndrome
m. organ dysfunction syndrome
(MODS)
m. organ failure
m. pregnancy
m. pterygium syndrome
m. resistant cell lines
m. sclerosis
m. sulfatase deficiency (MSD)
m. synostoses
m. synostoses syndrome
multiple-punch resection
multiples
m. of the appropriate gestational
median (MoM)
m. of the median (mom)
multiplex
arthrogryposis m.
dysostosis m.
myodysplasia fibrosa m.
synostosis m.
multiplexing
**MultiPRO 2000 disposable biopsy
needle guide for ultrasound**
multisite
m. BRACA
m. lower genital tract involvement
Multispatula cervical sampling device
multivariant analysis
multivitamin
MultiVysion PB assay test
multocida
Pasteurella m.
Mulvihill-Smith syndrome
mummified fetus
mumps
m. orchitis
m. vaccine
m. virus
Mumpsvax
**Münchausen syndrome by proxy
(MSBP)**
Munro
M. and Parker classification for
laparoscopic hysterectomy
M. point
Munro-Kerr maneuver

M

NOTES

345

mupirocin
mural pregnancy
MURCS
    measurable undesirable respiratory
      contaminants
    müllerian, renal, cervicothoracic, somite
      abnormalities
    MURCS syndrome
Murine Plus Ophthalmic
Murless
    M. head extractor
    M. head retractor
murmur
    continuous m.
    diamond-shaped m.
    diastolic m.
    functional m.
    Graham Steell m.
    holosystolic m.
    machinery m.
    mill wheel m.
    pansystolic m.
    pulmonary m.
    pulmonic m.
    Still m.
    systolic m.
    systolic ejection m.
Murphy hole
muscarinic
muscle
    abdominis m.
    m. atrophy-contracture-oculomuscle
      apraxia syndrome
    bulbocavernosus m.
    coccygeus m.
    detrusor m.
    gracilis m.
    Guthrie m.
    hypertrophic bundle of smooth m.
    hypoplasia of anguli oris
      depressor m. (HAODM)
    m. hypotonia
    iliococcygeal m.
    ischiocavernosus m.
    levator ani m.
    pectoralis major m.
    pubococcygeal m.
    puborectal m.
    puboviseral m.
    rectus abdominis m.
    m. relaxant
    smooth m.
    m. splitting
    striated m.
    striated circular m.
    thyroarytenoid m.
muscle-eye-brain (MEB)

    m.-e.-b. disease
    m.-e.-b. syndrome (MEBS)
muscle, liver, brain, eye (mulibrey)
muscular
    m. dystrophy (MD)
    m. hypertrophy syndrome
    m. infantilism
    m. injury
musculoskeletal
    m. disorder
    m. system
mushrooming of fimbria
mustard
    nitrogen m.
    phenylalanine m.
    ʟ-phenylalanine m. (L-PAM)
Mustard procedure
Mustargen
    M. Hydrochloride
musty odor
mutagen
mutagenesis
    insertional m.
Mutamycin
mutant
    m. allele
    m. cell
    m. gene
mutation
    BRCA1 gene m.
    BRCA2 gene m.
    cystic fibrosis transmembrane
      conductance regulator m.
    factor V Leiden m.
    frameshift m.
    function-enhancing m.
    gene m.
    genetic m.
    germ-line m.
    G-to-T transversion m.
    HBS m.
    heparin-binding site m.
    hew m.
    Leiden m.
    missense m.
    mutator m.
    new m.
    nonsense m.
    point m.
    reactive site m.
    RS m.
    trinucleotide repeat expansion m.
mutational dysostosis
mutator mutation
Mutchinick syndrome
mutilans
    keratoma hereditarium m.
mutilating keratoderma

mutilation
female genital m. (FMG)
female genital tract m. (FGTM)
Muzsani procedure
muzzled sperm
mV, mv
millivolt
MX2-300 xenon quality light source
myalgia
Myambutol
myasthenia gravis
Mycelex
M.-G
M.-G Topical
M. Troche
mycelium, pl. **mycelia**
Mycifradin
M. Sulfate
Myciguent
Myclo-Derm
Myclo-Gyne
mycobacterial
Mycobacterium
M. *avium-intracellulare*
M. *leprae*
M. *tuberculosis*
Mycogen II Topical
Mycolog-II Topical
Myconel Topical
Mycoplasma
M. *hominis*
M. *pneumoniae*
M. *ureaplasma*
mycoplasma
genital m.
T strain m.
mycoplasmal
mycosis
Mycostatin
mycotic
m. vaginosis
Myco-Triacet II
Mydfrin Ophthalmic
myelacephalus
myelencephalon
myelitis
transverse m.
myeloblastic leukemia
myelocele
myelodysplasia
myelodystrophy
myelofibrosis

myelogenous leukemia
myelography
myeloid/erythrocyte (M/E)
myeloma
multiple m.
m. protein
myelomeningocele
myelophthisis
myeloproliferative
m. disorder
myeloschisis
myelosuppression
myelosuppressive agent
myenteric plexus
Myhre syndrome
myiasis
Myleran
Mylicon
M. drops
myoblastoma
granular cell m.
myocardial
m. blood flow
m. function
m. hypertrophy
m. infarction
m. ischemia
myocarditis
myocardium
Myochrysine
myoclonia
myoclonic
m.-astatic seizure
m. epilepsy with ragged red fibers
(MERRF)
m. seizure
myoclonus
Baltic m.
m. epilepsy
jaw m.
myocolpitis
myocutaneous
m. flap
transverse rectus abdominis m.
(TRAM)
myodysplasia
m. fetalis deformans
m. fibrosa multiplex
myodystrophia fetalis deformans
myodystrophy
Duchenne m.
myoepithelial cell

M

NOTES

myoglobin
myoglobinuria
myognathus
myography
Myogyn II stimulator
myoma, pl. **myomata**
    asymptomatic m.
    calcified m.
    cervical m.
    degenerating m.
    m. fixation instrument
    intracavitary m.
    intraligamentous m.
    intramural m.
    parasitic m.
    retained m.
    m. screw
    submucosal m.
    submucous m.
    subserosal m.
    subserous m.
    myomata uteri
    uterine m.
myomectomy
    abdominal m.
    hysteroscopic m.
    vaginal m.
myometrial
    m. contraction
    m. coring
    m. neurofibroma
myometritis
myometrium
    hypotonic m.
    uterine m.
myomotomy
myonecrosis
myopathic limb-girdle syndrome
myopathy
    Batten-Turner congenital m.
    centronuclear m. (CNM)
    hypothyroid m.
    m.-lactic acidosis-sideroblastic
      anemia syndrome
    mitochondrial m.
    m.-myxedema syndrome
    nemaline m.
    ocular m.
    rod m.

    X-linked centronuclear m.
    X-linked myotubular m. (MTMX,
      XLMTM)
    X-linked recessive centronuclear m.
    X-linked recessive myotubular m.
myopia
myosalpingitis
myosalpinx
myosin
    m. filament
    m. light-chain kinase
    m. light-chain phosphorylation
myositis
    m. ossificans circumscripta
    m. ossificans progressiva
myotomy
    Heller m.
    Livadatis circular m.
myotonia
    chondrodystrophia m.
    m. chondrodystrophia
    chondrodystrophic m.
    m. congenita
    m. neonatorum
    spondyloepimetaphyseal dysphasia
      with m.
myotonic
    m. dystrophy (MD)
    m. muscular dystrophy
MyoTrac EMG
myringitis
    bullous m.
myrtiform caruncle
Mysoline
mystery syndrome
Mytelase
Mytrex F Topical
myxedema
    infantile m.
    m.-myotonic dystrophy syndrome
myxedematous infantilism
myxoid matrix
myxoma
    familial cardiac m.
myxorrhea
M-Zole
    M-Z. 7 Combination Pack
    M-Z. 3 combination pack
    M-Z. 7 dual pack

N-[2-hydroxyethyl]piperazine N′-[2-ethanesulfonic acid] (HEPES)
Nabi-HB hepatitis B immune globulin human
nabothian
    n. cyst
    n. follicle
    n. gland
    n. vesicle
N-acetylcysteine
N-acetyl-galactosamine
N-acetylgalactosamine-4-sulfatase deficiency
N-acetyl-glucosamine
N-acetylneuraminic
    N-a. acid (NANA)
    N-a. acid storage disease (NSD)
NACS
    Neurologic and Adaptative Capacity Score
Nadler forceps
nadolol
Nadopen-V
Nadostine
Nadrothyron-D
Naegele (var. of Nägele)
Naegeli
    chromatophore nevus of N.
    N. syndrome
naeslundii
    Actinomyces n.
nafarelin
    n. acetate
nafazatin
Nafcil
nafcillin
    n. sodium
naftifine
    n. hydrochloride
Naftin
Nafucci syndrome
Nägele, Naegele
    N. forceps
    N. obliquity
    N. pelvis
    N. rule
Nager
    N. acrofacial dysostosis
    N. anomaly
    N. sign
    N. syndrome
Nager-de Reynier syndrome
nail
    hypoplastic n.'s
    n.-patella syndrome

Najjar syndrome
nalbuphine
Nalcrom
Naldecon
Naldelate
Nalfon
Nalgest
Nal-Glu
nalidixic acid
Nallpen
nalorphine
naloxone
    n. hydrochloride
NALS
    neonatal adjuvant life support
Nalspan
naltrexone
NAME
    nevi, atrial myxoma, myxoid neurofibromas, ephilides
    NAME syndrome
NANA
    N-acetylneuraminic acid
NANBH
    non-A, non-B hepatitis
Nance-Horan syndrome (NHS)
nandrolone
nanism
    n.-constrictive pericarditis syndrome
    mulibrey n.
    Russell n.
    Seckel n.
nanocephalic dwarfism
nanocephaly
nanoid
nanomelia
nanosomia
nape nevus
NAPI
    Neurodevelopmental Assessment Procedure for Preterm Infants
nappy test
Naprosyn
naproxen
    n. sodium
Naqua
narasin
Narcan
narcolepsy
narcosis
narcotic
    n. analgesia
    n. analgesic
    n. antagonist
    n. depression

N

narcotic *(continued)*
    intrathecal n.
    n. withdrawal syndrome
Nardil
naris, pl. **nares**
Naropin
NARP
    neuropathy, ataxia, and retinitis
    pigmentosa
narrow pubic arch
NAS
    neonatal abstinence syndrome
Nasacort
    N. AQ
Nasahist B
nasal
    n. cannula
    n. CPAP
    Drixoral N.
    n. flaring
    n. hypoplasia, peripheral dysostosis,
     mental retardation syndrome
    Kondon's N.
    Otrivin N.
    Tyzine N.
Nasalcrom
Nasalide Nasal Aerosol
Nasarel Nasal Spray
nascentium
    trismus n.
nasi
    flaring of ala n.
nasoduodenal feeding
nasogastric (NG)
    n. tube
nasojejunal
nasolabial
    n. fold asymmetry
    n. reflex
nasolacrimal duct
Nasonex
nasopharyngeal
    n. airway obstruction
    n. aspirate
    n. suction
nasopharyngitis
nasotracheal
NASS
    Neonatal Abstinence Scoring System
natal
    n. cleft
    n. teeth
Natalins
    N. Rx
natality
natiform skull
natimortality

national
    N. Cancer Institute (NCI)
    N. Collaborative Diethylstilbestrol
     Adenosis Project (DESAD)
    N. Institute of Child Health and
     Human Development (NICHD)
    N. Surgical Adjuvant Breast
     Project (NSABP)
native squamous epithelium
natriuresis
natural
    n. antibody
    n. childbirth
    n. conception
    n. delivery
    n. fecundity
    n. killer cell
    n. killer lymphocyte
    n. penicillin
    n. selection
nature of specimen
Naturetin
nausea
    n. gravidarum
    n. and vomiting
Navajo brainstem syndrome
Navane
navel
    blue n.
    cutis n.
Navelbine
Navratil stirrups
Naxen
NBAS
    Neonatal Behavior Assessment Scale
NBCC
    nevoid basal cell carcinoma
NBCCS
    nevoid basal cell carcinoma syndrome
NBS
    nevoid basal cell carcinoma syndrome
    Nijmegen breakage syndrome
NC
    Histussin NC
NCAH
    nonclassical adrenal hyperplasia
NC-CAH
    nonclassic congenital adrenal hyperplasia
NCI
    National Cancer Institute
ND-Stat
Nd:YAG
    neodymium:yttrium-aluminum-garnet
    Nd:YAG laser
    Nd:YAG laser ablation
Neal insufflator
near-anhydramnios
near-infrared spectroscopy

near-term pregnancy
near-viable fetus
Nebcin
    N. Injection
nebulization ventilator
nebulizer
    Dura Neb portable n.
    Hudson T Up-Draft II
      disposable n.
    Micro-Mist disposable n.
    PulmoMate n.
    Schuco n.
NEC
    necrotizing enterocolitis
neck
    bladder n.
    n.-righting reflex
    vesical n.
    webbed n.
necrolysis
    toxic epidermal n. (TEN)
necrophorum
    Fusobacterium n.
necrosis
    acute tubular n. (ATN)
    cortical n.
    decidual fibrinoid n.
    fascial n.
    fat n.
    hepatic n.
    intrauterine facial n.
    miliary calcified n.
    periportal hemorrhagic n.
    pituitary n.
    postsurgical fat n.
    posttraumatic fat n.
    renal cortical n.
    renal tubular n.
    subcutaneous fat n.
    unilateral intrauterine facial n.
    uterine n.
    white matter n.
necrospermia
necrotic facial dysplasia
necrotizing
    n. enterocolitis (NEC)
    n. fasciitis
    n. funisitis
necrozoospermia
N.E.E. 1/35
needle
    Adair-Veress n.

    n. aspiration
    Bard Biopty cut n.
    n. biopsy
    Biopty cut n.
    butterfly n.
    Chiba n.
    coaxial sheath cut-biopsy n.
    Cobb-Ragde n.
    CUSALap accessory n.
    n. cystourethropexy
    Dieckmann intraosseous n.
    D-Tach removal n.
    Endopath Ultra Veress n.
    Ethiguard n.
    Euro-Med FNA-21 aspiration n.
    eXcel-DR Glasser laparoscopic n.
    FNA-21 n.
    GraNee n.
    Hawkins breast localization n.
    Howell biopsy aspiration n.
    insufflation n.
    J n.
    Kopan n.
    n. localization
    Pereyra n.
    Potocky n.
    scalp vein n.
    Schutt n.
    SonoVu US aspiration n.
    Stamey n.
    stereotactic breast biopsy n.
    n. suspension
    n. tip
    ULTRA-vue amniocentesis n.
    Veress n.
    Vim-Silverman n.
    Virginia n.
    Wolf-Veress n.
needle holder
    Crile-Wood n. h.
    Endo-Assist endoscopic n. h.
    Heaney n. h.
    Mayo-Hegar n. h.
    Wangensteen n. h.
    Wolf-Castroviejo n. h.
NEEP
    negative end-expiratory pressure
nefazodone
negative
    n. end-expiratory pressure (NEEP)
    n. interference

N

NOTES

negative *(continued)*
    n. π meson
    n. punch biopsy
negative-pressure
    n.-p. box
    n.-p. respirator
NegGram
negligence
Neill-Dingwall syndrome
Neisseria gonorrhoeae
neisserial infection
nelfinavir
Nellcor
    N. FS-10 oximeter sensor
    N. FS-14 oximeter sensor
    N. N-499 fetal oxygen saturation
    monitor
    N. N-400/FS system
Nellcor-Puritan-Bennett oxygen saturation monitor
Nelova
Nelson
    N. sign
    N. syndrome
nemaline myopathy
Nemasole
nem (breast milk nutritional unit)
Nembutal
neoadjuvant hormonal therapy (NHT)
Neo-Calglucon
Neocate formula
Neo-Codema
neocystostomy
neo-darwinism
Neo-Durabolic
neodymium:yttrium-aluminum-garnet (Nd:YAG)
Neo-Estrone
Neofed
neofetus
Neo-fradin
neogala
Neo-Gen screening
neomycin
    n. sulfate
neonatal
    N. Abstinence Scoring System (NASS)
    n. abstinence syndrome (NAS)
    n. acne
    n. adjuvant life support (NALS)
    n. alloimmune thrombocytopenia
    n. asphyxia
    N. Behavior Assessment Scale (NBAS)
    n. cholestatic hepatitis
    n. cholestatic jaundice
    n. condition

    n. conjunctivitis
    n. convulsion
    n. death
    n. diagnosis
    n. distress
    n. encephalopathy
    Exosurf N.
    N. Facial Coding System
    n. herpes
    n. HMD
    n. HSV-1, -2 encephalitis
    n. hyaline membrane disease
    n. hyperparathyroidism
    n. hypocalcemia
    n. hypoglycemia
    N. Infant Pain Scale (NIPS)
    n. infection
    n. intensive care unit (NICU)
    n. intubation
    n. line
    n. listeriosis
    n. lupus erythematosus (NLE)
    n. medicine
    n. metabolism
    n. monitor
    n. morbidity
    n. mortality
    n. mortality rate
    n. myasthenia gravis
    n. neutropenia
    n. ocular prophylaxis
    n. olivopontocerebellar atrophy (OPCA)
    n. osseous dysplasia
    n. outcome
    n. progeroid syndrome
    n. pseudohydrocephalic progeroid syndrome
    n. pustular melanosis
    n. respiratory distress syndrome
    n. resuscitation
    n. ring
    n. screening
    n. seizure
    n. sepsis
    n. small left colon syndrome
    n. teeth
    n. tetany
    n. thymectomy
neonate
    cola-colored n.
    early n.
    n. examination
    preterm n.
neonatologist
neonatology
neonatorum
    acne n.

adiponecrosis subcutanea n.
anemia n.
apnea n.
asphyxia n.
diabetes n.
eczema n.
edema n.
erythema toxicum n.
erythroblastosis n.
herpes n.
icterus gravis n.
impetigo n.
keratolysis n.
mastitis n.
melena n.
milia n.
myotonia n.
ophthalmia n.
sclerema n.
sepsis n.
tetania n.
tetanus n.
trismus n.
volvulus n.

Neopap
NeoPath
neoplasia
anal intraepithelial n. (AIN)
cervical epithelial n.
cervical intraepithelial n. (CIN)
endometrial n.
endometrial intraepithelial n. (EIN)
estrogen-dependent n.
gestational trophoblastic n. (GTN)
glandular n.
hematologic n.
intraepithelial n.
invasive n.
lobular n.
multicentric lower genital tract n.
multiple endocrine n., types I, II, III (MEN)
thyroid n.
trophoblastic n.
vaginal intraepithelial n. (VAIN)
vulvar intraepithelial n. (VIN)

neoplasm
adrenal n.
benign ovarian n.
bilateral ovarian n.
borderline epithelial ovarian n.
borderline malignant epithelial n.

cervix n.
endometrial n.
epithelial stromal ovarian n.
germ cell ovarian n.
hepatic n.
indigenous n.
isologous n.
large intestine n.
lipid cell n.
lipoid ovarian n.
malignant n.
malignant ovarian n.
mesenchymal n.
mucinous ovarian n.
neural n.
ovarian lipid cell n.
ovarian malignant epithelial n.
ovarian sex-cord stromal n.
posterior fossa n.
serous ovarian n.
sex-cord stromal n.
soft tissue ovarian n.
uterine n.

neoplastic
n. cyst
n. sequela
n. trophoblastic disease

neosalpingostomy
terminal n.

Neosar
Neo-Sert
N.-S. umbilical vessel catheter
N.-S. umbilical vessel catheter insertion set

Neosporin G.U. Irrigant
neostigmine
Neo-Synephrine
N.-S. 12 Hour Nasal Solution

Neo-Tabs
Neo-Therm neonatal skin temperature probe
Neo-trak 515A neonatal monitor
neoumbilicus
neovagina
gracilis flap n.

neovascularization
Nephramine
nephrectomy
nephritis, pl. **nephritides**
Alport syndrome-like n.
n. gravidarum
hemorrhagic familial n.

N

NOTES

nephritis *(continued)*
    hemorrhagic hereditary n.
    hereditary familial congenital n.
    interstitial n.
    lupus n.
nephroblastomatosis
Nephro-Calci
nephrocalcinosis
nephrogenic
    n. cord
    n. diabetes insipidus
nephrolithiasis
nephroma
    mesoblastic n.
nephron
    cortical n.
Nephronex
nephropathy
    diabetic n.
    reflux n.
    sickle cell n.
nephrophthisis
    familial juvenile n. (FJN)
nephrosclerosis
    malignant n.
nephrosis
    congenital n.
    n., microcephaly, hiatus hernia
      syndrome
nephrosis-microcephaly syndrome
nephrosis-neural dysmigration syndrome
nephrosis-neuronal dysmigration
  syndrome
nephrostomy
    percutaneous n.
nephrotic syndrome
nephrotoxicity
nephroureterectomy
nepiology
Neptazane
nerve
    n. block
    facial n.
    femoral n.
    genitofemoral n.
    n. growth factor (NGF)
    hemorrhoidal n.
    n. injury
    obturator n.
    n. palsy
    parasympathetic n.
    pelvic floor n.
    perineal n.
    presacral n.
    pudendal n.
    sacral n.
    saphenous n.
    sciatic n.

    n. sprouting
    thoracolumbar sympathetic n.
nervi erigentes
Nervine
Nervocaine
nervosa
    anorexia n.
    bulimia n.
nervous system
nesidioblastosis
nest
    cancer n.
    Walthard n.
Nestabs FA
NET-EN
    norethindrone enanthate
Netherton syndrome
netilmicin
    n. sulfate
Netromycin
Nettleship-Falls ocular albinism
Nettleship syndrome
network
    American College of Obstetrics &
      Gynecology n.
    Human Genome Epidemiology N.
      (HuGE Net)
    SEER n.
    Surveillance, Epidemiology and End
      Results n.
Neubauer
    N. chamber
    N. hemocytometer
Neugebauer-LeFort procedure
Neuhauser syndrome
neu/HER2 oncogene
Neu-Laxova syndrome (NLS)
Neupogen
neural
    n. arch
    n. crest
    n. crest-derived cell lineage
    n. crest tissue
    n. groove
    n. neoplasm
    n. plate
    n. tube
    n. tube defect (NTD)
neuralgia
    mammary n.
Neuramate
neuraminidase
neurectomy
    presacral n.
neurenteric canal
neurinomatosis
    n. centralis
    n. universalis

neuritis
    retrobulbar n.
neuritogenesis
neuroangiomatosis
    encephalocraniofacial n.
    n. encephalofacialis
neuroarthromyodysplasia
neurobehavioral abnormality
neuroblastoma
neurocutaneous syndrome
neurodermatitis
neurodevelopment
neurodevelopmental
    N. Assessment Procedure for Preterm Infants (NAPI)
    n. disability
neuroectodermal
    n. hamartoma
    n. tumor
neuroendocrine system
neuroendocrinology
neurofaciodigitorenal (NFDR)
    n. syndrome
neurofibroma
    clitoral n.
    myometrial n.
    ovarian n.
    vaginal n.
    vulvar n.
neurofibromatosis (NF)
    bilateral acoustic n.
    central type n.
    n.-Noonan syndrome (NF-NS, NFNS)
    plexiform n.
    vaginal n.
    von Recklinghausen n.
    n. with Noonan phenotype
neurofilament
    n. antibody
neurogenic
    n. bladder
    n. tumor
neurogram
    pudendal n.
neurohypophysis
neuroichthyosis-hypogonadism syndrome
neuroid cell
neuroleptic
neurologic
    n. abnormality

    N. and Adaptative Capacity Score (NACS)
    n. adverse effect
    n. complication
    n. development
    n. disability
    n. disorder
    n. examination
    n. involvement
neurological
    n. disease syndrome
    n. injury
neuroma
    acoustic n.
neuromodulator
neuromuscular
    n. blockage
    n. maturity
neuronal
    n. ceroid lipofuscinosis (CLN)
    n. GM1 gangliosidosis
    n. heterotopia
neuronotmesis
neuron-specific enolase (NSE)
neurooculocutaneous angiomatosis
neuropathic
neuropathy
    n., ataxia, and retinitis pigmentosa (NARP)
    Charcot-Marie-Tooth n., X-linked recessive, type II
    congenital sensory n.
    familial visceral n.
    femoral n.
    hereditary motor and sensory n. II (HMSN II)
    Leber hereditary optic n. (LHON)
    nutritional n.
    obstetric n.
    porphyric n.
    retractor n.
neuropharmacology
neurophysin I, II
neurophysiologic
neurophysiology
neuroplate
neuropsychopharmacology
neuroretinogangiomatosis
neurosecretion
neurosis, pl. **neuroses**
neurospongioblastosis diffusa
neurosteroid

NOTES

neurosyphilis
neurotensin
neurotoxin
neurotransmitter
>n. release
>substance P pain n.

neurovisceral lipidosis
neurtransmission
neurula, pl. **neurulae**
neutral
>n. protamine Hagedorn (NPH)
>n. protamine Hagedorn insulin

neutralization
neutron
>fast n.
>n. therapy

neutropenia
>congenital n.
>Kostmann n.
>neonatal n.

neutrophil
>n. apoptosis
>mature n.

neutrotrichocutaneous syndrome
nevi (*pl. of* nevus)
nevi, atrial myxoma, myxoid
neurofibromas, ephilides (NAME)
Neville-Barnes forceps
nevirapine
nevoid
>n. amentia
>n. basal cell carcinoma (NBCC)
>n. basal cell carcinoma syndrome (NBCCS, NBS)
>n. basal cell epithelioma, jaw cysts, bifid rib syndrome
>n. hyperkeratosis of the nipple and areola

Nevo syndrome
nevoxanthoendothelioma
nevus, pl. **nevi**
>n. anemicus
>bathing trunk n.
>benign n.
>blue n.
>cutaneous nevi
>n. depigmentosus
>n. elasticus of Lewandowsky
>faun tail n.
>n. flammeus
>halo n.
>n. of Ito
>Ito n.
>Jadassohn-Tieche n.
>linear verrucous epidermal n.
>melanocytic n.
>nape n.
>Ota n.
>n. of Ota
>n. pigmentosus systematicus
>port-wine n.
>n. sebaceous of Feuerstein and Mims
>n. sebaceous of Jadassohn (SNJ)
>n. sebaceus
>n. sebaceus linearis
>n. simplex
>spider n.
>Spitz n.
>strawberry n.
>n. unius lateris
>Unna n.
>n. varicosus osteohypertrophicus
>n. vasculosus osteohypertrophicus
>vulvar n.
>white sponge n.
>wooly-hair n.

new
>N. Decongestant
>n. mutation
>N. York Heart Association classification of heart disease

newborn
>congenital anemia of n.
>congenital epulis of the n.
>epulis of n.
>n. examination
>genital crisis of n.
>hemolytic disease of the n. (HDN)
>hemorrhagic disease of the n.
>jaundice of n.
>lamellar desquamation of n.
>n. nursery
>Parrot atrophy of n.
>persistent pulmonary hypertension of the n. (PPHN)
>physiologic jaundice of the n.
>respiratory distress syndrome of the n.
>n. respiratory distress syndrome
>n. resuscitation
>n. screening kit
>transient bullous dermolysis of the n.
>transient tachypnea of n. (TTN)

Newport medical instrument
newtonian aberration
Neyman-Pearson statistical hypothesis
Nezelof syndrome
Nezhat-Dorsey
>N.-D. aspirator
>N.-D. suction-irrigator

NF
>neurofibromatosis

NF1 gene

NFDR
 neurofaciodigitorenal
 NFDR syndrome
NF-NS, NFNS
 neurofibromatosis-Noonan syndrome
NG
 nasogastric
 NG tube
NGF
 nerve growth factor
N.G.T. Topical
NHS
 Nance-Horan syndrome
NHT
 neoadjuvant hormonal therapy
Niac
niacin
niacinamide
nialamide
Niaspan
nicardipine
NICHD
 National Institute of Child Health and
 Human Development
Nichols
 N. procedure
 N. sacrospinous fixation
**nicked free beta subunit of human
 chorionic gonadotropin**
nickel dermatitis
Nickerson
 N. Biggy vials
 N. medium
nick translation
Nico-400
Nicobid
Nicoderm
Nicolar
Nicolas-Favre disease
nicotinamide
nicotine
 n. patch therapy
Nicotinex
nicotinic acid
nicotinyl alcohol
nicoumalone
NICU
 neonatal intensive care unit
nidation
NIDDM
 non-insulin-dependent diabetes mellitus
Niemann-Pick disease

nifedipine
Niferex
 N.-PN Forte
niger
 Peptococcus n.
Nightingale examining lamp
nigra
 lingua n.
nigricans
 acanthosis n.
 hyperandrogenism, insulin resistance,
 acanthosis n. (HAIR-AN)
 n.-hyperinsulinemia syndrome
 n. syndrome
NIHF
 nonimmune hydrops fetalis
Niikawa-Kuroki syndrome
Nijmegen breakage syndrome (NBS)
Nikolsky sign
Nilandron
Nilstat
nilutamide
Nimbus
 N. test
nimodipine
Niplette device
nipple
 accessory n.
 n.-areola complex
 n. discharge
 Duckey n.
 extranumerary n.
 n. fed
 n. feeding
 inverted n.
 McGovern n.
 out-of-profile n.
 Paget disease of n.
 n. retraction
 n. shield
 n. stimulation
 n. stimulation test
 supernumerary n.
nipple-fed baby
nippling
Nipride
NIPS
 Neonatal Infant Pain Scale
Nisentil
Nissl bodies
nit
Nitabuch layer

NOTES

Nite
    N. Train'r Alarm
    N. Train-R enuresis conditioning
        device
nitidus
    lichen n.
nitrate
    butoconazole n. 2%
    miconazole n.
    silver n.
Nitrazine
    fern-positive N.
    N. test
nitric oxide
nitrite
    amyl n.
Nitro-Bid
nitroblue tetrazolium dye test
Nitrocap
nitrofurantoin
    n. monohydrate
    n. monohydrate macrocrystal
nitrogen
    blood urea n. (BUN)
    n. mustard
    n. partial pressure (PN_2)
    n. washout
nitrogenous base
nitroglycerin (NTG)
Nitroglyn
nitroimidazole compound
Nitrol
Nitrolingual
Nitronet
Nitrong
nitroprusside
    sodium n.
nitrosourea
Nitrospan
Nitrostat
nitrous oxide
nitrovasodilator
Nix Creme Rinse
Nizoral
NLE
    neonatal lupus erythematosus
NLGCLS
    Noonan-like giant cell lesion syndrome
NLS
    Neu-Laxova syndrome
NMGTD
    nonmetastatic gestational trophoblastic
        disease
NMR
    nuclear magnetic resonance
N-multistix
No.1
    Valertest No.1

Noack syndrome
Noble-Mengert perineal repair
nocardiosis
nociceptive pathway
Noctec
nocturia
nocturnal
    n. epilepsy
    n. micturition
nocturnus
    pavor n.
nodal involvement
nodding spasm
node
    aortic n.
    atrioventricular n.
    axillary lymph n.
    Cloquet n.
    femoral lymph n.
    gluteal lymph n.
    hypogastric lymph n.
    iliac n.
    inguinal lymph n.
    intramammary lymph n.
    lymph n.
    paraaortic n.
    parauterine lymph n.
    pelvic lymph n.
    periaortic lymph n.
    pericervical n.
    rectal n.
    retroperitoneal n.
    sacral lymph n.
    shotty n.
    signal n.
    subaortic lymph n.
    ureteral n.
    vulvar lymph n.
nodosa
    periarteritis n.
    polyarteritis n.
    trichorrhexis n.
nodular
    n. blueberry lips
    n. embryo
    n. melanoma
    n. renal blastoma
nodule
    Albini n.
    blueberry muffin n.
    Hoboken n.
    Lisch n.
    Lutz-Jeanselme n.
    pearly-white n.
    placental site n.
    rheumatoid n.
    Sister Mary Joseph n.

subserosal n.
thyroid n.
NOFT
nonorganic failure to thrive
Noguchi test
Nolahist
Nolvadex
noma
n. pudendi
n. vulvae
nomenclature
FIGO n.
TNM n.
nomogram
body mass index n.
gestation-specific n.
Lubchenco n.
Siggaard-Andersen n.
nomp C gene
nonalkylating agent
nonallele
non-A, non-B hepatitis (NANBH)
nonappendiceal carcinoid
noncalcified nodular mass
noncalculous
noncarbonic acid
noncarrier
nonclassic
n. congenital adrenal hyperplasia (NC-CAH)
n. 21-hydroxylase deficiency
nonclassical adrenal hyperplasia (NCAH)
noncommunicating uterine horn
noncornual placental location
noncosmetic panniculectomy
noncovalent
nondeciduous placenta
nondeletion
nondiagnostic culdocentesis
nondisjunction
Non-Drowsy
Drixoral N.-D.
nondysgerminomatous
nonestrogen drug
nonestrogen-regulated growth
nonfenestrated forceps
nonfollicular pustulosis
nonfrank breech
nonfrosted tip

nonfunctional
n. pituitary tumor
n. streak
nongenital pelvic organ
nongestational choriocarcinoma
nongonococcal
n. cervicitis
n. urethritis
nongranulomatous salpingitis
nonhemolytic aerobic organism
nonhistone
non-Hodgkin lymphoma
nonhomologous chromosome
nonimmune
n. fetal hydrops
n. hydrops fetalis (NIHF)
non-insulin-dependent
n.-i.-d. diabetes mellitus (NIDDM)
n.-i.-d. diabetic
noninvasive
n. detection of trisomy 18
n. management
nonionizing nonthermal application
nonketotic
n. hyperglycemia
n. hyperglycinemia
nonlactating breast
nonleukocytospermic
nonlinkage
nonlocking closure
nonmaternal death
nonmetastatic gestational trophoblastic disease (NMGTD)
nonmonogamous partner
nonmotile sperm
Nonne-Milroy-Meige syndrome
nonnucleoside analog reverse transcriptase inhibitor
nonnutritive sucking
nonorganic failure to thrive (NOFT)
nonoxynol-9
n./octoxynol-9
nonpalpable abnormality
nonparous
nonpigmented endometriosis
nonpregnant endometrium
nonprimary
n. first episode
n. infection
nonprogressive cerebellar disorder with mental retardation
nonradioactive immunoassay

N

NOTES

nonrandom mating
nonreactive
    n. immunoassay
    n. NST
nonreassuring
    n. fetal heart beat pattern
    n. fetal status
    n. fetal testing
    n. FHR
non-REM sleep
nonsalt-losing adrenogenital syndrome
nonsecreting pituitary tumor
nonsense mutation
non-sex hormone-binding globulin bound
  testosterone
nonspecific
    n. immunotherapy
    n. mesenchyme
    n. urethritis (NSU)
    n. vaginitis
nonsteroidal antiinflammatory drug
  (NSAID)
nonstressed fetus
nonstress test (NST)
nonstructural gene
non-syncytium-inducing (NSI)
    n.-s.-i. variant of AIDS virus
nontreponemal
    n. serology
    n. test
non-*Treponema* titer
nontyphoid salmonella
nonvertex fetus
nonvertex-nonvertex
nonvertex-vertex
nonviable
    n. fetus
nonvolatile acid
Noonan-like giant cell lesion syndrome
  (NLGCLS)
Noonan syndrome
No Pain-HP
noradrenaline
Noradryl
Norcept-E 1/35
Norcuron
Nordette
Nordryl
norepinephrine
Norethin
    N. 1/35E
    N. 1/35M
    N. 1/50M
norethindrone
    n. acetate
    n. acetate and ethinyl estradiol
    n. acetate 1 mg/ethinyl estradiol 5
    mcg

    n. enanthate (NET-EN)
    n. and ethinyl estradiol
    ethinyl estradiol and n.
    mestranol and n.
norethisterone
norethynodrel
    mestranol and n.
Norflex
norfloxacin
norgestimate
    ethinyl estradiol and n.
    n. and ethinyl estradiol
    n. progestin
norgestrel
    ethinyl estradiol and n.
d-norgestrel
norgestrel/ethinyl estradiol combination
Norinyl
    N. 1+35
    N. 1+50
Norio syndrome
Norisodrine
Norlestrin
    N. 1/50
    N. 2.5/50
    N. Fe
Norlutate
Norlutin
normal
    n. blood loss
    n. ovariotomy
    n. placentation
    n. pressure hydrocephalus (NPH)
    n. saline
    n. temperature
    n. transformation zone
normally progressing pregnancy (NPP)
Norman-Landing syndrome
Norman Miller vaginopexy
Norman-Roberts lissencephaly syndrome
Norman-Wood syndrome
Normegon
Nor-Mil
normoactive bowel sounds
normocephalic
Normodyne
normospermic
normotensive intrauterine growth
  restriction
Normotest
Noroxin
Norpace
Norplant
    N. II
    N. implant
    N. system
Norpramin

Norrie
    N. disease
    N. syndrome
Norrie-Warburg syndrome
Norris-Carrol criteria
19-nortestosterone
    19-n. derivative
Nor-tet
    N.-t. Oral
Northern
    N. blot
    N. blot test
Northway staging
Norton operation
nortriptyline
Nortussin
Norvir
no-scalpel vasectomy
nose
    bifid n.
    saddle n.
nose-breather
    obligate n.-b.
nosocomial
    n. infection
nostril
    anteverted n.
    reverse n.'s
Nostrilla
notencephalia
notochord
notogenesis
notomelus
Novafed
Novafil
Novak curette
Novamoxin
Novantrone
Novapren
Nova Rectal
Novasome
novo
    N.-AZT
    N.-Cimetidine
    N.-Clobetasol
    N.-Cloxin
    N.-Cromolyn
    N.-Difenac
    N.-Difenac-SR
    N.-Diflunisal
    N.-Dipam
    N.-Doxylin

    N.-Famotidine
    N.-Fibrate
    N.-Flurazine
    N.-Flurprofen
    N.-Hydrazide
    N.-Keto-EC
    N.-Lexin
    N.-Naprox
    N.-Pen-VK
    N.-Piroxicam
    N.-Pramine
    N.-Profen
    N.-Ranidine
    N.-Reserpine
    N.-Ridazine
    N.-Rythro Encap
    N.-Seven
    N.-Soxazole
    stress incontinence de n.
    N.-Tamoxifen
    N.-Terfenadine
    N.-Tetra
    N.-Trimel
novobiocin
Novobutamide
Novochlorhydrate
Novofuran
Novolin
    N. 70/30
    N. insulin
    N. L
    N. N
    N. N PenFil
    N. 70/30 PenFil
    N. R
    N. R PenFil
Novomepro
Novonidazol
Novopoxide
Novopropamide
Novorythro
Novosecobarb
Novotriptyn
Noz
    Whoo N.
Nozinan
NP-27
NPH
    neutral protamine Hagedorn
    normal pressure hydrocephalus
    NPH Iletin I, II

N

NOTES

NPH *(continued)*
    NPH insulin
    NPH-N
NPP
    normally progressing pregnancy
NSABP
    National Surgical Adjuvant Breast
    Project
NSAID
    nonsteroidal antiinflammatory drug
NSD
    N-acetylneuraminic acid storage disease
NSE
    neuron-specific enolase
NSI
    non-syncytium-inducing
NST
    nonstress test
        antepartum fetal NST
        fetal NST
        nonreactive NST
NSU
    nonspecific urethritis
NT
    N-telopeptide
NTD
    neural tube defect
N-telopeptide (NT)
NTG
    nitroglycerin
NTT
    nuchal translucency thickness
NTZ Long Acting Nasal Solution
Nu
    Nu-Amoxi
    Nu-Ampi
    Nu-Cephalex
    Nu-Cimet
    Nu-Cloxi
    Nu-Cotrimox
    Nu-Diclo
    Nu-Diflunisal
    Nu-Doxycycline
    Nu-Famotidine
    Nu-Flurprofen
    Nu Gauze dressing
    Nu-Gel dressing
    Nu-Ibuprofen
    Nu-Ketoprofen
    Nu-Ketoprofen-E
    Nu-Naprox
    Nu-Pen-VK
    Nu-Pirox
    Nu-Ranit
    Nu-Tetra
Nubain
nuchal
    n. arm

n. cord
n. cystic hygroma
n. fold
n. translucency
n. translucency in a fetus
n. translucency measurement
n. translucency screening
n. translucency thickness (NTT)
nuchal translucency
Nuck
    canal of N.
    N. canal
    N. diverticulum
    N. hydrocele
nuclear
    n. agenesis
    n. antigen
    n. aplasia
    n. jaundice
    n. magnetic resonance (NMR)
    n. milk
    n. radiation
    n. sex
nuclease
nucleated red blood cell
nucleatum
    Fusobacterium n.
nuclei (*pl. of* nucleus)
nucleic
    n. acid
    n. acid amplification technique
    n. acid probe
nucleoid
    Landovski n.
nucleolar
    n. chromosome
nucleolus, pl. **nucleoli**
nucleoside
    n. pair
nucleosome
nucleotide
nucleotidylexotransferase
    DNA n.
nucleotidyltransferase
    DNA n.
    RNA n.
nucleus, pl. **nuclei**
    arcuate n.
    steroid n.
    X-linked congenital recessive
        muscle hypotrophy with central
        nuclei
Nugent criteria
Nu-Knit
    Surgical N.-K.
null
    n. cell tumor
    n. zone

nulligravida
nullipara
nulliparity
nulliparous
number
    n. of needle passes
    ovulation n.
    Reynolds n.
nummular eczema
Numorphan
Nunn engorged corpuscles
Nupercainal cream
Nuprin
nurse
    wet n.
nursery
    newborn n.
Nursoy formula
nutans
    eclampsia n.
    spasmus n.
Nutracort Topical
Nutramigen formula
Nutr-E-Sol
nutrient requirement
nutrition
    diet and n.
    fetal n.
    fluids, electrolytes, n. (FEN)
    maternal n.
    parenteral n. (PN)
    PediaSure liquid n.

    ProBalance liquid n.
    total parenteral n. (TPN)
    total peripheral parenteral n.
      (TPPN)
nutritional
    n. neuropathy
    n. problem
    n. surveillance
Nutropin
Nyaderm
nyctalopia
Nydrazid
nylidrin hydrochloride
nylon suture material
nympha, pl. **nymphae**
nymphectomy
nymphitis
nymphomania
nymphoncus
nymphotomy
Nyquist frequency
nystagmus
    congenital n.
nystatin
    n. and triamcinolone cream
    n. and triamcinolone ointment
Nystat-Rx
Nystex
Nystop topical powder
Nytilax
Nytone enuretic control unit

NOTES

N

O₂

oxygen
hood O₂
THb O₂
total oxyhemoglobin

OA

occipitoanterior
occiput anterior

OAE

otoacoustic emission test

oat cell carcinoma

OATS

oligoasthenoteratozoospermia syndrome

OAV

oculoauriculovertebral

OAVS

oculoauriculovertebral spectrum

OB

obstetrician
obstetrics
Chemstrip 4 The OB
Chromagen OB
OB Gees maternity orthotic
Sil-K OB

Obalan

Obephen

obesity

android o.
o. in endometrial sarcoma
exogenous o.
experimental o.
gynecoid o.
hypotonia, hyperphagia,
hypogonadism, o. (HHHO)
o., short stature, mental deficiency,
hypogonadism, micropenis, finger
contracture, cleft-lip palate
syndrome

obesity-hypotonia syndrome

Obeval

Obezine

OBG

OBG Clinical Records Manager
software
OBG LabTrack software

OB/GYN

obstetrics and gynecology

objective probability

obligate

o. carrier
o. heterozygote
o. nose-breather

oblique

o. diameter

o. lie
o. presentation

obliquity

Litzmann o.
Nägele o.

Obrinsky syndrome

observation

observer variation

obstetric, obstetrical

o. accident
o. anesthesia
o. binder
o. brachial plexus palsy
o. care
o. complication
o. conjugate
o. conjugate of outlet
o. damage
o. forceps
o. history
o. hysterectomy
o. neuropathy
o. operation
o. outcome
o. paralysis
o. position
o. risk factor
o. traction injury
o. ultrasound

obstetrician (OB)

high-risk o.

obstetrician-gynecologist

obstetricopediatric

obstetrics (OB)

defensive o.
o. and gynecology (OB/GYN)
International Federation of
Gynecology and O. (FIGO)

Obstetrique

obstipation

obstruction

bilateral ureteral o. (BUO)
bladder neck o.
bowel o.
colonic o.
distal tubal o.
duodenal o.
extramural upper airway o.
fimbrial o.
gastrointestinal o.
iatrogenic urethral o.
intestinal o.
intraluminal upper airway o.
intramural upper airway o.
intrathoracic airway o.

O

obstruction *(continued)*
    nasopharyngeal airway o.
    pelviureteric junction o.
    tubal o.
    ureteral o.
    ureteropelvic junction o.
    ureterovesical o.
    urinary outlet o.
    urinary tract o.

obstructive
    o. azoospermia
    o. dysmenorrhea
    o. jaundice
    o. uropathy

obturator
    o. artery
    Endopath Optiview optical
      surgical o.
    o. fascia
    o. nerve
    o. nerve damage
    Optiview optical surgical o.

OC
    oral contraceptive

OC-125

OCC
    oculocerebrocutaneous
    OCC syndrome

occipital horn syndrome

occipitoanterior (OA)
    left o. position (LOA)
    o. position
    right o. position (ROA)

occipitofrontal
    o. circumference
    o. diameter

occipitomental diameter

occipitoposterior (OP)
    left o. position (LOP)
    o. position
    right o. position (ROP)

occipitotransverse (OT)
    left o. position (LOT)
    o. position
    right o. position (ROT)

occiput
    o. anterior (OA)
    o. posterior (OP)
    o. presentation
    o. transverse (OT)

Occlucort

occludens
    zonula o.

occlusion
    distal o.
    isthmic o.
    proximal o.
    roller o.

    salpingitis after previous tubal o.
      (SPOT)
    tracheal o.
    tubal o.
    tuboovarian abscess after previous
      tubal o. (TOAPOT)

occult
    o. cancer
    o. cord prolapse
    o. focus

occulta
    spina bifida o.

occultum
    cranium bifidum o.

occupational history

Ochoa syndrome

Ochsner forceps

O'Connor-O'Sullivan retractor

OCR
    oculocerebrorenal
    OCR syndrome

OCT
    ornithine carbamoyltransferase
    oxytocin challenge test
    OCT deficiency

Octamide

octaploidy

Octicair Otic

Octostim

octoxynol 9

OcuClear Ophthalmic

Ocugram

ocular
    o. albinism
    o. aspergillosis
    o. bobbing
    o. coloboma-imperforate anus
      syndrome
    o. hypertelorism
    o. hypotelorism
    o. hypotony
    o. involvement
    o. manifestation
    o. myopathy
    o. prophylaxis

oculoauricular dysplasia

oculoauricularis
    dysplasia o.

oculoauriculofrontonasal syndrome

oculoauriculovertebral (OAV)
    o. dysplasia
    o. spectrum (OAVS)

oculocephalic reflex

oculocephalogyric reflex

oculocerebral
    o. hypopigmentation syndrome

oculocerebrocutaneous (OCC)
    o. syndrome

oculocerebrofacial syndrome
oculocerebrorenal (OCR)
    o. dystrophy
    o. syndrome
oculocutaneous
    o. albinism
    o. tyrosinemia or tyrosinosis
oculodental syndrome
oculodentodigital (ODD)
    o. dysplasia
    o. syndrome
oculodentodigitalis
    dysplasia o.
oculodentoosseous dysplasia (ODOD)
oculodigitoesophagoduodental (ODED)
    o. syndrome
oculogenitolaryngeal syndrome
oculogyration
oculogyria
oculomandibular dyscephalia
oculomandibulodyscephaly (OMD)
oculomandibulodyscephaly-hypotrichosis
  syndrome
oculomandibulofacial (OMF)
    o. syndrome
oculomelic amyoplasia
oculopalatoskeletal syndrome
oculopharyngeal muscular dystrophy
Ocu-Sol
OD
    oocyte donation
OD$_{450}$
    delta OD$_{450}$
ODD
    oculodentodigital
    osteodental dysplasia
odd
    o. chromosome
    o.'s ratio (OR)
ODED
    oculodigitoesophagoduodental
    ODED syndrome
od gene
ODOD
    oculodentoosseous dysplasia
O'Donnell operation
odontogenic keratocytosis-skeletal
  anomalies syndrome
odontoonychodermal dysplasia
odor
    amine o.

    fishy o.
    musty o.
O'Driscoll report
OEC
    ovarian epithelial cancer
Oedipus complex
Oestrilin
oestrogens
OFD
    orofaciodigital
        OFD syndrome, type I–IV, VI–IX
office
    o. cystometrics
    o. laparoscopy under local
      anesthesia (OLULA)
offspring
ofloxacin
OG
    orogastric
OGCT
    oral glucose challenge test
Ogden vaginal cream
Ogen
    O. Oral
    O. Vaginal
Ogino-Knaus rule
Ogita test
OGTT
    oral glucose tolerance testing
Oguchi disease
Ohdo blepharophimosis syndrome
Ohio
    O. bed
    O. humidifier
    O. warmer
Ohmeda
    O. Care-Plus incubator
21-OH nonclassical adrenal hyperplasia
17-OHP
    17-hydroxyprogesterone
OHSS
    ovarian hyperstimulation syndrome
Ohtahara syndrome
OI
    osteogenesis imperfecta type I–IV
    oxygenation index
oil
    camphorated o.
    cod liver o.
    o. cyst
    o. enema
    ethiodized o.

O

NOTES

oil *(continued)*
    evening primrose o. (EPO)
    glycerin, lanolin and peanut o.
    MCT O.
    mineral o.
    peanut o.
    Progesterone O.
ointment
    A and D O.
    Cormax O.
    nystatin and triamcinolone o.
    Pazo hemorrhoid o.
    zinc oxide o.
Okazaki fragment
OKT8
17β-ol-dehydrogenase
oleandomycin phosphate
olfacto-ethmoidohypothalamic dysplasia
olfactoethmoidohypothalamic dysraphia
olfactogenital
    o. dysplasia
    o. syndrome
olfactory placode
oligiogenic inheritance
oligo
    allele specific o.
    o. man
oligoamnios
oligoasthenospermia
oligoasthenoteratozoospermia
    o. syndrome (OATS)
oligoclonal bands
oligodactyly
    postaxial o.
oligogalactia
oligohydramnios
oligomenorrhea
oligonucleotide
    antisense o.
oligo-ovulation
oligophrenia
    growth retardation, ocular
      abnormalities, microcephaly,
      brachydactyly, o. (GOMBO)
    o.-ichthyosis syndrome
oligospermia
    eugonadotropic o.
oligoteratoasthenozoospermia syndrome
oligozoospermia
oligozoospermic
oliguria
Oliver-McFarlane syndrome
Oliver syndrome
Ollier-Klippel-Trenaunay-Weber
  syndrome
Ollier syndrome
Olshausen
    O. procedure
    O. sign
    O. suspension
OLULA
    office laparoscopy under local anesthesia
Olympic Neostraint immobilizer
Olympus
    O. disposable cannula
    O. disposable trocar
    O. flexible hysterofiberscope
    O. hysteroscope
OMD
    oculomandibulodyscephaly
Omenn syndrome
omental
    o. adhesion
    o. biopsy
    o. cake
    o. cyst
omentectomy
omentum
    lymphocyst o.
omeprazole
OMF
    oculomandibulofacial
    OMF syndrome
OMI
    oocyte maturation inhibitor
OMM
    ophthalmomandibulomelic
Ommaya reservoir
Omnicef
Omnipen-N
Omniprobe test
Omni-Tract
    O.-T. vaginal retractor
omphal
omphaloangiopagous twins
omphaloangiopagus
omphalocele
omphalocele-cleft palate syndrome
omphalomesenteric
    o. artery
    o. cord
    o. cyst
    o. duct
omphalopagus
omphalorrhagia
omphalorrhea
omphalorrhexis
omphalotomy
omphalotripsy
OMPI colposcope
OMRON monitor
onanism
Onat syndrome
oncofetal
    o. antigen
    o. fibronectin

oncogene
> Bcl-2 o.
> c-*erb* B-2 o.
> c-*myc* o.
> o. expression
> H-*ras* o.
> human epidermal growth factor-
>   2 o.
> K-*ras* o.
> neu/HER2 o.

oncologist
> Society of Gynecologic O.'s (SGO)

oncology
> gynecologic o.
> Society of Gynecologic O.

On-Command catheter

oncoprotein
> c-*erb* B-2 o.

OncoScint
> O. CR103 monoclonal antibody
> O. CR/OV contrast medium
> O. test

oncotic

Oncovin

ondansetron
> o. HCl

Ondine
> curse of O.
> O. curse

Ondogyne

one-child sterility

one-egg twins

one-horned uterus

one-hour
> o.-h. glucose tolerance test
> o.-h. PG
> o.-h. prostaglandin

O'Neil cup

one-tail test

onlay patch anastomosis

onset
> optic atrophy 2 (OPA 2) with
>   early o.

ontogeny

ontology

onychia

**onychodystrophy-congenital deafness
  syndrome**

onychoosteodysplasia

Ony-Clear Spray

ooblast

oocyesis

oocyte
> aspiration of mature o.
> o. atresia
> o. collection
> o. cryopreservation
> o. culture
> o. donation (OD)
> o. donor
> o. extrusion
> GV o.'s
> o. maturation inhibitor (OMI)
> o. meiosis
> primary o.
> o. production
> o. recovery
> o. retrieval
> sibling o.

oogenesis

oogonia, sing. **oogonium**

oolemma

oophoralgia

oophorectomy
> prophylactic o.

oophoritic cyst

oophoritis
> autoimmune o.

oophorocystectomy

oophorocystosis

oophorohysterectomy

oophoroma

oophoropathy

oophoropeliopexy

oophoropexy
> laparoscopic o.

oophoroplasty

oophororrhaphy

oophorosalpingectomy

oophorosalpingitis

oophorostomy

oophorotomy

oophorrhagia

ooplasm

oozing
> venous o.

OP
> occipitoposterior
> occiput posterior

O&P
> ova and parasites
> stool culture for O&P

opacity
> corneal o.

O

NOTES

OPCA
  neonatal olivopontocerebellar atrophy
  X-linked olivopontocerebellar ataxia
OPD
  otopalatodigital
  OPD syndrome
open
  o. crib
  to keep o. (TKO)
  keep vein o. (KVO)
  o. reading frame
open-ended vasectomy
OpenGene automated DNA sequencing system
opening
  urethral o.
  vaginal o.
OPERA
  outpatient endometrial resection/ablation
  OPERA procedure
  OPERA Star morcellator
  OPERA Star SL hysteroscope
  OPERA Star tissue aspirating resectoscope
operation
  Alexander o.
  Bacon-Babcock o.
  Baldy o.
  Ball o.
  Band-Aid o.
  Baudelocque o.
  Blalock-Hanlon o.
  Blalock-Taussig o.
  Bozeman o.
  Brunschwig o.
  cesarean o.
  Cotte o.
  Counsellor-Davis artificial vagina o.
  Döderlein roll-flap o.
  Donald-Fothergill o.
  Doyle o.
  Emmet o.
  Estes o.
  Falk-Shukuris o.
  Fothergill o.
  Fothergill-Donald o.
  Fothergill-Hunter o.
  Frangenheim-Goebell-Stoeckel o.
  Fredet-Ramstedt o.
  Freund o.
  Gigli o.
  Gilliam o.
  Gilliam-Doleris o.
  Giordano o.
  Gittes o.
  Glenn o.
  Grant-Ward o.
  Halsted o.
  Haultain o.
  Heaney o.
  Heller-Belsey o.
  Heller-Nissen o.
  Kelly o.
  Kennedy-Pacey o.
  Lash o.
  LeFort o.
  LeFort-Neugebauer o.
  Madlener o.
  Manchester o.
  Manchester-Fothergill o.
  Marckwald o.
  Marshall-Marchetti-Krantz o.
  McIndoe o.
  morcellation o.
  Norton o.
  obstetric o.
  O'Donnell o.
  Pomeroy o.
  Porro o.
  Ramstedt o.
  Récamier o.
  Saenger o.
  Schauta vaginal o.
  Schroeder o.
  Schuchardt o.
  second-look o.
  sex change o.
  Shirodkar o.
  Spinelli o.
  Stamey o.
  Strap o.
  Strassman o.
  Sturmdorf o.
  suprapubic urethrovesical suspension o.
  suspensory sling o.
  switch o.
  TeLinde o.
  Tessier craniofacial o.
  transsphenoidal o.
  Urban o.
  Vecchietti o.
  Waters o.
  Way o.
  Webster o.
  Wertheim o.
  Wertheim-Schauta o.
operative
  o. cholangiogram
  o. laparoscopy
  o. site complication
  o. vaginal delivery
operator
  o. gene
  o. locus
operculum, pl. **opercula**

operon
Ophthaine
Ophthalgan
ophthalmia
    gonococcal o.
    gonorrheal o.
    o. neonatorum
ophthalmic
    Collyrium Fresh O.
    Eyesine O.
    Geneye O.
    Murine Plus O.
    Mydfrin O.
    OcuClear O.
    Optigene O.
    Tetrasine O.
    Tetrasine Extra O.
    Vira-A O.
    Viroptic O.
    Visine Extra O.
    Visine L.R. O.
ophthalmoacromelic syndrome
ophthalmologic
ophthalmomandibulomelic (OMM)
    o. dysplasia
ophthalmoplegia
    fibrotic o.
Ophthochlor
opiate
    endogenous o.
    o. receptor
opioid
    o. activity
    o. addiction
    epidural o.
    o. peptide
    o. receptor antagonist
opioidergic control impairment
opipramol hydrochloride
opisthotonic posturing
opisthotonos fetalis
opisthotonus
Opitz
    O. syndrome
    O. trigonocephaly syndrome
Opitz-Christian syndrome
Opitz-Frias syndrome
Opitz-Kaveggia syndrome
opium
OPMI drape
OPMILAS CO2 laser
opocephalus

opodidymus
Oppenheim
    O. congenital hypotonia
    O. disease
    O. syndrome
opposing wall
OPS
    osteoporosis-pseudoglioma syndrome
opsoclonus
opsomyoclonus
opsonin
opsonization
    o. system
optic
    o. atrophy-ataxia syndrome
    o. atrophy 2 (OPA 2) with early
    onset
    o. fundus
    o. glioma
optical
    o. density
    o. density measurement
Opticrom
Opti-Flow catheter
Optigene Ophthalmic
Optimal Observation Score
Optimine
Optimox
    O. C-500
    O. Mag 200
Optimyd
Optiphot-2UD microscope
Optiview optical surgical obturator
Optivite
    O. PMT
Optro
OPUS immunoassay system
OR
    odds ratio
ora (*pl. of* os)
Orabase
    O. HCA Topical
    Kenalog in O.
Oradexon
Oragrafin
oral
    o. administration
    Aller-Chlor O.
    AL-Rr O.
    Amen O.
    Ansaid O.
    Cataflam O.

O

NOTES

oral *(continued)*
    o. cavity
    Ceftin O.
    Chlo-Amine O.
    Chlorate O.
    Chlor-Trimeton O.
    Cleocin HCl O.
    Cleocin Pediatric O.
    o. conjugated estrogen
    o. contraception
    o. contraceptive (OC)
    o. contraceptive efficacy
    o. contraceptive use
    Curretab O.
    Cycrin O.
    Doxy O.
    Dynacin O.
    Estrace O.
    Estratest O.
    o. estrone
    o. glucose challenge test (OGCT)
    o. glucose tolerance testing
      (OGTT)
    o. hormone replacement therapy
    o. hypoglycemic
    Klorominr O.
    Mephyton O.
    Minocin O.
    o. misoprostol
    MSIR O.
    Nor-tet O.
    Ogen O.
    Ortho-Est O.
    Panmycin O.
    PediaCare O.
    Phenergan O.
    o. polio vaccine
    Provera O.
    Robitet O.
    o. sex
    o. steroid contraceptive
    Sumycin O.
    Telachlor O.
    Teldrin O.
    o. temperature
    Terramycin O.
    Tetracap O.
    o. transmucosal fentanyl citrate
    Valium O.
    Voltaren O.
    Voltaren-XR O.
oral-facial-digital
    o.-f.-d. syndrome
    o.-f.-d. syndrome with retinal
      abnormalities
orally administered estrogen
Oramide
Oramorph SR

orange
    Agent O.
Orap
OraSure
    O. device
    O. oral HIV-1 antibody testing
      system
    O. oral HIV test
Orbeli syndrome
Orbenin
orbit
    shallow o.'s
Orbit blade
orbitopagus
orchidectomy
orchidoblastoma
orchiometer
orchiopexy
    Fowler-Stephens o.
orchitis
    mumps o.
orciprenaline sulfate
order
    high fetal o.
Oregon-type tyrosinemia
Oretic
Oreton Methyl
organ
    internal generative o.
    nongenital pelvic o.
    o. of Rosenmüller
    sensory o.
    o. transplant
    o. transplantation
organic
    o. acid metabolism
    o. acid screen
organism
    Gram-negative o.
    Gram-positive o.
    nonhemolytic aerobic o.
    transgenic o.
organization
    health maintenance o. (HMO)
    managed care o. (MCO)
    preferred provider o. (PPO)
    World Health O. (WHO)
organogenesis
organoid nevus syndrome
organoleptic characteristic
organomegaly
Organon percutaneous E2 implant
orgasm
orgasmic
    o. flush
    o. plateau
orientation
    sexual o.

orifice
orificial
origin
    O. balloon
    congenital amaurosis of retinal o.
    fever of unknown o. (FUO)
    histoimmunological o.
    parental o.
    O. tacker
    O. trocar
Original Doan's
Orimune
Orinase
Ornade
ornithine
    o. carbamoyltransferase (OCT)
    o. carbamoyltransferase deficiency
    o. decarboxylase
    o. transcarbamylase (OTC)
    o. transcarbamylase deficiency
      (OTCD)
orocraniodigital syndrome
orodigitofacial
    o. dysostosis
    o. syndrome
orofacial cleft
orofaciodigital (OFD)
    o. syndrome types I–IV, VI–IX
    o. syndrome with tibial dysplasia
orogastric (OG)
orogenital
    o. sexual practices
    o. syndrome
oromandibular limb hypoplasia
oromandibuloauricular syndrome
oromandibulootic syndrome
oromotor function
oroorbital cleft
oropharyngeal colonization
oropharynx
orphan
    enteric cytopathogenic human o.
      (ECHO)
    enterocytopathogenic human o.
      (ECHO)
    respiratory enteric o. (REO)
orphenadrine
Orr rectal prolapse repair
orthagonal plane
Ortho
    O. All-Flex diaphragm
    O. diaphragm kit

    O. Personal Pak
    O. Tri-Cyclen
Ortho-Cept
Ortho-Chloram
Ortho-Creme
Ortho-Cyclen
Ortho-Dienestrol Vaginal
Ortho-Est Oral
orthogonal lead system
Ortho-Gynol
orthology
orthomyxovirus
Ortho-Novum
    O.-N. 1/35
    O.-N. 1/50
    O.-N. 7/7/7
    O.-N. 10/11
orthopaedic, orthopedic
    o. anomaly
    o. condition
Ortho-Prefest
orthostatic
    o. hypotension
    o. proteinuria
orthotic
    OB Gees maternity o.
orthotopic live transplantation
orthovoltage
    o. radiation
Orthoxine
Ortolani
    O. click
    O. maneuver
    O. sign
    O. test
Orudis
    O. KT
Oruvail
os, pl. ossa
    o. pubis
os, pl. ora
    cervical o.
    external o.
    internal o.
    parous o.
    Scanzoni second o.
Os-Cal 500
oscillation amplitude
oscillator
oscillatory
    high-frequency o. (HFO)
    o. ventilation

O

NOTES

OSD
> Profilate OSD

Osebold-Remondini syndrome
Osgood-Schlatter
> O.-S. disease
> O.-S. syndrome

Osler-Weber-Rendu disease
Osmitrol Injection
OsmoCyte pillow
Osmoglyn
osmolality
osmolarity
Osmolite formula
osmoregulation
osmotic
> o. dilator
> o. pressure

ossa (*pl. of* os)
osseous-oculodental dysplasia
ossicle
> Andernach o.
> anlagen of the auditory o.

ossicular
ossification
ossified ear cartilages, mental deficiency, muscle wasting, bony changes syndrome
OSSUM
> O. database
> O. database of skeletal dysplasias

osteitis
> o. condensans generalisata
> o. fibrosa
> o. pubis

OsteoAnalyzer densitometer
osteoangiohypertrophy
osteoarthropathy
osteoblast
osteoblast-like cell
osteoblastoma
osteocalcin
> serum o.

osteochondritis
> o. deformans juvenilis
> o. dissecans
> o. ischiopubica

osteochondrodysplasia
> hereditary o.
> hypertrichotic o.

osteochondrodystrophia deformans
osteochondrodystrophy
> familial o.

osteochondroma
osteochondromuscular dystrophy
osteochondrosis deformans tibiae
osteoclastic
osteocranium

osteocystoma
osteodental dysplasia (ODD)
osteodysplastic primordial dwarfism
osteodystrophia juvenilis
osteodystrophy
> Albright o.
> Albright hereditary o. (AHO)
> renal o.

osteogenesis
> o. imperfecta congenita syndrome
> o. imperfecta cystica
> o. imperfecta, optic atrophy, retinopathy, developmental delay syndrome
> o. imperfecta type I–IV (OI)

osteogenic
osteoglophonic
> o. dwarfism
> o. dysplasia

OsteoGram bone density test
osteohypertrophicus
> nevus varicosus o.
> nevus vasculosus o.

osteohypertrophic varicose syndrome
osteoid
> o. osteoma

osteoma
> osteoid o.

osteomalacia
> juvenile o.

Osteomark
> O. agent
> O. NTx assay
> O. NTx serum test

Osteomeasure computer-assisted image analyzer
osteomyelitis
> o. pubis

Osteopatch
osteopathia
> o. dysplastica familiaris
> o. striata
> o. striata, deafness, cranial osteopetrosis syndrome
> o. striata, macrocephaly, cranial sclerosis syndrome
> o. striata syndrome
> o. striata with cranial sclerosis

osteopedion
osteopenia
> relative o.

osteopenia-sparse hair-mental retardation syndrome
osteopenic
> o. halo

osteopetrosis
> infantile o.
> o. tarda

osteopoikilosis
osteopontin
osteoporosis
> postmenopausal o.

osteoporosis-pseudoglioma syndrome (OPS)
osteoporotic fracture
osteopsathyrosis
osteorhabdotosis
Osteosal test
osteosarcoma
osteotabes
OsteoView
> O. device
> O. 2000 system

ostium, pl. **ostia**
> abdominal o.
> o. primum defect
> o. secundum
> o. secundum defect

ostrich maneuver
Ostrum-Furst syndrome
O'Sullivan-O'Connor retractor
OT
> occipitotransverse
> occiput transverse

O.T.
> Mono-VaccTest (O.T.)

Ota
> O. nevus
> nevus of O.

otalgia
OTC
> ornithine transcarbamylase
> over-the-counter

OTCD
> ornithine transcarbamylase deficiency

Otic
> Acetasol HC O.
> AK-Spore H.C. O.
> AntibiOtic O.
> Cipro HC O.
> Cortatrigen O.
> Cortisporin O.
> Cortisporin-TC O.
> O. Domeboro
> Octicair O.
> Otic-Care O.
> Otocort O.
> Otosporin O.
> Pediotic O.
> UAD O.

> VoSol O.
> VoSol HC O.

otitis externa
otoacoustic emission test (OAE)
Otocalm Ear
otocyst
otofacial dysostosis
otofaciocervical syndrome
otomandibular
> o. dysostosis
> o. facial dysmorphogenesis
> o. syndrome

otomycosis
otopalatodigital (OPD)
> o. syndrome

otorrhea
otosclerosis
> o. syndrome

otoscope
> Siegel o.

otospondylomegaepliphyseal dysplasia
otospongiosis syndrome
Otosporin Otic
Ototemp 3000 thermometer
ototoxicity
Otovent autoinflation kit
Otrivin Nasal
Otto
> O. pelvis
> O. syndrome

ouabain
ounce (oz)
> calories per o. (cal/oz)

out
> toeing o.

outbreeding
outcome
> adverse o.
> fetal o.
> maternal o.
> neonatal o.
> obstetric o.
> perinatal o.
> poor pregnancy o.
> pregnancy o.
> teratogenic o.

outercourse
outlet
> conjugate of pelvic o.
> o. forceps delivery
> parous o.
> pelvic o.

O

NOTES

outlet *(continued)*
    vaginal o.
    vulvovaginal o.
outline
    fetal o.
out-of-phase endometrial biopsy
out-of-profile nipple
outpatient
    o. endometrial resection/ablation
      (OPERA)
    o. endometrial resection/ablation
      procedure
output
    cardiac o.
    elevated cardiac o.
    intake and o. (I&O)
    urine o.
out-toeing
ova (*pl. of* ovum)
ovale
    foramen o.
ovalocytary anemia
OvaRex
    O. MAb
Ovarex vaccine
ovaria (*pl. of* ovarium)
ovarialgia
ovarian
    o. ablation
    o. abnormality
    o. abscess
    o. activity
    o. agenesis
    o. amenorrhea
    o. antibody
    o. aplasia
    o. artery
    o. cancer
    o. cancer metastasis
    o. carcinoma antigen
    o. carcinoma debulking
    o. cautery
    o. clear cell adenocarcinoma
    o. colic
    o. cortex
    o. cycle
    o. cycle change
    o. cyst
    o. cystadenocarcinoma
    o. cystadenoma
    o. cystectomy
    o. cystic teratoma
    o. disorder
    o. dwarfism
    o. dysgenesis
    o. dysgenesis-sensorineural deafness
     syndrome
    o. dysgerminoma

    o. dysmenorrhea
    o. embryonal teratoma
    o. endometriosis
    o. epithelial cancer (OEC)
    o. estrogen synthesis
    o. factor
    o. failure
    o. fibroma
    o. fibromatosis
    o. follicle exhaustion
    o. function
    o. gametogenesis
    o. germinal epithelium
    o. gonadoblastoma
    o. hormone
    o. hyperandrogenism
    o. hyperstimulation
    o. hyperstimulation syndrome
     (OHSS)
    o. hypertrophy
    o. leiomyoma
    o. ligament
    o. lipid cell neoplasm
    o. lymphoma
    o. malignancy
    o. malignant epithelial neoplasm
    o. malignant germ cell tumor
    o. mass
    o. neurofibroma
    o. plexus
    o. pregnancy
    o. proliferative cystadenoma
    o. remnant syndrome
    o. retrieval
    o. seminoma
    o. sex-cord stromal neoplasm
    o. short stature syndrome
    o. small-cell carcinoma
    o. steroid
    o. steroidogenesis
    o. stimulation
    o. stroma
    o. stromal hyperthecosis
    o. thecoma
    o. torsion
    o. tubular adenoma
    o. varicocele
    o. vein
    o. vein syndrome
    o. vein thrombosis (OVT)
    o. wedge resection
ovarian-sparing irradiation
ovariectomy
ovarii
    hyperthecosis o.
    mesoblastoma o.
    struma o.
ovarioabdominal pregnancy

ovariocele
ovariocentesis
ovariocyesis
ovariodysneuria
ovariogenic
ovariohysterectomy
ovariolytic
ovarioncus
ovariopathy
ovariorrhexis
ovariosalpingectomy
ovariosalpingitis
ovarioscopy
ovariosteresis
ovariostomy
ovariotomy
    Beatson o.
    normal o.
ovaritis
ovarium, pl. ovaria
    o. bipartitum
    o. disjunctum
    o. gyratum
    o. lobatum
ovary
    accessory o.
    contralateral o.'s
    mulberry o.
    multifollicular o.
    oyster o.
    palpable postmenopausal o.
    polycystic o. (PCO)
    resistant o.
    sclerocystic o.
    Stein-Leventhal type of
      polycystic o.
    strumal carcinoid of o.
    supernumerary o.
    transposition of o.
    wandering o.
Ovation falloposcopy system
ovatus
    Bacteroides o.
O-Vax vaccine
Ovcon
    O. 35
    O. 50
over
    crossing o.
overdistention syndrome
overflow incontinence
overhead warmer

overlapping
    o. clones
    o. closure of peritoneum
overload
    fluid o.
    volume o.
overriding
    o. aorta
    o. of sutures
oversewing placental bed
overt
    o. insulin-dependent diabetes
      mellitus
over-the-counter (OTC)
    over-the-counter drug
    over-the-counter medication
Oves Cervical Cap
ovicidal
Ovide Topical
oviduct
    ampulla of o.
    angiomyoma of o.
ovigenesis
ovine
    o. trophoblast protein-1
ovoid
    Delclos o.
    Fleming o.
    Manchester o.
    tandem and o.'s (T&O)
ovotestis, pl. ovotestes
Ovral
Ovrette
O-V Staticin
OVT
    ovarian vein thrombosis
OvuDate hCG
OvuGen test kit
OvuKIT
    O. Self-Test kit
    O. test
ovular transmigration
ovulation
    contralateral o.
    incessant o.
    o. induction
    o. number
    paracyclic o.
    o. rate
    spontaneous o.
    o. stimulation
ovulational sclerosis

O

NOTES

ovulation-associated hemoperitoneum
ovulatory
    o. age
    o. defect
    o. disturbance
    o. menstrual cycle
    o. mucus
ovulocyclic
    o. porphyria
ovum, pl. **ova**
    blighted o.
    Bryce-Teacher o.
    o. capture
    o.-capture inhibitor
    o. donation
    fertilized o.
    o. forceps
    frozen o.
    hamster o.
    Hertig-Rock o.
    holoblastic o.
    Mateer-Streeter o.
    o. maturation
    Miller o.
    ova and parasites (O&P)
    Peters o.
    o. pickup mechanism
    primitive o.
    primordial o.
    o. transport
    trapped o.
Ovumeter
OvuQuick
    O. One-Step ovulation kit
    O. Self-Test
    O. Self-Test kit
    O. Self-Test ovulation predictor
OvuStick
owl's
    o. eye cell
    o. eye inclusion body
Owren disease
oxacillin
    o. sodium
Oxandrin
oxandrolone
oxazepam
Oxford Family Planning Association Contraceptive Study
oxiconazole
oxidase
    cholesterol o.
    diamine o.
    monoamine o. (MOA)
oxidative stress
oxide
    inhalation of nitrous o.
    magnesium o.

    nitric o.
    nitrous o.
    zinc o.
oximeter
    Accustat pulse o.
    Healthdyne o.
    INVOS 3100 cerebral o.
    OxyShuttle pulse o.
    Oxytrak pulse o.
    pulse o.
oximetry
    pulse o.
    reflectance pulse o.
Oxisensor
    O. fetal oxygen saturation monitor
    O. transducer
Oxistat Topical
Ox-Pam
oxprenolol
oxtriphylline
oxybutynin
    o. chloride
oxycardiorespirogram
Oxycel cautery
oxycephaly
oxychlorosone sodium
oxycodone
oxy-CR gram
Oxydess II
oxygen (O_2)
    blow-by o.
    o. concentration in pulmonary capillary blood (CcO_2)
    o. consumption
    o. delivery
    o. dissociation curve
    home o.
    hood o.
    hyperbaric o.
    partial pressure of o. (PO_2)
    partial pressure of arterial o. (PaO_2)
    o. saturation (SaO_2)
    o. tension
    tent o.
    o. toxicity lung disease
    transcutaneous partial pressure of o. ($tCpO_2$)
oxygenated fetal blood
oxygenation
    extracorporeal membrane o. (ECMO)
    fetal o.
    fetal scalp o.
    o. index (OI)
oxygenator
    Lilliput neonatal o.
oxygen-diffusing capacity

oxygen-free radical
Oxygent temporary blood substitute
oxyhemoglobin
    total o. (THb O_2, THb O_2)
Oxy-Hood oxygen hood
oxymetazoline
oxymorphone
oxyphenbutazone
oxyphencyclimine
oxyphenonium bromide
OxyShuttle pulse oximeter
oxytetracycline
    o. HCl
oxytoca
    Klebsiella o.
oxytocia
oxytocic
    o. stimulation

oxytocin
    o. analogue
    o. augmentation
    o. challenge test (OCT)
    o. stimulation of labor
    o. stress test
oxytocinase
Oxytrak pulse oximeter
Oyst-Cal 500
Oystercal 500
oyster ovary
oz
    ounce
ozaenae
    Klebsiella o.

NOTES

O

P
> partial pressure
> phosphorus
>> P value

³²P
> phosphorus-32

p
> short arm of chromosome

p53
> p. gene
> p. tumor suppressor gene

p24 antigen
P32 intraperitoneal treatment
p55 soluble tumor necrosis factor receptor
p75 soluble tumor necrosis factor receptor

PA
> alveolar partial pressure
> pernicious anemia
> plasminogen activator

Pa
> arterial partial pressure
> pascal unit of pressure measurement

p(A-a)O₂
> alveolar-arterial pressure difference

Paas disease
PAC
> papular acrodermatitis of childhood

PACE-2 test
PACE-2C DNA probe test
pacemaker
> artificial p.

Pacey technique
pachygyria
pachyonychia congenita
pachysalpingitis
pachysalpingoovaritis
pachytene
> p. phase of meiosis
> p. stage

pachyvaginitis
> p. cystica

Pacific yew
pacifier
pack
> Barrier p.
> Flents breast comfort p.
> M-Zole 3 combination p.
> M-Zole 7 Combination P.
> M-Zole 7 dual p.
> Peri-Cold P.
> Peri-Gel P.
> Peri-Warm P.

packaging RNA

packed
> p. erythrocyte
> p. red blood cells

packer
> Bernay uterine p.
> Kitchen postpartum gauze p.

packing
> Kaltostat p.
> uterine p.

paclitaxel
PaCO₂
> arterial carbon dioxide pressure (tension)
> partial pressure of arterial carbon dioxide

PAD
> pelvic adhesive disease

pad
> Aquaflex ultrasound gel p.
> buccal fat p.
> LePad breast exam training p.
> malar fat p.
> Padette interlabial p.
> sucking p.
> suctorial p.
> p. test

paddle forceps
Padette interlabial pad
PadKit sample collection system
Paget
> P. disease
> P. disease of anus
> P. disease of breast
> P. disease of nipple
> P. disease of vulva

pagetic
pagetoid
Pagon syndrome
pagophagia
PAI
> plasminogen activator inhibitor

PAI-1
> plasminogen-activator inhibitor-1

PAI-2
> plasminogen-activator inhibitor-2

pain
> acyclic pelvic p.
> after-p.'s
> bearing-down p.
> chronic pelvic p. (CPP)
> epigastric p.
> expulsive p.'s
> false p.'s
> flank p.
> hypogastric p.
> intermenstrual p.
> labor p.'s

P

pain *(continued)*
    low-back p.
    lower abdominal p.
    pelvic p.
    periumbilical p.
    pleuritic p.
    psychogenic pelvic p.
    recurrent abdominal p. (RAP)
    referred pelvic p.
    splanchnic pelvic p.
    suprapubic p.
Paine syndrome
Pain-HP
    No P.-HP
paint, painting
    chromosome p.
    wcp1 chromosome p.
    wcp7 chromosome p.
    wcp8 chromosome p.
    wcp10 chromosome p.
    wcp15 chromosome p.
    wcp18 chromosome p.
    wcp19 chromosome p.
    wcpX chromosome p.
    whole chromosome p. (wcp)
pair
    base p.
    chromosome p.
    kb p.
    kilobase p.
    nucleoside p.
    vertex-nonvertex p.
    vertex-vertex p.
paired
    p. allosome
    p. box
pairing
    chromosome p.
Pai syndrome
Pajot maneuver
Pak
    Ortho Personal P.
    Pedi-Boro Soak P.'s
    Trovan/Zithromax Compliance P.
Palant cleft palate syndrome
palatal-digital-oral syndrome
palate
    ankyloblepharon, ectodermal defects,
      cleft lip/p. (AEC)
    cleft p. (CP)
palatognathous
palatopharyngeal incompetence
palatoschisis
palivizumab
palliative surgery
pallid complexion
pallidum

Pallister
    P. mosaic aneuploidy
    P. mosaic syndrome
    P. syndrome 1
    P. W syndrome
Pallister-Hall syndrome
Pallister-Killian syndrome
pallor of skin
PALM
    premature accelerated lung maturation
palm
    p. leaf pattern
    tripe p.
palmar
    p. crease
    p. erythema
    p. and plantar keratosis and
      keratitis
palmaris
    tinea nigra p.
    xanthoma striata p.
palmatae
    plicae p.
palm-chin reflex
Palmer ovarian biopsy forceps
palmitate
    colfosceril p.
Palmitate-A 5000
palmomandibular sign
palmomental reflex
palmoplantaris
    keratosis p.
PalmVue system
palpable
    p. cord
    p. postmenopausal ovary
palpation
    spoke-wheel p.
palpebral fissure
palsy
    abducens p.
    abducent p.
    Bell p.
    brachial birth p.
    brachial plexus p.
    cerebral p. (CP)
    Duchenne p.
    Erb p.
    facial nerve p.
    familial congenital fourth cranial
      nerve p.
    familial congenital superior oblique
      oculomotor p.
    familial congenital trochlear
      nerve p.
    Klumpke p.
    Landry p.

nerve p.
obstetric brachial plexus p.
Pam
E P.
pamabrom
PAMBA
paraaminomethylbenzoic acid
Pamelor
Pamine
pamoate
pampiniform plexus
Pamprin IB
pancreas
pancreatic
p. agenesis
p. hypoplasia
p. insufficiency syndrome
p. oncofetal antigen (POA)
p. pseudocyst
pancreatitis
gallstone p.
pancreatoblastoma
panculture
pancuronium bromide
pancytopenia
aplastic p.
Pander islands
Panectyl
panencephalitis
Panex
Panhematin
panhysterectomy
panic disorder
Panmycin
P. Oral
Panner disease
panneuropathy
megalencephaly with hyaline p.
panniculectomy
noncosmetic p.
panniculitis
panniculus
PanoGauze
PanoPlex dressing
Panretin topical gel
pansystolic murmur
pantothenate
calcium p.
pantothenic acid
pants-over-vest technique
Panwarfin

PaO$_2$
arterial oxygen pressure (tension)
partial pressure of arterial oxygen
Pap
Papanicolaou
Pap plus speculoscopy (PPS)
Pap smear
ThinPrep Pap
Papanicolaou (Pap)
P. smear
PAPase
phosphatidic acid phosphohydrolase
papaverine
paper-doll fetus
Papette cervical collector
papilla, pl. **papillae**
columnar epithelium p.
p. of columnar epithelium
papillary
p. adenocarcinoma
p. dermis
p. endometrial carcinoma
p. hidradenoma
p. serous cervical carcinoma
p. serous cystadenocarcinoma
papilledema
papilloma
choroid plexus p.
ductal p.
intraductal p.
p. virus, polyoma virus, vacuolative virus (papova)
papillomatosis
juvenile p.
laryngeal p.
subareolar duct p.
vulvar p.
Papillomavirus
papillomavirus
p. hybridization
p. infection
transcriptionally active human p.
type 16 p.
Papillon-Léage-Psaume syndrome
Papillon-Léfevre syndrome
papillosis
introital p.
Pap-Kaps
PapNet
P. automated cervical cystology system
P. reader

NOTES

P

PapNet *(continued)*
 P. test
 P. testing
 P. testing system
papoose board
papova
 papilloma virus, polyoma virus,
 vacuolative virus
PAPPA
 pregnancy-associated plasma protein A
PAPPC
 pregnancy-associated plasma protein C
Pap-Perfect supply system
papular
 p. acrodermatitis of childhood
 (PAC)
 p. dermatitis of pregnancy (PDP)
 p. urticaria
papule
 pruritic urticarial p.
 urticarial p.
papulonecrotica
 tuberculosis p.
papulosis
 bowenoid p.
PAPVR
 partial anomalous pulmonary venous
 return
papyraceous fetus
papyraceus
 fetus p.
para
 p. I, II
paraaminomethylbenzoic acid (PAMBA)
paraaminosalicylate
paraaminosalicylic acid (PAS, PASA)
paraaortic
 p. lymphadenectomy
 p. node
 p. node irradiation
 p. positivity
parabasal cell
parabiosis
 dialytic p.
 vascular p.
parabola
 Keplerian p.
parabolic twins
paracentesis
paracentric inversion
paracervical
 p. anesthesia
 p. block
 p. blockade
 p. injection
 p. lymphatics
paracetamol

parachute
 p. mitral valve
 p. reflex
paracolic gutter
paracolpitis
paracolpium
paracrine
 p. communication
paracyclic ovulation
paracyesis
Paradione
paradoxical incontinence
paraduodenal hernia
Paraflex
ParaGard
 P. T380A intrauterine copper
 contraceptive
 P. T380 copper IUD
parainfluenza
parakeratosis
 multinucleation p.
paralogy
paralysis, pl. **paralyses**
 antigenic p.
 birth p.
 congenital abducens facial p.
 congenital oculofacial p.
 Duchenne p.
 Erb-Duchenne p.
 facial nerve p.
 hereditary abductor vocal cord p.
 hysterical p.
 immunologic p.
 Klumpke p.
 Klumpke-Dejerine p.
 laryngeal abductor p.
 laryngeal adductor p.
 obstetric p.
 parturient p.
 pseudohypertrophic muscular p.
 respiratory p.
 spastic spinal p.
 p. spinalis spastica
 unilateral partial facial p.
 Werdnig-Hoffmann p.
paralytic ileus
paramagnetic particle
paramedian incision
paramenia
paramesonephric
 p. duct
parameter
 fetal growth p.'s
paramethadione
 p. syndrome
parametrectomy
 radical p.
parametrial phlegmon

parametric abscess
parametritic
    p. abscess
parametritis
parametrium
paranasal sinus
paraneoplastic
    p. amyloidosis
    p. syndrome
paraneoplastica
    acrokeratosis p.
paraovarian cyst
parapertussis
paraphimosis
Paraplatin
paraplegia
    congenital spastic p.
    infantile spastic p.
    spastic p. (SP)
    p. spastica
parapsoriasis
    guttata p.
    p. guttata
pararectal
parasalpingitis
parasite
    intestinal p.
    ova and p.'s (O&P)
parasitic
    p. fetus
    p. leiomyoma
    p. myoma
    p. pregnancy
parasiticus
    dipygus p.
    janiceps p.
paraspadias
parastremmatic dwarfism
parasympathetic
    p. nerve
    p. nervous system
paratesticular
parathyroid
    p. adenoma
    p. disease
    p. disorder
    p. gland
    p. hormone (PTH)
    p. hormone-related peptide (PTHrP)
parathyromatosis
paratubal cyst

paratyphoid fever
paraurethral
    p. duct
    p. gland
parauterine lymph node
paravaginal
    p. cystocele
    p. cystocele repair
    p. defect
    p. hysterectomy
    p. soft tissue
    p. suspension
paravaginitis
paraventricular tachycardia
paraxial
    p. mesoderm
parchment skin
Par Decon
paregoric
parencephalia
parencephalous
parenchyma
    placental p.
    renal p.
parenchymal
parenchymatous mastitis
parent
    Coping Health Inventory for P.'s
    (CHIP)
    rearing p.'s
    social p.'s
parentage
    coefficient of p.
    p. determination
parental
    p. karyotype
    p. origin
parenteral
    p. administration
    p. alimentation
    p. hyperalimentation
    p. medication
    p. nutrition (PN)
    p. progesterone
Parenti-Fraccaro syndrome
paresis
    cochleovestibular p.
    congenital suprabulbar p.
paresthesia
    Berger p.
pargyline

NOTES

P

parietal
>p. foramina, brachymicrocephaly, mental retardation syndrome
>p. peritoneum
>p. shunt

parity
>vaginal p.

Parkes-Weber and Dimitri syndrome
Parkinson disease
Parkland
>P. Hospital technique
>P. tubal ligation

Parlodel
Parnate
parodynia
>p. perversa

paromomycin sulfate
paronychia
paronychial infection
paroophoritis
paroöphoron
parotid
>p. enlargement
>p. gland

parous
>p. introitus
>p. os
>p. outlet
>vaginally p.

parovarian
>p. tumor

parovariotomy
parovaritis
parovarium
paroxetine hydrochloride
paroxysmal
>p. nocturnal dyspnea (PND)
>p. nocturnal hemoglobinuria
>p. tachycardia

Parrot
>P. artery
>P. atrophy of newborn
>P. pseudoparalysis
>P. sign
>P. syndrome

parrot
>p. jaw

Parry-Jones vulvectomy
Parsidol
pars tuberalis
part
>fetal small p.'s
>presenting p.

parthenogenesis
partial
>p. agenesis of septa pellucida
>p. anomalous pulmonary venous return (PAPVR)

>p. breech extraction
>p. hydatidiform mole
>p. linkage
>p. monosomy 1p–22p
>p. monosomy XP21, XP22
>p. monosomy Xq
>p. pressure (P)
>p. pressure of arterial carbon dioxide (PaCO$_2$)
>p. pressure of arterial oxygen (PaO$_2$)
>p. pressure of carbon dioxide (PCO$_2$)
>p. pressure of oxygen (PO$_2$)
>p. rollerball endometrial ablation
>p. tetrasomy 10p
>p. thromboplastin time (PTT)
>p. trisomy 1p–22p
>p. trisomy Xq
>p. zona dissection (PZD)

particle
>alpha p.
>Dane p.
>paramagnetic p.

particulate
>p. matter
>p. radiation

Partington-Anderson syndrome
Partington syndrome (PRTS)
partner
>p. gene
>intimate p.
>nonmonogamous p.

partograph
parturient
>p. apoplexy
>p. canal
>p. fever
>p. paralysis

parturients
parturifacient
parturiometer
parturition
Parvolex
Parvovirus
>P. B19

PAS, PASA
>paraaminosalicylic acid
>periodic acid-Schiff
>>PAS stain

pascal unit of pressure measurement (Pa)
Pashayan-Pruzansky syndrome
Pashayan syndrome
Pasini variant
passage
>transplacental p.

passer
> MetraPass suture p.

passes
> number of needle p.

passive
> p. immunity
> p. incontinence

Passos-Bueno syndrome

paste
> butt p.
> Monsel p.

Pasteur
> P. Institute bacillus Calmette-Guérin vaccine
> P. pipette

Pasteurella multocida

Pastia sign

Patau syndrome

patch
> Alora p.
> blood p.
> Climara estradiol transdermal system p.
> Gore-Tex Soft Tissue P.
> Lyrelle p.
> Menorest hormone replacement p.
> moth p.
> Peyer p.
> salmon p.
> strawberry p.
> transdermal glyceryl trinitrate p.
> transdermal medication p.

patchy infiltrate

patella
> aplastic p.
> hypoplastic p.

patency
> probe p.
> tubal p.
> ureteral p.

patent
> p. anus
> p. ductus arteriosus (PDA)
> p. processus vaginalis peritonei

paternal
> p. age
> p. karyotype
> p. meiosis I (PMI)
> p. meiosis II (PMII)
> p. postnatal depression

paternity

Pathfinder DFA test

Pathilon

Pathocil

pathogen

pathogenesis

pathognomonic

pathologic
> p. amenorrhea
> p. finding
> p. retraction ring
> p. retraction ring of Bandl

pathological model

pathologist
> Cancer Committee of College of American P.'s
> perinatal/placental p.

pathology
> hilar cell p.
> International Society for Gynecologic P. (ISGP)
> intracranial p.
> perinatal p.
> Society for Gynecologic P.
> uterine p.

pathophysiologic
> p. mechanism

pathophysiology

pathway
> COX p.
> Embden-Meyerhof p.
> nociceptive p.

PATI
> Penetrating Abdominal Trauma Index

patient
> amenorrheic p.
> anovulatory p.
> antenatal p.
> gynecologic cancer p.
> high-risk p.
> infertile p.
> P. Outcomes Research Team (PORT)
> pregnant cardiac p.
> thalassemic p.

patient-controlled analgesia (PCA)

pattern
> atypical vessel colposcopic p.
> banding p.
> Christmas tree p.
> chromosomal p.
> dermatoglyphic p.
> dysfunctional labor p.
> fern leaf p.

NOTES

P

pattern *(continued)*
    fetal heart rate p.
    fleur de lis breast reconstruction p.
    four-leaf clover p.
    inactivation p.
    mixed p.
    mosaic p.
    nonreassuring fetal heart beat p.
    palm leaf p.
    prominent ductal p.
    silent fetal heart rate p.
    silent oscillatory p.
    snowflake p.
    startle p.
Patterson pseudoleprechaunism syndrome
Patterson-Stevenson-Fontaine syndrome
patty
    Cellolite p.
patulous rectal sphincter
paucicellular
paucity of interlobular bile duct (PILBD)
Paul-Bunnell-Davidsohn test
Pavabid
Pavacap
Pavacen
Pavarine
Pavatest
Pavlik harness
pavor
    p. diurnus
    p. nocturnus
Pavulon
PAW
    pulmonary artery wedge
Pawlik fold
PAWP
    pulmonary artery wedge pressure
paws
    Medevice surgical p.
Paxene
Paxil
Pazo hemorrhoid ointment
Pb4/27
    soft radiation grid Pb4/27
PBF
    pulmonary blood flow
PBL
    peripheral blood lymphocyte
PBMC
    peripheral blood mononuclear cell
PBS
    phosphate buffered saline
        isotonic PBS
PBZ
PBZ-SR
PCA
    patient-controlled analgesia

PCAinfuser-Model 310
PCC
    Poison Control Center
    postcoital contraception
PCE
    physical capacity evaluation
PCO
    polycystic ovary
PCO$_2$
    partial pressure of carbon dioxide
PCOD
    polycystic ovarian disease
PCOS
    polycystic ovary syndrome
PCP
    Pneumocystis carinii pneumonia
PCR
    polymerase chain reaction
    reverse transcription polymerase chain reaction
        PCR assay
        PCR test
PCT
    postcoital test
PCVC
    percutaneous central venous catheter
PCWP
    pulmonary capillary wedge pressure
PD
    Bromfenex PD
    Iofed PD
PDA
    patent ductus arteriosus
        bidirectional PDA
P.D. Access with Peel-Away needle introducer
PDD
    pervasive developmental disorder
PDHC
    pyruvate dehydrogenase complex deficiency
PDI
    psychomotor development index
PDP
    papular dermatitis of pregnancy
PE
    pleiotrophic functional defect
    preeclampsia
        AT3 type II PE
Peacock bromide
peak
    adaptive p.
    Bragg p.
    p. expiratory flow rate (PEFR)
    p. inspiratory pressure (PIP)
peak-and-trough levels

Pean
>   P. clamp
>   P. forceps

peanut oil

PEARL
>   physiologic endometrial
>   ablation/resection loop

Pearl index

pearls
>   Bohn epithelial p.
>   collar of p.
>   Epstein p.
>   perineal p.

pearly-white nodule

pear-shaped uterus

Pearson marrow-pancreas syndrome

peau
>   p. d'orange appearance
>   p. d'orange appearance of the
>   breast

pectoralis major muscle

pectus
>   p. carinatum
>   p. excavatum
>   p. gallinatum
>   p. recurvatum

peculiar facies

Pedameth

Pederson vaginal speculum

PediaCare
>   P. Oral

Pediacof

Pedialyte
>   P. formula
>   P. oral electrolyte maintenance
>   solution

PediaPatch

Pediapred

PediaProfen

PediaSure
>   P. formula
>   P. liquid nutrition

pediatric
>   p. bone rongeur
>   p. bulldog clamp
>   p. cocktail
>   P. Early Elemental Examination
>   (PEEX)
>   p. gynecology
>   p. ovarian teratoma
>   p. self-retaining retractor

>   P. Triban
>   p. vaginoscopy
>   p. vascular clamp

Pediazole

Pedi-Bath Salts

Pedi-Boro Soak Paks

pedicle
>   p. clamp
>   uterosacral ligament p.

Pedicran with Iron

pedicterus

pediculosis
>   p. capitis
>   p. corpus
>   p. pubis

pedigree
>   CEPH p.
>   p. chart

Pediotic Otic

Pedi PEG tube

Pedi-Pro Topical

pedis
>   tinea p.

Pedituss
>   P. Cough

pedologist

pedometer

Pedric

PedTE-Pak-4

Pedtrace-4
>   P. hyperalimentation

pedunculated polyp

PedvaxHIB

PEEP
>   positive end-expiratory pressure
>   PEEP ventilator

PEEX
>   Pediatric Early Elemental Examination

PEF-24 formula

PEFR
>   peak expiratory flow rate

pegademase bovine

Peganone

PEHO
>   progressive encephalopathy, edema,
>   hypsarrhythmia, optic atrophy
>   PEHO syndrome

Peiper reflex

Pelger-Huet anomaly

Pelizaeus-Merzbacher disease (PMD)

pellagra

NOTES

P

pellet
Testopel P.
YAG p.
Pelletier-Leisti syndrome
Pellizzi syndrome
pellucida
partial agenesis of septa p.
zona p. (ZP)
pellucidum
enlarged cavum p.
septum p.
Pelosi
P. hysterotomy
P. vaginal hysterectomy
pelves (*pl. of* pelvis)
pelvic
p. abscess
p. actinomycosis
p. adhesive disease (PAD)
p. architecture
p. arterial embolization
p. arteriogram
p. arteriovenous malformation
p. artery
p. axis
p. bleeding
p. boost radiotherapy
p. brim
p. cavity
p. cellulitis
p. congestion syndrome
p. contraction
p. diaphragm
p. direction
p. endometriosis
p. examination
p. exenteration
p. exercise
p. exostosis
p. fibromatosis
p. floor
p. floor electrical stimulation (PFS)
p. floor nerve
p. forceps
p. fracture
p. girdle relaxation (PGR)
p. gutter
p. hematocele
p. inclination
p. index
p. infection
p. inflammation
p. inflammatory disease (PID)
p. inlet
p. involvement
p. irradiation
p. joint
p. kidney

p. laparoscopy
p. lymphadenectomy
p. lymph node
p. malignancy in pregnancy
p. mass
p. nerve injury
p. node dissection
P. Organ Prolapse-Quantified system (POP-Q)
P. Organ Prolapse staging system
p. outlet
p. ovarian vein thrombosis (POVT)
p. pain
p. peritonitis
p. plane of greatest dimensions
p. plane of inlet
p. plane of least dimensions
p. plane of outlet
p. plexus
p. pole
p. presentation
p. reconstruction surgeon
p. recurrence
p. relaxation
p. rest
p. score
p. support defect
p. support index (PSI)
p. surgery
p. tenderness
p. vein congestion
p. vein thrombophlebitis
p. version
pelvicephalography
pelvicephalometry
pelvic-floor surgery
pelvicocleidocranial dysostosis
pelvifixation
pelvigraph
pelvimeter
Hanely-McDermitt p.
pelvimetry
computed tomographic p.
CT p.
manual p.
planographic p.
stereoscopic p.
x-ray p.
pelvioperitonitis
pelvioplasty
pelvioscopy, pelviscopy, pelvoscopy
pelviotomy
pelviperitonitis
pelvis, pl. **pelves**
android p.
anthropoid p.
arcus tendineus fasciae p.
assimilation p.

axis of p.
beaked p.
bifid p.
bowl of p.
breech location out of p.
p. capsular dysplasia
concrete p.
contracted p.
cordate p.
cordiform p.
Deventer p.
dwarf p.
elephant p.
p. enlargement
flat p.
flying-T p.
frozen p.
funnel-shaped p.
gynecoid p.
heart-shaped p.
inverted p.
p. justo major
p. justo minor
juvenile p.
Kilian p.
kyphotic p.
longitudinal oval p.
masculine p.
mesatipellic p.
Nägele p.
p. nana
Otto p.
p. plana
platypellic p.
Prague p.
reniform p.
Robert p.
Rokitansky p.
round p.
sweep the p.
transverse oval p.
true p.
well engaged in p.
pelviscope
pelviscopic intrafascial hysterectomy
pelviscopy (*var. of* pelvioscopy)
pelvis-shoulder dysplasia
pelvitherm
pelviureteric junction obstruction
pelvocephalography
pelvoscopy (*var. of* pelvioscopy)
pemoline

pemphigoid
bullous p.
p. gestationis
pemphigus
benign familial chronic p.
p. vulgaris
pen
P. A/N
Humalog P.
Humulin P.
insulin p.
Pena-Shokeir
P.-S. phenotype
P.-S. syndrome
Penbritin
penciclovir cream
pencil
ConMed electrosurgical p.
Valleylab p.
pencil-handled laryngoscope
Pendred syndrome
pendulous abdomen
Penecort Topical
Penetrak test
penetrance
penetrant
p. gene
p. trait
Penetrating Abdominal Trauma Index (PATI)
penetration
Penetrex
PenFil
Novolin 70/30 P.
Novolin N P.
Novolin R P.
penicillamine
penicillin
benzathine p.
extended-spectrum p.
p. G
p. G benzathine
p. G procaine
natural p.
penicillinase-resistant p.
phenoxymethyl p.
procaine p.
p. V
p. V potassium
penile
p. prosthesis

NOTES

P

391

penile *(continued)*
    p. urethra
    p. vibrator
penis
    corona of p.
    hypoplastic p.
penischisis
PEN-Kera moisturizing cream
Penlon infant resuscitator
Pennington
    P. clamp
    P. forceps
Penn pouch procedure
penoscrotal
    p. hypospadias
    p. transposition
Penrose drain
Pentacef
pentaerythritol tetranitrate
pentagastrin
pentalogy
    p. of Cantrell syndrome
    p. of Fallot
pentamidine isethionate
Pentamycetin
pentasomy
    chromosome X p.
    p. X syndrome
penta-X
    p.-X chromosomal aberration
    p.-X syndrome
Pentazine
pentazocine
Penthrane
Pentids
Pentids-P AS
pentobarbital
pentobarbitone
pentosifylline
pentosuria
Pentothal Sodium
pentoxifylline
Pentritol
penumbra
Pen.Vee K
PEP
    progestogen-dependent endometrial protein
Pepcid
    P. AC Acid Controller
    P. RPD
PEPI
    postmenopausal estrogen and progestin intervention
Pepper syndrome
Peptavlon

peptic
    p. ulcer
    p. ulcer disease
peptide
    amino-terminal p.
    atrial natriuretic p. (ANP)
    brain p.
    carboxyl terminal p. (CTP)
    corticotropin-like intermediate lobe p.
    endogenous opioid p.
    gonadotropin-releasing hormone-associated p. (GAP)
    gonadotropin-releasing hormone-like p.
    helix termination p.
    opioid p.
    parathyroid hormone-related p. (PTHrP)
    vasoactive intestinal p. (VIP)
Peptococcus
    P. anaerobius
    P. asaccharolyticus
    P. niger
Peptostreptococcus
Perceived Stress Scale
perchloroethylene
Percocet
Percodan
Percoll technique
percreta
    placenta p.
Percuflex Plus stent
percussion
    abdominal p.
    p. therapy
percutaneous
    p. blood sampling
    p. central venous catheter (PCVC)
    p. drainage
    p. epididymal sperm aspiration (PESA)
    p. fetal transfusion
    p. line
    p. nephrostomy
    p. nephrostomy catheter
    P. Stoller Afferent Nerve Stimulation System
    p. suprapubic teloscopy
    p. transluminal angioplasty (PTA)
    p. umbilical blood sampling (PUBS)
Pereyra
    P. needle
    P. needle suspension
    P. procedure
Pereyra-Lebhertz modification of Frangenheim-Stoeckel procedure

Perez
  P. reflex
  P. sign
Perez-Castro forceps
perflubron emulsion temporary blood substitute
perforated bowel
perforation
  ileal p.
  uterine p.
perforator
  Baylor amniotic p.
Performa
  P. Acoustic Imaging system
  P. diagnostic ultrasound imaging system
  P. ultrasound
performance
  reproductive p.
perfused twins
perfusion
  fallopian tube sperm p. (FTSP)
  placental p.
  renal p.
  twin reverse arterial p. (TRAP)
  uteroplacental p.
pergolide
Pergonal
Perheentupa syndrome
Periactin
perianal aphthosis
periaortic lymph node
periappendicitis
  p. decidualis
periarteritis nodosa
pericanalicular fibroadenoma
pericardial
  p. constriction-growth failure syndrome
  p. puncture
  p. tamponade
pericarditis
pericardium
pericentric inversion
pericentromeric
  p. marker
pericervical node
Peri-Cold Pack
pericolpitis
periconceptional rubella
periderm
Peridin-C

Peridol
peridural
  p. analgesia
  p. anesthesia
Peri-Gel Pack
perihepatitis
perihilar streaking
periimplantation
perilobular connective tissue
perimenopausal
  p. woman
perimenopause
perimenstrual tenesmus
perimesonephric cyst
perimetritic
perimetritis
perimortem delivery
perinatal
  p. acidosis
  p. asphyxia
  p. death
  p. distress prediction
  p. effect
  p. hypoxia
  p. medicine
  p. morbidity
  p. mortality
  p. mortality rate (PMR, PNMR)
  p. outcome
  p. pathology
  p. risk factor
  p. transmission
  p. trauma
perinatal/placental pathologist
perinate
perinatologist
perinatology
perineal
  p. analgesia
  p. anesthesia
  p. body
  p. defect
  p. hygiene
  p. laceration
  p. massage
  p. nerve
  p. pearls
  p. repair
  p. scar
  p. sphincter disruption
  p. surgical apron

NOTES

P

perineometer
Gynos p.
perineoplasty
perineorrhaphy
Emmet-Studdiford p.
vaginal p.
perineotomy
perineovaginal fistula
perinephric
p. abscess
p. phlegmon
perinephritis
perineum
dashboard p.
period
blastogenic p.
canalicular p.
effective refractory p.
embryonic p.
fertile p.
intrapartum p.
last menstrual p. (LMP)
last normal menstrual p. (LNMP)
menstrual p. (MP)
missed p.
postpartum p.
previous menstrual p. (PMP)
pseudoglandular p.
puerperal p.
saccular p.
terminal saccular p.
periodic
p. acid-Schiff (PAS, PASA)
p. auscultation
p. breathing
p. fever
p. patient assessment
periodicity
perioophoritis
perioophorosalpingitis
perioral cyanosis
periorificial
p. dermatitis
p. rhagades
periostitis
pubic symphysis p.
periovarian adhesion
periovaritis
peripartum
p. cardiomyopathy
p. symphysis separation
peripheral
p. androgen activity
p. androgen activity marker
p. arterial line
p. blood lymphocyte (PBL)
p. blood mononuclear cell (PBMC)

p. dysostosis, nail hypoplasia, mental retardation (PNM)
p. dysostosis, nail hypoplasia, mental retardation syndrome (PMN)
p. hormone level
p. hyperalimentation
p. jaundice
p. mononeuropathy
p. placental separation
p. vascular shock
peripherally inserted central catheter (PICC)
periportal hemorrhagic necrosis
perirectal fascia
perisalpingitis
perisalpingoovaritis
perissodactylous
peristalsis
intestinal p.
peritomy
peritoneal
p. biopsy
p. button
p. catheter
p. cavity
p. cytology
p. dialysis
p. dissemination
p. envelope
p. fluid
p. hernia
p. implant
p. insufflation
p. lavage
p. macrophage
p. mouse
p. oocyte sperm transfer (POST)
p. reflection
p. serosa
p. serous papillary carcinoma
p. studding
p. washing
peritonei
gliomatosis p.
patent processus vaginalis p.
peritoneoscopy
peritoneum
abdominal p.
Blake closure of p.
overlapping closure of p.
parietal p.
visceral p.
peritonitis
meconium p.
pelvic p.
spontaneous bacterial p. (SBP)
tuberculous p.

Peritrate
peritubal adhesion
periumbilical
    p. incision
    p. pain
periurethral
    p. collagen injection
    p. gland
perivaginitis
perivascular emphysema
periventricular
    p. leukomalacia
    p. nodular gray matter heterotopia
periventricular-intraventricular
  hemorrhage
perivitelline space
Peri-Warm Pack
Perlman nephroblastomatosis syndrome
Perlutal
Permapen
PermCath catheter
permeability
    vascular p.
permethrin
permission, limited information, specific
  suggestions and intensive therapy
  (PLISSIT)
Permitil
permutation
pernicious
    p. anemia (PA)
    p. anemia of pregnancy
perocormus
perodactyly
peromelia
peroneal
    p. atrophy, X-linked recessive
    p. muscle atrophy
    p. sign
peropus
perosomum
perosplanchnia
peroxidase
    glutathione p.
    streptavidin p.
peroxidation
    plasma lipid p.
peroxide
    hydrogen p.
    lipid p.
    zinc p.
peroxyoxalate

perphenazine
Per-Q-Cath catheter
PerQ SANS
Perrault syndrome
Persantine
persistens
    ductus arteriosus p.
persistent
    p. anovulation
    p. ectopic pregnancy
    p. estrogen secretion
    p. fetal circulation (PFC)
    p. hyperinsulinemic hypoglycemia
    of infancy (PHHI)
    p. müllerian duct syndrome
    p. occiput posterior position
    p. occiput posterior presentation
    p. ovarian mass
    p. postmolar gestational
    trophoblastic tumor
    p. pulmonary hypertension of the
    newborn (PPHN)
    p. right umbilical vein
personal probability
Persutte and Lenke study
Pertofrane
pertubation
pertussis
    Bordetella p.
    Haemophilus p.
pervasive developmental disorder (PDD)
perversa
    parodynia p.
perversus
    situs p.
per vias naturales
pes
    p. cavus
    p. equinovarus
PESA
    percutaneous epididymal sperm aspiration
pessary
    Albert-Smith p.
    Biswas Silastic vaginal p.
    blue ring p.
    cube p.
    diaphragm p.
    doughnut p.
    Dumontpallier p.
    Dutch p.
    Emmett-Gelhorn p.
    Findley folding p.

NOTES

P

pessary *(continued)*
    Gariel p.
    Gehrung p.
    Gelhorn p.
    Hodge p.
    p. injection
    Mayer p.
    Menge p.
    PGE_2 p.
    Prentif p.
    Prochownik p.
    prostaglandin p.
    ring p.
    Smith p.
    Zwanck p.
pestis
    Yersinia p.
PET
    preeclamptic toxemia
petechia, pl. **petechiae**
Peters
    P. anomaly
    P. anomaly-corneal clouding-growth
    and mental retardation syndrome
    P. anomaly-short limb dwarfism
    syndrome
    P. ovum
    P.-plus syndrome
pethidine
    intramuscular injection of p.
Petit
    P. ligament
petit
    p. mal
    p. mal epilepsy
petroleum distillate poisoning
petrositis
Pettigrew syndrome (PGS)
peudoprogeria syndrome
Peutz-Jeghers syndrome
Peyer patch
Peyronie disease
Pezzer catheter
PF
    Astramorph PF
Pfannenstiel incision
Pfaundler-Hurler syndrome
PFC
    persistent fetal circulation
Pfeiffer syndrome
P-fimbriae
Pfizerpen
    P. AS
Pflüger
    cord of P.
PFS
    pelvic floor electrical stimulation
    Adriamycin PFS

    Folex PFS
    Idamycin PFS
    Tarabine PFS
    Vincasar PFS
PG
    phosphatidylglycerol
    prostaglandin
        one-hour PG
pg
    picogram
PGD
    preimplantation genetic diagnosis
PGE
    prostaglandin E
PGE_2
    prostaglandin E_2
        PGE_2 pessary
PGF
    placental growth factor
$PGF_{2\alpha}$
    prostaglandin $F_{2\alpha}$
PGG
    prostaglandin G
PGH
    placental growth hormone
    prostaglandin H
PGI_2
    prostacyclin
PGK
    phosphoglycerate kinase
        PGK deficiency
        PGK hereditary nonspherocytic
        anemia
P-glycoprotein
PGR
    pelvic girdle relaxation
        symptom-giving PGR
PgR
    progesterone receptor
PGS
    Pettigrew syndrome
PGSI
    prostaglandin synthetase inhibitor
phacomatosis
    fifth p.
    fourth p.
Phaedra complex
phagedenic ulcer
phage phenotype
phagocytosis
phakomata
phakomatosis
    Jadassohn nevus p. (JNP)
phalloplasty
phallus
Phaneuf-Graves repair
Phaneuf uterine artery forceps

phantom
  p. pregnancy
  Schultze p.
pharaonic circumcision
pharmacokinetics
  quinolone p.
pharmacologic treatment
Pharmaseal
  P. disposable cervical dilator
  P. disposable uterine sound
pharyngeal
  p. diverticulum
  p. gonococcal infection
  p. pouch syndrome
pharyngitis
  lymphonodular p.
  purulent p.
  streptococcal p.
  viral p.
phase
  active p.
  aqueous p.
  catagen p.
  early proliferative p.
  end-expiratory p. (EEP)
  equilibrium p.
  fertile p.
  follicular p.
  hair growth p.
  hyperprolactinemia-associated
    luteal p.
  implantation p.
  inadequate luteal p.
  latent p.
  luteal p.
  menses p.
  menstrual p.
  mitosis p.
  proliferative p.
  secretory p.
  telogen p.
Phazyme infant drops
PHC
  primary hepatocellular carcinoma
Ph1 chromosome
  Philadelphia chromosome
phenacetin
phenanthrene
Phenaphen
Phenazine
  P. Injection
phenazocine

Phenazodine
phenazone
phenazopyridine
  p. HCl
  p. hydrochloride
phencyclidine
phendimetrazine
phenelzine
Phenerbel-S
Phenergan
  P. Injection
  P. Oral
  P. Rectal
Phenetron
phenindamine
phenindione
pheniramine
phenobarbital
phenobarbitone
phenocopy
phenogenetics
phenolsulfonphthalein (PSP)
  p. test
phenomenon, pl. phenomena
  all-or-none p.
  Arias-Stella p.
  dawn p.
  Hata p.
  Köbner p.
  lip p.
  Marcus Gunn p.
  memory p.
  piezoelectric p.
  rebound p.
  recall p.
  Rumpel-Leede p.
  Somogyi p.
  Strassman p.
  stuck-twin p.
  Wenckebach p.
  X-linked p.
phenothiazine
  p. poisoning
phenotype
  Bombay p.
  neurofibromatosis with Noonan p.
  Pena-Shokeir p.
  phage p.
  XX and XY Turner p.
phenotypic
  p. sex

NOTES

P

phenotypic *(continued)*
    p. threshold
    p. variance
Phenoxine
phenoxybenzamine
phenoxymethyl
    p. penicillin
phenprocoumon
phensuximide
phentermine
phentolamine
phenylalanine
    p. hydroxylase
    p. 4-monooxygenase
    p. mustard
phenylalaninemia
phenylbutazone
Phenyldrine
phenylephrine
phenylhydantoin
phenylketonuria (PKU)
    maternal p.
    preconceptionally treated p.
    p. test
phenylpiperazine
phenylpropanolamine
phenylpyruvate
phenylpyruvic acid
phenyl salicylate
phenyltoloxamine
phenytoin
pheochromocytoma
Pheryl E
PHHI
    persistent hyperinsulinemic hypoglycemia
    of infancy
pH (hydrogen ion concentration)
    fetal blood pH
    vaginal pH
Philadelphia chromosome (Ph1
  chromosome, Ph1 chromosome)
Philip gland
Phillips' Milk of Magnesia
philtrum
    hypoplastic p.
phimosis, pl. **phimoses**
    p. clitoridis
    p. vaginalis
pHisoHex soap
phlebarteriectasis
phlebectasia
phlebitis
    puerperal p.
    string p.
    suppurative p.
phlebography
phlebometritis
phlebothrombosis

phlegmasia
    p. alba dolens
    cellulitic p.
    p. dolens
    thrombotic p.
phlegmon
    parametrial p.
    perinephric p.
phlegmonous mastitis
Phocas
    P. disease
    P. syndrome
phocomelia
phonocardiography
phorbol ester
phosphatase
    acid p.
    alkaline p.
    bone alkaline p. (BAP)
    placental alkaline p. (PLAP)
phosphate
    p. buffered saline (PBS)
    chromic p.
    Cleocin P.
    clindamycin p.
    etoposide p.
    polyestradiol p.
phosphatidic acid phosphohydrolase
  (PAPase)
phosphatidylcholine
    dipalmitoyl p. (DPPC)
    disaturated p.
phosphatidylglycerol (PG)
phosphatidylinositol (PI)
phosphoglycerate
    p. kinase (PGK)
    p. kinase deficiency hereditary
    nonspherocytic anemia
phosphohydrolase
    phosphatidic acid p. (PAPase)
phosphokinase
    creatine p. (CPK)
phospholipase
    p. A_2
    p. activity
    p. C
phospholipid
    p. antibody
phospholipidosis
phosphorus (P)
    p.-32 (^{32}P)
phosphorylase kinase deficiency
phosphorylation
    myosin light-chain p.
    protein p.
    sperm tail protein p.
photochemotherapy

photocoagulation
    laser p.
PhotoDerm
    P. PL handpiece
photodynamic therapy
Photofrin
photometer
    HemoCue hemoglobin p.
    reflectance p.
photomicrograph
photon
photophoresis
photoplethysmography (PPG)
photoretinopathy
photosensitive epilepsy
photostethoscope
phototherapy
    double-blanket p.
photothermal sclerosis
photovaporization
    laser p.
PHS
    pseudoprogeria/Hallermann-Streiff
    PHS syndrome
phthalylsulfacetamide
phthalylsulfathiazole
phycomycosis
pHydrion strip
phygogalactic
Phyllocontin
phyllodes
    cystosarcoma p.
    p. tumor
phylloides
phylloquinone
phylogenesis
phylon
physiatrist
physical
    p. capacity evaluation (PCE)
    p. map
    p. maturity
    p. therapy (PT)
    p. victimization
physician
    p.-applied force
    p. payment review commission
    (PPRC)
physiognomy
physiologic
    p. amenorrhea
    p. anemia

    p. childbirth
    p. endometrial ablation/resection
    loop (PEARL)
    p. follicular regulation
    p. hypogammaglobulinemia
    p. icterus
    p. jaundice
    p. jaundice of the newborn
    p. replacement dose
    p. retraction ring
    p. saline
    p. sclerosis
    p. third-stage management
physiologist
    fetal p.
physiology
    maternal p.
physiotherapy (PT)
physometra
physopyosalpinx
physostigmine
phytanic acid storage disease
phytoagglutinin
phytobezoar
phytoestrogen
phytomenadione
phytonadione
PI
    phosphatidylinositol
    premature infant
    protease inhibitor
    pulsatility index
        alpha-1 PI
pi
    p. cone monochromatism
    p. dimeric protein
piano-wire adhesion
PICA
    posterior inferior cerebellar artery
    posterior inferior communicating artery
pica
    p. craving
PICC
    peripherally inserted central catheter
pickup
    Adson p.'s
    rat tooth p.'s
pickwickian syndrome
picogram (pg)
picornavirus
PID
    pelvic inflammatory disease

NOTES

P

PIE
> postinfectious encephalomyelitis
> preimplantation embryo
> pulmonary infiltrate with eosinophilia
> pulmonary interstitial emphysema

piebald
Pie Medical ultrasound
Pierre Robin syndrome
Piersol point
piezoelectric phenomenon
PIF
> prolactin inhibiting factor

pigeon
> p. breast
> p. toe

pigeon-toed
pigmentary
> p. retinopathy, hypogonadism, mental retardation, nerve deafness, glucose intolerance
> p. retinopathy, hypogonadism, mental retardation, nerve deafness, glucose intolerance syndrome
> p. retinopathy, mental retardation syndrome

pigmentation
pigmented lesion
pigmenti
> incontinentia p.

pigmentosa
> autosomal-dominant retinitis p. (adRP)
> congenital bullous urticaria p.
> congenital retinitis p.
> retinitis p.
> urticaria p.

pigmentosum
> xeroderma p.

PIH
> pregnancy-induced hypertension
> PIH symptom

pilaris
> keratosis p.
> pityriasis rubra p.

PILBD
> paucity of interlobular bile duct

pileum
pileus
pili
> p. annulati
> p. torti

pill
> birth control p. (BCP)
> combined birth control p.
> emergency contraceptive p. (ECP)
> morning-after p.
> progestin-only p. (POP)

pillar
> bladder p.
> rectal p.
> Usko p.

pillow
> Crescent p.
> OsmoCyte p.

Pilocar
pilocarpine
> p. iontophoresis method

pilonidal
> p. cyst
> p. dimple
> p. sinus

Piloptic
pilosebaceous unit
Pilot
> P. audiometer
> P. suturing guide

Pima
> P. population

pimozide
pin
> Beath p.
> Surgin hemorrhage occluder p.

Pinard
> P. maneuver
> P. sign

pincer grasp
pindolol
pineal
> p. gland
> p. tumor

pinealoma
pinguecula, pinguicula
pink
> p. disease
> p. tetralogy of Fallot

pinked up
pinna, pl. pinnae
> anteverted p.

pinopode
pinworm
> p. vaginitis

pion
PIP
> peak inspiratory pressure

Pipelle
> P. biopsy
> P. endometrial suction curette
> Unimar P.

Piper
> P. fatigue scale
> P. forceps

piperacetazine
piperacillin
> p. sodium/tazobactam sodium

piperazine
    p. dione
    p. estrone sulfate
piperidolate hydrochloride
piperonyl butoxide
pipestem urethra
Pipet Curette
pipette
    Pasteur p.
Pipracil
Pirie syndrome
piroxicam
Piskacek sign
Piso test
Pistofidis cervical biopsy forceps
Pisum sativum **agglutinin (PSA)**
pit
    commissural lip p.
Pitocin
    P. augmentation
    P. induction
Pitressin
Pitrex
Pitt-Rogers-Danks syndrome (PRDS)
Pitt syndrome
Pitt-Williams brachydactyly
pituitary
    p. adenoma
    p. axis
    p. cell
    p. desensitization
    p. disorder
    p. dwarfism
    p. forceps
    p. gigantism
    p. gland
    p. gland function
    p. gland transplantation
    p. gland tumor
    p. gonadotropin
    p. gonadotropin effect
    p. gonadotropin inhibition
    p. gonadotropin regulation
    p. gonadotropin secretion
    p. gonadotropin suppression
    p. hormone
    p. hormone release
    p. necrosis
pituitary-hypothalamic circulation
pituitary-ovarian function
pityriasis
    p. alba

    p. lichenoides
    p. lichenoides et varioliformis acuta
    (PLEVA)
    p. rosea
    p. rubra pilaris
pityrosporum yeast
pivampicillin
Piver type II procedure
PIVKA-II assay
pivot lock
PIXI bone densitometer
Pixie minilaparoscope
PK
    pyruvate kinase
Pk
    Synsorb Pk
PKC
    protein kinase C
PKD
    polycystic kidney disease
PKU
    phenylketonuria
    PKU test
placement
    extraovular p.
placenta, pl. **placentae**
    ablatio placentae
    abruptio placentae
    accessory p.
    p. accreta
    p. accreta vera
    adherent p.
    amotio placentae
    annular p.
    battledore p.
    bidiscoidal p.
    p. biloba
    p. bipartita
    central p. previa
    chorioallantoic p.
    chorioamnionic p.
    choriovitelline p.
    circummarginate p.
    p. circumvallata
    circumvallate p.
    cotyledonary p.
    deciduate p.
    diamniotic dichorionic p.
    dichorionic p.
    dichorionic-diamniotic p.
    p. diffusa
    p. dimidiata

NOTES

P

placenta *(continued)*
    disperse p.
    Duncan p.
    p. duplex
    endotheliochorial p.
    endothelio-endothelial p.
    epitheliochorial p.
    p. extrachorales
    p. fenestrata
    hemochorial p.
    horseshoe p.
    incarcerated p.
    p. increta
    labyrinthine p.
    p. marginata
    p. membranacea
    monochorionic p.
    monochorionic diamniotic p.
    monochorionic monoamniotic p.
    p. multiloba
    nondeciduous p.
    p. panduraformis
    p. percreta
    p. previa
    p. previa centralis
    p. previa creta
    p. previa marginalis
    p. previa partialis
    p. reflexa
    p. reniformis
    retained p.
    Schultze p.
    p. in situ
    p. spuria
    Stallworthy p.
    succenturiate p.
    supernumerary p.
    total p. previa
    p. triloba
    p. tripartita
    p. triplex
    twin p.
    p. velamentosa
    villous p.
    zonary p.

placental
    p. abruption
    p. adaptive angiogenesis
    p. alkaline phosphatase (PLAP)
    p. barrier
    p. bed
    p. bleeding
    p. bleeding site
    p. bruit
    p. cotyledon
    p. disk
    p. dysfunction
    p. dysfunction syndrome

p. dystocia
p. extrusion
p. fragment
p. giant cell
p. grade
p. grading
p. growth
p. growth factor (PGF)
p. growth hormone (PGH)
p. hemangioma syndrome
p. hemorrhage
p. hormone
p. hydrops
p. immunology
p. implantation
p. index
p. infarction
p. insufficiency
p. lactogen
p. leakage
p. lobule
p. localization
p. membrane
p. metastasis
p. migration
p. parenchyma
p. perfusion
p. polyp
p. presentation
p. protein
p. respiration
p. secretion
p. separation
p. septa
p. sign
p. site gestational trophoblastic disease
p. site nodule
p. site trophoblastic tumor (PSTT)
p. site tumor
p. souffle
p. stage
p. stage of labor
p. steroid
p. sulfatase deficiency
p. thickness
p. thrombosis
p. tissue
p. tissue transplant
p. transfer
p. transfusion
p. transfusion syndrome
p. vascular anastomosis
p. vasopressin
p. villus
p. weight
placenta, ovary, uterus (POU)
placentascan

placentation
  abnormal p.
  p. abnormality
  fundal p.
  hemochorioendothelial p.
  monochorionic p.
  normal p.
placentitis
placentography
  indirect p.
placentology
placentoma
placentotherapy
Placidyl
placing reflex
placode
  olfactory p.
plagiocephaly
plague vaccine
plan
  P. B
  cleavage p.
  exit p.
plana
  vertebra p.
plane
  equatorial p.
  first parallel pelvic p.
  fourth parallel pelvic p.
  p. of midpelvis
  orthagonal p.
  p. of pelvic canal
  pelvic p. of greatest dimensions
  pelvic p. of inlet
  pelvic p. of least dimensions
  pelvic p. of outlet
  second parallel pelvic p.
  third parallel pelvic p.
  wide p.
planning
  family p.
planographic pelvimetry
planovalgus
  talipes p.
plant alkaloid
plantar
  p. crease
  p. dermatosis
  p. reflex
  p. response
  p. wart

plantaris
  keratosis palmaris et p.
planus
  lichen p.
  lichen ruber p. (LRP)
  talipes p.
PLAP
  placental alkaline phosphatase
plaque
  pruritic urticarial papules and p.'s
    (PUPP)
  pulmonary p.
plasma
  p. albumin
  p. cell balanitis
  p. cell mastitis
  p. cell pneumonia
  p. cell vulvitis
  p. cholesterol
  p. fibrinogen
  p. fraction
  fresh frozen p. (FFP)
  frozen p.
  p. iron-binding capacity
  p. iron concentration
  p. level
  p. lipid peroxidation
  postingestion p.
  p. prolactin
  p. prorenin
  p. prorenin increase
  p. protein
  seminal p.
  p. testosterone
  p. thromboplastin antecedent (PTA)
  p. thromboplastin antecedent factor
  p. volume
  p. volume expansion
  p. volume regulation
plasmacellularis
  balanitis circumscripta p.
plasmacrit test
Plasmanate
plasmapheresis
plasmatic
  progesterone p.
plasmid
  p. DNA
  p. fingerprinting
plasmin

NOTES

P

403

plasminogen
    p. activator (PA)
    p. activator inhibitor (PAI)
plasminogen-activator
    p.-a. inhibitor-1 (PAI-1)
    p.-a. inhibitor-2 (PAI-2)
Plasmodium
Plastibell
plastica
    linitis p.
    rectal linitis p. (RLP)
plastic cup vacuum extractor
plastid
plate
    basal p.
    breast p.
    chocolate agar p.
    chorionic p.
    Mancini p.
    neural p.
    trigonal p.
    vaginal p.
plateau
    orgasmic p.
platelet
    p. adhesion
    p. aggregation
    p. antigen
    p. cofactor I
    p. concentrate
    p. count
    p. cyclooxygenase
    p. disorder
    p. function test
    p. neutralization procedure
    single donor apheresis p. (SDAP)
    p. transfusion
platelet-activating factor
platelet-associated antibody
platelet-derived growth factor
plate-like atelectasis
Platelin Plus Activator
platform
    Aspen ultrasound p.
Platinol
    P.-AQ
    bleomycin, Eldisine, mitomycin, P. (BEMP)
    bleomycin, etoposide, P. (BEP)
    bleomycin, ifosfamide, P. (BIP)
platinum
platinum-based regimen
platypellic, platypelloid
    p. pelvis
platyspondyly
Pleatman
    P. pouch
    P. sac

Plegine
pleiotrophic
    p. cytokine
    p. functional defect (PE)
pleiotropic gene
pleiotropy
pleomastia, pleomazia
pleomorphism
pleonosteosis
    Leri p.
plethora hydremica
plethoric
plethysmography
    impedance p.
    jugular occlusion p.
pleural effusion
Pleur-evac chest catheter
pleuritic pain
pleurodesis
    tetracycline p.
pleurodynia
pleuromelus
pleuroperitoneal
    p. foramen
    p. hernia
pleuropulmonic
pleurosomus
PLEVA
    pityriasis lichenoides et varioliformis acuta
plexiform neurofibromatosis
plexus
    Auerbach p.
    brachial p.
    branchial p.
    cavernous p.
    choroid p.
    Frankenhäuser p.
    hypogastric p.
    p. lesion
    Meissner p.
    myenteric p.
    ovarian p.
    pampiniform p.
    pelvic p.
    submucosal p.
    superior mesenteric p.
    vaginal venous p.
plica, pl. **plicae**
    plicae palmatae
plicamycin
plicate
plication
    Kelly p.
    Kelly-Kennedy p.
PLISSIT
    permission, limited information, specific suggestions and intensive therapy

ploidy
      DNA p.
Plott syndrome
PLUG
      plug the lung until it grows
            FETENDO PLUG
plug
      Avina female urethral p.
      copulation p.
      epithelial p.
      meconium p.
      mucous p.
      silicone p.
plug the lung until it grows (PLUG)
plumbism
plume
      electrosurgical p.
      laser p.
      smoke p.
plural pregnancy
plus
      Answer P.
      Duramist P.
      Fact P.
      Riopan P.
PM
      PMI
            paternal meiosis I
      PMII
            paternal meiosis II
PM-60/40
      P. formula
      Similac P.
PMB
      postmenopausal bleeding
            PMB 200
            PMB 400
PMD
      Pelizaeus-Merzbacher disease
PMDD
      premenstrual dysphoric disorder
PMN
      peripheral dysostosis, nail hypoplasia,
            mental retardation syndrome
PMP
      previous menstrual period
PMR
      perinatal mortality rate
PMS
      premenstrual syndrome
            PMS-Cyproheptadine
            PMS-Diazepam

PMS-Erythromycin
PMS-Imipramine
PMS-Ketoprofen
PMS-Lindane
Lurline PMS
PMS-Methylphenidate
PMS-Nystatin
PMS-Progesterone
PMS-Pseudoephedrine
PMS-Sodium Cromoglycate
PMT
      postmenstrual tension
            Optivite PMT
PMVP
      pulmonary microvascular permeability to
            protein
PN
      parenteral nutrition
      pronucleus
PN$_2$
      nitrogen partial pressure
PND
      paroxysmal nocturnal dyspnea
pneogaster
pneumatic
      p. compression
      p. compression stocking
pneumatocele
pneumatosis cystoides intestinalis
pneumococcal
      p. pneumonia
      p. vaccine
pneumococcus, pl. pneumococci
pneumocystiasis
Pneumocystis carinii **pneumonia (PCP)**
pneumocystography
pneumocystosis
pneumoencephalogram
pneumogram
pneumography
      impedance p.
pneumohydrometra
pneumomediastinum
pneumonia
      acute interstitial p.
      adenoviral p.
      p. alba
      p. alba of Virchow
      aspiration p.
      bacterial p.
      bronchiolitis obliterans organizing p.
      chemical p.

NOTES

P

pneumonia *(continued)*
    chickenpox p.
    chlamydial p.
    congenital p.
    eosinophilic p.
    giant cell p.
    Gram-negative p.
    Hecht p.
    hypostatic p.
    interstitial plasma cell p.
    intrauterine p.
    Kaufman p.
    lipoid p.
    lobar p.
    plasma cell p.
    pneumococcal p.
    Pneumocystis carinii p. (PCP)
    rheumatic p.
    staphylococcal p.
    streptococcal p.
    thrush p.
    varicella p.
    viral p.
pneumoniae
    Chlamydia p.
    Diplococcus p.
    Klebsiella p.
pneumonitis
    aspiration p.
    chemical p.
    viral p.
pneumonocyte
    type II p.
pneumopericardium
pneumoperitoneum
pneumosalpingography
pneumotachography
pneumothorax
    tension p.
Pneumovax 23
Pneumo-Wrap
PNM
    peripheral dysostosis, nail hypoplasia,
      mental retardation
PNMR
    perinatal mortality rate
Pnu-Imune 23
PO$_2$
    partial pressure of oxygen
POA
    pancreatic oncofetal antigen
POADS
    postaxial acrofacial dysostosis syndrome
POC
    products of conception
pocket
    amniotic fluid p.
Pocket-Dop 3 monitor

Pocketpeak peak flow meter
podalic
    p. extraction
    p. version
Pod-Ben-25
podencephalus
podofilox
podophyllin
podophyllotoxin
podophyllum
    p. resin
POEMS
    polyneuropathy, organomegaly,
      endocrinopathy, M protein, skin
      changes
    POEMS syndrome
POF
    premature ovarian failure
poikiloderma
poikiloploidy
point
    p. A, subspinale
    p. B, supramentale
    cardinal p.'s
    Halle p.
    Legat p.
    McBurney p.
    Munro p.
    p. mutation
    Piersol p.
    trigger p.
pointer syndrome
pointes
    torsade de p.
Poiseville-Hagen relationship
Poiseville law
Poison Control Center (PCC)
poisoning
    barbiturate p.
    carbon monoxide p.
    ciguatera p.
    lead p.
    petroleum distillate p.
    phenothiazine p.
    salicylate p.
    scopolamine p.
    strychnine p.
    thallium p.
    zinc p.
pokeweed mitogen (PWM)
Poladex
Poland
    P. anomaly
    P. malformation sequence
    P. syndrome
polar
    p. body

p. body in-situ hybridization
p. presentation
Polaramine
Polaris
    P. reusable cutters
    P. reusable dissector
    P. reusable forceps
    P. reusable grasper
    P. reusable instrument
polarity
polarization
    amniotic fluid fluorescence p.
    fluorescent p.
pole
    caudal p.
    cephalic p.
    fetal p.
    germinal p.
    head p.
    pelvic p.
poliodystrophia cerebri progressiva infantilis
poliodystrophy
polioencephalitis
    bulbar p.
poliomyelitis
Poliovirus hominis
polish
    Sklar p.
Politano-Leadbetter ureteroneocystostomy
pollination
Pollitt syndrome
Pollock forceps
Poloxamer 407 barrier material
poly
    p. A RNA
    P. CS device
    p. roticulator 55 stapler
polyamine
polyarteritis nodosa
polyarthritis
    juvenile p.
polyarticular juvenile rheumatoid arthritis
polybutester
polycarbophil
    calcium p.
polycheiria
polychlorinated biphenyl
polychondritis
Polycillin
    P.-N

polyclonal antibody
polyclonal-monoclonal antibody
Polycose formula
polycyesis
polycystic
    p. fibrous dysplasia
    p. kidney disease (PKD)
    p. ovarian disease (PCOD)
    p. ovary (PCO)
    p. ovary syndrome (PCOS)
    p. renal disease
polycythemia
    p. rubra vera
polydactylism
polydactyly
    p.-chondrodystrophy syndrome
    p.-craniofacial anomalies syndrome
    p.-craniofacial dysmorphism syndrome
    p.-imperforate anus syndrome
    p., imperforate anus, vertebral anomalies syndrome
    postaxial p.
    preaxial p.
    short rib-p.
polydimethylsiloxane
polydioxanone
polydipsia
polydysplasia
    hereditary ectodermal p.
polydysspondylism
polydystrophic dwarfism
polydystrophy
    pseudo-Hurler p.
polyembryoma
polyembryony
polyester suture material
polyestradiol
    p. phosphate
polyethylene suture material
polygalactia
Polygam
    P. S/D
polygene
polygenic
    p. disorder
    p. inheritance
polyglactic acid suture
polyglactide suture
polyglactin
    p. 910 suture
polyglandular syndrome

NOTES

P

polyglycolic acid
polyglycol suture
polyglyconate suture
polygnathus
PolyHeme blood substitute
Poly-Histine
    P.-H. CS
    P.-H. D
    P.-H.-D Capsule
    P.-H. DM
polyhydramnios
polyhypermenorrhea
polyhypomenorrhea
polymastia, polymazia
PolyMem dressing
polymenorrhea
polymerase
    p. chain reaction (PCR)
    DNA p.
    DNA-directed RNA p.
    RNA p.
    RNA-directed DNA p.
polymetacarpia
polymetatarsia
polymicrobial
    p. bacteremia
    p. pelvic infection
polymicrogyria
polymorphic
    p. light eruption of pregnancy
polymorphism
    amplified fragment length p.
    genetic p.
    restriction fragment length p.
     (RFLP)
    single nucleotide p.
    single stranded conformational p.
    TGFA p.
polymorphonuclear
polymorphonucleocyte
Polymox
polymyositis
    primary idiopathic p.
polymyxin
    p. B
    p. B sulfate
polyneuritis
    infantile p.
polyneuropathy
    p.-cataract-deafness syndrome
    erythredema p.
    p., organomegaly, endocrinopathy,
     M protein, skin changes
     (POEMS)
polyoncosis
    hereditary cutaneomandibular p.
    multiple hereditary
     cutaneomandibular p.

polyostotic fibrous dysplasia
polyotia
polyp
    adenomatous p.
    cervical p.
    colonic p.
    endocervical p.
    endometrial p.
    fibrinous p.
    fibroepithelial p.
    hydatid p.
    hyperplastic p.
    pedunculated p.
    placental p.
    retention p.
    sessile p.
polypectomy
polypeptide
    factor V p.
    S-methionine-labeled p.
    vasoactive intestinal p. (VIP)
polyphosphatidylinositide
polyphospholinositide
polyploid
polyploidy
polypoid hyperplasia
polyposis
    colonic p.
polypropylene
    p. fascial strip
    p. suture material
polysaccharide group-specific antigen
polyscelia
polyserositis
    idiopathic p.
polysomia
polysomnogram
    p. with flow-volume loop
polysomy
    sex chromosomal p.
Polysonic ultrasound lotion
Polysorb suture
polyspermy, polyspermia
polysplenia
    p. syndrome
polysymbrachydactyly
polysyndactyly
    crossed p.
    p.-dyscrania syndrome
    p.-peculiar skull syndrome
polysynostoses syndrome
Polytef
polytene
    p. chromosome
polytetrafluoroethylene graft
polythelia
polythiazide

Polytrim
P. eye drops
polyuria
polyvalent
p. immunoglobulin therapy
p. pneumococcal vaccine
Poly-Vi-Flor vitamins
polyvinylpyrrolidone (PVP)
Poly-Vi-Sol vitamins
PolyWic
POMARD anthropomorphic measurement reference chart
POMC
proopiomelanocortin
Pomeroy
P. method
P. operation
P. tubal ligation
Pompe
P. disease
P. syndrome
POMS
Profile of Mood States
POMS score
ponderal index
Pondimin
Pondocillin
Ponstan
Ponstel
Pontén fasciocutaneous flap
pontine
p. glioma
p. tumor
pontobulbar
p. palsy with deafness
p. palsy with neurosensory deafness
Pontocaine
PONV
postoperative nausea and vomiting
pool
gene p.
vaginal p.
poorly marginated mass
poor pregnancy outcome
POP
progestin-only pill
pop
Revital-Ice rehydrating freezer p.
popcorn-like calcification
popliteal
p. cyst

p. pterygium syndrome
p. web syndrome
POP-Q
Pelvic Organ Prolapse-Quantified system
population
Pima p.
Porak-Durante syndrome
porcine surfactant
porcupine skin
porencephalic cyst
porencephaly
p., cerebellar hypoplasia, internal malformations syndrome
porfimer
p. sodium
Porges-Meier test
pork insulin
porokeratosis
porphyria
acute intermittent p.
congenital erythropoietic p.
congenital photosensitive p.
p. cutanea tarda
p. cutanea tarda hereditaria
p. cutanea tarda symptomatica
cutaneous hepatic p.
erythrohepatic p.
erythropoietic p.
hepatic p.
hepatoerythropoietic p.
intermittent p.
mixed p.
ovulocyclic p.
South African genetic p.
Swedish p.
symptomatic p.
p. variegata
variegate p.
porphyric neuropathy
porphyrin metabolism disorder
Porro
P. cesarean section
P. hysterectomy
P. operation
PORT
Patient Outcomes Research Team
port
laparoscopy p.
10-mm umbilical p.
p. site hernia
portable respirator
Port-A-Cath catheter

NOTES

P

Portagen formula
portal hypertension
Porteous syndrome
Porter-Silber reaction
portio, pl. **portiones**
    p. vaginalis
Portland
    hemoglobin P.
Portuguese disease
port-wine
    p.-w. hemangioma
    p.-w. mark
    p.-w. nevus
    p.-w. stain
Posiero effect
position (*See also* presentation)
    arm p.
    asynclitic p.
    back-up p.
    batrachian p.
    Bozeman p.
    brow p.
    brow-anterior p.
    brow-down p.
    brow-posterior p.
    brow-up p.
    chin p.
    coaxial p.
    decubitus p.
    dorsal birthing p.
    dorsal lithotomy p.
    dorsal supine p.
    Duncan p.
    Edebohl p.
    en face p.
    English p.
    face-to-pubes p.
    fetal p.
    fetal head p.
    fetal spine p.
    Fowler p.
    frogleg p.
    frontoanterior p.
    frontoposterior p.
    frontotransverse p.
    genucubital p.
    genupectoral p.
    jackknife p.
    knee-chest p.
    knee-elbow p.
    Kraske p.
    lateral recumbent p.
    left frontoanterior p. (LFA)
    left frontoposterior p. (LFP)
    left frontotransverse p. (LFT)
    left lateral p.
    left mentoanterior p. (LMA)
    left mentoposterior p. (LMP)

    left mentotransverse p. (LMT)
    left occipitoanterior p. (LOA)
    left occipitoposterior p. (LOP)
    left occipitotransverse p. (LOT)
    left sacroanterior p. (LSA)
    left sacroposterior p. (LSP)
    left sacrotransverse p. (LST)
    left scapuloanterior p. (LScA)
    left scapuloposterior p. (LScP)
    lithotomy p.
    maternal birthing p.
    mentoanterior p.
    mentoposterior p.
    mentotransverse p.
    mentum anterior p.
    mentum posterior p.
    mentum transverse p.
    missionary p.
    obstetric p.
    occipitoanterior p.
    occipitoposterior p.
    occipitotransverse p.
    persistent occiput posterior p.
    right acromiodorsoposterior p.
    right frontoanterior p. (RFA)
    right frontoposterior p. (RFP)
    right frontotransverse p. (RFT)
    right lateral p.
    right mentoanterior p. (RMA)
    right mentoposterior p. (RMP)
    right mentotransverse p. (RMT)
    right occipitoanterior p. (ROA)
    right occipitoposterior p. (ROP)
    right occipitotransverse p. (ROT)
    right sacroanterior p. (RSA)
    right sacroposterior p. (RSP)
    right sacrotransverse p. (RST)
    right scapuloanterior p. (RScA)
    right scapuloposterior p. (RScP)
    sacroanterior p.
    sacroposterior p.
    sacrotransverse p.
    semi-Fowler p.
    semilithotomy p.
    semiprone p.
    Simon p.
    Sims p.
    sitting p.
    standing p.
    supine p.
    Trendelenburg p.
    Valentine p.
    vertex p.
    Walcher p.
positional cloning
positive
    p. end-expiratory pressure (PEEP)
    Hematest p.

heme p.
p. interference
P. Reinforcement In the
  Menopausal Experience patient-
  support program
P. Reinforcement in the
  Menopausal Experience (PRIME)
positivity
lymph node p.
paraaortic p.
possible migrational abnormality
POSSUM
P. database
P. database of genetic syndromes
POST
peritoneal oocyte sperm transfer
postadolescence
postaugmentation
postaxial
p. acrofacial dysostosis syndrome
  (POADS)
p. oligodactyly
p. polydactyly
postcesarean hemorrhage
postcoital
p. bleeding
p. contraception (PCC)
p. estrogen
p. test (PCT)
postconceptional age
postcordocentesis bradycardia
postdate
p. pregnancy
post dates
postdatism
postembolization syndrome
posterior
anterior and p. (A&P)
p. asynclitism
p. commissure
p. cul-de-sac
dysplasia marginalis p.
p. fontanel
p. fornix
p. fossa
p. fossa cyst
p. fossa neoplasm
p. inferior cerebellar artery (PICA)
p. inferior communicating artery
  (PICA)
p. lie

occiput p. (OP)
p. pelvic exenteration
p. probability
p. repair
p. sagittal diameter
p. urethral valve (PUV)
p. urethritis
p. uterosacral ligament
p. vagina
p. vaginal cuff
p. vaginismus
posterolateral fontanel
posthetomy
postinfectious encephalomyelitis (PIE)
postinflammatory adenopathy
postingestion plasma
postirradiation syndrome
postischemic stenosis
postlumpectomy
postmature
p. infant
postmaturity
p. syndrome
postmeiotic segregation
postmembrane
p. pressure
p. rupture
postmenarchal
p. bleeding
p. cyclicity
postmenopausal
p. amenorrhea
p. atrophy
p. bleeding (PMB)
p. body mass
p. bone loss
p. estrogen and progestin
  intervention (PEPI)
p. estrogen replacement therapy
p. hirsutism
p. level
p. osteoporosis
p. palpable ovary syndrome
p. woman
postmenopause
postmenstrual
p. stress
p. tension (PMT)
**postmolar persistent gestational
  trophoblastic tumor**
postmortal pregnancy

NOTES

P

postmortem
 p. delivery
 p. study
postnatal
 p. year
postnatally
postoperative
 p. bladder dysfunction
 p. care
 p. chemotherapy
 p. complication
 p. congestive epididymitis
 p. cuff cellulitis
 p. cystitis
 p. nausea and vomiting (PONV)
 p. pelvic radiation
 p. radiotherapy
 p. sepsis
 p. seroma
 p. shock
 p. symptom analysis
 p. voiding dysfunction
postovulatory age
postpartum
 p. amenorrhea
 p. attitude
 p. blues
 p. cardiomyopathy
 p. care
 p. depression
 p. endometritis
 p. endomyometritis
 p. febrile morbidity
 p. hemolytic uremic syndrome
 p. hemorrhage (PPH)
 p. hypertension
 p. incontinence
 p. infection
 p. metritis
 p. period
 p. pituitary necrosis syndrome
 p. pleuropulmonary and cardiac
  disease
 p. psychosis
 p. tetanus
 p. thyroiditis
postperfusion syndrome
postpill amenorrhea
postpoartum partial salpingectomy
postponing sexual involvement (PSI)
postprandial glucose
postpuberal, postpubertal
postpuberty
postpubescent
postradiation
 p. cystitis
 p. fistula
 p. periureteral fibrosis

postreduction
 p. mammaplasty
postrubella syndrome
postsurgical fat necrosis
post-term
 p.-t. AGA
 p.-t. infant
 p.-t. LGA
 p.-t. pregnancy
 p.-t. SGA
posttransfusion hematocrit
posttraumatic fat necrosis
posttubal ligation syndrome
postulate
 Koch p.'s
postural
 p. deformity
 p. proteinuria
 p. version
posture
 fetal p.
posturing
 opisthotonic p.
posturography
 dynamic p.
postvoid
 p. residual urine test
potassium
 p. bromide
 p. chloride
 p. citrate
 clavulanate p.
 diclofenac p.
 p. gluconate
 p. hydroxide (KOH)
 p. iodide
 p. iodide Enseals
 penicillin V p.
potato-like mass
potbelly
potential
 auditory evoked p. (AEP)
 brainstem auditory evoked p.
  (BAEP)
 fetal brain stem auditory evoked p.
 low malignant p. (LMP)
 resting membrane p.
 stage IIIc papillary tumor of low
  malignant p.
 vertex p.
 visual evoked p. (VEP)
Potocky needle
Potter
 P. disease
 P. facies
 P. syndrome
 P. version

POU
placenta, ovary, uterus
POU theory of Ishihara
pouch
Broca p.
continent urinary p.
copulating p.
Denis Browne p.
Douglas p.
Florida p.
Indiana p.
intravaginal p.
kangaroo p.
Kock p.
labioperineal p.
Mainz p.
Marsupial P.
Morison p.
Pleatman p.
Rathke p.
Reality vaginal p.
rectouterine p.
Rowland p.
Seessel p.
utriculovaginal p.
vaginal p.
wallaby p.
pouchogram
Poupart ligament
Pourcelot index
povidone-iodine
p.-i. douche
p.-i. gel
p.-i. wipe
POVT
pelvic ovarian vein thrombosis
powder
Absorbine Antifungal Foot P.
p. burn spot
Lotrimin AF P.
Lotrimin AF Spray P.
Nystop topical p.
p. pseudocalcification
Secretin-Ferring P.
Sklar Kleen p.
Zeasorb-AF P.
powder-burn endometrial lesion
powder-free glove
power
p. of attorney
p. Doppler sonography
Pozzi procedure

PPC
primary peritoneal carcinoma
PPG
photoplethysmography
PPH
postpartum hemorrhage
PPHN
persistent pulmonary hypertension of the newborn
PPHP
pseudopseudohypoparathyroidism
PPO
preferred provider organization
PPRC
physician payment review commission
PPROM
preterm premature rupture of membranes
PPROM UCI
PPS
Pap plus speculoscopy
PPTT
prepubertal testicular tumor
PR
progesterone receptor
practice
Advisory Committee on Immunization P.'s (ACIP)
orogenital sexual p.'s
practitioner
Prader-Gurtner syndrome
Prader-Labhart-Willi-Fanconi syndrome
Prader-Labhart-Willi syndrome
Prader-Willi
P.-W. habitus, osteopenia, camptodactyly syndrome
P.-W. habitus, osteoporosis, hand contracture syndrome
P.-W. syndrome
praecox
dentia p.
icterus p.
macrogenitosomia p.
pubertas p.
Prague
P. maneuver
P. pelvis
pralidoxime chloride
PrameGel
Pramet FA
Pramilet FA
Pramosone
pramoxine

NOTES

P

413

PRAMS
Pregnancy Risk Assessment Monitoring
System
Pratt
P. dilator
sigmoid pouch of P.
Prax
prazepam
prazosin
pRb tumor suppressor gene
PRDS
Pitt-Rogers-Danks syndrome
preadolescent vaginal bleeding
preantral follicle
preaxial
p. acrofacial dysostosis
p. hexadactyly
p. mandibulofacial dysostosis
p. polydactyly
precancerous lesion
PreCare Conceive
Precef
precipitable
precipitant
precipitate
p. labor
p. labor and delivery
precipitatory factor
precipitin
precipitous
p. delivery
p. labor
precise
P. disposable skin stapler
P. pregnancy test
precision
preclinical carcinoma
Preclude peritoneal membrane
precocious
p. adrenarche
p. pseudopuberty
p. pubarche
p. puberty
p. teeth
precocity
**preconceptionally treated
phenylketonuria**
preconception care
Precose
predecidual
predeciduous teeth
predecondensed spermhead injection
prediction
perinatal distress p.
prenatal risk p.
scar p.
predictive value of test

predictor
ClearPlan Easy ovulation p.
Conceive Ovulation P.
First Response ovulation p.
morbidity p.
mortality p.
OvuQuick Self-Test ovulation p.
Q-test ovulation p.
predisposition
genetic p.
prednicarbate
prednisolone
prednisone
cyclophosphamide, methotrexate, 5-
fluorouracil, and p. (CMFP)
Cytoxan, methotrexate,
fluorouracil, p. (CMFP)
Cytoxan, methotrexate, fluorouracil,
vincristine, p. (CMFVP)
vincristine, cyclophosphamide,
and p. (VCP)
preeclampsia (PE)
early-onset p.
p. headache
superimposed p.
preeclamptic toxemia (PET)
preembryo
triploid p.
preemie
preferred provider organization (PPO)
Prefill
Femstat P.
preformation
pregenesis
Pregestimil formula
PregnaGen test kit
pregnancy
abdominal p.
aborted ectopic p.
accidental p.
acute fatty liver of p. (AFLP)
adolescent p.
ampullar p.
anaphylactoid syndrome of p.
anembryonic p.
p. associated hypertension
at-risk p.
Besnier prurigo of p.
bichorial p.
bigeminal p.
bilateral ectopic p.
bilateral simultaneous tubal p.'s
biochemical p.
broad ligament p.
p. category B, C, X
p. cell
cervical p.
cervical ectopic p.

chemical p.
cholestatic hepatosis of p.
clinical p.
combined p.
p. complication
compound p.
cornual p.
p. dating
direct agglutination p. (DAP)
ectopic p. (EP)
epulis of p.
euploid p.
extraamniotic p.
extrachorial p.
extramembranous p.
extrauterine p.
fallopian p.
false p.
fatty liver of p.
p. fear
fimbrial ectopic p.
first-cycle clinical p.
full-term p.
gemellary p.
grand p.
hepatic p.
heterotopic p.
high-risk p.
p. hormone
hydatid p.
iatrogenic multiple p. (IMP)
idiopathic cholestasis of p.
inherited thrombophilia in p.
interstitial p.
intrahepatic cholestasis of p.
intraligamentary p.
intramural p.
intraperitoneal p.
intrauterine p. (IUP)
isoimmunization in p.
IVF-induced abdominal p.
late p.
ligamentary ectopic p.
ligamentous ectopic p.
linear IgM disease of p.
p. loss
p. luteoma
luteoma of p.
p. management
mask of p.
mesometric p.
molar p.

monochorionic twin p.
multifetal p.
multiple p.
mural p.
near-term p.
normally progressing p. (NPP)
p. outcome
ovarian p.
ovarioabdominal p.
papular dermatitis of p. (PDP)
parasitic p.
pelvic malignancy in p.
pernicious anemia of p.
persistent ectopic p.
phantom p.
plural p.
polymorphic light eruption of p.
postdate p.
postmortal p.
post-term p.
previous p.
prolonged p.
prurigo of p.
pruritic folliculitis of p.
pruritic urticarial papules and
  plaques of p. (PUPPP)
recurrent jaundice of p.
recurrent molar p.
p. reduction
refractory anemia of p.
P. Risk Assessment Monitoring
  System (PRAMS)
sarcofetal p.
secondary abdominal p.
singleton p.
Spangler papular dermatitis of p.
splenic p.
spurious p.
successful p.
SureCell rapid test kit for p.
term p.
p. test
time to p. (TIP)
toxemia of p.
toxemic rash of p.
treatment-associated p.
treatment-independent p.
tubal p.
tuboabdominal p.
tuboovarian p.
tubouterine p.
p. tumor

NOTES

P

415

pregnancy *(continued)*
    twin p.
    unplanned p.
    uterine p.
    uteroabdominal p.
    voluntary interruption of p. (VIP)
    p. wastage
    p. zone protein
pregnancy-associated
    p.-a. globulin
    p.-a. hypoplastic anemia
    p.-a. plasma protein
    p.-a. plasma protein A (PAPPA)
    p.-a. plasma protein C (PAPPC)
    p.-a. thrombosis
pregnancy-induced
    p.-i. hypertension (PIH)
    p.-i. hypertension symptom
pregnancy-specific protein
**pregnancy, uterine, not delivered
  (PUND)**
pregnane
pregnanediol
5β-pregnane-3,20-dione
pregnanetriol
pregnant
    p. cardiac patient
    p. diabetic
pregnenolone
**Preg-Net computerized information
  system**
Pregnosis
Pregnyl
pregranulosa cell
pre-hCG estradiol
preimplantation
    p. diagnosis
    p. embryo (PIE)
    p. genetic diagnosis (PGD)
preinvasive
    p. cervical disease
    p. lesion
prelabor
    p. membrane rupture
    p. rupture of the membranes
    (PROM)
Prelu-2
premalignant
    p. disease
    p. lesion
premammary abscess
preMA prenatal drink mix
Premarin
    P. With Methyltestosterone
    P. With Methyltestosterone Oral
premature
    p. accelerated lung maturation
    (PALM)

    p. adrenarche
    p. airway closure
    p. amnion rupture
    p. birth
    p. centromere division
    p. cervical dilation
    p. delivery
    p. ductus arteriosus closure
    p. ejaculation
    p. infant (PI)
    p. labor
    p. luteal regression
    p. membrane rupture
    p. menopause
    p. ovarian failure (POF)
    p. pubarche
    p. rupture of membranes (PROM)
    P. Special Care formula
    p. thelarche
    p. uterine contraction inhibition
    p. ventricular contraction (PVC)
prematurity
    anemia of p.
    apnea of p.
    pulmonary insufficiency of p.
    retinopapillitis of p.
    retinopathy of p. (ROP)
    sequelae of extreme p.
Pre-Meal
    Dexatrim P.-M.
premembrane
    p. pressure
    p. rupture
premenarchal vulvovaginitis
premenstrual
    p. dysphoria
    p. dysphoric disorder (PMDD)
    p. edema
    p. salivary syndrome
    p. symptoms
    p. syndrome (PMS)
    p. tension
    p. tension syndrome
premenstruum
PremesisRx
premie Simpson forceps
Premphase
Prempro
premutation allele
prenatal
    p. care
    p. detection
    p. diethylstilbestrol exposure
    p. genetic diagnosis
    p. genetics
    p. magnetic resonance imaging
    p. mortality

p. placement of thoracoamniotic shunt
p. risk factor
p. risk prediction
p. screening
p. selection
p. sex determination
p. tocolysis
p. treatment
p. ultrasound

Prenate
P. 90
P. Ultra

Prentif
P. cavity-rim cervical sap
P. pessary

preovulatory
p. follicle
p. LH surge

prep
preparation
KOH prep
wet prep

Pre-Par

preparation (prep)
bowel p.
International Reference P. (IRP)
recombinant FSH p.

Preparation-H hydrocortisone cream

prepared semen

preperitoneal fat

Prepidil
P. Gel cervical ripener
P. Vaginal Gel

prepregnancy care

prepubertal, prepuberal
p. testicular tumor (PPTT)

prepubescent

prepuce

prerenal azotemia

prereproductive

presacral
p. nerve
p. neurectomy
p. sympathectomy

prescription
p. drug
P. Strength Desenex

presentation (*See also* position)
acromion p.
arm p.
breech p.

brow p.
brow-down p.
cephalic p.
complete breech p.
compound p.
p. of cord
double-breech p.
double-footling p.
Duncan p.
face p.
face/chin p.
p. of fetus
foot p.
footling breech p.
frank breech p.
full-breech p.
funic p.
hand and head p.
head p.
incomplete breech p.
incomplete foot p.
incomplete knee p.
knee p.
longitudinal p.
mentoanterior p.
mentoposterior p.
oblique p.
occiput p.
pelvic p.
persistent occiput posterior p.
placental p.
polar p.
right occipitoposterior p.
shoulder p.
sincipital p.
single-breech p.
single-footling p.
singleton breech p.
torso p.
transverse lie p.
trunk p.
umbilical p.
vertex p.
vertex-breech twin p.
vertex-transverse twin p.

presenting part

PreservCyt

presomite embryo

Press-Mate model 8800T blood pressure monitor

pressor
p. agent

NOTES

P

417

pressor *(continued)*
 p. medication
 p. response
pressure
 arterial blood p.
 arterial carbon dioxide p. (tension)
  (PaCO$_2$)
 blood p.
 central venous p. (CVP)
 colloid oncotic p.
 colloid osmotic p. (COP)
 continuous distending airway p.
  (CDAP)
 continuous negative airway p.
  (CNAP)
 continuous positive airway p.
  (CPAP)
 cricoid p.
 end-expiratory p. (EEP)
 esophageal p.
 high-frequency positive p. (HFPP)
 intraabdominal p.
 intravesical p.
 jugular venous p. (JVP)
 leak-point p.
 maternal abdominal p.
 maximum urethral closure p.
  (MUCP)
 mean arterial p. (MAP)
 negative end-expiratory p. (NEEP)
 nitrogen partial p. (PN$_2$)
 osmotic p.
 partial p. (P)
 peak inspiratory p. (PIP)
 positive end-expiratory p. (PEEP)
 postmembrane p.
 premembrane p.
 pulmonary artery wedge p.
  (PAWP)
 pulmonary capillary wedge p.
  (PCWP)
 suprapubic p.
 p. transmission
 tubal perfusion p. (TPP)
 urethral p.
 zero end-expiratory p.
pressure-cycled ventilator
pressure-preset ventilator
pressure-separator tubing
Pressyn
preterm
 p. birth
 p. delivery
 p. infant
 p. labor (PTL)
 p. labor arrest
 p. neonate

 p. premature rupture of membranes
  (PPROM)
 p. rupture of membranes
pretransfusion Hct
Pretz-D
Prevacid
prevalence rate
Preven emergency contraception kit
preventive antioxidant
Preveon
Prevex
 P. Baby Diaper Rash
 P. HC
previa
 central placenta p.
 complete placenta p.
 placenta p.
 vasa p.
previable fetus
previllous embryo
previous
 p. maternal immunity
 p. menstrual period (PMP)
 p. pregnancy
 p. transfundal uterine surgery
Prevnar
Prevotella
 P. biviua
 P. disiens
prezygotic
PRF
 prolactin releasing factor
Pribnow box
prickly heat
Prieto syndrome (PRS)
prilocaine
Prilosec
Primagedine
primaquine
primary
 p. amenorrhea
 p. atelectasis
 p. bubo
 P. Care Evaluation of Mental
  Disorders (PRIME-MD)
 p. cesarean section
 p. congenital glaucoma
 p. dysfunctional labor
 p. dysmenorrhea
 p. embryonic cell
 p. empty sella syndrome
 p. follicle
 p. hepatocellular carcinoma (PHC)
 p. hyperuricemia syndrome
 p. hypogonadism
 p. idiopathic dermatomyositis
 p. idiopathic polymyositis
 p. infection

p. infertility
p. oocyte
p. ovarian insufficiency
p. peritoneal carcinoma (PPC)
p. syphilis
p. thrombocythemia (PT)
p. uterine inertia
Prima Series LEEP speculum
Primatene
P. Mist
Primaxin
PRIME
Positive Reinforcement in the
Menopausal Experience
PRIME patient support program
primed in situ labeling (PRINS)
PRIME-MD
Primary Care Evaluation of Mental
Disorders
primer
allele specific associated p.
arbitrarily p.
degenerate consensus p.
degenerate oligonucleotide p. (DOP)
p. pair system
primidone
primigravida
elderly p.
priming
cervical p.
primipara
primiparity
primiparous
primitive
p. (fetal) hemoglobin (HbP)
p. groove
p. knot
p. ovum
p. reflex
p. streak
primordial
p. dwarfism
p. follicle
p. germ cell
p. ovum
p. pluripotent stem cell
vesicourethral p.
primordium
Primrose syndrome
primum
foramen p.
septum p.

Principen
principle
ALARA p.
as low as reasonably achievable p.
Doppler p.
Fick p.
Frank-Starling p.
Mitrofanoff p.
Pringle disease
PRINS
primed in situ labeling
prion
prior
p. complete mole
p. low transverse uterine incision
p. low vertical uterine incision
p. partial mole
p. probability
Priscoline
Pritchard intramuscular regimen
private blood group
PRL
prolactin
PRO
P. 2000 Gel
P. infusion catheter
Pro-Amox
Pro-Ampi
probability
conditional p.
joint p.
objective p.
personal p.
posterior p.
prior p.
subjective p.
Probalan
ProBalance liquid nutrition
proband
Pro-Banthine
probe
bacterial artificial chromosome p.
BICAP p.
BiLAP bipolar laparoscopic p.
biopsy p.
biplane intracavitary p.
Bipolar Circumactive P.
blunt p.
Bruehl-Kjaer transvaginal
ultrasound p.
convex p.
DNA dual color p.

NOTES

P

probe *(continued)*
    Endopath needle tip
      electrosurgery p.
    endovaginal p.
    Envision endocavity p.
    gene p.
    IntraDop p.
    locus-specific p.
    MEVA p.
    multiplane intracavitary p.
    Neo-Therm neonatal skin
      temperature p.
    nucleic acid p.
    p. patency
    ribonucleic acid p.
    p. sheath
    Spencer p.
    p. system
    transrectal p.
    transvaginal transducer p.
    TrueVision transvaginal p.
    Universal vaginal p.
    ViraType p.
    V33W Endocavity p.
    YSI neonatal temperature p.
probenecid
problem
    dietary p.
    growth p.
    intersex p.
    nutritional p.
    psychosexual p.
proboscis, pl. **proboscides, proboscises**
    disruptive p.
    holoprosencephalic p.
    p. lateralis
    lateral nasal p.
    supernumerary p.
procainamide
procaine
    p. penicillin
    penicillin G p.
procarbazine
Procardia
procedure
    Abbe-McIndoe p.
    Abbe-McIndoe-Williams p.
    Abbe-Wharton-McIndoe p.
    Aldridge sling p.
    Altemeier p.
    Aries-Pitanguy p.
    Baden p.
    Baldy-Webster p.
    Ball-Burch p.
    Bastiaanse-Chiricuta p.
    Blair-Brown p.
    Blalock-Hanlon p.
    Blalock-Taussig p.

Bricker p.
Burch p.
buried vaginal island p.
Castaneda p.
Chassar Moir-Sims p.
Chassar Moir sling p.
Coblation-Channeling surgical p.
Cole intubation p.
Cyclops p.
Dall-Miles cable grip p.
Danus-Fontan p.
Davydov p.
diagnostic p.
Donald p.
dot-blot p.
Duckett p.
Estes p.
Evans-Steptoe p.
Everard Williams p.
fascial sling p.
Fontan p.
Frank p.
Fredet-Ramstedt p.
Gepfert p.
Gittes p.
Goebell p.
Goebell-Stoeckel-Frangenheim p.
hemi-Fontan p.
Heyman-Herndon clubfoot p.
Hood p.
Ilizarov p.
Ingelman-Sundberg gracilis
  muscle p.
interventional p.
intestinal bypass p.
Jatene p.
Kaliscinski ureteral p.
Kasai p.
Kelly plication p.
Kennedy p.
Kono p.
Ladd p.
laparoscopic unipolar coagulation p.
Lash p.
Ledbetter-Politano p.
LETZ p.
loop electrosurgical excision p.
  (LEEP)
lyodura sling p.
Marshall-Marchetti p.
Marshall-Marchetti-Krantz p.
Martius p.
Mayo-Fueth inversion p.
McCall-Schumann p.
McDonald p.
McIndoe-Hayes p.
Meigs-Okabayashi p.
Mitrofanoff neourethral p.

MMK p.
Moschowitz p.
Mustard p.
Muzsani p.
Neugebauer-LeFort p.
Nichols p.
Olshausen p.
OPERA p.
outpatient endometrial
resection/ablation p.
Penn pouch p.
Pereyra p.
Pereyra-Lebhertz modification of
Frangenheim-Stoeckel p.
Piver type II p.
platelet neutralization p.
Pozzi p.
psoas hitch p.
pubovaginal sling p.
Ramstedt p.
Raz-Leach p.
Raz sling p.
Récamier p.
retropubic suspension p.
Richter and Albrich p.
Schauffler p.
selective embolization p.
selective tubal occlusion p. (STOP)
Shauta-Aumreich p.
Shirodkar p.
sling p.
Soave abdominal pull-through p.
Spence p.
Stamey modification of Pereyra p.
Stanley Way p.
Swenson pull-through p.
Tompkins p.
triangular vaginal patch sling p.
Turner-Warwick vagino-obturator
shelf p.
Uchida p.
UPLIFT p.
urinary unidiversion p.
vaginal wall sling p.
valvulotomy p.
Whipple p.
W-stapled urinary reservoir p.
procedure-related pyrexia
Pro-Ception
process
binary p.
CytoRich p.

freezing p.
immune p.
quality assurance p.
processed blood product
processor
ThinPrep p.
processus
prochlorperazine
Prochownik
P. method
P. pessary
procidentia uteri
procollagenase
PRO/Covers ultrasound probe sheath
procreation
assisted medical p. (AMP)
proctitis
proctocolpoplasty
ProctoCream-HC
proctoelytroplasty
proctoepisiotomy
ProctoFoam-HC
ProctoFoam NS
proctography
evacuation p.
proctosigmoiditis
proctosigmoidoscopy
proctotomy
procyclidine
Procytox
Pro-Depo
prodromal labor
Prodrox
prodrug
product
P. 80056
alpha-1 thymosin p.
blood p.
clearance of fetal p.
p.'s of conception (POC)
fibrin degradation p.
gene p.
processed blood p.
Repliform alternative p.
production
oocyte p.
Pro-Duosterone
ProDynamic monitor
prodynorphin
proencephalus
proenkephalin
p. A, B

NOTES

P

proenzyme
Profasi HP
profenamine
proficiency
 Bruininks-Oscretsky Test of
  Motor P.
Profilate
 P. OSD
 P. SD
profile
 biometric p.
 biophysical p. (BPP)
 coagulation p.
 fetal biophysical p.
 lung p.
 P. of Mood States (POMS)
 P. of Mood States score
 protein p.
 urethral pressure p.
 urethral pressure cough p.
 velocity flow p.
 P. viral probe test
profiling
 multiple arbitrary amplicon p.
Profilnine SD
Profilomat monitor
profilometry
 urethral p.
 urethral pressure p.
profunda
 miliaria p.
PRO/Gel ultrasound transmission gel
progenitor
 fetal erythroid p.
progeny
progeria
progeria-like syndrome
progeroid
 p. short stature-pigmented nevi
  syndrome
Progestasert
 P. intrauterine device
progestational
 p. activity
 p. agent
 p. challenge
 p. protection
 p. therapy
progesterone
 p. antagonist
 p. breakthrough bleeding
 p. challenge test
 p. dermatitis
 first-generation p.
 p. metabolism
 micronized p.
 p. myometrial level

P. Oil
parenteral p.
p. plasmatic
P. Radioimmunoassay Kit
p. receptor (PgR, PR)
p.-releasing T-shaped device
second-generation p.
p. secretion
serum p.
p. synthesis
third-generation p.
urinary free p.
p. withdrawal bleeding
progestin
 C_{21} p.
 p.-impregnated vaginal ring
 norgestimate p.
 p.-only pill (POP)
 p. oral contraceptive
progestogen
 C_{21} p.
 p. support therapy
progestogen-dependent endometrial
 protein (PEP)
Proglycem
prognathism
prognosis
 clinical p.
prognostic
 p. factor
 p. indicator
 p. scoring system
prognostication
prognosticator
Prognosticon Dri-Dot
program
 Early and Periodic Screening,
  Diagnosis, and Treatment p.
  (EPSDT)
 FRIA muscle training device & p.
 Metropolitan Atlanta Congenital
  Defects P. (MACDP)
 Positive Reinforcement In the
  Menopausal Experience patient-
  support p.
 PRIME patient support p.
 PSI p.
 Restore p.
 STEPS p.
progress
 failure to p.
progression
 intraepithelial disease p.
 tumor p.
progressiva
 fibrodysplasia ossificans p. (FOP)
 myositis ossificans p.

progressive

    p. bulbar palsy with perceptive deafness

    p. encephalopathy, edema, hypsarrhythmia, optic atrophy (PEHO)

    p. encephalopathy, edema, hypsarrhythmia, optic atrophy syndrome

    p. familial scleroderma

    p. muscular dystrophy of childhood

    p. systemic sclerosis

project

    Breast Cancer Detection Demonstration P. (BCDDP)

    Diana P.

    genome p.

    Human Genome P.

    National Collaborative Diethylstilbestrol Adenosis P. (DESAD)

    National Surgical Adjuvant Breast P. (NSABP)

projectile vomiting

prokaryote

prolactin (PRL)

    p. deficiency

    p. inhibiting factor (PIF)

    p. level

    plasma p.

    p. regulation

    p. releasing factor (PRF)

    p. secretion

    serum p.

    p. stimulation

    p. suppression

prolactinoma

    bromocriptine-resistant p.

    estrogen-induced p.

prolactin-secreting

    p.-s. adenoma

    p.-s. macroadenoma

prolamine

prolan

prolapse

    cervical p.

    p. of corpus luteum

    fallopian tube p.

    fimbrial p.

    first degree p.

    genital p.

    intrapartum cord p.

    massive genital p.

    mitral valve p.

    occult cord p.

    second degree p.

    p. stage I–IV

    third degree p.

    umbilical cord p.

    p. of umbilical cord

    urethral p.

    uterine p.

    uterovaginal p.

    p. of uterus

    vaginal p.

    vaginal stump p.

    vaginal vault p.

    vault p.

prolapsed uterus

Prolastin

Prolene suture

proliferation

    bFGF-stimulated cell p.

    endometrial p.

proliferative

    p. histology

    p. phase

proline

    p. aminopeptidase activity

prolinemia

    encephalopathy with p.

Prolixin

Proloid

prolonged

    p. delivery hospitalization

    p. gestation

    p. labor

    p. pregnancy

    p. Q-T syndrome

    p. rupture

    p. rupture of membranes (PROM)

Proloprim

PROM

    prelabor rupture of the membranes

    premature rupture of membranes

    prolonged rupture of membranes

Promensil

prometaphase

promethazine

Prometh Injection

Prometrium

prominence

    Rokitansky p.

NOTES

P

prominent
    p. ductal pattern
    p. incisors-obesity-hypotonia
      syndrome
    p. quadrigeminal plate cistern
promontory
    sacral p.
promoter
pronate
pronatis
Pronema domesticum
pronephros
Pronestyl
Pronto Shampoo
pronuclear embryo
pronucleate
    p. stage embryo transfer (PROST)
    p. stage tubal transfer (PROST)
pronuclei
pronucleus (PN)
PROOF
proopiomelanocortin (POMC)
prooxyphysin
Propacil
Propaderm
Propagest
propantheline bromide
proparacaine
prophase
prophylactic
    p. antibiotic
    p. aspirin use
    p. chemotherapy
    p. culdoplasty
    p. heparin
    p. immunization
    p. isoniazid
    p. mastectomy
    p. medication
    p. oophorectomy
    p. red-cell transfusion
    p. tetracycline
prophylaxis
    antibiotic p.
    aspiration p.
    intrapartum antibiotic p.
    neonatal ocular p.
    ocular p.
    SBE p.
    silver nitrate eye p.
    vitamin K p.
propidium iodide
propionate
    beclomethasone p.
    clobetasol p.
propionica
    Actinomyces p.
Pro-Piroxicam

Proplex T
propositus
propoxyphene
propping reflex
propranolol hydrochloride
propressophysin
propria
    lamina p.
proptosis
propylene glycol diacetate
propylthiouracil
Propyl-Thyracil
prorenin
    plasma p.
prorenin-renin-angiotensin system
Prorex Injection
Proscar
Prosed/DS
ProSobee formula
prosogaster
prosopoanoschisis
prosopopagus
prosoposchisis
prosoposternodymus
prosopothoracopagus
PROST
    pronucleate stage embryo transfer
    pronucleate stage tubal transfer
prostacyclin (PGI$_2$)
    p. assay
    p. deficiency
prostaglandin (PG)
    p. biosynthesis
    p. E (PGE)
    p. E$_2$ (PGE$_2$)
    p. E analogue
    p. E$_2$ gel
    p. F$_2$
    p. F$_{2\alpha}$ (PGF$_{2\alpha}$)
    p. G (PGG)
    p. H (PGH)
    p. I$_2$
    intravaginal p.
    p. metabolism
    15-methyl p.
    one-hour p.
    p. pessary
    p. suppository
    p. synthase
    p. synthesis inhibition
    p. synthetase inhibitor (PGSI)
    vaginal p.
    vasoactive p.
prostanoid
    cyclooxygenase-2-derived p.
Prostaphlin
Prostasert
prostate-specific antigen (PSA)

prostatic utricle
Pro-Step hCG
prosthesis, pl. **prostheses**
    Becker breast p.
    breast p.
    Introl bladder neck support p.
    penile p.
    vaginal prolapse p.
    valvular p.
Prostigmin
Prostin
    P. E2 Vaginal Suppository
    P. F2 alpha
    P./15M
protamine
    p. insulin zinc suspension
protease
    p. inhibitor (PI)
    vitamin K-dependent serine p.
ProtectaCap cap
Protectaid
    P. contraceptive sponge
    P. contraceptive sponge with F-5
      gel
protection
    progestational p.
proteiform syndrome
protein
    alpha dimeric p.
    binding p.
    p. binding
    carrier p.
    p. C deficiency
    cow's milk p. (CMP)
    C-reactive p. (CRP)
    dimeric p.
    E-cadherin p.
    endometrial p.
    feto-neonatal estrogen-binding p.
    follicle regulatory p. (FRP)
    G p.
    gonadotrophin-releasing
      hormonelike p.
    H-*ras* p21 p.
    I IFG-binding p.
    KAL p.
    p. kinase C (PKC)
    lipoprotein receptor-related p. (LRP)
    p. marker
    mu dimeric p.
    myeloma p.
    ovine trophoblast p.-1

    p. phosphorylation
    pi dimeric p.
    placental p.
    plasma p.
    pregnancy-associated plasma p.
    pregnancy-specific p.
    pregnancy zone p.
    p. profile
    progestogen-dependent
      endometrial p. (PEP)
    pulmonary microvascular
      permeability to p. (PMVP)
    receptor-associated p. (RAP)
    Schwangerschafts p. 1
    p. S deficiency
    p. solder
    p. standard
    Tamm-Horsfall p.
    theta dimeric p.
    thrombus precursor p. (T_pP^T)
    thyroxine-binding p. (TBP)
    transport-associated p. (TAP)
    vitamin D-binding p.
    zona p.
proteinase
proteinosis
    lipoid p.
proteinuria
    gestational p.
    glomerular p.
    orthostatic p.
    postural p.
    tubular p.
proteoglycans
proteolysis
proteolytic enzyme
Proteus
    P. mirabilis
    P. morganii
    P. syndrome (PS)
Prothazine
prothrombin[G20210A]
prothrombin time (PT)
prothrombokinase
ProTime microcoagulation system
protoblast
protocol
    Bagshawe p.
    modified Bagshawe p. (MBP)
    rape p.
    wean-and-feed p.
Protocult stool sampling device

NOTES

P

protogaster
proton
protooncogene
    c-*erb* B-2 p.
    c-fms p.
    fes p.
    fgr p.
    fyn p.
    hck p.
    HER-2/neu p.
    lck p.
    lyn p.
    mos p.
    raf p.
    RET p.
    scr p.
    yes p.
Protopam
protoporphyria
Protostat
protraction
Pro-Trin
protriptyline
Protropin
protuberans
    dermatofibrosarcoma p.
Proud syndrome
Proval
Proventil
Provera
    P. Oral
    P.-Testpac
Providencia rettgeri
provocative stress test
proXeed
proxetil
    cefpodoxime p.
proximal
    p. femoral focal deficiency
    p. occlusion
    p. outflow tract
    p. tubal blockage
    p. urethra
proxy
    Münchausen syndrome by p.
    (MSBP)
Prozac
    P. pulvule
Prozine-50
PRS
    Prieto syndrome
PRTS
    Partington syndrome
prune-belly syndrome
prurigo
    p. gestationis
    p. of pregnancy

pruritic
    p. folliculitis of pregnancy
    p. urticarial papule
    p. urticarial papules and plaques
    (PUPP)
    p. urticarial papules and plaques
    of pregnancy (PUPPP)
pruritus
    p. ani
    p. gravidarum
    p. vulvae
    vulvar p.
PS
    Proteus syndrome
PSA
    Pisum sativum agglutinin
    prostate-specific antigen
    PSA test
psammoma body
P450scc
    cytochrome P.
pseudencephalus
pseudoacephalus
pseudoachondroplasia
    p. syndrome
pseudoachondroplastic dysplasia
pseudoallele
pseudoaminopterin syndrome
pseudoaneurysm
    uterine artery p.
pseudocalcification
    powder p.
pseudochromosome
pseudo-Crouzon disease
pseudocyesis
    monosymptomatic delusional p.
pseudocyst
    pancreatic p.
pseudodeciduosis
pseudodiastrophic dysplasia
pseudoephedrine
    p. hydrochloride
pseudofemale
    hairless p.
pseudogene
pseudogestational sac
pseudoglandular period
pseudoglioma congenita
pseudohermaphrodite
pseudohermaphroditism
    female p.
    male p.
pseudo-Hurler
    p.-H. polydystrophy
    p.-H. syndrome
pseudohypertrophic
    p. adult muscular dystrophy

p. muscular paralysis
p. progressive muscular dystrophy
pseudohypha, pl. **pseudohyphae**
pseudohypoparathyroidism
p. syndrome
pseudointraligamentous
pseudoleukemia
pseudolymphoma
pseudomembranous
p. colitis
p. enterocolitis
pseudomenopause
pseudomenstruation
Pseudomonas
P. aeruginosa
P. exotoxin
pseudomonoamniotic cavity
pseudomosaicism
pseudomucinous tumor
pseudomyxoma peritonei
pseudonuchal infantilism
pseudoovulation
pseudoparalysis
Parrot p.
pseudopolyp
pseudoprecocious puberty
pseudopregnancy
pseudoprogeria/Hallermann-Streiff (PHS)
p.-S. syndrome
pseudoprogeria syndrome
pseudopseudohypoparathyroidism (PPHP)
pseudopuberty
precocious p.
pseudosarcoma
botryoid p.
pseudotoxemia
pseudotoxoplasmosis syndrome
pseudotrisomy 13 syndrome
pseudotuberculosis
Yersinia p.
pseudotumor
p. cerebri
trophoblastic p.
pseudo-Turner syndrome
pseudo-Ullrich-Turner syndrome
pseudovagina
PSI
pelvic support index
postponing sexual involvement
PSI program
psittaci
Chlamydia p.

psittacosis
psoas hitch procedure
Psorcon
psoriasis
guttata p.
vulvar p.
psoriatic arthritis
Psorion Topical
PSP
phenolsulfonphthalein
PSTT
placental site trophoblastic tumor
psychiatric
p. drug
p. illness
psychoactive drug
psychogenic
p. impotence
p. pelvic pain
psychological
p. causes, excessive urine production, restricted mobility, stool impaction
p. effect
p. sex
p. stress
p. trauma
psychometric assessment
psychomotor
p. development index (PDI)
p. epilepsy
psychoprophylaxis
psychoses (*pl. of* psychosis)
psychosexual problem
psychosine lipidosis
psychosis, pl. **psychoses**
climacteric p.
gestational p.
involutional p.
postpartum p.
puerperal p.
symbiotic p.
psychosocial
p. adjustment
p. support
psychosomatic disease
psychotherapy
psychotropic drug
PT
physical therapy
physiotherapy

NOTES

P

PT *(continued)*
    primary thrombocythemia
    prothrombin time
PTA
    percutaneous transluminal angioplasty
    plasma thromboplastin antecedent
pterygium
    p. colli
    p. colli, mental retardation, digital
      anomalies syndrome
    p. colli syndrome
    congenital p.
    p. universale
pterygo arthromyodysplasia congenita
PTH
    parathyroid hormone
Pthirus pubis
PTHrP
    parathyroid hormone-related peptide
PTL
    preterm labor
ptosis
    congenital p.
    p., downslanting palpebral fissures,
      hypertelorism, seizures, mental
      retardation syndrome
    p. of eyelids-diastasis recti-hip
      dysplasia
PTT
    partial thromboplastin time
ptyalism
pubarche
    precocious p.
    premature p.
pubertal, puberal
    p. aberrancy
    p. growth
pubertas praecox
puberty
    complete precocious p.
    constitutional precocious p.
    delayed p.
    factitious precocious p.
    heterosexual precocious p.
    iatrogenic precocious p.
    idiopathic precocious p.
    incomplete precocious p.
    p. initiation
    isosexual idiopathic precocious p.
    precocious p.
    pseudoprecocious p.
    true precocious p.
pubescence
pubescent
pubic
    p. arch
    p. hair
    p. ramus

    p. symphysis
    p. symphysis periostitis
    p. triangle
    p. tubercle
pubiotomy
pubis
    arcuate ligament of p.
    mons p.
    os p.
    osteomyelitis p.
    pediculosis p.
    Pthirus p.
    ruptured symphysis p.
    symphysis p.
pubocervical fascia
pubococcygeal muscle
pubococcygeus
puborectal muscle
pubourethral
pubovaginal sling procedure
puboviseral muscle
PUBS
    percutaneous umbilical blood sampling
puddle sign
pudenda (*pl. of* pudendum)
pudendal
    p. anesthesia
    p. apron
    p. artery
    p. block
    p. canal
    p. hematocele
    p. nerve
    p. nerve terminal motor latency
    p. neurogram
pudendi
pudendum, pl. **pudenda**
    frenulum labiorum pudendi
    labium majus pudendi
    noma pudendi
Pudenz
    P. reservoir
    P. shunt
puerorum
    hydroa p.
puerpera, pl. **puerperae**
puerperal
    p. convulsion
    p. eclampsia
    p. endometritis
    p. fever
    p. hematoma
    p. infection
    p. inversion
    p. mastitis
    p. morbidity
    p. period
    p. phlebitis

p. psychosis
p. pyemia
p. sepsis
p. septicemia
p. tetanus
puerperant
puerperium, pl. **puerperia**
pug nose-peripheral dysostosis syndrome
Pulec and Freedman classification
pull-to-sit reflex
Pulmo-Aid ventilator
Pulmocare
    P. diet
    P. formula
PulmoMate nebulizer
pulmonale
    cor p.
pulmonary
    p. agenesis
    p. alveolus
    p. angiography
    p. anthrax
    p. arborization
    p. artery atresia
    p. artery banding
    p. artery catheterization
    p. artery/ductus view
    p. artery sling
    p. artery wedge (PAW)
    p. artery wedge pressure (PAWP)
    p. aspergillosis
    p. blood flow (PBF)
    p. capillary wedge pressure (PCWP)
    p. complication
    p. diffusing capacity
    p. disorder
    p. dysmaturity syndrome
    p. edema
    p. ejection click
    p. emboli
    p. embolism
    p. endometriosis
    p. eosinophilia
    p. fibrosis
    p. function
    p. function tests
    p. gangrene
    p. hemorrhage
    p. hemosiderosis
    p. histoplasmosis
    p. hypertension

p. hypoplasia
p. infarction
p. infiltrate with eosinophilia (PIE)
p. injury
p. insufficiency of prematurity
p. interstitial emphysema (PIE)
p. maturation
p. metastasis
p. microvascular permeability to protein (PMVP)
p. mucormycosis
p. murmur
p. parenchymal disease
p. plaque
p. recurrence
p. resection
p. sequestration
p. stenosis
p. suppuration
p. surfactant
p. thromboembolism
p. time constant
p. toilet
p. tuberculosis
p. valve
p. vascular bed
p. venous drainage
pulmonic
    p. murmur
    p. regurgitation
    p. stenosis/café-au-lait spots syndrome
Pulmozyme
pulsatile
    p. discharge
    p. GnRH administration
    p. human menopausal gonadotropin
    p. release
pulsatility
    p. index (PI)
    intrinsic p.
pulse
    cutaneous pressure p.
    p. frequency
    maternal p.
    p. oximeter
    p. oximetry
    sharp p.
pulsed
    p. Doppler
    p. Doppler ultrasound
    p. electromagnetic wave

NOTES

P

pulsed-wave ultrasound
pulseless disease
pulsion enterocele
pulvule
    Ilosone P.'s
    Prozac p.
pump
    Advanced Collection breast p.
    Barron p.
    Basis breast p.
    battery-operated breast p.
    bilateral breast p.
    breast p.
    Chicco breast p.
    Chid breast p.
    Clarus model 5169 peristaltic p.
    Egnell breast p.
    electric breast p.
    Elmed peristaltic irrigation p.
    Grafco breast p.
    insulin p.
    Kangaroo infusion p.
    Kendall McGaw Intelligent p.
    KM-1 breast p.
    Lactina breast p.
    manual breast p.
    Mary Jane breast p.
    Medela Dominant vacuum
      delivery p.
    Medela manual breast p.
    Medfusion 1001 syringe infusion p.
    Mityvac reusable vacuum p.
    Salem p.
    servocontrolled ventilation p.
    suction p.
    p. twins
    Unicare breast p.
    Zyklomat infusion p.
punch
    baby Tischler biopsy p.
    Baker p.
    p. biopsy
    p. biopsy forceps
    Eppendorfer biopsy p.
    Keyes dermatologic p.
    Keyes vulvar p.
    Miltex disposable biopsy p.
    Tischler-Morgan biopsy p.
    Townsend biopsy p.
    Wittner biopsy p.
puncta (*pl. of* punctum)
punctata
    chondrodysplasia p.
    chondrodystrophia congenita p.
    dysplasia epiphysealis p.
punctate
    p. epiphyseal dysplasia

punctation
punctum, pl. **puncta**
puncture
    bone marrow p.
    cisternal p.
    lumbar p. (LP)
    pericardial p.
    p. punch site
    subdural p.
    transvaginal amniotic p. (TAP)
PUND
    pregnancy, uterine, not delivered
Punnett square
PUPP
    pruritic urticarial papules and plaques
puppet children
puppetlike syndrome
PUPPP
    pruritic urticarial papules and plaques of
      pregnancy
pure
    p. gonadal dysgenesis
    p. random drift
pure-breeding
Puregon
Puri-Clens
purified
    p. hormone
    p. human factor IX
    p. urinary FSH
purine
    p. metabolism disorder
Purinethol
Puritan swab
Purkinje
    P. cell
    P. fibers
puromycin
purple toes syndrome
purpura
    alloimmune neonatal
      thrombocytopenic p.
    alloimmune thrombocytopenia p.
    anaphylactoid p.
    autoimmune thrombocytopenic p.
    p. fulminans
    p. hemorrhagica
    Henoch-Schönlein p.
    idiopathic thrombocytopenic p.
      (ITP)
    immune thrombocytopenic p. (ITP)
    Schönlein-Henoch p.
    thrombocytopenic p.
    thrombotic p.
    thrombotic thrombocytopenic p.
      (TTP)
purse
    shepherd's p.

purse-string
    p.-s. mouth
    p.-s. suture
Purtilo X-linked lymphoproliferative
  syndrome
purulent pharyngitis
Pusey emulsion
pusher
    Endo-Assist endoscopic knot p.
    heliX knot p.
    knot p.
    MetraTie knot p.
    Ranfac knot p.
pus tube
pustular varicella
pustulosa
    miliaria p.
    varicella p.
pustulosis
    Juliusberg p.
    Malassezia furfur p.
    nonfollicular p.
putrefaciens
    Alteromonas p.
putrescence
putrescine
PUV
    posterior urethral valve
Puzo method
PVC
    premature ventricular contraction
P.V. Carpine Liquifilm
PVF K
PVP
    polyvinylpyrrolidone
    PVP solution
PWM
    pokeweed mitogen
pycnodysostosis
pyelectasis
    fetal p.
pyelitis
pyelocaliectasis
pyelogram
    dragon p.
    intravenous p. (IVP)
    retrograde p.
    washout p.
pyelography
    intravenous p. (IVP)
pyelonephritis
    antepartum p.

    p. in exenteration
    hereditary interstitial p.
pyemia
    puerperal p.
pygoamorphus
pygodidymus
pygomelus
pygopagus
pyknocytosis
pyknodysostosis syndrome
Pyle
    bone age standard of Greulich
      and P.
    P. syndrome
pyloric
    p. atresia
    p. stenosis
    p. string sign
pyloromyotomy
    Ramstedt p.
pyloroplasty
    Heineke-Mikulicz p.
pylorospasm
pyocolpocele
pyocolpos
pyogenes
    Streptococcus p.
pyogenic
    p. abscess
    p. arthritis
    p. granuloma
    p. salpingitis
pyometra
pyometritis
pyomyoma
    uterine p.
pyoovarium
Pyopen
pyophysometra
pyosalpingitis
pyosalpingo-oophoritis
pyosalpingo-oothecitis
pyosalpinx
pyramidal
pyrantel pamoate
pyrazinamide
pyrethrin
    p.'s and piperonyl butoxide
pyrexia
    maternal p.
    procedure-related p.
Pyribenzamine

NOTES

P

431

Pyridiate
Pyridium
pyridostigmine
    p. bromide
pyridoxine
pyridoxine-responsive anemia
pyrilamine
    p. maleate
Pyrilinks-D
    P.-D. assay
pyrimethamine
pyrimidine
Pyrinex Pediculicide Shampoo
Pyrinyl
    P. II Liquid
    P. Plus Shampoo

pyrogen
    endogenous p.
pyroglutamic acidemia
pyropoikilocytosis
pyrosis
pyruvate
    p. decarboxylase
    p. dehydrogenase complex
      deficiency (PDHC)
    p. kinase (PK)
    p. kinase deficiency
pyrvinium pamoate
PYtest for *Helicobacter pylori*
pyuria
PZD
    partial zona dissection

Q
  quotient
  Q band
  Q fever
Q506
  VF Q506
Q
  blood flow
q
  long arm of chromosome
21q
  tetrasomy 21q
Qa (series of loci)
QCT
  quantitative computed tomography
QDR
  quantitative digital radiography
Qf
  rate of fluid filtration
QIAamp Tissue kit
QNS
  quantity not sufficient
qr
  quadriradial
QRS
  Q. complex
  Q. morphology
Qs/Qt
  intrapulmonary shunt ratio
  right-to-left shunt ratio
Q-test ovulation predictor
Qtest Strep test
Q-T interval
Q-tip test
quadrantectomy
  q., axillary dissection, radiation
    therapy (QUART)
quadratum
  caput q.
quadriplegia
quadriradial (qr)
quadruplet
quality assurance process
Quan-Smith syndrome
Quantikine human IL-6 Immunoassay
quantitation
  amniotic fluid q.
quantitative
  q. computed tomography (QCT)
  q. digital radiography (QDR)
  q. inheritance
  q. trait locus
quantity not sufficient (QNS)
quarantine

QUART
  quadrantectomy, axillary dissection,
    radiation therapy
quarter-strength formula
Quarzan
quasicontinuous inheritance
quasidiploid
quasidominance
Queensland fever
Quest Diagnostics
questionnaire
  Family Apgar Q.
  Incontinence Impact Q.-Revised
    (IIQ-R)
Questran
Quetelet
  Q. body mass index
Queyrat erythroplasia
Quibron
  Q.-T
  Q.-T/SR
quickening
Quickpac-II OneStep hCG pregnancy test
Quick test
QuickVue
  Q. *Chlamydia* test
  Q. One-Step hCG-Combo test
  Q. One-Step hCG-urine test
  Q. One-Step *H. pylori* test
Quidel Group B Strep Test
quiescence
  uterine q.
QuietTrak monitor
Quik Connect fetal monitor
quinacrine
  q. banding
  q. hydrochloride
Quinaglute
quinalbarbitone sodium
Quinamm
quinate
quinestrol
quinethazone
quingestanol acetate
Quinidex Extentabs
quinidine
quinine
quinolone
  q. pharmacokinetics
Quinones-Neubüser uterine-grasping forceps
Quinones uterine-grasping forceps
Quinora

quintana
    Bartonella q.
quintessence
quintipara
Quinton dual-lumen catheter
quintuplet

quintuple-X syndrome
Quips genetic imaging system
Quisling hammer
quotidian fever
quotient (Q)
    ventilation/perfusion q.

R
radius
roentgen
  R band
  R & C Shampoo
r
ring chromosome 1–22
  r (1)–(22) syndrome
RA
rheumatoid arthritis
^{226}Ra
radium-226
rabeprazole
rabies
  r. immune globulin
  r. vaccine
Rabson-Mendenhall syndrome
racemic
  r. ephedrine
  r. epinephrine
Racephedrine
rachiopagus
rachischisis
  thoracolumbar r.
rachitic
  r. metaphysis
  r. rosary
rachitis fetalis
racial difference
RAD
reactive airway disease
rad
radiation-absorbed dose
radial
  r. aplasia-thrombocytopenia
  syndrome
  r. arterial catheter
  r. arterial line
  r. artery
  r. clubhand
  r. curettage
  r. jaw biopsy forceps
  r. ray defects, triangular face,
  telecanthus, sparse hair, dwarfism,
  mental retardation syndrome
  r. scar
  r. sclerosing lesion
radial-renal syndrome
radiant warmer
radiata
  corona r.
radiation
  adaptive r.
  r. cystitis
  r. danger

diagnostic r.
r. dose
r. dose limit
r. dysplasia
electromagnetic r.
intraoperative r.
ionizing r.
nuclear r.
orthovoltage r.
particulate r.
postoperative pelvic r.
state-of-the-art r.
supervoltage r.
r. therapy
tissue tolerance to r.
r. tolerance
whole abdominal r.
radiation-absorbed dose (rad)
radical
  r. excision
  free hydroxyl r.
  r. hysterectomy
  r. mastectomy
  oxygen-free r.
  r. parametrectomy
  r. surgery
  r. vaginal trachelectomy
  r. vulvectomy
radioactive
  r. applicator
  r. cobalt
  r. colloid
  r. gold
  r. immunoassay
  r. implant
  r. isotope
  r. microsphere
  r. ribbons
  r. seed
  r. seed implantation
  r. tracer
  r. uptake
radioallergosorbent test (RAST)
radiocontrast material
radiocurable
radiofibrinogen uptake scan
radiographic absorptiometry
radiography
  chest r.
  digital r.
  quantitative digital r. (QDR)
radioimmunoassay (RIA)
  solid phase r.
radioimmunodetection (RAID)

radioimmunoscintigraphy
radioimmunosorbent test (RIST)
radioiodination
radioiodine
radioisotope
radiolabeled
radiological examination
radiolucent
radiomutation
radionuclide
    r. venography
radioreceptor assay
radiorenoocular syndrome
radioresistant yolk sac tumor
radiosensitivity
radiosensitization
radiotherapy
    abdominal strip r.
    adjuvant r.
    pelvic boost r.
    postoperative r.
radioulnar
    r. synostosis
    r. synostosis-developmental
      retardation-hypotonia syndrome
    r. synostosis, short stature,
      microcephaly, scoliosis, mental
      retardation syndrome
radium
    r.-226 (^{226}Ra)
    r. implant
    intracavitary r.
RADIUS
    routine antenatal diagnostic imaging with
      ultrasound
    RADIUS trial
radius (R)
    thrombocytopenia-absent r. (TAR)
radon
    r.-222 (^{222}Rn)
Radovici
    R. reflex
    R. sign
raf protooncogene
RAID
    radioimmunodetection
Raimondi catheter
Raine syndrome
Raji cell
rale
    coarse r.
    crackling r.
    crepitant r.
    wet r.
raloxifene
    Study of Tamoxifen and R.
      (STAR)

RALPH
    renal, anus, lung, polydactyly,
      hamartoblastoma
    RALPH syndrome
      renal, anus, lung, polydactyly,
        hamartoblastoma
Rambam-Hasharon syndrome
rami (*pl. of* ramus)
ramipril
Ramon syndrome
RAMP hCG assay
Ramses
Ramsey-Hunt syndrome
Ramstedt
    R. operation
    R. procedure
    R. pyloromyotomy
ramus, pl. **rami**
    infrapubic r.
    ischiopubic r.
    pubic r.
Randall
    R. stone forceps
    R. suction curette
random
    r. mating
    r. sampling
    r. variable
randomization
randomly amplified polymorphic DNA
Ranfac knot pusher
range
    cervical tissue impedance r.
    chromatofocusing pH r.
range-gated Doppler (RGD)
ranitidine
    r. hydrochloride
RAP
    receptor-associated protein
    recurrent abdominal pain
rape
    r. evidence
    r. evidence kit
    r. protocol
    statutory r.
    r. trauma syndrome
raphe
    anococcygeal r.
    median r.
    sternoumbilical r.
rapid
    r. bone loss
    r. descent
    r. eye movement (REM)
    r. plasma reagin (RPR)
    r. plasma reagin card test
    r. slide test

Rapp-Hodgkin
   R.-H. ectodermal dysplasia
   R.-H. syndrome
rare-cutter enzyme
Rasch sign
rash
   butterfly r.
   diaper r.
   ECM r.
   malar r.
   monilial r.
   morbilliform r.
   Prevex Baby Diaper R.
   slapped cheek r.
Rashkind balloon
Rasmussen encephalitis
RAST
   radioallergosorbent test
rat-bite fever
rate
   aldosterone excretion r. (AER)
   baseline fetal heart r.
   beat-to-beat variability of fetal
      heart r.
   birth r.
   contraceptive failure r.
   cumulative conception r. (CCR)
   erythrocyte sedimentation r. (ESR)
   fertility r.
   fetal death r.
   fetal heart r. (FHR)
   flow r.
   r. of fluid filtration (Qf)
   glomerular filtration r. (GFR)
   heart r.
   infant mortality r.
   maternal death r.
   maternal mortality r.
   metabolic clearance r. (MCR)
   mortality r.
   neonatal mortality r.
   ovulation r.
   peak expiratory flow r. (PEFR)
   perinatal mortality r. (PMR,
      PNMR)
   prevalence r.
   recurrence r.
   sedimentation r.
   seroprevalence r.
   sinusoidal fetal heart r.
   sinusoidal heart r. (SHR)

   survival r.
   ventricular response r.
Rathke pouch
rating
   Apgar r.
ratio
   A/G r.
      albumin-globulin ratio
   albumin-globulin r. (A/G ratio, A/G
      ratio)
   amylase-creatinine clearance r.
   cerebral-placental r.
   cough-pressure transmission r.
   crude risk r.
   estrogen/progesterone r.
   fetal head:abdominal
      circumference r.
   head:body r.
   head circumference/abdominal
      circumference r.
   I/E r.
   intrapulmonary shunt r. (Qs/Qt)
   I/T r.
   lecithin and sphingomyelin r.
   lecithin/sphingomyelin r.
   left atrial to aortic r. (LA:A)
   L/S r.
   M/E r.
   milk-plasma r.
   odds r. (OR)
   right-to-left shunt r. (Qs/Qt)
   RVPEP/RVET r.
   S/D r.
   sex r.
   standardized incidence r. (SIR)
   systolic/diastolic r.
   testosterone/dihydrotestosterone r.
   umbilical velocity r.
   urinary lactate:creatinine r.
   vaginal pool L/S r.
   variance r.
   ventilation/perfusion r.
   V̇/Q̇ r.
   waist:hip r.
rat tooth pickups
Rauber layer
Raussly disease
Raxar
ray
   beta r.
   gamma r.
Ray-Tec sponge

NOTES

R

Raz
    R. bladder neck suspension
    R. double-prong ligature carrier
    R. sling
    R. sling procedure
Raz-Leach procedure
Razoxane
RBM **gene**
RBRVS
    resource-based relative value scale
RCA test
RD1000 resuscitator
RDF
    Adriamycin RDF
RDS
    respiratory distress syndrome
Reactine
reaction
    acetowhite r.
    acrosome r.
    allele polymerase chain r.
    anorectic r.
    arbitrarily primed polymerase
        chain r.
    Arias-Stella r.
    Arthus r.
    Chlamydia trachomatis ligase
        chain r.
    cortical r.
    decidual r.
    graft-versus-host r.
    harlequin r.
    intrusive stress r.
    inverse polymerase chain r.
    Jarisch-Herxheimer r.
    laser r.
    mixed agglutination r. (MAR)
    polymerase chain r. (PCR)
    Porter-Silber r.
    Shwartzman r.
    single primer amplification r.
    startle r.
    Staudinger r.
    vagal r.
    wheal-and-flare r.
    zona r.
reactive
    r. airway disease (RAD)
    r. oxygen species
    r. site (RS)
    r. site mutation
reactivity
    fetal heart rate r.
    vascular r.
reader
    AutoPap r.
    PapNet r.
reading frame

reagin
    rapid plasma r. (RPR)
Reality vaginal pouch
real-time
    r.-t. B-scanner
    r.-t. imaging on ultrasound
    r.-t. sonography
    r.-t. ultrasonography
    r.-t. ultrasound
reanastomosis
    tubocornual r.
rearing parents
rearrangement
    balanced chromosome r.
    de novo balanced chromosome r.
REAR syndrome
rebound
    r. dermatitis
    r. phenomenon
recalcitrant condyloma
recall phenomenon
Récamier
    R. operation
    R. procedure
recanalization
    tubal r.
receptor
    activated estrogen r.
    adrenergic r.
    alpha-adrenergic r.
    androgen r.
    r. assay
    beta-adrenergic r.
    calcitonin r.
    class I, II r.
    r. coupling
    r. cross talk
    endogenous opiate r.
    endometrial r.
    E-rosette r.
    estradiol r.
    estrogen r. (ER)
    fibroblast growth factor r. (FGF-R)
    G-protein-coupled r.
    hormone complex r.
    human EP1 r.
    r. internalization
    J pulmonary r.
    opiate r.
    progesterone r. (PgR, PR)
    p55 soluble tumor necrosis
        factor r.
    p75 soluble tumor necrosis
        factor r.
    steroid hormone r.
    type A IL-8 r.
    type B IL-8 r.

tyrosine kinase r.
vitamin D r. (VDR)
receptor-associated protein (RAP)
receptor-mediated endocytosis
recessive
r. allele
autosomal r. (AR)
r. deafness-onychodystrophy
  syndrome
r. disorder
r. enhanced S-cone syndrome
r. fundus albipunctatus
r. gene
r. inheritance
r. Leber congenital amaurosis
peroneal atrophy, X-linked r.
r. trait
r. Usher syndrome
rechallenge
reciprocal
r. cross
r. gene
r. translocation
Recklinghausen
R. disease type I, II
R. tumor
Reclomide
Reclus disease
recoil
arm r.
Recombigen assay
recombinant
r. antihemophilic factor
r. chromosome 8 syndrome
r. clone
r. DNA
r. DNA molecule
r. DNA technique
r. DNA technology
r. factor VIIa
r. follicle-stimulating hormone
(rFSH)
r. FSH preparation
r. hepatitis B immunization series
r. human growth hormone
r. immunosorbent assay (RIBA)
r. inbred strain
r. interleukin 2 (rIL-2)
r. substitution line
r. tissue type plasminogen activator
(rt-PA)
recombinase

Recombinate
recombination
r. fraction
r. frequency
homologous r.
r. map
Recombivax HB
recommendation
screening r.
recommended dietary allowance
reconstruction
Abbe-McIndoe vaginal r.
Dibbell cleft lip-nasal r.
transvaginal ultrasound-guided
  urethral r.
uterine r.
Young-Dees-Leadbetter bladder
  neck r.
reconstructive mammaplasty
record
bring-your-own medical r.
daily fetal movement r.
Infant Behavior R. (IBR)
women-held antenatal r.
recorder
Dopcord r.
multichannel r.
recovery
bacterial r.
labor, delivery, and r. (LDR)
oocyte r.
r. score
r. time
ultrasonic egg r.
recreational drug
rec (8) syndrome
rectal
r. cancer
r. examination
r. injury
r. linitis plastica (RLP)
r. node
Phenergan R.
r. pillar
r. postradiation ulcer
r. suppository
r. temperature
recti
diastasis r.
rectocele
r. repair
rectolabial fistula

NOTES

rectosacral space
rectoscopic endometrial ablation
rectosigmoid
    r. colon
    r. colon endometriosis
rectosigmoidectomy
    Altemeier perineal r.
rectouterine
    r. cul-de-sac
    r. fold
    r. pouch
rectovaginal
    r. examination
    r. fistula
    r. septum
rectovestibular fistula
rectovulvar fistula
rectrovaginal space
rectum
rectus
    r. abdominis muscle
recurrence
    pelvic r.
    pulmonary r.
    r. rate
    r. risk
recurrent
    r. abdominal pain (RAP)
    r. anaphylaxis
    r. aneuploidy
    r. candida
    r. carcinoma
    r. euploidic abortion
    r. infection
    r. jaundice of pregnancy
    r. lymphoma
    r. miscarriage
    r. molar pregnancy
    r. pregnancy loss
    r. premenstrual lithium serum
      concentration decline
    r. spontaneous abortion (RSA)
    r. vaginitis
recurvatum
    genu r.
    pectus r.
red
    r. blood cell
    r. cell antigen
    r. cell enzyme deficiency
    r. cell phosphoglycerate kinase
      deficiency
    r. cell volume
    r. degeneration
    r. reflex
    r. rubber catheter
Reddick-Saye forceps

Redi-kit
    LEEP R.-k.
Redisol
RediTab
    Claritin R.
reduced hemoglobin
reducible hernia
reductase
    5α-r.
    dihydrofolate r.
    17-ketosteroid r. deficiency
    methylene tetrahydrofolate r.
      (MTHFR)
reduction
    chromosome r.
    embryo r.
    fetal r.
    funic r.
    Lejour-type modified breast r.
    limb r.
    r. mammaplasty
    multifetal pregnancy r.
    pregnancy r.
    selective transvaginal embryo r.
    weight r.
redundant hymen
Redux capsule
Reed syndrome
reel foot
referred pelvic pain
Refetoff syndrome
reflectance
    r. photometer
    r. pulse oximetry
reflection
    peritoneal r.
reflex
    acoustic blink r.
    anal r.
    asymmetric tonic neck r. (ATNR)
    autonomic walking r.
    Babinski r.
    Babkin r.
    bowing r.
    bregmocardiac r.
    Breuer-Hering inflation r.
    bulbocavernous r.
    cochleopalpebral r.
    crossed adductor r.
    crossed extension r.
    darwinian r.
    dazzle r.
    deep tendon r. (DTR)
    diving r.
    doll's eye r.
    embrace r.
    fencing r.
    Ferguson r.

forced grasping r.
gag r.
Gallant r.
Gamper bowing r.
gastrocolic r.
Gerace r.
grasp r.
Grünfelder r.
Head r.
Hering-Breuer r.
r. HPV test
r. irritability
labial r.
labyrinthine r.
Landau r.
letdown r.
lip r.
Magnus and de Kleijn tonic
  neck r.
milk ejection r.
Moro r.
nasolabial r.
neck-righting r.
oculocephalic r.
oculocephalogyric r.
palm-chin r.
palmomental r.
parachute r.
Peiper r.
Perez r.
placing r.
plantar r.
primitive r.
propping r.
pull-to-sit r.
Radovici r.
red r.
rooting r.
sacral r.
snout r.
startle r.
stepping r.
sucking r.
swallow r.
r. sympathetic dystrophy (RSD)
tonic neck r.
white pupillary r.
reflexic apnea
reflexology
reflux
esophageal r.
r. esophagitis

gastroesophageal r. (GER)
intrarenal r.
menstrual r.
r. nephropathy
ureterovesical r.
vesicoureteral r.
refractile body
refractory anemia of pregnancy
Refsum
R. disease
R. syndrome
Regenbogen-Donnai syndrome
regeneration
tissue r.
regimen
downregulation r.
feeding r.
monophasic r.
platinum-based r.
Pritchard intramuscular r.
Yuzpe r.
Zuspan r.
region
flanking r.
regulatory r.
sequence characterized amplified r.
sex-determining r. (SRY)
regional
r. analgesia
r. anesthesia
r. enteritis
register linkage study
registry
Herbst r.
Teratogen R.
Regitine
Reglan
Regonol
regression
change-point r.
Galton law of r.
r. of the mean
premature luteal r.
trophoblast in r. (TIR)
tumor r.
regressive infantilism
regular
R. (Concentrated) Iletin II U-500
R. Iletin I, II
r. purified pork insulin
regulation
Baby Doe r.'s

NOTES

regulation *(continued)*
gonadotropin r.
menstrual cycle r.
physiologic follicular r.
pituitary gonadotropin r.
plasma volume r.
prolactin r.
regulator
r. gene
Medela membrane r.
regulatory
autoimmune r. (AIRE)
r. region
r. sequence
regurgitation
aortic r.
mitral r.
pulmonic r.
tricuspid r.
Regutol
Rehydralyte
R. formula
Reichel cloacal duct
Reid sleeve
Reifenstein syndrome
Reiki
reimplantation
Reiner-Beck snare
Reiner-Knight forceps
reinfusion
autologous bone marrow r.
Reinke
crystalloids of R.
R. crystalloids
reinsemination
reinsufflated
Reis-Wertheim vaginal hysterectomy
Reiter syndrome
rejection
fetal r.
ReJuveness
relapsing fever
related
alcohol r. (AR)
relation
Frank-Starling r.
relationship
avuncular r.
Poiseville-Hagen r.
relative
r. osteopenia
r. risk
r. sterility
relaxant
muscle r.
relaxation
pelvic r.

pelvic girdle r. (PGR)
uterine r.
relaxin
intracervical purified porcine r.
r. serum level
vaginal recombinant human r.
release
R. catheter
colonoscopic r.
GnRH-facilitated FSH r.
GnRH-facilitated LH r.
growth hormone r.
neurotransmitter r.
pituitary hormone r.
pulsatile r.
reliance
R. urinary control insert
R. urinary control insert catheter
relief
Fleet Pain R.
REM
rapid eye movement
REM PolyHesive II patient return
electrode
remission
complete r.
induced r.
spontaneous r.
remnant
branchial cleft r.
removal
extracorporeal CO_2 r. (ECOR)
Renaissance spirometry system
renal
r. acidification
r. agenesis
r., anus, lung, polydactyly,
hamartoblastoma (RALPH
syndrome, RALPH, RALPH
syndrome)
r., anus, lung, polydactyly,
hamartoblastoma syndrome
r. artery stenosis
r. biopsy
r. calculi
r. cell carcinoma
r. clearance test
r. colic
r. cortex
r. cortical necrosis
r. disease
r. duplication
r. dysgenesis
r. dysplasia
r. ectopia
r. failure
r. fistula
r. function

r. function test
r. hyperechogenicity
r. hypertension
r. infarction
r. insufficiency
r. malrotation
r. manifestation
r. mesangial sclerosis-eye defects
  syndrome
r. osteodystrophy
r. parenchyma
r. perfusion
r. plasma flow
r. stone
r. transplantation
r. tuberculosis
r. tubular acidosis (RTA)
r. tubular function
r. tubular necrosis
r. vascular thrombosis
r. vein thrombosis
Rendu-Osler-Weber syndrome
Renese
reniform pelvis
renin
  r.-angiotensin-aldosterone system
  r.-angiotensin system
  r. substrate
Renografin
Renoquid
Renpenning syndrome
REO
respiratory enteric orphan
  REO virus
Reovirus
repair
  abdominal paravaginal r.
  anterior and posterior r.
  A&P r.
  cystocele r.
  Danus-Stanzel r.
  defect-specific r.
  episiotomy r.
  LeFort-Wehrbein-Duplay
    hypospadias r.
  Macer abdominal cystocele r.
  Magpi hypospadius r.
  Noble-Mengert perineal r.
  Orr rectal prolapse r.
  paravaginal cystocele r.
  perineal r.
  Phaneuf-Graves r.

posterior r.
rectocele r.
Senning r.
sphincter r.
surgical r.
vaginal wall r.
vesicovaginal r.
York-Mason r.
Zancolli clawhand deformity r.
repeat
  r. cesarean section
  r. curettage
  inter-simple sequence r.
  inverted r.
  r.-loop excision
  simple sequence r.
repeated
  r. abortion
  r. complete mole
repermeabilization
repetitive
  r. DNA
  r. pregnancy loss
replacement
  cephalic r.
  estrogen r.
  fluid r.
  frozen embryo r. (FER)
  heart valve r.
replenishment
Replens
  R. lubricant
replication
  DNA r.
  r. fork
replicon
Repliform alternative product
Replogle tube
report
  Janus r.
  O'Driscoll r.
  Walton r.
representative section
repressed gene
repressor gene
reproduction
  assisted r.
  vegetative r.
reproductive
  r. cycle
  r. disorder
  r. endocrinology

NOTES

reproductive *(continued)*
    r. failure
    r. function
    r. genetics
    r. history
    r. liberty
    r. mortality
    r. performance
    r. system
    r. system development
    r. technology
    r. toxin
    r. tract
    r. tract abnormality
    r. tract embryology
    r. wastage
Repronex
repulsion
requirement
    iron r.
    mineral r.
    nutrient r.
    sodium r.
    vitamin r.
Rescriptor
rescue
    abdominal r.
    ablative therapy with bone
      marrow r.
    r. cervical cerclage
    r. surfactant
    trisomic r.
research
    International Committee for
      Contraceptive R. (ICCR)
resectability
resecting intrapartum uterine wall
resection
    cystoscopic transurethral tumor r.
    en bloc r.
    endocervical r. (ECR)
    endometrial r.
    endomyometrial r. (EMR)
    laparoscopic multiple-punch r.
    laterally extended endopelvic r.
    late uterine wedge r.
    low rectal r.
    multiple-punch r.
    ovarian wedge r.
    pulmonary r.
    segmental r.
    surgical r.
    Torpin cul-de-sac r.
    transcervical r.
    transsphenoidal microsurgical r.
    wedge r.
resection/ablation
    outpatient endometrial r. (OPERA)

resectoscope
    OPERA Star tissue aspirating r.
    specialized tissue aspirating r.
      (STAR)
    USA Elite System GYN rotating
      continuous flow r.
Resercen
reserpine
reservatus
    coitus r.
reservoir
    Camey r.
    double-bubble flushing r.
    Mainz pouch urinary r.
    Ommaya r.
    Pudenz r.
    sperm r.
residronate
residual
    r. ductal tissue
    r. ovary syndrome
    r. in situ
    r. urine
    r. volume
residue
    methanol extraction r. (MER)
resin
    podophyllum r.
resistance
    activated protein C r.
    airway r.
    androgen r.
    antibiotic r.
    bromocriptine r.
    dicumarol r.
    differential vascular r.
    r. index (RI)
    insulin r.
    systemic vascular r.
    thyroid hormone r.
    total pulmonary r.
    vascular r.
resistant
    r. condyloma
    r. ovary
    r. ovary syndrome
resolution
    axial r.
    lateral r.
resonance
    nuclear magnetic r. (NMR)
resorption
    bone r.
    fetal r.
resource-based relative value scale
  (RBRVS)
Respbid

respiration
- agonal r.'s
- Bouchut r.
- Cheyne-Stokes r.
- grunting r.'s
- Kussmaul r.
- placental r.
- vicarious r.

respirator
- Ambu r.
- Bath r.
- Bear r.
- Bennett r.
- Bird Mark 8 r.
- Bourns infant r.
- Bragg-Paul r.
- Breeze r.
- Clevedan positive pressure r.
- cuirass r.
- Dann r.
- Drinker r.
- Emerson r.
- Engstrom r.
- Gill r.
- Huxley r.
- Kirmisson r.
- mechanical r.
- Med-Neb r.
- Merck r.
- Monaghan r.
- Morch r.
- Morsch-Retec r.
- Moynihan r.
- negative-pressure r.
- portable r.
- Sanders jet r.

respiratory
- r. acidosis
- r. alkalosis
- r. arrest
- r. complication
- r. depression
- r. distress
- r. distress syndrome (RDS)
- r. distress syndrome of the newborn
- r. enteric orphan (REO)
- r. enteric orphan virus
- r. insufficiency
- r. paralysis
- r. sinus arrhythmia
- r. syncytial virus (RSV)

- r. system
- r. toilet
- r. tract
- r. tract infection

response
- abnormal r.
- amnestic r.
- anamnestic r.
- antibody r.
- auditory brainstem r. (ABR)
- auditory evoked r.
- automated brainstem auditory evoked r. (ABAER)
- baroreflex r.
- brainstem auditory evoked r. (BAER)
- cortisol r.
- fetal startle r.
- immune r.
- infantile breath-holding r.
- lactation letdown r.
- maternal immune r.
- maternal inflammatory r.
- maternal vascular r.
- menstrual cycle hemodynamic r.
- plantar r.
- pressor r.
- Rh immune r.
- righting r.
- startle r.
- target organ r.
- visual evoked r. (VER)

rest
- bed r.
- cystic Walthard r.
- Marchand r.
- mesonephric r.
- pelvic r.
- Walthard cell r.
- wolffian r.

resting membrane potential
restless legs syndrome
Restore program
Restoril
restriction
- r. endonuclease
- r. endonuclease analysis
- r. enzyme
- r. enzyme cutting site
- fetal growth r. (FGR)
- r. fragment

NOTES

restriction *(continued)*
r. fragment length polymorphism (RFLP)
intrauterine growth r. (IUGR)
r. landmark genomic scanning
r. map
normotensive intrauterine growth r.
salt r.
sodium r.
restrictive dermopathy
resumption
menstrual cycle r.
resuscitation
cardiopulmonary r. (CPR)
intrauterine r.
mouth-to-mouth r.
neonatal r.
newborn r.
resuscitator
Ambu infant r.
Fisher and Paykel RD1000 r.
Laerdal r.
Penlon infant r.
RD1000 r.
Resyl
RETA
rete testis aspiration
retained
r. bladder syndrome
r. hygroscopic cervical dilator
r. menstruation
r. myoma
r. placenta
r. products of conception (RPC)
retardation
alopecia mental r. (AMR)
alopecia universalis with mental r.
alpha-thalassemia mental r. (ATR)
aniridia, ambiguous genitalia, mental r. (AGR)
r., aphasia, shuffling gait, adducted thumbs syndrome
basal ganglion disorder-mental r. (BGMR)
Belgian type mental r.
Buenos Aires type mental r.
cataract, hypertrichosis, mental r. (CAHMR)
Charcot-Marie-Tooth syndrome, X-linked type II with deafness and mental r.
congenital progressive muscular dystrophy with mental r.
r., deafness, microgenitalism syndrome
deafness, onychodystrophy, osteodystrophy, r.

deafness, onychoosteodystrophy, mental r. (DOOR)
fetal growth r.
fragile site mental r. 1 (FMR1)
fragile site mental r. 2 (FMR2)
fragile X mental r.
FRAXE-associated mental r.
growth r.
hemoglobin H related mental r.
hereditary motor sensory neuropathy II-deafness-mental r. (MSN)
infantile spasms with mental r.
intrauterine growth r. (IUGR)
mental r.
motor-sensory neuropathy, X-linked Type II, with deafness and mental r.
multiple exostosis mental r. (MEMR)
nonprogressive cerebellar disorder with mental r.
peripheral dysostosis, nail hypoplasia, mental r. (PNM)
Wilms tumor, aniridia, gonadoblastoma, mental r. (WAGR)
X-linked mental r. (MRX1, XLMR)
X-linked mental r.-fragile site 1 (FRAXE1)
X-linked mental r.-fragile site 2 (FRAXE2)
X-linked mental r. 8 (MRX8, XLMR8)
retardation-growth
retarded fetal growth
rete
r. cord
r. cyst of ovary
r. testis
r. testis aspiration (RETA)
retention
r. enema
fetal r.
r. polyp
r. suture
urinary r.
Rethoré syndrome
reticular dermis
reticulare
magma r.
reticularis
adrenal r.
reticulocyte count
reticulocytosis
reticuloendothelial system
reticuloendotheliosis

reticulum
    endoplasmic r. (ER)
Retin-A
retinae
    dysgenesis neuroepithelialis r.
    hereditary epithelial dysplasia of
      the r.
retinal
    r. angioma
    r. aplasia
    r. change
    r. detachment
    r. pigmentary degeneration,
      microcephaly, mental retardation
      syndrome
retinitis
    gravidic r.
    neuropathy, ataxia, and r.
      pigmentosa (NARP)
    r. pigmentosa
    r. pigmentosa-congenital deafness
      syndrome
retinoblastoma
retinoblastoma-mental retardation
  syndrome
retinoic
    r. acid
    r. acid embryopathy
retinoid
retinol
retinopapillitis of prematurity
retinopathy
    diabetic r.
    eclamptic r.
    gravidic r.
    r. of prematurity (ROP)
    r. punctata albescens
    toxemic r. of pregnancy
retinopathy-mental retardation syndrome
retinoschisis
retocolysis
retort-shaped sactosalpinx
RET protooncogene
retraction
    intercostal r.'s
    nipple r.
    r. ring
    subcostal r.'s
    r. syndrome
retractor
    Airlift balloon r.
    Allport r.

Army-Navy r.
Aufricht nasal r.
Balfour r.
Berkeley-Bonney r.
bladder r.
Bookwalter r.
Breisky-Navratil r.
Brown uvula r.
Cer-View lateral vaginal r.
Cottle-Neivert r.
Deaver r.
DeLee Universal r.
Eccentric Y r.
eXpose r.
Ferris Smith-Sewall r.
Gazayerlie endoscopic r.
Gelpi-Lowrie r.
Gelpi perineal r.
Gott malleable r.
Goulet r.
Haight baby r.
Harrington r.
Heaney-Hyst r.
Heaney-Simon r.
Iron Intern r.
Jackson right-angle r.
Kelly r.
lateral wall r.
Latrobe r.
long atraumatic r.
Luer r.
Magrina-Bookwalter vaginal r.
malleable r.
Murless head r.
r. neuropathy
O'Connor-O'Sullivan r.
Omni-Tract vaginal r.
O'Sullivan-O'Connor r.
pediatric self-retaining r.
Richardson r.
right-angle r.
Roberts thumb r.
Schuknecht r.
self-retaining r.
Senn-Dingman r.
Shambaugh r.
Sims r.
thumb r.
vaginal r.
Weitlaner r.
Wullstein r.

R

NOTES

retraining
 bladder r.
 r. diary
retrieval
 egg r.
 oocyte r.
 ovarian r.
 transvaginal ultrasound-directed
  oocyte r. (TUDOR)
 ultrasound-directed egg r.
retrobulbar neuritis
retrocecal hernia
retrocervical
retrocession
retroconversion
 chemotherapeutic r.
retrodeviation
retrodisplacement
retroesophageal abscess
retroflexion, retroflection
 uterine r.
retrognathia
retrograde
 r. ejaculation
 r. intussusception
 r. menstrual flow
 r. menstruation
 r. pyelogram
retrolental fibroplasia
retromammary mastitis
retroorbital lymphoma
retroperitoneal
 r. drain
 r. endometriosis
 r. fetus
 r. hematoma
 r. infection
 r. lymphadenectomy
 r. node
 r. soft tissue
 r. soft tissue sarcoma
retroperitoneum
retropharyngeal abscess
retroposed
retroposition
retropubic
 r. colpourethrocystopexy
 r. cystourethropexy
 r. fibrosis
 r. sling
 r. suspension procedure
 r. urethropexy
 r. vesicourethrolysis
retrosternal hernia
retrotonsillar abscess
retroversioflexion
retroversion
 uterine r.

retroverted
 r. uterus
Retrovir
retroviral disease
retrovirus
rettgeri
 Providencia r.
Rett syndrome
return
 partial anomalous pulmonary
  venous r. (PAPVR)
 total anomalous pulmonary
  venous r.
returning-soldier effect
Retzius
 space of R.
Reuter tube
ReVele
reversal
 dosage-sensitive sex r. (DSS)
 gender r.
 hyperandrogenism r.
 sex r.
 vasectomy r.
reverse
 r. banding
 r. chylous syndrome
 r. dot blot sequence-specific
  oligonucleotide method
 r. FISH cytogenetic technique
 r. form McRoberts maneuver
 r. genetics
 r. nostrils
 r. transcriptase
 r. transcription
 r. transcription polymerase chain
  reaction (PCR)
 r. triiodothyronine
reverse-dot-blot-hybridization
 colorimetric r.-d.-b.-h.
reverse-last shoes
ReVia
revision
 Dibbell cleft lip-nasal r.
Revital-Ice rehydrating freezer pop
Reye syndrome
Reynell Development Scale
Reynolds number
Rezulin
RF
 White class, B through RF
RFA
 right frontoanterior position
RFLP
 restriction fragment length polymorphism
RFP
 right frontoposterior position

rFSH
recombinant follicle-stimulating hormone
RFT
right frontotransverse position
RGD
range-gated Doppler
R-Gel
Rh
rhesus
Rh antibody
Rh antigen
Rh blood group
Rh blood group system
Rh disease
Rh factor
Gamulin Rh
Rh hemolytic disease
Rh immune globulin (RhIg)
Rh immune response
Rh immunization
Rh immunoglobulin
Rh incompatibility
Rh isoimmunization
Mini-Gamulin Rh
Rh-negative
Rh negative antibody
Rh-positive
Rh positive antibody
Rh sensitization
rhabdocrania
rhabdomyoblast
rhabdomyoma
cardiac r.
rhabdomyosarcoma
embryonal r.
rhabdosphincter
rhagades
periorificial r.
RhC
rhesus antigen C
RhCE gene
rhesus CE gene
RhD
rhesus antigen D
RhD gene
$Rh_0(D)$ imaging
RhE
rhesus antigen E
rheometer
rhesus (Rh)
r. antibody
r. antigen

r. antigen C (RhC)
r. antigen D (RhD)
r. antigen E (RhE)
r. CE gene (RhCE gene, RhCE gene)
r. D disease
r. D gene (RhD gene)
r. hemolytic disease
r. isoimmunization
rheumatic
r. disease
r. fever
r. heart disease
r. pneumonia
rheumatica
rheumaticosis
rheumatoid
r. arthritis (RA)
r. factor
r. nodule
r. vasculitis
rheumatologic disorder
Rheumatrex
RhIg
Rh immune globulin
Rhinall
Rhindecon
rhinitis
r. medicamentosa
syphilitic r.
rhinocephaly
Rhinocort
rhinoprobe
Rhino Triangle brace
rhinovirus
rhizomelia syndrome
rhizomelic
r. brachymelia
r. chondrodysplasia punctata syndrome
r. dwarfism
r. limb shortening
r. shortening
Rh-null syndrome
Rho
R. imaging
R. immune globulin
rho(D)
r. human immune globulin
Rhodis
R.-EC

NOTES

R

449

RhoGAM
rhonchus, pl. **rhonchi**
Rhoprolene
Rhoprosone
Rh-positive
    R.-p. infant
    R.-p. red cell stroma
Rh-sensitized fetus
rhythm
    circadian r.
    diurnal r.
    junctional r.
    r. method
RI
    resistance index
RIA
    radioimmunoassay
Rias-Stella change
rib
    supernumerary r.
    wavy r.
RIBA
    recombinant immunosorbent assay
ribavirin
ribbon
    radioactive r.'s
rib-gap defect-micrognathia syndrome
riboflavin
    r. deficiency
ribonuclease (RNase)
ribonucleic
    r. acid (RNA)
    r. acid probe
ribonucleotide
ribosomal RNA
ribosome
Ricelyte
rich
    Rolaids Calcium R.
Richardson retractor
Richards-Rundle syndrome
Richner-Hanhart syndrome
Richner syndrome
Richter
    R. and Albrich procedure
    R. hernia
ricin
Ricinus communis **agglutinin**
rickets
    hypophosphatemic r.
    vitamin D-resistant r.
Rickettsia
rickettsial
Rid
ridge
    cervicovaginal r.
    genital r.
    germ r.

    gonadal r.
    mesonephric r.
    transverse r.
    wolffian r.
ridging
    metopic r.
Ridpath curette
RID Shampoo
Riechert-Mundinger stereotactic system
Rieger
    R. anomaly
    R. malformation
    R. syndrome
riesedronate
Rifadin
rifampicin
rifampin
rifamycin, rifomycin
Riga-Fede disease
right
    r. acromiodorsoposterior position
    r. to be well-born
    Breathe R.
    r. frontoanterior position (RFA)
    r. frontoposterior position (RFP)
    r. frontotransverse position (RFT)
    r. lateral position
    left to r. (L-R)
    R. Light examination light
    r. mentoanterior position (RMA)
    r. mentoposterior position (RMP)
    r. mentotransverse position (RMT)
    r. occipitoanterior position (ROA)
    r. occipitoposterior position (ROP)
    r. occipitoposterior presentation
    r. occipitotransverse position (ROT)
    r. ovarian vein syndrome
    r. sacroanterior position (RSA)
    r. sacroposterior position (RSP)
    r. sacrotransverse position (RST)
    r. scapuloanterior position (RScA)
    r. scapuloposterior position (RScP)
    r. ventricular hypertrophy (RVH)
right-angle
    r.-a. retractor
    r.-a. scissors
righting response
right-sided
    r.-s. arch, mental deficiency, facial dysmorphism syndrome
    r.-s. stomach
right-to-left
    r.-t.-l. shunt
    r.-t.-l. shunt ratio (Qs/Qt)
rigid cystoscopy
rigidity
    congenital articular r.
    multiple articular r.

rIL-2
  recombinant interleukin 2
Riley-Day syndrome
Riley-Schwachman syndrome
Riley-Smith syndrome
Rimactane
rind
ring
  amnion r.
  r. applicator
  Bandl r.
  r. chromosome 1–22 (r)
  r. chromosome 1–22 syndrome
  constriction r.
  continence r.
  contraceptive r.
  double r.
  estradiol-releasing silicone vaginal r.
  Estring estradiol vaginal r.
  Falope r.
  r. forceps
  gestational r.
  Graefenberg r.
  hymenal r.
  Imlach r.
  Kayser-Fleischer r.
  neonatal r.
  pathologic retraction r.
  r. pessary
  physiologic retraction r.
  progestin-impregnated vaginal r.
  retraction r.
  Silastic r.
  r. 1–22 syndrome
  trigonal r.
  T-shaped constriction r.
  tubal r.
  Yoon r.
  zipper r.
Ringer
  heparinized lactated R.
  lactated R.
  R. solution
ringworm
Rinman sign
Rinne test
rinse
  Nix Creme R.
Riopan
  R. Plus
ripener
  Prepidil Gel cervical r.

ripening
  Bishop score of cervical r.
  cervical r.
risk
  age-related r.
  elevated obstetric r.
  r. factor
  fetal r.
  r. management
  maternal age-related r.
  recurrence r.
  relative r.
  teratogenic r.
RIST
  radioimmunosorbent test
ristocetin cofactor
Ritalin
Ritalin-SR
Rite Time
Ritgen maneuver
ritodrine
  r. hydrochloride
ritonavir
Ritscher-Schinzel syndrome
Ritter disease
Rivotril
RLP
  rectal linitis plastica
RMA
  right mentoanterior position
RMP
  right mentoposterior position
RMSS
  Ruvalcaba-Myhre-Smith syndrome
RMT
  right mentotransverse position
^{222}Rn
  radon-222
RNA
  ribonucleic acid
    complementary RNA
    RNA-directed DNA polymerase
    messenger RNA
    RNA nucleotidyltransferase
    packaging RNA
    poly A RNA
    RNA polymerase
    ribosomal RNA
    soluble RNA
    RNA splicing
    transfer RNA (tRNA)

R

NOTES

RNase
  ribonuclease
ROA
  right occipitoanterior position
Robert pelvis
Roberts
  R. pseudothalidomide syndrome
  R. tetraphocomelia syndrome
  R. thumb retractor
robertsonian
  r. fusion
  r. translocation
Robicillin VK
Robidrine
Robimycin
Robin
  R. anomalad
  R. syndrome
Robinow
  R. dwarfism
  R. mesomelic dysplasia
  R. syndrome
Robinow-Silverman-Smith syndrome
Robinow-Sorauf syndrome
Robinul
Robitet
  R. Oral
Robitussin
ROC
  R. XS suture fastener
Rocaltrol
Rocephin
Roche Diagnostics
Rocher-Sheldon syndrome
Rochester-Ochsner forceps
Rochester-Pean forceps
rockerbottom foot
rocket dilator
Rockey-Davis incision
Rock-Mulligan hood
Rocky Mountain spotted fever
rod
  Gram-negative r.'s
  Gram-positive r.'s
  r. myopathy
Rodeck method
Rodrigues syndrome
Rodriquez lethal acrofacial dysostosis
Roeder
  R. knot
  R. loop slipknot
  modified R.
  R. slipknot
roentgen (R)
roentgenography
Roe v. Wade
rofecoxib
Roferon-A

Rofsing test
Rogaine
Roger
  R. forceps
  maladie de R.
rogletimide
Rohr stria
Rokitansky
  R. pelvis
  R. prominence
  R. tubercle
Rokitansky-Küster-Hauser syndrome
Rolaids Calcium Rich
rolandic
  r. epilepsy
  r. sharp waves
rolandometer
role
  gender r.
Rolland-Desbuquois syndrome
roller
  r.-barrel electrode
  r. occlusion
rollerball
  r. electrode
  r. endometrial ablation
  r. technique
rollerbar
  r. electrode
  r. endometrial ablation
rollover test
Rolon spatula
Rolserp
Romaña sign
Ronase
Rondomycin
rongeur
  pediatric bone r.
  Tobey ear r.
room
  birthing r.
  LDR r.
  r. temperature
rooming-in
rooting reflex
ROP
  retinopathy of prematurity
  right occipitoposterior position
ropivacaine
  r. HCl
  r. hydrochloride
rosary
  rachitic r.
Rosch-Thurmond fallopian tube catheterization set
rosea
  pityriasis r.
Rose equation

Rosenmüller
    organ of R.
Rosenthal-Kloepfer syndrome
roseola infantum
rosette test
Rosewater syndrome
ROSNI
    round spermatid nuclei injection
Ro (SS-A) autoantigen
Rossavik growth model
Rosselli-Gulienetti syndrome
Ross growth chart
Rossi syndrome
ROT
    right occipitotransverse position
rotation
    r. and descent
    external r.
    forceps r.
    internal r.
    Kielland r.
    manual r.
rotational delivery
rotavirus
Rotazyme test
Rothmann-Makai syndrome
Rothmund
    R. syndrome
Rothmund-Thomson syndrome
Rothmund-Werner syndrome
roticulating endograsper
Roticulator 55 stapler
Rotor syndrome
Rotunda treatment
Roubac
Rouget bulb
round
    r. ligament
    r. ligament syndrome
    r. pelvis
    r. spermatid nuclei injection
    (ROSNI)
round-headed acrosomeless spermatozoa
roundworm
Rous sarcoma virus
Roussy-Lévy syndrome
routine
    r. antenatal diagnostic imaging
    ultrasound study
    r. preoperative test
routine antenatal diagnostic imaging
with ultrasound (RADIUS)

Roux-en-Y choledochojejunostomy
Rovamycine
Rowasa enema
Rowden uterine manipulator injector
    (RUMI)
Rowland pouch
Roxanol
    R. SR
Royal College of General Practioners'
    Oral Contraception Study
RPC
    retained products of conception
RPD
    Pepcid RPD
RPR
    rapid plasma reagin
RPS4X
RPS4Y
RP with progressive sensorineural
    hearing loss
RS
    reactive site
      AT3 type II RS
      RS mutation
      RS virus
RSA
    recurrent spontaneous abortion
    right sacroanterior position
RScA
    right scapuloanterior position
RScP
    right scapuloposterior position
RSD
    reflex sympathetic dystrophy
RSH/SLO syndrome
RSH/Smith-Lemli-Opitz
    R.-L.-O. syndrome
RSH syndrome
RSP
    right sacroposterior position
RST
    right sacrotransverse position
RSV
    respiratory syncytial virus
RTA
    renal tubular acidosis
rt-PA
    recombinant tissue type plasminogen
    activator
RU 486
rubella
    congenital r.

R

NOTES

rubella *(continued)*
    r. embryopathy
    r. immune
    measles, mumps, r. (MMR)
    periconceptional r.
    r. scarlatinosa
    r. syndrome
    r. vaccine
    r. virus
rubella-immune mother
rubella-negative mother
Rubens flap
rubeola
    r. scarlatinosa
Rubex
Rubin
    R. cannula
    R. maneuver
    R. test
Rubinstein syndrome
Rubinstein-Taybi syndrome
rubor
rubra
    miliaria r.
ruddy
Rudiger syndrome
rudimentary
    r. testis syndrome
    r. uterine horn
Rud syndrome
ruga, pl. **rugae**
    rugae of vagina
    rugae vaginales
rugal fold
rugation
rule
    Arey r.
    Budin r.
    four-hour r.
    Haase r.
    His r.
    informed consent disclosure r.'s
    Lossen r.
    Mittendorf-Williams r.
    Nägele r.
    r. of 60s
    Ogino-Knaus r.
    r. of outlet
    Sandberg r.
    r. of threes
    Weinberg r.
RUMI
    Rowden uterine manipulator injector
    RUMI uterine manipulator
rumination
Rum-K
Rumpel-Leede phenomenon
Runeberg anemia

running
    r. imbricating stitch
    r. locked stitch
runt disease
runting
rupture
    amnion r.
    angiomyolipoma r.
    follicle aspiration, sperm injection, and assisted r. (FASIAR)
    hepatic r.
    marginal sinus r.
    membrane r.
    postmembrane r.
    prelabor membrane r.
    premature amnion r.
    premature membrane r.
    premembrane r.
    prolonged r.
    splenic r.
    total perineal r.
    tubal r.
    uterine r.
ruptured
    r. capsule
    r. cerebral aneurysm
    r. episiotomy
    r. symphysis pubis
    r. uterus
Rusconi
    anus of R.
Russell
    R. dwarf
    R. nanism
    R. syndrome
    R. viper venom time
Russell-Silver
    R.-S. dwarfism
    R.-S. syndrome
Russian tissue forceps
Rutherfurd syndrome
Rutledge
    R. classification of extended hysterectomy
    R. lethal multiple congenital anomaly syndrome
Ruvalcaba-Myhre-Smith syndrome (RMSS)
Ruvalcaba-Myhre syndrome
Ruvalcaba-Reichert-Smith syndrome
Ruvalcaba syndrome
RVH
    right ventricular hypertrophy
RVPEP/RVET ratio
Rx
    Mission Prenatal Rx
    Natalins Rx

RxFISH DNA probe and analysis
system

Rynacrom

NOTES

R

S-100 antibody
SA
    sacroanterior
SAB
    spontaneous abortion
Sabbagha formula
saber shin
Sabinas brittle hair syndrome
Sabin-Feldman dye test
Sabin vaccine
Sabouraud medium
sabre shin deformity
Sabril
sac
    allantoic s.
    amniotic s.
    chorionic s.
    embryonic s.
    gestational s. (GS)
    Lap S.
    Pleatman s.
    pseudogestational s.
    vitelline s.
    widened thecal s.
    yolk s.
Saccharomyces cerevisiae
saccular
    s. aneurysm
    s. period
sacculation
    s. of uterus
sacral
    s. agenesis
    s. colpopexy
    s. dimple
    s. lymph node
    s. nerve
    s. neural tube defect
    s. promontory
    s. reflex
sacroanterior (SA)
    left s.
    s. position
sacrococcygeal teratoma
sacrocolpopexy
    s. graft
    laparoscopic s.
sacroiliac joint
sacropexy
    abdominal s.
sacroposterior (SP)
    left s. position (LSP)
    s. position
    right s. position (RSP)
sacrosidase

sacrospinous
    s. colpopexy
    s. ligament
    s. ligament suspension
    s. vaginal vault suspension
sacrotransverse (ST)
    left s. position (LST)
    s. position
    right s. position (RST)
sacrotuberous
sacrum
    hollow of s.
    hypoplastic s.
sactosalpinx
    retort-shaped s.
SAD
    source-to-axis distance
saddle
    s. block
    s. block anesthesia
    s. nose
saddlebag flap
Saenger
    S. operation
    S. ovum forceps
Saethre-Chotzen syndrome
SAFE
    sexual assault forensic evidence
    SAFE kit
safety
    system for thalidomide education
      and prescription s. (STEPS)
sagittal
    s. fontanel
    s. suture
    s. suture line
sagrada
    cascara s.
sail sign
Sakati-Nyhan syndrome
SAL
    suction-assisted lipoplasty
salaam
    s. convulsion
    infantile s.
    s. seizures
sal ammoniac
Salazopyrin
salbutamol
Saldino-Noonan
    S.-N. dwarfism
    S.-N. syndrome
Salem pump
SalEst
    S. immunoassay

S

SalEst *(continued)*
  S. preterm labor test system
  S. system test
Saleto-200
Saleto-400
Salflex
Salgesic
salicylate
  choline s.
  magnesium s.
  phenyl s.
  s. poisoning
  sodium s.
salicylism
salicylsalicylic acid
saline
  s. abortion
  s. cathartic
  s. drop test
  Dulbecco phosphate buffered s.
  hypertonic s.
  s. implant
  s. infusion sonography (SIS)
  s. infusion sonohysterography (SIS)
  normal s.
  phosphate buffered s. (PBS)
  physiologic s.
  s. solution
  Tris[hydroxymethyl]aminomethane-
    buffered s.
  s. wet smear
salivary
  s. estriol
  s. estriol test
  s. gland
salivation, lacrimation, urination,
  defecation, gastrointestinal distress
  and emesis (SLUDGE)
Salk vaccine
Salla disease
salmeterol xinafoate
salmon
  s. patch
Salmonella
  S. enteritidis
  S. typhi
salmonella
  nontyphoid s.
salmonellosis
Salmon sign
Salpha-reductase
salpingectomy
  postpoartum partial s.
salpingemphraxis
salpinges (*pl. of* salpinx)
salpingioma
salpingitic

salpingitis
  s. after previous tubal occlusion
    (SPOT)
  chronic interstitial s.
  foreign body s.
  gonorrheal s.
  granulomatous s.
  s. isthmica nodosa (SIN)
  leperous s.
  nongranulomatous s.
  s. in previously occluded tubes
    (SPOT)
  pyogenic s.
  tuberculous s.
salpingocele
salpingocentesis
salpingocyesis
salpingography
  transcervical selective s.
salpingolysis
salpingoneostomy
salpingo-oophorectomy
  abdominal s.-o.
  bilateral s.-o. (BSO)
  unilateral s.-o.
salpingo-oophoritis
salpingo-oophorocele
salpingoovariectomy
salpingoovariolysis
salpingoperitonitis
salpingopexy
salpingoplasty
salpingorrhagia
salpingorrhaphy
salpingoscopy
salpingostomatomy
salpingostomy
  linear s.
salpingotomy
  abdominal s.
salpinx, pl. **salpinges**
salsalate
Salsitab
salt
  gold s.
  Pedi-Bath S.'s
  s. and pepper fundus
  s. restriction
  s. wasting
salt-losing adrenogenital syndrome
  (SLAS)
salt-wasting congenital adrenal
  hyperplasia (SW-CAH)
Saluron
salutary effect
salvage
  s. cesarean

s. intervention
s. therapy
sample
arterial blood s. (ABS)
venous blood s. (VBS)
sampler
Cervex-Brush cervical cell s.
Cordguard umbilical cord s.
Cytobrush Plus endocervical cell s.
SelectCells Mini endometrial s.
Wallach Endocell endometrial
cell s.
sampling
biological s.
blood s.
capillary blood s.
chemical s.
chorion s.
chorionic villus s. (CVS)
endocervical s.
endometrial s.
fetal scalp blood s.
fetal scalp platelet s.
fetal skin s.
Mucat cervical s.
percutaneous blood s.
percutaneous umbilical blood s.
(PUBS)
random s.
scalp blood s.
transabdominal chorionic villus s.
transcervical chorionic villus s.
trophoblast s.
ultrasound-directed percutaneous
umbilical blood s.
umbilical blood s.
Sampson cyst
Sanchez-Cascos syndrome
Sanchez-Corona syndrome
Sanchez-Salorio syndrome
Sandberg rule
Sanders jet respirator
Sandhoff disease
Sandifer syndrome
Sandimmune
Sandle gap foot deformity
Sandoglobulin
Sandostatin
sandwich
s. assay
solid phase s.

Sanfilippo
S. disease A, B, C, D
S. syndrome
Sanger incision
sanguinolentis
fetus s.
sanguinopurulent
Sani-Spec vaginal speculum
Sanjad-Sakati syndrome
Sanorex
SANS
PerQ S.
Sansert
Santavuori-Haltia syndrome
Santavuori syndrome
SaO$_2$
oxygen saturation
Sao Paulo MCA/MR syndrome
SAP-35
sap
Prentif cavity-rim cervical s.
saphenous nerve
saponification
saquinavir
sarcofetal pregnancy
sarcoidosis
sarcoma
alveolar soft part s.
botryoid s.
s. botryoides
cervical s.
clear cell s.
embryonal s.
endometrial s.
endometrial stromal s. (ESS)
Ewing s.
heterologous uterine s.
homologous uterine s.
Kaposi s.
Kaposi varicelliform s.
mesodermal s.
mixed mesodermal s. (MMS)
mixed müllerian s.
mixed ovarian mesodermal s.
obesity in endometrial s.
retroperitoneal soft tissue s.
secretory s.
uterine s.
uterine müllerian s.
vulvar s.
sarcomatous myoma degeneration
sarcomere

S

NOTES

Sarcoptes scabiei
sarcosinemia
SART
    Sexual Assault Response Team
    Society for Assisted Reproductive
    Technology
sartorial elegance
S.A.S.-500
SASPP
    septum pellucidum with porencephalia
    syndrome
Sassone score
satellite
    s. DNA
    s. lesion
Sato syndrome
satumomab pendetite imaging agent
saturated solution of potassium iodide
  (SSKI)
saturation
    s. analysis
    fetal arterial oxygen s. (FS_pO_2)
    fetal oxygen s.
    oxygen s. (SaO_2)
    s. strip
satyr ear
satyriasis
Saunders
    S. disease
    S. sign
Savage syndrome
Save-A-Tooth
saver
    Cell S.
saw
    Gigli s.
Saxonius
    coitus S.
Saxtorph maneuver
Say-Gerald syndrome
Say-Meyer syndrome
Say syndrome
SB-6 antiserum
SBE
    self-breast examination
    subacute bacterial endocarditis
    SBE prophylaxis
SBM
    selective broth medium
SBP
    spontaneous bacterial peritonitis
SC
    hemoglobin SC
    SC phocomelia syndrome
    SC syndrome
ScA
    scapuloanterior
scabies

scalded-skin syndrome
scale
    Albert Einstein Neonatal
      Developmental S. (AENNS)
    Anger Expression S.
    Apgar s.
    Borg Perceived Exertion S.
    Borg Physical Activity S.
    Brazelton Neonatal Behavioral
      Assessment S. (BNBAS)
    Capute s.
    Cattell Infant Intelligence S.
    Conners S.
    Cooke-Medley Hostility S.
    Cranley Maternal-Fetal
      Attachment S.
    Dyadic Adjustment S.
    Early Neonatal Neurobehavioural S.
    Family Adaptability and
      Cohesion S.-III (FACES-III)
    Family Environment S. (FES)
    Flint Infant Security S. (FISS)
    Gesell Developmental S.
    Graham-Rosenblith s.
    HSC S.
    Impact of Events S. (IES)
    Kent Infant Development S.
      (KIDS)
    Maternal Attitude S. (MAS)
    Neonatal Behavior Assessment S.
      (NBAS)
    Neonatal Infant Pain S. (NIPS)
    Perceived Stress S.
    Piper fatigue s.
    resource-based relative value s.
      (RBRVS)
    Reynell Development S.
    Toddler Temperament S.
    Vineland Adaptive Behavior S.'s
scalloping
    frontal bones s.
scalp
    s. blood sampling
    cutis aplasia of s.
    s. laceration
    s. pH determination
    s. vein needle
scalpel
    Bowen double-bladed s.
    Endo-Assist retractable s.
    Harmonic s.
    Shaw I, II s.
scan
    abdominopelvic s.
    gallium s.
    iodine 125-labeled fibrinogen s.
    longitudinal s.
    MUGA s.

multiple gated acquisition s.
  (MUGA)
radiofibrinogen uptake s.
time position s.
transverse s.
S. ultrasound gel
ventilation/perfusion s.
$\dot{V}/\dot{Q}$ s.
Scanlon Assessment
scanner
  Aloka 650 s.
  Aloka SSD-720 real-time s.
  BladderManager portable
    ultrasound s.
  EUB-405 ultrasound s.
scanning
  duplex s.
  s. electron microscopy (SEM)
  iodomethyl-norcholesterol s.
  isotope s.
  MEVA Probe for endovaginal s.
  restriction landmark genomic s.
Scanzoni
  S. forceps
  S. maneuver
  S. second os
Scanzoni-Smellie maneuver
scaphocephaly
scaphoid fontanel
scapula
  congenital elevation of the s.
scapuloanterior (ScA)
  left s. position (LScA)
  right s. position (RScA)
scapuloposterior (ScP)
  left s. position (LScP)
  right s. position (RScP)
scar
  lower-segment s.
  perineal s.
  s. prediction
  radial s.
  s. tissue
SCARF
  skeletal abnormalities, cutis laxa,
    craniostenosis, psychomotor retardation,
    facial abnormalities
scarf
  s. maneuver
  s. sign
scarification
scarlatina

scarlatinosa
  rubella s.
  rubeola s.
scarlet fever
SCARMD
  severe childhood autosomal recessive
    muscular dystrophy
Scarpa fascia
scarred womb
scatoma
scattered echo
SCCD
  Schnyder crystalline corneal dystrophy
SCCMS
  slow-channel congenital myasthenic
    syndrome
S-C disease
  sickle cell-hemoglobin C disease
SCE
  split hand-cleft lip/palate and ectodermal
    dysplasia
SCF
  somatic cell-derived growth factor
Schafer syndrome
Schatz maneuver
Schauffler procedure
Schaumann body
Schauta vaginal operation
scheduled feeding
Scheie syndrome
Scheuermann disease
Scheuthauer-Marie-Sainton syndrome
Schick
  S. sign
  S. test
Schilder
  S. disease
  S. encephalitis
Schiller
  S. solution
  S. test
  S. tumor
Schiller-Duvall body
Schilling test
Schimmelbusch
  S. disease
  S. syndrome
Schimmelpenning-Feuerstein-Mims
  syndrome
Schinzel
  S. acrocallosal syndrome

NOTES

S

Schinzel-Giedion
    S.-G. midface-retraction syndrome
    S.-G. syndrome (SGS)
Schirmer syndrome
schistocelia
schistocephalus
schistocormia
schistocystis
schistoglossia
schistomelia
schistoprosopia
schistorrachis
schistosomia
schistosomiasis
schistosomus
schistosternia
schistothorax
schistotrachelus
schizencephaly
schizocyte
schizocytosis
schizophrenia
Schlesinger solution
Schlusskoagulum
Schmid-Fraccaro syndrome
Schmidley syndrome
Schmidt syndrome
Schmorl jaundice
Schneckenbecken dysplasia
Schnyder crystalline corneal dystrophy (SCCD)
Scholz disease
Schönlein-Henoch purpura
Schroeder
    S. operation
    S. tenaculum forceps
    S. tenaculum loop
    S. uterine tenaculum
    S. vulsellum forceps
Schubert uterine biopsy forceps
Schuchardt
    S. incision
    S. operation
Schuco nebulizer
Schüffner dot
Schuknecht
    S. classification
    S. retractor
Schultze
    S. mechanism
    S. phantom
    S. placenta
Schutt needle
Schwachman-Diamond syndrome
Schwachman syndrome
Schwangerschafts protein 1
Schwann cell
schwannoma

Schwartz-Jampel-Aberfeld syndrome
Schwartz-Jampel syndrome
Schwartz syndrome
sciatic nerve
SCID
    severe combined immunodeficiency disease
    severe combined immunodeficiency disorder
SciMed-Kolobow membrane lung
scimitar syndrome
scintigraphy
    ventilation s.
scintimammography (SMM)
    s. prone breast cushion
scirrhous carcinoma
scissoring
scissors
    Adson ganglion s.
    Aslan endoscopic s.
    bandage s.
    Braun episiotomy s.
    curved Mayo s.
    Electroscope disposable s.
    Evershears bipolar laparoscopic s.
    fine Metzenbaum s.
    Jorgenson s.
    Lister s.
    Mayo s.
    Metzenbaum s.
    right-angle s.
    Seilor s.
    Smellie s.
    Spencer stitch s.
    straight s.
    umbilical s.
    Z-Scissors hysterectomy s.
sclera, pl. sclerae
    blue s.
scleral hemorrhage
sclerema neonatorum
scleroatonic muscular dystrophy
sclerocystic
    s. disease of the ovary
    s. ovary
scleroderma
    progressive familial s.
sclero-oophoritis
sclerosing
    s. adenitis
    s. adenosis
    s. agent
    s. lesion
sclerosis, pl. scleroses
    diffuse globoid body s.
    diffuse globoid cell cerebral s.
    dominant choroidal s.
    familial centrolobal s.

globoid cell cerebral s.
glomerular s.
menstrual s.
multiple s.
osteopathia striata with cranial s.
ovulational s.
photothermal s.
physiologic s.
progressive systemic s.
Sholz s.
s. tuberosa
tuberous s.
sclerosteosis
sclerotherapy
SCMC
sperm-cervical mucus contact
scoliosis
Adams test for s.
congenital thoracic s.
s. with dural ectasia
Scopemaster contact hysteroscope
Scopette device
scopolamine
s. methylbromide
s. poisoning
score
abstinence s.
Amiel-Tison s.
Apgar s.
Beck Depression Inventory s.
biophysical profile s.
Bishop s.
BPP s.
cervical s.
Charlson s.
Dubowitz s.
Ferriman-Gallwey hirsutism s.
Johnson s. 1–10
LOD s.
Manning s.
Manning s. of fetal activity
Neurologic and Adaptative
  Capacity S. (NACS)
Optimal Observation S.
pelvic s.
POMS s.
Profile of Mood States s.
recovery s.
Sassone s.
Silverman s.
T bone density s.
Wood-Downes asthma s.

Yale Optimal Observation S.
Zatuchni-Andros s.
Z bone density s.
scoring
follicular s.
Scotch-tape test
Scott
S. cannula
S. craniodigital syndrome
Scott-Taor syndrome
ScP
scapuloposterior
SC-pseudothalidomide syndrome
scrape
s. cytology
s. and smear
screen
antigen s.
Glucola s.
organic acid s.
toxicology s.
triple s.
triple-biochemical s.
triple-marker s.
urine toxicology s.
screener
Algo newborn hearing s.
screening
Amniostat fetal lung maturity s.
antenatal s.
antibody s.
colposcopic s.
cytologic s.
genetic s.
hepatitis s.
s. laboratory test
mammographic s.
maternal s.
multiple marker s.
Neo-Gen s.
neonatal s.
nuchal translucency s.
prenatal s.
s. recommendation
sickle cell s.
triple serum marker s.
ultrasound s.
uterine s.
screw
myoma s.
screwdriver teeth
scrofula

S

NOTES

scrofuloderma
scrofulosorum
    lichen s.
scroll ear
scrotal tongue
scrotum
    bifid s.
    shawl s.
scr protooncogene
scrub
    Sklar s.
Scully tumor
scultetus binder
scurvy
    hemorrhagic s.
    infantile s.
SD
    standard deviation
    AlphaNine SD
    Profilate SD
    Profilnine SD
    WinRho SD
S/D
    systolic/diastolic
    Gammagard S/D
    Polygam S/D
    S/D ratio
SDAP
    single donor apheresis platelet
SDAT
    senile dementia of the Alzheimer type
S-D disease
SDF
    WinRho SDF
SDU-400 EchoView ultrasound machine
SDYS
    Simpson dysmorphia syndrome
sea-blue histiocyte syndrome
Seabright bantam syndrome
seal fingers
search
    MEDLINE s.
seat
    Hospital Recliner s.
    Ingram bicycle s.
sebaceous
    s. cyst
    s. hyperplasia
    s. miliaria
    s. nevus syndrome
sebaceum
    adenoma s.
sebaceus
    nevus s.
seborrheic
    s. dermatitis
    s. keratosis

sebum
    s. preputiale
Sechrist neonatal ventilator
Seckel
    bird-headed dwarf of S.
    S. bird head syndrome
    S. dwarfism
    S. nanism
secobarbital
Seconal
second
    cycles per s. (cps)
    s. degree prolapse
    S. International Standard (SIS)
    s. messenger
    s. parallel pelvic plane
    s. stage of labor
    s. trimester
    s. trimester acute gestosis
    s. twin
secondary
    s. abdominal pregnancy
    s. amenorrhea
    s. atelectasis
    s. dysmenorrhea
    s. infertility
    s. sex characteristic
    s. syphilis
    s. uterine inertia
    s. vestibular dyspareunia
second-degree laceration
second-generation progesterone
second-hand smoke
second-look
    s.-l. laparoscopy
    s.-l. laparotomy
    s.-l. operation
secretin
Secretin-Ferring Powder
secretion
    abnormal cortisol s.
    adrenal androgen s.
    androgen s.
    cervicovaginal s.
    expressed prostatic s. (EPS)
    follicle-stimulating hormone s.
    follicular phase gonadotropin s.
    gonadotropin s.
    impaired s.
    luteinizing hormone s.
    melatonin s.
    persistent estrogen s.
    pituitary gonadotropin s.
    placental s.
    progesterone s.
    prolactin s.
    steroid s.

syndrome of inappropriate antidiuretic hormone s.

vaginal s.

secretory

s. adenocarcinoma

s. carcinoma

s. disease

s. IgA

s. phase

s. sarcoma

section

cesarean s. (C-section)

cut s.

frozen s.

lower uterine segment transverse cesarean s.

Porro cesarean s.

primary cesarean s.

repeat cesarean s.

representative s.

vaginal birth after cesarean s. (VBAC)

sectioning

serial s.

Sectral

secundigravida

secundina, pl. **secundinae**

secundines

secundipara

secundum

s. atrial septal defect

foramen s.

ostium s.

septum s.

secundum-type

Sedabamate

sedation

sedative

s. effect

Sedatuss

sedentary

cyclic s. (CS)

sediment

spun urine s.

urinary s.

sedimentation rate

Sedlacková syndrome

seed

gold s.

radioactive s.

Seemanová-Lesny syndrome

Seemanová syndrome 1, 2

SEER network

Seessel pouch

segment

chromosomal s.

lower uterine s.

mesodermal dysgenesis of anterior s.

uterine s.

segmental

s. epidural analgesia

s. resection

segregation

postmeiotic s.

segregation distortion

SEH

subependymal hemorrhage

Seilor scissors

Seip-Lawrence syndrome

Seip syndrome

Seitelberger disease

Seitleis syndrome

Seitzinger

S. device

S. tripolar cutting forceps

seizure

absence s.

s.'s, acquired microcephaly, agenesis of corpus callosum syndrome

atonic s.

autonomic s.

benign familial neonatal s.

s. control

s. disorder

eclamptic s.

epileptic s.

febrile s.

fetal s.

gelastic s.

s.'s, hypotonic cerebral palsy, megalocornea, mental retardation syndrome

hysterical s.

infantile myoclonic s.

jackknife s.

lightning s.

multifocal clonic s.

myoclonic s.

myoclonic-astatic s.

neonatal s.

salaam s.'s

sylvian s.

S

NOTES

465

seizure *(continued)*
    tonic s.
    tonic-clonic s.
    vertiginous s.
Seldane
SelectCells Mini endometrial sampler
selection
    bulk s.
    family s.
    fecundity s.
    gametic s.
    natural s.
    prenatal s.
    truncate s.
selective
    s. abortion
    s. broth medium (SBM)
    s. embolization procedure
    s. estrogen receptor modulator
      (SERM)
    s. feticide
    s. inguinal node dissection
    s. no-fault system
    s. termination
    s. transvaginal embryo reduction
    s. tubal assessment to refine
      reproductive therapy (STARRT)
    s. tubal occlusion procedure
      (STOP)
    s. tubal occlusion procedure system
self-breast examination (SBE)
self-catheter
    Mentor female s.-c.
self-catheterization
    clean intermittent s.-c.
self-examination
    BD Sensability breast s.-e.
    breast s.-e. (BSE)
self-incompatibility
selfing
self-monitoring
self-pollination
self-priming action
self-retaining retractor
self-sterility
self-test
    OvuQuick S.-t.
sellar enlargement
sella turcica
Sellheim incision
Sellick maneuver
SEM
    scanning electron microscopy
    standard error of the mean
semantic memory
SEMD
    spondyloepimetaphyseal dysplasia

semen
    s. analysis
    frozen s.
    s. liquefaction
    prepared s.
    viscous s.
    s. volume
SEMG
    surface electromyography
Semicid
semi-conservative
semi-Fowler position
semilithotomy position
seminal
    s. fluid
    s. fluid analysis (SFA)
    s. plasma
    s. vesicle
semination
seminiferous
    s. tubule
    s. tubule dysgenesis
seminoma
    ovarian s.
semiprone position
semisolid
Semken forceps
Semm
    S. hysterectomy
    S. Pelvi-Pneu insufflator
    S. RX morcellator
    S. uterine vacuum cannula
    S. Z technique
Sengstaken-Blakemore tube
senile
    s. dementia of the Alzheimer type
      (SDAT)
    s. vaginitis
senna
    s. concentrate/docusate sodium
    S. X-Prep
Senn-Dingman retractor
Senning repair
Senokot
    S.-S
senology
Sensability breast self-examination aid
Sensenbrenner-Dorst-Owens syndrome
Sensenbrenner syndrome
sensimotor induction in disturbed
    equilibrium syndrome
sensitivity
    assay s.
    culture and s. (C&S)
    insulin s.
    microbial s.
sensitization
    Kell s.

latex s.
Rh s.
sensitizer
hypoxic cell s.
sensor
anal EMG PerryMeter s.
BreastAlert differential
temperature s.
differential temperature s. (DTS)
Nellcor FS-10 oximeter s.
Nellcor FS-14 oximeter s.
Sensorcaine
Sensorimedics Horizon Metabolic Cart
sensorineural
s. change
s. deafness, imperforate anus,
hypoplastic thumbs syndrome
sensory
s. deficit
s. loss
s. organ
s. stimulation
SensoScan mammography system
Senter syndrome
SEPA
superficial external pudendal artery
separation
amnion-chorion s.
blastomere s.
peripartum symphysis s.
peripheral placental s.
placental s.
sperm s.
symphyseal s.
tripus s.
uterine scar s.
separator
Benson baby pylorus s.
Sephadex binding test
Seprafilm bioresorbable membrane
sepsis, pl. **sepses**
Chlamydia s.
Escherichia coli s.
neonatal s.
s. neonatorum
postoperative s.
puerperal s.
sepsis-pneumonia syndrome
septa (*pl. of* septum)
septal hypertrophy

septate
s. hymen
s. uterus
septation
internal s.
septectomy
balloon s.
septic
s. abortion
s. arthritis
s. pelvic thrombophlebitis (SPT)
s. pelvic vein thrombophlebitis
s. shock
septicemia
puerperal s.
streptococcal s.
septimetritis
septi pellucidi agenesis
septooptic dysplasia
septostomy
atrial s.
balloon s.
Septra
S. DS
septum, pl. **septa**
aortopulmonary s.
atrioventricular s.
enlarged cavum s.
s. pellucidum
s. pellucidum with porencephalia
syndrome (SASPP)
placental septa
s. primum
rectovaginal s.
s. secundum
supravaginal s.
transverse vaginal s.
urogenital s.
uterine s.
septuplet
sequela, pl. **sequelae**
sequelae of extreme prematurity
long-term sequelae
neoplastic s.
sequence
amniotic band s.
autonomous replication s.
base s.
blepharophimosis s.
s. characterized amplified region
cleaved amplified polymorphic s.
complementary s.

S

NOTES

sequence *(continued)*
    consensus s.
    conserved s.
    DiGeorge s.
    DNA s.
    fetal akinesia deformation s.
      (FADS)
    fetal brain disruption s.
    FLAIR s.
    Goldenhar s.
    half-Fourier acquisition single-shot
      turbo spin-echo s.
    insertion s.
    laterality s.
    mental retardation-overgrowth s.
    Möbius s.
    Poland malformation s.
    regulatory s.
    sirenomelia s.
    Sotos s.
    s. tagged microsatellite
    s. tagged site
    tandem repeat s.
    TRAP s.
    twin reverse arterial perfusion s.
    X-linked hydrocephalus-stenosis of
      aqueduct of Sylvius s.
    Y chromosome-specific DNA s.
sequencing
    chromosome s.
    gene s.
sequential
    s. administration
    s. delivery
    s. hormone therapy
    s. multiple analysis (SMA)
    s. oral contraceptive
sequestered lung
sequestration
    extralobar s.
    fetal pulmonary s.
    pulmonary s.
Sequoia Acuson system
sera (*pl. of* serum)
Serax
Sereen
Serentil
Serevent
serial sectioning
series
    gastrointestinal s.
    recombinant hepatitis B
      immunization s.
serine protease inhibitor (SERPIN)
serine-threonine kinase
seriography
    biplane s.

SERM
    selective estrogen receptor modulator
sermorelin acetate
serologic
    s. test
    s. test for syphilis (STS)
serology
    nontreponemal s.
    treponemal s.
serology-negative mother
seroma
    postoperative s.
Seroma-Cath system
seromuscular intestinal patch graft
Seromycin
seronegative
Serono
    S. SR1 FSH analyzer
    S. test
Serophene
seropositive
    cytomegalovirus s.
seroprevalence rate
serosa
    lochia s.
    peritoneal s.
serosal
    s. adhesion
    s. surface
serosanguineous fluid
serostatus
serotonin
    s. reuptake inhibitor
serous
    s. adenocarcinoma
    s. carcinoma
    s. cystadenoma
    s. ovarian neoplasm
    s. tumor
serovar-specific immunoglobulin IgG
serovar-specified immunoglobulin IgM
Serpalan
Serpasil
serpiginosa
    elastosis performans s.
SERPIN
    serine protease inhibitor
Sertoli
    S. cell
    S. cell tumor
Sertoli-cell-only syndrome
Sertoli-Leydig
    S.-L. cell
    S.-L. cell tumor
sertraline
    s. HCl
    s. hydrochloride
serum, pl. **sera**

s. albumin
s. amylase
s. analyte
s. antibody
s. assay
s. estrogen
fetal s.
s. free hemoglobin
s. hepatitis
s. iron
s. level
s. lithium concentration
maternal s.
s. müllerian inhibiting
s. osteocalcin
s. parathyroid hormone
s. pregnancy assay cartridge
s. progesterone
s. prolactin
s. sickness
stored sera
s. testosterone
s. transaminase
services
Child Protective S. (CPS)
Department of Children and Youth S.
Guidelines for Adolescent Preventive S. (GAPS)
servocontrolled ventilation pump
Servo 900C ventilator
servomechanism
sessile polyp
SEST
supine empty stress test
set
Embryon GIFT transfer catheter s.
haploid s.
Janacek reimplantation s.
Mi-Mark endocervical curette s.
Mi-Mark endometrial curette s.
Neo-Sert umbilical vessel catheter insertion s.
Rosch-Thurmond fallopian tube catheterization s.
Setleis syndrome
setting-sun sign
severe
s. childhood autosomal recessive muscular dystrophy (SCARMD)
s. combined immunodeficiency disease (SCID)

s. combined immunodeficiency disorder (SCID)
s. ovarian hyperstimulation syndrome (SOHS)
sex
s. assignment
s. cell
s. change operation
s. chromatin
chromosomal s.
s. chromosomal abnormality
s. chromosomal anomaly
s. chromosomal polysomy
s. chromosome
s. cord
s. determination
endocrinologic s.
genetic s.
gonadal s.
s. hormone
s. hormone-binding globulin (SHBG)
illicit s.
morphological s.
nuclear s.
oral s.
phenotypic s.
psychological s.
s. ratio
s. reversal
social s.
s. steroid
s. steroid add-back therapy
s. steroid modulation
s. surrogate
sex-conditioned gene
sex-cord
s.-c. stromal germ cell tumor
s.-c. stromal neoplasm
sex-determining region (SRY)
sex-influenced gene
sex-limited gene
sex-linked
s.-l. chromosome
s.-l. gene
s.-l. heredity
s.-l. inheritance
s.-l. neurodegenerative disease with monilethrix
sextuplet
sexual
s. abuse

NOTES

S

sexual (*continued*)
    s. activity
    s. ambiguity
    s. asphyxia
    s. assault
    s. assault forensic evidence (SAFE)
    S. Assault Response Team (SART)
    s. debut
    s. derivation
    s. deviation
    s. differentiation
    s. dimorphism
    s. dwarfism
    s. dysfunction
    s. function
    s. habit
    s. hair
    s. history
    s. infantilism
    s. initiation
    s. intercourse
    s. molestation
    S. Opinion Survey
    s. orientation
    s. response curve
    s. response cycle
    s. transmission
    s. victimization
sexually transmitted disease (STD)
SFA
    seminal fluid analysis
SFMS
    Smith-Fineman-Myers syndrome
SG
    Chemstrip 10 with SG
SGA
    small for gestational age
    post-term SGA
    term SGA
Sgambati test
SGB
    Simpson-Golabi-Behmel
SGBS
    Simpson-Golabi-Behmel syndrome
SGO
    Society of Gynecologic Oncologists
    SGO classification of cancer
SGS
    Schinzel-Giedion syndrome
shadow
    acoustic s.
    cardiothymic s.
    thymic s.
shaggy heart border
Shah-Waardenburg syndrome
shaken
    s. baby syndrome
    s. impact syndrome

shake test
shallow
    s. acetabular fossae
    s. orbits
    s. orbits, ptosis, coloboma, trigonocephaly, gyral malformations, mental and growth retardation syndrome
Shambaugh retractor
Shampoo
    A-200 S.
    Lice-Enz S.
    Pronto S.
    Pyrinex Pediculicide S.
    Pyrinyl Plus S.
    R & C S.
    RID S.
    Tisit S.
Shapleigh curette
sharp
    s. curettage
    s. pulse
Sharplan USA ultrasonic surgical aspirator
Shauta-Aumreich procedure
Shaw
    S. I, II scalpel
shawl
    s. scrotum
    s. scrotum syndrome
SHBG
    sex hormone-binding globulin
Shea forceps
Shearer forceps
shears
    ADC Medicut s.
    LaparoSonic coagulating s. (LCS)
sheath
    Bakelite cystoscopy s.
    ERA resectoscope s.
    fibrin s.
    Hemaflex s.
    Insul-Sheath vaginal speculum s.
    MicroSpan s.
    probe s.
    PRO/Covers ultrasound probe s.
    Vimule permanent s.
Sheathes ultrasound probe cover
shedding
    asymptomatic viral s.
    s. domain
    endometrial s.
    s. syndrome
    viral s.
Sheehan syndrome
sheet
    amniotic s.

impervious s.
Ioban 2 iodophor cesarean s.
shelf
Blumer s.
Shenton line
Shepard equation
shepherd's purse
Shereshevskii-Turner syndrome
**Sherwood intrascopic suction/irrigation
system**
SHG
sonohysterography
Shiatsu therapeutic massage
shield
Dalkon s.
FACE-IT protective s.
Fuller s.
Lea S.
nipple s.
Surety S.
shift
Doppler s.
luteoplacental s.
Shigella
shigellosis
Shimadzu
S. IIQ ultrasound
S. SDU-400 ultrasound
S. ultrasound system
shin
saber s.
shingles
Shirley wound drain
Shirodkar
S. cervical cerclage
S. operation
S. procedure
shock
anaphylactic s.
bacteremic s.
cardiogenic s.
endotoxic s.
hemorrhagic s.
hypovolemic s.
insulin s.
peripheral vascular s.
postoperative s.
septic s.
toxic s.
shoe
reverse-last s.'s
shoelace technique

Shohl solution
Shokeir syndrome
Sholz sclerosis
Shone anomaly
Shorr stain
SHORT
short stature, hyperextensibility of joints
and/or inguinal hernia, ocular
depression, Rieger anomaly, teething
delay
SHORT syndrome
short
s. arm of chromosome (p)
s. course
s. frenulum linguae
s. limb
s. limb dwarfism, saddle nose,
spinal alterations, metaphyseal
striation syndrome
s. rib-polydactyly
s. rib-polydactyly syndrome (SRPS)
s. stature, characteristic facies,
mental retardation, macrodontia,
skeletal anomalies syndrome
s. stature, hyperextensibility of
joints and/or inguinal hernia,
ocular depression, Rieger
anomaly, teething delay (SHORT)
s. stature, microcephaly, mental
retardation, multiple epiphyseal
dysplasia syndrome
s. stature, microcephaly, syndactyly,
dysmorphic face, mental
retardation syndrome
s. tandem repeat typing (STR
typing)
short-axis view
short-bowel syndrome
shortened cervix
shortening
rhizomelic s.
rhizomelic limb s.
uterosacral s.
shorthand vertical mattress stitch
short-increment sensitivity index (SISI)
short-limb
s.-l. dwarfism
s.-l. dystrophy
short-rib dwarfism
Shoshin beriberi
shotgun method
shotty node

S

NOTES

shoulder
s. dystocia
s. presentation
show
bloody s.
Shprintzen-Goldberg craniosynostosis syndrome
Shprintzen syndrome
SHR
sinusoidal heart rate
SHS
Sutherland-Haan syndrome
Shug male contraceptive device
shunt, shunting
aortopulmonary s.
arteriovenous s.
atrioventricular s.
AV s.
Blalock-Taussig s.
Codman Accu-Flow s.
Denver hydrocephalus s.
fetoamniotic s.
H-H neonatal s.
intrapulmonary s.
Kasai peritoneal venous s.
left-to-right s.
LeVeen s.
parietal s.
prenatal placement of thoracoamniotic s.
Pudenz s.
right-to-left s.
thoracoamniotic s.
ventriculoamniotic s.
ventriculoperitoneal s.
VP s.
Waterston s.
Shur-Clens
Shur-Seal
Shur-Strip
Shute forceps
Shutt suture punch system
Shwachman-Bodian syndrome
Shwachman-Diamond syndrome
Shwartzman reaction
Shy-Drager syndrome
Shy-Magee syndrome
SIADH
syndrome of inappropriate secretion of antidiuretic hormone
sialic
s. acid
s. acid storage disease
sialidase
sialidosis
sialorrhea
sialuria, Finnish type
sialyl Tn antigen

sialyted Lewis A antigen
Siamese
S. twins
sibling oocyte
sicca syndrome
sicchasia
sickle
s. cell
s. cell anemia
s. cell-beta-thalassemia disease
s. cell crisis
s. cell dactylitis
s. cell disease
s. cell hemoglobin (HbS)
s. cell hemoglobin C (HbsC)
s. cell-hemoglobin C disease (S-C disease)
s. cell-hemoglobin D disease
s. cell-hemoglobin S disease
s. cell nephropathy
s. cell screening
s. cell thalassemia
s. cell-thalassemia disease
s. cell trait
s. thalassemia (HbS-Thal)
Sickledex test
sicklemia
sickness
acute mountain s. (AMS)
morning s.
serum s.
SID
sudden infant death
side effect
sideroblastic anemia
siderophagic cyst
sidewalls
convergent s.
SIDS
sudden infant death syndrome
sulfoiduronate sulfatase deficiency
Siegel otoscope
Siegert sign
Siemens
S. Servo 900C ventilator
S. Servo 300 ventilator
S. SI 400 ultrasound
S. Sonoline SI-400 ultrasound system
S. Vision MRI
Siemens-Bloch pigmented dermatosis
Siemens-Elema Servo 900C ventilator
Siemerling-Creutzfeldt syndrome
Sierra-Sheldon tracheotome
SieScape imaging
sievert (Sv)
SIFT
transvaginal intrafallopian sperm transfer

Siggaard-Andersen nomogram
sigma tumor marker
sigmoid
    s. colon
    s. pouch of Pratt
sigmoiditis
sigmoidoscopy
sign
    Ahlfeld s.
    Alstrom s.
    Arnoux s.
    Babinski s.
    banana s.
    Beccaria s.
    Béclard s.
    Biederman s.
    Blumberg s.
    Bolt s.
    Borsieri s.
    Braxton Hicks s.
    brim s.
    Brudzinski s.
    Calkins s.
    Chadwick s.
    chandelier s.
    cherub s.
    Chvostek s.
    Comby s.
    Coopernail s.
    cranial s.
    crenation s.
    Cullen s.
    Dalrymple s.
    Danforth s.
    Darier s.
    double-bleb s.
    double-bubble s.
    dovetail s.
    dragon s.
    Elliot s.
    fadir s.
    Federici s.
    flag s.
    Foerster s.
    fontanel s.
    Galeazzi s.
    Gauss s.
    Golden s.
    Goldstein s.
    Goodell s.
    Gottron s.
    Gowers s.

    Granger s.
    Grisolle s.
    groove s.
    Hahn s.
    halo s.
    halo s. of hydrops
    harlequin s.
    Hartman s.
    Hegar s.
    Hellendall s.
    Higoumenakia s.
    Hoehne s.
    Homans s.
    Hutchinson s.
    Jacquemier s.
    jello s.
    Kanter s.
    Kantu s.
    Kergaradec s.
    Kernig s.
    Kleppinger envelope s.
    Krisovski s.
    Küstner s.
    Ladin s.
    lemon s.
    MacDonald s.
    Macewen s.
    Mayor s.
    Metenier s.
    Mirchamp s.
    Nager s.
    Nelson s.
    Nikolsky s.
    Olshausen s.
    Ortolani s.
    palmomandibular s.
    Parrot s.
    Pastia s.
    Perez s.
    peroneal s.
    Pinard s.
    Piskacek s.
    placental s.
    puddle s.
    pyloric string s.
    Radovici s.
    Rasch s.
    Rinman s.
    Romaña s.
    sail s.
    Salmon s.
    Saunders s.

S

NOTES

sign (*continued*)
scarf s.
Schick s.
setting-sun s.
Siegert s.
Simon s.
Sisto s.
Spalding s.
square window s.
Stellwag s.
Sumner s.
Tanyoz s.
Tenney-Parker s.
Toriello-Carey s.
Tresilian s.
Trousseau s.
turtle s.
twin peak s.
Vipond s.
vital s.
Von Fernwald s.
von Graefe s.
Weill s.
Wimberger s.
Wreden s.
Zaufal s.
SignaDress dressing
signal
abnormal feedback s.
centromeric s.
s. node
signet ring cell carcinoma
significance
atypical glandular cells of
uncertain s. (AGCUS, AGUS)
atypical glandular cells of
unknown s.
atypical squamous cells of
undetermined s. (ASCUS)
Siker laryngoscope
SIL
squamous intraepithelial lesion
SIL/ASCUS lesion
Silastic
S. band
S. cup extractor
S. ring
S. silo reduction of gastroschisis
Silc extractor
silent
s. allele
s. amnionitis
s. fetal heart rate pattern
s. gene
s. oscillatory pattern
s. pelvic inflammatory disease
Silfedrine
Children's S.

silicone
s. band application
s. implant
s. implant leakage
s. injection
s. microimplant
s. plug
silicosis
Sil-K
S.-K OB
S.-K OB barrier
silk suture
Silon
S. tent
S. wound dressing
Silver
S. dwarfism
S. syndrome
silver
s. cell
s. nitrate
s. nitrate conjunctivitis
s. nitrate eye prophylaxis
s. wire suture
Silverman-Anderson index
Silverman score
Silver-Russell
S.-R. dwarfism
S.-R. syndrome
Silverskiöld syndrome
Sim
S. SC-20 formula
S. SC-24 formula
S. SC-40 formula
simethicone
simian
s. crease
s. immunodeficiency virus (SIV)
s. lines
Similac
S.-20
S.-24-LBW with whey and iron
S.-24 with iron
S. PM-60/40
S. Special Care-24
S. Special Care formula
similar
s. twins
Simmonds
S. disease
S. syndrome
Simon
S. position
S. sign
Simonton technique
simple
s. colloid goiter
s. cyst

s. mastectomy
s. sequence repeat
s. urethritis
s. virilizing congenital adrenal hyperplasia (SV-CAH)
s. vulvectomy

simplex
epidermolysis bullosa s.
herpes s.
lichen s.
nevus s.
toxoplasmosis, other agents, rubella, cytomegalovirus, herpes s. (TORCH)

Simpson
S. dysmorphia syndrome (SDYS)
S. dysplasia syndrome
S. forceps
S. uterine sound

Simpson-Golabi-Behmel (SGB)
S.-G.-B. syndrome (SGBS)

Sims
S. curette
S. position
S. retractor
S. uterine sound
S. vaginal speculum

Sims-Huhner test
SIMV
synchronized intermittent mandatory ventilation

Simview 3000
SIN
salpingitis isthmica nodosa

Sinai System
Sinarest 12 Hour Nasal Solution
sincipital presentation
Sine-Aid IB
Sinequan
Sinex Long-Acting
single
s. cell biopsy
s. donor apheresis platelet (SDAP)
s. gene defect
s. intrauterine death
s. kidney
s. nucleotide polymorphism
s. photon emission computed tomography (SPECT)
s. primer amplification reaction
s. shot fast spin echo (SSFSE)
s. site BRACA

s. stranded conformational polymorphism

single-breech presentation
single-dose methotrexate therapy
single-energy photon absorptiometry
single-field hyperthermia combined with radiation therapy
single-film cholangiography
single-footling presentation
single-gene disorder
single-photon absorptiometry
singleton
breech s.
s. breech presentation
s. fetus
s. infant
s. pregnancy

single-tooth tenaculum
single-use diagnostic system (SUDS)
Singley forceps
sinistrocardia
sinistrocerebral
sinistrotorsion
sinobronchitis
Sinografin
sinovaginal bulb
Sinubid
Sinufed
sinus, pl. **sinus, sinuses**
s. abruption
s. arrest
s. arrhythmia
s. bradycardia
s. histiocytosis
Motrin IB S.
paranasal s.
pilonidal s.
s. tachycardia
urogenital s.
uterine s.
uteroplacental s.
s. of Valsalva
Valsalva s.
s. venosus defect

sinusoid
sinusoidal
s. fetal heart rate
s. heart rate (SHR)

Sioux alarm
Sipple syndrome
Sippy diet

NOTES

S

475

SIR
standardized incidence ratio
Sirecust 404N neonatal monitoring system
sireniform fetus
sirenomelia
s. sequence
sirenomelic fetus
SIRS
systemic inflammatory response syndrome
SIS
saline infusion sonography
saline infusion sonohysterography
Second International Standard
SISI
short-increment sensitivity index
SISI test
sister
s. chromatid
s. chromatid exchange
S. Mary Joseph nodule
Sisto sign
site
antibody reaction s.
antigen binding s.
binding s.
bleeding s.
fragile chromosome s. (FRA, fra)
placental bleeding s.
puncture punch s.
reactive s. (RS)
restriction enzyme cutting s.
sequence tagged s.
transcription start s.
sitting position
situ, in situ
adenocarcinoma in s.
carcinoma in s. (CIS)
ductal carcinoma in s. (DCIS)
placenta in s.
residual in s.
vulvar carcinoma in s.
situs
s. abnormality
s. ambiguus
s. inversus
s. inversus totalis
s. inversus totalis syndrome
s. inversus viscerum
s. perversus
s. solitus
s. transversus
sitz bath
SIV
simian immunodeficiency virus
six-fingered dwarfism
sixth disease

size
corpus luteum s.
focal spot s.
gestational sac s. (GSS)
maternal s.
tumor s.
uterine s.
size-date discrepancy
Sjögren-Larsson syndrome
Sjögren syndrome
SK-Amitriptyline
skate flap technique
skein
craniotubular dysplasia, growth retardation, mental retardation, ectodermal dysplasia, loose s.
skeletal
s. abnormalities, cutis laxa, craniostenosis, psychomotor retardation, facial abnormalities (SCARF)
s. abnormality
s. calcium deficiency
s. and cardiac malformations-thrombocytopenia syndrome
s. dysplasia
s. dysplasial, joint laxity, mental retardation syndrome
s. dysplasia, sparse hair, dental anomalies syndrome
s. growth
s. maturation
skeleton
gill arch s.
skeleton-skin-brain syndrome
Skene
Bartholin, urethral, S. (BUS)
S. duct
S. gland
skin
alligator s.
s. atrophy
s. calcification
collodion s.
congenital localized absence of s. (CLAS)
crocodile s.
s. disease
dusky s.
fish s.
s. fold
gelatinous s.
India rubber s.
s. lesion
s. mastocytosis-hearing loss-mental retardation syndrome
meconium-stained s.
mottling of s.

pallor of s.
parchment s.
porcupine s.
s. staple
s. tag
s. temperature
s. test
s. thickening
s. trigger theory
vulvar s.
skin-eye-brain syndrome
skinning
s. colpectomy
s. vulvectomy
skinny needle biopsy
skip area
Sklar
S. aseptic germicidal cleaner
S. aseptic germicidal disinfectant
S. cream
S. foam
S. Kleen liquid
S. Kleen powder
S. lube
S. polish
S. scrub
Sklarasol
Sklarsoak disinfectant
4S knot
SK-pramine
skull
cloverleaf s.
coronal suture line of s.
s. fracture
hot cross bun s.
lacunar s.
maplike s.
natiform s.
strawberry shaped s.
sutures of s.
tower s.
West-Engstler s.
SKY
spectral karyotype
SKY epidural pain control system
Sky-Boot stirrup system
SLA
superficial linear array
SLA transducer
slapped cheek rash
SLAS
salt-losing adrenogenital syndrome

Slavianski membrane
SLE
systemic lupus erythematosus
sleep
s. apnea
hour of s. (hs)
non-REM s.
twilight s.
sleeping habit
sleeve
Reid s.
SLI
subdermal levonorgestrel implant
slick-gut syndrome
slide
Testsimplets prestained s.
sliding
s. hernia
s. lock
sling
Aldridge rectus fascia s.
s. arm
fascia lata suburethral s.
levator s.
Martius flap and fascial s.
Mersilene mesh s.
modified s.
s. procedure
pulmonary artery s.
Raz s.
retropubic s.
suburethral s.
two-team s.
slipknot
Duncan s.
Roeder s.
Roeder loop s.
Weston s.
SLO
Smith-Lemli-Opitz syndrome
SLO syndrome
Slo-bid
Slo-Niacin
Slo-Phyllin
Slo-Salt
Slo-Salt-K
Slotnick-Goldfarb syndrome
sloughed urethra syndrome
slow
S. Fe
S. Fe with folic acid

NOTES

S

**slow-channel congenital myasthenic
syndrome (SCCMS)**
Slow-K
Slow-Mag
slow-release sodium fluoride
Slow-Trasicor
SLUDGE
> salivation, lacrimation, urination,
> defecation, gastrointestinal distress and
> emesis

Sly
> S. disease
> S. syndrome

SMA
> sequential multiple analysis

S-M-A formula
small
> s. bowel
> s. bowel endometriosis
> s. bowel strangulation
> s. cell carcinoma
> s. chromosome
> s. for gestational age (SGA)
> s. intestine decompression
> s. left colon syndrome
> s. patella syndrome
> s. single copy

small-for-dates
small-for-gestational-age infant
smallpox vaccine
SMART
> surgical myomectomy as reproductive
> therapy

SMC
> supernumerary marker chromosome

Smead-Jones
> S.-J. closure
> S.-J. closure of peritoneum and
> fascia

smear
> anal Pap s.
> ASCUS s.
> cervical s.
> cytologic s.
> low-grade positive s.
> LSIL Pap s.
> Pap s.
> Papanicolaou s.
> saline wet s.
> scrape and s.
> squash and s.
> ThinPrep s.
> vaginal s.
> vaginal irrigation s. (VIS)
> wet s.

smegma
> s. clitoridis

> s. embryonum
> s. preputii

Smellie
> S. method
> S. scissors

Smellie-Veit method
S-methionine-labeled polypeptide
SMG22
smiling
> s. facies
> s. incision

Smith
> S. pessary
> S. syndrome

Smith-Fineman-Myers syndrome (SFMS)
Smith-Lemli-Opitz syndrome (SLO)
Smith-Magenis syndrome (SMS)
Smith-McCort dwarfism
Smith-Theiler-Schachenmann syndrome
SMM
> scintimammography

smoke
> s. evacuator
> s. inhalation
> s. plume
> s. removal tube (SRT)
> second-hand s.

smooth
> s. muscle
> s. muscle contraction

SMS
> Smith-Magenis syndrome

SMZ-TMP
> trimethoprim-sulfamethoxazole

snapshot GRASS technique
snare
> Reiner-Beck s.

SNJ
> nevus sebaceous of Jadassohn

Sn-mesoporphyrin
snout reflex
snowflake pattern
snowstorm appearance
Sn-protoporphyrin
snuffles
Snyder-Robinson syndrome (SRS)
soak
> Cidex s.

soap
> Basis s.
> pHisoHex s.
> TLC antiseptic s.

soapsuds enema
Soave abdominal pull-through procedure
SOC
> surgical overhead canopy

social
> s. drinking

s. drug
s. factor effect
s. issue
s. parents
s. sex
S. Support Scale for Children test (SSSC)
society
American Fertility S. (AFS)
American Urogynecologic S.
S. for Assisted Reproductive Technology (SART)
S. of Gynecologic Oncologists (SGO)
S. of Gynecologic Oncology
S. for Gynecologic Pathology
sociobiologic
socioeconomic status
Soderstrom-Corson electrode
sodium
alendronate s.
ampicillin sodium/sulbactam s.
aqueous penicillin s.
s. balance
s. bicarbonate
s. bromide
cefazolin s.
cefoperazone s.
cefotaxime s.
cefoxitin s.
ceftriaxone s.
cephalothin s.
s. chloride
s. citrate with citric acid
cromolyn s.
s. cyclamate
1D s. dodecyl sulfate gel
s. diatrizoate
diclofenac s.
Diphenylan s.
docusate s.
s. equilin sulfate
estramustine phosphate s.
s. estrone sulfate
s. etidronate
s. excretion
fluorescein s.
s. fluoride
heparin s.
imipenem-cilastaten s.
s. iodide
s. iodide I-125

s. iodide I-131
methicillin s.
mezlocillin s.
nafcillin s.
naproxen s.
s. nitroprusside
oxacillin s.
oxychlorosone s.
Pentothal S.
piperacillin sodium/tazobactam s.
porfimer s.
s. requirement
s. restriction
s. salicylate
senna concentrate/docusate s.
tazobactam s.
s. tetradecyl sulfate
thiopental s.
s. thiosulfate
s. valproate
zobactam s.
zomepirac s.
sodium-free formula
sodomize
sodomy
Soehendra dilator
soft
s. hands syndrome
s. radiation grid Pb4/27
s. seal catheter
s. tissue abnormality
s. tissue ovarian neoplasm
S. Torque uterine catheter
S.-Wand atraumatic tissue manipulator balloon
Soft-Cell catheter
softener
stool s.
Softgels
Vita-Plus E S.
softness
variable s. (VS)
Softpatch
Impress S.
software
Medical Manager s.
OBG Clinical Records Manager s.
OBG LabTrack s.
SOHS
severe ovarian hyperstimulation syndrome
Sohval-Soffer syndrome

S

NOTES

Solar Beam medical examination light
Solatene
Solazine
solder
    protein s.
sole crease
Solfoton
Solganal
solid
    s. phase radioimmunoassay
    s. phase sandwich
solid-phase enzyme immunoassay
solitary
    s. dilated duct
    s. kidney
solitus
    situs s.
Solium
Solomon-Fretzin-Dewald syndrome
Solomon syndrome
Solos
    S. disposable cannula
    S. disposable trocar
soluble
    s. antigen excess
    s. gas technique
    s. intercellular adhesion molecule 1
    s. RNA
solution
    Afrin Nasal S.
    Allerest 12 Hour Nasal S.
    Bouin s.
    Burow s.
    Chlorphed-LA Nasal S.
    clindamycin phosphate topical s.
    colloid s.
    Cornoy s.
    crystalloid s.
    Dakin antibacterial s.
    Denhardt s.
    Dey-Drop Ophthalmic S.
    Dianeal dialysis s.
    Dristan Long Lasting Nasal S.
    DuraPrep surgical s.
    Duration Nasal S.
    Earle balanced salt s.
    Fungoid AF Topical S.
    Hartmann s.
    hetastarch s.
    lacmoid staining s.
    lactated Ringer s.
    Locke s.
    Lugol iodine s.
    Melanex topical s.
    modified Ham F-10 s.
    Monsel s.
    Neo-Synephrine 12 Hour Nasal S.
    NTZ Long Acting Nasal S.

    Pedialyte oral electrolyte
      maintenance s.
    PVP s.
    Ringer s.
    saline s.
    Schiller s.
    Schlesinger s.
    Shohl s.
    Sinarest 12 Hour Nasal S.
    sperm viability staining s.
    Transeptic cleansing s.
    Twice-A-Day Nasal S.
    Tyrode s.
    4-Way Long Acting Nasal S.
somaclonal variation
somamammotropin
    chorionic s.
SomaSensor
somatic
    s. cell
    s. cell-derived growth factor (SCF)
    s. chromosome
    s. differentiation
    s. hybrid
    s. nervous system feedback loop
somatoliberin
somatomedin
    s. C
    s. level
Somatom Plus computed tomography
somatopagus
somatoschisis
somatostatin
somatotridymus
somatotrope
somatotropinoma
somatrem
    s. growth hormone
somatropin
    s. growth hormone
Somer uterine elevator
somite
    s. embryo
    s. formation
Sommer syndrome
somnambulism
somnogram
Somogyi phenomenon
Somophyllin
Sonicaid
    S. Axis monitor
    S. SYSTEM 8000 fetal monitor
    S. Vasoflow Doppler system
Sonoclot
    S. coagulation analyzer
    S. test
Sonoda syndrome
Sono-Gram fetal ultrasound image card

sonographic finding
sonography
    Acuson computed s.
    color Doppler s. (CDS)
    laparoscopic s.
    power Doppler s.
    real-time s.
    saline infusion s. (SIS)
    transvaginal s. (TVS)
    transvaginal color Doppler s. (TV-CDS)
    vaginal s.
sonohysterogram
sonohysterography (SHG)
    saline infusion s. (SIS)
Sonoline
    S. Prima ultrasound
    Sonoline Sienna ultrasound
      system S.
sonolucency
sonolucent
    s. tissue
sonomicroscopy
SonoMix ultrasound gel
Sonopsy ultrasound-guided breast biopsy system
SonoSite 180
SonoVu US aspiration needle
S.O.P.
    Genoptic S.O.P.
Sopamycetin
Sopher ovum forceps
Sorbitrate
Soriatane
sorivudine
Sorsby syndrome
sorter
    fluorescence-activated cell s. (FACS)
    magnetically activated cell s. (MACS)
sorting
Sotos
    S. sequence
    S. syndrome
Sotradecol
souffle
    fetal s.
    funic s.
    funicular s.
    mammary s.
    placental s.

    umbilical s.
    uterine s.
Soules intrauterine insemination catheter
sound
    active bowel s.'s
    bowel s.'s
    cracked pot s.
    fetal heart s.'s
    grating s.
    heart s.'s
    hyperactive bowel s.'s
    hypoactive bowel s.'s
    Korotkoff s.
    normoactive bowel s.'s
    Pharmaseal disposable uterine s.
    Simpson uterine s.
    Sims uterine s.
    urethral s.
    uterine s.
    Waring blender s.
sound-stimulated fetal movement
source
    cesium s.
    dummy s.
    MX2-300 xenon quality light s.
source-to-axis distance (SAD)
source-to-skin distance (SSD)
south
    S. African genetic porphyria
    S. African tick fever
Southern
    S. blot
    S. blot technique
    S. blot test
Soyacal
    S. IV fat emulsion
Soyalac formula
SP
    sacroposterior
    spastic paraplegia
    Cordran SP
SPA
    sperm penetration assay
space
    anechoic s.
    Bogros s.
    extraembryonic celomic s. (EECS)
    intervillous s.
    lymphovascular s.
    perivitelline s.
    rectosacral s.
    rectrovaginal s.

S

NOTES

481

space *(continued)*
    s. of Retzius
    s. of Retzius bleeding
    subchorial s.
    vesicocervical s.
    volume of dead s.
    yolk s.
spacer
    dummy s.
Spalding sign
span
    fertilizable life s.
**Spangler papular dermatitis of
pregnancy**
Spanish fly
sparfloxacin
Sparine
sparing
    brain s.
spasm
    flexion s.
    greeting s.
    infantile s.
    jackknife s.
    levator ani s.
    nodding s.
    urethral s.
    X-linked infantile s.
spasmodic dysmenorrhea
spasmus nutans
spastic
    s. ataxia
    s. diplegia
    s. diplegia syndrome
    s. paraplegia (SP)
    s. quadriplegia, congenital
      ichthyosiform erythroderma,
      oligophrenia syndrome
    s. quadriplegia, retinitis pigmentosa,
      mental retardation syndrome
    s. spinal paralysis
spastica
    paralysis spinalis s.
    paraplegia s.
spasticity
spatula
    Aylesbury s.
    Ayre s.
    Cytobrush s.
    s. foot
    Milex s.
    Rolon s.
Spearman-Brown prediction formula
Spearman correlation coefficient
Special Care formula
**specialized tissue aspirating resectoscope
(STAR)**
speciation

species
    reactive oxygen s.
species-specific antibody
specific
    s. immunotherapy
    s. phosphodiesterase inhibitor
    s. urethritis
specificity
    assay s.
specimen
    catheter s.
    clean-catch urine s.
    hemolyzed s.
    lost surgical s.
    midstream urine s.
    nature of s.
    xanthochromic s.
speckled irides
Speck test
SPECT
    single photon emission computed
    tomography
Spect-Align laser system
Spectazole
spectinomycin
spectra (*pl. of* spectrum)
spectral
    s. Doppler
    s. karyotype (SKY)
Spectranetics catheter
Spectrobid
    S. Tablet
spectrometer
    Digilab FTS 40A s.
spectrometry
    mass s.
spectrophotometer
spectrophotometry
    atomic absorption s.
spectroscopy
    atomic absorption s.
    infrared s.
    near-infrared s.
spectrum, pl. **spectra**
    Spectra-Diasonics ultrasound
    Doppler shift spectra
    electromagnetic s.
    Spectra 400 extended surveillance
      and alert system
    oculoauriculovertebral s. (OAVS)
    S. stethoscope
specula (*pl. of* speculum)
specular echo
Speculite chemiluminescent light
speculoscopy
    Pap plus s. (PPS)
speculum, pl. **specula**
    Amko vaginal s.

Auvard s.
bivalve s.
blackened s.
duckbill s.
s. examination
Graves bivalve s.
Halle infant nasal s.
Holinger infant esophageal s.
Huffman infant vaginal s.
illuminated vaginal s.
Kogan endocervical s.
long weighted s.
Pederson vaginal s.
Prima Series LEEP s.
Sani-Spec vaginal s.
Sims vaginal s.
SRT vaginal s.
Vu-Max vaginal s.
weighted s.
speech reception threshold (SRT)
Spee embryo
spell
A&B s.
Spemann induction
Spence
axillary tail of S.
S. axillary tail
S. and Duckett marsupialization
S. procedure
Spencer
S. probe
S. stitch scissors
sperm
acrosome-intact s.
s. agglutination test
s. allergy
anonymous donor s. (ADS)
s. attrition
s. bank
s. capacitation
s. capacitation medium
s. chromatin decondensation
s. count
s. donation
donor s.
s. donor
epididymal s.
frozen s.
s. function test
s. granuloma
haploid s.
s. immobilization test

microsurgical extraction of
ductal s. (MEDS)
motile s.
s. motility
muzzled s.
nonmotile s.
s. penetration assay (SPA)
s. progression scale 0 through 4
s. reservoir
S. Select sperm recovery system
s. separation
s. surface antibody
s. tail protein phosphorylation
s. transport
s. viability staining solution
washed s.
s. washing insemination method
(SWIM)
Spermac stain
spermagglutination
sperm-aster
spermatic
s. cord
spermatid
spermatin
spermatocele
artificial s.
spermatocide
spermatogenesis
spermatogonia
spermatotoxin
spermatozoon, pl. **spermatozoa**
haploid s.
round-headed acrosomeless
spermatozoa
washed spermatozoa
sperm-cervical mucus contact (SCMC)
sperm-counting fluid
sperm-egg adhesion
sperm-free ejaculate
spermicide
spermidine
spermine
spermiogenesis
**SpermMAR mixed antiglobulin reaction
test**
sperm-mediated
sperm-mucus interaction
sperm-oocyte interaction
sperm-zona pellucida binding
Spersacarpine
S phase fraction

NOTES

S

sphenocephaly
sphenoid fontanel
sphenopagus
spherocytosis
    hereditary s.
sphincter
    AMS 800 artificial urethral s.
    anal s.
    artificial urethral s. (AUS)
    s. deficiency
    esophageal s.
    external anal s.
    internal anal s.
    lower esophageal s. (LES)
    patulous rectal s.
    s. repair
    urethral s.
sphingolipidosis
sphingolipodystrophy
    Gm3 hematoside s.
sphingomyelin
spiculated lesion
spider
    s. angioma
    s. finger
    s. nevus
    vascular s.
Spiegel
    S. criteria
    S. method
Spiegelberg criteria
Spielberger State Anxiety Inventory
Spielmeyer-Vogt disease
spigelian hernia
spike
    Monoscopy locking trocar with Woodford s.
    s. and wave complex
spilled
    filled and s.
spina
    s. bifida
    s. bifida cystica
    s. bifida occulta
spinal
    s. analgesia
    s. anesthesia
    s. angioma
    s. blockade
    s. bone loss
    s. compression fracture
    s. cord
    s. cord injury
    s. headache
    s. meningocele
    s. muscular atrophy-mental retardation syndrome

    s. muscular atrophy, microcephaly, mental retardation syndrome
spindle cell
spine
    cleft s.
    cloven s.
    dysraphia of s.
Spinelli operation
spinnbarkeit
spinning-top deformity
spinocerebellar
    s. ataxia
    s. ataxia-dysmorphism syndrome
    s. degenerative disease
spinosa
    ichthyosis s.
spinulosus
    lichen s.
spiral
    s. artery
    s. electrode
spiramycin
Spirette
    CCD S.
spirochete
spironolactone
Spitz nevus
splanchnic
    s. fold
    s. pelvic pain
splanchnocystica
    dysencephalia s.
splanchnopathy
splanchnopleuric
spleen
    accessory s.
splenectomy
splenic
    s. artery aneurysm
    s. flexure
    s. pregnancy
    s. rupture
    s. sequestration syndrome
    s. tissue
    s. torsion
splenium
    absent s.
splenomegaly
splenosis
splicing
    gene s.
    RNA s.
splint
    Denis Browne clubfoot s.
    Freidman s.
    Ilfeldt s.
    talipes hobble s.

split

cricoid s.

s. foot, microphthalmia, cleft lip/palate-mental retardation syndrome

s. hand-cleft lip/palate and ectodermal dysplasia (SCE)

s. hand/feet syndrome

s. sheath catheter

split-flap technique

split-foot deformity

split-thickness graft

splitting

blastocyst s.

embryo s.

muscle s.

spoke-wheel palpation

SPONASTRIME

spondylar changes-nasal anomaly-striated-metaphyses

SPONASTRIME dysplasia

spondylar

s. changes-nasal anomaly-striated-metaphyses (SPONASTRIME)

s. and nasal alterations-striated metaphyses syndrome

spondylitis

ankylosing s.

spondylocostal dysplasia syndrome

spondyloepimetaphyseal

s. dysphasia with myotonia

s. dysplasia (SEMD)

spondyloepiphyseal

s. dysplasia

s. dysplasia congenita syndrome

s. dysplasia-diabetes mellitus syndrome

s. dysplasia tarda-mental retardation syndrome

spondylohumerofemoral hypoplasia

spondylolisthesis

spondylolysis

spondylometa-epiphyseal dysplasia-extreme short stature syndrome

spondylometaphyseal

s. dysplasia

s. dysplasia-short limb-abnormal calcification syndrome

s. dysplasia, X-linked

spondyloperipheral dysplasia

spondylothoracic

s. dysplasia

s. dysplasia syndrome

sponge

absorbable gelatin s.

contraceptive s.

s. forceps

Incert bioabsorbable s.

intravaginal s.

Lapwall s.

Protectaid contraceptive s.

Ray-Tec s.

s. stick

Today vaginal contraceptive s.

vaginal s.

Weck-cel s.

sponge-holding forceps

spongiform encephalopathy

spongioblastoma

spongiosum

stratum s.

spongy degeneration of infancy

spontaneous

s. abortion (SAB)

s. abortion material

s. amputation

s. apoptosis

s. bacterial peritonitis (SBP)

s. breech

s. breech extraction

s. cephalic delivery

s. evolution

s. gangrene of newborn

s. involution

s. labor

s. menstrual cycle

s. miscarriage

s. ovulation

s. preterm birth (SPTB)

s. preterm delivery

s. preterm labor with intrapartum demise

s. remission

s. rupture of membranes (SROM)

s. vaginal delivery (SVD)

s. version

s. vertex

spoon forceps

sporadic

s. chromosome abnormality

s. nonfamilial clear cell carcinoma

Sporanox

S

NOTES

spore
sporotrichosis
SPOT
    salpingitis after previous tubal occlusion
    salpingitis in previously occluded tubes
spot
    blood s.
    blue s.
    blueberry muffin s.
    Brushfield s.
    café au lait s.
    s. compression
    s. compression view
    Fordyce s.
    Koplik s.
    s. magnification
    mongolian s.
    powder burn s.
    strawberry s.
spotlight
    KDC-Healthdyne nonfluorescent s.
spotted fever
Sprangeler-Wiedemann syndrome
spray
    butorphanol tartrate nasal s.
    CaldeCort Anti-Itch Topical S.
    DDAVP nasal s.
    ipratropium bromide nasal s.
    Itch-X s.
    Nasarel Nasal S.
    Ony-Clear S.
SprayGel Adhesion Barrier System
spread
    halstedian concept of tumor s.
    lymphatic s.
    transcelomic s.
    vessel s.
Sprengel
    S. anomaly
    S. deformity
spring clip application
sprout
    syncytial s.
sprouting
    nerve s.
sprue
    celiac s.
SPT
    septic pelvic thrombophlebitis
SPTB
    spontaneous preterm birth
SPTL vascular lesion laser
spud dissector
spun
    s. hematocrit
    s. urine
    s. urine sediment

spuria
    melena s.
    placenta s.
spurious pregnancy
Spurway syndrome
sputum
    s. culture
    s. cytology
squalamine
squama
squamocolumnar junction
squamous
    s. cell
    s. cell carcinoma
    s. dysplasia
    s. epithelium
    s. intraepithelial lesion (SIL)
    s. metaplasia
    s. metaplasia of amnion
square
    s. knot
    s. matrix
    Punnett s.
    s. window
    s. window sign
squash and smear
squirming Valsalva
SR
    Indocin SR
    Oramorph SR
    Roxanol SR
Srb syndrome
SRI automated immunoassay analyzer
SROM
    spontaneous rupture of membranes
SRPS
    short rib-polydactyly syndrome
SRS
    Snyder-Robinson syndrome
SRT
    smoke removal tube
    speech reception threshold
    SRT vaginal speculum
SRY
    sex-determining region
SS
    hemoglobin SS
    Uroplus SS
SSC-20 formula
SSC-24 formula
SSCVD
    sterile, spontaneous, controlled vaginal
    delivery
SSD
    source-to-skin distance
SSFSE
    single shot fast spin echo

SSKI
saturated solution of potassium iodide
SSSC
Social Support Scale for Children test
SSVD
sterile, spontaneous vaginal delivery
ST
sacrotransverse
St.
St. Clair-Thompson curette
St. Louis encephalitis
St. Vitus dance
stabilization
stable access cannula
stab wound
staccato cough
stadiometer
Harpenden s.
Stadol
stage
band s.
cleavage s.
cortical supremacy s.
delayed first s.
developmental s.
diakinesis s.
dictyate s.
germinal vesicle s.
s. IIIc papillary tumor of low
malignant potential
indifferent gonadal s.
International Continence Society
prolapse s. I, II
s. IV epithelial ovarian cancer
Jirasek s.
s.'s of labor
leptotene s.
pachytene s.
placental s.
prolapse s. I–IV
Tanner s.
trophectoderm s.
two-part nuclear s.
zygotene s.
staging
clinical s.
FIGO s.
genital prolapse s.
Marshall-Tanner pubertal s.
Northway s.
surgical s.
Tanner s. (Grade I–IV)

stagnation mastitis
STAI-I
State-Trait-Anxiety Index-I
stain, staining
Betke s.
blood pigment s.
Bryan-Leishman s.
DA-DAPI s.
eosin-4 s.
Feulgen s.
Giemsa s.
Gram s.
Gram-Weigert s.
Grimelius s.
iodine s.
Kinyoun s.
Kleihauer-Betke s.
KOH s.
Leder s.
Masson-Fontana s.
meconium s.
Movat s.
PAS s.
port-wine s.
Shorr s.
Spermac s.
TUNEL s.
Warthin-Starry s.
Wright s.
stainless steel suture
stalk
allantoic s.
body s.
infundibular s.
yolk s.
stalking
celery s.
Stallworthy placenta
Stamey
S. catheter
S. modification of Pereyra
procedure
S. needle
S. operation
Stamey-Malecot catheter
Stamey-Pereyra needle suspension
stammering
s. bladder
stance
Buddha s.
standard
s. curve

NOTES

standard *(continued)*
    s. deviation (SD)
    s. error of the mean (SEM)
    MapMarkers fluorescent DNA
       sizing s.
    protein s.
    Second International S. (SIS)
standardized incidence ratio (SIR)
standing position
Stanley Way procedure
Staphcillin
staphylococcal pneumonia
staphylococcal-scalded skin syndrome
Staphylococcus
    S. aureus
    S. epidermidis
    S. saprophyticus
staphylococcus, pl. **staphylococci**
    coagulase-positive s.
staple
    absorbable s.
    metallic skin s.
    skin s.
    titanium s.
stapler
    Auto Suture Multifire Endo GIA
       30 s.
    EEA s.
    Endo-GIA s.
    Endo-GIA30 suture s.
    Endopath endoscopic articulating s.
    GIA 60 s.
    GIA 80 s.
    poly roticulator 55 s.
    Precise disposable skin s.
    Roticulator 55 s.
    TA 55 s.
    thoracoabdominal 55 s.
    30-V-3 s.
    Vista disposable skin s.
STAR
    specialized tissue aspirating resectoscope
    Study of Tamoxifen and Raloxifene
star
    s. effect
    S. ventilator
Stargardt disease
Stargate falloposcopy catheter
Starling
    S. equation
    S. equilibrium
    S. law of transcapillary exchange
STARRT
    selective tubal assessment to refine
    reproductive therapy
       STARRT falloscopy system
startle
    s. disease

    s. pattern
    s. reaction
    s. reflex
    s. response
starvation
    accelerated s.
    s. ketoacidosis
    s. ketosis
stasigenesis
stasis
    urinary s.
state
    accompanying mood s.
    fugue s.
    gradient recalled acquisition in the
       steady s. (GRASS)
    menstrual s.
    mood s.
    Profile of Mood S.'s (POMS)
state-of-the-art radiation
State-Trait-Anxiety Index-I (STAI-I)
static
    s. B-scanner
    s. immersion
Staticin
    O-V S.
station
    0 s.
    complete/complete/+ s.
    fetal s.
statistical model
Statobex
stature
    brittle hair, intellectual impairment,
       decreased fertility, short s.
       (BIDS)
    constitutional short s.
    goniodysgenesis, mental retardation,
       short s. (GMS)
    idiopathic short s.
    maternal s.
status
    acid-base s.
    s. asthmaticus
    s. degenerativus amstelodamensis
    s. dysmyelinisatus
    s. epilepticus
    Karnofsky performance s.
    s. lymphaticus
    s. marmoratus
    nonreassuring fetal s.
    socioeconomic s.
    s. thymicolymphaticus
    s. thymicus
statutory rape
Staudinger reaction
stavudine (d4T)
S-T Cort Topical

STD
   sexually transmitted disease
steatorrhea
   idiopathic s.
steely-hair syndrome
Steiner
   S. canal
   S. electromechanical morcellator
   S. tumor
Steinert
   S. disease
   S. myotonic dystrophy
   S. syndrome
Steinfeld syndrome
Stein-Leventhal
   S.-L. syndrome
   S.-L. type of polycystic ovary
Stelazine
stellate mass
Stellwag sign
stem
   s. cell
   s. cell assay
Stemetil
stenogyria
   agenesis of corpus callosum
      with s.
stenosis
   anorectal s.
   antral s.
   aortic s.
   aqueductal s.
   bladder neck s.
   cervical s.
   cholestasis-peripheral pulmonary s.
   congenital tubular s.
   esophageal s.
   hypertrophic s.
   idiopathic hypertrophic subaortic s.
   infundibular s.
   mitral s.
   postischemic s.
   pulmonary s.
   pyloric s.
   renal artery s.
   subaortic s.
   tracheal s.
   tubular s.
   urethral s.
   valvular pulmonic s.
   X-linked aqueductal s. (XLAS)
Stensen duct

stent
   double-J s.
   Fader Tip ureteral s.
   Lubri-Flex s.
   Percuflex Plus s.
   urinary s.
step-down cannula
Step laparoscopic trocar
stepping reflex
STEPS
   system for thalidomide education and
      prescription safety
   STEPS program
2-step testing
stercoroma
stereocolpogram
stereocolposcope
stereoscopic pelvimetry
stereotactic
   s. breast biopsy
   s. breast biopsy needle
stereotaxis
Steri-Drape 2
sterile
   s. isolation bag
   s. pyuria syndrome
   s. specimen trap
   s., spontaneous, controlled vaginal
      delivery (SSCVD)
   s., spontaneous vaginal delivery
      (SSVD)
   s. vaginal examination (SVE)
sterility
   absolute s.
   adolescent s.
   one-child s.
   relative s.
sterilization
   intermittent s.
   involuntary s.
   microlaparoscopic s.
   tubal s.
   voluntary s.
sterilize
Steri-Strip skin closure
sternocleidomastoid hemorrhage
sternodymus
sternopagus
sternoschisis
sternoumbilical raphe
sternoxiphopagus

S

NOTES

steroid
    adrenal s.
    adrenocortical s.
    anabolic s.
    s. biosynthesis
    s. cell
    s. concentration
    s. conjugate hydrolysis
    s. contraceptive
    endogenous s.
    gonadal s.
    s. hormone
    s. hormone receptor
    17-ketogenic s.
    long-acting contraceptive s.
    low-dose s.'s
    s. metabolism
    s. metabolite
    s. nucleus
    ovarian s.
    placental s.
    s. secretion
    s. secretion inhibition
    sex s.
    s. sulfatase
    s. sulfatase deficiency
    s. therapy
steroidogenesis
    adrenal s.
    adrenocortical s.
    fetal-placental s.
    follicle s.
    ovarian s.
    testicular s.
steroidogenic
    s. aberration
stethoscope
    Allen fetal s.
    Doptone fetal s.
    Medasonic first beat ultrasound s.
    Spectrum s.
    ultrasound s.
Stevens-Johnson syndrome
Stewart-Treves syndrome
S-thalassemia
    hemoglobin S-t. (HbS-Thal)
stick
    sponge s.
    Universalindicator s.
Sticker disease
Stickler syndrome
stiff-baby syndrome
stigma, pl. **stigmata**
stilbestrol
Still
    S. disease
    S. murmur
stillbirth

stillborn infant
Stilling-Türk-Duane syndrome
Stillman cleft
Stilphostrol
Stimate
Stimmler syndrome
stimulant
stimulation
    ACTH s.
    cervical carcinoma s.
    cranial electrical s. (CES)
    endogenous estrogenic s.
    endometrial s.
    exogenous estrogenic s.
    exogenous gonadotropin s.
    follicle maturation s.
    nipple s.
    ovarian s.
    ovulation s.
    oxytocic s.
    pelvic floor electrical s. (PFS)
    prolactin s.
    sensory s.
    tactile s.
    s. test
    vibratory acoustic s. (VAS)
    vibroacoustic s. (VAS)
    visual s.
stimulator
    adrenergic s.
    alpha-adrenergic s.
    hematopoietic system s.
    Innova pelvic floor s.
    long-acting thyroid s. (LATS)
    luteinization s.
    Myogyn II s.
stimulus, pl. **stimuli**
    tactile s.
stippling
    bone s.
stirrups
    Allen laparoscopic s.
    candy-cane s.
    high s.
    Lloyd-Davies s.
    Navratil s.
stitch
    s. abscess
    baseball s.
    Endo S.
    McCall s.
    running imbricating s.
    running locked s.
    shorthand vertical mattress s.
Stocco dos Santos syndrome
stockinette
    impervious s.

stocking
    antiembolism s.
    elastic s.
    Juzo-Hostess two-way stretch compression s.
    leg-compression s.
    pneumatic compression s.
    TED s.
Stock-Spielmeyer-Vogt syndrome
Stoll syndrome
Stolte forceps
stomach
    s. cancer metastasis
    leather-bottle s.
    right-sided s.
stomatitis
    aphthous s.
    herpetic s.
stomatocytosis
stomatomy
stomatoschisis
stomatotomy
stomocephalus
stone
    kidney s.
    renal s.
    womb s.
stool
    acholic s.
    s. colonization
    s. culture for O&P
    currant jelly s.
    s. examination
    s. impaction
    s. softener
STOP
    selective tubal occlusion procedure
stopcock
    Acel s.
stop codon
storage disease
STORCH
    syphilis, toxoplasmosis, rubella, cytomegalovirus, and herpes
    STORCH test
stored sera
stork bite
storm
    thyroid s.
Stormby brush

Stortz
    S. disposable cannula
    S. disposable trocar
Storz
    S. endoscope
    S. infant bronchoscope
    S. laparoscope
Stoxil
strabismus
straddle injury
straight
    s. catheter test
    s. line velocity (VSL)
    s. scissors
straight-back syndrome
Straight-In surgical system
strain
    recombinant inbred s.
strait
    inferior s.
    superior s.
strand
    antisense s.
strangulated hernia
strangulation
    small bowel s.
strangury
S-transferase
    glutathione S.-t.
strap
    Montgomery s.
Strap operation
Strassman
    S. metroplasty
    S. operation
    S. phenomenon
    S. technique
    transverse fundal incision of S.
strata (*pl. of* stratum)
StrataSorb dressing
stratification
stratified squamous epithelium
Stratton-Parker syndrome
stratum, pl. **strata**
    s. basale
    s. compactum
    s. functionale
    s. spongiosum
strawberry
    s. appearance
    s. cervix
    s. hemangioma

S

NOTES

strawberry *(continued)*
- s. mark
- s. nevus
- s. patch
- s. shaped skull
- s. spot
- s. tongue

streak
- gonadal s.
- s. gonads
- intraabdominal s.
- nonfunctional s.
- primitive s.

streaking
- perihilar s.

streaky infiltrate
streblodactyly
Streeter
- S. band
- S. dysplasia
- S. horizon

Strema
strength
- Clocort Maximum S.
- Tums Extra S.

Streptase
streptavidin peroxidase
Streptex rapid strep test
streptococcal
- s. group
- s. infection
- s. pharyngitis
- s. pneumonia
- s. septicemia
- s. vaginitis

Streptococcus
- *S. agalactiae*
- *S. milleri*
- *S. mitis*
- *S. pneumoniae*
- *S. pyogenes*
- *S. viridans*

streptococcus, pl. **streptococci**
- beta-hemolytic s.
- fecal streptococci
- group A s.
- group B s. (GBS)
- group D s.

streptogenes
- erythema s.

streptokinase
- s.-urokinase

streptomycin
streptozocin
streptozocin-induced diabetes
stress
- s. incontinence
- s. incontinence de novo

life s.
- maternal s.
- oxidative s.
- postmenstrual s.
- psychological s.
- s. reaction in exenteration
- s. test
- s. urinary incontinence (SUI)
- visual analog scale for s.

stressed fetus
stretching
- brachial plexus s.

stria, pl. **striae**
- abdominal s.
- striae atrophicae
- striae cutis distensae
- striae gravidarum
- Langhans s.
- Rohr s.

striata
- osteopathia s.

striated
- s. circular muscle
- s. muscle

striation
- hyperostosis generalisata with s.

striatus
- lichen s.

stridor
- inspiratory s.
- laryngeal s.

Stringer technique
string phlebitis
strip
- ColorpHast Indicator S.'s
- Cover-Strip wound closure s.
- fascial s.
- fetal monitoring s.
- leucocyte detection s.
- lung s.
- Mersilene fascial s.
- pHydrion s.
- polypropylene fascial s.
- saturation s.

stripe
- endometrial s.

stripping
- capsular s.
- membrane s.

Stroganoff method
stroke
stroma
- cervical s.
- fibromuscular cervical s.
- gonadal s.
- ovarian s.
- Rh-positive red cell s.
- uterine endolymphatic s.

uterine endometrial s.
uterine epithelial s.
stromal
    s. adenomyosis
    s. cell
    s. development
    s. endometriosis
    s. hyperplasia
    s. hyperthecosis
    s. luteoma
    s. microinvasion
stromal-epithelial interaction
stromatosis
stromelysin
strongyloidiasis
strontium bromide
strophocephaly
strophulus
STR typing
structural
    s. change
    s. gene
    s. heart defect
structure
    genetic fine s.
Strudwick syndrome
struma, pl. **strumae**
    s. ovarii
strumal carcinoid of ovary
Strumpell-Lorrain disease
strutural brain defect
strychnine poisoning
STS
    serologic test for syphilis
Stuart
    S. factor
    S. index
    S. Prenatal vitamins
Stuartnatal 1+1
Stuart-Prower factor
stub thumb
stuck-twin phenomenon
stuck twins
studding
    peritoneal s.
Student-Newman-Keuls test
study
    acoustic stimulation s.
    acute-phase serum s.
    barium s.
    biochemical s.
    CASH s.

cohort s.
cytogenetic s.
cytologic s.
fetal blood s.
Heart and Estrogen/Progestin
    Replacement S. (HERS)
histopathological s.
Oxford Family Planning Association
    Contraceptive S.
Persutte and Lenke s.
postmortem s.
register linkage s.
routine antenatal diagnostic imaging
    ultrasound s.
Royal College of General
    Practioners' Oral Contraception S.
S. of Tamoxifen and Raloxifene
    (STAR)
tissue s.
transesophageal electrophysiologic s.
urodynamic s.
stunted
    s. embryo
    s. fetus
Sturge
    S. disease
    S. syndrome
Sturge-Kalischer-Weber syndrome
Sturge-Weber
    S.-W. angiomatosis
    S.-W. anomalad
    S.-W. syndrome
Sturge-Weber-Dimitri syndrome
Sturge-Weber-Krabbe syndrome
Sturmdorf
    S. hemostatic suture
    S. operation
stuttering
Stüve-Wiedemann (SW)
    S.-W. syndrome (SWS)
sty, stye, pl. **styes**
stylopodium
stype
subacute bacterial endocarditis (SBE)
subaortic
    s. lymph node
    s. stenosis
    s. stenosis-short stature syndrome
subarachnoid
    s. block
    s. hemorrhage

S

NOTES

subareolar
- s. abscess
- s. duct papillomatosis

subcapsular cyst

subchorial
- s. lake
- s. space

subchorionic
- s. hematoma
- s. hemorrhage

subclinical hypothyroidism

subcortical band heterotopia

subcostal retractions

subcutanea
- lipogranulomatosis s.

subcutaneous
- s. fat necrosis
- s. mastectomy

subdermal
- s. contraceptive system
- s. implant
- s. levonorgestrel implant (SLI)

subdural
- s. hematoma
- s. hemorrhage
- s. puncture

subependymal
- s. hemorrhage (SEH)
- s. tuber

subfecundity

subfertility

subfragment-1

subgaleal hemorrhage

subinvolution

subjective probability

sublethal gene

Sublimaze

sublingual
- s. gland
- s. hematoma

submammary mastitis

submaxillary gland

submental hematoma

submetacentric chromosome

submucosa

submucosal
- s. myoma
- s. plexus
- s. urethral augmentation

submucous
- s. fibroid
- s. leiomyoma
- s. myoma

subnormality
- mental s.

subnormal temperature

suboccipitobregmatic diameter

suboptimal surgery

subpanicular area

subpectoral implant

subphrenic
- s. abscess

Sub-Q-Set

subseptate uterus

subsequent fertility

subserosal
- s. myoma
- s. nodule

subserous
- s. fascia
- s. fibroid
- s. myoma

subset
- T-cell s.

substance
- müllerian inhibiting s. (MIS)
- s. P pain neurotransmitter
- vasoactive s.
- s. X

substitute
- Oxygent temporary blood s.
- perflubron emulsion temporary blood s.
- PolyHeme blood s.

substrate
- renin s.

subtotal hysterectomy

subunit
- inhibin s.
- inhibin-A s.

suburethral sling

subzonal
- s. insemination (SUZI)
- s. insertion (SUZI)

succedaneum
- caput s.

succenturiate placenta

successful pregnancy

succinate
- sumatriptan s.

succinylcholine

succinylsulfathiazole

sucking
- s. cushion
- nonnutritive s.
- s. pad
- s. reflex

suckle

suckling

Sucraid

sucralfate

Sucrets Sore Throat

sucrose-free formula

suction
- airway s.
- bulb s.

s. catheter
s. curettage
s. drainage
nasopharyngeal s.
s. pump
Tis-u-Trap endometrial s.
Vabra s.
vacuum s.
suction-assisted lipoplasty (SAL)
suctioning
bulb s.
DeLee s.
suction-irrigator
Nezhat-Dorsey s.-i.
suctorial pad
Sudafed
S. 12 Hour
sudanophilic leukodystrophy
sudden
s. infant death (SID)
s. infant death syndrome (SIDS)
Sudrin
SUDS
single-use diagnostic system
SUDS HIV-1 antibody test
Sufedrin
sufentanil citrate
sufficient
quantity not s. (QNS)
sugar
blood s.
fasting blood s.
Sugarman
S. brachydactyly
S. syndrome
SUI
stress urinary incontinence
suicide
s. gene
suis
Herpesvirus s.
suit
MAST s.
sulbactam
sulconazole
Sulf-10
sulfa
sulfabenzamide
sulfacarbamide
sulfacetamide
sulfachlorpyridazine
sulfacytine

sulfadiazine
sulfadimethoxine
sulfadimidine
sulfadoxine
sulfaethidole
sulfafurazole, sulphafurazole
sulfaguanidine
Sulfa-Gyn
sulfalene
sulfamerazine
sulfameter
sulfamethazine
sulfamethizole
Sulfamethoprim
sulfamethoxazole
sulfamethoxazole/phenazopyridine hydrochloride
sulfamethoxydiazine
sulfamethoxypyridazine
Sulfamylon
sulfanilamide
sulfaphenazole
sulfapyridine
sulfasalazine
sulfatase
steroid s.
sulfate
abacavir s.
amikacin s.
anhydrous magnesium s.
bleomycin s.
dehydroepiandrosterone s. (DHEAS)
dehydroisoandrosterone s.
dextran s.
dextrin s.
DHEA s.
ephedrine s.
estrone s.
ferric s.
ferrous s. ($FeSO_4$)
gentamicin s.
hexoprenaline s.
hyoscyamine s.
magnesium s. ($MgSO_4$)
metaproterenol s.
morphine s.
Mycifradin S.
neomycin s.
netilmicin s.
orciprenaline s.
piperazine estrone s.
polymyxin B s.

S

NOTES

sulfate *(continued)*
    sodium equilin s.
    sodium estrone s.
    sodium tetradecyl s.
    terbutaline s.
    trimethoprim s.
    vincristine s.
sulfathiazole
sulfathiourea
sulfatidosis
    juvenile s.
Sulfatrim
    S. DS
Sulfa-Trip
sulfaturia
    keratan s.
sulfisomidine
sulfisoxazole
sulfisoxazole/phenazopyridine
    hydrochloride
Sulfizole
sulfoiduronate sulfatase deficiency
    (SIDS)
sulfonamide
sulfonate
    2-mercaptoethane s. (MESNA)
sulfonylurea
sulfotransferase
    estrogen s.
sulfur-deficient brittle hair syndrome
sulfur granule
sulindac
sulphafurazole *(var. of* sulfafurazole)
sulpiride
sulprostone
Sultrin
sumatriptan succinate
Sumner sign
Sumycin
    S. Oral
sunken fontanel
Sunna circumcision
sunny-side up delivery
superfecundation
super female
superfetation
superficial
    s. external pudendal artery (SEPA)
    s. linear array (SLA)
    s. spreading melanoma
    s. thrombophlebitis
superimposed
    s. eclampsia
    s. preeclampsia
superimpregnation
superinvolution
superior
    s. mesenteric artery

    s. mesenteric plexus
    s. strait
    s. vena cava
    s. vena caval syndrome
superlactation
supernumerary
    s. breast
    s. chromosome
    s. digit
    s. mamma
    s. marker chromosome (SMC)
    s. nipple
    s. ovary
    s. placenta
    s. proboscis
    s. rib
superovulation
    s. induction
supervoltage
    s. radiation
supine
    s. empty stress test (SEST)
    s. hypotensive syndrome
    s. position
    s. pressor test
supplement, supplementation
    Aminosyn-PF s.
    calcium s.
    Fer-In-Sol s.
    vitamin s.
supplementary
    s. gene
    s. menstruation
supply
    iodine s.
support
    s. catheter
    s. group
    luteal phase s.
    neonatal adjuvant life s. (NALS)
    psychosocial s.
    vaginal vault s.
    ventilator s.
suppository
    AVC s.
    glycerin s.
    Monistat-3 vaginal s.
    prostaglandin s.
    Prostin E2 Vaginal S.
    rectal s.
    Terazol vaginal s.
    triple sulfa s.
    vaginal s.
Supprelin
suppressed menstruation
suppression
    antibody-mediated immune s.
    (AMIS)

endogenous gonadotropin activity s.
estradiol s.
gonadal steroid s.
immunologic s.
pituitary gonadotropin s.
prolactin s.
testosterone s.

suppressor
s. cell
s. gene
s. T cell

Supprettes
Aquachloral S.

suppuration
pulmonary s.

suppurativa
vulvar hidradenitis s.

suppurative
s. appendicitis
s. mastitis
s. phlebitis

supracervical hysterectomy
supraciliary tap
suprapubic
s. catheter
s. cystotomy
5-mm s. trocar
s. pain
s. pressure
s. stab wound
s. urethrovesical suspension
operation

supraumbilical incision
supravaginal septum
supraventricular
s. tachyarrhythmia (SVT)
s. tachycardia (SVT)
s. tachydysrhythmia

Supravital stain test
Suprax
SureCell
S. Chlamydia Test kit
S. Herpes (HSV) Test
S. rapid test kit
S. rapid test kit for pregnancy

SurePress
S. dressing
S. wrap

SureSite dressing
Surety Shield
surface
adaptive s.

s. antigen
s. antigen subtype ayw1
s. antigen subtype ayw2
s. antigen subtype ayw3
s. antigen subtype ayw4
antimesenteric s.
denuded s.
s. electromyography (SEMG)
s. epithelium
s. epithelium vascular channel
external s.
s. immunoglobulin
s. irradiation
serosal s.
s. tension

surfactant
beractant s.
bovine s.
s. deficiency syndrome
heterologous s.
Human Surf s.
Infrasurf s.
porcine s.
pulmonary s.
rescue s.
Survanta s.

Surfaxin
surf test
surge
estrogen s.
LH s.
midcycle s.
preovulatory LH s.

surgeon
pelvic reconstruction s.

surgeon's knot
surgery
antivesicoureterel reflux s.
bypass s.
conservative s.
cytoreductive s.
feminizing s.
fetal s.
FETENDO s.
gastric reduction s.
hysteroscopic s.
intraabdominal s.
laser s.
palliative s.
pelvic s.
pelvic-floor s.
previous transfundal uterine s.

NOTES

surgery *(continued)*
    radical s.
    suboptimal s.
    tubal reconstruction s.
    vaginal s.
    zero gravity s.
surgical
    s. debulking
    s. infection
    s. management
    s. myomectomy as reproductive
      therapy (SMART)
    S. Nu-Knit
    s. overhead canopy (SOC)
    s. repair
    s. resection
    s. staging
    s. weight loss
surgically-induced abortion
Surgicel
Surgicenter 40 CO2 laser
Surgidac suture
Surgilase 55W laser
Surgilene
Surgin hemorrhage occluder pin
Surgiport
Surgi-Prep
Surgiview laparoscope
Surmontil
surrogacy
    gestational s.
    traditional s.
surrogate
    gestational s.
    s. gestational motherhood
    s. mother
    sex s.
Survanta surfactant
surveillance
    antepartum fetal s.
    S., Epidemiology and End Results
      network
    fetal s.
    immunologic s.
    maternal s.
    nutritional s.
    s. technique
    s. tracheal aspirate
    ultrasound s.
survey
    Sexual Opinion S.
survival
    actuarial s.
    allograft s.
    life table s.
    long-term s.
    s. rate

susceptibility
    genetic s.
Susp
    Megacillin S.
suspected pituitary adenoma
suspension
    Aldridge-Studdefort urethral s.
    Alexander-Adams uterine s.
    Baldy-Webster uterine s.
    bladder neck s.
    Children's Advil S.
    Children's Motrin S.
    Coffey s.
    Cortisporin-TC Otic S.
    endoscopic bladder neck s. (EBNS)
    Gilliam-Doleris uterine s.
    Gittes urethral s.
    minimal-incision pubovaginal s.
    needle s.
    Olshausen s.
    paravaginal s.
    Pereyra needle s.
    protamine insulin zinc s.
    Raz bladder neck s.
    sacrospinous ligament s.
    sacrospinous vaginal vault s.
    Stamey-Pereyra needle s.
    transvaginal bladder neck s.
    uterine s.
    uterosacral ligament s.
suspensory
    s. ligaments of Cooper
    s. sling operation
Sus-Phrine
Sustacal formula
Sustagen formula
Sustaire
Sustiva
Sutherland-Haan syndrome (SHS)
sutural calcification
suture
    absorbable s.
    apposition of skull s.
    catgut s.
    coated Vicryl Rapide s.
    continuous running monofilament s.
    cranial s.
    Dexon II s.
    Dexon Plus s.
    DG Softgut s.
    Endoloop s.
    Ethibond polybutilate-coated
      polyester s.
    frontal s.
    Gambee s.
    s. grasper forceps
    gut s.
    interrupted s.

inverted subcuticular s.
Investa s.
s. ligated
s. material
Maxon delayed-absorbable s.
Mersilene s.
Monocryl s.
overriding of s.'s
polyglactic acid s.
polyglactide s.
polyglactin 910 s.
polyglycol s.
polyglyconate s.
Polysorb s.
Prolene s.
purse-string s.
retention s.
sagittal s.
silk s.
silver wire s.
s.'s of skull
stainless steel s.
Sturmdorf hemostatic s.
Surgidac s.
Sutureloop colposuspension s.
Vicryl Rapide s.
suture-ligation
suture-ligature
Sutureloop colposuspension suture
Suture-Mate
SUZI
subzonal insemination
subzonal insertion
Sv
sievert
SV-CAH
simple virilizing congenital adrenal
hyperplasia
SVD
spontaneous vaginal delivery
SVE
sterile vaginal examination
SVT
supraventricular tachyarrhythmia
supraventricular tachycardia
SW
Stüve-Wiedemann
swab
s. examination
Puritan s.
s. test

swallowing
fetal s.
swallow reflex
Swan-Ganz catheter
SW-CAH
salt-wasting congenital adrenal
hyperplasia
sweat
s. chloride determination
s. chloride level
s. chloride test
sweaty feet syndrome
Swedish porphyria
Sween Cream
sweep
s. gas
s. the pelvis
The Cell S.
Sweet syndrome
swelling
hypoosmotic s. (HOS)
labioscrotal s.
Swenson pull-through procedure
Swift disease
SWIM
sperm washing insemination method
swim-up technique
Swiss
S. cheese endometrium
S. cheese hyperplasia
switch
genetic s.
s. operation
swivel-arm system
swordfish test
SWS
Stüve-Wiedemann syndrome
Swyer-James-Macleod syndrome
Swyer-James syndrome
Swyer syndrome
Sydenham chorea
Sydney
S. crease
S. line
**Syed-Neblett dedicated vulvar plastic
template**
sylvian seizure
Sylvius
aqueduct of S.
hydrocephalus due to congenital
stenosis of aqueduct of S.
(HSAS, HYCX)

NOTES

symbiotic psychosis
symbrachydactyly
symmelia
Symmetrel
symmetrical
  s. conjoined twins
symmetric communicating uterus
symmetros
  duplicitas s.
sympathectomy, sympathetectomy
  presacral s.
sympathetic nervous system
sympathomimetic
symphalangism
symphocephalus
symphyseal
  s. separation
  s. wall
symphyses (*pl. of* symphysis)
symphysiotome, symphyseotome
symphysiotomy, symphyseotomy
symphysis, pl. symphyses
  pubic s.
  s. pubis
symphysis-fundus height
symphysodactyly
sympodia
symptom
  PIH s.
  pregnancy-induced hypertension s.
  premenstrual s.'s
symptomatic
  s. infection
  s. porphyria
symptomatica
  porphyria cutanea tarda s.
symptom-giving PGR
symptothermal method
sympus
Synacort Topical
synadelphus
synagiosis
  encephalomyoarterial s.
Synagis
Synalar
  S. Topical
Synalar-HP Topical
Synalgos-DC
synangiosis
  encephaloduroarterial s.
synapsis
Synapton
synaptonemal complex
synaptophysin
Synarel
Synasal
syncephalus
  craniothoracopagus s.

syncheilia
synchondrotomy
synchronic
synchronized intermittent mandatory
  ventilation (SIMV)
synchronous breathing
synclitic
synclitism
syncope
syncytia (*pl. of* syncytium)
syncytial
  s. bud
  s. cell
  s. knot
  s. sprout
syncytiotrophoblast
  malignant s.
syncytium, pl. syncytia
syndactyly
  s.-anophthalmos syndrome
  s.-cataracts-mental retardation
    syndrome
  Cenani-Lenz s.
  s.-microcephaly-mental retardation
    syndrome
syndesis
syndrome
  Aarskog s.
  Aarskog-Scott s. (ASS)
  Aase s.
  Aase-Smith s.
  Abderhalden-Fanconi s.
  abdominal compartment s. (ACS)
  abdominal muscle deficiency s.
  abdominal musculature aplasia s.
  abducted thumbs s.
  Aberfeld s.
  ablepharon-macrostomia s.
  absence of abdominal muscle s.
  Abt-Letterer-Siwe s.
  abuse dwarfism s.
  Accutane dysmorphic s.
  ACD s.
  ACD mental retardation s.
  achalasia-microcephaly s.
  Achard s.
  Achard-Thiers s.
  achondrogenesis s.
  achondroplasia s.
  acid aspiration s.
  acquired immune deficiency s.
    (AIDS)
  acquired immunodeficiency s.
    (AIDS)
  acral-renal-mandibular s.
  acrocallosal s. (ACS)
  acrodysgenital s.
  acrodysostosis s.

acrodysplasia-dysostosis s.
acrofacial dysostosis with postaxial
  defects s.
acrofrontofacionasal dysostosis s.
  types 1, 2
acromegaloid-cutis verticis gyrata-
  leukoma s.
acromegaloid facial appearance s.
acroosteolysis s.
acrorenal s.
acrorenoocular s.
acute urethral s.
Adair-Dighton s.
Adams-Oliver s.
Adams-Stokes s.
Addison disease-cerebral sclerosis s.
Addison disease-spastic
  paraplegia s.
addisonian s.
Addison-Schilder s.
adducted thumb-clubfoot s.
adducted thumbs s.
adducted thumbs-mental
  retardation s.
Adie s.
adiposogenital s.
adrenal virilizing s.
adrenocortical atrophy-cerebral
  sclerosis s.
adrenogenital s. (AGS)
adult-onset polyglandular s.
adult respiratory distress s. (ARDS)
AEC s.
AFA s.
agenesis of corpus callosum-mental
  retardation-osseous lesions s.
aglossia-adactylia s.
agonadism, mental retardation, short
  stature, retarded bone age s.
agyria-pachygyria s.
Aicardi s.
Aicardi-Goutieres s.
air leak s.
Alagille s.
Alagille-Watson s. (AWS)
Alajouanine s.
Albers-Schönberg s.
albinism-deafness s.
Albright s.
ALCAPA s.
aldosteronism-normal blood
  pressure s.

Aldred s.
Aldrich s.
Ale-Calo s.
Alexander s.
Allemann s.
Allen-Herndon s.
Allen-Herndon-Dudley s. (AHDS)
Allen-Masters s.
alopecia, anosmia, deafness,
  hypogonadism s.
alopecia, contracture, dwarfism s.
alopecia, contracture, dwarfism,
  mental retardation s.
alopecia, epilepsy, oligophrenia s.
alopecia mental retardation s.
alopecia, mental retardation,
  epilepsy, microcephaly s.
alpha-thalassemia mental
  retardation s.
alpha-thalassemia mental
  retardation s. deletion type
alpha-thalassemia mental
  retardation s. nondeletion type
Alport s.
Ambras s.
ameloonychohypohidrotic s.
amenorrhea-galactorrhea s.
aminopterin embryopathy s.
aminopterin-like embryopathy s.
Amish brittle hair s.
amniotic band s.
amniotic banding s.
amniotic fluid embolism s.
amniotic fluid embolus s.
amniotic infection s.
anal-ear-renal radial malfunction s.
Andermann s.
Andersen s.
Anderson s.
Andogsky s.
androgen insensitivity s.
androgen resistance s.
Angelman s.
angiomatosis, oculoorbito, thalamo-
  encephalic s.
angioosteohypertrophy s.
aniridia, ambiguous genitalia,
  mental retardation triad s.
aniridia, cerebellar ataxia-
  oligophrenia s.
aniridia, Wilms tumor
  association s.

S

NOTES

syndrome *(continued)*
    aniridia, Wilms tumor,
        gonadoblastoma s.
    ankyloblepharon, ectodermal
        dysplasia, clefting s.
    ankyloglosson superius s.
    anomalous left coronary artery
        from pulmonary artery s.
    anophthalmia, hand-foot defects-
        mental retardation s.
    anophthalmia-Waardenburg s.
    anophthalmos-limb anomalies s.
    anophthalmos-syndactyly s.
    anterior chamber cleavage s.
    antiphospholipid s. (APS)
    Antley-Bixler s.
    anus-hand-ear s.
    aortic arch anomaly-peculiar facies
        mental retardation s.
    aortic stenosis, corneal clouding,
        growth and mental retardation s.
    Apak s.
    Apert s.
    Apert-Crouzon s.
    aplastic abdominal muscle s.
    Appelt-Gerkin-Lenz s.
    apraxia-ataxia-mental deficiency s.
    apraxia-oculomotor contracture-
        muscle atrophy s.
    aprosencephaly s.
    aprosencephaly-atelencephaly s.
    ARCS s.
    Argonz-Del Castillo s.
    Arkawa s. 1, 2
    Arkless-Graham s.
    Arnold-Chiari s.
    arrhinia, choanal atresia,
        microphthalmia s.
    ARSB s.
        arylsulfatase B
    arthrogryposis, ectodermal dysplasia,
        cleft lip/palate developmental
        delay s.
    Arts s.
    arylsulfatase B s.
    Ascher s.
    Asherman s.
    asphyxiating thoracic dysplasia s.
    asphyxiating thoracodystrophy s.
    asplenia s.
    association with hydrocephalus s.
    asymmetric short stature s.
    ataxia-deafness s.
    ataxia-deafness-retardation s.
    ataxia-microcephaly-cataract s.
    ataxia, myoclonic encephalopathy,
        macular degeneration, recurrent
        infections s.

    ataxia-telangiectasia s.
    atelencephalic s.
    Atkin-Flaitz s.
    Atkin-Flaitz-Patil s.
    ATR s.
    ATRX s.
        X-linked alpha-thalassemia/mental
        retardation syndrome
    Austin s.
    autism, dementia, ataxia, loss of
        purposeful hand use s.
    autism-fragile X s. (AFRAX)
    autoimmune polyglandular s.
    autosomal dominant
        macrocephaly s.
    autosomal dominant Opitz s.
        (ADOS)
    Axenfeld s.
    Axenfeld-Rieger s.
    Babinski-Fröhlich s.
    BADS s.
    Ballantyne-Runge s.
    Ballantyne-Smith s
    Baller-Gerold s. (BGS)
    Ballinger-Wallace s.
    Bamforth s.
    Banki s.
    Bannayan s.
    Bannayan-Riley-Ruvalcaba s.
        (BRRS)
    Bannayan-Zonna s. (BZS)
    Banti s.
    Baraitser-Burn s.
    Baraitser-Winter s.
    Barber-Say s.
    Bardet-Biedl s. (BBS)
    bare lymphocyte s.
    Barlow s.
    Bart s.
    Bartholin-Patau s.
    Bartsocaas-Papas s.
    Bartter s. (BS)
    basal cell nevus s. (BCNS)
    basal ganglion disorder-mental
        retardation s.
    Bassen-Kornzweig s.
    battered buttock s.
    battered child s.
    battered fetus s.
    battered wife s.
    Bazex s.
    Bazex-Dupré-Christol s.
    BBB s.
    BCD s.
    BD s.
    Beare s.
    Beare-Stevenson cutis gyrata s.
    Beckwith s.

Beckwith-Wiedemann s.
Beemer-Langer s.
Beemer lethal malformation s.
Begeer s.
Behçet s.
Behr s.
Benjamin s.
Berardinelli s.
Berardinelli-Seip s.
Berardinelli-Seip-Lawrence s.
Berdon s.
Bergia s.
Berlin breakage s.
Bernard-Soulier s.
Berry s.
Berry-Kravis and Israel s.
Berry-Treacher Collins s.
Bertini s.
Beuren s.
BGMR s.
Bianchine-Lewis s.
Bickers-Adams s.
BIDS s.
Bielschowsky s.
Biemond s. 1, 2
Biglieri s.
bile-plug s.
Binder s.
bird-headed dwarf s.
birdlike face s.
bitemporal forceps marks s.
Bixler s.
Blackfan-Diamond s.
black lock-albinism-deafness s.
bladder outlet s.
blepharocheilodontic s.
blepharonasofacial malformation s.
blepharophimosis-ptosis s.
blepharophimosis, ptosis, epicanthus
  inversus s. (BPEIS)
blepharophimosis, ptosis, epicanthus
  inversus, primary amenorrhea s.
blepharophimosis, ptosis, syndactyly,
  short stature s.
blepharoptosis, blepharophimosis,
  epicanthus inversus, telecanthus s.
blind loop s.
Bloch-Siemens s.
Bloch-Sulzberger s.
Bloodgood s.
Bloom s.
blue diaper s.

blue dome s.
blue rubber-bleb nevus s.
BOD s.
Bodian-Schwachman s.
Bohring s.
Bonneau s.
Bonnevie-Ullrich s.
boomerang s.
BOR s.
Börjeson s.
Börjeson-Forssman-Lehmann s.
  (BFLS)
Bourneville s.
Bourneville-Pringle s.
Bowen-Conradi s.
Bowen Hutterite s.
brachioskeletogenital s.
Brachmann-Cornelia de Lange s.
  (BDLS)
Brachmann-de Lange s. (BDLS)
brachycephaly, deafness, cataract,
  microstomia, mental retardation s.
brachydactyly-distal
  symphalangism s.
brachydactyly, dwarfism, hearing
  loss, microcephaly, mental
  retardation s.
brachydactyly, mesomelia, mental
  retardation, aortic dilation, mitral
  valve prolapse, characteristic
  facies s.
brachydactyly, nystagmus, cerebellar
  ataxia s.
brachymesomelia-renal s.
brachymetacarpalia, cataract,
  mesiodens s.
brachymorphism, onychodysplasia,
  dysphalangism s.
Brailsford s.
brain-death s.
branchial arch s.
branchial clefts-lip pseudocleft s.
branchiooculofacial s. (BOFS)
branchiootic s.
branchiootorenal s.
Brandt s.
Brentano s.
Briard-Evans s.
Brissaud s.
brittle hair, intellectual impairment,
  decreased fertility, short stature s.
brittle hair-mental deficit s.

S

NOTES

syndrome *(continued)*
    broad thumb-hallux s.
    broad thumb-mental retardation s.
    bronze baby s.
    Brooks s.
    Brooks-Wisniewski-Brown s.
    brown baby s.
    Brown vertical retraction s.
    Brown-Vialetto-Van Laere s.
    Bruck-de Lange s.
    Brunner s.
    Brusa-Toricelli s.
    Brushfield-Wyatt s.
    BSG s.
        branchioskeletogenital
    bubbly lung s.
    Budd-Chiari s.
    bulldog s.
    burning vulva s.
    Burn-McKeown s.
    C s.
    3C s.
    Caffey pseudo-Hurler s.
    Caffey-Silverman s.
    Caffey-Smyth-Roske s.
    CAHMR s.
    Calabro s.
    Calvé-Legg-Perthes s. ,
    CAMAK s.
    CAMFAK s.
    camptomelic s.
    Camurati-Englemann s.
    Cantrell s.
    Cantú s.
    carbohydrate-deficient
        glycoprotein s. (CDGS)
    carcinoid s.
    cardiac-limb s.
    cardiocranial s.
    cardiofacial s.
    cardiofaciocutaneous s. (CFC)
    cardiogenital s.
    cardiovertebral s.
    Carey-Fineman-Ziter s.
    Carmi s.
    Carnevale s.
    Carney s.
    Carpenter s.
    cataract, ataxia, deafness,
        retardation s.
    cataract-dental s.
    cataract, hypertrichosis, mental
        retardation s.
    cataract, mental retardation,
        hypogonadism s.
    cataract, microcephaly,
        arthrogryposis, kyphosis s.

    cataract, microcephaly, failure to
        thrive, kyphoscoliosis s.
    cataract, motor system disorder,
        short stature, learning difficulty,
        skeletal abnormalities s.
    cataract-oligophrenia s.
    CATCH-22 s.
    Catel-Manzke s.
    cat eye s. (CES)
    cat's cry s.
    caudal appendage, short terminal
        phalanges, deafness,
        cryptorchidism, mental
        retardation s.
    caudal dysplasia s.
    caudal regression s. (CRS)
    cavum septum pellucidum, cavum
        vergae, macrocephaly, seizures,
        mental retardation s.
    Cayler s.
    CCC s.
    centromeric instability-
        immunodeficiency s.
    cephalopolysyndactyly s.
    cerebral malformations, seizures,
        hypertrichosis, overlapping
        fingers s.
    cerebral, ocular, dental, auricular,
        skeletal s. (CODAS)
    cerebral palsy-hypotonic seizures-
        megalocornea s.
    cerebroarthrodigital s.
    cerebrocostomandibular s.
    cerebrofacioarticular s. (CFA)
    cerebrohepatorenal s.
    cerebroocular
        dentoauriculoskeletal s. (CODAS)
    cerebroocular dysgenesis-muscular
        dystrophy s.
    cerebrooculomuscular s. (COMS)
    cerebrooculonasal s.
    cerebroosteonephrosis s.
    Chapple s.
    characteristic face-hypogenitalism-
        hypotonia-pachygyria s.
    Charcot-Marie-Tooth s. (CMTS)
    Charcot-Marie-Tooth-Hoffmann s.
    Charcot-Marie-Tooth s. X-linked
        recessive type II
    CHARGE s.
    Cheadle s.
    Chédiak-Higashi s.
    Chemke s.
    Cheney s.
    cherubism, gingival fibromatosis,
        epilepsy, mental deficiency s.
    Chiari-Arnold s.
    Chiari-Frommel s.

Chilaiditi s.
CHILD s.
cholestasis, pigmentary retinopathy, cleft palate s.
chondrodysplasia-pseudohermaphrodism s.
chondroectodermal dysplasia-like s.
chorioretinal anomalies, corpus callosum agenesis, infantile spasms s.
Chotzen s.
Christian s. 1. 2
Christian-Andrews-Conneally-Muller s.
Christian-Opitz s.
Christ-Siemens-Touraine s.
chromosomal breakage-immunodeficiency s.
chromosome diploid/tetraploid mixoploidy s.
chromosome diploid/triploid mixoploidy s.
chromosome GI deletion s.
chromosome 9 inversion s.
chromosome 1–22 monosomy s.
chromosome 22 monosomy s.
chromosome 11p detention s.
chromosome 1p–22p deletion s.
chromosome 1q–22q deletion s.
chromosome 1q–22q duplication s.
chromosome 1q–22q tetrasomy s.
chromosome 1q–22q triplication s.
chromosome 8 recombinant s.
chromosome 1–22 ring s.
chromosome tetraploidy s.
chromosome triploidy s.
chromosome 1–22 trisomy s.
chromosome 14 uniparental disomy s.
chromosome X autosome translocation s.
chromosome X fragility s.
chromosome X inversion s.
chromosome XO s.
chromosome Xp21 deletion s.
chromosome Xp22 deletion s.
chromosome Xq deletion s.
chromosome Xq duplication s.
chromosome XXX s.
chromosome 47,XXX s.
chromosome XXXXX s.
chromosome XXXXY s.

chromosome XXY s.
chromosome Y;18 translocation s.
Chudley s. 1, 2
Chudley-Lowry-Hoar s.
Cianchetti s.
circumferential skin creases-psychomotor retardation s.
Clarke Hadfield s.
clasped thumbs-mental retardation s.
clefting, ocular anterior chamber defect, lid anomalies s.
cleft lip, cleft palate, lobster claw deformity s.
cleft palate, diaphragmatic hernia, coarse facies, acral hypoplasia s.
cleft palate-lateral synechia s. (CPLS)
cleft palate, microcephaly, large ears, short stature s.
cleidocranial dysplasia s.
cleidorhizomelic s.
Clifford s.
climacteric s.
clitoris tourniquet s. (CTS)
clomiphene-resistant polycystic ovary s.
Clouston s.
cloverleaf skull s.
Cockayne s.
cocktail party s.
COD-MD s.
Coffin s. 1, 2
Coffin-Lowry s.
Coffin-Siris s.
Coffin-Siris-Wegienka s.
Cohen s.
Cole s.
Cole-Carpenter s.
Cole-Rauschkolb-Toomey s.
coloboma-anal atresia s.
coloboma, clefting, mental retardation s.
coloboma cleft lip/palate-mental retardation s.
coloboma, heart defects, ichthyosiform dermatosis, mental retardation, ear defects s. (CHIME)
coloboma, mental retardation, hypogonadism, obesity s.
coloboma-microphthalmos s.

S

NOTES

505

syndrome *(continued)*

coloboma, microphthalmos, hearing loss, hematuria, cleft lip/palate s.

coloboma, obesity, hypogenitalism, mental retardation s.

complete androgen insensitivity s. (CAIS)

complete androgen resistance s.

complete feminizing testes s.

congenital acromicria s.

congenital anosmia-hypogonadotropic hypogonadism s.

congenital arthromyodysplastic s.

congenital cataracts, sensorineural deafness, Down syndrome facial appearance, short stature, mental retardation s.

congenital central hypoventilation s.

congenital clasped thumbs-mental retardation s.

congenital emphysema, cryptorchidism, penoscrotal web, deafness, mental retardation s.

congenital hydantoin s.

congenital hypertrichosis-osteochondrodysplasia-cardiomegaly s.

congenital hypocupremia s.

congenital hypothyroidism s.

congenital ichthyosis-mental retardation-spasticity s.

congenital ichthyosis-trichodystrophy s.

congenital microcephaly-hiatus hernia-nephrotic s.

congenital muscular hypertrophy-cerebral s.

congenital pseudohydrocephalic progeroid s.

congenital rubella s. (CRS)

congenital thrombocytopenia, Robin sequence, agenesis of corpus callosum, distinctive facies, developmental delay s.

congenital warfarin s.

congestive cardiomyopathy-hypergonadotropic hypogonadism s.

Conn s.

conotruncal anomaly face s. (CTAF)

Conradi s.

Conradi-Hünermann s.

constrictive pericarditis-dwarfism s.

contiguous gene s.

contractural arachnodactyly s.

contracture, muscle atrophy, oculomotor apraxia s.

Cooks s.

Cooper s.

Cornelia de Lange s. (CLS)

corpus callosum agenesis, chorioretinal abnormality s.

corpus callosum agenesis, chorioretinopathy, infantile spasms s.

corpus callosum agenesis, facial anomalies, salaam seizures s.

corpus callosum hypoplasia, retardation, adjusted thumbs, spastic paraparesis, hydrocephalus s. (CRASH)

corpus luteum deficiency s.

Costello s.

coumarin s.

Cowchock s.

Cowchock-Fischbeck s.

coxoauricular s.

Crane-Heise s.

cranial sclerosis, osteopathia striata, macrocephaly s.

cranioacrofacial s.

craniocarpotarsal s.

craniocerebellocardiac s.

craniofacial anomalies, polysyndactyly s.

craniofacial-deafness-hand s.

craniofacial dysmorphism, absent corpus callosum, iris colobomas, connective tissue dysplasia s.

craniofacial dysmorphism-polysyndactyly s.

craniofrontonasal s. (CNFS)

cranioorodigital s.

craniosynostosis, arachnodactyly, abdominal hernia s.

craniosynostosis, arthrogryposis, cleft palate s.

craniosynostosis, ataxia, trigeminal anesthesia, parietal anesthesia and pons, vermis fusion s.

craniosynostosis-lid anomalies s.

craniosynostosis-radial aplasia s.

CREST s.

cretinism-muscular hypertrophy s.

Creutzfeldt-Jakob s.

cri du chat s.

Crigler-Najjar s.

Crisponi s.

Crome s.

crooked fingers s.

Cross s.

Cross-McKusick-Breen s.

Crouzon s.

CRST s.

crying cat s.

cryptomicrotia-brachydactyly s.
cryptophthalmia s.
cryptophthalmia-syndactyly s.
cryptophthalmos s.
cryptophthalmos-syndactyly s.
Curran s.
Curry-Jones s.
Curtis s.
Cushing s.
cushingoid s.
cutis verticis gyrata, thyroid aplasia, mental retardation s.
Cypress facial neuromusculoskeletal s.
Dandy-Walker s. (DWS)
Dandy-Walker-like s.
Dandy-Walker malformation-basal ganglia disease-seizures s.
Danlos s.
Darrow-Gamble s.
David-O'Callaghan s.
dead fetus s.
deafness-craniofacial s.
deafness, femoral epiphyseal dysplasia, short stature, developmental delay s.
deafness, hypogonadism, hypertrichosis, short stature s.
deafness, imperforate anus, hypoplastic thumbs s.
deafness-nephritis s.
Debré-Fibiger s.
Debré-Sémélaigne s.
De Crecchio s.
defective abdominal wall s.
de Grouchy s. 1, 2
Dejerine-Klumpke s.
del (1)–(22) s.
de Lange s. 1, 2
delay s.
del Castillo s.
deletion 1–22 s.
deletion 1p–22p s.
deletion 1q–22q s.
deletion Xp21 s.
deletion Xp22 s.
deletion Xq s.
Delleman s.
del (1p)–(22p) s.
del (1q)–(22q) s.
del (Xp21) s.
del (Xq) s.

Demons-Meigs s.
de Morsier s.
de Morsier-Gauthier s.
Dennie-Marfan s.
Denys-Drash s.
depressor anguli oris muscle hypoplasia s.
dermotrichic s.
Derry s.
De Sanctis-Cacchione s.
Desbuquois s.
Desmons s.
De Toni-Fanconi s.
De Toni-Fanconi-Debré s.
De Vaal s.
developmental delay-multiple strawberry nevi s.
dextrocardia/situs inversus s.
diabetes-deafness s.
diabetes insipidus, diabetes mellitus, optic atrophy s. (DIDMO)
diabetes mellitus, mental retardation, lipodystrophy, dysmorphic traits s.
Diamond-Blackfan s.
diaphragmatic hernia, abnormal face, distal limb anomalies s.
diaphragmatic hernia-distal digital hypoplasia s.
diaphragmatic hernia-exophthalmos-hypertelorism s.
diaphragmatic hernia-myopia-deafness s.
Dickinson s.
diencephalic s.
DiFerrante s.
diffuse mesangial sclerosis-ocular abnormalities s.
DiGeorge microdeletion s.
Dighton-Adair s.
digital anomalies, short palpebral fissures, atresia of esophagus or duodenum s.
digitoorofacial s. I–V
digitooropalatal s.
digitorenocerebral s. (DCR)
Dilantin s.
DiSala s.
disequilibrium s.
dislocated elbow, bowed tibiae, scoliosis, deafness, cataract,

S

NOTES

syndrome *(continued)*
    microcephaly, mental
      retardation s.
    distal arthrogryposis,
      hypopituitarism, mental retardation,
      facial anomalies s.
    distal arthrogryposis, mental
      retardation, characteristic facies s.
    distal limb deficiency-mental
      retardation s.
    distal transverse limb defects-mental
      retardation-spasticity s.
    disturbed equilibrium s.
    Donahue-Uchida s.
    Donohue s.
    DOOR s.
    Down s.
    Drash s.
    Duane s.
    Dubin-Johnson s.
    Dubowitz s.
    Duchenne s.
    Duchenne-Griesinger s.
    duplication-deficiency s.
    duplication 1p–22p s.
    duplication 1q–22q s.
    duplication Xq s.
    dup (10p)/del (10q) s.
    dup (1p)–(22p) s.
    dup (9q/del (9p) s.
    dup (1q)–(22q) s.
    dup (Xq) s.
    dwarf s.
    dwarfism, cerebral atrophy,
      keratosis follicularis s.
    dwarfism, congenital medullary
      stenosis s.
    dwarfism, eczema, peculiar
      facies s.
    dwarfism, ichthyosiform
      erythroderma, mental deficiency s.
    dwarfism, onychodysplasia s.
    dwarfism, pericarditis s.
    dwarfism, polydactyly, dysplastic
      nails s.
    Dyggve-Melchior-Clausen s.
    Dyke-Davidoff s.
    dyscephaly-congenital cataract-
      hypotrichosis s.
    dysequilibrium s. (DES)
    dysmaturity s.
    dysmorphic s.
    dysostotic-idiocy-gargoylism-
      lipochondrodystrophy s.
    dystocia-dystrophia s.
    dystonia-deafness s.
    dystrophia retinae-dysacousis s.

    dystrophia retinae pigmentosa-
      dysostosis s. (DRD)
    dysuria-sterile pyuria s.
    Eagle-Barrett s.
    early-onset diabetes mellitus-
      epiphyseal dysplasia s.
    early onset Parkinsonism-mental
      retardation s.
    ear, patella, short stature s. (EPS)
    Eastman-Bixler s.
    ectodermal dysplasia, cleft lip and
      palate, hand and foot deformity,
      mental retardation s.
    ectodermal dysplasia, cleft lip and
      palate, mental retardation,
      syndactyly s. I, II
    ectodermal dysplasia, mental
      retardation, syndactyly s.
    ectrodactyly-cleft lip/palate s.
    ectrodactyly, ectodermal dysplasia,
      clefting s.
    ectrodactyly, ectodermal dysplasia
      and cleft lip/palate s.
    ectrodactyly-ectodermal dysplasia,
      cleft palate s.
    ectrodactyly-mandibulo facial
      dysostosis s.
    ectrodactyly-spastic paraplegia-mental
      retardation s.
    Eddowes s.
    Edinburgh malformation s.
    Edwards s.
    EEC s.
    Ehlers-Danlos s. (EDS)
    Ehlers-Danlos s. type VII
    Ehlers-Danlos s. type VIIA1
    Ehlers-Danlos s. type VIIA2
    Ehlers-Danlos s. type VIIB
    Ehlers-Danlos s. type VIII–XI
    Eisenmenger s.
    Elejalde s.
    elfin facies hypercalcemia s.
    Ellis-Sheldon s.
    Ellis-van Creveld s.
    ElSahy-Waters s.
    embryofetal alcohol s. (EFAS)
    embryonic testicular regression s.
    EMG s.
    empty scrotum s.
    empty sella s.
    encephalotrigeminal s.
    Engman s.
    epidermal nevus s.
    epiphyseal dysplasia-microcephaly-
      nystagmus s.
    epiphyseal dysplasia, short stature,
      microcephaly, nystagmus s.
    Erb s.

Erb-Charcot s.
Erb-Goldflam s.
Erlacher-Blount s.
Eronen s.
Escalante s.
Escobar s.
ethmocephaly s.
exomphalos, macroglossia, and gigantism s.
extended rubella s.
eye defects-diffuse renal mesangial sclerosis s.
facial-digital-genital s.
facial dysplasia, hyperextensibility of joints, clinodactyly, growth retardation, mental retardation s.
faciocardiorenal s.
faciocerebroskeletocardiac s.
faciocutaneoskeletal s.
faciodigitogenital s.
faciogenital s.
faciooculoacousticorenal s.
faciopalatoosseous s.
Fadhil s.
failure-to-thrive s.
Fairbank-Keats s.
Fallot s.
familial aortic ectasia s.
familial ataxia-hypogonadism s.
familial atypical multiple mole melanoma s.
familial cardiac myxoma s.
familial congenital alopecia, mental retardation, epilepsy, unusual EEG s.
familial endocrine-neuroectodermal abnormalities s.
familial macroglossia-omphalocele s.
familial polysyndactyly-craniofacial anomalies s.
familial pterygium s.
familial third and fourth pharyngeal pouch s.
familial Turner s.
Fanconi s.
Fanconi-Albertini Zellweger s.
Fanconi-Petrassi s.
Fanconi-Prader s.
Fanconi-Schlesinger s.
FCS s.
Feingold s.
Feinmesser-Zelig s.

Felty s.
female pseudo-Turner s.
feminization s.
feminizing testes s.
femoral-facial s.
femur-fibula-ulna s.
fetal Accutane s.
fetal akinesia s.
fetal alcohol s. (FAS)
fetal aminopterin s.
fetal aminopterin like s.
fetal anticoagulant s.
fetal aspiration s.
fetal cocaine s.
fetal Dilantin s.
fetal distress s.
fetal face s.
fetal facies s.
fetal gigantism-renal hamartoma-nephroblastomatosis s.
fetal hydantoin s. (FHS)
fetal isotretinoin s.
fetal methotrexate s.
fetal paramethadione-trimethadione s.
fetal phenytoin s.
fetal trimethadione s.
fetal valproate s. (FVS)
fetal varicella s. (FVS)
fetal warfarin s.
fetofetal transfusion s.
Feuerstein-Mims s.
Fèvre-Languepin s.
FFU s.
FG s.
FHUF s.
fibrinogen-fibrin conversion s.
fifth digit s.
Filippi s.
Fine-Lubinsky s.
first arch s.
first and second branchial arch s.
Fishman s.
Fitz-Hugh and Curtis s.
Fitzsimmons s.
floating harbor s. (FHS)
floppy infant s.
FOAR s.
focal dermal hypoplasia s.
follicular atrophoderma-basal cell carcinoma s.

S

NOTES

509

syndrome *(continued)*
follicular atrophoderma-basocellular proliferation-hypotrichosis s.
Fontaine s.
Forbes-Albright s.
formiminotransferase deficiency s.
Fountain s.
four-day s.
FPO s.
fragile X s.
fragile X-mental retardation s.
fragile Xq s.
Franceschetti s.
Franceschetti-Goldenhar s.
Franceschetti-Jadassohn s.
Franceschetti-Klein s.
Franceschetti-Zwahlen s.
Franceschetti-Zwahlen-Klien s.
Francois dyscephalic s.
Fraser s.
Fraser-Francois s.
Fraser-like s.
fra(X) s.
fra(X)(28) s.
fra(X)(q27) s.
Freeman-Sheldon s.
Friend s.
Fritsch s.
Fritsch-Asherman s.
Fröhlich s.
frontodigital s.
Fryns s. 1–3
Fryns-Moerman s.
Fryns-van den Berghe s.
Fuhrmann s.
Fukuyama s.
Fuller Albright s. 1
functional prepubertal castrate s.
Funston s.
G s.
Gailliard s.
galactorrhea-amenorrhea s.
Galloway s.
Galloway-Mowat s.
Gamble-Darrow s.
Garcia-Lurie s.
Gardner s.
Gardner-Silengo-Wachtel s.
Gareis-Mason s.
Gasser s.
GBBB s.
Gee-Herter-Heubner s.
gender dysphoria s.
Genée-Wiedemann s.
generalized hypertrichosis terminals-gingival hyperplasia s.
generalized hypotonia, congenital hydronephrosis, characteristic face s.
genital anomaly-cardiomyopathy s.
genital ulcer s.
genitopalatocardiac s.
Genoa s.
Gerhardt s.
German s.
Gerstmann s.
Gianotti-Crosti s.
giant platelet s.
Gilbert-Dreyfus s.
Gilbert-Lereboullet s.
Gilles de la Tourette s.
Gillespie s. 1, 2
gingival fibromatosis, hypertrichosis, cherubism, mental retardation, epilepsy s.
gingival fibromatosis, hypertrichosis, mental retardation, epilepsy s.
gingival hyperplasia, hirsutism, convulsions s.
gingival hypertrophy-corneal dystrophy s.
Glanzmann s.
Glanzmann-Riniker s.
glossopalatine ankylosis s.
GMS s.
goiter-deafness s.
Golabi-Ito-Hall s.
Golabi-Rosen s. (GRS)
Goldenhar s.
Goldenhar-Gorlin s.
Goldston s.
Goltz s.
Goltz-Gorlin s.
Goltz-Peterson-Gorlin-Ravitz s.
GOMBO s.
Gomez and López-Hernández s.
gonadal agenesis s.
gonadal dysgenesis s.
gonadal failure, short stature, mitral valve prolapse, mental retardation s.
gonadotrophin-resistant ovary s.
goniodysgenesis, mental retardation, short stature s.
Goodman s.
Goodpasture s.
Gordan-Overstreet s.
Gordon s.
Gorlin s. 1, 2
Gorlin-Goltz s.
Gorlin-Psaume s.
Gougerot-Carteaud s.
Gradenigo s.
Graefe-Usher s.

Graham s.
granddad s.
Grant s.
gray baby s.
Greig s.
Greig cephalopolysyndactyly s.
(GCPS)
Griscelli s.
growth failure-pericardial
constriction s.
growth retardation, ocular
abnormalities, microcephaly,
brachydactyly, oligophrenia s.
growth retardation, small and puffy
hands, eczema s.
Grubben s.
Gruber s.
Guerin-Stein s.
Guillain-Barré s.
Gurrieri s.
Gustavson s.
HAIR-AN s.
hair-brain s.
Hajdu-Cheney s.
Hakim s.
Hakim-Adams s.
Halban s.
Halbrecht s.
Hall s. 1, 2
Hallermann s.
Hallermann-Streiff s.
Hallermann-Streiff-François s.
Hallervorden-Spatz s.
Hallgen s.
Hallopeau-Siemens s.
Hall-Pallister s.
Hall-Riggs s.
Halpern s.
hamartoneoplastic s.
hamartopolydactyly s.
Hamel s.
Hamman-Rich s.
hand-foot s.
hand-foot-genital s.
hand-foot-mouth s.
hand-foot-uterus s.
Hand-Schüller-Christian s.
Hanhart s.
happy puppet s.
HARD s.
Hardikar s.
Harrod s.

Hart s.
Hay-Wells s.
HbH disease-mental retardation s.
hearing loss, mental deficiency,
growth retardation, clubbed digits,
EEG abnormalities s.
hearing-loss nephritis s.
heart-hand s.
HELLP s.
hemangioma-thrombocytopenia s.
hemangiomatous branchial clefts-lip
pseudocleft s.
hematuria-nephropathy-deafness s.
hemignathia and microtia s.
hemoglobin H disease-mental
retardation s.
hemolytic-uremic s.
Hennekam lymphangiectasia-
lymphedema s.
hepatic ductular hypoplasia-multiple
malformations s.
hepatofacioneurocardiovertebral s.
hereditary benign intraepithelial
dyskeratosis s.
hereditary blepharophimosis, ptosis,
epicanthus inversus s.
hereditary dysplastic nevus s.
hereditary hematuria s.
hereditary motor sensory neuropathy
II-deafness-mental retardation s.
hereditary nephritis-deafness s.
hereditary nephritis deafness-
abnormal thrombogenesis s.
Hermansky-Pudlak s.
Hernandez s.
heterotaxia s.
HHHO s.
hiatus hernia, microcephaly,
nephrosis s.
Hirschsprung disease, microcephaly,
mental retardation, characteristic
facies s.
hirsutism, skeletal dysplasia, mental
retardation s.
HLHS s.
HMC s.
H2O s.
Holt-Oram s.
Holzgreve s.
Hootnick-Holmes s.
Horner s.
Hoyeraal-Hreidarsson s.

S

NOTES

syndrome *(continued)*
Hughes s.
Hunter s.
Hunter-Fraser s.
Hunter-MacMurray s.
Hunter-McAlpine s.
Hunter-McAlpine craniosynostosis s.
Hurler s.
Hurler-like s.
Hurler-Pfaundler s.
Hurler-Scheie s.
Hurst s.
Hutchinson s.
Hutchinson-Gilford s.
hyaline membrane s.
hydantoin s.
Hyde-Forster s.
hydrocephalus, agyria, retinal
  dysplasia with or without
  encephalocele s. (HARD+/-E)
hydrocephalus-cerebellar agenesis s.
hydrocephalus, skeletal anomalies,
  mental disturbances s.
hydrolethalis s.
hydronephrocolpos,
  postaxialpolydactyly, congenital
  heart disease s.
17-hydroxylase deficiency s.
21-hydroxylase deficiency s.
hyperammonemic s.
hyperandrogenism, insulin resistance,
  acanthosis nigricans s.
hypercalcemia, peculiar facies,
  supravalvular aortic stenosis s.
hypercalcemia/Williams-Beuren s.
hyper-IgE s.
hyperlucent lung s.
hypertelorism-hypospadias s.
hypertelorism, microtia, clefting s.
hypertrichosis, coarse face,
  brachydactyly, obesity, mental
  retardation s.
hyperviscosity s.
hypocalcemia, dwarfism, cortical
  thickening s.
hypochondroplasia s.
hypogenital dystrophy with diabetic
  tendency s.
hypoglossia-hypodactyly s.
hypogonadism-anosmia s.
hypogonadotropic hypogonadism-
  anosmia s. (HHA)
hypogonadotropic hypogonadism,
  mental retardation,
  microphthalmia s.
hypohidrotic ectodermal dysplasia-
  hypothyroidism-agenesis of corpus
  callosum s.

hypomelia, hypotrichosis, facial
  hemangioma s.
hypoparathyroidism, stature, mental
  retardation, seizures s.
hypoplasia, endocrine disturbances,
  tracheostenosis s.
hypoplastic congenital anemia s.
hypoplastic left heart s. (HLHS)
hypoplastic right heart s. (HRHS)
hyposmia-hypogonadotropic
  hypogonadism s.
hypospadias-dysphagia s.
hypospadias-mental retardation s.
hypothalamic hamartoblastoma s.
hypothalamic hamartoblastoma,
  hypopituitarism, imperforate anus,
  postaxial polydactyly s.
hypothyroid-large muscle s.
hypotonia, hypogonadism, obesity s.
hypotonia, hypopigmentia,
  hypogonadism, obesity s.
hypotonia, obesity, hypogonadism,
  mental retardation s.
hypotonia, obesity, prominent
  incisors s.
IADH s.
ICF s.
ichthyosiform erythroderma, corneal
  involvement, deafness s.
ichthyosiform erythroderma, hair
  abnormality, mental and growth
  retarding s.
ichthyosis, alopecia, ectropion,
  mental retardation s.
ichthyosis, brittle hair, impaired
  intelligence, decreased fertility,
  short stature s. (IBIDS)
ichthyosis, characteristic appearance,
  mental retardation s.
ichthyosis, cheek, eyebrow s.
ichthyosis, follicularis, atrichia (or
  alopecia), photophobia s. (IFAP)
ichthyosis, hypogonadism, mental
  retardation, epilepsy s.
ichthyosis, male hypogonadism s.
ichthyosis, mental retardation,
  dwarfism, renal impairment s.
ichthyosis, mental retardation,
  epilepsy, hypogonadism s.
ichthyosis, oligophrenia, epilepsy s.
ichthyosis, spastic neurologic
  disorder, oligophrenia s.
ichthyosis, split hair,
  aminoaciduria s.
Idaho s.
idiopathic hypercalcemia-
  supravalvular aortic stenosis s.
idiopathic infantile hypercalcemia s.

idiopathic respiratory distress s.
(IRDS)
IFAP s.
Illum s.
Imerslund s.
Imerslund-Graesback s.
immotile cilia s.
immunodeficiency, centromeric
heterochromatin instability, facial
anomalies s.
immunodeficiency, centromeric
instability, facial anomalies s.
(ICF)
imperforate anus-hands and foot
anomalies s.
imperforate anus-polydactyly s.
s. of inappropriate antidiuretic
hormone secretion
s. of inappropriate secretion of
antidiuretic hormone (SIADH)
infancy-onset diabetes mellitus,
multiple epiphyseal dysplasia s.
infantile bilateral striatal necrosis s.
(IBSN)
infantile optic atrophy-ataxia s.
infantile respiratory distress s.
infantile spasms, hypsarrhythmia,
mental retardation s.
infant respiratory distress s. (IRDS)
infection-associated
hemophagocytic s. (IAHS)
insensitive ovary s.
inspissated bile s.
inspissated milk s.
insulin-resistant diabetes, acanthosis
nigricans, hypogonadism
pigmentary retinopathy, deafness,
mental retardation s.
intestinal lymphangiectasia,
lymphedema, mental retardation s.
intrauterine growth retardation-
microcephaly-mental retardation s.
intrauterine parabiotic s.
inversion 9 s.
inversion duplication (15)
chromosome s.
inversion duplication (8p) s.
Ionasescu s.
iris, coloboma, ptosis,
hypertelorism, mental
retardation s.
irritable bowel s. (IBS)

Isaac s.
isochromosome 10p s.
isochromosome 12p s.
isolated autosomal dominant s.
isotretinoin dysmorphic s.
isotretinoin teratogenic s.
Ito s.
Ivemark s.
Jabs s.
Jackson-Weiss s. (JWS)
Jacob s.
Jacobsen s.
Jacobsen-Brodwall s.
Jadassohn-Lewandowski s.
Jaeken s.
Jaffe-Campanacci s.
Jaffe-Gottfried-Bradley s.
Jaffe-Lichtenstein s.
Jahnke s.
Jaksch s.
Jancar s.
Jansen s.
Jansky-Bielschowsky s.
Jarcho-Levin s.
jaw cysts, basal cell tumors,
skeletal anomalies s.
Jensen s.
Jervell-Lange-Nielson s.
Jervis s.
Jessner-Cole s.
Jeune s.
Job s.
Johanson-Blizzard s.
Johnie Mel s.
Johnson-McMillin s.
Johnson neuroectodermal s.
Joseph s.
Josephs-Blackfan-Diamond s.
Joubert s.
Joubert-Boltshauser s.
Juberg-Hayward s.
Juberg-Holt s.
Juberg-Marsidi s. (JMS)
Junius-Kuhnt s.
juvenile cataract, cerebellar atrophy,
mental retardation, myopathy s.
juvenile hyperuricemia s.
juxtaglomerular hyperplasia s.
Kabuki s. (KS)
Kabuki makeup s. (KMS)
Kalischer s.
Kallmann s.

S

NOTES

syndrome *(continued)*

Kallmann-de Morsier s.
Kanner s.
Kaplan s.
Kapur-Toriello s.
Kartagener s.
Kasabach-Merritt s.
Kasnelson s.
Kaufman s. 3
Kaufman-McKusick s.
Kaufman oculocerebrofacial s.
Kaveggia s.
KBG s.
Kearns-Sayre s.
Keipert s.
Keller s.
Kelly s.
Kenny s.
Kenny-Caffey s.
Kenny-Linarelli s.
Kenny-Linarelli-Caffey s.
keratitis, ichthyosis, deafness s.
keratosis palmaris et plantaris-
  corneal dystrophy s.
keratosis palmoplantaris-corneal
  dystrophy s.
Kesaree-Wooley s.
ketoaciduria-mental deficiency s.
Keutel s. 1 & 2
KID s.
Killian s.
Kimmelstiel-Wilson s.
kinky-hair s.
Kinsbourne s.
KIO s.
kleeblattschädel s.
Klein-Waardenburg s.
Klinefelter s.
Klinefelter-Reifenstein s.
Klinefelter-Reifenstein-Albright s.
Klippel-Feil s.
Klippel-Trenaunay s.
Klippel-Trenaunay-Parkes-Weber s.
Klippel-Trenaunay-Weber s.
Kloepfer s.
Klotz s.
Klüver-Bucy s.
Kniest s.
Kobberling-Dunnigan s.
Koby s.
Kocher-Debré-Sémélaigne s.
Koerber-Salus-Elschnig s.
Kosenow-Sinios s.
Kostmann s.
Kowarski s.
Kramer s.
Krause s.
Krause-Kivlin s.

Krause-van Schooneveld-Kivlin s.
Laband s.
lacrimoauriculodentodigital s.
Ladd s.
Lambert s.
Lambotte s.
Landing s.
Landry-Guillain-Barré s.
Langdon Down s.
Lange-Akeroyd s.
Langer s.
Langer-Giedion s.
Langer-Petersen-Spranger s.
Langer-Saldino s.
Laron s.
Larsen s.
late embryonic testicular
  regression s.
late-onset local junctional
  epidermolysis bullosa-mental
  retardation s.
Launois s.
Launois-Cléret s.
Laurence-Moon s.
Laurence-Moon-Biedl s.
Laurence-Moon-Biedl-Bardet s.
  (LMBBS)
Läwen-Roth s.
Lawford s.
Lawrence s.
lazy leukocyte s.
Leigh s.
Lejeune s.
Lennox s.
Lennox-Gastaut s.
lentigines (multiple),
  electrocardiographic abnormalities,
  ocular hypertelorism, pulmonary
  stenosis, abnormalities of
  genitalia, retardation of growth,
  and deafness (sensorineural) s.
Lenz dysmorphogenic s.
Lenz-Majewski s.
Lenz-Majewski-like s.
Lenz microphthalmia s.
LEOPARD s.
Leri s.
Leri-Weill s.
Leroy s.
Leschke s.
Lesch-Nyhan s.
lethal multiple pterygium s.
leukoerythroblastic s.
Levin s.
Levy-Hollister s.
Li-Fraumeni cancer s.
Lightwood-Albright s.
limb-girdle s. (LGS)

limp infant s.
linear nevus sebaceous s.
linear sebaceous nevus s.
Lin-Gettig s.
lipodystrophy-acromegaloid
  gigantism s.
lip-palate s.
lip pseudocleft-hemangiomatous
  branchial cyst s.
Lison s.
Lobstein s.
lobster-claw with ectodermal
  defects s.
lobulation-polydactyly s.
Löffler s.
Louis-Bar s.
Lowe s. (LS)
Lowe oculocerebrorenal s.
Lowe-Terry-MacLachlan s.
Lowry s.
Lowry-Maclean s.
Lowry-Wood s. (LWS)
Lub s.
Lucey-Driscoll s.
Lujan-Fryns s.
lupus obstetric s.
luteinized unruptured follicle s.
  (LUFS)
Lutembacher s.
Lyell s.
lymphoproliferative s.
Lynch s.
lysine malabsorption s.
MacDermot-Winter s.
Macleod s.
macrocephaly, cutis marmorata,
  telangiectatica congenita s.
macrocephaly, facial abnormalities,
  disproportionate tall stature mental
  retardation s.
macrocephaly-hamartomas s.
macrocephaly, hypertelorism, short
  limbs, hearing loss, developmental
  delay s.
macrocephaly, multiple lipomas,
  hemangiomata s.
macrocephaly, pseudoepithelioma,
  multiple hemangiomas s.
macroglossia-omphalocele s.
macroglossia-omphalocele-
  visceromegaly s.
macroorchidism marker X s.

macrosomia-mental retardation s.
macrosomia, obesity, macrocephaly,
  ocular abnormality s. (MOMO)
Maestre de San Juan-Kallmann-de
  Morsier s.
Maestre-Kallmann-de Morsier s.
Maffucci s.
Majewski s.
malabsorption s.
male pseudohermaphroditism-
  persistent müllerian structures-
  mental retardation s.
male Turner s.
Mallory-Weiss s.
Malouf s.
Malpuech facial clefting s.
mandibulofacial dysostosis with
  epibulbar dermoids s.
mandibulofacial dysostosis with
  limb malformations s.
Marañón s.
Marden-Walker s.
Marfan s.
marfanoid craniosynostosis s.
marfanoid habitus-mental
  retardation s.
marfanoid habitus-microcephaly-
  glomerulonephritis s.
Marie s.
Marie-Sainton s.
Marinesco-Garland s.
Marinesco-Sjögren s.
Marinesco-Sjögren-Garland s.
Marinesco-Sjögren-like s.
marker X s.
Maroteaux-Lamy s.
Maroteaux-Malamut s.
Marshall s.
Marshall-Smith s. (MSS)
Martin-Bell s. (MBS)
Martin-Bell-Renpenning s.
Martsolf s.
marX s.
MASA s.
Masters-Allen s.
maternal Bernard-Soulier s.
maternal deprivation s.
maternal hydrops s.
Mauriac s.
Mayer-Rokitansky-Küster-Hauser s.
McCune-Albright s.
McDonough s.

S

NOTES

syndrome *(continued)*
>McKusick-Kaufman s.
Meadows s.
Meckel s.
Meckel-Gruber s.
meconium aspiration s. (MAS)
meconium blockage s.
meconium plug s.
median cleft upper lip, mental retardation, pugilistic facies s.
median facial cleft s.
megacystis-megaureter s.
megacystis, microcolon, intestinal hypoperistalsis s.
megalencephaly, cranial sclerosis, osteopathia striata s.
megalocornea, developmental retardation, dysmorphic s.
megalocornea-macrocephaly-mental and motor retardation s. (MMMM)
megalocornea-mental retardation s. (MMR)
Meier-Gorlin s.
Meigs s.
Meigs-Kass s.
Meinecke-Peper s.
Melinck-Needles s.
Melkersson-Rosenthal s.
Melnick-Fraser s.
Melnick-Needles s.
Mendelson s.
Mendenhall s.
Mengert shock s.
Menkes s.
Menkes-Kaplan s.
Menkes kinky hair s. (MKHS)
menopausal s.
mental deficiency, spasticity, congenital ichthyosis s.
mental and growth retardation-amblyopia s.
mental and physical retardation, speech disorders, peculiar facies s.
mental retardation-absent nails of hallux and pollex s.
mental retardation-adducted thumbs s.
mental retardation, ataxia, hypotonia, hypogonadism, retinal dystrophy s.
mental retardation, blepharonasofacial abnormalities, hand malformations s.
mental retardation-clasped thumb s.
mental retardation, coarse face, microcephaly, epilepsy, skeletal abnormalities s.
mental retardation, coarse facies, epilepsy, joint contracture s.
mental retardation, congenital contracture, low fingertip arches s.
mental retardation-distal arthrogryposis s.
mental retardation, dysmorphism, cerebral atrophy s.
mental retardation, dystonic movements, ataxia, seizures s.
mental retardation, epilepsy, short stature, skeletal dysplasia s.
mental retardation, facial anomalies, hypopituitarism, distal arthrogryposis s.
mental retardation, gynecomastia, obesity s.
mental retardation, hearing impairment, distinct facies, skeletal anomalies s.
mental retardation, hip luxation, G6PD variant s.
mental retardation, macroorchidism s.
mental retardation, microcephaly, blepharochalasis s.
mental retardation, mitral valve prolapse, characteristic face s.
mental retardation, optic atrophy, deafness, seizures s.
mental retardation-overgrowth s.
mental retardation, pre-and postnatal overgrowth, remarkable face, acanthosis nigricans s.
mental retardation-psoriasis s.
mental retardation, retinopathy, microcephaly s.
mental retardation, scapuloperoneal muscular dystrophy, lethal cardiomyopathy s.
mental retardation, short stature, hypertelorism s.
mental retardation, short stature, obesity, hypogonadism s.
mental retardation, skeletal dysplasia, abducens palsy s. (MRSD)
mental retardation-sparse hair s.
mental retardation, spasticity, distal transverse limb defects s.
mental retardation-spastic paraplegia s.

mental retardation, spastic paraplegia, palmoplantar hyperkeratosis s.
mental retardation, typical facies, aortic stenosis s.
mesiodens-cataracts s.
mesoaxial hexadactyly-cardiac malformation s.
mesomelic dwarfism-small genitalia s.
metabolic acidosis s.
methionine malabsorption s.
Meyer-Schwickerath and Weyers s.
Meyer-Schwinkerath s.
Michelin tire baby s.
Michels s.
microcephalic primordial dwarfism-cataracts s.
microcephaly-calcification of cerebral white matter s.
microcephaly-cardiomyopathy s.
microcephaly-chorioretinopathy s.
microcephaly-deafness s.
microcephaly-digital anomalies s.
microcephaly, hiatus hernia, nephrotic s.
microcephaly, hypergonadotropic hypogonadism, short stature s.
microcephaly, infantile spasm, psychomotor retardation, nephrotic s.
microcephaly, mental retardation, cataract, hypogonadism s.
microcephaly, mental retardation, retinopathy s.
microcephaly, mesobrachyphalangy, tracheoesophageal fistula s. (MMT)
microcephaly, microphthalmia, ectrodactyly, prognathism s. (MMEP)
microcephaly, mild developmental delay, short stature, distinctive face s.
microcephaly, mild mental retardation, short stature, skeletal anomalies s.
microcephaly, muscular build, rhizomelia-cataracts s.
microcephaly-oculo-digito-esophageal-duodenal s. (MODED)

microcephaly, sparse hair, mental retardation, seizures s.
microcephaly-spastic diplegia s.
microdeletion s.
microdontia-microcephaly-short stature s.
micrognathia-glossoptosis s.
microphthalmia, dermal aplasia, sclerocornea s. (MIDAS)
microphthalmia-mental deficiency s.
microtia-absent patellae-micrognathia s.
midfetal testicular regression s.
midline cleft s.
Miescher s.
Mietens s.
Mietens-Weber s.
Mikity-Wilson s.
Miles s.
Miles-Carpenter s. (MCS)
Miller s.
Miller-Dieker s.
Miller-Dieker lissencephaly s. (MDLS)
Minkowski-Chauffard s.
Minot-von Willebrand s.
Mirhosseini-Holmes-Walton s.
mixed sclerosing bone dysplasia, small stature, seizures, mental retardation s.
MMIH s.
MNBCC s.
Möbius s.
Mohr s.
Mohr-Claussen s.
Mohr-Tranebjaerg s. (MTS)
Mollica s.
Mollica-Pavone-Anterer s.
monosomy 7 s.
monosomy G s.
Montefiore s.
Moore-Federman s.
Morgagni-Turner s.
Morgagni-Turner-Albright s.
morning glory s.
Morquio s.
Morquio-Brailsford s.
Morquio-Ullrich s.
mosaic tetrasomy 8p s.
mosaic Turner s.
Moynahan alopecia s.
MSN s.

S

NOTES

syndrome *(continued)*

mucocutaneous lymph node s. (MLNS)

mucosal neuroma s.

Müller s.

müllerian, renal, cervicothoracic, somite abnormalities s.

multiple basal cell carcinoma syndrome multiple basal cell nevus s.

multiple epiphyseal dysplasia-early onset diabetes mellitus s. (MED-IDDM)

multiple epiphyseal dysplasia tarda s.

multiple exostosis mental retardation s.

multiple lentigines s.

multiple neuroma s.

multiple nevoid-basal cell carcinoma s.

multiple nevoid, basal cell epithelioma, jaw cysts, bifid rib s.

multiple organ dysfunction s. (MODS)

multiple pterygium s.

multiple synostoses s.

Mulvihill-Smith s.

MURCS s.

muscle atrophy-contracture-oculomuscle apraxia s.

muscle-eye-brain s. (MEBS)

muscular hypertrophy s.

Mutchinick s.

Myhre s.

myopathic limb-girdle s.

myopathy-lactic acidosis-sideroblastic anemia s.

myopathy-myxedema s.

mystery s.

myxedema-myotonic dystrophy s.

Naegeli s.

Nafucci s.

Nager s.

Nager-de Reynier s.

nail-patella s.

Najjar s.

NAME s.

Nance-Horan s. (NHS)

nanism-constrictive pericarditis s.

nasal hypoplasia, peripheral dysostosis, mental retardation s.

Navajo brainstem s.

Neill-Dingwall s.

Nelson s.

neonatal abstinence s. (NAS)

neonatal progeroid s.

neonatal pseudohydrocephalic progeroid s.

neonatal respiratory distress s.

neonatal small left colon s.

nephrosis-microcephaly s.

nephrosis, microcephaly, hiatus hernia s.

nephrosis-neural dysmigration s.

nephrosis-neuronal dysmigration s.

nephrotic s.

Netherton s.

Nettleship s.

Neuhauser s.

Neu-Laxova s. (NLS)

neurocutaneous s.

neurofaciodigitorenal s.

neurofibromatosis-Noonan s. (NF-NS, NFNS)

neuroichthyosis-hypogonadism s.

neurological disease s.

neutrotrichocutaneous s.

Nevo s.

nevoid basal cell carcinoma s. (NBCCS, NBS)

nevoid basal cell epithelioma, jaw cysts, bifid rib s.

newborn respiratory distress s.

Nezelof s.

NFDR s.

nigricans s.

nigricans-hyperinsulinemia s.

Niikawa-Kuroki s.

Nijmegen breakage s. (NBS)

Noack s.

Nonne-Milroy-Meige s.

nonsalt-losing adrenogenital s.

Noonan s.

Noonan-like giant cell lesion s. (NLGCLS)

Norio s.

Norman-Landing s.

Norman-Roberts lissencephaly s.

Norman-Wood s.

Norrie s.

Norrie-Warburg s.

obesity-hypotonia s.

obesity, short stature, mental deficiency, hypogonadism, micropenis, finger contracture, cleft lip-palate s.

Obrinsky s.

OCC s.

occipital horn s.

Ochoa s.

OCR s.

ocular coloboma-imperforate anus s.

oculoauriculofrontonasal s.

oculocerebral hypopigmentation s.
oculocerebrocutaneous s.
oculocerebrofacial s.
oculocerebrorenal s.
oculodental s.
oculodentodigital s.
oculodigitoesophagoduodenal s.
oculogenitolaryngeal s.
oculomandibulodyscephaly-
  hypotrichosis s.
oculomandibulofacial s.
oculopalatoskeletal s.
ODED s.
odontogenic keratocytosis-skeletal
  anomalies s.
OFD s., type I–IV, VI–IX
Ohdo blepharophimosis s.
Ohtahara s.
olfactogenital s.
oligoasthenoteratozoospermia s.
  (OATS)
oligophrenia-ichthyosis s.
oligoteratoasthenozoospermia s.
Oliver s.
Oliver-McFarlane s.
Ollier s.
Ollier-Klippel-Trenaunay-Weber s.
Omenn s.
OMF s.
omphalocele-cleft palate s.
Onat s.
onychodystrophy-congenital
  deafness s.
OPD s.
ophthalmoacromelic s.
Opitz s.
Opitz-Christian s.
Opitz-Frias s.
Opitz-Kaveggia s.
Opitz trigonocephaly s.
Oppenheim s.
optic atrophy-ataxia s.
Orbeli s.
organoid nevus s.
orocraniodigital s.
orodigitofacial s.
orofaciodigital s. type I–IV, VI–IX
orogenital s.
oromandibuloauricular s.
oromandibulootic s.
Osebold-Remondini s.
Osgood-Schlatter s.

ossified ear cartilages, mental
  deficiency, muscle wasting, bony
  changes s.
osteogenesis imperfecta congenita s.
osteogenesis imperfecta, optic
  atrophy, retinopathy, developmental
  delay s.
osteohypertrophic varicose s.
osteopathia striata s.
osteopathia striata, deafness, cranial
  osteopetrosis s.
osteopathia striata, macrocephaly,
  cranial sclerosis s.
osteopenia-sparse hair-mental
  retardation s.
osteoporosis-pseudoglioma s. (OPS)
Ostrum-Furst s.
otofaciocervical s.
otomandibular s.
otopalatodigital s.
otosclerosis s.
otospongiosis s.
Otto s.
ovarian dysgenesis-sensorineural
  deafness s.
ovarian hyperstimulation s. (OHSS)
ovarian remnant s.
ovarian short stature s.
ovarian vein s.
overdistention s.
Pagon s.
Pai s.
Paine s.
Palant cleft palate s.
palatal-digital-oral s.
Pallister-Hall s.
Pallister-Killian s.
Pallister mosaic s.
Pallister W s.
pancreatic insufficiency s.
Papillon-Léage-Psaume s.
Papillon-Léfevre s.
paramethadione s.
paraneoplastic s.
Parenti-Fraccaro s.
parietal foramina,
  brachymicrocephaly, mental
  retardation s.
Parkes-Weber and Dimitri s.
Parrot s.
Partington s. (PRTS)
Partington-Anderson s.

S

NOTES

519

syndrome *(continued)*

Pashayan s.
Pashayan-Pruzansky s.
Passos-Bueno s.
Patau s.
Patterson pseudoleprechaunism s.
Patterson-Stevenson-Fontaine s.
Pearson marrow-pancreas s.
PEHO s.
Pelletier-Leisti s.
Pellizzi s.
pelvic congestion s.
Pena-Shokeir s.
Pendred s.
pentalogy of Cantrell s.
pentasomy X s.
penta-X s.
Pepper s.
Perheentupa s.
pericardial constriction-growth
  failure s
peripheral dysostosis, nail
  hypoplasia, mental retardation s.
  (PMN)
Perlman nephroblastomatosis s.
Perrault s.
persistent müllerian duct s.
Peters anomaly-corneal clouding-
  growth and mental retardation s.
Peters anomaly-short limb
  dwarfism s.
Peters-plus s.
Pettigrew s. (PGS)
peudoprogeria s.
Peutz-Jeghers s.
Pfaundler-Hurler s.
Pfeiffer s.
pharyngeal pouch s.
Phocas s.
PHS s.
pickwickian s.
Pierre Robin s.
pigmentary retinopathy,
  hypogonadism, mental retardation,
  nerve deafness, glucose
  intolerance s.
pigmentary retinopathy, mental
  retardation s.
Pirie s.
Pitt s.
Pitt-Rogers-Danks s. (PRDS)
placental dysfunction s.
placental hemangioma s.
placental transfusion s.
Plott s.
POEMS s.
pointer s.
Poland s.

Pollitt s.
polycystic ovary s. (PCOS)
polydactyly-chondrodystrophy s.
polydactyly-craniofacial anomalies s.
polydactyly-craniofacial
  dysmorphism s.
polydactyly-imperforate anus s.
polydactyly, imperforate anus,
  vertebral anomalies s.
polyglandular s.
polyneuropathy-cataract-deafness s.
polysplenia s.
polysyndactyly-dyscrania s.
polysyndactyly-peculiar skull s.
polysynostoses s.
Pompe s.
popliteal pterygium s.
popliteal web s.
Porak-Durante s.
porencephaly, cerebellar hypoplasia,
  internal malformations s.
Porteous s.
POSSUM database of genetic s.'s
postaxial acrofacial dysostosis s.
  (POADS)
postembolization s.
postirradiation s.
postmaturity s.
postmenopausal palpable ovary s.
postpartum hemolytic uremic s.
postpartum pituitary necrosis s.
postperfusion s.
postrubella s.
posttubal ligation s.
Potter s.
Prader-Gurtner s.
Prader-Labhart-Willi s.
Prader-Labhart-Willi-Fanconi s.
Prader-Willi s.
Prader-Willi habitus, osteopenia,
  camptodactyly s.
Prader-Willi habitus, osteoporosis,
  hand contracture s.
premenstrual s. (PMS)
premenstrual salivary s.
premenstrual tension s.
Prieto s. (PRS)
primary empty sella s.
primary hyperuricemia s.
Primrose s.
progeria-like s.
progeroid short stature-pigmented
  nevi s.
progressive encephalopathy, edema,
  hypsarrhythmia, optic atrophy s.
prolonged Q-T s.
prominent incisors-obesity-
  hypotonia s.

proteiform s.
Proteus s. (PS)
Proud s.
prune-belly s.
pseudoachondroplasia s.
pseudoaminopterin s.
pseudo-Hurler s.
pseudohypoparathyroidism s.
pseudoprogeria s.
pseudoprogeria/Hallermann-Streiff s.
pseudotoxoplasmosis s.
pseudotrisomy 13 s.
pseudo-Turner s.
pseudo-Ullrich-Turner s.
pterygium colli s.
pterygium colli, mental retardation,
  digital anomalies s.
ptosis, downslanting palpebral
  fissures, hypertelorism, seizures,
  mental retardation s.
pug nose-peripheral dysostosis s.
pulmonary dysmaturity s.
pulmonic stenosis/café-au-lait
  spots s.
puppetlike s.
purple toes s.
Purtilo X-linked
  lymphoproliferative s.
pyknodysostosis s.
Pyle s.
13q-deletion s.
Quan-Smith s.
quintuple-X s.
r (1)–(22) s.
Rabson-Mendenhall s.
radial aplasia-thrombocytopenia s.
radial ray defects, triangular face,
  telecanthus, sparse hair, dwarfism,
  mental retardation s.
radial-renal s.
radiorenoocular s.
radioulnar synostosis-developmental
  retardation-hypotonia s.
radioulnar synostosis, short stature,
  microcephaly, scoliosis, mental
  retardation s.
Raine s.
RALPH s.
  renal, anus, lung, polydactyly,
    hamartoblastoma
Rambam-Hasharon s.
Ramon s.

Ramsey-Hunt s.
rape trauma s.
Rapp-Hodgkin s.
REAR s.
rec (8) s.
recessive deafness-
  onychodystrophy s.
recessive enhanced S-cone s.
recessive Usher s.
recombinant chromosome 8 s.
Reed s.
Refetoff s.
Refsum s.
Regenbogen-Donnai s.
Reifenstein s.
Reiter s.
renal, anus, lung, polydactyly,
  hamartoblastoma s.
renal mesangial sclerosis-eye
  defects s.
Rendu-Osler-Weber s.
Renpenning s.
residual ovary s.
resistant ovary s.
respiratory distress s. (RDS)
respiratory distress s. of the
  newborn
restless legs s.
retained bladder s.
retardation, aphasia, shuffling gait,
  adducted thumbs s.
retardation, deafness,
  microgenitalism s.
Rethoré s.
retinal pigmentary degeneration,
  microcephaly, mental
  retardation s.
retinitis pigmentosa-congenital
  deafness s.
retinoblastoma-mental retardation s.
retinopathy-mental retardation s.
retraction s.
Rett s.
reverse chylous s.
Reye s.
rhizomelia s.
rhizomelic chondrodysplasia
  punctata s.
Rh-null s.
rib-gap defect-micrognathia s.
Richards-Rundle s.
Richner s.

S

NOTES

syndrome *(continued)*
Richner-Hanhart s.
Rieger s.
right ovarian vein s.
right-sided arch, mental deficiency,
  facial dysmorphism s.
Riley-Day s.
Riley-Schwachman s.
Riley-Smith s.
ring 1–22 s.
ring chromosome 1–22 s.
Ritscher-Schinzel s.
Roberts pseudothalidomide s.
Roberts tetraphocomelia s.
Robin s.
Robinow s.
Robinow-Silverman-Smith s.
Robinow-Sorauf s.
Rocher-Sheldon s.
Rodrigues s.
Rokitansky-Küster-Hauser s.
Rolland-Desbuquois s.
Rosenthal-Kloepfer s.
Rosewater s.
Rosselli-Gulienetti s.
Rossi s.
Rothmann-Makai s.
Rothmund s.
Rothmund-Thomson s.
Rothmund-Werner s.
Rotor s.
round ligament s.
Roussy-Lévy s.
RSH s.
RSH/SLO s.
RSH/Smith-Lemli-Opitz s.
rubella s.
Rubinstein s.
Rubinstein-Taybi s.
Rud s.
Rudiger s.
rudimentary testis s.
Russell s.
Russell-Silver s.
Rutherfurd s.
Rutledge lethal multiple congenital
  anomaly s.
Ruvalcaba s.
Ruvalcaba-Myhre s.
Ruvalcaba-Myhre-Smith s. (RMSS)
Ruvalcaba-Reichert-Smith s.
Sabinas brittle hair s.
Saethre-Chotzen s.
Sakati-Nyhan s.
Saldino-Noonan s.
salt-losing adrenogenital s. (SLAS)
Sanchez-Cascos s.
Sanchez-Corona s.

Sanchez-Salorio s.
Sandifer s.
Sanfilippo s.
Sanjad-Sakati s.
Santavuori s.
Santavuori-Haltia s.
Sao Paulo MCA/MR s.
Sato s.
Savage s.
Say s.
Say-Gerald s.
Say-Meyer s.
SC s.
scalded-skin s.
Schafer s.
Scheie s.
Scheuthauer-Marie-Sainton s.
Schimmelbusch s.
Schimmelpenning-Feuerstein-Mims s.
Schinzel acrocallosal s.
Schinzel-Giedion s. (SGS)
Schinzel-Giedion midface-
  retraction s.
Schirmer s.
Schmid-Fraccaro s.
Schmidley s.
Schmidt s.
Schwachman s.
Schwachman-Diamond s.
Schwartz s.
Schwartz-Jampel s.
Schwartz-Jampel-Aberfeld s.
scimitar s.
Scott craniodigital s.
Scott-Taor s.
SC phocomelia s.
SC-pseudothalidomide s.
sea-blue histiocyte s.
Seabright bantam s.
sebaceous nevus s.
Seckel bird head s.
Sedlacková s.
Seemanová s. 1, 2
Seemanová-Lesny s.
Seip s.
Seip-Lawrence s.
Seitleis s.
seizures, acquired microcephaly,
  agenesis of corpus callosum s.
seizures, hypotonic cerebral palsy,
  megalocornea, mental
  retardation s.
Sensenbrenner s.
Sensenbrenner-Dorst-Owens s.
sensimotor induction in disturbed
  equilibrium s.
sensorineural deafness, imperforate
  anus, hypoplastic thumbs s.

Senter s.
sepsis-pneumonia s.
septum pellucidum with
    porencephalia s. (SASPP)
Sertoli-cell-only s.
Setleis s.
severe ovarian hyperstimulation s.
    (SOHS)
Shah-Waardenburg s.
shaken baby s.
shaken impact s.
shallow orbits, ptosis, coloboma,
    trigonocephaly, gyral
    malformations, mental and growth
    retardation s.
shawl scrotum s.
shedding s.
Sheehan s.
Shereshevskii-Turner s.
Shokeir s.
SHORT s.
short-bowel s.
short limb dwarfism, saddle nose,
    spinal alterations, metaphyseal
    striation s.
short rib-polydactyly s. (SRPS)
short stature, characteristic facies,
    mental retardation, macrodontia,
    skeletal anomalies s.
short stature, microcephaly, mental
    retardation, multiple epiphyseal
    dysplasia s.
short stature, microcephaly,
    syndactyly, dysmorphic face,
    mental retardation s.
Shprintzen s.
Shprintzen-Goldberg
    craniosynostosis s.
Shwachman-Bodian s.
Shwachman-Diamond s.
Shy-Drager s.
Shy-Magee s.
sicca s.
Siemerling-Creutzfeldt s.
Silver s.
Silver-Russell s.
Silverskiöld s.
Simmonds s.
Simpson dysmorphia s. (SDYS)
Simpson dysplasia s.
Simpson-Golabi-Behmel s. (SGBS)
Sipple s.

situs inversus totalis s.
Sjögren s.
Sjögren-Larsson s.
skeletal and cardiac malformations-
    thrombocytopenia s.
skeletal dysplasial, joint laxity,
    mental retardation s.
skeletal dysplasia, sparse hair,
    dental anomalies s.
skeleton-skin-brain s.
skin-eye-brain s.
skin mastocytosis-hearing loss-
    mental retardation s.
slick-gut s.
SLO s.
Slotnick-Goldfarb s.
sloughed urethra s.
slow-channel congenital
    myasthenic s. (SCCMS)
Sly s.
small left colon s.
small patella s.
Smith s.
Smith-Fineman-Myers s. (SFMS)
Smith-Lemli-Opitz s. (SLO)
Smith-Magenis s. (SMS)
Smith-Theiler-Schachenmann s.
Snyder-Robinson s. (SRS)
soft hands s.
Sohval-Soffer s.
Solomon s.
Solomon-Fretzin-Dewald s.
Sommer s.
Sonoda s.
Sorsby s.
Sotos s.
spastic diplegia s.
spastic quadriplegia, congenital
    ichthyosiform erythroderma,
    oligophrenia s.
spastic quadriplegia, retinitis
    pigmentosa, mental retardation s.
spinal muscular atrophy-mental
    retardation s.
spinal muscular atrophy,
    microcephaly, mental
    retardation s.
spinocerebellar ataxia-
    dysmorphism s.
splenic sequestration s.
split foot, microphthalmia, cleft
    lip/palate-mental retardation s.

S

NOTES

syndrome *(continued)*
    split hand/feet s.
    spondylar and nasal alterations-striated metaphyses s.
    spondylocostal dysplasia s.
    spondyloepiphyseal dysplasia congenita s.
    spondyloepiphyseal dysplasia-diabetes mellitus s.
    spondyloepiphyseal dysplasia tarda-mental retardation s.
    spondylometa-epiphyseal dysplasia-extreme short stature s.
    spondylometaphyseal dysplasia-short limb-abnormal calcification s.
    spondylothoracic dysplasia s.
    Sprangeler-Wiedemann s.
    Spurway s.
    Srb s.
    staphylococcal-scalded skin s.
    steely-hair s.
    Steinert s.
    Steinfeld s.
    Stein-Leventhal s.
    sterile pyuria s.
    Stevens-Johnson s.
    Stewart-Treves s.
    Stickler s.
    stiff-baby s.
    Stilling-Türk-Duane s.
    Stimmler s.
    Stocco dos Santos s.
    Stock-Spielmeyer-Vogt s.
    Stoll s.
    straight-back s.
    Stratton-Parker s.
    Strudwick s.
    Sturge s.
    Sturge-Kalischer-Weber s.
    Sturge-Weber s.
    Sturge-Weber-Dimitri s.
    Sturge-Weber-Krabbe s.
    Stüve-Wiedemann s. (SWS)
    subaortic stenosis-short stature s.
    sudden infant death s. (SIDS)
    Sugarman s.
    sulfur-deficient brittle hair s.
    superior vena caval s.
    supine hypotensive s.
    surfactant deficiency s.
    Sutherland-Haan s. (SHS)
    sweaty feet s.
    Sweet s.
    Swyer s.
    Swyer-James s.
    Swyer-James-Macleod s.
    syndactyly-anophthalmos s.
    syndactyly-cataracts-mental retardation s.
    syndactyly-microcephaly-mental retardation s.
    systemic inflammatory response s. (SIRS)
    TAR s.
    Tariverdian s.
    tarsal-carpal coalition s.
    Taussig-Bing s.
    Tay s.
    Taybi s.
    Taybi-Linder s.
    Teebi s.
    telecanthus-hypospadias s.
    Temtamy s.
    ter Haar s.
    Terry s.
    Teschler-Nicola and Killian s.
    testicular feminization s.
    tethered cord s.
    tetra amelia s.
    tetrahydrofolate-methyltransferase deficiency s.
    tetralogy of Fallot s.
    tetraphocomelia-cleft lip-palate s.
    tetraploidy s.
    tetrasomy 15p s.
    tetra-X s.
    thalidomide teratogenicity s.
    Thiemann s.
    third and fourth pharyngeal pouch s.
    Thompson s.
    Thomsen s.
    Thomson s.
    thoracic compression s.
    thrombocytopenia-absent radius s.
    thymic aplasia s.
    thymic and parathyroid agenesis s.
    thyrohypophysial s.
    tibial aplasia-ectrodactyly s.
    Tietze s.
    tired housewife s.
    tooth anomalies, skeletal dysplasia, sparse hair s.
    TORCH s.
    Toriello s. 1, 2
    Toriello-Carey s.
    Torsten Sjögren s.
    Tourette s.
    Townes s.
    Townes-Brocks s.
    toxemia s.
    toxic shock s. (TSS)
    tracheal agenesis s.

tracheoesophageal fistula, esophageal atresia, multiple congenital anomaly s.
Tranebjaerg s. 1, 2
transient respiratory distress s. (TRDS)
translocation Down s.
trapezoidocephaly-synostosis s.
Treacher Collins s.
Treacher Collins-Franceschetti s.
trichodental dysplasia-microcephaly-mental retardation s.
trichorhinophalangeal s.
trichorrhexis nodosa s.
trichothiodystrophy-congenital ichthyosis s.
trichothiodystrophy-neurocutaneous s.
trichothiodystrophy-xeroderma pigmentosum s.
Tridione s.
trigonocephaly s.
trimethadione s.
triple-X s.
triploidy s.
triplo-X s.
trip (15q) s.
trip-X chromosome s.
trismus-pseudocamptodactyly s.
trisomy 1–22 s.
trisomy 11q s.
trisomy C,D,E,G s.
trisomy 18-like s.
Troyer s. (TS)
tubular stenosis, hypocalcemia, convulsions, dwarfism s.
Turner s.
Turner-Albright s.
Turner-Kieser s.
Turner-like s.
twin-peak s.
twin-to-twin transfusion s. (TTS, TTTS)
twin transfusion s.
twin-twin transfusion s. (TTTS)
t (Y;18) s.
tyrosinemia-palmar and plantar keratosis-ocular keratitis s.
Ullrich s.
Ullrich-Bonnevie s.
Ullrich-Feichtiger s.
Ullrich and Fremerey-Dohna s.
Ullrich-Noonan s.

Ullrich-Turner s.
ulnar hypoplasia-club feet-mental retardation s.
ulnar-mammary s.
umbilical cord s.
unilateral fibular aplastic s.
universal joint s.
unusual facies-mental retardation-intrauterine growth retardation s.
Unverricht-Lundborg s.
Urban s.
Urban-Rogers-Meyer s.
urethral s.
urofacial s.
Usher s. (US)
Usher s. type II
uterine hernia s.
uveal coloboma, cleft lip/palate, mental retardation s.
VACTERL s.
VACTERL-H s.
valproic acid s.
Van Buchem s.
van den Bosch s.
van der Hoeve s.
Van der Woude s.
Van Haldergem s.
vanishing testes s.
vanishing twin s.
Van Maldergem s.
Váradi s.
Váradi-Papp s.
vascular ring s.
VATER s.
velocardiofacial s.
Verner-Morrison s.
Virchow-Seckel s.
viscous s.
Vles s.
Vogt s.
Vohwinkel s.
von Willebrand s.
Voorhoeve s.
vulvar vestibulitis s. (VVS)
W s.
Waardenburg-Klein s.
Waardenburg recessive anophthalmia s.
Wagner s.
WAGR s.
Waisman s.
Waisman-Laxova s.

S

NOTES

syndrome *(continued)*
Walker-Clodius s.
Walker lissencephaly s.
Walker-Warburg s.
Walton s.
Warburg s.
warfarin s.
Waring blender s.
Warkany s. 1, 2
Waterhouse-Friderichsen s.
Watson s.
Watson-Alagille s.
Watson-Miller s.
Weaver s.
Weaver-Smith s. (WSS)
Weaver-Williams s.
Weber s.
Weber-Christian s.
Weber-Dimitri s.
Weill-Marchesani s.
Weismann-Netter s.
Weissenbacher-Zweymuller s.
Went s.
Werdnig-Hoffmann s.
Wermer s.
Werner s.
West s.
wet lung s.
Weyers oligodactyly s.
Whelan s.
Whipple s.
whistling face s.
whistling face-windmill vane
  hand s.
Wieacker s.
Wieacker-Wolff s.
Wiedeman-Beckwith-Combs s.
Wiedemann s.
Wiedemann-Rautenstrauch s. (WR)
Wildervanck s.
Wildervanck-Smith s.
Wilkins s.
Willebrand-Jurgens s.
Williams s.
Williams-Barratt s.
Williams-Beuren s.
Williams-Campbell s.
Wilms tumor-aniridia s.
Wilms tumor, aniridia
  gonadoblastoma, mental
  retardation s.
Wilson-Mikity s.
Wilson-Turner s. (WTS)
Winter s.
Wisconsin s.
Wiskott-Aldrich s.
Wittwer s.
Wolcott-Rallison s.

Wolf s.
Wolff mental retardation s.
Wolff-Parkinson-White s.
Wolf-Hirschhorn s.
Wolfram s.
Woods s.
Worster-Drought s.
wrinkly skin s. (WSS)
Wyburn-Mason s.
s. X
45,X s.
XK s.
XK-aprosencephaly s.
X-linked alpha-thalassemia/mental
  retardation s. (ATRX syndrome,
  ATRX syndrome)
X-linked cataract-dental s.
X-linked congenital cataracts-
  microcornea s.
X-linked dominant s.
X-linked dysplasia-gigantism s.
  (DGSX)
X-linked Hurler s.
X-linked lymphoproliferative s.
X-linked mental deficiency-
  megalotestes s.
X-linked mental handicap-retinitis
  pigmentosa s.
X-linked mental retardation s. 1–6
  (MRXS1–6)
X-linked mental retardation-
  aphasia s. (MRXA)
X-linked mental retardation-
  blindness-deafness-multiple
  congenital anomalies s.
X-linked mental retardation-fragile
  site s. 2
X-linked mental retardation-growth
  hormone deficiency s.
X-linked mental retardation-
  hypogenitalism-cerebral anomaly s.
X-linked mental retardation-
  marfanoid habitus s.
X-linked mental retardation,
  microphthalmia, microcornea,
  cataract, hypogenitalism, mental
  retardation-spasticity s.
X-linked mental retardation-
  psoriasis s.
X-linked mental retardation-spastic
  diplegia s.
X-linked mental retardation, thin
  habitus, osteoporosis,
  kyphoscoliosis s.
X-linked mental retardation with
  fragile X s.
X-linked Opitz s. (XLOS)
X-linked recessive deafness s.

X-linked seizures, acquired
micrencephaly, agenesis of corpus
callosum s.
XO s.
Xq+ s.
Xq- s.
Xq Klinefelter s.
XX male s.
46,XX male s.
XXX s.
47,XXX s.
XXXX s.
XXXXX s.
XXXXY s.
49XXXXY s.
XXY s.
47,XXY s.
XYY s.
yellow vernix s.
Young s.
Young-Hughes s.
Young-Madders s.
Yunis-Varon s.
YY s.
Zellweger s.
Zellweger cerebrohepatorenal s.
Zerres s.
Ziehen-Oppenheim s.
Zimmermann-Laband s. (ZLS)
Zinsser s.
Zinsser-Engman-Cole s.
Zipokowski-Margolis s.
Zlotogora-Ogür s.
Zollinger-Ellison s.
Zollino s.
Zunich s.
Zwahlen s.
synechia, pl. **synechiae**
intrauterine s.
s. vulvae
Synemol Topical
synencephalocele
Synevac vacuum curettage system
Synflex
syngamy
syngeneic tissue
syngnathia
s. congenita
syngraft
Synkayvite
synkinesia
mouth-and-hand s.

Syn-Minocycline
Synodroy
synophthalmia
Synophylate
synorchidism
synoscheos
synostosis, pl. **synostoses**
cranial s.
humeroradial s.
multiple synostoses
s. multiplex
radioulnar s.
tribasilar s.
synovitis
Synphasic
Synsorb Pk
syntenic gene
synteny
synthase
methionine s.
prostaglandin s.
synthesis, pl. **syntheses**
decidual prolactin s.
estrogen s.
ovarian estrogen s.
progesterone s.
synthetic
s. conjugated estrogen
s. prostaglandin E_1, E_2
s. suture material
Synthroid
Syntocinon
syphilis
biological false-positive serologic
test for s. (BF-STS)
congenital s.
s. hereditaria tarda
latent s.
primary s.
secondary s.
serologic test for s. (STS)
tertiary s.
s., toxoplasmosis, rubella,
cytomegalovirus, and herpes
(STORCH)
syphilitic rhinitis
syphilotherapy
Syprine
syringe
Asepto s.
Auto S.
bulb s.

S

NOTES

syringe *(continued)*
    s. feeding
    FNA-21 s.
    Luer-Lok s.
    tuberculin s.
syringes (*pl. of* syrinx)
syringocele
syringoma
syringomeningocele
syringomyelocele
syrinx, pl. **syringes**
syrup
    Bromfed S.
    Decofed S.
    Drixoral S.
    Karo s.
system
    Abbott LifeCare PCA Plus II
      infusion s.
    ABI model 373 sequencing gel s.
    ABI model 377 sequencing gel s.
    ABO blood group s.
    ACCESS immunoassay s.
    adnexal adhesion classification s.
    AEGIS sonography management s.
    Affirm VP microbial
      identification s.
    Affymetrix GeneChip s.
    AFS adhesion scoring s.
    AI 5200 S Open Color Doppler
      imaging s.
    Aloka SD ultrasound s.
    alternative s.
    Apogee 800 ultrasound s.
    AquaSens FMS 1000 Fluid
      Monitoring S.
    ASG s.
      Adhesion Scoring Group
    AspenVAC smoke evacuation s.
    ATL HDI 3000 ultrasound s.
    Aurora MR breast imaging s.
    AutoCyte S.
    autonomic nervous s.
    AutoPap 300 QC s.
    Autoread centrifuge hematology s.
    Auto Suture ABBI s.
    Aviva mammography s.
    BABE OB ultrasound reporting s.
    Babyflex heated ventilation s.
    Bair Hugger patient warming s.
    BDProbeTec ET s.
    Bethesda classification s.
    Bethesda II s.
    bicarbonate-carbonic acid s.
    BiliBlanket Plus Phototherapy s.
    Biogel Reveal puncture
      indication s.
    Bishop pelvic scoring s.

    Bishop Prelabor Scoring S.
    breast leakage inhibitor s. (BLIS)
    Breslow microstaging s.
    bursa-dependent s.
    CADD-Prizm pain control s.
    Capasee diagnostic ultrasound s.
    cardiovascular s.
    CatsEye digital camera s.
    CDE blood group s.
    Cell Recovery S. (CRS)
    central nervous s. (CNS)
    Chung microstaging s.
    Cineloop image review
      ultrasound s.
    Clark microstaging s.
    Climara estradiol transdermal s.
    community health management
      information s. (CHMIS)
    Companion 318 Nasal CPAP S.
    Conceptus fallopian tube
      catheterization s.
    continuous distention irrigation s
      (CDIS)
    Cooper Surgical Monopolor ELSG
      LEEP S.
    cotyledon perfusion s.
    cre-loxP s.
    CRYOcare cryoablation s.
    Cryomedics electrosurgery s.
    CrystalEyes endoscopic video s.
    CS-5 cryosurgical s.
    digestive s.
    digital mammography s.
    Dolphin hysteroscopic fluid
      management s.
    Duffy s.
    dynamic optical breast imaging s.
      (DOBI)
    Dynamite mattress s.
    Eccocee ultrasound s.
    Eklund positioning s.
    electroshield monitoring s.
    EnAbl thermal ablation s.
    End-Flo laparoscopic irrigating s.
    endocrine s.
    Endodermologie LPG s.
    EndoMed LSS laparoscopy s.
    Endotek urodynamics s.
    Entree II trocar and cannula s.
    Entree Plus trocar and cannula s.
    Esclim estradiol transdermal s.
    Esclim transderm s.
    Estraderm transdermal s.
    estradiol transdermal s.
    17β-E2 transdermal drug-delivery s.
    EUB-405 ultrasound s.
    Exact-Touch Saccomanno Pap
      smear collection s.

E-Z-EM BioGun automated biopsy s.
Ferriman-Gallwey hirsutism scoring s.
fibrinolytic and clotting s.
Force GSU argon-enhanced electrosurgery s.
Gleeson FloVAC Hi-Flo laparoscopic suction/irrigation s.
Glucometer Elite diabetes care s.
Guardian DNA s.
Gynecare Thermachoice uterine balloon therapy s.
Gynecare Verascope Hysteroscopy S.
HabitEX smoking cessation s.
HemoCue blood glucose s.
HemoCue blood hemoglobin s.
Histofreezer cryosurgical s.
Hitachi EUB 405 imaging s.
Hitachi UB 420 digital ultrasound s.
hybrid capture s.
hypothalamic-pituitary s.
Hysteroser s.
Illumina Pro Series CO2 surgical laser s.
image recording s.
IMEXLAB vascular diagnostic s.
immune s.
In-Fast bone screw s.
InfraGuide delivery s.
Innova electrotherapy s.
Innova feminine incontinence treatment s.
International Neuroblastoma Staging S. (INSS)
International Staging S.
JustVision diagnostic ultrasound s.
Kagan staging s.
KOH colpotomizer s.
Laparolift s.
Lapro-Clip ligating clip s.
Lectromed urinary investigation s.
lymphatic s.
male reproductive s.
Mammomat C3 mammography s.
Mammoscan digital imaging s.
Mammotest breast biopsy s.
Maturna bra s.
MED-1/InfoChart paperless medical record s.

MicroLap Gold s.
MicroSpan microhysterescopy s.
MicroSpan minihysteroscopy s.
mini Vidas automated immunoassay s.
Mityvac vacuum delivery s.
Mullin s.
musculoskeletal s.
Nellcor N-400/FS s.
Neonatal Abstinence Scoring S. (NASS)
Neonatal Facial Coding S.
nervous s.
neuroendocrine s.
Norplant s.
OpenGene automated DNA sequencing s.
opsonization s.
OPUS immunoassay s.
OraSure oral HIV-1 antibody testing s.
orthogonal lead s.
OsteoView 2000 s.
Ovation falloposcopy s.
PadKit sample collection s.
PalmVue s.
PapNet automated cervical cystology s.
PapNet testing s.
Pap-Perfect supply s.
parasympathetic nervous s.
Pelvic Organ Prolapse-Quantified s. (POP-Q)
Pelvic Organ Prolapse staging s.
Percutaneous Stoller Afferent Nerve Stimulation S.
Performa Acoustic Imaging s.
Performa diagnostic ultrasound imaging s.
Pregnancy Risk Assessment Monitoring S. (PRAMS)
Preg-Net computerized information s.
primer pair s.
probe s.
prognostic scoring s.
prorenin-renin-angiotensin s.
ProTime microcoagulation s.
Quips genetic imaging s.
Renaissance spirometry s.
renin-angiotensin s.
renin-angiotensin-aldosterone s.

S

NOTES

system *(continued)*
    reproductive s.
    respiratory s.
    reticuloendothelial s.
    Rh blood group s.
    Riechert-Mundinger stereotactic s.
    RxFISH DNA probe and
      analysis s.
    SalEst preterm labor test s.
    selective no-fault s.
    selective tubal occlusion
      procedure s.
    SensoScan mammography s.
    Sequoia Acuson s.
    Seroma-Cath s.
    Sherwood intrascopic
      suction/irrigation s.
    Shimadzu ultrasound s.
    Shutt suture punch s.
    Siemens Sonoline SI-400
      ultrasound s.
    Sinai S.
    single-use diagnostic s. (SUDS)
    Sirecust 404N neonatal
      monitoring s.
    Sky-Boot stirrup s.
    SKY epidural pain control s.
    Sonicaid Vasoflow Doppler s.
    Sonopsy ultrasound-guided breast
      biopsy s.
    Spect-Align laser s.
    Spectra 400 extended surveillance
      and alert s.
    Sperm Select sperm recovery s.
    SprayGel Adhesion Barrier S.
    STARRT falloscopy s.
    Straight-In surgical s.
    subdermal contraceptive s.
    swivel-arm s.
    sympathetic nervous s.
    Synevac vacuum curettage s.
    Technos ultrasound s.
    Teratogen Information S. (TERIS)
    Tesla s.
    s. for thalidomide education and
      prescription safety (STEPS)
    ThermaChoice uterine balloon
      therapy s.
    thermal balloon s.
    ThermoChem-HT s.
    TLX alloantigen s.
    TroGARD electrosurgical blunt
      trocar s.
    1.5 T superconductive s.
    T-TAC s.
    Tylok high-tension cerclage
      cabling s.
    UD 2000 urodynamic
      measurement s.
    ULPA/charcoal filtration s.
    Ultramark ultrasound s.
    UPS 2020 ambulatory
      measurement s.
    urinary s.
    Urocyte diagnostic cytometry s.
    UroVive s.
    Vacutainer s.
    Vakutage suction s.
    Valleylab REM s.
    Valley Vac smoke evacuation s.
    VestaBlate s.
    VIDAS automated immunoassay s.
    VIDAS immunoanalysis testing s.
    Vivelle-Dot estradiol transdermal s.
    WAVE DNA Fragment Analysis S.

systematicus
    nevus pigmentosus s.

systematisata
    melanoblastosis cutis linearis
      sive s.

systemic
    s. arteriovenous fistula
    s. illness
    s. inflammatory response syndrome
      (SIRS)
    s. lupus erythematosus (SLE)
    s. vascular resistance

systemic-active nonspecific
  immunotherapy

systolic
    s. ejection murmur
    s. murmur

systolic/diastolic (S/D)
    s./d. ratio

Syva test

Sztehlo umbilical clamp

T

    testosterone
      T band
      T bone density score
      T cell
      T strain mycoplasma
      1.5 T superconductive system
      T, T_3, T_4 lymphocyte

T_2

    diiodothyronine

T_4

    thyroxine

t

    t. (Y;18) syndrome

T6 antigen

TA

    thoracoabdominal
      TA 55 stapler

^{182}Ta

    tantalum-182

TAA

    tumor-associated antigen

TAB

    therapeutic abortion

Tabb curette

tabes

    t. infantum
    t. mesenterica

table

    Bayley-Pinneau t.
    contingency t.
    cross t.
    height t.

tablet

    Actifed Allergy T.
    Benadryl Decongestant Allergy T.
    Bromfed T.
    Bromphen T.
    Cenestin t.'s
    Dimaphen T.'s
    Dimetapp T.
    Histalet Forte T.
    Hista-Vadrin T.
    hormonal pregnancy test t.
    Levlite t.
    Materna T.'s
    Mircette t.
    Spectrobid T.
    Vicks DayQuil Allergy Relief 4
      Hour T.

Tabs

    Apo-Doxy T.

TACE

tachyarrhythmia

    supraventricular t. (SVT)

tachycardia

    atrial t.
    atrioventricular reciprocating t.
    paraventricular t.
    paroxysmal t.
    sinus t.
    supraventricular t. (SVT)
    ventricular t.

tachydysrhythmia

    supraventricular t.

tachyphylaxis

tachypnea

    transient t.

tachysystole

    uterine t.

tacker

    Origin t.

Tacozin

tactile

    t. fever
    t. sensory monitoring
    t. stimulation
    t. stimulus
    t. temperature

taeniasis

TAF

    tumor angiogenesis factor

tag

    expressed sequence t.
    hymenal t.
    skin t.

Tagamet

    T.-HB

TAGO diagnostic kit

TAH

    total abdominal hysterectomy

tail

    axillary t.
    t. bud
    Spence axillary t.

tailgut

talc

talipes

    t. calcaneovalgus
    t. calcaneovarus
    t. calcaneus
    t. cavovalgus
    t. cavus
    t. equinovalgus
    t. equinovarus
    t. equinus
    t. hobble splint
    t. planovalgus
    t. planus

T

talipes *(continued)*
    t. valgus
    t. varus
talipomanus
talk
    receptor cross t.
Talwin
Tambocor
Tamine
Tamm-Horsfall
    T.-H. mucoprotein
    T.-H. protein
Tamofen
Tamone
tamoxifen
    t. citrate
    Cytoxan, methotrexate, fluorouracil,
      prednisone, t. (CMFPT)
tampon
    Genupak t.
    vaginal t.
tamponade
    cardiac t.
    pericardial t.
Tanac
tandem
    afterload t.
    Fleming afterloading t.
    Fletcher-Suit afterloading t.
    T. Icon II hCG
    t. and ovoids (T&O)
    t.-repeat
    t. repeat sequence
    T.-R Ostase osteoporosis test
    variable-number t.-repeat
Tangier disease
Tanner
    T. stage
    T. stages of development
    T. staging (Grade I–IV)
tantalum
    t.-182 (^{182}Ta)
tanycyte
Tanyoz sign
Tao
TAP
    transport-associated protein
    transvaginal amniotic puncture
tap
    supraciliary t.
Tapar
Tapazole
tape
    lap t.
    Medipore H soft cloth surgical t.
    tension-free vaginal t. (TVT)
tapir mouth
tapiroid

TAR
    thrombocytopenia-absent radius
    TAR syndrome
Tarabine PFS
Taractan
Tarasan
tarda
    chondrodystrophia t.
    chondrodystrophia congenita t.
    Edwardsiella t.
    osteopetrosis t.
    porphyria cutanea t.
Tardieu test
target
    t. cell
    molybdenum t.
    t. organ response
    tungsten t.
targeted ultrasound
targeting
    gene t.
Tariverdian syndrome
Tarkowski method
Tarnier axis-traction forceps
Taro-Ampicillin
Taro-Cloxacillin
Taro-Sone
tarry cyst
tarsal-carpal coalition syndrome
tartrate
    butorphanol t.
    ergotamine t.
    metoprolol t.
    zolpidem t.
TAS/TVS
    transabdominal/transvaginal ultrasound
TAT
    tray agglutination test
    tyrosine aminotransferase
TATA box
TATD
    tyrosine aminotransferase deficiency
taurine
taurodontism
Taussig-Bing
    T.-B. disease
    T.-B. syndrome
tautomenial
Tavist
    T.-1
    T.-D
taxis
    bipolar t.
Taxol
Taxotere
Taybi-Linder syndrome
Taybi syndrome

Tay-Sachs
  T.-S. disease
  T.-S. disease with visceral
    involvement
Tay syndrome
Tazicef
Tazidime
tazobactam
  t. sodium
Tazol
TB
  tuberculosis
TBG
  thyroid-binding globulin
TBLC
  term birth, living child
TBP
  thyroxine-binding protein
3TC
TC7 adhesion barrier
TCA
  trichloroacetic acid
  tricyclic antidepressant
T-cell subset
TCIFTT
  transcervical intrafallopian tube transfer
tCpO$_2$
  transcutaneous partial pressure of oxygen
T.D.
  Diamine T.D.
TDF
  testis-determining factor
TDI
  therapeutic donor insemination
TDLU
  terminal duct lobular unit
TDxFLM
  TDxFLM Assay
  TDxFLM Assay test
TDxFLx Assay
tdy gene
team
  Patient Outcomes Research T.
    (PORT)
  Sexual Assault Response T.
    (SART)
Tebamide
Tebrazid
technetium
technical artifact
technique
  agar gel precipitation t.

Ayre spatula-Zelsmyr Cytobrush t.
balloon catheter t.
Ball pelvimetry t.
Beverly-Douglas lip-tongue
  adhesion t.
Brockenbrough t.
Brown-Wickham t.
Bruhat t.
clip t.
clonogenic t.
cobalt-60 moving strip t.
Colcher-Sussman t.
contraceptive t.
Counsellor-Flor modification of
  McIndoe t.
Culcher-Sussman t.
2-diameter pocket t.
Döderlein t.
dot-blot t.
double-freeze t.
Dufourmentel t.
Dyban t.
Eklund t.
enzyme-multiplication
  immunoassay t. (EMIT)
evoked potential t.
ferning t.
fetal surveillance t.
Frank t.
Gittes t.
Goebell-Frangenheim-Stoeckel t.
Gomco t.
gracilis flap t.
GRASS MRI t.
Hamou t.
Heaney t.
hemisection uterine morcellation t.
hold t.
hook traction t.
hybridoma t.
hyperinsulinemic-euglycemic
  clamp t.
immune monitoring t.
immune separation t.
immunoperoxidase t.
insemination swim-up t.
Jones and Jones wedge t.
Kehr t.
Kety-Schmidt t.
Kidde cannula t.
Kleihauer t.
Krönig t.

T

NOTES

technique *(continued)*
 Lapides t.
 Lazarus-Nelson t.
 Leboyer t.
 Lich t.
 Limberg t.
 loss-of-resistance t.
 Madden t.
 March t.
 marsupialization t.
 M-FISH cytogenetic t.
 Millen t.
 minimally invasive surgical t.
  (MIST)
 Miyazaki t.
 modified Pomeroy t.
 molecular genetic t.
 multicolor FISH cytogenetic t.
 nucleic acid amplification t.
 Pacey t.
 pants-over-vest t.
 Parkland Hospital t.
 Percoll t.
 recombinant DNA t.
 reverse FISH cytogenetic t.
 rollerball t.
 Semm Z t.
 shoelace t.
 Simonton t.
 skate flap t.
 snapshot GRASS t.
 soluble gas t.
 Southern blot t.
 split-flap t.
 Strassman t.
 Stringer t.
 surveillance t.
 swim-up t.
 three-point t.
 Tompkins median bivalving t.
 tubal ligation band t.
 U t.
 Wallace t.
technology
 assisted reproduction t.
 assisted reproductive t. (ART)
 Center for Health T. (CHT)
 genetic engineering t.
 recombinant DNA t.
 reproductive t.
 Society for Assisted
  Reproductive T. (SART)
 t. transfer
Technos ultrasound system
tectocephaly
TED
 thromboembolic disease
 TED stocking

Teebi syndrome
Teejel
teeth
 Fournier t.
 Hutchinson t.
 mental retardation, congenital heart
  disease, blepharophimosis,
  blepharoptosis, hypoplastic t.
 milk t.
 Moon t.
 natal t.
 neonatal t.
 precocious t.
 prediciduous t.
 screwdriver t.
 X-linked cataract with
  hutchinsonian t.
TEF
 tracheoesophageal fistula
Teflon periurethral injection
Tegison
Tegopen
Tegretol
Tegrin-HC Topical
Teilum tumor
Telachlor Oral
Teladar Topical
telangiectasia
 ataxia t.
 calcinosis cutis, Raynaud
  phenomenon, esophageal motility
  disorders, sclerodactyly, t.
  (CREST)
 calcinosis cutis, Raynaud
  phenomenon, sclerodactyly, t.
  (CRST)
 hemorrhagic t.
telangiectatic granuloma
Teldrin
 T. Oral
telecanthus
telecanthus-hypospadias syndrome
telefetal monitoring
telemammography
telemedicine
teleologic theory
telephase
telepsychiatry
teleradiology
telescopy
 percutaneous suprapubic t.
teletherapy
TeLinde operation
Teline
telocentric chromosome
telogen
 t. effluvium
 t. phase

telomere
telophase
TEM
> transanal endoscopic microsurgery

Temaril
temazepam
Temovate
temperature
> absolute t.
> artificial t.
> aseptic t.
> aural t.
> axillary t.
> basal body t. (BBT)
> body t.
> core t.
> critical t.
> ephemeral t.
> eruptive t.
> maximum t. (T-max)
> normal t.
> oral t.
> rectal t.
> room t.
> skin t.
> subnormal t.
> tactile t.
> tympanic t.

template
> Syed-Neblett dedicated vulvar plastic t.

temporal lobe epilepsy
temporary diverting colostomy
Tempra
Temtamy syndrome
TEN
> toxic epidermal necrolysis

tenaculum, pl. **tenacula**
> Braun-Schroeder single-tooth t.
> cervical t.
> double-tooth t.
> Emmett cervical t.
> t. hook
> t. hook loop
> Jacobs t.
> Schroeder uterine t.
> single-tooth t.
> uterine t.

Tenckhoff catheter
tendency
> familial t.

tenderness
> cervical motion t. (CMT)
> costovertebral angle t. (CVAT)
> lower abdominal t.
> pelvic t.

Tender-Touch
> T.-T. extractor
> T.-T. vacuum birthing cup

tendineus
> arcus t.

tendinous arch
tendon forceps
tenesmus
> perimenstrual t.

Tenex
teniposide
Tenney-Parker sign
Tenol
Tenoretic
Tenormin
Tensilon
tension
> carbon dioxide t.
> t. cyst
> t. hydrothorax
> oxygen t.
> t. pneumothorax
> postmenstrual t. (PMT)
> premenstrual t.
> surface t.
> vaginal tape ' t.

tension-free vaginal tape (TVT)
tensor fasciae latae (TFL)
tent
> CAM t.
> Hypan t.
> intracervical t.
> laminaria t.
> t. mist
> mist t.
> t. oxygen
> Silon t.

tentorial laceration
Tenuate
tenuous
Tenzel calipers
Tepanil
teranthanasia
teras, pl. **terata**
teratism
teratoblastoma
teratocarcinoma

NOTES

T

535

teratogen
    t.-induced malformation
    T. Information System (TERIS)
    T. Registry
teratogenesis
teratogenic
    t. agent
    t. effect
    t. medication
    t. outcome
    t. risk
teratogenicity
teratology
teratoma, pl. **teratomata**
    benign cystic t. (BCT)
    benign cystic ovarian t.
    germ cell t.
    immature ovarian t.
    malignant ovarian t.
    mature cystic ovarian t.
    ovarian cystic t.
    ovarian embryonal t.
    pediatric ovarian t.
    sacrococcygeal t.
teratomata
teratophobia
teratospermia
teratozoospermia
Terazol
    T. Vaginal
    T. 3 vaginal cream
    T. 7 vaginal cream
    T. vaginal suppository
terbinafine
terbutaline
    t. sulfate
terconazole
terfenadine
Terfluzine
ter Haar syndrome
TERIS
    Teratogen Information System
term
    t. AGA
    t. birth, living child (TBLC)
    t. delivery
    t. gestation
    t. infant
    t. infants, premature infants,
      abortions, living children (TPAL)
    t. LGA
    t. pregnancy
    t. SGA
Term-Guard
terminal
    t. cardiotocogram
    t. deletion
    t. ductal-lobular unit

    t. duct lobular unit (TDLU)
    t. hair
    t. ileitis
    t. motor latency
    t. neosalpingostomy
    t. saccular period
    t. transverse acheiria defect
    t. transverse limb defect
termination
    t. codon
    selective t.
terminator
terminus
    vaginal t.
Ter-Pogossian cervical applicator
Terramycin
    T. I.M. Injection
    T. Oral
Terry syndrome
tertiary syphilis
TESA
    testicular sperm aspiration
Teschler-Nicola and Killian syndrome
TESE
    testicular sperm extraction
Teslac
Tesla system
Tessier craniofacial operation
TEST
    tubal embryo stage transfer
test
    Accu-Chek t.
    AccuStat hCG pregnancy t.
    acid elution t.
    acoustic stimulation t. (AST)
    ACTH stimulation t.
    Affirm VPIII t.
    agglutination inhibition t.
    Alcohol Use Disorders
      Identification T. (AUDIT)
    Allen-Doisy t.
    ambulatory uterine contraction t.
    Amiel-Tison t.
    Amniostat-FLM t.
    AneuVysion Assay prenatal
      genetic t.
    antiglobulin t.
    antitreponemal t.
    Apt t.
    arginine tolerance t. (ATT)
    Aschheim-Zondek t.
    A.-Z. t.
    Ballard t.
    Barlow hip dysplasia t.
    Bernstein t.
    Betke-Kleihauer t.
    Biocept-G pregnancy t.
    Biocept-5 pregnancy t.

BioStar strep A 1A t.
bitterling t.
bladder muscle stress t.
bladder neck elevation t.
blood-type t.
Bonney blue stress incontinence t.
bovine mucus penetration t.
BRACA gene t.
branching snowflake t.
breast stimulation contraction t. (BSCT)
breath hydrogen excretion t.
Brodie-Trendelenburg t.
Brouha t.
BTA stat t.
bubble stability t.
CAGE t.
cancer antigen 125 t.
caramel t.
Children's Depression Inventory t.
Chlamydiazyme t.
chlortetracycline fluorescence t.
Clearview hCG pregnancy t.
clomiphene citrate challenge t. (CCCT, C3T)
coagulation t.
Collins t.
complementation t.
complement fixation t.
Concise Plus hCG urine t.
contraction stress t. (CST)
Coombs t.
Corner-Allen t.
Cortrosyn stimulation t.
cosyntropin stimulation t.
cotton swab t.
cough t.
cumulative sum t.
cytochrome oxidase t.
DAP t.
dehydroepiandrosterone sulfate loading t.
Denver Developmental Screening T. (DDST)
dexamethasone suppression t.
DFA t.
Dienst t.
Digene Hybrid Capture II HPV T.
diiodothyronine t.
direct agglutination pregnancy t.
direct fluorescent antigen t.
dot-blot HPV hybridization t.

Eagle t.
early pregnancy t. (EPT)
Eastern blot t.
Elispot t.
enzyme-linked antiglobulin t.
estrogen-progestin t.
factor V Leiden mutation t.
family-based t.
Farber t.
Farris t.
fern t.
fetal acoustic stimulation t.
fetal activity t.
fetal fibronectin t.
Fetal Lung Maturity t.
fetal surveillance t.
Fibrindex t.
Fisher-Yates t.
five-hour glucose tolerance t.
FLM t.
Fluhmann t.
fluorescein-conjugated monoclonal antibody t.
fluorescein treponema antibody t.
fluorescent antimembrane antibody t. (FAMA)
fluorescent treponemal antibody-absorption t.
foam stability t.
Fortel ovulation t.
Franklin-Dukes t.
free beta t.
Frei t.
Friberg microsurgical agglutination t.
Friedman-Lapham t.
Friedman rabbit t.
FTA t.
Galli-Mainini t.
gelatin agglutination t.
genetic t.
Gen-Probe amplified CT t.
germ tube t.
Gesell t.
glucose challenge t.
glucose tolerance t. (GTT)
gonadotropin agonist stimulation t. (GAST)
Gono Kwik t.
Gonozyme t.
Gravindex t.

NOTES

test *(continued)*
    growth hormone stimulation t. (GHST)
    guaiac t.
    Guthrie t.
    HABA binding t.
    hamster t.
    hanging-drop t.
    HealthCheck One-Step One Minute pregnancy t.
    Heller t.
    hemagglutination treponemal t. (HATT)
    Hematest t.
    Hemoccult II t.
    HemoCue glucose t.
    HemoCue hemoglobin t.
    heparin challenge t.
    Heprofile ELISA t.
    Heritage Panel genetic screening t.
    Herp-Check t.
    Hi-Gonavis t.
    Hinton t.
    Hirschberg t.
    HIV t.
    HIVAGEN t.
    Hogben t.
    home pregnancy t.
    Hooker-Farbes t.
    Huhner t.
    human immunodeficiency virus t.
    human ovum fertilization t.
    hydrogen breath t.
    hypoosmotic swelling t.
    Icon serum pregnancy t.
    Icon strep B t.
    Icon urine pregnancy t.
    IgA HIV antibody t.
    immunobead t. (IBT)
    immunofluorescent antibody t.
    immunofluorescent *Chlamydia* t.
    immunologic pregnancy t.
    Indiclor t.
    inhibin t.
    InSight prenatal t.
    insulin tolerance t.
    InTray CCD t.
    Isojima t.
    Isojima-Koyama t.
    Jadassohn t.
    Kahn t.
    Kapeller-Adler t.
    Kell t.
    Kibrick t.
    Kibrick-Isojima infertility t.
    Kleihauer t.
    Kleihauer-Betke t.
    Kline t.

    Knobloch-Gesell t.
    Kodak hCG serum t.
    Kodak SureCell Chlamydia T.
    Kodak SureCell hCG-Urine T.
    Kodak SureCell Herpes (HSV) T.
    Kodak SureCell LCH in-office pregnancy t.
    Kodak SureCell Strep A t.
    KOH t.
    Kolmer t.
    Kolmer-Kline-Kahn t.
    Korotkoff t.
    Kremer penetration t.
    Kruskal-Wallis t.
    Kupperman t.
    Kurzrok-Miller t.
    Kurzrok-Ratner t.
    Kveim t.
    laboratory t.
    Landau t.
    Lange t.
    latex agglutination inhibition t. (LAIT)
    latex fixation· t.
    LCx Probe System t.
    leak-point pressure t.
    Leiter t.
    Lendersloot version t.
    leucine tolerance t.
    levothyroxine t.
    LH Color t.
    Liddle t.
    t. of linkage disequilibrium
    liver function t.'s
    Locke-Wallace Marital Adjustment t.
    Lumadex-FSI t.
    lysoPC diagnostic ovarian cancer t.
    Mantel-Cox t.
    Mantel-Haenszel t.
    MAR t.
    Marchetti t.
    Marshall t.
    Matritech NMP22 bladder cancer t.
    Mazzini t.
    McCaman-Robins t.
    Meinicke t.
    Metopirone t.
    metyrapone t.
    Micral urine dipstick t.
    MicroTrak t.
    Miraluma t.
    Miyazaki-Bonney t.
    monoclonal antibody coagglutination t.
    Monospot t.
    Monosticon Dri-Dot t.
    MultiVysion PB assay t.

nappy t.
Nimbus t.
nipple stimulation t.
Nitrazine t.
nitroblue tetrazolium dye t.
Noguchi t.
nonstress t. (NST)
nontreponemal t.
nontreponemal t.
Northern blot t.
Ogita t.
Omniprobe t.
OncoScint t.
one-hour glucose tolerance t.
one-tail t.
oral glucose challenge t. (OGCT)
OraSure oral HIV t.
Ortolani t.
OsteoGram bone density t.
Osteomark NTx serum t.
Osteosal t.
otoacoustic emission t. (OAE)
OvuKIT t.
OvuQuick Self-T.
oxytocin challenge t. (OCT)
oxytocin stress t.
PACE-2 t.
PACE-2C DNA probe t.
T. Pack hCG
pad t.
PapNet t.
Pathfinder DFA t.
Paul-Bunnell-Davidsohn t.
PCR t.
Penetrak t.
phenolsulfonphthalein t.
phenylketonuria t.
Piso t.
PKU t.
plasmacrit t.
platelet function t.
Porges-Meier t.
postcoital t. (PCT)
postvoid residual urine t.
Precise pregnancy t.
predictive value of t.
pregnancy t.
Profile viral probe t.
progesterone challenge t.
provocative stress t.
PSA t.
pulmonary function t.'s

Qtest Strep t.
Q-tip t.
Quick t.
Quickpac-II OneStep hCG
  pregnancy t.
QuickVue *Chlamydia* t.
QuickVue One-Step hCG-Combo t.
QuickVue One-Step hCG-urine t.
QuickVue One-Step *H. pylori* t.
Quidel Group B Strep T.
radioallergosorbent t. (RAST)
radioimmunosorbent t. (RIST)
rapid plasma reagin card t.
rapid slide t.
RCA t.
reflex HPV t.
renal clearance t.
renal function t.
Rinne t.
Rofsing t.
rollover t.
rosette t.
Rotazyme t.
routine preoperative t.
Rubin t.
Sabin-Feldman dye t.
SalEst system t.
saline drop t.
salivary estriol t.
Schick t.
Schiller t.
Schilling t.
Scotch-tape t.
screening laboratory t.
Sephadex binding t.
serologic t.
Serono t.
Sgambati t.
shake t.
Sickledex t.
Sims-Huhner t.
SISI t.
skin t.
Social Support Scale for
  Children t. (SSSC)
Sonoclot t.
Southern blot t.
Speck t.
sperm agglutination t.
sperm function t.
sperm immobilization t.

T

NOTES

test (*continued*)

SpermMAR mixed antiglobulin reaction t.
stimulation t.
STORCH t.
straight catheter t.
Streptex rapid strep t.
stress t.
Student-Newman-Keuls t.
SUDS HIV-1 antibody t.
supine empty stress t. (SEST)
supine pressor t.
Supravital stain t.
SureCell Herpes (HSV) T.
surf t.
swab t.
sweat chloride t.
swordfish t.
Syva t.
Tandem-R Ostase osteoporosis t.
Tardieu t.
TDxFLM Assay t.
Tes-Tape urine glucose t.
TestPackChlamydia t.
tetraiodothyronine t.
Thayer-Martin gonorrhea t.
ThinPrep Pap t.
Thomas t.
Thorn t.
thrombin clot t.
Thrombostat platelet function t.
Thrombo-Wellco t.
thrombus precursor protein t.
thymol turbidity t.
thyroid function t. (TFT)
thyrotropin-releasing hormone stimulation t.
thyroxine t.
tilt t.
tine t.
tissue thromboplastin-inhibition t.
toluidine blue t.
TPI t.
transmission/disequilibrium t.
tray agglutination t. (TAT)
Treponema pallidum immobilization t.
triiodothyronine t.
triple screen t.
triple swab t.
T2, T3, T4 t.
Tuttle t.
two-tail t.
Tzanck t.
UCG-Slide T.
UNIprobe t.
Uniscreen urine t.
Uri-Check t.

urinary concentration t.
urine CIE t.
urine latex t.
Uriscreen t.
Urispec GP+A t.
Urispec 9-Way t.
vaginal cornification t.
vaginal mucification t.
van den Bergh t.
VDRL t.
Venereal Disease Research Laboratory t.
Venning-Brown t.
Vernes t.
ViraPap HPV DNA t.
ViraType t.
Visscher-Bowman t.
von Poehl t.
Wampole t.
Wasserman t.
water t.
Weber t.
Wepman t.
Western blot t.
wet mount t.
wheat sperm agglutination t.
whiff amine t.
Whittaker t.
withdrawal bleeding t.
Woodcock-Johnson T.
Xenopus t.
zona-free hamster egg penetration t.

Tes-Tape urine glucose test

test/assay

Berkson-Gage t.

test-cross

tester

testes (*pl. of* testis)

testicle

testicular

t. cancer
t. differentiation
t. feminization
t. feminization syndrome
t. sperm aspiration (TESA)
t. sperm extraction (TESE)
t. steroidogenesis

testing

carrier t.
couple t.
fecal occult blood t. (FOBT)
fragile X chromosome t.
nonreassuring fetal t.
oral glucose tolerance t. (OGTT)
PapNet t.
2-step t.

testis, pl. **testes**

t. determination
testes down
gonadotropin-resistant t.
maldescensus t.
rete t.
undescended t.
yolk sac tumor of t.
testis-determining factor (TDF)
testolactone
Testopel Pellet
testosterone (T)
bound t.
t. cypionate
t. enanthate
ethinyl t.
free t.
t. index
non-sex hormone-binding globulin
bound t.
plasma t.
serum t.
t. suppression
topical t.
testosterone/dihydrotestosterone ratio
testosterone-estrogen-binding globulin
testosterone-secreting adrenal adenoma
TestPackChlamydia test
Testred
Testsimplets prestained slide
test-tube baby
TET
tubal embryo transfer
tetania
t. gravidarum
t. neonatorum
tetanic
t. contraction
t. uterine contraction
t. uterus
tetanism
tetanus
diphtheria, pertussis, t. (DPT)
t. immune globulin
t. neonatorum
postpartum t.
puerperal t.
t. toxoid
t. toxoid booster
uterine t.
tetany
neonatal t.

tethered
t. cord
t. cord syndrome
tetotoxicity
tetra amelia syndrome
tetrabenazine
tetrabrachius
tetracaine
Tetracap
T. Oral
tetrachirus
tetracycline
t. analogue
t. hydrochloride
t. pleurodesis
prophylactic t.
Tetracyn
tetrad
tetradactyly
9-tetrahydrocannabinol (THC)
tetrahydrocortisol
tetrahydrofolate (THF)
tetrahydrofolate-methyltransferase
deficiency syndrome
tetrahydrozoline
tetraiodothyronine test
Tetralan
tetralogy
t. of Fallot (TF, TOF)
t. of Fallot syndrome
Tetram
tetramastia
tetramelus
tetramonodactyly
tetranitrate
erythrityl t.
pentaerythritol t.
tetranophthalmos
tetraotus
tetraparesis
tetraphocomelia-cleft lip-palate syndrome
tetraplegia
tetraploid
t. distribution
t. embryo
tetraploidy
t. syndrome
tetrascelus
Tetrasine
T. Extra Ophthalmic
T. Ophthalmic

NOTES

tetrasomy
> chromosome 8p mosaic t.
> partial t. 10p
> t. 15p syndrome
> t. 21q

tetra-X
> t.-X chromosomal aberration
> t.-X syndrome

tetrazolium dye assay
Texacort Topical
TF
> tetralogy of Fallot

TFL
> tensor fasciae latae

TFPI
> tissue factor pathway inhibitor

TFT
> thyroid function test

TFX catheter stylet
TGA
> transposition of great arteries

TGEF
> transabdominal thin-gauge
> embryofetoscopy

T-Gen
TGF
> transforming growth factor

TGFalpha
> transforming growth factor alpha

TGFA polymorphism
TGFbeta
> transforming growth factor beta

TGV
> transposition of great vessels

thalassemia
> alpha t.
> beta t.
> beta-t. major
> beta-t. minor
> t. intermedia
> Lepore t.
> t. major
> t. minor
> sickle t. (HbS-Thal)
> sickle cell t.

thalassemic patient
thalidomide
> fetal t.
> t. teratogenicity syndrome

thallium poisoning
THAM
> tromethamine

thanatophoric
> t. dwarfism
> t. dysplasia

thawing
> embryo t.

Thayer-Martin
> T.-M. gonorrhea test
> T.-M. medium

THb O$_2$
> total oxyhemoglobin

THC
> 9-tetrahydrocannabinol

theca
> t. cell
> t. cell tumor
> t. externa
> t. interna
> t. lutein cell
> t. lutein cyst

theca-granulosa cell cooperativity
theca-interstitial cell
thecal interstitial cell
The Cell Sweep
thecoma
> luteinized t.
> ovarian t.

Theelin
thelarche
> premature t.

theleplasty
theloncus
thelorrhagia
Theo-24
Theobid
Theochron
Theoclear
> T. L.A.

Theo-Dur
Theolair
Theon
theophyllinate
> choline t.

theophylline
> t. level

theory
> clonal selection t.
> Cohnheim t.
> endorphin, dopamine, and
> prostaglandin t.
> Freud t.
> ganglion trigger t.
> gate control t.
> grandmother t.
> implantation t.
> Ishihara t.
> Larmarck t.
> skin trigger t.
> teleologic t.
> Trivers-Willard t.

Theospan-SR
Theostat
Theovent
Theo-X

TherAblator
Hydro T.
therapeutic
t. abortion (TAB)
t. donor insemination (TDI)
t. husband insemination (THI)
t. insemination
t. touch
t. window
therapeutics
biofield t.
therapy
add-back t.
adjuvant t.
adjuvant chemoradiation t.
aerosol t.
aldosterone replacement t.
antenatal corticosteroid t.
antibacterial t.
anticoagulant t.
antidepressant t.
antiemetic t.
antifungal drug t.
antihelminthic t.
antihypertensive t.
antimicrobial t.
antiparasitic drug t.
antiretroviral t.
antithyroid drug t.
antiviral t.
axillary irradiation t.
balloon heating t.
belly bath t.
BEP t.
biofeedback t.
bisphosphonate t.
blood component t.
breast-conserving t. (BCT)
breast-covering t.
breast-preserving t.
broad-spectrum antibiotic t.
bromocriptine t.
caffeine t.
cancer t.
chest physical t.
dexamethasone t.
dinitrochlorobenzene t.
DNCB t.
electroshock t.
embolization t.
endocavitary radiation t.
enterostomal t.

estrogen add back t. (EABT)
estrogen-progestin replacement t.
estrogen replacement t. (ERT)
extended field irradiation t.
external radiation t.
external x-ray t.
ex vivo liver-directed gene t.
fetal drug t.
fluid t.
fractionated radiation t.
frappage t.
Functional Assessment of
Cancer T. (FACT)
gene t.
glucocorticosteroid t.
gold t.
hormonal antineoplastic t.
hormone t.
hormone replacement t. (HRT)
human gene t.
hyperbaric oxygen t.
immunosuppressive t.
internal radiation t.
InterStim t.
interstitial t.
intraperitoneal radiation t.
iodide t.
laser t.
LH-releasing hormone agonist t.
maternal blood clot patch t.
menopausal estrogen replacement t.
monoclonal antibody t.
neoadjuvant hormonal t. (NHT)
neutron t.
nicotine patch t.
oral hormone replacement t.
percussion t.
permission, limited information,
specific suggestions and
intensive t. (PLISSIT)
photodynamic t.
physical t. (PT)
polyvalent immunoglobulin t.
postmenopausal estrogen
replacement t.
progestational t.
progestogen support t.
quadrantectomy, axillary dissection,
radiation t. (QUART)
radiation t.
salvage t.

NOTES

T

543

therapy *(continued)*
>> selective tubal assessment to refine reproductive t. (STARRT)
>> sequential hormone t.
>> sex steroid add-back t.
>> single-dose methotrexate t.
>> single-field hyperthermia combined with radiation t.
>> steroid t.
>> surgical myomectomy as reproductive t. (SMART)
>> ThermaChoice uterine balloon t.
>> thrombolytic t.
>> L-thyroxine t.
>> tocolytic t.
>> transdermal hormone replacement t.
>> transdermal nicotine replacement t.
>> uterine balloon t. (UBT)
>> vaginal estrogen t.
>> vasopressin t.
>> in vivo gene t.
>> xanthochromia t.
>> x-ray t.

Theratope vaccine
Therex
ThermaChoice
>> T. uterine balloon
>> T. uterine balloon therapy
>> T. uterine balloon therapy system

Therma Jaw hot urologic forceps
thermal
>> t. balloon ablation
>> t. balloon system
>> t. gel gradient electrophoresis
>> t. injury

Thermasonic gel warmer
thermistor thermometer
ThermoChem-HT system
Thermodigital thermometer
thermodilution method
thermogenesis
>> brown fat nonshivering t.

thermography
thermolability
>> methylene tetrahydrofolate reductase t.

thermometer
>> basal body t.
>> LighTouch Neonate t.
>> Ototemp 3000 t.
>> thermistor t.
>> Thermodigital t.
>> Thermoscan tympanic instant t.

thermoplasty
>> balloon t.

thermoregulation
Thermoscan tympanic instant thermometer

theta dimeric protein
thetaiotaomicron
>> *Bacteroides t.*

THF
>> tetrahydrofolate

THI
>> therapeutic husband insemination
>> transient hypogammaglobulinemia of infancy

thiabendazole
thiamine
>> t. hypovitaminemia

thiazide
>> t. diuretic

thickening
>> decidual mural t.
>> skin t.

thickness
>> endometrial t.
>> fetal nuchal translucency t.
>> nuchal translucency t. (NTT)
>> placental t.

Thiemann
>> T. disease
>> T. syndrome

Thiersch-Duplay urethroplasty
thin-layer chromatography
ThinPrep
>> T. Pap
>> T. Pap test
>> T. processor
>> T. smear

Thinsite dressing
thioglycollate broth medium
thioguanine
thiomalate
>> gold sodium t.

thiopental sodium
Thioplex
thiopropazate
thiosulfate
>> sodium t.

Thiosulfil
thiotepa
thiothixene
thiphenamil hydrochloride
third
>> t. degree prolapse
>> t. disease
>> t. and fourth pharyngeal pouch syndrome
>> t. parallel pelvic plane
>> t. stage of labor
>> t. trimester
>> t. trimester bleeding

third-degree laceration
third-generation progesterone

THL
    true histiocytic lymphoma
thlipsencephalus
Thomas
    T. curette
    T. test
Thomas-Gaylor biopsy forceps
Thompson syndrome
Thomsen
    T. disease
    T. myotonia congenita
    T. syndrome
Thomson
    T. complex
    T. disease
    T. syndrome
thonzonium
thoracentesis
thoracic
    t. asphyxiant dystrophy
    t. cavity
    t. clamp
    t. compression syndrome
thoracic-pelvic-phalangeal dystrophy
thoracoabdominal (TA)
    t. 55 stapler
thoracoamniotic shunt
thoracoceloschisis
thoracodelphus
thoracodidymus
thoracogastrodidymus
thoracogastroschisis
thoracolumbar
    t. gibbus
    t. rachischisis
    t. sympathetic nerve
thoracomelus
thoracopagus
    t. conjoined twins
thoracoparacephalus
thoracoschisis
thoracoscopy
thoracotomy
thoradelphus
thorax
    amazon t.
    t. compression
Thorazine
Thorn
    T. maneuver
    T. test
threadworm infestation

threatened
    t. abortion
    t. miscarriage
three-day measles
three-dimensional
    t.-d. ultrasound
    t.-d. videoendoscope
three-point technique
three-quarter strength formula
three-vessel cord
threshold
    phenotypic t.
    speech reception t. (SRT)
    t. trait
thrive
    failure to t. (FTT)
    nonorganic failure to t. (NOFT)
throat
    Sucrets Sore T.
thrombasthenia
    Glanzmann t.
thrombectomy
thrombi (*pl. of* thrombus)
thrombin
    t. clot test
    t. time
thrombocythemia
    primary t. (PT)
thrombocytopenia
    alloimmune t.
    immune t.
    isoimmune fetal t.
    neonatal alloimmune t.
    transfusion-induced t.
thrombocytopenia-absent
    t.-a. radius (TAR)
    t.-a. radius syndrome
thrombocytopenic purpura
thrombocytosis
thromboembolic disease (TED)
thromboembolism
    idiopathic venous t.
    pulmonary t.
    venous t. (VTE)
thrombolytic therapy
thrombomodulin
thrombophilia
    factor V Leiden t.
    familial t.
thrombophlebitis
    deep vein t.
    pelvic vein t.

T

NOTES

thrombophlebitis *(continued)*
    septic pelvic t. (SPT)
    septic pelvic vein t.
    superficial t.
thromboplastin
thrombosis, pl. **thromboses**
    arterial t.
    cavernous sinus t.
    cerebral t.
    coronary t.
    decidual fibrin t.
    deep vein t. (DVT)
    deep venous t. (DVT)
    maternal cortical vein t.
    ovarian vein t. (OVT)
    pelvic ovarian vein t. (POVT)
    placental t.
    pregnancy-associated t.
    renal vascular t.
    renal vein t.
    venous t.
thrombospondin
Thrombostat platelet function test
thrombotic
    t. microangiopathy
    t. phlegmasia
    t. purpura
    t. thrombocytopenic purpura (TTP)
Thrombo-Wellco test
thromboxane
    t. A_2
    t. dominance
thrombus, pl. **thrombi**
    intramural t.
    t. precursor protein (T_pP^T)
    t. precursor protein test
thrush
    t. pneumonia
thrust
    tongue t.
thumb
    congenital clasped t.'s
    hitchhiker t.
    t. retractor
    stub t.
    triphalangeal t.
thumbsucking
thymectomy
    neonatal t.
thymic
    t. agenesis
    t. aplasia
    t. aplasia syndrome
    t. asthma
    t. dysplasia
    t. lymphocyte antigen (TL)

    t. and parathyroid agenesis
      syndrome
    t. shadow
thymic-dependent deficiency
thymic-parathyroid aplasia
thymine
thymol turbidity test
thymoma
thymopoietin
thymosin
thymus gland
thyroarytenoid muscle
Thyro-Block
thyroglobulin
thyroglossal
thyrohypophysial syndrome
thyroid
    t. autoantibody
    t. cancer
    t. crisis
    t. deficiency
    t. disease
    t. dysfunction
    t. function
    t. function test (TFT)
    t. gland
    t. hormone
    t. hormone resistance
    t. hormone unresponsiveness
    t. index
    t. Lahey clamp
    t. neoplasia
    t. nodule
    t. storm
thyroid-binding globulin (TBG)
thyroidectomy
thyroiditis
    Hashimoto t.
    postpartum t.
thyroid-stimulating
    t.-s. hormone (TSH)
    t.-s. hormone assay
    t.-s. immunoglobulin
Thyrolar
thyrotoxicosis
    gestational t.
thyrotrope
thyrotropic, thyrotrophic
    t. hormone
thyrotropin
    chorionic t.
    t. deficiency
thyrotropin-releasing
    t.-r. hormone (TRH)
    t.-r. hormone stimulation test
thyroxine, thyroxin (T_4)
    t. test

L-thyroxine
 L-t. therapy
thyroxine-binding
 t.-b. globulin
 t.-b. protein (TBP)
Thytropar
TIA
 transient ischemic attack
Tiamol
TIBC
 total iron-binding capacity
tibiae
 osteochondrosis deformans t.
tibial
 t. aplasia-ectrodactyly syndrome
 t. torsion
tibolone
tic
 maladie des t.'s
Ticar
ticarcillin
 t. disodium
tick fever
Ticon
ticonazole
TID
 tubal inflammatory damage
tidal volume
Tietze syndrome
Tigan
Tillaux disease
tilt test
tiludronate
time
 activated clotting t. (ACT)
 activated partial thromboplastin t.
  (APTT)
 bleeding t.
 capillary refill t.
 cell generation t.
 circulation t.
 doubling t.
 gastric emptying t.
 inspiration t. (I-time)
 Ivy bleeding t.
 kaolin clotting t.
 Lee-White clotting t.
 partial thromboplastin t. (PTT)
 t. position scan
 t. to pregnancy (TIP)
 prothrombin t. (PT)
 recovery t.

 Rite T.
 Russell viper venom t.
 thrombin t.
timed intercourse
Timentin
timer
 Apgar t.
Timolide
timolol
Timoptic
Tinactin
tincture
 Fungoid T.
Tindal
tinea
 t. capitis
 t. corpus
 t. cruris
 t. nigra palmaris
 t. pedis
 t. versicolor
tine test
tinnitus
tioconazole
TIP
 time to pregnancy
 Marlow Primus handle, shaft, and
  TIP
tip
 Corometrics Gold Quik Connect
  Spiral electrode t.
 Corometrics Quik Connect Spiral
  electrode t.
 needle t.
 nonfrosted t.
tiptoeing
TIR
 trophoblast in regression
tired housewife syndrome
Tischler cervical biopsy forceps
Tischler-Morgan
 T.-M. biopsy punch
 T.-M. uterine biopsy forceps
Tisit
 T. Blue Gel
 T. Liquid
 T. Shampoo
tissue
 t. activator-induced fibrinolysis
 adipose t.
 anechoic t.
 attenuating t.

T

NOTES

tissue *(continued)*
> breast biopsy t.
> choriodecidual t.
> connective t.
> echogenic t.
> ectopic endometrial t.
> t. expansion vaginoplasty
> t. factor
> t. factor pathway inhibitor (TFPI)
> fibrous connective t.
> t. forceps
> glandular t.
> granulation t.
> t. homogeneity
> t. inhibitors of metalloproteinase
> intralobular connective t.
> lymphoid t.
> maternal t.
> neural crest t.
> paravaginal soft t.
> perilobular connective t.
> t. pH monitoring
> placental t.
> t. plasminogen activator (t-PA)
> t. regeneration
> residual ductal t.
> retroperitoneal soft t.
> scar t.
> sonolucent t.
> T. Specific imaging
> splenic t.
> t. study
> syngeneic t.
> t. thromboplastin-inhibition test
> t. tolerance to radiation
> t. transplant
> trophoblastic t.
> xenogeneic t.

tissue-specific antibody
Tis-U-Trap
> T.-U.-T. endometrial suction

titanium staple
titer
> anti-RHO-D t.
> antistreptolysin t.
> HI t.
> maternal t.
> non-*Treponema* t.
> viral t.

Titralac
> T. Plus Liquid

titrate
TKO
> to keep open

TL
> thymic lymphocyte antigen
> tubal ligation

TLC
> total lung capacity
> > TLC antiseptic soap

TLX alloantigen system
T-max
> maximum temperature

TNDM
> transient neonatal diabetes mellitus

TNF
> tumor necrosis factor
> > TNF-alpha
> > TNF-alpha converting enzyme

TNM
> tumor, node, metastasis
> > TNM classification
> > TNM nomenclature

T&O
> tandem and ovoids

TOA
> tuboovarian abscess

TOAPOT
> tuboovarian abscess after previous tubal occlusion

toast
> bananas, rice cereal, applesauce, and t. (BRAT)

Tobey ear rongeur
tobramycin
Tobrex
TOC
> tuboovarian complex

tococardiography
tocodynagraph
tocodynamometer
tocodynamometry
tocograph
tocography
tocology
tocolysis
> acute t.
> prenatal t.

tocolytic
> t. agent
> t. therapy

tocometer
tocophobia
Today vaginal contraceptive sponge
Todd-Hewitt broth
Toddler Temperament Scale
toe
> catheter t.'s
> hammer t.
> pigeon t.

toeing
> t. in
> t. out

toe-in gait
Toesen

toewalking
TOF
  tetralogy of Fallot
Tofranil
  T.-PM
togavirus
toilet
  pulmonary t.
  respiratory t.
Toitu MT-810 cardiographic monitor
Tokos monitor
Tolamide
tolazamide
tolazoline
  t. hydrochloride
tolbutamide
Toldt
  line of T.
Tolectin
tolerance
  antigen t.
  carbohydrate t.
  glucose t.
  immunologic t.
  impaired glucose t. (IGT)
  radiation t.
tolfenamic acid
Tolinase
tolmetin
tolnaftate
Toloxan
toluene
toluidine
  t. blue
  t. blue test
Tolzol
Tom Jones closure
tomography
  automated computerized axial t.
    (ACAT)
  computed t. (CT)
  computed axial t. (CAT)
  hypocycloidal t.
  quantitative computed t. (QCT)
  single photon emission computed t.
    (SPECT)
  Somatom Plus computed t.
Tompkins
  T. median bivalving technique
  T. procedure
tone
  uterine t.

tongue
  black t.
  t. crib
  crocodile t.
  fern leaf t.
  geographic t.
  hairy t.
  scrotal t.
  strawberry t.
  t. thrust
  t.-tie
tonic
  t. neck reflex
  t. seizure
tonic-clonic, tonoclonic
  t.-c. movements
  t.-c. seizure
tonicoclonic seizure activity
tonsillectomy
tonsillitis
  white t.
tonsillopharyngitis
tonus
  baseline t.
tool
  Adolescent and Pediatric Pain T.
    (APPT)
tooth anomalies, skeletal dysplasia,
  sparse hair syndrome
Topactin
Topaz-UPS
topical
  Aclovate T.
  Acticort T.
  Aeroseb-HC T.
  Ala-Cort T.
  Ala-Scalp T.
  Alphatrex T.
  Anusol HC-1 T.
  Anusol HC-2.5% T.
  Aristocort T.
  Aristocort A T.
  Betalene T.
  Betatrex T.
  Beta-Val T.
  CaldeCort T.
  Caldesene T.
  Canesten T.
  Carmol-HC T.
  Cetacort T.
  Cloderm T.
  CortaGel T.

T

NOTES

topical *(continued)*
    Cortaid Maximum Strength T.
    Cortaid with Aloe T.
    Cort-Dome T.
    Cortef Feminine Itch T.
    Cortizone-5 T.
    Cortizone-10 T.
    Cyclocort T.
    Delcort T.
    Delta-Tritex T.
    Dermacort T.
    Dermarest Dricort T.
    Derma-Smoothe/FS T.
    Dermolate T.
    Dermtex HC with Aloe T.
    DesOwen T.
    Diprolene T.
    Diprolene AF T.
    Efudex T.
    Eldecort T.
    Eurax T.
    Exelderm T.
    Fluonid T.
    Fluoroplex T.
    Flurosyn T.
    Flutex T.
    FS Shampoo T.
    Gynecort T.
    Hi-Cor-1.0 T.
    Hi-Cor-2.5 T.
    Hycort T.
    Hydrocort T.
    Hydro-Tex T.
    Hytone T.
    t. iodine application
    Kenalog T.
    LactiCare-HC T.
    Lanacort T.
    Locoid T.
    Lotrimin T.
    Lotrimin AF T.
    Maxivate T.
    Micatin T.
    Monistat-Derm T.
    Mycelex-G T.
    Mycogen II T.
    Mycolog-II T.
    Myconel T.
    Mytrex F T.
    N.G.T. T.
    Nutracort T.
    Orabase HCA T.
    Ovide T.
    Oxistat T.
    Pedi-Pro T.
    Penecort T.
    Psorion T.
    S-T Cort T.

    Synacort T.
    Synalar T.
    Synalar-HP T.
    Synemol T.
    Tegrin-HC T.
    Teladar T.
    t. testosterone
    Texacort T.
    t. treatment
    Triacet T.
    Tridesilon T.
    Tri-Statin II T.
    U-Cort T.
    Valisone T.
    Vytone T.
    Westcort T.
Topicort
    T -LP
Topilene
Topisone
topographic cervical
topoisomerase I inhibitor
Toposar Injection
topotecan
    t. hydrochloride
Topsyn
Toradol
TORCH
    toxoplasmosis, other agents, rubella,
        cytomegalovirus, herpes simplex
    TORCH syndrome
toremifene
    t. citrate
Toriello-Carey
    T.-C. sign
    T.-C. syndrome
Toriello syndrome 1, 2
Torpin cul-de-sac resection
Torpin-Waters-McCall culdoplasty
torr
torsade de pointes
torsion
    adnexal t.
    cord t.
    t. dystonia
    ovarian t.
    splenic t.
    tibial t.
torso presentation
Torsten Sjögren syndrome
torti
    pili t.
torticollis
Torueopsis glabrata
torulosis
Totacillin
    T.-N

total
t. abdominal hysterectomy (TAH)
t. anomalous pulmonary venous return
t. ascertainment
t. body water
t. breech extraction
t. cavopulmonary connection
t. iron-binding capacity (TIBC)
t. lung capacity (TLC)
t. mastectomy
t. oxyhemoglobin (THb O$_2$, THb O$_2$)
t. parenteral nutrition (TPN)
t. pelvic exenteration
t. perineal rupture
t. peripheral parenteral nutrition (TPPN)
t. placenta previa
t. pulmonary resistance
t. quality management (TQM)
t. testosterone index
totalis
situs inversus t.
totipotent cell
totipotential cell
in toto
toto
in t.
touch
t. imprint
therapeutic t.
Tourette syndrome
TOVA ADD/ADHD assessment
Towako method
towel
t. clip
DisCide disinfecting t.
tower skull
Townes-Brocks syndrome
Townes syndrome
Townsend
T. biopsy punch
T. endocervical biopsy curette
toxemia, toxicemia
florid t.
preeclamptic t. (PET)
t. of pregnancy
t. syndrome
toxemic
t. rash of pregnancy
t. retinopathy of pregnancy

toxic
t. epidermal necrolysis (TEN)
t. megacolon
t. shock
t. shock syndrome (TSS)
t. shock syndrome toxin 1
toxicemia (*var. of* toxemia)
toxicity
bone marrow t.
lidocaine t.
toxicology screen
toxicum
erythema neonatorum t.
toxin
bacterial t.
environmental t.
reproductive t.
toxic shock syndrome t. 1
toxocariasis
toxoid
tetanus t.
Toxoplasma gondii
toxoplasmosis
congenital t.
fetal t.
t., other agents, rubella, cytomegalovirus, herpes simplex (TORCH)
TP10
T$_p$P^T
thrombus precursor protein
t-PA
tissue plasminogen activator
TPAL
term infants, premature infants, abortions, living children
TPHA
Treponema pallidum hemagglutination
TPI
Treponema pallidum immobilization
TPI test
TPN
total parenteral nutrition
TPP
tubal perfusion pressure
TPPN
total peripheral parenteral nutrition
TQM
total quality management
trabeculate
trace metal

NOTES

tracer
    radioactive t.
trachea
tracheal
    t. agenesis syndrome
    t. compression
    t. intubation
    t. occlusion
    t. stenosis
    t. web
tracheal-aspirate culture
trachelectomy
    radical vaginal t.
trachelitis
trachelobregmatic diameter
trachelopanus
trachelopexia, trachelopexy
tracheloplasty
    ex utero intrapartum t. (EXIT)
trachelorrhaphy
tracheloschisis
trachelotomy
tracheobronchial
tracheobronchitis
tracheobronchomegaly
tracheoesophageal
    t. fistula (TEF)
    t. fistula, esophageal atresia,
      multiple congenital anomaly
      syndrome
tracheolaryngomalacia
tracheomalacia
tracheostomy
    t. tube
tracheotome
    Sierra-Sheldon t.
tracheotomy
trachoma
    t. inclusion conjunctivitis (TRIC)
trachomatis
    Chlamydia t. (CT)
tract
    gastrointestinal t.
    genital t.
    intestinal t.
    proximal outflow t.
    reproductive t.
    respiratory t.
traction
    t. atrophy
    axis t.
    Bryant t.
    t. enterocele
trade
    Lact-Aid STARTrainer Nursing
      System & t.

traditional surrogacy
tragus, pl. **tragi**
    accessory t.
TRAIDS
    transfusion-related AIDS
training
    Lovas t.
trait
    autosomal dominant t.
    autosomal recessive t.
    cytoplasmic t.
    dominant lethal t.
    familial t.
    galtonian t.
    hereditary t.
    mendelian t.
    multifactorial t.
    penetrant t.
    recessive t.
    sickle cell t.
    threshold t.
    X-linked t.
TRAM
    transverse rectus abdominis
    myocutaneous
    TRAM flap
TRAMPE
    trichorhinophalangeal multiple exostosis
TRAMP flap
Trandate
Tranebjaerg syndrome 1, 2
tranexamic acid
tranquilizer
trans
transabdominal
    t. amnioinfusion
    t. cervicoisthmic cerclage
    t. chorionic villus sampling
    t. needle transfer
    t. thin-gauge embryofetoscopy
      (TGEF)
    t. transducer
    t. ultrasound
**transabdominal/transvaginal ultrasound
(TAS/TVS)**
transaminase
    serum t.
**transanal endoscopic microsurgery
(TEM)**
transanimation
transcapillary
    t. fluid
    t. fluid balance
transcarbamylase
    ornithine t. (OTC)
transcatheter
    t. uterine artery embolization

transcelomic
　t. spread
transcephalic impedance
transcervical
　t. balloon tuboplasty
　t. chorionic villus sampling
　t. division
　t. intrafallopian tube transfer
　　(TCIFTT)
　t. resection
　t. selective salpingography
　t. tubal access
　t. tubal access catheter (T-TAC)
　t. ultrasound
transcobalamine deficiency
transcortin
transcranial Doppler ultrasonography
transcript
　X inactive, specific t. (XIST)
transcriptase
　reverse t.
transcription
　gene t.
　helix-loop-helix t.
　reverse t.
　t. start site
transcriptionally active human
　papillomavirus
transcutaneous
　t. measurement
　t. monitor
　t. oxygen tension monitoring
　t. partial pressure of oxygen
　　(tCpO$_2$)
transcystoscopically
transdermal
　t. administration
　Alora T.
　Climara T.
　17β-E2 t. drug-delivery system
　Esclim T.
　Estraderm T.
　t. estrogen
　t. glyceryl trinitrate patch
　t. hormone replacement therapy
　t. medication patch
　t. nicotine replacement therapy
　Vivelle T.
transdiaphragmatic
transducer
　annular-array t.
　endovaginal t.

　Oxisensor t.
　SLA t.
　transabdominal t.
　ultrasound t.
　Voluson sector t.
transduction
Transeptic cleansing solution
transesophageal electrophysiologic study
transfection
transfer
　blastocyst t.
　direct oocyte t. (DOT)
　direct oocyte sperm t. (DOST)
　donor oocyte t.
　embryo t. (ET)
　embryo intrafallopian t. (EIFT)
　embryo thawing with t.
　t. factor
　frozen-thawed embryo t.
　gamete intrafallopian t. (GIFT)
　gas t.
　gene t.
　intrafallopian t.
　in vitro fertilization-embryo t.
　　(IVF-ET)
　linear energy t. (LET)
　t. medium
　peritoneal oocyte sperm t. (POST)
　placental t.
　pronucleate stage embryo t.
　　(PROST)
　pronucleate stage tubal t. (PROST)
　t. ribonucleic acid (tRNA)
　t. RNA (tRNA)
　technology t.
　transabdominal needle t.
　transcervical intrafallopian tube t.
　　(TCIFTT)
　transvaginal intrafallopian sperm t.
　　(SIFT)
　tubal embryo t. (TET)
　tubal embryo stage t. (TEST)
　zygote intrafallopian t. (ZIFT)
transferase
　gamma glutamyl t. (GGT)
transferrin
transformation
　t. zone
transforming
　t. growth factor (TGF)
　t. growth factor alpha (TGFalpha)
　t. growth factor beta (TGFbeta)

T

NOTES

transfusion
- acute intrapartum t.
- acute perinatal t.
- antenatal fetofetal t.
- autologous t.
- blood t.
- t. controversy
- double-volume exchange t.
- erythrocyte t.
- exchange t.
- fetal t.
- fetofetal t.
- fetomaternal t.
- fetoplacental t.
- HLA-matched platelet t.
- intrapartum fetoplacental t.
- intraperitoneal blood t.
- intraperitoneal fetal t.
- intrauterine intraperitoneal fetal t.
- intravascular t.
- leukocyte t.
- massive t.
- percutaneous fetal t.
- placental t.
- platelet t.
- prophylactic red-cell t.
- twin-twin t.
- umbilical cord t.

transfusion-induced thrombocytopenia
transfusion-related AIDS (TRAIDS)
transgenesis
- mammalian t.

transgenic
- t. organism

transient
- t. bullous dermolysis of the newborn
- t. congenital hypothyroidism
- t. erythroid hypoplasia
- t. fetal distress
- t. hematuria
- t. hypertension
- t. hypogammaglobulinemia
- t. hypogammaglobulinemia of infancy (THI)
- t. ischemic attack (TIA)
- t. neonatal diabetes mellitus (TNDM)
- t. neonatal pustular melanosis
- t. respiratory acidosis
- t. respiratory distress syndrome (TRDS)
- t. tachypnea
- t. tachypnea of newborn (TTN)

transillumination
- t. of head

transition
- fetal-neonatal t.

translabial ultrasound
translation
- nick t.

translocation
- autosome t.
- balanced t.
- t. carrier
- centric fusion t.
- chromosomal t.
- t. chromosome
- t. of chromosome 22
- de novo balanced t.
- t. Down syndrome
- mosaic t.
- reciprocal t.
- robertsonian t.
- unbalanced t.
- X 19 t.
- X-autosome t.

translucency
- nuchal t.

transmesenteric hernia
transmigration
- ovular t.

transmission
- maternal-fetal t.
- t. medium
- perinatal t.
- pressure t.
- sexual t.
- vertical HIV t.
- viral t.

transmission/disequilibrium test
transnasal administration
Transorbent dressing
transpeptidase
- glutamyl t. (GTP)

transperineal
- t. implant
- t. ultrasonography
- t. ultrasound

transperitoneal cesarean
transplacental
- t. hemorrhage
- t. infection
- t. passage

transplant
- fetal tissue t.
- fetus-to-fetus t.
- liver t.
- organ t.
- placental tissue t.
- tissue t.

transplantation
- t. antigen
- bone marrow t.
- cord blood t.
- hematopoietic stem cell t.

high-dose chemotherapy with
autologous bone marrow t.
(HDC-ABMT)
t. immunology
kidney t.
liver t.
organ t.
orthotopic live t.
pituitary gland t.
renal t.

transport
t. disorder
egg t.
ovum t.
sperm t.

transport-associated protein (TAP)
transposable element
transposase
transposed adnexa
transposition
t. of great arteries (TGA)
t. of great vessels (TGV)
lateral ovarian t.
t. of ovary
penoscrotal t.
t. of viscera
Watkins t.

transposon
transpyloric enteral feeding
transrectal
t. approach
t. probe
t. surgical treatment
t. ultrasound (TRUS)

transsexual
transsexualism
transsphenoidal
t. microsurgical resection
t. operation

transudate
mucosal t.
vaginal t.

**transumbilical breast augmentation
(TUBA)**
transurethral
t. catheter
t. collagen injection
t. electrocautery
t. marsupialization

transvaginal
t. amniotic puncture (TAP)
t. bladder neck suspension

t. color Doppler sonography (TV-
CDS)
t. cone
t. fine-needle biopsy
t. implant
t. intrafallopian sperm transfer
(SIFT)
t. sacrospinous colpopexy
t. sonography (TVS)
t. transducer probe
t. tubal catheterization
t. ultrasonography
t. ultrasound (TVU, TV-UST)
t. ultrasound-directed oocyte
retrieval (TUDOR)
t. ultrasound-guided urethral
reconstruction

transverse
t. arrest
t. cervical ligament
t. colon
t. diameter
t. fetal lie
t. fundal incision of Strassman
t. hemimelia
t. incision
t. lie
t. lie presentation
lower uterine segment t. (LUST)
t. myelitis
occiput t. (OT)
t. oval pelvis
t. rectus abdominis myocutaneous
(TRAM)
t. rectus abdominis myocutaneous
flap
t. ridge
t. scan
t. vaginal septum

transversion
transversus
t. and rectus musculoperitoneal flap
situs t.

Tranxene
tranylcypromine
TRAP
twin reverse arterial perfusion
TRAP sequence

trap
filtered specimen t.
sterile specimen t.

trapezoidocephaly-synostosis syndrome

T

NOTES

trapped ovum
Trasicor
trastuzumab
trauma, pl. **traumata, traumas**
  acoustic t.
  birth t.
  blunt t.
  childhood genital t.
  fetal t.
  genital tract t.
  maternal t.
  perinatal t.
  psychological t.
trauma-related acute pelvic hemorrhage
traumatic
  t. amenorrhea
  t. birth injury
  t. injury
  t. vaginitis
Travamine
Travamulsion IV fat emulsion
tray
  t. agglutination test (TAT)
  E-Z-EM PercuSet amniocentesis t.
  HSG t.
  Unimar HSG t.
trazodone
TRDS
  transient respiratory distress syndrome
Treacher
  T. Collins-Franceschetti syndrome
  T. Collins mandibulofacial
    dysostosis
  T. Collins syndrome
treatment
  add-back t.
  ambulatory antibiotic t.
  antenatal t.
  antenatal phenobarbital t.
  biologic t.
  continuous/combined t.
  Fliess t.
  hormonal t.
  hyperbaric oxygen t.
  infertility t.
  laser t.
  pharmacologic t.
  P32 intraperitoneal t.
  prenatal t.
  Rotunda t.
  topical t.
  transrectal surgical t.
  updraft t.
  ureteral surgical t.
  zidovudine t.
treatment-associated pregnancy
treatment-independent pregnancy
Trecator-SC

Treitz
  ligament of T.
Trendar
Trendelenburg position
Trental
trephine, trepan
Treponema
  microhemagglutination assay for
    antibodies to T. pallidum (MHA-
    TP)
  T. *pallidum* antibody
  T. *pallidum* antigen
  T. *pallidum* hemagglutination
    (TPHA)
  T. *pallidum* immobilization (TPI)
  T. *pallidum* immobilization test
treponemal
  t. serology
treponematosis
Tresilian sign
tretinoin
Trevor disease
Trexan
TRH
  thyrotropin-releasing hormone
Triacet Topical
triacetyloleandomycin
triad
  aniridia, genitourinary abnormalities,
    mental retardation t.
  Currarino t.
  Hutchinson t.
  Hutchison t.
  Wilms tumor, aniridia genitourinary
    abnormalities, mental retardation t.
trial
  Bernoulli t.
  Cetus t.
  chemotherapy phase t.
  t. forceps
  labor t.
  RADIUS t.
  voiding t.
triamcinolone
  t. acetonide
Triaminic
  T. AM Decongestant Formula
  T. Oral Infant Drops
triamterene
triangle
  Burger t.
  Einthoven t.
  Hesselbach t.
  pubic t.
  Ward t.
triangular
  t. ligament

t. uterus
t. vaginal patch sling procedure
Triasox
triatriatum
cor t.
Triazole
Triban
Pediatric T.
tribasilar synostosis
tribrachius
TRIC
trachoma inclusion conjunctivitis
tricephalus
tricheiria
trichinosis
Trichlorex
trichlormethiazide
trichloroacetic acid (TCA)
trichloroethylene
2,4,5-trichlorophenoxyacetic acid
trichobezoar
trichodental dysplasia-microcephaly-mental retardation syndrome
trichoepithelioma
trichomonad
Trichomonas
T. infection
T. vaginalis
T. vaginalis vaginitis
trichomoniasis
t. vaginitis
trichopoliodystrophy
trichorhinoauriculophalangeal multiple exostoses dysplasia
trichorhinophalangeal
t. dysplasia
t. multiple exostosis (TRAMPE)
t. multiple exostosis dysplasia
t. syndrome
trichorionic
trichorrhexis
t. blastysis
t. invaginata
t. nodosa
t. nodosa syndrome
trichoschisis
trichothiodystrophy
t. 2 (TTD 2)
t.-congenital ichthyosis syndrome
t.-neurocutaneous syndrome
t.-xeroderma pigmentosum syndrome
trichotillomania

trichuriasis
tricuspid
t. atresia
t. insufficiency
t. regurgitation
Tri-Cyclen
Ortho T.-C.
tricyclic antidepressant (TCA)
tridactylism
tridermogenesis
Tridesilon Topical
tridihexethyl chloride
Tridione
T. syndrome
tridymus
triencephalus
trientine
triethnic
triethylenethiophosphoramide
trifluoperazine hydrochloride
triflupromazine hydrochloride
trifluridine
trigeminoencephaloangiomatosis
trigger
environmental t.
t. point
triglyceride
medium chain t. (MCT)
trigonal
t. plate
t. ring
t. urothelium
trigone
urinary t.
trigonitis
trigonocephaly
t. syndrome
Trihexane
Trihexy
trihexyphenidyl
trihydrate
ampicillin t.
trihydroxycoprostanic acidemia
triiniodymus
triiodothyronine
reverse t.
t. test
Trikacide
Trilafon
trilaminar blastoderm
Tri-Levlen
Trilisate

T

NOTES

triloba
    placenta t.
trilogy of Fallot
trilostane
Trilucent breast
Trimazide
trimegestone
trimeprazine
trimester
    first t.
    second t.
    third t.
trimethadione
    t. embryopathy
    t. syndrome
trimethaphan camsylate
trimethoprim
    t.-sulfamethoxazole (SMZ-TMP)
    t. sulfate
trimipramine
Trimox
Trimpex
Trimstat
Trimtabs
trinitrate
    glyceryl t. (GTN)
Tri-Norinyl
Trinsicon
trinucleotide repeat expansion mutation
triocephalus
triodurin
triophthalmos
triose phosphate isomerase
triotus
tripartita
    placenta t.
Tripedia
tripelennamine
    t. hydrochloride
tripe palm
triphalangeal thumb
triphasic oral contraceptive
Triphasil
Tri-Phen-Chlor
triphenylethylene selective estrogen receptor modulator
triphosphate
    adenosine t. (ATP)
    digoxigenin-labeled deoxyuridine t.
    guanosine t. (GTP)
5'-triphosphate
    guanosine 5'-t. (GTP)
triple
    t. bromide
    t. diapers
    t. dye
    t. screen
    t. screen test

    t. serum marker screening
    t. sulfa cream
    t. sulfa suppository
    t. swab test
triple-biochemical screen
triple-lumen catheter
triple-marker screen
triplet
triple-X
    t.-X chromosomal aberration
    t.-X female
    t.-X syndrome
Triple X Liquid
triplication
    ureteral t.
triploid
    t. embryo
    t. fetus
    t. preembryo
triploidy
    t. syndrome
triplo-X syndrome
tripodia
tripoding
trip (15q) syndrome
triprolidine hydrochloride
Triptil
triptorelin
tripus
    t. conjoined twins
    ischiopagus t. separation
    ischiopagus t. twins
    t. limb
    t. separation
    t. twins
trip-X chromosome syndrome
trisalicylate
    choline magnesium t.
Tris[hydroxymethyl]aminomethane-buffered saline
trismus
    t. nascentium
    t. neonatorum
trismus-pseudocamptodactyly syndrome
trisomic
    t. fetus
    t. rescue
trisomy
    t. 1–22
    autosomal t.
    t. C,D,E,G syndrome
    chromosome 1–22 t.
    chromosome 1p–22p t.
    chromosome 1q–22q t.
    chromosome Xq t.
    t. 18-like syndrome
    mosaic t. 14
    t. 8 mosaicism

t. 1p–22
partial t. 1p–22p
t. 11q syndrome
t. 1–22 syndrome
t. X
X t.
t. Xq
trisphosphate
inositol t.
1,4,5-trisphosphate
inositol 1,4,5-t.
TriStar trocar
Tri-Statin II Topical
Trisulfa
T.-S
tritodrine hydrochloride
Trivagizole
Trivers-Willard theory
Tri-Vi-Flor vitamins
Tri-Vi-Sol vitamins
Trizol RNA extractor
trizygotic
tRNA
transfer ribonucleic acid
transfer RNA
Trobicin
trocar
Bluntport disposable t.
Cabot t.
Circon-ACMI t.
Core Dynamics disposable t.
Dexide disposable t.
Endopath TriStar t.
Ethicon disposable t.
t. guide
t. implantation metastasis
Jarit disposable t.
laparoscopic t.
Marlow disposable t.
10-mm t.
5-mm suprapubic t.
Olympus disposable t.
Origin t.
t. site ecchymosis
Solos disposable t.
Step laparoscopic t.
Stortz disposable t.
TriStar t.
Visiport optical t.
Weck disposable t.
Wisap disposable t.

Wolf disposable t.
Ximed disposable t.
troche
Cepacol T.
Cepacol Anesthetic T.
Mycelex T.
trochocephaly
TroGARD electrosurgical blunt trocar system
troglitazone
troleandomycin
tromethamine (THAM)
carboprost t.
dinoprost t.
fosfomycin t.
ketorolac t.
TrophAmine
T. hyperalimentation
trophectoderm
t. biopsy
t. stage
trophoblast
extravillous t.
t. in regression (TIR)
t. sampling
trophoblastic
t. cell
t. disease
t. embolus
t. invasion
t. neoplasia
t. neoplastic disease
t. pseudotumor
t. tissue
t. tumor
trophospongia
trophotropism
tropical bubo
tropicamide
tropic hormone
troponin
t. T
trousers
antishock t.
military antishock t. (MAST)
Trousseau sign
Trovan/Zithromax Compliance Pak
Troyer syndrome (TS)
true
t. conjugate
t. hermaphroditism
t. histiocytic lymphoma (THL)

NOTES

true *(continued)*
    t. incontinence
    t. knot of umbilical cord
    t. labor
    t. pelvis
    t. precocious puberty
    t. twins
    t. uterine inertia
true-breeding
TrueVision transvaginal probe
trumpet
    t. cannula
    Iowa t.
truncate selection
truncation
    uterine positioning via ligament
      investment fixation t. (UPLIFT)
truncus
    t. arteriosus
    t. arteriosus communis
trunk presentation
Truphylline
TRUS
    transrectal ultrasound
Tru-Trax
Trymegen
trypanosomiasis
tryptophan
    t. malabsorption
tryptophanuria
tryptorelin
Trysul
TS
    Troyer syndrome
TSH
    thyroid-stimulating hormone
T-shaped
    T.-s constriction ring
    T.-s uterus
TSS
    toxic shock syndrome
TSTA
    tumor-specific transplantation antigen
T-TAC
    transcervical tubal access catheter
    T-TAC catheter
    T-TAC system
TTD 2
    trichothiodystrophy 2
TTN
    transient tachypnea of newborn
TTP
    thrombotic thrombocytopenic purpura
TTS
    twin-to-twin transfusion syndrome
    Estraderm TTS

TTTS
    twin-to-twin transfusion syndrome
    twin-twin transfusion syndrome
T2, T3, T4 test
TUBA
    transumbilical breast augmentation
tubage
tubal
    t. abortion
    t. banding
    t. colic
    t. diverticulum
    t. dysmenorrhea
    t. embryo stage transfer (TEST)
    t. embryo transfer (TET)
    t. endometriosis
    t. endometrium
    t. factor
    t. gestation
    t. inflammatory damage (TID)
    t. insufflation
    t. ligation (TL)
    t. ligation band technique
    t. mass
    t. metaplasia
    t. microsurgery
    t. obstruction
    t. occlusion
    t. patency
    t. perfusion pressure (TPP)
    t. pregnancy
    t. recanalization
    t. reconstruction surgery
    t. ring
    t. rupture
    t. sterilization
tubatorsion *(var. of* tubotorsion)
tube
    bilateral myringotomy t.'s (BMT)
    Cantor t.
    Cole endotracheal t.
    Cole orotracheal t.
    Dobbhoff nasogastric feeding t.
    double-focus t.
    embryonic neural t.
    endotracheal t.
    eustachian t.
    fallopian t.
    fimbriated end of fallopian t.
    follicle aspiration t.
    Keofeed t.
    knuckle of t.
    Miller-Abbott t.
    molybdenum rotating anode x-ray t.
    Moss t.
    nasogastric t.
    neural t.

NG t.
Pedi PEG t.
pus t.
Replogle t.
Reuter t.
salpingitis in previously
  occluded t.'s (SPOT)
Sengstaken-Blakemore t.
smoke removal t. (SRT)
tracheostomy t.
tympanostomy t.
uterine t.

tubectomy
tuber
cortical t.
subependymal t.

tubercle
genital t.
Ghon t.
Montgomery t.
Morgagni t.
Müller t.
pubic t.
Rokitansky t.

tuberculin syringe
tuberculoma
tuberculosis (TB)
abdominal t.
endometrial t.
extrathoracic t.
gastrointestinal t.
genital t.
intrathoracic t.
miliary t.
Mycobacterium t.
t. papulonecrotica
pulmonary t.
renal t.

tuberculous
t. colitis
t. dactylitis
t. keratoconjunctivitis
t. meningitis
t. peritonitis
t. salpingitis

tuberosa
sclerosis t.

tuberous
t. breast abnormality
t. mole
t. sclerosis
t. sclerosis complex

t. subchorial hematoma of the
  decidua
Tubex
tubing
pressure-separator t.
tuboabdominal pregnancy
tubocornual
t. anastomosis
t. microsurgery
t. reanastomosis
tubocurarine
t. chloride
tuboendometrial cell
tuboovarian
t. abscess (TOA)
t. abscess after previous tubal
  occlusion (TOAPOT)
t. complex (TOC)
t. pregnancy
t. varicocele
tuboovariectomy
tuboovaritis
tuboplasty
balloon t.
transcervical balloon t.
ultrasound-guided transcervical t.
ultrasound transcervical t.
tubotorsion, tubatorsion
tubouterine
t. implantation
t. pregnancy
tubovaginal fistula
tubular
t. carcinoma
t. proteinuria
t. stenosis
t. stenosis, hypocalcemia,
  convulsions, dwarfism syndrome
tubule
annular t.
mesonephric t.
seminiferous t.
tubulointerstitial disease
tuck
tummy t.
Tucker-McLane forceps
Tucker-McLane-Luikart forceps
TUDOR
transvaginal ultrasound-directed oocyte
  retrieval
tularemia vaccine
tummy tuck

T

NOTES

tumor

adenomatoid oviduct t.
adnexal t.
adrenal cell rest t.
adult granulosa cell t. (AGCT)
androgen-producing t.
t. angiogenesis factor (TAF)
angiomatoid t.
t. antigen
t. antigenicity
t. ascites
autochthonous t.
benign t.
bladder t.
bone t.
borderline epithelial ovarian t.
Brenner t.
t. burden
carcinoid t.
central nervous system t.
cervical stump t.
colorectal t.
debulking of t.
desmoid t.
endocervical sinus t.
endodermal sinus t. (EST)
endometrioid t.
epithelial stromal t.
Ewing t.
feminizing adrenal t.
genital tract t.
germ cell testicular t.
gestational trophoblastic t. (GTT)
Glazunov t.
glcyoprotein-producing t.
gonadal stromal ovarian t.
t. grading
granulosa cell t.
granulosa-stromal cell t.
granulosa-theca cell t.
hilar cell t.
t. immunology
t. immunotherapy
intracranial t.
islet cell t.
Krukenberg t.
Leydig cell t.
lipid cell ovarian t.
lipoid ovarian t.
liver t.
malignant mixed müllerian t. (MMMT)
malignant ovarian germ cell t.
t. marker
t. mass
mesonephroid t.
mixed germ cell t.
mixed mesodermal t.

mixed müllerian t. (MMT)
mixed uterine t.
monodermal t.
mucinous t.
t. necrosis factor (TNF)
t. necrosis factor-alpha
neuroectodermal t.
neurogenic t.
t., node, metastasis (TNM)
t., node, metastasis classification
nonfunctional pituitary t.
nonsecreting pituitary t.
null cell t.
ovarian malignant germ cell t.
parovarian t.
persistent postmolar gestational trophoblastic t.
phyllodes t.
pineal t.
pituitary gland t.
placental site t.
placental site trophoblastic t. (PSTT)
pontine t.
postmolar persistent gestational trophoblastic t.
pregnancy t.
prepubertal testicular t. (PPTT)
t. progression
pseudomucinous t.
radioresistant yolk sac t.
Recklinghausen t.
t. regression
Schiller t.
Scully t.
serous t.
Sertoli cell t.
Sertoli-Leydig cell t.
sex-cord stromal germ cell t.
t. size
Steiner t.
t. suppression gene
t. suppressor gene
Teilum t.
theca cell t.
trophoblastic t.
ulcerative t.
uterine corpus t.
virilizing adrenal t.
vitelline t.
Wilms t.
yolk sac t.
tumor-associated antigen (TAA)
tumor-cloning assay
tumorigenesis
tumorigenicity
tumor-limiting factor

tumorous hyperprolactinemia
tumor-specific transplantation antigen
(TSTA)
Tums Extra Strength
TUNEL stain
tungsten target
tunica
    t. albuginea
    t. vaginalis
Turbinaire
    Dexacort Phosphate T.
turcica
    sella t.
Turco posteromedial release of clubfoot
turgescence
turgescent
turmschädel
Turner
    T. mosaicism
    T. phenotype with normal
     karyotype
    T. syndrome
    T. syndrome in female with X
     chromosome
Turner-Albright syndrome
Turner-Kieser syndrome
Turner-like syndrome
Turner-Warwick
    T.-W. urethroplasty
    T.-W. vagino-obturator shelf
     procedure
turnover
    bone t.
    iron t.
turricephaly
turtle sign
Tusstat
Tuttle test
TV-CDS
    transvaginal color Doppler sonography
TVS
    transvaginal sonography
TVT
    tension-free vaginal tape
TVU
    transvaginal ultrasound
TV-UST
    transvaginal ultrasound
TWAR agent
Tween
twenty-nail dystrophy
Twice-A-Day Nasal Solution

twilight sleep
twin
    acardiac t.'s
    allantoidoangiopagous t.'s
    asymmetrical conjoined t.'s
    binovular t.'s
    t. birth
    t. birth weight discordance
    breech first t.
    concordant t.'s
    conjoined t.'s
    t. delivery
    t. demise
    diamniotic t.'s
    dicephalic t.'s
    dichorionic t.'s
    dichorionic-diamniotic t.'s
    diovular t.'s
    discordant t.'s
    dissimilar t.'s
    dizygotic t.'s
    enzygotic t.
    equal conjoined t.'s
    false t.'s
    false heteroovular t.'s
    fraternal t.'s
    t. gestation
    growth-discordant t.'s
    heterologous t.'s
    heteroovular t.'s
    identical t.'s
    impacted t.'s
    incomplete conjoined t.'s
    janiceps t.'s
    locked t.'s
    locking t.'s
    membranous t.'s
    t. method
    monoamniotic t.'s
    monochorial t.'s
    monochorionic t.'s
    monovular t.'s
    monozygotic t.'s
    omphaloangiopagous t.'s
    one-egg t.'s
    parabolic t.'s
    t. peak sign
    perfused t.'s
    t. placenta
    t. pregnancy
    pump t.'s
    t. reverse arterial perfusion (TRAP)

NOTES

T

twin *(continued)*
>t. reverse arterial perfusion sequence
>second t.
>Siamese t.'s
>similar t.'s
>stuck t.'s
>symmetrical conjoined t.'s
>thoracopagus conjoined t.'s
>t. transfusion syndrome
>tripus t.'s
>tripus conjoined t.'s
>true t.'s
>two-egg t.'s
>unequal conjoined t.'s
>uniovular t.'s
>unlike t.'s
>vanishing t.'s
>xiphoomphaloischiopagus tripus conjoined t.'s

twinning
twin-peak syndrome
twin-to-twin transfusion syndrome (TTS, TTTS)
twin-twin
>t.-t. transfusion
>t.-t. transfusion syndrome (TTTS)

Two-Cal HN
two-cell
>t.-c. embryo
>t.-c. mechanism
>t.-c. zygote

two-dimensional echocardiogram
two-egg twins
two-hit hypothesis
two-part nuclear stage
two-tail test
two-team sling
two-vessel cord
Tygon catheter
Tylenol
Tylok high-tension cerclage cabling system
Tylox
tympanic temperature
tympanites
>uterine t.

tympanogram
tympanostomy tube
type
>accelerated skeletal maturation, Marshall-Smith t.
>t. A IL-8 receptor
>alpha-thalassemia/mental retardation syndrome, deletion t.
>alpha-thalassemia/mental retardation syndrome, nondeletion t.
>Amsterdam t.

>t. B IL-8 receptor
>breech t.
>dementia of the Alzheimer t. (DAT)
>t. 1, 2 diabetes mellitus
>D-mosaic blood t.
>D^U variant blood t.
>epidermolysis bullosa, macular t. (EBM)
>facial clefting syndrome, Gypsy t.
>gangliosidosis GM1 juvenile t.
>generalized gangliosidosis GM1 adult t.
>t. 7 glycogenosis
>hereditary bullous skin dystrophy, macular t.
>t. I collagen C-telopeptide
>t. I glycogen storage disease
>t. I H/S mucopolysaccharidosis
>t. II AT3 deficiency
>t. I, II AT3 deficiency
>t. II, III, VI–VIII mucopolysaccharidosis
>t. III incontinence
>t. II pneumonocyte
>t. IS mucopolysaccharidosis
>t. IVA B mucopolysaccharidosis
>mating t.
>t. 16 papillomavirus
>senile dementia of the Alzheimer t. (SDAT)
>sialuria, Finnish t.
>wild t.

typhlitis
typhoid
>t. autoantibody
>t. fever
>t. vaccine

typhus fever
typing
>blood t.
>DNA t.
>short tandem repeat t. (STR typing)
>STR t.
>>short tandem repeat typing

typus
>t. amstelodamensis
>t. ediburgensis

Tyrode solution
tyropanoate sodium
tyrosinase-negative oculocutaneous albinism
tyrosinase-positive oculocutaneous albinism
tyrosine
>t. aminotransferase (TAT)
>t. aminotransferase deficiency (TATD)

t. kinase
t. kinase receptor
t. transaminase deficiency
tyrosinemia
Oregon-type t.
tyrosinemia II
**tyrosinemia-palmar and plantar
keratosis-ocular keratitis syndrome**

tyrosinosis
oculocutaneous tyrosinemia or t.
tyrosinosis II
Tyzine Nasal
Tzanck test

NOTES

T

U46619
U-500
  Regular (Concentrated) Iletin II U.
UA
 umbilical artery
UAC
 umbilical artery catheter
UAD Otic
UAE
 uterine artery embolization
UbetaCF
 urinary beta-core fragment
UBT
 uterine balloon therapy
UCAC
 uterine cornual access catheter
UCG
 urinary chorionic gonadotropin
UCG-Slide Test
Uchida
 U. fimbriectomy
 U. method
 U. procedure
 U. tubal ligation
UCI
 umbilical coiling index
  PPROM UCI
U-Control
U-Cort Topical
UDI
 urinary diagnostic index
 Urogenital Distress Inventory
UDP
 uridine diphosphate
UDP-galactose-4-epimerase deficiency (GALE)
UD 2000 urodynamic measurement system
U-elevator
 Gregersen U-e.
Uendex
UGA
 urogenital atrophy
u-hFSH
 urinary-derived human follicle-stimulating hormone
Uhl anomaly
UICC
 International Union Against Cancer
ulcer
 aphthous u.
 Curling u.
 Cushing-Rokitansky u.
 duodenal u.
 genital u.

Hunner u.
kissing u.
Lipschütz u.
peptic u.
phagedenic u.
rectal postradiation u.
vaginal u.
ulcerative
 u. colitis
 u. tumor
ulcus, pl. **ulcera**
 u. vulvae acutum
uLH
 urinary luteinizing hormone
Ullrich
 U. and Fremerey-Dohna syndrome
 U. syndrome
Ullrich-Bonnevie syndrome
Ullrich-Feichtiger syndrome
Ullrich-Noonan syndrome
Ullrich-Turner syndrome
ulnar hypoplasia-club feet-mental retardation syndrome
ulnar-mammary syndrome
ULPA/charcoal filtration system
ultra
 U. Cover transducer cover
 Prenate U.
Ultracef
ultrafast
 u. magnetic resonance imaging
 u. MRI
ultrafiltration
 u. membrane
 u. virus clearance
Ultralente
 Humulin U U.
 U. insulin
 U. U
Ultramark
 U. 4 ultrasound
 U. ultrasound system
Ultrascope obstetrical Doppler
ultrasonic
 u. cephalometry
 u. egg recovery
 u. endovaginal finding
 u. fetometry
ultrasonography
 color Doppler u.
 compression u.
 Doppler u.
 endoanal u.
 endovaginal u.
 gray-scale u.

U

ultrasonography *(continued)*
    high-resolution u.
    level III u.
    real-time u.
    transcranial Doppler u.
    transperineal u.
    transvaginal u.
ultrasound
    Acuson 128 Doppler u.
    Acuson 128XP u.
    Acuson 128XP-10 u.
    Advantage u.
    Aloka 650 CL u.
    Aloka OB/GYN u.
    A-mode u.
    Ansaldo AU560 u.
    antenatal u.
    u. assessment
    ATL Ultramark 4,8,9 u.
    B-mode u.
    Cineloop U.
    cranial u.
    3-D u.
    u. diagnosis
    digital u.
    Doppler u.
    duplex u.
    dynamic image on u.
    Elscint ESI-3000 u.
    endoanal u.
    Endosound endoscopic u.
    endovaginal u. (EVUS)
    fetal u.
    u. fetometry
    General Electric Model RT-3200 u.
    GE RT 3200 Advantage II u.
    u. guidance
    HDI 3000 u.
    head u.
    high-resolution u.
    Hitachi EUB 420 digital u.
    MAGGI disposable biopsy needle
      guide for u.
    M-mode u.
    MultiPRO 2000 disposable biopsy
      needle guide for u.
    obstetric u.
    Performa u.
    Pie Medical u.
    prenatal u.
    pulsed Doppler u.
    pulsed-wave u.
    real-time u.
    real-time imaging on u.
    u. screening
    Shimadzu IIQ u.
    Shimadzu SDU-400 u.
    Siemens SI 400 u.

    Sonoline Prima u.
    Spectra-Diasonics u.
    u. stethoscope
    u. surveillance
    targeted u.
    three-dimensional u.
    transabdominal u.
    transabdominal/transvaginal u.
      (TAS/TVS)
    transcervical u.
    u. transcervical tuboplasty
    u. transducer
    translabial u.
    transperineal u.
    transrectal u. (TRUS)
    transvaginal u. (TVU, TV-UST)
    Ultramark 4 u.
    vaginal probe u.
ultrasound-directed
    u.-d. egg retrieval
    u.-d. percutaneous umbilical blood
      sampling
**ultrasound-guided transcervical
tuboplasty**
Ultratard
    Actrapid insulin with U.
Ultravate
ULTRA-vue amniocentesis needle
umbilical
    u. artery (UA)
    u. artery catheter (UAC)
    u. artery catheterization
    u. artery Doppler velocimetry
    u. artery line
    u. artery waveform notching
      abnormality
    u. blood flow
    u. blood sampling
    u. circulation
    u. clamp
    u. coiling index (UCI)
    u. cord
    u. cord accident
    u. cord anomaly
    u. cord blood bank
    u. cord hematoma
    u. cord leptin
    u. cord mass
    u. cord prolapse
    u. cord syndrome
    u. cord transfusion
    u. cyst
    u. fistula
    u. fungus
    u. hernia
    u. ligament
    10-mm u. port
    u. presentation

u. scissors
u. souffle
u. vein
u. vein catheter (UVC)
u. vein catheterization
u. vein varix
u. velocity ratio
u. vessel
umbilicalis
arteritis u.
umbilicoplacental vessel
umbilicus
furrowlike u.
Umbilicutter
umbrella
vascular u.
Unasyn
unavoidable hemorrhage
unbalanced translocation
unborn child
unconjugated estriol (E3)
undecylenic acid
underlying disease
undernourishment
maternal u.
underwater delivery
Underwood disease
undescended testis
undifferentiated gonad
Undritz anomaly
undulant fever
unequal
u. conjoined twins
unexplained
u. fever
u. infertility
u. recurrent miscarriage
unfractionated heparin
UNG
uracil-N-glycosylase
Unicare breast pump
unicellular
unicolic uterus
unicollis
uterus bicornis u.
unicornous
unicornuate uterus
unicorn uterus
Uni-Decon
unilateral
u. facial microsomia
u. fibular aplastic syndrome

u. intrauterine facial necrosis
u. mandibulofacial dysostosis
u. microtia
u. partial facial paralysis
u. salpingo-oophorectomy
unilocular ovarian cyst
Unimar
U. HSG tray
U. Pipelle
uninterrupted estrogen
uniovular twins
uniparental
u. disomy
u. disomy of chromosome 14
(UPD)
Unipath
Unipen
Uniphyl
Uniplant
unipolar electrode
Uni-Pro
UNIprobe test
Uniscreen urine test
Unisom
unit
Alexander u.
BiliBed phototherapy u.
Bovie u.
bubble isolation u.
calf compression u.
Ferrold-Hisaw u.
fetal-placental u.
Flowtron DVT prophylaxis u.
human *neu* u. (HNU)
u. inheritance
Magnetrode cervical u.
maternal-placental u.
maternal-placental-fetal u.
Montevideo u.
mouse uterine u.
nem (breast milk nutritional u.)
neonatal intensive care u. (NICU)
Nytone enuretic control u.
pilosebaceous u.
terminal ductal-lobular u.
terminal duct lobular u. (TDLU)
Unitensen
Unitrol
universal
u. joint syndrome
U. reducer cap
U. vaginal probe

U

NOTES

universale
pterygium u.
Universalindicator stick
universalis
neurinomatosis u.
unlike twins
Unna
U. mark
U. nevus
unopposed estrogen
unpaired chromosome
unplanned pregnancy
unresponsiveness
immunologic u.
thyroid hormone u.
unruptured tubal gestation
unsaturated fat
unstable
u. hemoglobin
u. lie
unusual
u. facies
u. facies-mental retardation-
intrauterine growth retardation
syndrome
Unverricht disease
Unverricht-Lundborg syndrome
UOAC
uterine ostial access catheter
up
pinked u.
u-PA
urokinase plasminogen activator
UPD
uniparental disomy of chromosome 14
updraft treatment
UPLIFT
uterine positioning via ligament
investment fixation truncation
UPLIFT procedure
upper
u. contractile portion of uterus
u. genital tract infection
u. respiratory infection (URI)
UPS 2020 ambulatory measurement
system
upslanting palpebral fissure
upstream
uptake
maximum oxygen u. (VO$_2$max)
minute oxygen u.
radioactive u.
Uracel
urachal
u. cyst
u. fistula
urachus
Uracid

Uracil
uracil-N-glycosylase (UNG)
uranoschisis
uranostaphyloschisis
urate
u. calculus
u. clearance
Urban
U. operation
U. syndrome
Urban-Rogers-Meyer syndrome
urea
u. clearance
urealyticum
Ureaplasma u.
Ureaphil
Ureaplasma
U. culture
U. urealyticum
Urea urealyticum
Urecholine
uremia
ureter
double u.
ectopic u.
u. fistula
ureteral
u. duplication
u. dysmenorrhea
u. ectopia
u. jet
u. node
u. obstruction
u. patency
u. surgical treatment
u. triplication
ureteric bud
ureterocele
ureterocervical
ureteroileostomy
Bricker u.
ureteroneocystostomy
Glen Anderson u.
Politano-Leadbetter u.
ureteropelvic
u. junction obstruction
ureteropyelocaliectasis
ureteropyeloscope
Karl Storz flexible u.
ureterotubal anastomosis
ureteroureteral anastomosis
ureteroureterostomy
ureterouterine
ureterovaginal fistula
ureterovesical
u. junction (UVJ)
u. obstruction
u. reflux

ureterovesicoplasty
  Leadbetter-Politano u.
urethra
  drain-pipe u.
  imperforate u.
  lead pipe u.
  low-pressure u.
  penile u.
  pipestem u.
  proximal u.
urethral
  u. angle
  u. candle
  u. caruncle
  u. coaptation
  u. discharge
  u. diverticulectomy
  u. diverticulum
  u. function
  u. gland
  u. hypermobility
  u. lumen
  u. meatus
  u. opening
  u. pressure
  u. pressure cough profile
  u. pressure profile
  u. pressure profilometry
  u. profilometry
  u. prolapse
  u. sound
  u. spasm
  u. sphincter
  u. stenosis
  u. syndrome
urethritis
  anterior u.
  chlamydial u.
  follicular u.
  granular u.
  nongonococcal u.
  nonspecific u. (NSU)
  u. petrificans
  posterior u.
  simple u.
  specific u.
  u. venerea
urethrocele
urethrocystometry
urethrocystoscopy
urethrography
urethrolysis

urethropexy
  Burch retropubic u.
  retropubic u.
urethroplasty
  Badenoch u.
  Thiersch-Duplay u.
  Turner-Warwick u.
urethroscopy
urethrotome
  Huffman-Huber infant u.
urethrotomy
  direct vision internal u. (DVIU)
urethrovaginal fistula
urethrovesical
  u. angle
Urex
urge incontinence
urgency
  urinary u.
URI
  upper respiratory infection
uric
  u. acid
  u. acid infarction
Uri-Check test
URICULT culture
uridine diphosphate (UDP)
Uridon
uridyltransferase
  galactose 1-phosphate u. (GALT)
urinalysis
  clean-catch u.
urinary
  u. beta-core fragment (UbetaCF)
  u. bladder
  u. bladder dysfunction
  u. calculus
  u. catheterization
  u. chorionic gonadotropin (UCG)
  u. concentrating capacity
  u. concentration test
  u. conduit
  u. diagnostic index (UDI)
  u. diversion
  u. exertional incontinence
  u. fistula
  u. free progesterone
  u. frequency
  u. glucose
  u. lactate:creatinine ratio
  u. luteinizing hormone (uLH)
  u. menopausal gonadotropin

U

NOTES

urinary *(continued)*
  u. outlet obstruction
  u. ovulation predictor kit
  u. retention
  u. sediment
  u. stasis
  u. stent
  u. steroid conjugate
  u. stress incontinence
  u. system
  u. tract abnormality
  u. tract disorder
  u. tract endometriosis
  u. tract infection (UTI)
  u. tract obstruction
  u. trigone
  u. trypsin inhibitor
  u. unidiversion procedure
  u. urgency
urinary-derived human follicle-stimulating hormone (u-hFSH)
urination
  fetal u.
urine
  u. CIE test
  u. culture
  u. cytology
  delirium, infection, atrophic urethritis/vaginitis, pharmaceuticals, psychological, excess u.
  u. flow
  u. latex test
  maple syrup u.
  u. output
  residual u.
  spun u.
  u. toxicology screen
urinoma
Uriscreen test
Urispas
Urispec
  U. GP+A test
  U. 9-Way test
Uristat
Uri-Three urine culture kit
Uri-Two petri dish
Urobak
Urobiotic-250
Urocyte diagnostic cytometry system
urocytogram
urodynamically
urodynamic study
urofacial syndrome
uroflometry
uroflowmetry
urofollitropin
urogenital
  u. anomaly

  u. atrophy (UGA)
  u. diaphragm
  U. Distress Inventory (UDI)
  u. epithelium
  u. fistula
  u. fold
  u. hiatus
  u. septum
  u. sinus
urogram
  intravenous u. (IVU)
urography
  magnetic resonance u. (MRU)
urogynecologic
urogynecologist
urogynecology
urokinase
  u. plasminogen activator (u-PA)
Urolene Blue
urolithiasis
urological
  u. evaluation
  u. injury
urologic history
urology
Uro-Mag
  U.-M. capsule
uropathy
  bilateral u.
  fetal u.
  fetal lower obstructive u.
  lower obstructive u. (LOU)
  obstructive u.
Uroplus
  U. DS
  U. SS
urorectal
urosepsis
urothelium
  trigonal u.
UroVive system
Urozide
urticaria
  papular u.
  u. pigmentosa
urticarial papule
US
  Usher syndrome
  US 1005 uroflow meter
USA Elite System GYN rotating continuous flow resectoscope
u-score method
use
  conservative drug u.
  u. effectiveness
  illicit drug u.
  maternal cocaine u.

oral contraceptive u.
prophylactic aspirin u.
Usher
  U. syndrome (US)
  U. syndrome type II
Usko pillar
U technique
uterectomy
uteri (*pl. of* uterus)
uterine
  u. abnormality
  u. absence
  u. access
  u. action
  u. activity
  u. activity alteration
  u. activity monitor
  u. angiosarcoma
  u. anomaly
  u. arteriovenous malformation
  u. artery
  u. artery embolization (UAE)
  u. artery hemodynamic adaptation
  u. artery pseudoaneurysm
  u. atony
  u. balloon therapy (UBT)
  u. ballottement
  u. bleeding
  u. blood flow
  u. calculus
  u. carcinosarcoma
  u. cast
  u. cavity
  u. chondrosarcoma
  u. colic
  u. compression
  u. contraction
  u. coring
  u. cornu
  u. cornual access catheter (UCAC)
  u. corpus
  u. corpus carcinoma
  u. corpus tumor
  u. cough
  u. cramping
  u. cry
  u. curette
  u. decompression
  u. dysfunction
  u. dysmenorrhea
  u. elevator
  u. endolymphatic stroma

u. endometrial stroma
u. epithelial stroma
u. epithelium
u. evacuator
u. evaluation
U. Explora Curette endometrial
  sampling device
u. exteriorization
u. factor
u. fibroid
u. fibromyoma
u. flora
u. fragmentation
u. gland
u. hematoma
u. hemodynamics
u. hernia syndrome
u. horn
u. hyperstimulation
u. hypertonus
u. hypotonia
u. incarceration
u. incision
u. inertia
u. infection
u. insufficiency
u. inversion
u. lateral fusion defect
u. leiomyomata
u. lysosome level
u. malposition
u. manipulator
u. mass
u. massage
u. milk
u. morcellation
u. müllerian sarcoma
u. myoma
u. myometrium
u. necrosis
u. neoplasm
u. ostial access catheter (UOAC)
u. packing
u. papillary serous carcinoma
u. pathology
u. perforation
u. positioning via ligament
  investment fixation truncation
  (UPLIFT)
u. pregnancy
u. prolapse
u. pyomyoma

U

NOTES

uterine *(continued)*
- u. quiescence
- u. reconstruction
- u. relaxation
- u. retroflexion
- u. retroversion
- u. rupture
- u. sarcoma
- u. sarcoma metastasis
- u. scar separation
- u. screening
- u. segment
- u. septum
- u. sinus
- u. size
- u. souffle
- u. sound
- u. suspension
- u. tachysystole
- u. tenaculum
- u. tenaculum forceps
- u. tetanus
- u. tone
- u. tube
- u. tympanites
- u. vein
- u. vessel

uterinus
- vagitus u.

uterismus

uteritis

utero
- fetal death in u. (FDIU)
- fetal version in u.
- in u.

uteroabdominal pregnancy

Uterobrush endometrial sample collector

uterocystostomy

uterofixation

uterolith

uterometer

uteroovarian
- u. circulation
- u. ligament
- u. varicocele

uteroperitoneal fistula

uteropexy

uteroplacental
- u. apoplexy
- u. blood flow
- u. circulation
- u. insufficiency
- u. perfusion
- u. sinus
- u. vessel

uteroplasty

uterosacral
- u. complex

- u. ligament
- u. ligament pedicle
- u. ligament suspension
- u. nerve ablation
- u. shortening

uterosalpingography

uteroscope

uteroscopy

uterotomy

uterotonic

uterotonin

uterotropin

uterotubal junction (UTJ)

uterotubography

uterovaginal
- u. canal
- u. prolapse

uterus, pl. **uteri**
- u. acollis
- anomalous u.
- arcuate u.
- u. arcuatus
- AV/AF u.
- benign nonprolapsed u.
- u. bicameratus vetularum
- u. bicornis
- u. bicornis unicollis
- bicornuate u.
- bifid u.
- u. bifidus
- biforate u.
- u. biforis
- u. bilocularis
- bipartite u.
- u. bipartitus
- bivalving of the u.
- boggy u.
- capped u.
- communicating u.
- cordiform u.
- u. cordiformis
- corpus of u.
- Couvelaire u.
- Credé maneuver of u.
- u. didelphys
- double-mouthed u.
- duplex u.
- u. duplex
- gravid u.
- heart-shaped u.
- hourglass u.
- hypoplastic u.
- impacted u.
- incarcerated gravid u.
- incudiform u.
- u. incudiformis
- large-for-dates u.
- myomata uteri

one-horned u.
u. parvicollis
pear-shaped u.
placenta, ovary, u. (POU)
prolapsed u.
retroverted u.
ruptured u.
sacculation of u.
septate u.
u. septus
subseptate u.
u. subseptus
symmetric communicating u.
tetanic u.
triangular u.
u. triangularis
T-shaped u.
unicolic u.
unicorn u.
u. unicornis
unicornuate u.
upper contractile portion of u.

UTI
  urinary tract infection
UTJ
  uterotubal junction
Utrata forceps
utricle
    prostatic u.
utriculoplasty
utriculovaginal pouch
UVC
  umbilical vein catheter
uveal coloboma, cleft lip/palate, mental retardation syndrome
uveitis
uveokeratitis
UVJ
  ureterovesical junction
uvula
    bifid u.
uvulitis

NOTES

U

V
>venous
>ventral
>V line

V^{Leiden}

>factor V.

V33W Endocavity probe

VA
>venoarterial
>ventriculoatrial

Vabra
>V. cannula
>V. catheter
>V. cervical aspirator
>V. suction
>V. suction curette

VAC
>vincristine, actinomycin D,
>cyclophosphamide

vaccination
>varicella v.

vaccinatum
>eczema v.

vaccine
>ActHIB *H. influenzae* type B v.
>autogenous v.
>Bacille bilié de Calmette-Guérin v.
>bacillus Calmette-Guérin v.
>BCG v.
>cholera v.
>DPT v.
>Engerix-B hepatitis B v.
>*Escherichia coli* v.
>*Haemophilus pertussis* v. (HPV)
>hepatitis B v. (HBV)
>HIB polysaccharide v.
>inactivated poliovirus v. (IPV)
>influenza v.
>killed virus v.
>live poliovirus v.
>live-virus v.
>measles v.
>meningococcal polysaccharide v.
>*Meningococcus* v.
>MMR v.
>mumps v.
>oral polio v.
>Ovarex v.
>O-Vax v.
>Pasteur Institute bacillus Calmette-
>   Guérin v.
>plague v.
>pneumococcal v.
>polyvalent pneumococcal v.
>rabies v.

>rubella v.
>Sabin v.
>Salk v.
>smallpox v.
>Theratope v.
>tularemia v.
>typhoid v.
>varicella virus v.
>yellow fever v.

vaccinia of the vulva

vacciniforme
>hydroa v.

vache
>coitus la v.

VACTERL
>vertebral, anal, cardiac, tracheal,
>   esophageal, renal, limb
>VACTERL anomaly
>VACTERL association
>VACTERL syndrome

VACTERL-H syndrome

Vacu-Irrigator
>Vozzle V.-I.

vacuolated

vacuole

vacuolization

Vacurette
>Berkeley V.

vacuronium

Vacutainer system

vacuum
>v. aspiration
>v. aspirator
>v. cannula
>Egnell v.
>v. extraction
>v. extractor
>v. extractor delivery
>v. suction

vacuum-assisted delivery

vagal reaction

vagina
>anterior v.
>apex of v.
>artificial v.
>atretic v.
>blind-ending v.
>dry v.
>posterior v.
>rugae of v.

vaginal
>v. absence
>v. acidification
>v. adenocarcinoma
>v. adenosis

V

vaginal *(continued)*
- v. administration
- v. advancement
- v. agenesis
- v. anomaly
- v. apex
- v. artery
- v. atresia
- v. atrophy
- v. birth after cesarean (VBAC)
- v. birth after cesarean section (VBAC)
- v. birth after cesarean—trial of labor (VBAC-TOL)
- v. bleeding
- v. cancer
- v. candidiasis
- v. candle
- Canesten V.
- v. carcinoma
- v. celiotomy
- v. clear cell adenocarcinoma
- v. colonization
- v. condom
- v. cone
- v. contraceptive
- v. contraceptive film (VCF)
- v. cornification test
- v. cream
- v. cuff
- v. cuff adhesion
- v. cystourethropexy
- v. cytology
- v. decompression
- v. delivery
- v. dilator
- v. dimple
- v. discharge
- v. douche
- v. dysmenorrhea
- v. dysontogenetic cyst
- v. ectopic anus
- v. embryonic cyst
- v. epithelial abnormality
- v. estrogen therapy
- v. eversion
- v. evisceration
- v. examination
- v. flora
- v. fluid arborization
- v. fluid ferning
- v. fornix
- v. GIFT
- Gyne-Lotrimin V.
- v. hand
- v. hematoma
- v. hood
- v. hysterectomy
- v. hysterotomy
- v. inclusion cyst
- v. infection
- v. injury
- v. interruption of pregnancy with dilatation and curettage (VIP-DAC)
- v. intraepithelial neoplasia (VAIN)
- v. irrigation smear (VIS)
- v. laceration
- v. lithotomy
- v. lubricant
- v. microflora
- v. misoprostol
- Monistat V.
- v. morcellation
- v. mucification test
- v. mucosa
- v. myomectomy
- v. neurofibroma
- v. neurofibromatosis
- Ogen V.
- v. opening
- Ortho-Dienestrol V.
- v. outlet
- v. parity
- v. perineorrhaphy
- v. pH
- v. plate
- v. pool
- v. pool L/S ratio
- v. pouch
- v. probe ultrasound
- v. prolapse
- v. prolapse prosthesis
- v. prostaglandin
- v. recombinant human relaxin
- v. retractor
- v. ring contraception
- v. secretion
- v. smear
- v. smear intermediate cell
- v. smear parabasal cell
- v. smear superficial cell
- v. sonography
- v. speculum loop
- v. sponge
- v. squamous metaplasia
- v. stump prolapse
- v. suppository
- v. surgery
- v. tampon
- v. tape tension
- Terazol V.
- v. terminus
- v. transudate
- v. ulcer
- Vagistat-1 V.

v. vault
v. vault prolapse
v. vault support
v. venous plexus
v. wall flap
v. wall repair
v. wall sling procedure
v. wash
v. yeast
vaginales
    rugae v.
vaginalis
    Haemophilus v.
    Trichomonas v.
vaginalis
    portio v.
    tunica v.
    vagitus v.
vaginally parous
vaginam
vaginapexy
vaginectomy
vaginism
vaginismus
    posterior v.
vaginitis, pl. vaginitides
    v. adhesiva
    adhesive v.
    amebic v.
    atrophic v.
    chlamydial v.
    Corynebacterium v.
    v. cystica
    desquamative inflammatory v.
    v. emphysematosa
    Gardnerella v.
    Haemophilus v.
    Mobiluncus v.
    monilial v.
    nonspecific v.
    pinworm v.
    recurrent v.
    senile v.
    v. senilis
    streptococcal v.
    traumatic v.
    Trichomonas vaginalis v.
    yeast v.
vaginocele
vaginodynia
vaginofixation
vaginography

vaginohysterectomy
vaginometer
vaginomycosis
vaginopathy
vaginoperineoplasty
vaginoperineorrhaphy
vaginoperineotomy
vaginopexy
    Norman Miller v.
vaginoplasty
    Fenton v.
    tissue expansion v.
vaginoscope
    Huffman-Huber infant v.
    Huffman infant v.
vaginoscopy
    pediatric v.
vaginosis
    anaerobic v.
    bacterial v. (BV)
    mycotic v.
vaginotomy
    anterior v.
Vagisil intimate moisturizer
Vagistat-1
    V.-1 Vaginal
Vagi-TEST
Vagitrol
vagitus
    v. uterinus
    v. vaginalis
VAIN
    vaginal intraepithelial neoplasia
Vakutage suction system
valacyclovir
    v. HCl
    v. hydrochloride
Valadol
Valchev uterine manipulator
Valentine position
valerate
    betamethasone v.
    estradiol v.
Valergen
Valertest No.1
valga
    coxa v.
valgum
    genu v.
valgus
    cubitus v.

NOTES

V

valgus *(continued)*
    hallux v.
    talipes v.
validity
Valisone
    V. Topical
Valium
    V. Injection
    V. Oral
Valle hysteroscope
Valley
    V. fever
    V. Vac smoke evacuation system
Valleylab
    V. ball electrode
    V. Force IC electrosurgical
      generator
    V. loop electrode
    V. pencil
    V. REM system
Valorin
Valpin 50
valproate
    sodium v.
valproic
    v. acid
    v. acid embryopathy
    v. acid syndrome
Valrelease
valrubicin
Valsalva
    V. maneuver
    V. sinus
    sinus of V.
    squirming V.
valsartan
Valstar
Valtrex
value
    acid-base v.
    adaptive v.
    Astrup blood gas v.
    baseline v.
    complement v.
    fetal blood v.
    maturation v.
    P v.
valve
    bicuspid aortic v.
    mitral v.
    parachute mitral v.
    posterior urethral v. (PUV)
    pulmonary v.
valvotomy
valvular
    v. heart disease
    v. prosthesis
    v. pulmonic stenosis

valvuloplasty
    balloon v.
valvulotomy
    balloon v.
    v. procedure
Van
    V. Buchem syndrome
    V. der Woude syndrome
    V. Haldergem syndrome
    V. Hoorn maneuver
    V. Maldergem syndrome
van
    v. Bogaert disease
    v. den Bergh test
    v. den Bosch syndrome
    v. der Hoeve syndrome
Vancaillie uterine cannula
Vancenase
    V. AQ
    V. AQ Inhaler
    V. AQ 84 mcg
    V. Nasal Inhaler
Vanceril
Vancocin
Vancoled
vancomycin
    v. hydrochloride
vanillylmandelic acid (VMA)
Vaniqa
vanishing
    v. fetus
    v. testes syndrome
    v. twins
    v. twin syndrome
Vanquin
Vantin
Vantos vacuum extractor
VAP
    average path velocity
vapocauterization
Vapo-Iso
Vaponefrin
vaporization
    bipolar v.
    endometrial v.
    laser v.
vara
    coxa v.
Váradi-Papp syndrome
Váradi syndrome
variabilis
    erythrokeratodermia v.
variability
    baseline v. of fetal heart rate
    beat-to-beat v.
    fetal heart rate v.
    interpretation v.

variable
- continuous random v.
- v. deceleration
- discrete random v.
- mixed discrete-continuous random v.
- random v.
- v. softness (VS)

variable-number tandem-repeat

variance
- additive genetic v.
- dominance v.
- environmental v.
- genetic v.
- phenotypic v.
- v. ratio

variant
- albopapuloid v.
- v. hemoglobin
- Hurler v.
- Klinefelter v.
- Pasini v.

variation
- beat-to-beat v. of fetal heart rate
- interobserver v.
- observer v.
- somaclonal v.

varicella
- v. bullosa
- v. gangrenosa
- v. infection
- v. inoculata
- v. pneumonia
- pustular v.
- v. pustulosa
- v. vaccination
- v. virus vaccine

varicellae
- Herpesvirus v.

varicella-zoster
- v.-z. immune globulin (VZIG)
- v.-z. virus (VZV)
- v.-z. virus infection

varicelliform

varices (*pl. of* varix)

varicocele
- ovarian v.
- tuboovarian v.
- uteroovarian v.

varicocelectomy

varicomphalus

varicosity
- vulvar v.

variegata
- porphyria v.

variegate
- v. porphyria

variety

variola

varioliform

Varivax

varix, pl. **varices**
- esophageal v.
- gelatinous v.
- umbilical vein v.
- vulvar v.

varus
- metatarsus v.
- talipes v.

VAS
- vibratory acoustic stimulation
- vibroacoustic stimulation

vas, pl. **vasa**
- v. deferens
- v. deferens aplasia
- vasa previa

vasal

Vas-Cath catheter

vascular
- v. anomaly
- v. calcification
- v. cell adhesion molecule
- v. channel
- v. communication
- v. complication
- v. disease
- v. endothelial growth factor (VEGF)
- v. engorgement
- v. headache
- v. injury
- v. leiomyoma
- v. malformation
- v. metastasis
- v. parabiosis
- v. permeability
- v. reactivity
- v. resistance
- v. ring syndrome
- v. spider
- v. umbrella

vascularization
- chorionic v.

V

NOTES

vasculature
    atypical v.
vasculitis
    cutaneous v.
    rheumatoid v.
vasculopathy
    fetal thrombotic v. (FTV)
vasectomy
    no-scalpel v.
    open-ended v.
    v. reversal
vasitis
vasoactive
    v. intestinal peptide (VIP)
    v. intestinal polypeptide (VIP)
    v. prostaglandin
    v. substance
vasocongestion
Vasodilan
vasoepididymostomy (VE)
vasogram
vasomotor flush
vasopressin
    arginine v. (AVP)
    v. infusion
    placental v.
    v. therapy
vasopressor
Vasospan
vasospasm
    cerebral v.
    coronary v.
Vasotec
vasotocin
    arginine v.
vasovagal
    v. reflex apnea
vasovasostomy
Vasoxyl
VATER
    vertebral defects, imperforate anus,
    tracheoesophageal fistula, renal defects
    VATER complex
    hydrocephaly with features of
        VATER
    VATER syndrome
vault
    v. cap
    v. prolapse
    vaginal v.
VBAC
    vaginal birth after cesarean
    vaginal birth after cesarean section
VBAC-TOL
    vaginal birth after cesarean—trial of
    labor
VBG
    venous blood gas

VBP
    vinblastine, bleomycin, cisplatin
VBS
    venous blood sample
V-Cath catheter
VCF
    vaginal contraceptive film
    VCF vaginal contraceptive film
V-Cillin K
VCL
    curvilinear velocity
VCP
    vincristine, cyclophosphamide, and
    prednisone
VCU
    videocystourethrography
    voiding cystourethrogram
VCUG
    vesicoureterogram
    voiding cystourethrogram
VD
    venereal disease
VDCC
    voltage-dependent calcium channel
VDR
    vitamin D receptor
VDRL
    Venereal Disease Research Laboratories
    VDRL test
VE
    vasoepididymostomy
    volume of expired gas
Vecchietti
    V. method
    V. operation
vectis
vector
    cloning v.
Vectrin
vecuronium
Veetids
vegan
vegetable
    cruciferous v.
vegetarian
    v. diet
vegetative reproduction
VEGF
    vascular endothelial growth factor
Veillonella
vein
    v. compression
    dilated collateral v.
    v. of Galen
    iliac v.
    maternal cortical v.
    ovarian v.
    persistent right umbilical v.

umbilical v.
uterine v.
Veingard dressing
vela (*pl. of* velum)
velamen, pl. **velamina**
v. vulvae
velamentosa
placenta v.
velamentous
v. insertion
v. vessel
velamina (*pl. of* velamen)
Velban
Velbe
vellus
v. hair
v. hypertrichosis
velocardiofacial syndrome
velocimetry
Doppler v.
fetal aortic Doppler v.
umbilical artery Doppler v.
velocity
average path v. (VAP)
curvilinear v. (VCL)
v. flow profile
straight line v. (VSL)
velofacial hypoplasia
Velosef
Velosulin Human
Veltane
velum, pl. **vela**
vena
v. cava
v. caval filter
v. caval interruption
venenata
dermatitis v.
venereal
v. bubo
v. collar
v. disease (VD)
V. Disease Research Laboratories (VDRL)
V. Disease Research Laboratory test
v. lymphogranuloma
v. wart
venereology
venereum
lymphogranuloma v. (LGV)
venipuncture

venlafaxine
v. HCl
Venning-Brown test
venoarterial (VA)
Venodyne pneumatic compressive device
Venoglobulin-I
Venoglobulin-S
venography
ascending v.
radionuclide v.
venous (V)
v. blood gas (VBG)
v. blood sample (VBS)
v. catheter
v. collateral
v. line
v. oozing
v. thromboembolism (VTE)
v. thrombosis
venovenous (VV)
venter
v. propendens
vent gleet
ventilation
bag-and-mask v.
bagged mask v.
controlled mechanical v. (CMV)
hand v.
HFJ v.
HFO v.
HFPP v.
high-frequency jet v. (HFJV)
high-frequency oscillatory v. (HFOV)
high-frequency positive pressure v. (HFPPV)
intermittent mandatory v. (IMV)
intermittent mechanical v. (IMV)
intermittent positive pressure v. (IPPV)
v. by mask
oscillatory v.
v. scintigraphy
synchronized intermittent mandatory v. (SIMV)
ventilation/perfusion (V/Q)
v./p. mismatch
v./p. quotient
v./p. ratio
v./p. scan
ventilator
Amsterdam infant v.

V

NOTES

ventilator *(continued)*
    BABYbird II v.
    Babyflex v.
    Bear Cub infant v.
    Bennett PR-2 v.
    Bourns LS104-150 infant v.
    Breeze v.
    CPAP v.
    Healthdyne v.
    HFJ v.
    HFO v.
    HFPP v.
    high-frequency v.
    humidification v.
    HVF v.
    infant Star high-frequency v.
    nebulization v.
    PEEP v.
    pressure-cycled v.
    pressure-preset v.
    Pulmo-Aid v.
    Sechrist neonatal v.
    Servo 900C v.
    Siemens-Elema Servo 900C v.
    Siemens Servo 300 v.
    Siemens Servo 900C v.
    Star v.
    v. support
    Vickers Neovent v.
    Vix infant v.
    volume-limited v.
    Wave v.
Ventolin
ventouse
ventral (V)
    v. hernia
ventricle
    dilation of v.
    double-inlet left v.
    double-inlet right v.
    isolated double outlet right v.
    left v. (LV)
ventricular
    v. function
    v. response rate
    v. septal defect (VSD)
    v. tachycardia
ventriculoamniotic shunt
ventriculoatrial (VA)
ventriculography
ventriculomegaly
    bilateral cerebral v.
    fetal v.
ventriculoperitoneal (VP)
    v. shunt
ventriculoseptal defect (VSD)
ventroposterior (VP)
ventrosuspension

Venturi mask
Venus
    collar of V.
VEP
    visual evoked potential
VePesid
VER
    visual evoked response
vera
    placenta accreta v.
    polycythemia rubra v.
Veralba
verapamil
***Veratrum* alkaloid**
Veress needle
vergeture
Veriloid
vermiculation
vermiform lesion
vermilion
vermina cyst
Vermox
Verner-Morrison syndrome
Vernes test
vernix
    v. caseosa
    v. membrane
Vero cell
verruca
    v. acuminata
    v. plana juvenilis
Verruca-Freeze
verruciformis
    epidermodysplasia v.
verrucous
    v. carcinoma
VersaLap
versicolor
    tinea v.
version
    bimanual v.
    bipolar v.
    Braxton Hicks v.
    cephalic v.
    combined v.
    external v.
    external cephalic v. (ECV)
    Hicks v.
    internal podalic v.
    pelvic v.
    podalic v.
    postural v.
    Potter v.
    spontaneous v.
    Wigand v.
    Wright v.
vertebra, pl. **vertebrae**
    butterfly vertebrae

cleft vertebrae
v. plana
vertebral
v., anal, cardiac, tracheal, esophageal, renal, limb (VACTERL)
v., anal, cardiac, tracheal, esophageal, renal, limb anomaly
v. artery
v. bone loss
v. bone mass
v. compression fracture
v. defects, imperforate anus, tracheoesophageal fistula, renal defects (VATER)
vertebrodidymus
vertex, pl. **vertices**
v. delivery
external cephalic version and spontaneous v.
instrumental v.
v. position
v. potential
v. presentation
spontaneous v.
vertex-breech twin presentation
vertex-nonvertex
v.-n. pair
vertex-transverse twin presentation
vertex-vertex
v.-v. pair
vertical
v. HIV transmission
v. hymen
v. transmission
vertically-acquired infection
vertices (*pl. of* vertex)
vertiginous seizure
vertigo
very
v. low birth weight (VLBW)
v. low-density lipoprotein (VLDL)
very-low-birth-weight infant
vesical neck
Vesica sling kit
vesicle
chorionic v.
germinal v.
graafian v.
nabothian v.
seminal v.
vesicobullous

vesicocele
vesicocentesis
vesicocervical
v. space
vesicofixation
vesicomyectomy
vesicomyotomy
vesicoureteral
v. reflux
vesicoureterogram (VCUG)
vesicourethral
v. canal
v. primordial
vesicourethrolysis
retropubic v.
vesicouterine fistula
vesicovaginal
v. fistula
v. repair
vesicovaginorectal
v. fistula
vesicular
v. mole
vessel
chorioallantoic v.
corkscrew v.
great v.
hairpin v.
infundibulopelvic v.
laser photocoagulation of the communicating v. (LPCV)
v. spread
transposition of great v.'s (TGV)
umbilical v.
umbilicoplacental v.
uterine v.
uteroplacental v.
velamentous v.
VestaBlate system
vestibular
v. adenitis
v. anus
v. bulb
v. cyst
v. duct
v. dyspareunia
v. gland
vestibule
vestibulectomy
Woodruff v.
vestibulitis
vulvar v.

V

NOTES

vestibulodynia
vestigial
VF Q506
viability
    fetal v.
viable
    v. endometrial cell
    v. fetus
    v. infant
vial
    Nickerson Biggy v.'s
Viasorb dressing
Vibramycin
Vibra-Tabs
vibrator
    penile v.
vibratory acoustic stimulation (VAS)
Vibrio fetus
vibroacoustic-induced fetal movement
vibroacoustic stimulation (VAS)
vicarious
    v. menstruation
    v. respiration
Vickers Neovent ventilator
Vicks DayQuil Allergy Relief 4 Hour Tablet
Vicodin
Vicryl Rapide suture
victimization
    physical v.
    sexual v.
Victor Gomel method
vidarabine
VIDAS
    V. automated immunoassay system
    V. Estradiol II assay kit
    V. immunoanalysis testing system
    V. varicella zoster assay
Vi-Daylin vitamins
videocystourethrography (VCU)
videoendoscope
    three-dimensional v.
videolaparoscope
Video Overlay Method
VideoZoomscope
Videx
Vi-Drape bowel bag
view
    axillary v.
    craniocaudal v.
    exaggerated craniocaudal v.
    four-chamber v.
    frogleg v.
    lateral oblique v.
    lateromedial oblique v.
    long-axis v.
    medial oblique v.
    mediolateral v.

    pulmonary artery/ductus v.
    short-axis v.
    spot compression v.
vigabatrin
Vignal cell
vigorous infant
villi (*pl. of* villus)
villitis
    focal v.
villoglandular configuration
villositis
villous placenta
villus, pl. **villi**
    chorionic villi
    hydropic chorionic v.
    placental v.
vimentin
Vim-Silverman needle
Vimule
    V. cap
    V. permanent sheath
VIN
    vulvar intraepithelial neoplasia
vinblastine
    v., bleomycin, cisplatin (VBP)
    cisplatin, methotrexate, v. (CMV)
    methotrexate, cisplatin, v. (MCV)
Vinca **alkaloid**
Vincasar PFS
vincristine
    v., actinomycin D, cyclophosphamide (VAC)
    v., cyclophosphamide, and prednisone (VCP)
    v. sulfate
Vineland Adaptive Behavior Scales
violaceous lesion
violence
    domestic v.
violet
    crystal v.
    gentian v.
Vioxx
VIP
    vasoactive intestinal peptide
    vasoactive intestinal polypeptide
    voluntary interruption of pregnancy
VIP-DAC
    vaginal interruption of pregnancy with dilatation and curettage
Vipond sign
viprynium
Vira-A
    V.-A. Ophthalmic
Viracept
viral
    cellular v.
    v. encephalitis

v. hepatitis
v. infection
v. pharyngitis
v. pneumonia
v. pneumonitis
v. shedding
v. titer
v. transmission
Viramune
ViraPap
V. HPV DNA test
ViraType
V. HPV DNA typing assay
V. probe
V. test
Virazole
V. Aerosol
Virchow
pneumonia alba of V.
Virchow-Seckel syndrome
viremia
Viresolve ultrafiltration membrane
virgin
virginal
v. breast hypertrophy
v. hymen
v. introitus
Virginia needle
virginity
viridans
Streptococcus v.
virilescence
virilism
adrenal v.
virilization
virilizing
v. adenoma
v. adrenal tumor
v. 3α-androstanediol glucuronide
Virilon
V. capsule
virion
intranuclear v.
Viroptic Ophthalmic
virtual labor monitor (VLM)
virulent bubo
virus
acquired immunodeficiency
syndrome-related v. (ARV)
v. clearance
ECHO v.
Epstein-Barr v. (EBV)

hepatitis A v.
hepatitis B v. (HBV)
hepatitis C v. (HCV)
herpes simplex v. (HSV)
herpes simplex v. type 1 (HSV-1)
herpes simplex v. type 2 (HSV-2)
herpes zoster v. (HZV)
horizontal transmission of v.
human immunodeficiency v. (HIV)
human papilloma v. (HPV)
human T-cell leukemia v. (HTLV)
human T-cell leukemia v. type I
(HTLV-I)
human T-cell leukemia v. type II
(HTLV-II)
human T-cell leukemia v. type III
(HTLV-III)
influenza v.
Kirsten murine sarcoma v.
lymphadenopathy-associated v.
(LAV)
molluscum contagiosum v. (MCV)
mouse mammary tumor v.
(MMTV)
mumps v.
non-syncytium-inducing variant of
AIDS v.
papilloma virus, polyoma virus,
vacuolative v. (papova)
REO v.
respiratory enteric orphan v.
respiratory syncytial v. (RSV)
Rous sarcoma v.
RS v.
rubella v.
simian immunodeficiency v. (SIV)
varicella-zoster v. (VZV)
VIS
vaginal irrigation smear
viscera
inversion of v.
transposition of v.
visceral
v. cleft
v. heterotaxy
v. peritoneum
viscerale
cranium v.
visceromegaly
viscerum
situs inversus v.
viscoelastic

NOTES

587

viscosity
viscous
> v. semen
> v. syndrome

visible cortical mantle
Visicath
Visine
> V. Extra Ophthalmic
> V. L.R. Ophthalmic

Visiport optical trocar
Visken
Visscher-Bowman test
Vista disposable skin stapler
Vistaril
visual
> v. analog scale for stress
> v. change
> v. disturbance
> v. evoked potential (VEP)
> v. evoked response (VER)
> v. field
> v. stimulation

visualization
visuomotor integration
Vitabee 6
vital
> v. capacity
> v. sign

vitamin
> v. A
> v. B_6
> v. B_{12}
> v. B complex
> B complex v.'s
> Bronson chewable prenatal v.'s
> v. C
> v. D
> v. D-binding protein
> v. D receptor (VDR)
> v. D-resistant rickets
> v. E
> Fer-In-Sol v.'s
> v. K
> v. K-dependent serine protease
> v. K prophylaxis
> Materna prenatal v.
> v. metabolism
> Poly-Vi-Flor v.'s
> Poly-Vi-Sol v.'s
> v. requirement
> Stuart Prenatal v.'s
> v. supplement
> Tri-Vi-Flor v.'s
> Tri-Vi-Sol v.'s
> Vi-Daylin v.'s

Vita-Plus E
> V. P. E. Softgels

Vitec

Vite E Cream
vitelliform degeneration
vitelline
> v. cord
> v. duct
> v. duct cyst
> v. fistula
> v. membrane
> v. sac
> v. tumor

vitellinum
> mesoblastoma v.

vitellointestinal
> v. cyst
> v. duct

vitiligo
Vitrasert
vitreoretinopathy
> familial exudative v.

vitreous
vitro
> in v. fertilization (IVF)
> in v.

vitronectin
Vivactil
Vivelle
> V. Transdermal

Vivelle-Dot estradiol transdermal system
Vivigen diagnostics
viviparity
viviparous
vivo
> ex v.
> in v.

Vivol
Vix infant ventilator
VK
> Apo-Pen VK

VLBW
> very low birth weight

VLDL
> very low-density lipoprotein

Vles syndrome
V-line
VLM
> virtual labor monitor

VM-26
VMA
> vanillylmandelic acid

Vogt
> V. cephalodactyly
> V. syndrome

Vogt-Spielmeyer disease
Vohwinkel syndrome
voiceprint
voiding
> v. cystourethrogram (VCU, VCUG)
> v. diary

v. dysfunction
v. trial
volatile
v. acid
v. anesthetic
Volkmann
V. deformity
V. disease
volt
electron-v. (eV, ev)
megaelectron v. (MeV)
voltage-dependent calcium channel (VDCC)
Voltaren
V. Oral
Voltaren-XR Oral
Voltolini disease
volume
amniotic fluid v. (AFV)
blood v.
closing v.
v. of dead space
v. expansion
v. of expired gas (VE)
fetal blood v.
fetoplacental blood v.
v. flow
forced expiratory v. (FEV)
intrauterine v.
lung v.
mean corpuscular v. (MCV)
minute ventilatory v.
v. overload
plasma v.
red cell v.
residual v.
semen v.
tidal v.
volume-limited ventilator
voluntary
v. interruption of pregnancy (VIP)
v. sterilization
Voluson sector transducer
volvulus
gastric v.
midgut v.
v. neonatorum
VO$_2$max
maximum oxygen uptake
vomiting
nausea and v.

postoperative nausea and v. (PONV)
v. of pregnancy
projectile v.
vomitus
von
V. Fernwald sign
v. Gierke disease
v. Gierke glycogenolysis
v. Graefe sign
v. Hippel-Lindau disease
V. Jaksch anemia
v. Poehl test
v. Recklinghausen disease
v. Recklinghausen neurofibromatosis
v. Willebrand disease
v. Willebrand disease type IIB, III
v. Willebrand factor (vWF)
v. Willebrand factor antigen
v. Willebrand syndrome
Voorhees bag
Voorhoeve
V. dyschondroplasia
V. syndrome
vorozole
VoSol
V. HC Otic
V. Otic
Vozzle Vacu-Irrigator
VP
ventriculoperitoneal
ventroposterior
VP shunt
VP-16
V/Q
ventilation/perfusion
V̇/Q̇ ratio
V̇/Q̇ scan
Vrolik disease
VS
variable softness
VSD
ventricular septal defect
ventriculoseptal defect
VSL
straight line velocity
VTE
venous thromboembolism
vu
jamais v.
vulgaris
acne v.

V

NOTES

589

vulgaris *(continued)*
    ichthyosis v.
    lupus v.
    pemphigus v.
Vulpe Assessment Battery
vulsellum clamp
vulva, pl. **vulvae**
    noma vulvae
    Paget disease of v.
    synechia vulvae
    vaccinia of the v.
vulvar
    v. adenoid cystic adenocarcinoma
    v. algesiometer
    v. angiokeratoma
    v. atrophy
    v. atypia
    v. biopsy
    v. carcinoma
    v. carcinoma in situ
    v. colposcopy
    v. congenital dysplastic angiopathy
    v. dermatitis
    v. dermatosis
    v. dystrophy
    v. edema
    v. endometriosis
    v. fibroma
    v. hemangioma
    v. hematoma
    v. hidradenitis suppurativa
    v. hydatidiform mole
    v. inclusion cyst
    v. infection
    v. intraepithelial neoplasia (VIN)
    v. lipoma
    v. lymph node
    v. malignancy
    v. melanoma
    v. neurofibroma
    v. nevus
    v. papillomatosis
    v. pigmented lesion
    v. pruritus
    v. psoriasis
    v. sarcoma
    v. skin
    v. squamous hyperplasia
    v. varicosity
    v. varix
    v. vestibulitis
    v. vestibulitis syndrome (VVS)
vulvectomy
    Basset radical v.
    Parry-Jones v.
    radical v.

    simple v.
    skinning v.
vulvismus
vulvitis
    adhesive v.
    chronic atrophic v.
    chronic hypertrophic v.
    creamy v.
    cyclic v.
    erosive v.
    focal v.
    follicular v.
    leukoplakic v.
    plasma cell v.
vulvodynia
    cyclic v.
    dysesthetic v.
    essential v.
    idiopathic v.
vulvoperianal
vulvoplasty
vulvovaginal
    v. anus
    v. candidiasis (VVC)
    v. carcinoma
    v. cystectomy
    v. disorder
    v. lesion
    v. outlet
    v. pouch of Williams
    v. premenarchal infection
vulvovaginitis
    candidal v.
    cyclic v.
    premenarchal v.
vulvovaginoplasty
    Williams v.
Vu-Max vaginal speculum
Vumon
VV
    venovenous
VVC
    vulvovaginal candidiasis
VVS
    vulvar vestibulitis syndrome
V.V.S.
vWF
    von Willebrand factor
    vWF antigen
Vytone
    V. Topical
VZIG
    varicella-zoster immune globulin
VZV
    varicella-zoster virus

W

W chromosome

W syndrome

Waardenburg

W. recessive anophthalmia
syndrome

Waardenburg-Klein syndrome

Wachendorf membrane

waddling gait

Wade

Roe v. W.

Wadia elevator

Wagner-Missner body

Wagner syndrome

WAGR

Wilms tumor, aniridia, gonadoblastoma,
mental retardation

WAGR syndrome

Waisman-Laxova syndrome

Waisman syndrome

waist:hip ratio

Walcher position

Waldenstrom macroglobinemia

Waldeyer

W. fossae

germinal epithelium of W.

W. preurethral ligament

Walker

W. chart

W. lissencephaly syndrome

Walker-Clodius syndrome

Walker-Warburg

W.-W. malformation

W.-W. syndrome

walking

automatic w.

chromosome w.

w. epidural anesthetic

wall

opposing w.

resecting intrapartum uterine w.

symphyseal w.

wallaby pouch

Wallace

W. catheter

W. technique

Wallach

W. Endocell collection device

W. Endocell endometrial cell
sampler

W. LL100 cryosurgical cryogun

**Wallach-Papette disposable cervical cell
collector**

Walt Disney dwarfism

Walthard

W. cell rest

W. nest

Walther dilator

Walton

W. report

W. syndrome

Wampole test

wandering ovary

Wangensteen needle holder

Warburg syndrome

Wardill four-flap method

**Wardill-Kilner advancement flap
method**

Ward-Mayo vaginal hysterectomy

Ward triangle

Ware Short Form-35

warfarin

w. embryopathy

w. syndrome

Warfilone

Waring

W. blender sound

W. blender syndrome

Warkany syndrome 1, 2

warmer

Echowarm gel w.

Kreiselman infant w.

Ohio w.

overhead w.

radiant w.

Thermasonic gel w.

Warren flap

wart

anogenital w.

flat w.

genital w.

plantar w.

venereal w.

water w.

Warthin-Starry stain

wash

vaginal w.

washed

w. intrauterine insemination

w. sperm

w. spermatozoa

washer

Gravlee jet w.

washing

cytologic w.

peritoneal w.

washout

nitrogen w.

w. pyelogram

W

Wassel classification
Wasserman test
wastage
    early pregnancy w.
    fetal w.
    pregnancy w.
    reproductive w.
wasting
    salt w.
water
    w. aerobics
    bag of w.'s (BOW)
    dextrose in w.
    w. excretion
    false w.'s
    w. intoxication
    w. loss
    w. metabolism
    w. test
    total body w.
    w. wart
Waterhouse-Friderichsen syndrome
waterseal drainage
Waters operation
Waterston shunt
Watkins transposition
Watson
    W. method
    W. syndrome
Watson-Alagille syndrome
Watson-Crick
    W.-C. helix
    W.-C. method
Watson-Miller syndrome
WAVE
    WAVE DNA Fragment Analysis
    System
wave
    fluid w.
    pulsed electromagnetic w.
    rolandic sharp w.'s
    W. ventilator
waveform
    aortic blood flow velocity w.
    arterial w.
    discordant artery flow velocity w.
    Doppler w.
    flow velocity w.
wavy rib
4-Way Long Acting Nasal Solution
Way operation
WBC
    AF WBC
wcp
    whole chromosome paint
    wcp1 chromosome paint
    wcp7 chromosome paint
    wcp8 chromosome paint

    wcp10 chromosome paint
    wcp15 chromosome paint
    wcp18 chromosome paint
    wcp19 chromosome paint
    wcpX chromosome paint
weakness
    collagen w.
wean
wean-and-feed protocol
weaning
    w. brash
Weaver-Smith syndrome (WSS)
Weaver syndrome
Weaver-Williams syndrome
web
    w. cerclage
    laryngeal w.
    tracheal w.
webbed
    w. finger
    w. neck
Weber
    W. syndrome
    W. test
Weber-Christian syndrome
Weber-Dimitri syndrome
webspace
Webster operation
Wechsler
    W. Intelligence Scale for Children-
    Revised (WISC-R)
    W. Preschool and Primary Scale
    of Intelligence-Revised (WPPSI-R)
Weck
    W. disposable cannula
    W. disposable trocar
Weck-cel sponge
wedge
    pulmonary artery w. (PAW)
    w. resection
weekend drug holiday
Weerda laparoscope
Wegener granulomatosis
Wegner disease
weight
    birth w. (BW)
    body w.
    estimated fetal w. (EFW)
    extremely low birth w. (ELBW)
    fetal w.
    w. gain
    w. loss
    low birth w. (LBW)
    maternal w.
    placental w.
    w. reduction
    very low birth w. (VLBW)
weightbearing bone

weighted speculum
Weightrol
Weil disease
Weill-Marchesani syndrome
Weill sign
Weinberg rule
Weismann-Netter syndrome
Weissenbacher-Zweymuller syndrome
Weitlaner retractor
well-being
    fetal w.-b.
well-born
    right to be w.-b.
well-circumscribed
well-circumscribed carcinoma
Wellcovorin
well-defined mass
well engaged in pelvis
well-hydrated baby
well-oxygenated infant
well-perfused baby
Wenckebach
    W. heart block
    W. phenomenon
Went syndrome
Wepman test
Werdnig-Hoffmann
    W.-H. disease
    W.-H. muscular atrophy
    W.-H. paralysis
    W.-H. syndrome
Werlhof disease
Wermer syndrome
Werner syndrome
Wernicke aphasia
Wertheim operation
Wertheim-Schauta operation
West
    W. Haven-Yale Multidimensional
    Pain Inventory
    W. Nile fever
    W. syndrome
Westcort Topical
West-Engstler skull
Western
    W. blot
    W. blot test
Weston
    W. knot
    W. slipknot
wet
    w. burp

    w. lung syndrome
    w. mount
    w. mount test
    w. nurse
    w. prep
    w. rale
    w. smear
Weyers oligodactyly syndrome
Wharton jelly
wheal
wheal-and-flare reaction
Wheaton Pavlik harness
wheat sperm agglutination test
Whelan syndrome
whiff
    w. amine test
Whipple
    W. disease
    W. procedure
    W. syndrome
whipworm
whistle-tip catheter
whistling
    w. face syndrome
    w. face-windmill vane hand
    syndrome
white
    w. blood cell
    W. class, B through RF
    W. classification
    w. coat hypertension
    w. epithelium
    w. grape juice
    w. infarction
    w. leg
    w. matter necrosis
    w. pupillary reflex
    w. sponge nevus
    w. tonsillitis
Whittaker test
Whitten medium
Whittingham medium
WHO
    World Health Organization
whole
    w. abdominal radiation
    w. blood
    w. chromosome paint (wcp)
whole-abdomen irradiation
whole-body irradiation
whole-pelvis irradiation
Wholey balloon occlusion catheter

W

NOTES

Whoo Noz
whooping cough
whorl
WI-38 cell
wick
        gauze w.
widened thecal sac
wide plane
width
        funnel w.
Wieacker syndrome
Wieacker-Wolff syndrome
Wiedeman-Beckwith-Combs syndrome
Wiedemann-Rautenstrauch syndrome
    (WR)
Wiedemann syndrome
Wiegernick culdocentesis
Wigand
        W. maneuver
        W. version
Wildermuth ear
Wildervanck-Smith syndrome
Wildervanck syndrome
wild-type
        w.-t. allele
        w.-t. gene
wild type
Wilkins
        W. disease
        W. syndrome
Willebrand-Jurgens syndrome
Willett
        W. clamp
        W. forceps
Williams
        W. disease
        W. syndrome
        vulvovaginal pouch of W.
        W. vulvovaginoplasty
Williams-Barratt syndrome
Williams-Beuren syndrome
Williams-Campbell syndrome
Willis
        circle of W.
Willy Meyer mastectomy
Wilms
        W. tumor
        W. tumor, aniridia genitourinary
          abnormalities, mental retardation
          triad
        W. tumor, aniridia, gonadoblastoma,
          mental retardation (WAGR)
        W. tumor, aniridia gonadoblastoma,
          mental retardation syndrome
        W. tumor-aniridia syndrome
        W. tumor suppression gene
Wilson disease

Wilson-Mikity syndrome
Wilson-Turner syndrome (WTS)
Wimberger sign
Winckel disease
window
        aortopulmonary w.
        square w.
        therapeutic w.
wink
        anal w.
Winkler body
Winkler-Waldeyer
        closing ring of W.-W.
WinRho
        W. DS
        W. SD
        W. SDF
Winston cervical clamp
Winter
        W. placental forceps
        W. syndrome
Wintrobe index
wipe
        povidone-iodine w.
wire
        electrosurgical w.
        iridium w.
Wisap
        W. disposable cannula
        W. disposable trocar
Wisconsin syndrome
WISC-R
        Wechsler Intelligence Scale for Children-
        Revised
WISH
        Wistar Institute Susan Hayflick
        WISH cell
Wiskott-Aldrich syndrome
Wistar
        W. Institute Susan Hayflick
          (WISH)
        W. Institute Susan Hayflick cell
witch's milk
withdrawal
        w. bleeding
        w. bleeding test
        estrogen w.
withdrawal-like activity
Witness
        Jehovah's W.
Wittner biopsy punch
Wittwer syndrome
Wolcott-Rallison syndrome
Wolf
        W. disposable cannula
        W. disposable trocar
        W. laparoscope
        W. syndrome

Wolf-Castroviejo needle holder
Wolfe classification of breast cancer
wolffian
- w. duct
- w. duct carcinoma
- w. rest
- w. ridge

wolffian-remnant cyst
Wolff mental retardation syndrome
Wolff-Parkinson-White syndrome
Wolf-Hirschhorn syndrome
Wolfram syndrome
Wolf-Veress needle
Wolman disease
woman
- amenorrheic w.
- androgenized w.
- battered w.
- eumenorrheic w.
- euprolactinemic w.
- formula-feeding w.
- hirsute w.
- hypoestrogenic w.
- hypogonadal w.
- hypomenorrheic w.
- infertile w.
- lactating w.
- perimenopausal w.
- postmenopausal w.

womb
- falling of the w.
- scarred w.
- w. stone

women-held antenatal record
Wood
- W. lamp
- W. light

Woodcock-Johnson Test
Wood-Downes asthma score
Woodruff vestibulectomy
Woods
- W. screw maneuver
- W. syndrome

wooly-hair
- w.-h. disease
- w.-h. nevus

Word
- W. Bartholin gland catheter
- W. bladder catheter

work factor
World Health Organization (WHO)

wormian bones
Worster-Drought syndrome
wort
- mother w.

wound
- w. breakdown
- w. dehiscence
- w. infection
- stab w.
- suprapubic stab w.

WPPSI-R
- Wechsler Preschool and Primary Scale of Intelligence-Revised

WR
- Wiedemann-Rautenstrauch syndrome

Wramsby hypothesis
wrap
- CircPlus compression w./dressing
- SurePress w.

Wreden sign
Wright
- W. peak flow meter
- W. stain
- W. version

wrinkly skin syndrome (WSS)
wrist
- gymnast's w.

written consent
wrongful
- w. birth
- w. birth and life
- w. conception

wryneck
WSS
- Weaver-Smith syndrome
- wrinkly skin syndrome

W-stapled urinary reservoir procedure
WTS
- Wilson-Turner syndrome

Wullstein retractor
Wyamycin S
Wyanoids
Wyburn-Mason syndrome
Wycillin
Wydase
Wygesic
Wymox
Wytensin

W

NOTES

X

    X chromatin
    X chromosome
    X inactivation
    X inactivation center (XIC)
    X inactive, specific transcript (XIST)
    X 19 translocation
    X trisomy

45,X

    45,X karyotype
    45,X syndrome

(X)

    inv (X)

Xact
Xanax
xanthelasmas
xanthine
xanthinuria, xanthiuria
xanthochromia

    x. therapy

xanthochromic

    x. specimen

xanthogranuloma

    juvenile x. (JXG)

xanthoma striata palmaris
xanthomatosis
xanthosis cutis
xanthous
xanthurenic aciduria
X-autosome translocation
X-chromosome abnormality
Xe

    xenon

xenogamy
xenogeneic

    x. antibody
    x. tissue

xenograft
xenon (Xe)
Xenopus test
xeroderma pigmentosum
xerodermic idiocy
xerography
xeromammography
xeromenia
xerophthalmia
xerosis

    x. conjunctiva
    x. cornea

xerostomia
xerotocia
XIC

    X inactivation center

Ximed

    X. disposable cannula
    X. disposable trocar

xinafoate

    salmeterol x.

xiphoid

    bifid x.

xiphoomphaloischiopagus tripus conjoined twins
xiphopagus
XIST

    X inactive, specific transcript

XK-aprosencephaly syndrome
XK syndrome
XL

    Ditropan XL

XLAS

    X-linked aqueductal stenosis

XLCM

    X-linked cardiomyopathy
    X-linked dilated cardiomyopathy

X-linked

    X-l. agammaglobulinemia
    X-l. alpha-thalassemia mental retardation (ATRX)
    X-l. alpha-thalassemia/mental retardation syndrome (ATRX syndrome, ATRX syndrome)
    X-l. aqueductal stenosis (XLAS)
    X-l. cardiomyopathy (XLCM)
    X-l. cataract-dental syndrome
    X-l. cataract with hutchinsonian teeth
    X-l. centronuclear myopathy
    X-l. cerebral ataxia (CLA)
    X-l. cerebral hypoplasia/hydrocephalus
    X-l. congenital cataracts-microcornea syndrome
    X-l. congenital glycerol kinase deficiency
    X-l. congenital recessive muscle hypotrophy with central nuclei
    X-l. dilated cardiomyopathy (XLCM)
    X-l. disease
    X-l. disorder
    X-l. dominant condition
    X-l. dominant disease
    X-l. dominant inheritance
    X-l. dominant syndrome
    X-l. dysplasia-gigantism syndrome (DGSX)
    X-l. first site of fragility (FRAXA)
    X-l. gene

X

X-linked *(continued)*
- X-l. heredity
- X-l. Hurler syndrome
- X-l. hydrocephalus
- X-l. hydrocephalus-stenosis of aqueduct of Sylvius sequence
- X-l. hypogammaglobulinemia
- X-l. ichthyosis
- X-l. infantile spasm
- X-l. lymphoproliferative syndrome
- X-l. mental deficiency-megalotestes syndrome
- X-l. mental handicap-retinitis pigmentosa syndrome
- X-l. mental retardation (MRX1, XLMR)
- X-l. mental retardation-aphasia syndrome (MRXA)
- X-l. mental-retardation-bilateral clasp thumb anomaly
- X-l. mental retardation-blindness-deafness-multiple congenital anomalies syndrome
- X-l. mental retardation-fragile site 1 (FRAXE1)
- X-l. mental retardation-fragile site 2 (FRAXE2)
- X-l. mental retardation-fragile site syndrome 2
- X-l. mental retardation-growth hormone deficiency syndrome
- X-l. mental retardation-hypogenitalism-cerebral anomaly syndrome
- X-l. mental retardation-marfanoid habitus syndrome
- X-l. mental retardation, microphthalmia, microcornea, cataract, hypogenitalism-mental retardation-spasticity syndrome
- X-l. mental retardation 8 (MRX8, XLMR8)
- X-l. mental retardation/multiple congenital anomaly (XLMR/MCA)
- X-l. mental retardation-psoriasis syndrome
- X-l. mental retardation-seizures-acquired microcephaly-agenesis of corpus callosum
- X-l. mental retardation-spastic diplegia syndrome
- X-l. mental retardation syndrome 1–6 (MRXS1–6)
- X-l. mental retardation, thin habitus, osteoporosis, kyphoscoliosis syndrome
- X-l. mental retardation with fragile X syndrome
- X-l. monoamine oxidase deficiency
- X-l. myotubular myopathy (MTMX, XLMTM)
- X-l. olivopontocerebellar ataxia (OPCA)
- X-l. Opitz syndrome (XLOS)
- X-l. phenomenon
- X-l. primary hyperuricemia
- X-l. recessive centronuclear myopathy
- X-l. recessive condition
- X-l. recessive deafness syndrome
- X-l. recessive disease
- X-l. recessive dysgenesis
- X-l. recessive inheritance
- X-l. recessive muscular dystrophy
- X-l. recessive myotubular myopathy
- X-l. second site of fragility (FRAXE)
- X-l. seizures, acquired micrencephaly, agenesis of corpus callosum syndrome
- spondylometaphyseal dysplasia, X-l.
- X-l. trait
- X-l. uric aciduria enzyme defect

XLMR
- X-linked mental retardation

XLMR/MCA
- X-linked mental retardation/multiple congenital anomaly

XLMTM
- X-linked myotubular myopathy

XLOS
- X-linked Opitz syndrome

XO
- XO chromosome
- XO syndrome

XomaZyme-H65

X-Prep
- Senna X-P.

XP Xcelerator ultrasound enhancer

Xq
- Xq Klinefelter syndrome
- monosomy Xq
- partial trisomy Xq
- trisomy Xq

Xq+ syndrome

Xq- syndrome

x-ray
- x-r. absorptiometry
- chest x-r. (CXR)
- x-r. mammogram
- x-r. mammography
- x-r. pelvimetry
- x-r. therapy

X-tra
- AFP X-t.

XX
    XX chromosome
    XX hermaphroditism
    XX karyotype
    XX male syndrome
    XX and XY Turner phenotype
46,XX
    46,XX karyotype
    46,XX male
    46,XX male syndrome
47,XX karyotype
XX-type gonadal dysgenesis
XXX
    XXX karyotype
    XXX syndrome
47,XXX
    47,XXX syndrome
XXXX syndrome
XXXXX syndrome
XXXXY
    XXXXY aneuploidy
    XXXXY syndrome

49XXXXY syndrome
XXY
    XXY male
    XXY syndrome
47,XXY
    47,XXY karyotype
    47,XXY syndrome
69,XXY
XY
    XY gonadal dysgenesis
    XY karyotype
46,XY
    46,XY karyotype
46,XY/47,XY karyotype
47,XY karyotype
Xylocaine
xylometazoline
xylulose dehydrogenase deficiency
XYY
    XYY male
    XYY syndrome
X-zone

NOTES

X

Y

Y chromatin
Y chromosome
Y chromosome-specific DNA sequence

YAC

yeast artificial chromosome

YAG

yttrium-aluminum-garnet
YAG laser
YAG pellet

Yale Optimal Observation Score
Yankauer curette
yaws
y-cystathionase deficiency
year

y. of birth (YOB)
postnatal y.

yeast

y. artificial chromosome (YAC)
y. infection
pityrosporum y.
vaginal y.
y. vaginitis

Yellen clamp
yellow

y. fever
y. fever vaccine
y. vernix syndrome

Yeoman forceps
Yersinia

Y. enterocolitica
Y. pestis
Y. pseudotuberculosis

yes protooncogene
yew

English y.
Pacific y.

Y-incision
Y-linkage
Y-linked

Y.-l. character
Y.-l. gene

YOB

year of birth

Yocon
Yodoxin
yogurt douche
yohimbine
Yohimex
yolk

accessory y.
y. cell
formative y.
y. membrane
y. sac
y. sac carcinoma
y. sac tumor
y. sac tumor of testis
y. space
y. stalk

Yom Kippur effect
Yoon ring
York-Mason repair
young

maturity-onset diabetes of the y. (MODY)

Young-Dees-Leadbetter bladder neck reconstruction
Young-Hughes syndrome
Young-Madders syndrome
Young syndrome
Yp53.3
Y-plasty
Yq

AZFa region of Yq
AZFb region of Yq
AZFc region of Yq

YSI neonatal temperature probe
Y-specific

Y-s. DNA
Y-s. DNA amplification

yttrium-aluminum-garnet (YAG)

y.-a.-g. laser

Yunis-Varon syndrome
Yutopar
Yuzpe regimen
YY

YY syndrome

Y

Z

Z bone density score
Z chromosome
Z degree of contraction
Z sampler endometrial sampling
device
Zagam
zalcitabine (ddC)
Zancolli clawhand deformity repair
Zanosar
Zantac
Z. 75
Zarontin
Zaroxolyn
Zartan
Zatuchni-Andros score
Zaufal sign
Zavanelli maneuver
Z-Clamp hysterectomy forceps
ZD
zona drilling
Z′ degree of contraction
Z″ degree of contraction
ZDV
zidovudine
Zeasorb-AF
Z.-AF Powder
Zefazone
Zeiss colposcope
Zellweger
Z. cerebrohepatorenal syndrome
Z. syndrome
Zelsmyr Cytobrush
Zenate
Zephrex LA
Zeppelin clamp
Zerit
zero
z. end-expiratory pressure
z. gravity surgery
Zerres syndrome
Zetran
Ziagen
zidovudine (ZDV)
z. monotherapy
z. treatment
Ziehen-Oppenheim syndrome
ZIFT
zygote intrafallopian transfer
ZIG
zoster immune globulin
zigzagplasty
Zimmermann arch
Zimmermann-Laband syndrome (ZLS)
Zinacef Injection

Zinaderm
zinc
z.-dependent enzyme
z. oxide
z. oxide ointment
z. peroxide
z. poisoning
Z-incision
Zincofax
Zinnanti uterine manipulator-injector (ZUMI)
Zinsser-Engman-Cole syndrome
Zinsser syndrome
Zipokowski-Margolis syndrome
zipper ring
Zithromax
Zixoryn
Z-line
Z-linked gene
Zlotogora-Ogür syndrome
ZLS
Zimmermann-Laband syndrome
ZM-1 coloscope
zobactam sodium
Zofran
Zoladex
Z. Implant
Zolicef
Zollinger-Ellison syndrome
Zollino syndrome
Zoloft
zolpidem tartrate
zomepirac sodium
zona, pl. **zonae**
z. basalis
z. compacta
z. drilling (ZD)
z.-free hamster egg penetration test
z. functionalis
z. pellucida (ZP)
z. protein
z. reaction
z. spongiosa
zonary placenta
zone
Barnes z.
basement membrane z. (BMZ)
beta-hCG discriminatory z.
cervical transformation z.
interthreshold z.
ipsilon z.
large loop excision of
transformation z. (LLETZ)
loop excision of the
transformation z. (LETZ)

zone *(continued)*
    normal transformation z.
    null z.
    transformation z.
Zone-A Forte
zonoskeleton
zonula, pl. **zonulae**
    z. adherens
    z. occludens
zoogonous
zoogony
Zoomscope colposcope
Zoon erythroplasia
zoosperm
zoster
    herpes z.
    z. immune globulin (ZIG)
zosteriform
Zostrix
    Z.-HP
Zosyn
Zovia
Zovirax
ZP
    zona pellucida
Z-plasty
    four-flap Z.-p.
Z-Sampler endometrial suction curette
Z-Scissors hysterectomy scissors

Z-stitch
zuclopenthixol
ZUMI
    Zinnanti uterine manipulator-injector
Zunich syndrome
Zuska disease
Zuspan regimen
Zwahlen syndrome
Zwanck pessary
Zweifel-DeLee cranioclast
Zyclast
Zyderm
zygodactyly
zygomycosis
zygopodium
zygosity
zygote
    frozen z.
    z. intrafallopian transfer (ZIFT)
    two-cell z.
zygotene
    z. phase of meiosis
    z. stage
Zyklomat infusion pump
zymogen
    coagulation factor z.
Zynergy Zolution catheter
Zyrtec
ZZ male

Appendix 1
Anatomical Illustrations

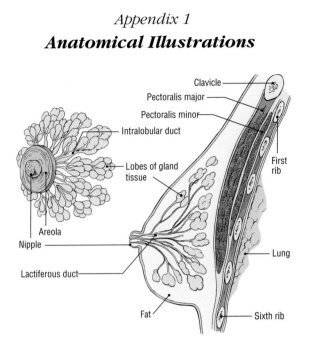

Figure 1. Breast. Glandular tissue and ducts of the mammary gland.

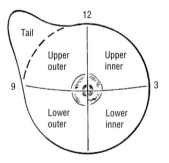

Figure 2. Breast. Schematic of breast as clock with nipple at center to assist reference.

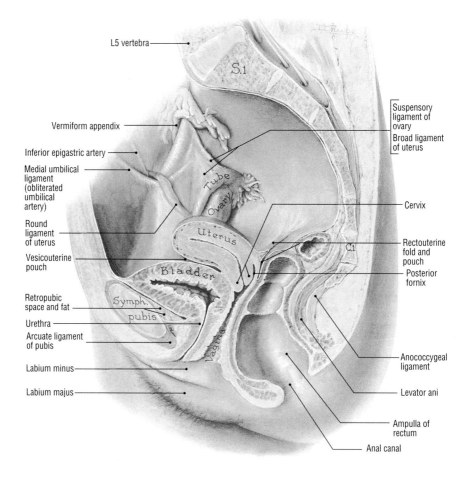

L5 vertebra

S.1

Suspensory ligament of ovary

Broad ligament of uterus

Vermiform appendix

Inferior epigastric artery

Medial umbilical ligament (obliterated umbilical artery)

Round ligament of uterus

Vesicouterine pouch

Retropubic space and fat

Urethra

Arcuate ligament of pubis

Labium minus

Labium majus

Tube

Ovary

Uterus

Bladder

Symph. pubis

Vagina

C.1

Cervix

Rectouterine fold and pouch

Posterior fornix

Anococcygeal ligament

Levator ani

Ampulla of rectum

Anal canal

Figure 3. Female pelvis, median section.

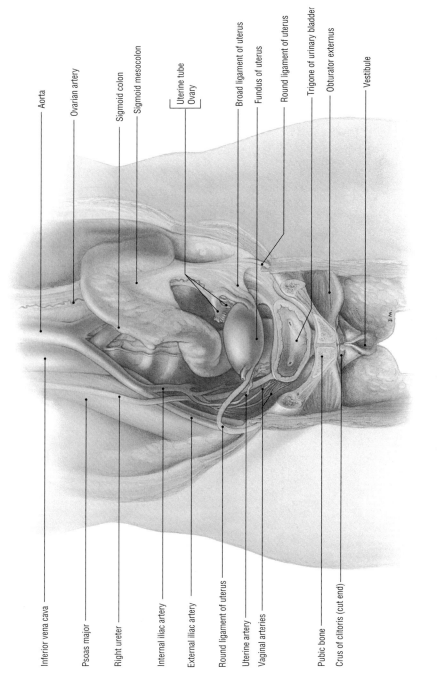

Aorta

Ovarian artery

Sigmoid colon

Sigmoid mesocolon

Uterine tube
Ovary

Broad ligament of uterus

Fundus of uterus

Round ligament of uterus

Trigone of urinary bladder

Obturator externus

Vestibule

Inferior vena cava

Psoas major

Right ureter

Internal iliac artery

External iliac artery

Round ligament of uterus

Uterine artery

Vaginal arteries

Pubic bone

Crus of clitoris (cut end)

Figure 4. Female genital organs, anteroposterior view.

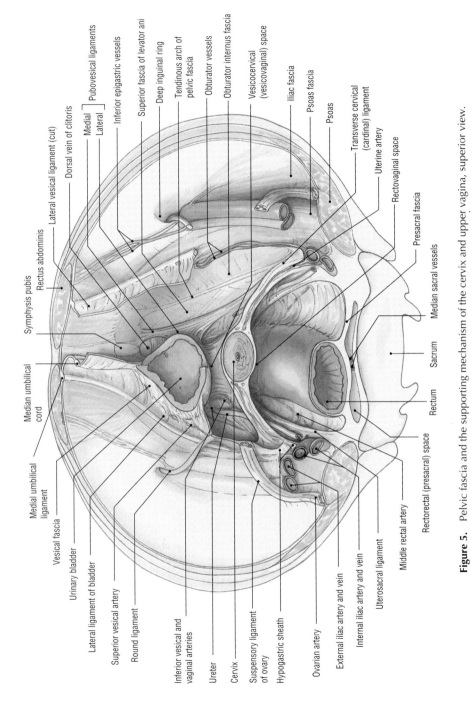

Figure 5. Pelvic fascia and the supporting mechanism of the cervix and upper vagina, superior view.

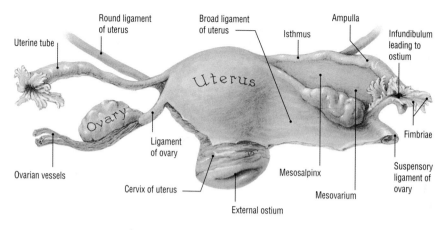

Figure 6. Uterus and adnexa, posterior view.

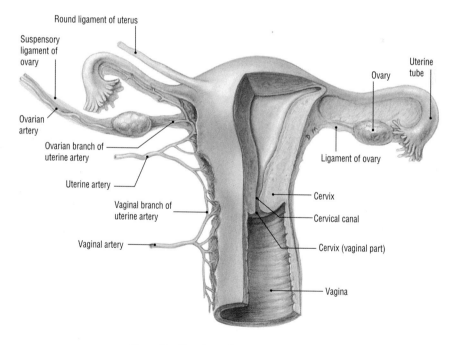

Figure 7. Blood supply to uterus and adnexa.

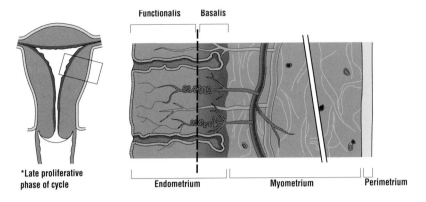

Functionalis | Basalis

*Late proliferative phase of cycle

Endometrium | Myometrium | Perimetrium

Figure 8. Layers of uterus.

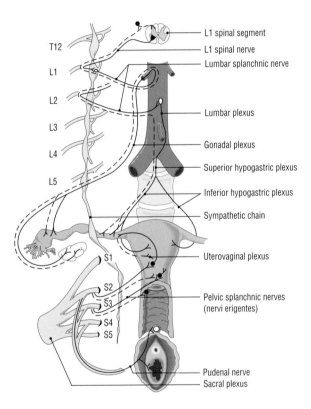

T12

L1 spinal segment
L1 spinal nerve
Lumbar splanchnic nerve

L1
L2
L3
L4
L5

Lumbar plexus

Gonadal plexus

Superior hypogastric plexus

Inferior hypogastric plexus

Sympathetic chain

S1

Uterovaginal plexus

S2
S3
S4
S5

Pelvic splanchnic nerves
(nervi erigentes)

Pudenal nerve
Sacral plexus

Figure 9. Innervation of the female reproductive tract and genitalia. The sympathetic pathways arise from the lower thoracic and upper lumbar spinal levels (*black triangle*). There are no white rami below L2. These reach the aortic plexus via thoracic and lumbar splanchnic nerves. Synapse occurs in the aortic plexus (*white circles*). The postsynaptic neurons reach the pelvic viscera via hypogastric plexuses. The parasympathetic pathways arise from the midsacral spinal levels and reach the pelvic viscera via the splanchnic nerves. Synapse occurs in the walls of the viscera (*white circles*). Visceral afferent fibers (*dashed*) from the pelvic viscera travel specifically along either one or the other autonomic pathways, have their cells bodies in the dorsal root ganglia, and produce specific patterns of referred pain. The pudenal nerve provides somatic innervation to and from the perineum.

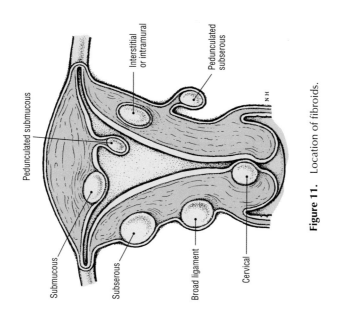

Figure 11. Location of fibroids.

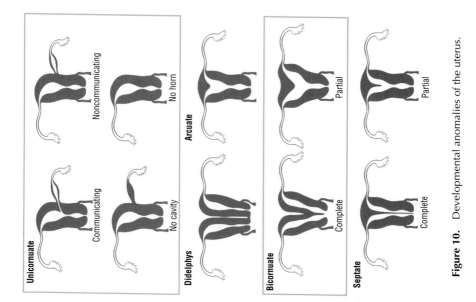

Figure 10. Developmental anomalies of the uterus.

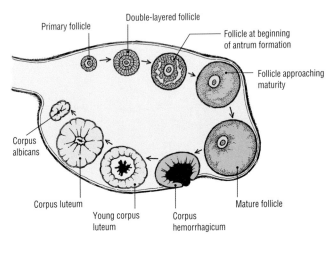

Figure 12. Ovulation.

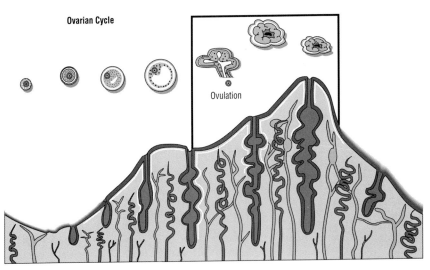

Figure 13. Menstrual cycle.

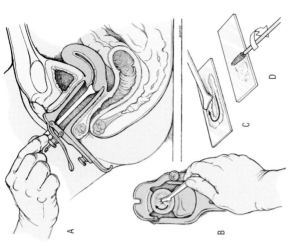

A

B

C

D

Figure 16. Pap smear. (A) Speculum in place and Ayre spatula in position at cervical os, (B) tip of spatula placed in the cervical os and rotated 360 degrees, (C) cellular material clinging to spatula is then smeared smoothly on glass slide, which is promptly placed in fixative solution, (D) cytobrush is rotated in cervical os and rolled onto glass slide.

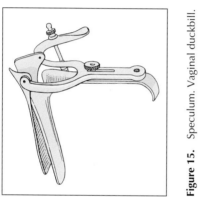

Figure 15. Speculum. Vaginal duckbill.

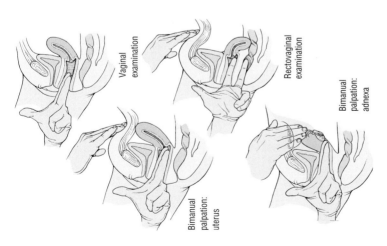

Vaginal examination

Rectovaginal examination

Bimanual palpation: uterus

Bimanual palpation: adnexa

Figure 14. Pelvic examination.

A9

Appendix 1

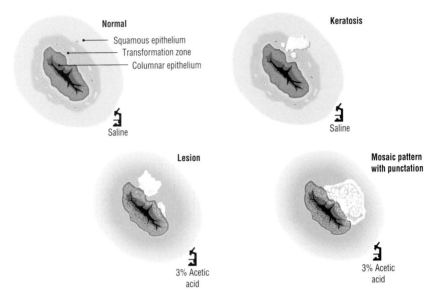

Figure 17. Colposcopy findings.

Figure 18. Cervical lesions.

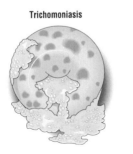

Figure 19. Vaginitis.

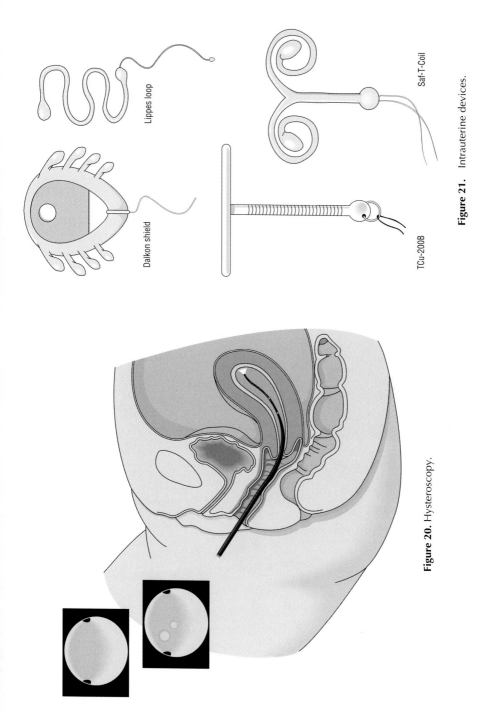

Lippes loop

Saf-T-Coil

Dalkon shield

TCu-200B

Figure 21. Intrauterine devices.

Figure 20. Hysteroscopy.

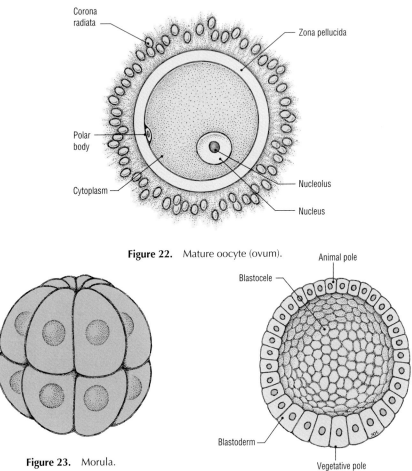

Figure 22. Mature oocyte (ovum).

Figure 23. Morula.

Figure 24. Blastula, hemisected.

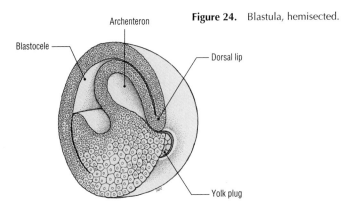

Figure 25. Gastrula.

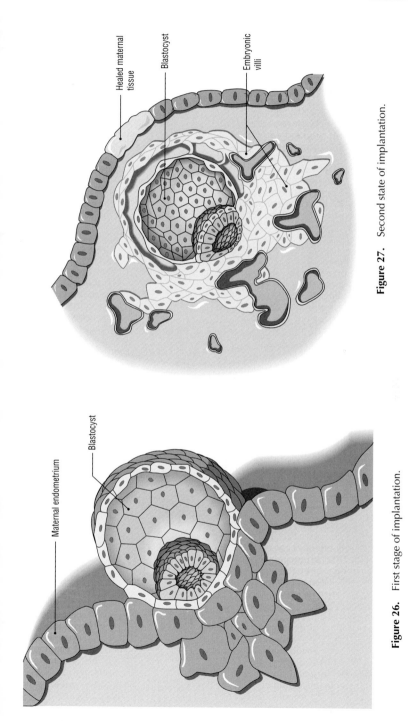

Figure 27. Second state of implantation.

Figure 26. First stage of implantation.

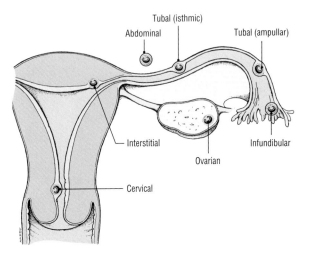

Figure 28. Sites of ectopic pregnancy.

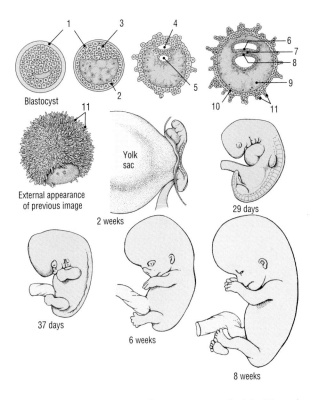

Figure 29. Development; from blastocyst to fetus. (1) Zona pellucida, (2) trophectoderm, (3) inner cell mass, (4) amniotic cavity, (5) yolk sac, (6) ectoderm, (7) mesoderm, (8) entoderm, (9) mesoderm, (10) trophectoderm, (11) chorionic villi.

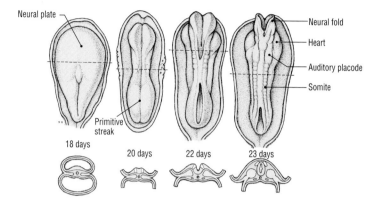

Neural plate

Primitive streak

18 days

20 days

22 days

23 days

Neural fold

Heart

Auditory placode

Somite

Figure 30. Blastoderm. Top, dorsal views. Bottom, cross-sectional views.

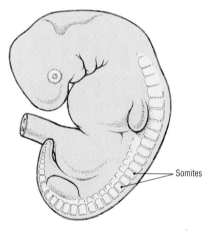

Somites

Figure 31. Somites in a 29-day human embryo.

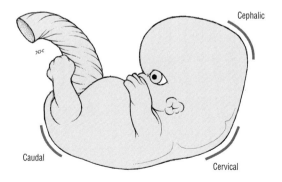

Cephalic

Caudal

Cervical

Figure 32. Flexures seen in a 6-week old embryo.

A15

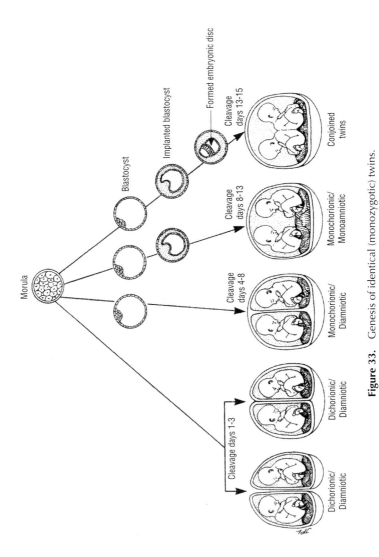

Figure 33. Genesis of identical (monozygotic) twins.

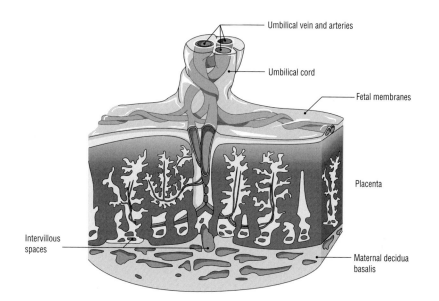

Figure 34. Placental circulation.

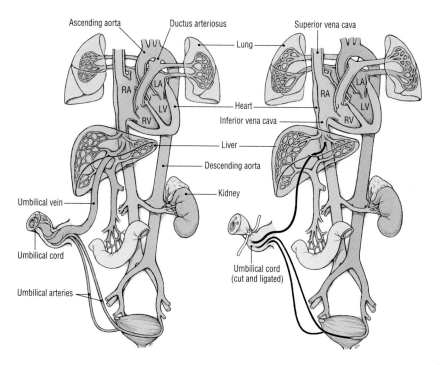

Figure 35. Fetal circulation. *Left,* during pregnancy, oxygen diffuses from the maternal circulation to the fetal circulation in the placenta; oxygenated blood returns to the fetus through the umbilical vein. *Right,* after birth, umbilical cord is cut and blood is oxygenated as it passes through the lungs. (*RA*) Right atrium, (*LA*) left atrium, (*LV*) left ventricle, (*RV*) right ventricle.

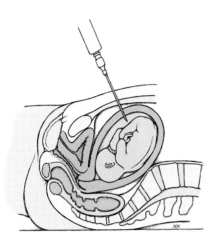

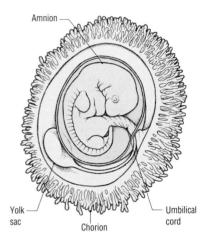

Figure 36. Amniocentesis.

Figure 37. Amnion and related structures showing 5-week embryo.

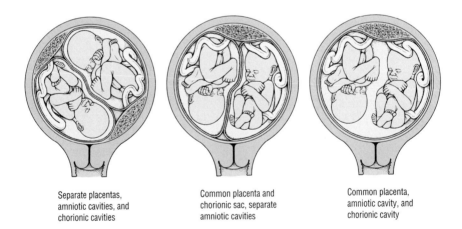

Figure 38. Twins. Schematic diagrams showing the possible relations of the fetal membranes in monozygotic twins.

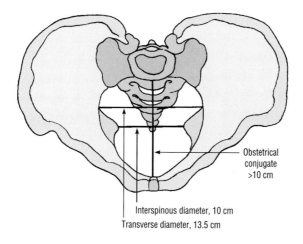

Figure 39. Pelvic diameters. Superior view of female pelvis, indicating normal distances between structures.

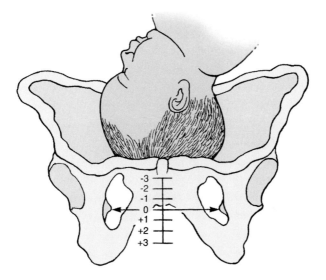

Figure 40. Presentation, vertex, station estimation. Estimation of station by traditional three-station system. Station is estimated by palpation of the bony segment of the presenting part during a vaginal examination and determining the distance from the plane of the ischial spines.

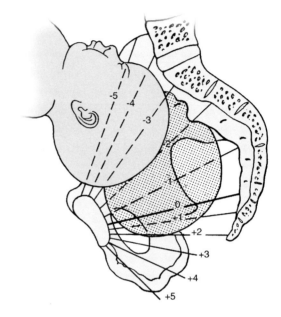

Figure 41. Presentation, vertex, station estimation. Illustration showing the estimation of station by current ACOG centimeter system.

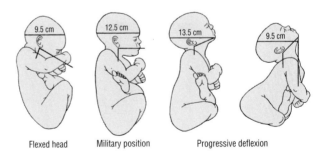

Figure 42. Presentation, breech, cranial diameters. Importance of cranial flexion is emphasized by noting the increased diameters presented to the birth canal with progressive deflexion.

ROA–Right occiput anterior

ROP–Right occiput posterior

ROT–Right occiput transverse

LOA–Left occiput anterior

LOP–Left occiput posterior

LOT–Left occiput transverse

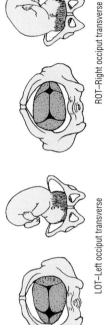

Figure 44. Vertex presentation. Fetal head positions within the pelvic girdle in a vertex presentation.

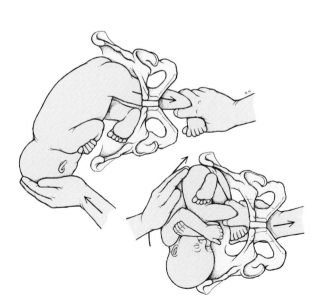

Figure 43. Internal podalic version. Conversion from dorsoposterior transverse lie to breech. *Left,* obstetrician's right hand grasps fetal foot within uterus while left hand applies pressure externally to rotate breech toward pelvic inlet. *Right,* obstetrician maneuvers fetus into longitudinal orientation by applying traction to foot while externally directing head into fundus, so that delivery can proceed as in breech presentation.

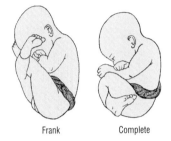

Frank   Complete

Figure 45. Full breech presentations refer to the relationships at the hip and knee joints. Frank breech: both hip joints are flexed, and both knee joints are extended. Complete breech: both hip joints and knee joints are flexed.

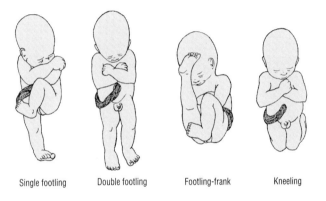

Single footling   Double footling   Footling-frank   Kneeling

Figure 46. Varieties of incomplete breech presentations refer to incomplete flexion at either the hip or knee joints.

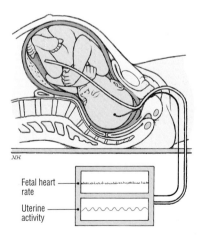

Fetal heart
rate

Uterine
activity

Figure 47. Electronic fetal monitoring.

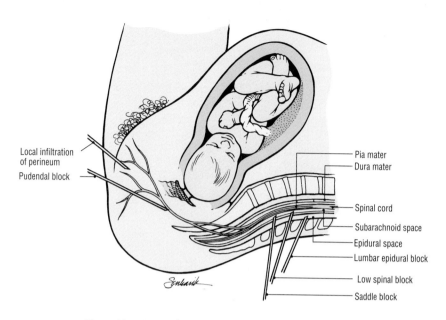

Figure 48. Regional anesthesia for childbirth. Sites of injection.

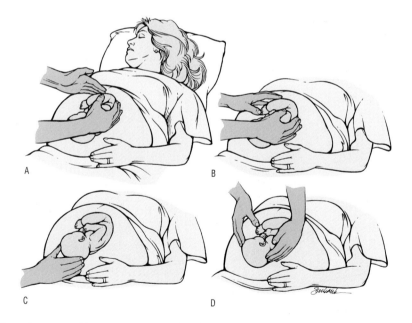

Figure 49. Leopold maneuvers. (A) First maneuver, palpate superior surface of fundus, (B) second maneuver, palpate sides of uterus to determine which direction fetal back is facing, (C) third maneuver, palpate to discover what is at inlet of pelvis, (D) fourth maneuver, assuming fetus has been found to be in a cephalic presentation, fetal attitude should then be determined (degree of flexion).

A23

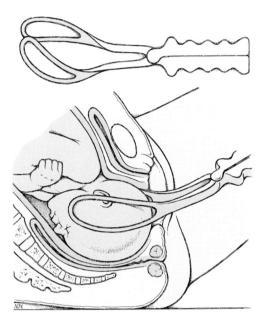

Figure 50. Obstetrical forceps.

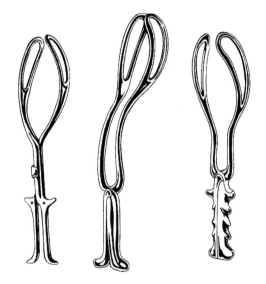

Figure 51. Obstetrical forceps. *Left,* Kjelland. *Middle,* Piper. *Right,* Simpson.

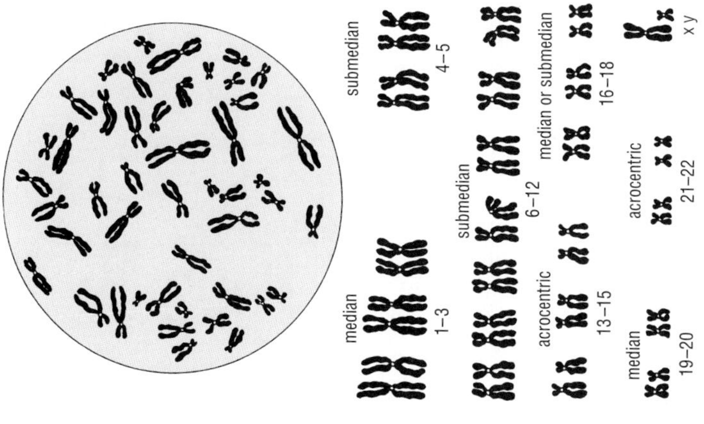

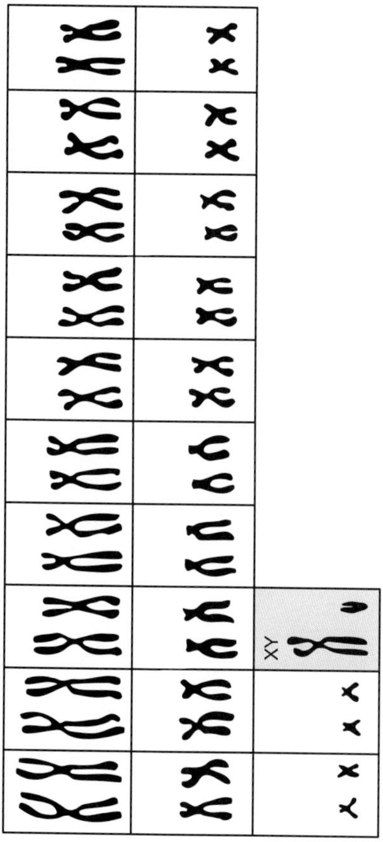

Figure 53. Normal male karyotype.

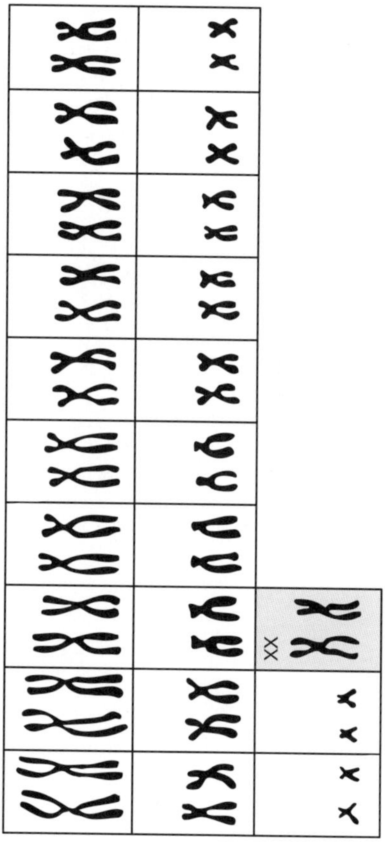

Figure 54. Normal female karyotype.

Figure 52. Karyotype of a normal human cell.

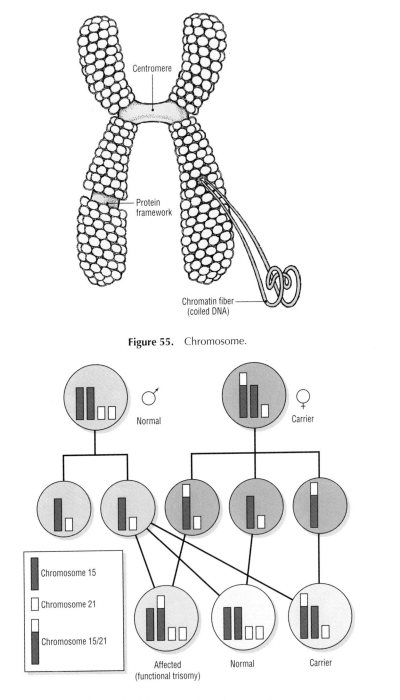

Figure 55. Chromosome.

Figure 56. Inheritance of translocation trisomy.

Appendix 2
Genetic Symbols

□   male

○   female

◇   sex unspecified

□○   normal individuals

■●◆   affected individual (with ≥ 2 conditions, the symbol is partitioned and shaded with different fill defined in a key or legend)

⑤⑤⑤   multiple individuals, number known (number of siblings written inside symbol)

ⓝⓝⓝ   multiple individuals, number unknown ("n" used in place of specific number

□○   mating

□○   consanguinity

(+)   uncommon or uncertain mode of inheritance

□○ / □○   parents and offspring, in generations

□○   dizygotic twins

□○   monozygotic twins

④③   number of children of sex indicated

□○   adopted individuals

□○   individual died without leaving offspring

□○   no issue

■●   affected individuals

■● proband or propositus (first affected family member coming to medical attention)

⊞   examined professionally · normal for trait

⊡   not examined · dubiously reported to have trait

⊡   not examined · reliably reported to have trait

■●   heterozygotes for autosomal recessive

⊙   carrier of sex-linked recessive

□⊘   death

⊘⊘⊘ SB 28wk SB 30wk SB 34wk   stillbirth (SB)

P Ⓟ Ⓟ LMP 7/1/94 20wk   pregnancy (P); gestational age and karotype (if known) below symbol

□○   consultand (individual seeking genetic counseling/testing)

△△△ male female ECT   spontaneous abortion (SAB); ECT below symbol indicates ectopic pregnancy

▲▲▲ male female 16wk   affected SAB (gestational age, if known, below symbol, and key or legend used to define shading)

△△△ male female   termination of pregnancy (TOP)

▲▲▲ male female   affected TOP (key or legend used to define shading)

Source: Genetic symbols are public domain. We credit and gratefully acknowledge the *American Journal of Human Genetics* (56:746–747, 1995) as our source for these symbols.

A27

Normal Lab Values

Tests	Conventional Units	SI Units
alpha1-antitrypsin, serum	78–200 mg/dL	0.78–2.00 g/L
alpha-fetoprotein (AFP), serum	<15 ng/mL	<15 μg/L
bilirubin, serum, adult		
conjugated	0.0–0.3 mg/dL	0–5 μmol/L
unconjugated	0.1–1.1 mg/dL	1.7–1.9 μmol/L
delta	0–0.2 mg/dL	0–3 μmol/L
total	0.2–1.3 mg/L	3–22 μmol/L
CA 125, serum	<35 U/mL	<35 kU/L
CA 19–9, serum	<37 U/mL	<37 kU/L
CEA, serum, nonsmokers	<5.0 ng/mL	<5.0 μg/L
cholinesterase, serum	4.9–11.9 U/mL	4.9–11.9 kU/L
chorionic gonadotropin,		
intact serum or plasma		
male and nonpregnant female	<5.0 mIU/mL	<5.0 IU/L
pregnant female	varies with gestational age	
urine, qualitative		
male and nonpregnant female	negative	negative
pregnant female	positive	positive
coagulation tests		
antithrombin III (synthetic substrate)	80–120% of normal	0.8–1.2 of normal
bleeding time (Duke)	0–6 min	0–6 min
bleeding time (Ivy)	1–6 min	1–6 min
bleeding time (template)	2.3–9.5 min	2.3–9.5 min
clot retraction, qualitative	50–100% in 2 h	0.5–1.0/2 h
complement components		
total hemolytic complement activity, plasma (EDTA)	75–160 U/mL	75–160 kU/L
total complement decay rate (functional), plasma (EDTA)	10–20%	fraction decay rate: 0.10–0.20
	deficiency: >50%	>0.50
C1q, serum	14.9–22.1 mg/dL	149–221 mg/L
C1r, serum	2.5–10.0 mg/dL	25–100 mg/L
C1s (C1 esterase), serum	5.0–10.0 mg/dL	50–100 mg/L
C2, serum	1.6–3.6 mg/dL	16–36 mg/L

continued

Tests	Conventional Units	SI Units
C3, serum	90–180 mg/dL	0.9–1.8 g/L
C4, serum	10–40 mg/dL	0.1–0.4 mg/L
C5, serum	5.5–11.3 mg/dL	55–113 mg/L
C6, serum	17.9–23.9 mg/dL	179–239 mg/L
C7, serum	2.7–7.4 mg/dL	27–74 mg/L
C8, serum	4.9–10.6 mg/dL	49–106 mg/L
C9, serum	3.3–9.5 mg/dL	33–95 mg/L
Coombs test		
direct	negative	negative
indirect	negative	negative
creatine kinase (CK), serum		
male	15–105 U/L (30°C)	0.26–1.79 μkat/L (30°C)
female	10–80 U/L (30°C)	0.17–1.36 μkat/L (30°C)
cryoglobulins, serum	0	0
C-reactive protein, serum	<0.5 mg/dL	<5 mg/L
dehydroepiandrosterone (DHEA), serum		
male	180–1250 ng/dL	6.2–43.3 nmol/L
female	130–980 ng/dL	4.5–34.0 nmol/L
dehydroepiandrosterone sulfate (DHEAS), serum or plasma		
male	59–452 μg/mL	1.6–12.2 μmol/L
female		
premenopausal	12–379 μg/mL	0.8–10.2 μmol/L
postmenopausal	30–260 μg/mL	0.8–7.1 μmol/L
delta aminolevulinic acid, urine	1.3–7.0 mg/24 h	10–53 μmol/24 h
estradiol, serum		
adult male	10–50 pg/mL	37–184 pmol/L
adult female	varies with menstrual cycle	
ferritin, serum		
male	20–250 ng/mL	20–250 μg/L
female	10–120 ng/mL	10–120 μg/L
follicle-stimulating hormone (FSH), serum and plasma		
male	1.4–15.4 mIU/mL	1.4–15.4 IU/L
female		
follicular phase	1–10 mIU/mL	1–10 IU/L
mid-cycle	6–17 mIU/mL	6–17 IU/L
luteal phase	1–9 mIU/mL	1–9 IU/L
postmenopausal	19–100 mIU/mL	19–100 IU/L

continued

Appendix 3

Tests	Conventional Units	SI Units
glucose-6-phosphate dehydrogenase (G6PD) in erythrocytes, whole blood	12.1 ± 2.1 U/g Hb (SD) 351 ± 60.6 U/10^{12} RBC 4.11 ± 0.71 U/mL RBC	0.78 ± 0.13 mU/mol Hb 0.35 ± 0.06 nU/RBC 4.11 ± 0.71 kU/L RBC
haptoglobin, serum	30–200 mg/dL	0.3–2.0 g/L

hematocrit
  males
  females
	42–52%
	37–47%
		0.42–0.52
		0.37–0.47

hemoglobin (Hb)
  males
  females
	14.0–18.0 g/dL
	12.0–16.0 g/dL
		2.17–2.79 mmol/L
		1.86–2.48 mmol/L

hemoglobin, fetal
  0–6 mo
  6 mo to adult
	<75% of total Hb
	<2% of total Hb
		<0.75% of total Hb
		<0.02% of total Hb

immunoglobulins, serum
  IgG
  IgA
  IgM
  IgD
  IgE
	700–1600 mg/dL
	70–400 mg/dL
	40–230 mg/dL
	0–8 mg/dL
	3–423 IU/mL
		7–16 g/L
		0.7–4.0 g/L
		0.4–2.3g/L
		0–80 mg/L
		3–423 kIU/L

immunoglobulin G (IgG), CSF	0.5–6.1 mg/dL	0.5–6.1 g/L

iron, serum
  males
  females
	65–175 μg/dL
	50–170 μg/dL
		11.6–31.3 μmol/L
		9.0–30.4 μmol/L

iron binding capacity, serum, total (TIBC)	250–425 μg/dL	44.8–71.6 μmol/L
lecithin-sphingomyelin (L/S) ratio, amniotic fluid	2.0–5.0 indicates probable fetal lung maturity; >3.5 in diabetics	same

luteinizing hormone (LH), serum or plasma
  male
  female
    follicular phase
    mid-cycle peak
    luteal phase
    postmenopausal
	1.24–7.8 mIU/mL
	
	1.68–15.0 mIU/mL
	21.9–56.6 mIU/mL
	0.61–16.3 mIU/mL
	14.2–52.5 mIU/mL
		1.24–7.8 IU/L
		
		1.68–15.0 IU/L
		21.9–56.6 IU/L
		0.61–16.3 IU/L
		14.2–52.5 IU/L

mean corpuscular hemoglobin (MCH)	27–31 pg	0.42–0.48 fmol

continued

Tests	Conventional Units	SI Units
mean corpuscular hemoglobin concentration (MCHC)	33–37 g/dL	330–370 g/L
mean corpuscular volume (MCV)		
male	80–94 μ^3	80–94 fL
female	81–99 μ^3	81–99 fL
metanephrines, total, urine	0.1–1.6 mg/24 h	0.5–8.1 μmol/24 h
5′-nucleotidase, serum	2–17 U/L	0.034–0.29 μkat/L
partial thromboplastin time, activated (APTT)	<35 sec	<35 sec
phosphatidylglycerol (PG), amniotic fluid		
fetal lung immaturity	absent	absent
fetal lung maturity	present	present
porphobilinogen, urine		
qualitative	negative	negative
quantitative	<2.0 mg/24 h	<9 μmol/24 h
porphyrins, urine		
coproporphyrin	34–230 μg/24 h	52–351 nmol/24 h
uroporphyrin	27–52 μg/24 h	32–63 nmol/24 h
progesterone, serum		
adult male	13–97 ng/dL	0.4–31 nmol/L
adult female		
follicular phase	15–70 ng/dL	0.5–2.2 nmol/L
luteal phase	200–2500 ng/dL	6.4–79.5 nmol/L
pregnancy	varies with gestational week	
protoporphyrin, total, WB	<60 μg/dL	<600 μg/L
sedimentation rate		
Wintrobe		
males	0–10 mm in 1 h	0–10 mm/h
females	0–20 mm in 1 h	0–20 mm/h
Westergren		
males (< 50 yr)	0–15 mm in 1 h	0–15 mm/h
females (< 50 yr)	0–20 mm in 1 h	0–20 mm/h
transferrin, serum		
adult	212–360 mg/dL	2.12–3.60 g/L
>60 y	190–375 mg/dL	1.9–3.75 g/L
urobilinogen, urine	0.1–0.8 Ehrlich unit/2 h	0.1–0.8 Ehrlich unit/2 h
	0.5–4.0 mg/24 h	0.5–4.0 mg/24 h
vanillylmandelic acid (VMA), urine (4-hydroxy-3-methoxymandelic acid)	1.4–6.5 mg/24 h	7–33 μmol/d
viscosity, serum	1.00–1.24 cP	1.00–1.24 cP
vitamin B_{12}, serum	110–800 pg/mL	81–590 pmol/L

Appendix 4

Amino Acids

Essential Amino Acids

histidine (essential to infants only)
isoleucine
leucine
lysine
methionine
phenylalanine
threonine
tryptophan
valine

Nonessential Amino Acids

alanine
arginine
asparagine
aspartic acid
cysteine
glutamine
glutamic acid
glycine
proline
serine
tyrosine

Storage Diseases

Glycogenoses

Disease	Eponym	Enzyme Deficiency	Stored or Accumulated Materials
hepatic glycogen storage disease	Pompe disease	alpha-1,4-glucosidase, lysosomal glucosidase, and acid maltase	hepatic, cardiac, skeletal muscle glycogen
myopathic glycogen storage disease	von Gierke disease	glucose-6-phosphatase	hepatorenal, intranuclear glycogen
generalized glycogen storage disease	McArdle disease	muscle phosphorylase	skeletal muscle accumulation of glycogen
glycogen storage disease*	Smith disease	acid maltase	glycogen
glycogen storage disease*	Engel disease (adult)	acid maltase	glycogen
glycogen storage disease*	Forbes disease	debrancher- or amylo-1,6-glucosidase	glycogen
glycogen storage disease*	Anderson disease	debrancher- or amylo-transglucosidase	glycogen
glycogen storage disease*	McArdle disease	myophosphorylase	glycogen
glycogen storage disease*	Hers disease	hepatophosphorylase	glycogen
glycogen storage disease*	Tarui disease	phosphofructokinase	glycogen
glycogen storage disease*	Hug disease	hepatic phosphorylase kinase	glycogen
glycogen storage disease*	Satoyoshiy disease	phosphohexose isomerase	glycogen
glycogen storage disease*	Thomson disease	phosphoglucomutase	glycogen
glycogen storage disease*		glycogen synthase	glycogen
glycogen storage disease*	Bresolin disease	phosphoglycerate kinase	glycogen
glycogen storage disease*	Tonin disease	phosphoglycerate mutase	glycogen

continued

Different classification schemes may assign different "type" numbers and eponyms.

Glycogenoses

Disease	Eponym	Enzyme Deficiency	Stored or Accumulated Materials
glycogen storage disease*	Tsujimo disease	lactate dehydrogenase	glycogen

Sphingolipidoses

Disease	Eponym	Enzyme Deficiency	Stored or Accumulated Materials
GM_1 gangliosidosis		beta-galactosidase	GM_1 ganglioside, galactose-containing oligosaccharides
GM_1 gangliosidosis type I infantile, generalized	Caffey pseudo-Hurler syndrome Caffey syndrome Hurler-like syndrome Hurler variant Landing syndrome Norman-Landing syndrome	beta-galactosidase	abnormal accumulation of GM_1 ganglioside in neurons; in hepatic, splenic and other histiocytes; in renal glomerular epithelium
GM_1 gangliosidosis type II juvenile gangliosidosis GM_1 juvenile type gangliosidosis GM_1 late onset without bony involvement generalized gangliosidosis juvenile type generalized juvenile gangliosidosis juvenile GM_1 gangliosidosis late infantile systemic lipidosis	Derry syndrome	B and C isoenzymes of beta-galactosidase	ganglioside GM_1 in the brain; large amounts of keratosulfate-like mucopolysaccharide in the visceral organs
GM_2 gangliosidosis		hexosaminidase	GM_2 ganglioside
GM_2 gangliosidosis type I	Tay-Sachs disease, infantile form	hexosaminidase-alpha subunit	GM_2 ganglioside

continued

*Different classification schemes may assign different "type" numbers and eponyms.

Sphingolipidoses

Disease	Eponym	Enzyme Deficiency	Stored or Accumulated Materials
GM_2 gangliosidosis type II	Tay-Sachs disease, juvenile form	hexosaminidase-alpha subunit	GM_2 ganglioside
GM_2 gangliosidosis type III	Tay-Sachs disease, adult form	hexosaminidase-alpha subunit	GM_2 ganglioside
GM_2 gangliosidosis: activator factor abnormality		hexosaminidase A deficiency due to activator factor dysfunction	GM_2 ganglioside
GM_2 gangliosidosis	Sandhoff disease	hexosaminidase-beta subunit hexosaminidase A and B	GM_2 ganglioside and globoside
GM_2 gangliosidosis variant AB hexosaminidase activator deficiency AB variant		ganglioside activator protein	GM_2 ganglioside
GM_3 gangliosiosis hematoside sphingolipodystrophy		UPD-Gal-NAC: Gm(3)-N-acetyl-galactosaminyl-transferase	GM_3 hematoside

Sulfatidoses

Disease	Eponym	Enzyme Deficiency	Stored or Accumulated Materials
metachromatic leukodystrophy		arylsulfatase A	sulfatide
multiple sulfatase deficiency mucosulfatidosis juvenile sulfatidosis	Austin sulfatidosis	arylsulfatases A, B, C; steroid sulfatase; iduronate sulfatase; heparan N-sulfatase	sulfatide, steroid sulfate, heparan sulfate, dermatan sulfate

Lipidoses

Disease	Eponym	Enzyme Deficiency	Stored or Accumulated Materials
globoid cell leukodystrophy	Krabbe disease	galactocerebrosidase galactosylceramidase	galactocerebroside galactosylceramide

continued

Lipidoses

Disease	Eponym	Enzyme Deficiency	Stored or Accumulated Materials
angiokeratoma hereditary dystopic lipidosis	Fabry disease; Anderson-Fabry disease	alpha-galactosidase A ceramide trihexosidase	globotriaosylcera-mide
ceramide deficiency	Farber disease	ceramidase	ceramide
cholesterol ester storage disease	Wolman disease	acid lipase	cholesterol esters and triglyceride

Cerebrosidoses

Disease	Eponym	Enzyme Deficiency	Stored or Accumulated Materials
cerebroside multi-systemic lipidosis type 1 adult nonneuro-nopathic type 2 infantile neuro-nopathic type 3 juvenile or Norrbotten form	Gaucher disease	glucose cerebrosi-dase	glucose cerebroside

Sphingomyelinoses

Disease	Eponym	Enzyme Deficiency	Stored or Accumulated Materials
sphingomyelinosis type A classic type B nonneurological variant type C childhood or adult	Niemann-Pick disease, type A Niemann-Pick disease, type B Niemann-Pick disease, type C	sphingomyelinase	sphingomyelin in CNS in type A; systemically in types B and C.
type D Nova Scotian variant	Niemann-Pick disease, type D, Nova Scotia variant	no defect identified yet	sphingomyelin, Niemann-Pick cells in bone marrow

Mucopolysaccharidoses (MPS)

Disease	Eponym	Enzyme Deficiency	Stored or Accumulated Materials
MPS I-H	Hurler disease or syndrome	alpha-L-iduronidase	dermatan sulfate, heparan sulfate

continued

Mucopolysaccharidoses (MPS)

Disease	Eponym	Enzyme Deficiency	Stored or Accumulated Materials
MPS I-S	Scheie disease or syndrome	alpha-L-iduronidase	dermatan sulfate, heparan sulfate
MPS I H/S	Hurler-Scheie disease or syndrome	alpha-L-iduronidase	dermatan sulfate, heparan sulfate
MPS II	Hunter disease or syndrome	iduronate-2-sulfatase	heparan sulfate, dermatan sulfate
MPS III A	Sanfilippo A disease or syndrome	heparan N-sulfatase (sulfamidase)	heparan sulfate
MPS III B	Sanfilippo B disease or syndrome	alpha-N-acetyl-glucosaminidase	heparan sulfate
MPS III C	Sanfilippo C disease or syndrome	N-acetyl-CoA: alpha glucosaminidase transferase	heparan sulfate
MPS III D	Sanfilippo D disease or syndrome	N-acetylglucosamine-6-sulfatase	heparan sulfate
MPS IV A	Morquio A disease or syndrome	galactose-6-sulfatase	keratan sulfate, chondroitin
MPS IV B	Morquio B disease or syndrome	beta-galactosidase	keratan sulfate
MPS VI	Maroteaux-Lamy disease or syndrome	arylsulfatase B	dermatan sulfate
MPS VII	Sly disease or syndrome	beta-glucoronidase	dermatan sulfate, heparan sulfate
MPS IX		hyaluronidase	hyaluronan
multiple sulfatase A, B, C		not known	dermatan sulfate, heparan sulfate, chondroitin sulfate

Mucolipidoses (ML)

Disease	Eponym	Enzyme Deficiency	Stored or Accumulated Materials
ML I sialidosis		neuraminidase	oligosaccharide, glycolipid

continued

A37

Mucolipidoses (ML)

Disease	Eponym	Enzyme Deficiency	Stored or Accumulated Materials
ML I type 2 lipomucosaccharidosis		beta-galactosidase and sialidase	oligosaccharide, glycolipid
I-cell disease ML II		mannose-6-phosphate transferase	mucopolysaccharide, glycolipid
ML III	pseudo-Hurler polydystrophy	N-acetylglucosamine-1-phosphotransferase	oligosaccharide, glycolipid
ML type IV sialipidosis	Berman mucolipidosis	not yet characterized	ganglioside, mucopolysaccharide (hyaluronic acid)

Other Diseases of Complex Carbohydrates

Disease	Eponym	Enzyme Deficiency	Stored or Accumulated Materials
fucosidosis		alpha-fucosidase	fucose-containing sphingolipids and glycoprotein fragments
mannosidosis		alpha-mannosidase	mannose-containing oligosaccharides
aspartylglycosaminuria		aspartylglycosamine amide hydrolase	spartyl-2-deoxy-2-acetamido-glycosylamine

Other Lysosomal Storage Diseases

Disease	Eponym	Enzyme Deficiency	Stored or Accumulated Materials
acid phosphate deficiency		lysosomal acid phosphatase	phosphate esters

Neuronal Ceroid Lipofuscinoses

Disease	Eponym	Enzyme Deficiency	Stored or Accumulated Materials
NCL1 juvenile chronic form	Batten disease	not known	lipopigments of ceroid, lipofuscin, and excessive

continued

Neuronal Ceroid Lipofuscinoses

Disease	Eponym	Enzyme Deficiency	Stored or Accumulated Materials
			amounts of protein subunit c.
NCL2 acute late infantile	Bielschowsky disease	not known	lipopigments of ceroid, lipofuscin, and excessive amounts of protein subunit c.
NCL3 acute infantile	Santavuori disease	not known	lipopigments of ceroid, lipofuscin and excessive amounts of protein subunit c.
NCL4 adult recessive	Kuf disease	not known	lipopigments of ceroid, lipofuscin, and excessive amounts of protein subunit c.
NCL5 adult dominant	Boehme disease	not known	lipopigments of ceroid, lipofuscin, and excessive amounts of protein subunit c.
NCL6 early juvenile	Lake disease	not known	lipopigments of ceroid, lipofuscin, and excessive amounts of protein subunit c.
NCL7 atypical			lipopigments of ceroid, lipofuscin, and excessive amounts of protein subunit c.

Appendix 6

Biophysical Profile Scoring:
Technique and Interpretation

Biophysical variable	Normal (score = 2)	Abnormal (score = 0)
Fetal breathing movements	≥1 episode of ≥30 sec in 30 min	Absent or no episode of ≥30 min
Gross body movements	≥3 discrete body-limb movements in 30 min (episodes of active continuous movement considered)	≤2 episodes of body-limb movements in 30 min as single movement
Fetal tone	≥1 episode of active extension with return to flexion of fetal limb(s) or trunk Opening and closing of hand considered normal tone	Either slow extension with return to partial flexion movement of limb in full extension or absent fetal movement
Reactive fetal heart rate	≥2 episodes of acceleration of ≥15 bpm and of >15 sec associated with fetal movement in 20 min	<2 episodes of acceleration fetal heart rate or acceleration of >15 bpm in 20 min
Qualitative amniotic fluid volume	≥1 pocket of fluid measuring 2 cm in vertical axis	Either no pockets or largest pocket <2 cm in vertical axis

From Reece EA, Hobbins JC, eds. Medicine of the Fetus and Mother, 2nd Edition. Philadelphia: Lippincott-Raven Publishers, 1999.

Appendix 7
Sample Reports

ASSISTED REPRODUCTIVE TECHNIQUE: EMBRYO TRANSFER

TITLE OF PROCEDURE: Embryo transfer.

PROCEDURE IN DETAIL: The patient was taken to the operating room, where she was placed in the dorsal lithotomy position on the operating table. Her vagina and perineum were draped in the usual sterile fashion, and a bivalved speculum was placed into the vagina. The cervix was visualized and cleansed with transfer medium. A Wallace catheter, which had been premeasured to appropriate depth, was preloaded with embryos, one at the two-part nuclear stage; the other three were multicells. The catheter was threaded into the cervix, and the embryos were transferred high in the endometrial cavity. The catheter was then slowly removed, inspected, and found to be free of all embryos. The speculum was then removed. The patient was moved to the recovery area in satisfactory condition.

ASSISTED REPRODUCTIVE TECHNIQUE: SONOGRAPHICALLY ASSISTED TRANSVAGINAL OOCYTE RETRIEVAL

TITLE OF PROCEDURE: Sonographically assisted transvaginal oocyte retrieval.

PROCEDURE IN DETAIL: The patient was placed on the operating room table in the supine position. After adequate analgesia and sedation, she was placed in the dorsal lithotomy position and prepped and draped in the standard sterile fashion. The ultrasound transducer was then placed into the vagina, and culdocentesis was performed, first on the right and then on the left, with aspiration of mature oocytes recovered. The pelvis was rescanned, and there was no increase in peritoneal fluid noted. The probe was removed. The vagina was inspected and found to be hemostatic. The patient tolerated the procedure well and left the operating room in good condition.

CONTROLLED VAGINAL DELIVERY

TITLE OF PROCEDURE: 1. Controlled vaginal delivery.
                           2. Repair of episiotomy.

PROCEDURE IN DETAIL: The patient is a primigravida 27-year-old white female who received prenatal care during her first trimester of pregnancy. She has remained normotensive throughout her pregnancy, and dipsticks remained negative. Maternal blood type is O negative, so RhoGAM was administered postdelivery.

The patient arrived in active labor with a good mechanism at 0430 hours. She was 80% effaced with a cervix dilated to 4 cm. The fetus was noted to be in a vertex presentation at a -2 station. She progressed rapidly in labor, and by 1045 hours, she was 100% effaced and dilated to 2 cm. An epidural was started by anesthesia at the patient's request. The fetus remained in the vertex presentation and had normal fetal monitoring strips throughout labor, with a heart rate ranging from 120 to 152.

At 1205 hours, the patient was moved to delivery. The epidural was continued. The patient's vagina and perineum were prepped, and drapes were applied after the patient was placed in Allen stirrups in the lithotomy position. It was felt necessary to do a midline episiotomy to prevent tearing. The infant's head was delivered, and the nose and oropharynx were suctioned with a bulb. The shoulders were gently rotated, and the infant was delivered and placed on the mother's abdomen. The infant cried spontaneously and vigorously. The mouth and nose were once again suctioned. The cord was clamped and cut. Cord blood was obtained from a three-vessel cord. The infant was handed off the field to the neonatologist in attendance. The infant's blood type will be determined, and the infant will be closely monitored for any signs of Rh incompatibility, but none was apparent at birth.

The patient delivered a viable male infant weighing 7 pounds 9 ounces with Apgars of 8 at one minute and 10 at five minutes. RhoGAM will be administered. The midline episiotomy was repaired without complications. The infant was sent to the newborn nursery, and the mother will be closely observed prior to returning to her room for recovery.

FRACTIONAL DILATATION AND CURETTAGE AND HYSTEROSCOPY

TITLE OF PROCEDURE: Fractional dilatation and curettage and diagnostic hysteroscopy.

PROCEDURE IN DETAIL: After informed consent was obtained, the patient was taken to the operating room, where she underwent general anesthesia with mask inhalation. The patient was prepped and draped in the normal sterile fashion in the dorsal lithotomy position in high stirrups. The patient was then placed in Trendelenburg. Her bladder was emptied of approximately 50 cc of urine. A weighted speculum was placed in the posterior vagina, and a retractor was placed in the anterior vagina. The anterior lip of the cervix was grasped with a single-tooth tenaculum. The cervix was serially dilated to a #10 dilator. A uterine sound was done and sounded to 8 cm. Endocervical curettage was then done. A hysteroscope was placed through the cervix into the uterus. The uterine cavity was examined. There was a questionable polyp versus just a tear of the superficial endometrium noted anteriorly. The remainder of the cavity seemed normal, with normal ostia. Sharp curettage was done, and a small amount of tissue was obtained.

The hysteroscope was placed back through the cervix and the uterus, and the previous area was noted to have been removed with the D&C. Both ostia, as stated, were identified. The cervix was examined, and the hysteroscope was removed. There was no evidence of any abnormalities in the cervical canal. The single-tooth tenaculum was then removed, and the cervix was noted to be hemostatic.

The patient was then taken out of high stirrups, awakened, and transferred to the recovery room in stable condition. Specimens were submitted to pathology. The patient tolerated the procedure well. All counts were correct times three. The patient had approximately 300 cc of hysteroscopy fluid in and 200 cc out. Estimated blood loss <100 cc.

GENETIC COUNSELING NOTE: ADVANCED MATERNAL AGE AND FAMILY BIRTH ANOMALIES

REASON FOR VISIT: Amniocentesis because of family birth anomalies.

The patient was seen today for genetic counseling and amniocentesis. She was referred by her obstetrician with an indication of advanced maternal age.

The patient is a gravida 3, para 1, SAB 1, healthy 39-year-old. Her previous miscarriage was a first trimester loss of unknown cause. Her current pregnancy has been complicated only by spotting earlier in the pregnancy. She denied exposure to medications, alcohol, drugs, tobacco, and x-rays during the pregnancy.

The patient's husband is a healthy 43-year-old. In review of the family history, the patient reported that her sister was born with spina bifida and died shortly after birth. Given the multifactorial nature of neural tube defects, the patient's children are at a slightly increased risk of approximately 1% for neural tube defects. The patient also reported that her aunt had a son born with muscular dystrophy. This boy died at age 11 and, given the description of his symptoms, was likely suffering from Duchenne muscular dystrophy (DMD). DMD, an X-linked disorder affecting males, may be caused by a new mutation or may be inherited from a carrier mother. Given this history, she may also be a carrier for DMD and may be at risk to have sons with this disorder. The option of carrier testing for DMD was offered to the patient, and she accepted testing. The results of this testing will be forwarded to her physician's office upon receipt.

The patient also reported that she has two paternal cousins, siblings who are affected with myotonic dystrophy (MD), a dominant disorder with extreme variability and evidence of anticipation. In other words, symptoms of MD may vary greatly in different family members, and symptoms often become more severe with future generations. Although the patient does not experience symptoms of MD, there is a possibility that she is a carrier and is not yet expressing symptoms. Therefore, carrier testing for MD was offered to her, and she accepted this testing. Results of this testing will be forwarded to her physician's office upon receipt. The remainder of the family history is negative for suspected genetic disorders or mental retardation. The patient and her husband are both of Bulgarian heritage.

The maternal age-related risk for chromosome abnormalities was reviewed. The procedure of amniocentesis and its associated risks were reviewed. The patient indicated that she wished to proceed with testing and signed our consent form. A complete obstetrical ultrasound was performed. Please see our ultrasound report for additional details. Amniocentesis was performed successfully, and 20 cc of fluid was drawn for alpha-fetoprotein and chromosome analysis.

GENETIC COUNSELING NOTE: CHORIONIC VILLI SAMPLING

REASON FOR VISIT: Followup for abnormal chorionic villi sampling (CVS) results.

The patient was seen today for genetic counseling and blood drawing because of an abnormal CVS result. She was seen previously for genetic counseling and CVS testing. A review of the family history at the time was remarkable for a niece with a craniosynostosis but was otherwise unremarkable for birth defects or suspected genetic disorders. CVS was successfully performed.

Results of CVS have since revealed a marker chromosome in all cells analyzed. Studies in the laboratory are currently under way to try to determine the nature of the marker, but laboratory specialists have suggested that it may be an additional piece of chromosome 10 (possible isochromosome 10p). They do comment that the marker is large and appears to have banding. This leads laboratory specialists to suspect that this is a clinically significant piece of chromosomal material that, if present in the fetus, would carry an abnormal prognosis for the baby. Specialists in the lab have also suggested that this marker may be confined to the placenta and have recommended that we consider the option of amniocentesis for confirmation. They have also requested parental chromosome analysis for comparison. There is a small possibility that this marker is inherited from the patient or her partner, or that either of them may have a balanced chromosome rearrangement that has led to the generation of this marker and may be significant for future pregnancies or other family members. The patient is here today for further discussion of this finding.

The above information was discussed with the patient. She indicated that at this time she did wish to pursue parental chromosome testing and would consider further options for amniocentesis pending these results. We did describe that the laboratory will be using fluorescent in situ hybridization (FISH) techniques, as well as another technique called "chromosome painting" to try to identify the origin of the marker chromosome. This may lead to more specific prognostic information. Blood was drawn from the patient today for the purpose of chromosome analysis. She was given a prescription for chromosome testing for her partner.

The patient and her partner return today for discussion of results of further lab analysis. Since that time, results of FISH analysis and chromosome painting have indicated that the marker chromosome is apparently an isochromosome 10p, consistent with partial tetrasomy 10p. The results indicated that if this is present in the fetus, it will indeed carry a poor prognosis. However, because this marker is fairly large, it does have a higher probability of being confined to the placenta, as demonstrated in CVS fairly commonly by the finding of trisomy 7, virtually never found in the fetus itself when demonstrated on CVS. Alternately, trisomy 21 in CVS is virtually 100% predictive of fetus status. These statistics indicate that the nature of the chromosome abnormalities seen in CVS is significant to presence in the fetus. The possibility that a severe chromosome abnormality is confined to the placenta is greater when ultrasound evaluation is within normal limits. In other words, when a chromosome abnormality inconsistent with viable pregnancy is noted in a CVS with a normal ultrasound, it is significantly less likely to be present in the fetus. This event is most likely to occur as a result of a phenomenon called "trisomic rescue." In trisomic rescue, the relatively small number of cells that continue on to develop the fetus lose the extra chromosome, resulting in a fetus with normal chromosomes and an abnormal placenta.

Amniocentesis should be considered when trisomic rescue is suspected. Unfortunately, amniocentesis cannot entirely eliminate the possibility that some percentages of cells in the fetus carry the extra chromosome or chromosome marker (chromosomal mosaicism in the fetus). However, normal amniocentesis following CVS with a suspicious chromosome finding would significantly improve the probability of a normal outcome. Followup studies by ultrasound evaluation remaining within normal limits would further increase the likelihood of favorable outcome.

This information was reviewed in detail with the patient and her partner. They indicated that they wished to proceed with amniocentesis. The patient was scheduled for that procedure. We encouraged them to contact us in the mean time if they have additional questions and will provide additional consultation and discussions on the day of testing. Careful attention will be given to the appearance of the fetus at amniocentesis in hopes that ultrasound evaluation may give some clue as to the potential for abnormal findings on amniocentesis.

Amniocentesis will be analyzed using FISH techniques in hopes of providing them with a rapid preliminary result. We hope to have final chromosome results 7–10 days after amniocentesis.

GENETIC COUNSELING NOTE: INCREASED RISK FOR DOWN SYNDROME

REASON FOR VISIT: Amniocentesis for increased risk of Down syndrome.

The patient and her husband were seen today for genetic counseling and discussion of prenatal diagnosis. The patient was referred by her obstetrician with an indication of an abnormal triple screen, indicating an increased risk for Down syndrome of 1/71.

The patient is a gravida 3, para 1, SAB 1, healthy 31-year-old. The cause of her previous miscarriage is unknown. Her current pregnancy has been uncomplicated. She denied exposure to medications, alcohol, drugs, tobacco, and x-rays during the pregnancy.

The patient's husband is a 28-year-old with glucose-6-phosphate dehydrogenase (G6PD) deficiency. G6PD deficiency is an X-linked recessive disorder, which is more common in African-American males. It is characterized by hemolytic anemia; however, many affected individuals are asymptomatic until exposed to environmental triggers. Triggers of a hemolytic crisis include oxidizing drugs, infections, diabetic acidosis, ingestion of fava beans, and occupational exposures. The husband's daughters are obligate carriers for the G6PD gene because they must inherit their X chromosomes from their father. Although they would likely not be at risk for health complications, their sons would be at risk for inheriting the gene for G6PD. The husband's sons would be at no risk for inheriting the G6PD gene because the gene for G6PD is found on the X chromosome, and fathers pass their Y chromosomes to their sons.

Also of significance is the family history. The husband reported that he had a son with a previous partner who was affected with Wilms tumor at the age of 9. Wilms tumor is a kidney tumor that is typically sporadic in nature; however, some cases are thought to be inherited in an autosomal dominant fashion. Therefore, there may be a small chance for the husband's future children to inherit the gene for Wilms tumor. This information was shared with the couple, and the importance of monitoring for kidney abnormalities following delivery was stressed. The remainder of the family history is negative for suspected genetic disorders, birth defects, or mental retardation.

Both the patient and husband are African-American. The patient is unaware whether she has had sickle cell screening. It may prove beneficial to screen her for the sickle cell trait, given the high carrier frequency in the African-American population, if the screen has not already been performed.

The results of the triple screen indicating an increased risk for Down syndrome of 1/71 were reviewed. A complete obstetrical ultrasound was performed which estimated the date of delivery to be October 25, 1999. Today's ultrasound dating introduces a discrepancy with the dating that was established by an earlier ultrasound, which was used to calculate the triple screen results. Therefore, a recalculation of the triple screen results was performed using current ultrasound dating. The recalculated triple screen indicated no increased risk for Down syndrome; however, it showed an elevated alpha-fetoprotein (AFP) level, increasing the risk for open neural tube defects. This information was reviewed with the couple. The discrepancy in ultrasound dating is difficult to interpret, and the patient was told that she could be at risk either for a baby with Down syndrome or with spina bifida. However, the ultrasound revealed no defects of the spine and no findings associated with Down syndrome. The procedure of amniocentesis and its associated risks were reviewed. The couple declined amniocentesis on the day of counseling.

A complete obstetrical ultrasound report will be forwarded to the office. If we can be of any further assistance in clarification of dating, we would be happy to do so. It is recommended that the patient have a repeat ultrasound in approximately 3–4 weeks to better establish dating as well as follow the growth of the fetus. If her AFP is truly elevated, she may be at an increased risk for pregnancy complications. Therefore, serial ultrasonography throughout the pregnancy may prove beneficial.

GENETIC COUNSELING NOTE: MATERNAL AGE-RELATED AND OTHER RISKS

REASON FOR VISIT: Donor's egg and husband's sperm being implanted into a surrogate; genetic testing relevant to the donor wife and husband's ethnic heritage.

The donor and surrogate were seen today for genetic counseling regarding the maternal age-related risk for Down syndrome in a current pregnancy. Donor is also accompanied by her mother.

The surrogate is a gravida 3, para 2, healthy 28-year-old. Her current pregnancy is the result of in vitro fertilization involving gametes from the donor and her husband. These egg cells were retrieved from the donor in 1997 following a diagnosis of breast cancer in the donor. Since that time, donor has successfully undergone treatment for breast cancer and remains in good health. Embryo thawing with transfer to the surrogate took place on May 21, 2000. Ultrasound evaluation has since revealed an apparently normal twin gestation which at the time of this report we estimate to be approximately 8–1/2 weeks' gestation. The surrogate is taking medications related to her in vitro fertilization. She denies exposure to other medications or alcohol, and reports no significant complications during the pregnancy.

A review of the family history for donor and husband reveals that they have experienced an ectopic pregnancy and then two early spontaneous miscarriages together. This child will be their fourth pregnancy. Both of the donor's spontaneous miscarriages were thought to be related to complications of a fibroid. The donor does have a niece through her sister who has unilateral microtia but is otherwise a healthy 15-year-old. The donor's father died of Lou Gehrig disease. The donor's father-in-law died of chronic lymphocytic leukemia. The remainder of family history is negative for suspected genetic disorders, birth defects, or mental retardation. Both the donor and her husband are of eastern European (Ashkenazi) Jewish heritage. The donor's husband has been screened and apparently identified as noncarrier for Tay-Sachs disease. The family has not undertaken screening for Canavan disease.

The maternal age-related risk for this pregnancy is estimated by the age that the donor would have been at the time of delivery for a pregnancy that would have taken place at the time of egg retrieval in 1997. Therefore, the discussion was based on a maternal age of 37 at delivery. This risk, approximately 1/230, is doubled in a twin pregnancy. Therefore the risk for Down syndrome in this pregnancy at birth is approximately 1/115. The risk in the midtrimester is slightly higher.

Options for prenatal testing include amniocentesis and chorionic villus sampling (CVS). Each of these procedures and their associated risks were reviewed with the donor and the surrogate. The significance of a twin gestation and options for pregnancy management with particular regard to the possibility of the presence of Down syndrome in one twin were reviewed in detail. The donor indicated that she felt that CVS testing would be more helpful to the family, and the surrogate was scheduled to undergo that procedure.

We did suggest that the donor consider carrier testing for the common recessive conditions in the Jewish population. Her husband has already been screened for Tay-Sachs disease, which takes away concern about perhaps the most important of these. However, we did offer to provide carrier screening for Canavan disease as well as cystic fibrosis. Donor indicated that this testing would be helpful to her, and blood was drawn for a panel including Tay-Sachs disease, Canavan disease, Gaucher disease, cystic fibrosis, and Fanconi anemia. If this testing indicates that the donor is a carrier for any of these conditions, then carrier testing on the donor will be pursued. We anticipate having these results in approximately two weeks.

We have encouraged donor and the surrogate to call if they have additional questions in the near future. Additional consultation will be provided on the day of CVS.

GENETIC COUNSELING NOTE: MULTIPLE MISCARRIAGES

REASON FOR VISIT: Genetic counseling for history of multiple miscarriages.

The patient is a gravida 7, para 3, SAB 3, 32-year-old. Her first two pregnancies were shared with a previous partner. The first pregnancy resulted in a miscarriage at 11 weeks for which the cause was unknown. The second pregnancy resulted in a learning-disabled, dyslexic daughter who has been diagnosed with obsessive-compulsive disorder. The cause for her difficulties has yet to be identified. The remainder of the patient's pregnancies was conceived with her current husband. The first of their pregnancies resulted in a healthy daughter. The second pregnancy was a miscarriage of a male fetus at approximately 16 weeks in pregnancy. An autopsy performed on the products of conception revealed a twisted umbilical cord but otherwise no known established cause for the miscarriage. The third pregnancy of the patient and her husband resulted in a healthy daughter. The fourth pregnancy resulted in a miscarriage of a male fetus at again approximately 16 weeks of pregnancy. No testing was performed on the products of conception. The final pregnancy is the current pregnancy, which has been uncomplicated thus far. The patient reported that she experienced a bladder infection earlier in the pregnancy for which she was treated with antibiotics. She denied exposure to alcohol, drugs, tobacco, and x-rays during the pregnancy.

The various etiologies of miscarriage were discussed with the patient. Common causes of miscarriage, especially first trimester losses, are sporadic chromosome abnormalities; approximately 50%-60% of first trimester miscarriages are caused by such abnormalities. Other causes for miscarriage are obstetrical issues, such as maternal health conditions, uterine abnormalities, and blood incompatibility. Some women have circulating antibodies, such as antiphospholipids and anticardiolipins, that may interfere with pregnancy. Another cause for miscarriage is a translocation in the chromosomes of the fetus. In couples who have experienced two or more pregnancy losses, there is an approximate 6% chance that one member of the couple is a carrier of a balanced translocation. Individuals who carry a balanced translocation are at an increased risk for miscarriages and also for liveborns with birth defects and mental retardation. Miscarriages can also be caused by single-gene disorders or multifactorial traits, such as neural tube defects. The above possibilities for the patient's pregnancy losses were reviewed.

The patient's husband is the healthy father of the pregnancy. A review of the family history is significant for a sister of the patient's husband who was described to be developmentally delayed with possible mental retardation; autistic, dependent, with little communication, and repetitive movements. The cause of her difficulties is unknown. The patient reported a maternal cousin (6-year-old female) who is developmentally delayed with unusual behaviors. An ultrasound during the pregnancy of this 6-year-old girl revealed abnormalities; however, the patient was unable to clarify the types of abnormalities that were visualized.

The couple was offered fragile X carrier testing, given the history in both of their families of learning disabilities. Fragile X is an X-linked disorder that affects mostly males but can be found in milder forms in females. Affected individuals display learning disabilities or mental retardation, autism, unusual behaviors, and characteristic physical features. The couple was given lab slips for blood to be drawn for fragile X carrier testing.

The patient also reported a history in her family of psychiatric disorders, including bipolar disorder and obsessive-compulsive disorder. The multifactorial nature of these disorders and the tendency to occur in close relatives were reviewed. The patient is of English, Scottish, and German heritage, while the patient's husband is of Irish, Swedish, and English descent.

Appropriate testing for the current pregnancy was discussed. Blood chromosome analyses on the patient and her husband were offered to look for balanced translocations. They accepted this testing and were given lab slips to have their blood drawn for chromosome analysis. Testing for antiphospholipids and anticardiolipins, if not already performed, may prove beneficial for this couple. The benefits and limitations of level III ultrasonography to detect birth defects were discussed with the patient. She has scheduled an ultrasound appointment. Prenatal diagnosis via amniocentesis will be discussed with the couple in more detail if indicated by their blood chromosome results.

The report from the ultrasound will be forwarded to the office. Additional information will be provided upon the receipt of fragile X and blood chromosome testing.

GENETIC PERIPHERAL BLOOD REPORT: POSSIBLE MOSAICISM FOR MONOSOMY 22

CYTOGENETICS RESULTS:

METAPHASES ANALYZED:	40
METAPHASES KARYOTYPED:	3+ partials
STAINING METHOD USED:	G-banding
BAND STAGE:	650
KARYOTYPE ANALYSIS:	?mos 45,XY, -22 [3]/ 46,XY [47]

KARYOTYPE INTERPRETATION: Male karyotype with possible low-grade mosaicism for monosomy 22. Three of 50 cells (6%) examined were missing chromosome 22. Forty-seven cells (94%) were of normal male karyotype. To clarify these findings, a chromosome analysis on a skin biopsy and/or a repeat blood sample is recommended.

GENETIC PERIPHERAL BLOOD REPORT: SUSPICION OF MOSAIC TURNER SYNDROME

CYTOGENETICS RESULTS:

METAPHASES ANALYZED:	50
METAPHASES KARYOTYPED:	4
STAINING METHOD USED:	G-banding
BAND STAGE:	500
KARYOTYPE ANALYSIS:	46,XX

KARYOTYPE INTERPRETATION: This karyotype is consistent with normal female. Two of 50 cells (4%) examined were monosomy X (45,X). The remaining 48 cells (96%) were of normal female karyotype (46,XX). Because of the clinical suspicion of mosaic Turner syndrome, a chromosome analysis on a skin biopsy is recommended to clarify whether the 45,X cells are due to technical/cultural artifact or represent true mosaicism.

LAPAROSCOPIC-ASSISTED HYSTERECTOMY

TITLE OF OPERATION: Laparoscopic-assisted hysterectomy and bilateral salpingo-oophorectomy.

PROCEDURE IN DETAIL: After informed consent was obtained, the patient was taken to the operating room where she underwent general endotracheal anesthesia. The patient was then prepped and draped in normal sterile fashion in the dorsal lithotomy position in Allen stirrups. A Foley catheter was placed, and a Rubin cannula was placed in the cervix. The infraumbilical area was injected with Xylocaine. A 1-cm incision was then made. The Veress needle was then placed. Two towel clips were then placed on the skin to elevate the abdomen. The Veress needle was placed through this incision with proper placement verified by saline drop test and an opening pressure of 3 mmHg. The abdomen was then insufflated. The pelvis was examined and noted to be grossly normal. Two additional 10-mm trocars were placed in the left and right lower quadrants under direct visualization. Each area was injected with lidocaine first. The pelvis was fully examined.

The uterus was grossly normal and retroverted. There were normal bilateral tubes and ovaries. There was a small amount of endometriosis along the posterior uterosacral ligaments which was ablated. Next, the left infundibulopelvic ligament was clamped and cut using the Endostapler. Prior to this being transected, care was taken to verify that the ureter was not in the clamp. The remainder of the broad ligament was then serially transected using the stapling device. Once again, prior to transection and stapling, each pedicle was checked to ensure that there was no ureter noted in the clamp. In a similar fashion, the right infundibulopelvic ligament was clamped and transected using the stapler. Once again, the ureter was identified prior to any transection. The remainder of the broad ligament was transected in a similar fashion. The bladder was elevated, and the peritoneum above the bladder was incised with the Endoshears. The space was developed using the Endoshears. All areas appeared to be dry.

Attention was then turned to the vagina. The patient was converted to high stirrups using the Allen stirrups. A weighted speculum was placed in the most posterior vagina with a Deaver retractor in the anterior vagina. The cervix was grasped with two thyroid Lahey clamps. The mucosa of the cervix was circumscribed with Bovie cautery.

The bladder was bluntly dissected off of the lower uterine segment, and the peritoneum was then entered. The posterior cul-de-sac was grasped with a Kocher clamp and elevated. Curved Mayo scissors were used to enter the posterior cul-de-sac. A speculum was then placed. A Heaney clamp was then used on the right side to clamp across the uterosacral complex. This was then transected and suture-ligated with 0 Vicryl. In a similar fashion, the left uterosacral was transected, cut, and suture-ligated with 0 Vicryl. The small amount of remaining broad ligament was then clamped with Heaney clamps on each side, transected, and suture-ligated with 0 Vicryl. The uterus was then easily delivered through the vagina.

The vaginal cuff and perineum were closed together using a 2–0 Vicryl from 12 o'clock to 6 o'clock, being tied in the midline. The uterosacral ligament pedicles had been tagged, and these were tied in the midline and reattached to the vaginal cuff. The vaginal cuff, once it was sewn, was noted to be hemostatic. The Foley catheter drained clear urine.

The patient was taken out of stirrups, awakened, extubated, and transferred to the recovery room in stable condition. The patient tolerated the procedure well. All counts were correct times three. The patient received Ancef for antibiotic prophylaxis.

LAPAROSCOPY AND HYSTEROSCOPIC POLYPECTOMY

TITLE OF OPERATION: 1. Laparoscopy with lysis of adhesions.
                            2. Hysteroscopic polypectomy.

PROCEDURE IN DETAIL: The patient was taken to the operating room and placed in the dorsal lithotomy position on the operating table. After adequate general anesthesia, endotracheal intubation was performed. The abdomen, vagina, and perineum were prepped and draped in the usual sterile fashion. A Foley catheter was inserted under sterile conditions. The cervix was dilated to 9 mm, and the hysteroscope was placed in the cavity. The above findings were noted. Using the graspers, the polyps were all resected without difficulty, leaving a completely normal cavity. We instilled 400 cc of glycine and recovered 400 cc, for a net usage of essentially zero.

The Kronner manipulator was then placed in the cavity, and attention was directed to the abdomen, where an incision was made in the umbilicus. The Veress needle was inserted. The abdomen was insufflated with 2.5 liters of carbon dioxide gas. The Veress needle was removed. The laparoscopic trocar was inserted, followed by introduction of the laparoscope. There was no evidence of trauma from insertion of the above instruments. A second puncture site was developed two fingerbreadths below the pubis on the left under direct visualization without difficulty.

Using the CO_2 laser at 18 watts with sharp pulse, the adhesions were all taken down and lysed. Chromotubation before and after lysis of adhesions revealed different findings such that after lysis of adhesions, both fallopian tubes filled and spilled. Again, there was no evidence of any pelvic endometriosis. The pelvis was then copiously irrigated and found to be completely hemostatic. Gas was allowed to escape and suctioned out with vacuum suction. The instruments were removed under direct visualization. We left approximately 150 cc of heparinized lactated Ringer in place. The incisions were cleaned with Betadine and sutured with inverted subcuticular sutures of 4–0 Vicryl. Marcaine 0.5% with epinephrine was instilled around the incisions. The patient was transferred to the recovery room stable, awake, extubated, and in good condition. There were no complications noted. Estimated blood loss for the entire procedure was less than 10 cc.

LAPAROSCOPY, OOPHORECTOMY, AND VAGINAL HYSTERECTOMY

TITLE OF OPERATION:
1. Laparoscopy.
2. Oophorectomy.
3. Peritoneal omental biopsies.
4. Cytologic washings.
5. Vaginal hysterectomy.

PROCEDURE IN DETAIL: The patient was taken to the operating room and placed in the semilithotomy position. Satisfactory anesthesia was achieved. She was prepped and draped in the usual fashion. A Foley catheter was placed in the bladder. A uterine manipulator was placed. A knife was used to make a 1-cm umbilical incision. The Veress needle was introduced, and pneumoperitoneum was achieved. A 5-mm suprapubic and an 11-mm left lower quadrant trocar were placed under direct visualization. The blunt aspirator was used to obtain fluid for cytology. She had a small cystic lesion in the pelvis. This was biopsied. Peritoneal biopsies were obtained. Cul-de-sac biopsy and omental biopsies were obtained. Visualization of the entire peritoneal cavity revealed nothing suspicious for malignancy.

The left round ligament was grasped with the Seitzinger device. It was then cauterized and cut. The left ureter was isolated. The left infundibulopelvic ligament was then clamped, coagulated, and cut with the Seitzinger. The peritoneum below the ovary was incised, and the ureter visualized. The peritoneum overlying the bladder flap was incised.

Attention was then turned to the vaginal portion of the procedure. The cautery was used to incise the vaginal mucosa around the cervix. Mayo scissors were used to enter the posterior cul-de-sac. The anterior peritoneum was isolated and incised. A long weighted speculum was placed in the posterior cul-de-sac, and a Deaver retractor was placed in the anterior cul-de-sac. Uterosacral ligaments were clamped, cut, and ligated bilaterally. Uterine vessels were clamped, cut, and ligated bilaterally. Vessels in the broad ligament were clamped, cut, and ligated. The uterus and left ovary were then removed. There was a little bit of bleeding from the posterior cuff, but not much. Otherwise, the pedicles were dry.

The peritoneum was closed with a running 0 Vicryl suture. The uterosacral ligaments were plicated together in the midline. The vaginal cuff was closed with figure-of-eight 0 Vicryl sutures. A pack was placed in the vagina. A Foley catheter was placed.

The pneumoperitoneum was then reachieved. Hemostasis was noted to be good. The pneumoperitoneum was evacuated. The 5-mm suprapubic and 11-mm left lower quadrant trocars, as well as the 10-mm umbilical port, were all removed under direct visualization. The fascial incisions were closed in the left lower quadrant, and the umbilicus was closed with 0 Vicryl sutures. The skin was reapproximated with 4–0 Vicryl sutures and Steri-Strips. The patient was awakened from anesthesia and transferred to the recovery room in stable condition. Blood loss was approximately 200 cc. Counts were correct.

LOW TRANSVERSE CESAREAN SECTION

TITLE OF OPERATION: Repeat low transverse cesarean section with upper midline vertical uterine extension.

PROCEDURE IN DETAIL: The patient was taken to the operating room after successful epidural anesthesia had been placed in labor and delivery. She was placed in the supine position on the operating room table with her right side elevated and supported with a bag. The patient's abdomen was then prepared and draped in a sterile manner. An Allis clamp was used to check for an appropriate level of anesthesia, and the patient's breathing was noted to be unlabored.

A scalpel was used to make a Pfannenstiel incision through the old scar, which was excised and removed. The incision was extended through the subcutaneous tissue with the scalpel, and the bleeding was controlled with an electrocautery device. The fascia of the abdominis muscle was identified and nicked transversely with the Mayo scissors. Kocher clamps were placed on the proximal fascial flap, which was bluntly and sharply dissected from the rectus abdominis muscle. Kocher clamps were then placed on the lower fascial flap, which was bluntly and sharply dissected free from the rectus abdominis muscle. The median raphe was then bluntly and sharply dissected in a careful manner. The parietal peritoneum was visualized, grasped with hemostats using a three-point technique and nicked with a scalpel, and the incision extended both cephalad and caudad.

There were omental adhesions present to the anterior abdominal wall, presumed due to the previous cesarean section. These were carefully lysed with electrocautery. A bladder blade was positioned, and the peritoneal serosa was grasped with pickups and nicked with the Metzenbaum scissors, and the incision extended transversely with the Metzenbaum scissors.

A scalpel was used to score a low transverse uterine incision, which was then extended utilizing blunt dissection. The surgeon's hand was placed into the uterus to remove the fetal head, but it was necessary to extend the incision through the rectus muscles, the fascia, and the skin to remove the fetal head safely. There was a nuchal cord times two, which was reduced, and the nares and oropharynx were bulb-suctioned prior to removing the remainder of the fetus. The infant cried spontaneously. The cord was clamped and cut, and the neonate was handed off the field to the waiting neonatologist in attendance. The infant weighed 8 pounds 3 ounces and had Apgars of 9 at one minute and 10 at five minutes. Cord gases were obtained, as was cord blood.

At this time, the uterus was exteriorized and wrapped in a moist towel. The vertical upper midline extension in the uterus was reapproximated using #1 chromic sutures in a running locked fashion. The second layer was reapproximated using #1 chromic sutures in a simple fashion. Hemostasis was excellent. The peritoneal serosa was reapproximated with 3–0 Vicryl in a running fashion. The low transverse uterine incision was closed in two layers, the first a #1 chromic running locked stitch, and the second, a #1 chromic running imbricating stitch. Cautery was used to assure complete hemostasis. After copious irrigation with normal saline, the bladder flap was reapproximated with 3–0 Vicryl in a running fashion.

The cul-de-sac was wiped free of all blood and clots, and the uterus was returned to the abdominal cavity. An Interceed barrier was placed over the upper vertical uterine incision, and the pelvic gutters were freed of all blood and clots. Once again, the uterine incision was inspected and found to have excellent hemostasis.

The parietal peritoneum was reapproximated with 3–0 Vicryl in a running fashion. The fascia was reapproximated with 2–0 chromic sutures in a running fashion. The area was irrigated once again with normal saline, and hemostasis was noted to be excellent. The rectus muscles were reapproximated using 3–0 Vicryl in a running fashion, and the skin was reapproximated utilizing metallic skin staples. A sterile, mildly compressive dressing was applied.

The Foley catheter was noted at this time to be draining clear urine. The sponge, needle, instrument, and lap counts were reported as being correct times two prior to the final closure.

The patient was taken to the recovery room in stable condition with an estimated blood loss of 700 cc. She tolerated the procedure well. The neonate was sent to the newborn nursery in excellent condition.

PATHOLOGY REPORT

NATURE OF SPECIMEN: Cervix, uterus, bilateral tubes, and ovaries.

GROSS: The specimen consists of a pear-shaped uterus, attached ovaries, and fallopian tubes. The uterus weighs 140 g and measures $10 \times 6 \times 5.3$ cm. The external surface is smooth and glistening. There is a small subserosal nodule protruding through the serosal surface located close to the left cornu. On cut section, the ecto-cervix is smooth, and the ecto-endocervical junction is poorly defined and granular, and shows a few nabothian cysts. The endocervical canal is patent and trabeculated. The endometrial cavity is pyramidal in shape and lined by a light brown, 0.3-cm thick endometrium. Serial sectioning through the myometrium shows six additional well-circumscribed pearly-white nodules ranging in diameter from 0.3 to 0.5 cm.

CASSETTE CODE: A. Cervix
                        B. Endomyometrium and nodules

The right ovary measures $6 \times 1.8 \times 1.4$ cm. The external surface is yellowish and smooth. On cut section, there are several subcapsular cysts ranging in diameter from 0.3 to 0.6 cm. The adjacent fallopian tube measures 8 cm in length by 0.7 cm in diameter. There are two paratubal cysts present, measuring 2.5 cm and 0.5 cm in diameter. Representative sections are submitted and labeled "C."

The left ovary measures 5.5 × 2.7 × 2 cm. The external surface is yellowish and cerebriform. On cut section, there are several small subcapsular cysts ranging in diameter from 0.3 to 1 cm. The adjacent fallopian tube measures 8.5 cm in length by 0.8 cm in diameter. The cut section is unremarkable. Sections labeled "D" and stock.

MICROSCOPIC: Sections of specimen A of uterine cervix reveal chronic inflammation.

Sections of specimen B of endometrium reveal a proliferative histology. There are nodules composed of hypertrophic bundles of smooth muscle and fibrous tissue with dilated capillaries.

Sections of specimens C and D show sections of ovaries with follicular cysts and serosal adhesions. Sections of fallopian tubes are unremarkable. Paratubal cysts are present.

DIAGNOSES

1. Endometrium of proliferative histology.
2. Uterine leiomyomata, hypercellular.
3. Follicular cysts and serosal adhesions of ovaries.
4. Fallopian tubes, unremarkable.
5. Paratubal cysts, bilateral.
6. Chronic cervicitis.

TENSION-FREE VAGINAL TAPE

TITLE OF OPERATION: Tension-free vaginal tape (TVT).

PROCEDURE IN DETAIL: After informed consent was obtained, the patient was taken to the operating room, and IV sedation with monitored anesthesia care was administered. She was prepped and draped in a sterile fashion while in the dorsal lithotomy position. A rigid cystoscopy was carried out with both a 30-degree and 70-degree lens. Then the bladder, ureteral orifices, bladder neck, and urethra were fully inspected and found to be completely normal. The patient's bladder was then drained with a 16-French catheter with 10 cc of water in the balloon. The anterior vaginal wall was infiltrated with 0.25% Marcaine with epinephrine. Injection was also directed laterally to the infrapubic ramus to the endopelvic fascia bilaterally. A marking pen was used to localize the area of the proximal and mid urethra.

A midline vaginal incision was made in the anterior vaginal wall with a #15 blade at the level of the mid and proximal urethra. Tissue forceps and fine Metzenbaum scissors were used to dissect a vaginal wall flap toward the infrapubic ramus. The mid urethra was palpated easily through the vaginal incision and with the catheter in place.

Using the trocar guide, the TVT was then tunneled under the infrapubic ramus, perforating the endopelvic fascia bilaterally and exiting through a counter stab wound incision that was made times two just above the pubic tubercle approximately 2 cm in width. The tape was then laid loosely at the midurethral level. The Foley catheter was deflated and removed. Again rigid cystoscopy was carried out and showed no evidence of bladder perforation or injury and no evidence of urethral injury.

The patient's bladder was then left full, the cystoscope was removed, and she was asked to cough and bear down, at which point stress incontinence was noted, and vaginal tape tension was then adjusted loosely to prevent any leakage with Valsalva-related maneuver. Once there was no leakage seen with Valsalva and/or cough, the plastic sheath was then removed from the synthetic tape, and it was cut flush at the level of the subcutaneous tissues from the suprapubic stab wound. A 16-French Foley catheter was then reinserted with 10 cc of water in the balloon. Copious antibiotic irrigation was used in the suprapubic stab wound areas as well as the vaginal incision.

The small stab wounds were then closed with a single 4–0 Vicryl suture in a subcuticular fashion, and a single Steri-Strip was applied to each one. The vaginal incision was then closed with interrupted 3–0 Vicryl sutures.

There was minimal blood loss. No intraoperative complications were encountered, and the patient tolerated the procedure well. She was transported back to the recovery room in stable condition with plans made to perform a voiding trial prior to her discharge today.

Appendix 8
Common Terms by Procedure

Assisted Reproductive Techniques
aspiration of mature oocyte
bivalved speculum
culdocentesis
dorsal lithotomy position
endometrial cavity
multicells
peritoneal fluid
transfer medium
two-part nuclear stage
ultrasound transducer
Wallace catheter

Controlled Vaginal Delivery
Allen stirrups
Apgar score
dipstick
effaced
epidural
fetal monitoring strip
lithotomy position
midline episiotomy
normotensive
perineum
primigravida
Rh incompatibility
RhoGAM
station
three-vessel cord
vertex presentation
viable infant

Fractional Dilatation and Curettage and Hysteroscopy
#10 dilator
anterior lip of the cervix
anterior vagina
cervical canal

dorsal position
endocervical curettage
endometrium
high stirrups
hysteroscope
hysteroscopy fluid
mask inhalation anesthesia
ostium, *pl.* ostia
posterior vagina
sharp curettage
single-tooth tenaculum
Trendelenburg position
uterine sound
weighted speculum

Genetic Counseling
advanced maternal age
age-related risk
alpha-fetoprotein (AFP) level
amniocentesis
anticardiolipin
antiphospholipid
Ashkenazi Jewish heritage
autosomal dominant inheritance
balanced chromosome rearrangement
balanced translocation
birth anomaly
blood chromosome analysis
Canavan disease
carrier testing
chorionic villus sampling (CVS)
chromosomal mosaicism
chromosome painting
circulating antibody
cystic fibrosis
diabetic acidosis
dominant disorder
donor egg
Down syndrome

Duchenne muscular dystrophy (DMD)
ectopic pregnancy
egg retrieval
egg cells
embryo thawing with transfer
environmental trigger
ethnic heritage
evidence of anticipation
Fanconi anemia
fluorescent in situ hybridization (FISH)
fragile X carrier testing
gamete
Gaucher disease
genetic counseling
glucose-6-phosphate dehydrogenase
   (G6PD) deficiency
hemolytic anemia
in vitro fertilization
isochromosome
level III ultrasonography
liveborn
Lou Gehrig disease
marker chromosome
maternal age-related risk
mosaicism
multifactorial trait
myotonic dystrophy (MD)
neural tube defect
new mutation
noncarrier
obligate carrier
obstetrical ultrasound
partial tetrasomy
sickle cell screening
sickle cell trait
single-gene disorder
spina bifida
spontaneous miscarriage
sporadic chromosome abnormality
surrogate
Tay-Sachs disease
translocation in the chromosome

triple screen
trisomic rescue
trisomy
unilateral microtia
Wilms tumor
X-linked disorder

Genetic Peripheral Blood/Cytogenetics

band stage
chromosome
chromosome analysis
cultural artifact
female karyotype
G-banding
karyotype
low-grade mosaicism
male karyotype
metaphases
monosomy
mosaic Turner syndrome
technical artifact

Laparoscopic-Assisted Hysterectomy

10-mm trocar
Allen stirrups
Ancef
anterior vagina
antibiotic prophylaxis
bilateral salpingo-oophorectomy
blunt dissection
Bovie cautery
broad ligament
cervical mucosa
curved Mayo scissors
Deaver retractor
endometriosis
Endoshears
Endostapler
Foley catheter
general endotracheal anesthesia

Heaney clamp
high stirrups
infraumbilical
infundibulopelvic ligament
Kocher clamp
lower uterine segment
pedicle
peritoneum
posterior cul-de-sac
posterior uterosacral ligament
posterior vagina
retroverted
Rubin cannula
saline drop test
suture ligated
thyroid Lahey clamp
towel clip
uterosacral complex
uterosacral ligament pedicle
vaginal cuff
weighted speculum

Laparoscopy and Hysteroscopic Polypectomy

Betadine
carbon dioxide gas
chromotubation
CO_2 laser
dorsal lithotomy position
endotracheal intubation
filled and spilled
fingerbreadth
Foley catheter
glycine
graspers
hemostatic
heparinized lactated Ringer
hysteroscope
inverted subcuticular suture
Kronner manipulator
laparoscopic trocar
lysis of adhesion

Marcaine with epinephrine
pelvic endometriosis
puncture site
sharp pulse
vacuum suction
Veress needle

Laparoscopy, Oophorectomy, and Vaginal Hysterectomy

10-mm umbilical port
11-mm left lower quadrant trocar
5-mm suprapubic trocar
bladder flap
blunt aspirator
broad ligament
cul-de-sac biopsy
cystic lesion
cytologic washings
Deaver retractor
infundibulopelvic ligament
long weighted speculum
Mayo scissors
omental biopsy
pedicle
peritoneal biopsy
plicated
pneumoperitoneum
posterior cul-de-sac
posterior vaginal cuff
round ligament
Seitzinger device
semilithotomy position
uterine manipulator
uterosacral ligament

Low Transverse Cesarean Section

#1 chromic
Allis clamp
Apgar score
bladder blade
bladder flap

blunt and sharp dissection
bulb suction
caudad
cephalad
compressive dressing
cord blood
cord gas
cul-de-sac
electrocautery
epidural anesthesia
fascial flap
Foley catheter
hemostat
in toto
Interceed barrier
Kocher clamp
lap count
Mayo scissors
median raphe
metallic skin staple
midline vertical uterine extension
newborn nursery
nicked
nuchal cord
omental adhesion
parietal peritoneum
pelvic gutter
peritoneal serosa
Pfannenstiel incision
pickup
rectus abdominis muscle
running imbricating stitch
running locked stitch
three-point technique

Pathology
cerebriform
chronic cervicitis
cornua
cut section
dilated capillary
ectocervix

ecto-endocervical junction
endocervical canal
endometrial cavity
endometrium
endomyometrium
external surface
follicular cyst
hypercellular uterine leiomyoma
hypertrophic bundle of smooth muscle
myometrium
nabothian cyst
nature of specimen
paratubal cyst
pearly-white nodule
pear-shaped uterus
proliferative histology
pyramidal
representative section
serial sectioning
serosal adhesion
serosal surface
subcapsular cyst
subserosal nodule
trabeculated
well circumscribed

Tension-Free Vaginal Tape (TVT)
#15 blade
30-degree lens
70-degree lens
copious antibiotic irrigation
counter stab wound incision
endopelvic fascia
fine Metzenbaum scissors
Foley catheter
infrapubic ramus
mid urethra
monitored anesthesia care
proximal urethra
pubic tubercle
rigid cystoscopy

Steri-Strip
stress incontinence
suprapubic stab wound
tissue forceps
trocar guide

vaginal tape tension
vaginal wall flap
Valsalva maneuver
voiding trial

Appendix 9
Drugs by Indication

ABDOMINAL DISTENTION (POSTOPERATIVE)
Hormone, Posterior Pituitary
  Pitressin®
  Pressyn® (Can)
  vasopressin

ABETALIPOPROTEINEMIA
Vitamin, Fat Soluble
  Amino-Opti-E® [OTC]
  Aquasol A®
  Aquasol E® [OTC]
  Del-Vi-A®
  E-Complex-600® [OTC]
  E-Vitamin® [OTC]
  Palmitate-A® 5000 [OTC]
  vitamin A
  vitamin E
  Vita-Plus® E Softgels® [OTC]
  Vitec® [OTC]
  Vite E® Creme [OTC]

ACQUIRED IMMUNODEFICIENCY SYNDROME (AIDS)
Antiviral Agent
  Apo®-Zidovudine (Can)
  Combivir®
  Crixivan®
  delavirdine
  didanosine
  Epivir®
  Epivir®-HBV™
  Fortovase®
  Hivid®
  indinavir
  Invirase®
  lamivudine
  nelfinavir

nevirapine
Norvir®
Novo-AZT® (Can)
Rescriptor®
Retrovir®
ritonavir
saquinavir
stavudine
3TC® (Can)
Videx®
Viracept®
Viramune®
zalcitabine
Zerit®
zidovudine
zidovudine and lamivudine
Nonnucleoside Reverse Transcriptase Inhibitor
  efavirenz
  Sustiva™
Nucleoside Analog Reverse Transcriptase Inhibitor
  abacavir
  Ziagen™
Protease Inhibitor
  Agenerase™
  amprenavir
Reverse Transcriptase Inhibitor
  adefovir
  Preveon®

ALPHA1-ANTITRYPSIN DEFICIENCY (CONGENITAL)
Antitrypsin Deficiency Agent
  alpha1-proteinase inhibitor
  Prolastin®

AMENORRHEA
Antihistamine

cyproheptadine
Periactin®
PMS-Cyproheptadine (Can)
Ergot Alkaloid and Derivative
Apo® Bromocriptine (Can)
bromocriptine
Parlodel®
Gonadotropin
Factrel®
gonadorelin
Lutrepulse®
Progestin
Amen® Oral
Aygestin®
Crinone™
Curretab® Oral
Cycrin® Oral
Depo-Provera® Injection
hydroxyprogesterone caproate
Hylutin®
Hyprogest® 250
medroxyprogesterone acetate
Micronor®
norethindrone
NOR-QD®
PMS-Progesterone (Can)
Progestasert®
progesterone
Progesterone Oil (Can)
Provera® Oral

AMYLOIDOSIS
Mucolytic Agent
acetylcysteine
Mucomyst®
Mucosil™
Parvolex® (Can)

APNEA (NEONATAL IDIOPATHIC)
Theophylline Derivative
aminophylline
theophylline

CANDIDIASIS
Antifungal Agent
Abelcet™
Absorbine® Antifungal [OTC]
Absorbine® Antifungal Foot Powder [OTC]
Absorbine® Jock Itch [OTC]
Absorbine Jr.® Antifungal [OTC]
Aftate® [OTC]
Amphotec®
amphotericin B cholesteryl sulfate complex
amphotericin B (conventional)
amphotericin B (lipid complex)
Ancobon®
Ancotil® (Can)
AVC™ Cream
AVC™ Suppository
Breezee® Mist Antifungal [OTC]
butoconazole
Candistatin® (Can)
Canesten® Topical (Can)
Canesten® Vaginal (Can)
ciclopirox
Clotrimaderm (Can)
clotrimazole
Desenex® [OTC]
Diflucan®
econazole
Ecostatin® (Can)
Exelderm® Topical
Femizol-M® [OTC]
fluconazole
flucytosine
Fungizone®
Fungoid® Creme
Fungoid® Tincture
Genaspor® [OTC]
Gyne-Lotrimin® Vaginal [OTC]
itraconazole
ketoconazole
Lamisil® Cream
Loprox®

Lotrimin® AF Powder [OTC]
Lotrimin® AF Spray Liquid [OTC]
Lotrimin® AF Spray Powder [OTC]
Lotrimin AF® Topical [OTC]
Lotrimin® Topical
Maximum Strength Desenex®
   Antifungal Cream [OTC]
Mestatin® (Can)
Micatin® Topical [OTC]
miconazole
Monistat-Derm™ Topical
Monistat i.v.™ Injection
Monistat™ Vaginal
Mycelex®-G Topical
Mycelex®-G Vaginal [OTC]
Mycelex® Troche
Myclo-Derm® (Can)
Myclo-Gyne® (Can)
Mycostatin®
M-Zole® 7 Dual Pack [OTC]
Nadostine® (Can)
naftifine
Naftin®
Nilstat®
Nizoral®
NP-27® [OTC]
Nyaderm (Can)
nystatin
Nystat-Rx®
Nystex®
Ony-Clear® Spray
O-V Staticin®
oxiconazole
Oxistat® Topical
Pitrex® (Can)
PMS-Nystatin (Can)
Prescription Strength Desenex®
   [OTC]
Spectazole™
Sporanox®
sulconazole
sulfanilamide
Terazol® Vaginal

terbinafine, topical
terconazole
Tinactin® [OTC]
tioconazole
tolnaftate
Vagistat®-1 Vaginal [OTC]
Vagitrol®
Zeasorb-AF® [OTC]
Zeasorb-AF® Powder [OTC]
Antifungal Agent, Systemic
AmBisome®
amphotericin B (liposomal)
Antifungal/Corticosteroid
Mycogen II Topical
Mycolog®-II Topical
Myconel® Topical
Myco-Triacet® II
Mytrex® F Topical
N.G.T.® Topical
nystatin and triamcinolone
Tri-Statin® II Topical

CARCINOMA

Antineoplastic Agent
Adriamycin PFS™
Adriamycin RDF®
Adrucil® Injection
Alkaban-AQ®
Alkeran®
Alpha-Tamoxifen® (Can)
altretamine
aminoglutethimide
Anandron® (Can)
anastrozole
Apo-Tamox® (Can)
Arimidex®
BiCNU®
Blenoxane®
bleomycin
Camptosar®
carboplatin
carmustine
CeeNU®

chlorambucil
cisplatin
Cosmegen®
cyclophosphamide
Cytadren®
cytarabine
Cytosar-U®
Cytoxan®
dacarbazine
dactinomycin
docetaxel
doxorubicin
Droxia™
DTIC-Dome®
Efudex® Topical
Emcyt®
estramustine
Etopophos®
etoposide
etoposide phosphate
Fareston®
floxuridine
Fludara® (Can)
Fluoroplex® Topical
fluorouracil
Folex® PFS
FUDR®
gemcitabine
Gemzar®
Herceptin®
Hexalen®
Hycamtin™
Hydrea®
hydroxyurea
Idamycin®
Idamycin® PFS
idarubicin
Ifex®
ifosfamide
irinotecan
Leukeran®
leuprolide acetate
lomustine

Lupron®
Lupron Depot®
Lupron Depot-3® Month
Lupron Depot-4® Month
Lupron Depot-Ped®
Lysodren®
mechlorethamine
Megace®
megestrol acetate
melphalan
methotrexate
Mithracin®
mitomycin
mitotane
mitoxantrone
Mustargen® Hydrochloride
Mutamycin®
Navelbine®
Neosar®
Nilandron™
nilutamide
Nolvadex®
Novantrone®
Novo-Tamoxifen (Can)
Oncovin®
paclitaxel
Paraplatin®
Paxene®
Photofrin®
Platinol®
Platinol®-AQ
plicamycin
porfimer
Procytox® (Can)
Rheumatrex®
Rubex®
streptozocin
Tamofen® (Can)
Tamone® (Can)
tamoxifen
Tarabine® PFS
Taxol®
Taxotere®

teniposide
Thioplex®
thiotepa
Toposar® Injection
topotecan
toremifene
trastuzumab
valrubicin
Valstar™
Velban®
Velbe® (Can)
VePesid®
vinblastine
Vincasar® PFS™
vincristine
vinorelbine
Vumon
Zanosar®
Antineoplastic Agent, Hormone
(Antiestrogen)
Femara™
letrozole
Estrogen and Androgen Combination
Estratest®
Estratest® H.S.
estrogens and methyltestosterone
Premarin® With Methyltestosterone
Estrogen Derivative
chlorotrianisene
diethylstilbestrol
estradiol
estrogens, conjugated (equine)
estrone
polyestradiol
Premarin®
Stilphostrol®
TACE®
Gonadotropin-Releasing Hormone
Analog
goserelin
Zoladex® Implant
Progestin
Amen® Oral

Androcur® (Can)
Androcur® Depot (Can)
Crinone™
Curretab® Oral
Cycrin® Oral
cyproterone Canada only
Depo-Provera® Injection
hydroxyprogesterone caproate
Hylutin®
Hyprogest® 250
medroxyprogesterone acetate
PMS-Progesterone (Can)
Progestasert®
progesterone
Progesterone Oil (Can)
Provera® Oral

CYSTINURIA
Chelating Agent
Cuprimine®
Depen®
penicillamine

CYSTITIS (HEMORRHAGIC)
Antidote
mesna
Mesnex™

DIAPER RASH
Antifungal Agent
Caldesene® Topical [OTC]
Fungoid® AF Topical Solution
[OTC]
Pedi-Pro Topical [OTC]
undecylenic acid and derivatives
Dietary Supplement
methionine
Pedameth®
Protectant, Topical
A and D™ Ointment [OTC]
Desitin® [OTC]
vitamin A and vitamin D
zinc oxide, cod liver oil, and talc

Topical Skin Product
  Diaparene® [OTC]
  methylbenzethonium chloride
  Prevex™ Baby Diaper Rash (Can)
  Puri-Clens™ [OTC]
  Sween Cream® [OTC]
  Zinaderm (Can)
  Zincofax® (Can)
  zinc oxide

DYSBETALIPOPROTEINEM-IA (FAMILIAL)

Antihyperlipidemic Agent,
    Miscellaneous
  Abitrate® (Can)
  Atromid-S®
  Claripex® (Can)
  clofibrate
  Novo-Fibrate® (Can)
Antilipemic Agent
  bezafibrate Canada only
  Bezalip® (Can)
Vitamin, Water Soluble
  niacin
  Niaspan®
  Nicobid® [OTC]
  Nicolar® [OTC]
  Nicotinex [OTC]
  Slo-Niacin® [OTC]

DYSMENORRHEA

Nonsteroidal Antiinflammatory Drug
    (NSAID)
  Aches-N-Pain® [OTC]
  Actiprofen® (Can)
  Actron® [OTC]
  Advil® [OTC]
  Aleve® [OTC]
  Anaprox®
  Ansaid® Oral
  Apo®-Diclo (Can)
  Apo-Diflunisal® (Can)
  Apo®-Ibuprofen (Can)

Apo-Keto-E® (Can)
Apo®-Naproxen (Can)
Apo®-Piroxicam (Can)
Apro-Flurbiprofen® (Can)
Cataflam® Oral
diclofenac
diflunisal
Dolobid®
Excedrin® IB [OTC]
Feldene®
flurbiprofen
Froben® (Can)
Froben-SR® (Can)
Genpril® [OTC]
Haltran® [OTC]
Ibuprin® [OTC]
ibuprofen
Ibuprohm® [OTC]
Ibu-Tab®
Junior Strength Motrin® [OTC]
ketoprofen
Medipren® [OTC]
mefenamic acid
Menadol® [OTC]
Midol® IB [OTC]
Motrin®
Motrin® IB [OTC]
Naprosyn®
naproxen
Naxen® (Can)
Novo-Difenac® (Can)
Novo-Difenac-SR® (Can)
Novo-Diflunisal (Can)
Novo-Flurprofen® (Can)
Novo-Keto-EC (Can)
Novo-Naprox® (Can)
Novo-Piroxicam® (Can)
Novo-Profen® (Can)
Nu-Diclo® (Can)
Nu-Diflunisal (Can)
Nu-Flurprofen® (Can)
Nu-Ibuprofen® (Can)
Nu-Ketoprofen (Can)

Nu-Ketoprofen-E (Can)
Nu-Naprox® (Can)
Nu-Pirox® (Can)
Nuprin® [OTC]
Orudis®
Orudis® KT [OTC]
Oruvail®
Pamprin IB® [OTC]
Pedia-Profen™
piroxicam
PMS-Ketoprofen (Can)
Ponstan® (Can)
Ponstel®
Pro-Piroxicam® (Can)
Rhodis® (Can)
Rhodis-EC® (Can)
Saleto-200® [OTC]
Saleto-400®
Synflex® (Can)
Trendar® [OTC]
Uni-Pro® [OTC]
Voltaren® Oral
Voltaren®-XR Oral
Selective Cyclooxygenase-2 Inhibitor
rofecoxib
Vioxx®

ECLAMPSIA
Barbiturate
Barbilixir® (Can)
Barbita®
Luminal®
phenobarbital
Solfoton®
Benzodiazepine
Apo®-Diazepam (Can)
Diazemuls® Injection
diazepam
E Pam® (Can)
Meval® (Can)
Novo-Dipam® (Can)
PMS-Diazepam (Can)
Valium® Injection

Valium® Oral
Vivol® (Can)

ENDOMETRIOSIS
Androgen
Cyclomen® (Can)
danazol
Danocrine®
Contraceptive, Oral
Alesse™
Brevicon®
Demulen®
Desogen®
Estrostep® 21
Estrostep® Fe
ethinyl estradiol and desogestrel
ethinyl estradiol and ethynodiol
diacetate
ethinyl estradiol and levonorgestrel
ethinyl estradiol and norethindrone
ethinyl estradiol and norgestimate
ethinyl estradiol and norgestrel
Genora® 0.5/35
Genora® 1/35
Genora® 1/50
Jenest-28™
Levlen®
Levlite®
Levora®
Loestrin®
Lo/Ovral®
mestranol and norethindrone
Modicon™
N.E.E.® 1/35
Nelova™ 0.5/35E
Nelova® 1/50M
Nelova™ 10/11
Nordette®
Norethin™ 1/35E
Norethin 1/50M
Norinyl® 1+35
Norinyl® 1+50
Ortho® 0.5/35 (Can)

Ortho-Cept®
Ortho-Cyclen®
Ortho-Novum® 1/35
Ortho-Novum® 1/50
Ortho-Novum® 7/7/7
Ortho-Novum® 10/11
Ortho Tri-Cyclen®
Ovcon® 35
Ovcon® 50
Ovral®
Preven™
Synphasic® (Can)
Tri-Levlen®
Tri-Norinyl®
Triphasil®
Zovia®
Contraceptive, Progestin Only
   Aygestin®
   Micronor®
   norethindrone
   norgestrel
   NOR-QD®
   Ovrette®
Gonadotropin Releasing Hormone
      Analog
   histrelin
   Supprelin™
Hormone, Posterior Pituitary
   nafarelin
   Synarel®
Progestin
   hydroxyprogesterone caproate
   Hylutin®
   Hyprogest® 250

ERYTHROPOIETIC PROTOPORPHYRIA (EPP)

Vitamin, Fat Soluble
   beta-carotene

FACTOR VIII DEFICIENCY

Blood Product Derivative
   antihemophilic factor (human)
   Hemofil® M

Humate-P®
Koate®-HP
Monoclate-P®
Profilate® OSD
Profilate® SD
Hemophilic Agent
   anti-inhibitor coagulant complex
   Autoplex® T
   Feiba VH Immuno®

FACTOR IX DEFICIENCY

Antihemophilic Agent
   AlphaNine® SD
   BeneFix™
   factor IX complex (human)
   Hemonyne®
   Konyne® 80
   Profilnine® SD
   Proplex® T

FIBROCYSTIC BREAST DISEASE

Androgen
   Cyclomen® (Can)
   danazol
   Danocrine®

FOLLICLE STIMULATION

Ovulation Stimulator
   Fertinex®
   Metrodin®
   urofollitropin

FUNGAL INFECTION (SEE CANDIDIASIS)

GALACTORRHEA

Antihistamine
   cyproheptadine
   Periactin®
   PMS-Cyproheptadine (Can)
Ergot Alkaloid and Derivative
   Apo® Bromocriptine (Can)

bromocriptine
Parlodel®

GAUCHER DISEASE
Enzyme
  alglucerase
  Ceredase®
  Cerezyme®
  imiglucerase

GENITAL HERPES
Antiviral Agent
  famciclovir
  Famvir™
  valacyclovir
  Valtrex®

GENITAL WART
Immune Response Modifier
  Aldara™
  imiquimod

GONOCOCCAL OPHTHALMIA NEONATORUM
Topical Skin Product
  Dey-Drop® Ophthalmic Solution
  silver nitrate

GONORRHEA
Antibiotic, Macrolide
  Rovamycine® (Can)
  spiramycin Canada only
Antibiotic, Miscellaneous
  spectinomycin
  Trobicin®
Antibiotic, Quinolone
  Raxar®
Cephalosporin (Second Generation)
  cefoxitin
  Ceftin® Oral
  cefuroxime
  Kefurox® Injection
  Mefoxin®

Zinacef® Injection
Cephalosporin (Third Generation)
  cefixime
  ceftriaxone
  Rocephin®
  Suprax®
Quinolone
  Cipro®
  ciprofloxacin
  Floxin®
  ofloxacin
Tetracycline Derivative
  Apo®-Doxy (Can)
  Apo®-Doxy Tabs (Can)
  Apo®-Tetra (Can)
  Bio-Tab®
  Doryx®
  Doxy-200®
  Doxy-Caps®
  Doxychel®
  Doxycin (Can)
  doxycycline
  Doxy-Tabs®
  Doxytec (Can)
  Dynacin®
  Monodox®
  Nor-tet® Oral
  Novo-Doxylin® (Can)
  Novo-Tetra® (Can)
  Nu-Doxycycline® (Can)
  Nu-Tetra® (Can)
  Panmycin® Oral
  Robitet® Oral
  Sumycin® Oral
  Tetracap® Oral
  tetracycline
  Tetracyn® (Can)
  Vibramycin®
  Vibra-Tabs®

GRANULOMATOUS DISEASE, CHRONIC
Biological Response Modulator

Actimmune®
interferon gamma-1b

HARTNUP DISEASE
Vitamin, Water Soluble
niacinamide

HEMOLYTIC DISEASE OF THE NEWBORN
Immune Globulin
Gamulin® Rh
HypRho®-D
HypRho®-D Mini-Dose
MICRhoGAM™
Mini-Gamulin® Rh
Rho(D) immune globulin
(intramuscular)
rho(D) immune globulin
(intravenous-human)
RhoGAM™
WinRho SD®
WinRho SDF®

HEMORRHAGE (POSTPARTUM)
Ergot Alkaloid and Derivative
Methergine®
methylergonovine
Oxytocic Agent
oxytocin
Pitocin®
Toesen® (Can)
Prostaglandin
carboprost tromethamine
Hemabate™

HERPES SIMPLEX
Antiviral Agent
acyclovir
Avirax™ (Can)
Cytovene®
famciclovir
Famvir™
foscarnet

Foscavir®
ganciclovir
trifluridine
vidarabine
Vira-A® Ophthalmic
Viroptic® Ophthalmic
Vitrasert®
Zovirax®

HERPES ZOSTER
Analgesic, Topical
capsaicin
Capsin® [OTC]
Capzasin-P® [OTC]
Dolorac™ [OTC]
No Pain-HP® [OTC]
R-Gel® [OTC]
Zostrix® [OTC]
Zostrix®-HP [OTC]
Antiviral Agent
acyclovir
Avirax™ (Can)
famciclovir
Famvir™
valacyclovir
Valtrex®
vidarabine
Vira-A® Ophthalmic
Zovirax®

HIV [SEE ACQUIRED IMMUNODEFICIENCY SYNDROME (AIDS)]

HOMOCYSTINURIA
Urinary Tract Product
betaine anhydrous
Cystadane®

HYDATIDIFORM MOLE (BENIGN)
Prostaglandin
Cervidil® Vaginal Insert
dinoprostone

Prepidil® Vaginal Gel
Prostin E2® Vaginal Suppository

HYPERMENORRHEA (TREATMENT)

Contraceptive, Oral
  Alesse™
  Brevicon®
  Demulen®
  Desogen®
  Estrostep® 21
  Estrostep® Fe
  ethinyl estradiol and desogestrel
  ethinyl estradiol and ethynodiol
    diacetate
  ethinyl estradiol and levonorgestrel
  ethinyl estradiol and norethindrone
  ethinyl estradiol and norgestimate
  ethinyl estradiol and norgestrel
  Genora® 0.5/35
  Genora® 1/35
  Genora® 1/50
  Jenest-28™
  Levlen®
  Levlite®
  Levora®
  Loestrin®
  Lo/Ovral®
  mestranol and norethindrone
  Modicon™
  N.E.E.® 1/35
  Nelova™ 0.5/35E
  Nelova® 1/50M
  Nelova™ 10/11
  Nordette®
  Norethin™ 1/35E
  Norethin 1/50M
  Norinyl® 1+35
  Norinyl® 1+50
  Ortho® 0.5/35 (Can)
  Ortho-Cept®
  Ortho-Cyclen®
  Ortho-Novum® 1/35
  Ortho-Novum® 1/50
  Ortho-Novum® 7/7/7
  Ortho-Novum® 10/11
  Ortho Tri-Cyclen®
  Ovcon® 35
  Ovcon® 50
  Ovral®
  Preven™
  Synphasic® (Can)
  Tri-Levlen®
  Tri-Norinyl®
  Triphasil®
  Zovia®
Contraceptive, Progestin Only
  norgestrel
  Ovrette®

HYPERPLASIA, VULVAR SQUAMOUS

Estrogen Derivative
  Alora® Transdermal
  Aquest®
  Cenestin™
  C.E.S.™ (Can)
  chlorotrianisene
  Climara® Transdermal
  Congest (Can)
  depGynogen® Injection
  Depo®-Estradiol Injection
  Depogen® Injection
  dienestrol
  Dioval® Injection
  DV® Vaginal Cream
  Esclim® Transdermal
  Estrace® Oral
  Estraderm® Transdermal
  estradiol
  Estra-L® Injection
  Estratab®
  Estring®
  Estro-Cyp® Injection
  estrogens, conjugated, a (synthetic)
  estrogens, conjugated (equine)

estrogens, esterified
estrone
estropipate
Femogen® (Can)
Gynogen L.A.® Injection
Kestrone®
Menest®
Neo-Estrone® (Can)
Oestrillin® (Can)
Ogen® Oral
Ogen® Vaginal
Ortho®-Dienestrol Vaginal
Ortho-Est® Oral
Premarin®
TACE®
Vivelle™ Transdermal

IDIOPATHIC APNEA OF PREMATURITY
Respiratory Stimulant
  caffeine, citrated

INFERTILITY
Antigonadotropic Agent
  Antagon™
  ganirelix
Ovulation Stimulator
  Follistim®
  follitropin alpha
  follitropin beta
  Gonal-F®

INFERTILITY (FEMALE)
Ergot Alkaloid and Derivative
  Apo® Bromocriptine (Can)
  bromocriptine
  Parlodel®
Gonadotropin
  A.P.L.®
  Chorex®
  chorionic gonadotropin
  Choron®
  Corgonject®

Follutein®
Glukor®
Gonic®
Humegon™
menotropins
Pergonal®
Pregnyl®
Profasi® HP
Repronex™
Ovulation Stimulator
  Clomid®
  clomiphene
  Milophene®
  Serophene®
Progestin
  Crinone™
  PMS-Progesterone (Can)
  Progestasert®
  progesterone
  Progesterone Oil (Can)

INFERTILITY (MALE)
Gonadotropin
  A.P.L.®
  Chorex®
  chorionic gonadotropin
  Choron®
  Corgonject®
  Follutein®
  Glukor®
  Gonic®
  Humegon™
  menotropins
  Pergonal®
  Pregnyl®
  Profasi® HP
  Repronex™

LABOR (PREMATURE)
Adrenergic Agonist Agent
  ritodrine
  terbutaline
  Yutopar®

LABOR INDUCTION

Oxytocic Agent
  oxytocin
  Pitocin®
  Toesen® (Can)
Prostaglandin
  carboprost tromethamine
  Cervidil® Vaginal Insert
  dinoprostone
  dinoprost tromethamine
  Hemabate™
  Prepidil® Vaginal Gel
  Prostin E2® Vaginal Suppository
  Prostin F2 Alpha®

LACTATION (SUPPRESSION)

Ergot Alkaloid and Derivative
  Apo® Bromocriptine (Can)
  bromocriptine
  Parlodel®

MARFAN SYNDROME

Rauwolfia Alkaloid
  Novo-Reserpine® (Can)
  reserpine
  Serpalan®

MECONIUM ILEUS

Mucolytic Agent
  acetylcysteine
  Mucomyst®
  Mucosil™
  Parvolex® (Can)

MENOPAUSE

Ergot Alkaloid and Derivative
  belladonna, phenobarbital, and
    ergotamine tartrate
  Bellergal-S®
  Bel-Phen-Ergot S®
  Phenerbel-S®
Estrogen and Androgen Combination
  Andro/Fem®
  Deladumone®
  depAndrogyn®
  Depo-Testadiol®
  Depotestogen®
  Duo-Cyp®
  Duratestrin®
  estradiol and testosterone
  Valertest No.1®
Estrogen and Progestin Combination
  Activelle™
  estradiol and norethindrone
  estrogens and medroxyprogesterone
  Premphase™
  Prempro™
Estrogen Derivative
  Alora® Transdermal
  Cenestin™
  C.E.S.™ (Can)
  chlorotrianisene
  Climara® Transdermal
  Congest (Can)
  depGynogen® Injection
  Depo®-Estradiol Injection
  Depogen® Injection
  diethylstilbestrol
  Dioval® Injection
  Esclim® Transdermal
  Estinyl®
  Estrace® Oral
  Estraderm® Transdermal
  estradiol
  Estra-L® Injection
  Estratab®
  Estring®
  Estro-Cyp® Injection
  estrogens, conjugated, a (synthetic)
  estrogens, conjugated (equine)
  estrogens, esterified
  ethinyl estradiol
  Gynogen L.A.® Injection
  Honvol® (Can)
  Menest®
  Premarin®

Stilphostrol®
TACE®
Vivelle™ Transdermal

MENORRHAGIA
Androgen
    Cyclomen® (Can)
    danazol
    Danocrine®

NIPPLE CARE
Topical Skin Product
    glycerin, lanolin, and peanut oil
    Masse® Breast Cream [OTC]

OVARIAN FAILURE
Estrogen and Progestin Combination
    estrogens and medroxyprogesterone
    Premphase™
    Prempro™
Estrogen Derivative
    Alora® Transdermal
    Aquest®
    Cenestin™
    C.E.S.™ (Can)
    Climara® Transdermal
    Congest (Can)
    depGynogen® Injection
    Depo®-Estradiol Injection
    Depogen® Injection
    Dioval® Injection
    Esclim® Transdermal
    Estrace® Oral
    Estraderm® Transdermal
    estradiol
    Estra-L® Injection
    Estratab®
    Estring®
    Estro-Cyp® Injection
    estrogens, conjugated, a (synthetic)
    estrogens, conjugated (equine)
    estrogens, esterified
    estrone
    estropipate

Femogen® (Can)
Gynogen L.A.® Injection
Kestrone®
Menest®
Neo-Estrone® (Can)
Oestrillin® (Can)
Ogen® Oral
Ogen® Vaginal
Ortho-Est® Oral
Premarin®
Vivelle™ Transdermal

OVULATION INDUCTION
Gonadotropin
    A.P.L.®
    Chorex®
    chorionic gonadotropin
    Choron®
    Corgonject®
    Follutein®
    Glukor®
    Gonic®
    Humegon™
    menotropins
    Pergonal®
    Pregnyl®
    Profasi® HP
    Repronex™
Ovulation Stimulator
    Fertinex®
    Metrodin®
    urofollitropin

OVULATION
Ovulation Stimulator
    Clomid®
    clomiphene
    Milophene®
    Serophene®

PAIN (ANOGENITAL)
Anesthetic/Corticosteroid
    Enzone®
    Pramosone®

pramoxine and hydrocortisone
Proctofoam®-HC
Zone-A Forte®
Local Anesthetic
  Americaine® [OTC]
  benzocaine
  Dermoplast® [OTC]
  dibucaine
  Dyclone®
  dyclonine
  Fleet® Pain Relief [OTC]
  Foille® [OTC]
  Itch-X® [OTC]
  Lanacane® [OTC]
  PrameGel® [OTC]
  pramoxine
  Prax® [OTC]
  ProctoFoam® NS [OTC]
  Tanac® [OTC]
  tetracaine

PELVIC INFLAMMATORY DISEASE (PID)

Aminoglycoside (Antibiotic)
  amikacin
  Amikin®
  Cidomycin® (Can)
  Garamycin®
  Garatec (Can)
  gentamicin
  Nebcin® Injection
  netilmicin
  Netromycin®
  Ocugram® (Can)
  tobramycin
Cephalosporin (Second Generation)
  cefmetazole
  Cefotan®
  cefotetan
  cefoxitin
  Mefoxin®
  Zefazone®
Cephalosporin (Third Generation)

Cefizox®
Cefobid®
cefoperazone
cefotaxime
ceftizoxime
ceftriaxone
Claforan®
Rocephin®
Macrolide (Antibiotic)
  AK-Mycin®
  Apo®-Erythro E-C (Can)
  azithromycin
  Diomycin (Can)
  E.E.S.®
  E.E.S.® Chewable
  E.E.S.® Granules
  E-Mycin®
  E-Mycin-E®
  Erybid™ (Can)
  Eryc®
  Ery-Tab®
  Erythro-Base® (Can)
  Erythrocin®
  erythromycin (systemic)
  Ilosone®
  Ilosone® Pulvules®
  Ilotycin®
  Ilotycin® (Can)
  Novo-Rythro Encap (Can)
  PCE®
  PMS-Erythromycin (Can)
  Wyamycin® S
  Zithromax™
Penicillin
  ampicillin and sulbactam
  Mezlin®
  mezlocillin
  piperacillin
  piperacillin and tazobactam sodium
  Pipracil®
  Tacozin® (Can)
  Ticar®
  ticarcillin

ticarcillin and clavulanate potassium
Timentin®
Unasyn®
Zosyn™
Quinolone
  Cipro®
  ciprofloxacin
  Floxin®
  ofloxacin
Tetracycline Derivative
  Apo®-Doxy (Can)
  Apo®-Doxy Tabs (Can)
  Apo®-Tetra (Can)
  Bio-Tab®
  Doryx®
  Doxy-200®
  Doxy-Caps®
  Doxychel®
  Doxycin (Can)
  doxycycline
  Doxy-Tabs®
  Doxytec (Can)
  Dynacin®
  Monodox®
  Nor-tet® Oral
  Novo-Doxylin® (Can)
  Novo-Tetra® (Can)
  Nu-Doxycycline® (Can)
  Nu-Tetra® (Can)
  Panmycin® Oral
  Robitet® Oral
  Sumycin® Oral
  Tetracap® Oral
  tetracycline
  Tetracyn® (Can)
  Vibramycin®
  Vibra-Tabs®

PERSISTENT FETAL CIRCULATION (PFC)

Alpha-Adrenergic Blocking Agent
  Priscoline®
  tolazoline

PERSISTENT PULMONARY HYPERTENSION OF THE NEWBORN (PPHN)

Alpha-Adrenergic Blocking Agent
  Priscoline®
  tolazoline

PREECLAMPSIA

Electrolyte Supplement, Oral
  magnesium sulfate

PREGNANCY (PROPHYLAXIS)

Contraceptive, Implant (Progestin)
  levonorgestrel
  Norplant® Implant
Contraceptive, Oral
  Alesse™
  Brevicon®
  Demulen®
  Desogen®
  Estrostep® 21
  Estrostep® Fe
  ethinyl estradiol and desogestrel
  ethinyl estradiol and ethynodiol
    diacetate
  ethinyl estradiol and levonorgestrel
  ethinyl estradiol and norethindrone
  ethinyl estradiol and norgestimate
  ethinyl estradiol and norgestrel
  Genora® 0.5/35
  Genora® 1/35
  Genora® 1/50
  Jenest-28™
  Levlen®
  Levlite®
  Levora®
  Loestrin®
  Lo/Ovral®
  mestranol and norethindrone
  Modicon™
  N.E.E.® 1/35
  Nelova™ 0.5/35E

Nelova® 1/50M
Nelova™ 10/11
Nordette®
Norethin™ 1/35E
Norethin 1/50M
Norinyl® 1+35
Norinyl® 1+50
Ortho® 0.5/35 (Can)
Ortho-Cept®
Ortho-Cyclen®
Ortho-Novum® 1/35
Ortho-Novum® 1/50
Ortho-Novum® 7/7/7
Ortho-Novum® 10/11
Ortho Tri-Cyclen®
Ovcon® 35
Ovcon® 50
Ovral®
Preven™
Synphasic® (Can)
Tri-Levlen®
Tri-Norinyl®
Triphasil®
Zovia®
Contraceptive, Progestin Only
Amen® Oral
Aygestin®
Curretab® Oral
Cycrin® Oral
Depo-Provera® Injection
levonorgestrel
medroxyprogesterone acetate
Micronor®
norethindrone
norgestrel
Norplant® Implant
NOR-QD®
Ovrette®
Provera® Oral
Spermicide
Because® [OTC]
Delfen® [OTC]
Emko® [OTC]

Encare® [OTC]
Gynol II® [OTC]
Koromex® [OTC]
nonoxynol 9
Ramses® [OTC]
Semicid® [OTC]
Shur-Seal® [OTC]

RESPIRATORY DISTRESS SYNDROME (RDS)

Lung Surfactant
beractant
calfactant
colfosceril palmitate
Exosurf® Neonatal™
Infasurf®
Survanta®

SYPHILIS

Antibiotic, Miscellaneous
chloramphenicol
Chloromycetin® Injection
Diochloram (Can)
Ortho-Chloram® (Can)
Pentamycetin® (Can)
Sopamycetin® (Can)
Penicillin
Ayercillin® (Can)
Bicillin® L-A
Crysticillin® A.S.
Megacillin® Susp (Can)
penicillin G benzathine
penicillin G, parenteral, aqueous
penicillin G procaine
Permapen®
Pfizerpen®
Wycillin®
Tetracycline Derivative
Apo®-Doxy (Can)
Apo®-Doxy Tabs (Can)
Apo®-Tetra (Can)
Bio-Tab®
Doryx®

Doxy-200®
Doxy-Caps®
Doxychel®
Doxycin (Can)
doxycycline
Doxy-Tabs®
Doxytec (Can)
Dynacin®
Monodox®
Nor-tet® Oral
Novo-Doxylin® (Can)
Novo-Tetra® (Can)
Nu-Doxycycline® (Can)
Nu-Tetra® (Can)
Panmycin® Oral
Robitet® Oral
Sumycin® Oral
Tetracap® Oral
tetracycline
Tetracyn® (Can)
Vibramycin®
Vibra-Tabs®

VAGINAL ATROPHY

Estrogen and Progestin Combination
Activelle™
estradiol and norethindrone

VAGINITIS

Antibiotic, Vaginal
Femguard®
Gyne-Sulf®
sulfabenzamide, sulfacetamide, and
sulfathiazole
Sulfa-Gyn®
Sulfa-Trip®
Sultrin™
Trysul®
V.V.S.®
Estrogen and Progestin Combination
estrogens and medroxyprogesterone
Premphase™
Prempro™

Estrogen Derivative
Alora® Transdermal
Aquest®
Cenestin™
C.E.S.™ (Can)
chlorotrianisene
Climara® Transdermal
Congest (Can)
depGynogen® Injection
Depo®-Estradiol Injection
Depogen® Injection
dienestrol
diethylstilbestrol
Dioval® Injection
DV® Vaginal Cream
Esclim® Transdermal
Estinyl®
Estrace® Oral
Estraderm® Transdermal
estradiol
Estra-L® Injection
Estring®
Estro-Cyp® Injection
estrogens, conjugated, a (synthetic)
estrogens, conjugated (equine)
estrone
ethinyl estradiol
Femogen® (Can)
Gynogen L.A.® Injection
Honvol® (Can)
Kestrone®
Neo-Estrone® (Can)
Oestrillin® (Can)
Ortho®-Dienestrol Vaginal
Premarin®
Stilphostrol®
TACE®
Vivelle™ Transdermal

VENEREAL WARTS

Biological Response Modulator
Alferon® N
interferon alfa-n3

WILSON DISEASE
Chelating Agent
  Syprine®
  trientine

ZOLLINGER-ELLISON SYNDROME (DIAGNOSTIC)
Diagnostic Agent
  secretin
  Secretin-Ferring Powder

ZOLLINGER-ELLISON SYNDROME
Antacid
  calcium carbonate and simethicone
  magaldrate
  magaldrate and simethicone
  magnesium hydroxide
  magnesium oxide
  Maox®
  Phillips'® Milk of Magnesia [OTC]
  Riopan® [OTC]
  Riopan Plus® [OTC]
  Titralac® Plus Liquid [OTC]
Antineoplastic Agent
  streptozocin
  Zanosar®
Diagnostic Agent
  pentagastrin
  Peptavlon®

Gastric Acid Secretion Inhibitor
  Aciphex™
  lansoprazole
  Losec® (Can)
  omeprazole
  Prevacid®
  Prilosec™
  rabeprazole
Histamine H2 Antagonist
  Apo®-Cimetidine (Can)
  Apo®-Famotidine (Can)
  Apo®-Ranitidine (Can)
  cimetidine
  famotidine
  Novo-Cimetidine® (Can)
  Novo-Famotidine® (Can)
  Novo-Ranidine® (Can)
  Nu-Cimet® (Can)
  Nu-Famotidine® (Can)
  Nu-Ranit® (Can)
  Pepcid®
  Pepcid® AC Acid Controller [OTC]
  Pepcid RPD®
  Peptol® (Can)
  ranitidine hydrochloride
  Tagamet®
  Tagamet-HB® [OTC]
  Zantac®
  Zantac® 75 [OTC]
Prostaglandin
  Cytotec®
  misoprostol